AF520908

Biotechnology of Plantation Crops

THE EDITORS

Dr. Pallem Chowdappa received M.Sc. in 1980 from Sri Venkateswara University, Tirupathi, Ph.D in 1985 from Mangalore University, Mangalore, Karnataka and post doctoral research at CABI Bioscience, U.K. He joined as Scientist-SI in 1985 at ICAR-Central Plantation Crops Research Institute, Kasaragod, Kerala and was elevated to Principal Scientist in 2006 at Indian Institute of Horticultural Research, Bangalore. Dr. Chowdappa served as Scientist-in-Charge, Central Plantation Crops Research Institute Research Centre, Hirehalli and Head, Central Horticultural Experimental Station, Hirehalli from December, 2000 till April, 2006. He became Director, Central Plantation Crops Research Institute, Kasaragod in September, 2014. Dr. Chowdappa is specialized in molecular plant pathology and has over 30 years of research experience in molecular characterization and management of *Alternaria, Colletotrichum* and *Phytophthora* associated with diseases of horticultural crops. He attended international training program on 'Oomycetes bioinformatics' at Virginia Tech, USA in 2014. Dr. Chowdappa was awarded DFID fellowship for Post-Doctoral research at CABI Bioscience, UK in 1998. Dr. Chowdappa has published more than 120 research papers in leading national and international journals, 12 books, 35 technical bulletins, 42 book chapters and 65 experimental manuals. He is a fellow of Scientific Academia and has won several awards of repute. He is also president of many scientific societies in India.

Dr. Anitha Karun completed her professional degree courses B.Sc. (Agri.), M.Sc. (Agri.) in Horticulture and Ph.D (Horticulture) from the University of Agricultural Sciences, Bangalore. She started her career as Assistant Professor in Horticulture at Kerala Agriculture University in 1988 and was subsequently selected to Agricultural Research Service in 1990. She was posted as Scientist at ICAR-Central Plantation Crops Research Institute, Kasaragod, Kerala. Since 2012, she is the Principal Scientist and Head of Crop Improvement Division of ICAR-CPCRI, Kasaragod, Kerala. Her areas of specialization are palm tissue/organ culture, cryopreservation and molecular biology. She had undergone international training programme on molecular biology at University of Adelaide, South Australia in 1995. She has been awarded 'ICAR Award for Team Research' for the biennium 1999-2000' for her major contribution in germplasm exchange of coconut in the form of embryos for the first time in the world. As a country expert in the field of embryo culture and cryopreservation, she has visited South Korea, the Philippines, Sri Lanka and El Salvador. She has published more than 100 research papers in leading national and international Journals, 12 book chapters and 3 technical bulletins and has guided 6 Ph. D students.

Dr. M.K. Rajesh has received his B.Sc. (Ag.), M.Sc. (Biotechnology) and Ph.D (Biotechnology) from Tamil Nadu Agricultural University, Coimbatore. He joined as Scientist at ICAR-Central Plantation Crops Research Institute (ICAR-CPCRI), Kasaragod, Kerala, in 1998. Presently, he is a Principal Scientist (Biotechnology) in the Division of Crop Improvement, ICAR-CPCRI. He has over 18 years experience in molecular characterization of coconut, cocoa and arecanut germplasm, use of molecular markers for hybrid authentication in coconut and arecanut, *in vitro* culture and *in vitro* conservation of coconut and transcriptome analysis and gene expression studies in coconut, with respect to somatic embryogenesis and host-pathogen interactions. He has published more than 75 research papers in peer reviewed journals, written over 20 book chapters and edited five books. He is also serving as Assistant Editor, Journal of Plantation Crops. His current area of research includes genome and transcriptome analysis of coconut.

Dr. S.V. Ramesh is serving as Scientist (Biotechnology) at ICAR-Central Plantation Crops Research Institute (ICAR-CPCRI), Kasaragod, Kerala. He has obtained his Masters and Doctorate degrees in Plant Biochemistry. His research interests include molecular biology of small non-coding RNAs (ncRNAs) during biotic and abiotic stresses of crop plants. Ramesh is a recipient of DBT-CTEP Travel Award (2013), DBT-CREST Fellowship-Washington State University, USA (2013-14), and Young Plant Biotechnologist Award by AABS (2015). He has authored over 20 peer-reviewed articles, including lead articles, in the journals of national and international repute. He has also presented invited talks in the arena of plant small RNAs mediated genetic engineering and implications, at international scientific gatherings *viz.*, World Soybean Research Conference, Durban, South Africa and International Symposium on Non-Coding Genomes, IBAB etc., to name a few.

Biotechnology of Plantation Crops

— *Editors* —

P. Chowdappa

Anitha Karun

M.K. Rajesh

S.V. Ramesh

2017

Daya Publishing House®

A Division of

Astral International Pvt. Ltd.

New Delhi – 110 002

ISBN: 978-93-86071-72-9 (International Edition)

Published by : **Daya Publishing House®**
A Division of
Astral International Pvt. Ltd.
– ISO 9001:2015 Certified Company –
4736/23, Ansari Road, Darya Ganj
New Delhi-110 002
Ph. 011-43549197, 23278134
E-mail: info@astralint.com
Website: www.astralint.com

त्रिलोचन महापात्र, पीएच.डी.
एफ एन ए, एफ एन ए एस सी, एफ एन ए ए एस
सचिव एवं महानिदेशक

TRILOCHAN MOHAPATRA, Ph.D.
FNA, FNASc, FNAAS
SECRETARY & DIRECTOR GENERAL

भारत सरकार
कृषि अनुसंधान और शिक्षा विभाग एवं
भारतीय कृषि अनुसंधान परिषद
कृषि एवं किसान कल्याण मंत्रालय, कृषि भवन, नई दिल्ली 110 001

GOVERNMENT OF INDIA
DEPARTMENT OF AGRICULTURAL RESEARCH & EDUCATION
AND
INDIAN COUNCIL OF AGRICULTURAL RESEARCH
MINISTRY OF AGRICULTURE AND FARMERS WELFARE
KRISHI BHAVAN, NEW DELHI 110 001
Tel.: 23382629; 23386711 Fax: 91-11-23384773
E-mail: dg.icar@nic.in

Foreword

Plantation crops are considered to be the major segment of the horticulture crops and the mainstay of agrarian economies of many countries in the tropics. They contribute a significant amount to the national exchequer and exports by way of excise and export earnings in countries where they are grown. The magnitude of direct and indirect employment provided by the sector, especially in the rural areas, makes it a vital cog in policy perspectives for overall economic development. The major plantation crops include coconut, arecanut, oil palm, cashew, tea, coffee, rubber, cocoa and spices. These crops have played a key role in the socio-economic development and transition from a subsistence agrarian economy to market oriented commercial cultivation due to their trade significance. Plantation crops also provide adequate interspaces for intercropping of seasonal crops and thus ensure food security to a great extent.

Cultivation of plantation crops also has a rich diversity and varied history with each crop having its own distinct historical and economic context of development. The area and productivity of the major plantation crops has witnessed significant positive trends during the last five decades. The rising population and the increase in purchasing power portends that the demand for the major agricultural produce from the plantation crops sector will continue its robust growth.

Given their importance, there exists a tremendous scope for improvement of plantation crops, especially in developing genotypes resistant to biotic and abiotic stresses, responsive to low input management and micropropagation, through biotechnological approaches. Biotechnology is one branch of modern sciences in which break-through information and fascinating discoveries come out virtually every day and it is an uphill task to keep abreast of the latest developments in one's field of specialization. The book 'Biotechnology of Plantation Crops' invites due attention of readers on all topics of current interest in the area of biotechnology of plantation crops. This book is more than a compendium of information; it is an integrated work written by experts who have experienced first-hand the intricacies and nuances associated with use of biotechnological tools in plantation crops. This book is designed in such a manner that it covers a broad range of areas of biotechnology, and more importantly, presents the subject in a crisp and concise manner with conceptual clarity, ensuring a rewarding reading experience.

(Dr. Trilochan Mohapatra)

Secretary, DARE and

Director General, ICAR, New Delhi

Preface

This book entitled 'Biotechnology of Plantation Crops' strives to provide a wide-ranging, up-to-date and methodical account of the application of biotechnological tools to the improvement of plantation crops. The greatest motivation behind the publication of this book is the desire we have had to compile and present a compendium of cutting edge biotechnological research in the arena of plantation crops. The thirty three chapters, organized into five sections, deal with the latest biotechnological achievements in the field of Regeneration Systems, Molecular Markers and Marker-Assisted Selection, In Vitro Conservation, Transgenics and 'Omics' Applications.

This book has been intended to serve as a comprehensive textbook as well as a wide-ranging reference book, designed to inform and inspire the next generation of plant biotechnologists in the plantation crops sector. The book is recommended for junior- and senior-level courses at the undergraduate and graduate levels and research scholars and post-doctoral students. It would also serve as an ideal reference book for practitioners, biotechnology scientists and industry researchers alike.

Biotechnology of plantation crops has been lagging behind other crops owing to various reasons viz., long juvenile phase, high heterozygosity, very little genomics information, lack of genome sequences and problems associated with regeneration and transformation protocols. In this context, the authors of the individual chapters have been chosen for their recognized expertise and their contributions to the various fields of biotechnology, especially in the field of plantation crops research. We intend that the first hand information from experts, who have been involved in plantation crops biotechnology, would therefore provide greater insights and deeper understanding of nuances to the readers. Their willingness to impart this knowledge, benefiting the plantation crops research community, is gratefully acknowledged. We really hope that the readers will appreciate the high scientific content of any chapter included in this book.

Finally, this work could not have been brought to fruition without the foresight and the constant and diligent support of the publisher, M/s Astral International Pvt. Ltd., New Delhi, for which we are indebted to them.

P. Chowdappa

Anitha Karun

M.K. Rajesh

S.V. Ramesh

Contents

List of Contributors

Afruza Zaman, *Tocklai Tea Research Institute, Jorhat, Assam, India*

Alexandre de Kochko, *Institut de Recherche pour le Développement (IRD), centre de Montpellier, BP64501 Montpellier Cedex 5, France*

Ambatkar M., *Ramniranjan Jhunjhunwala College, Mumbai, Maharashtra, India*

Ananda K. S., *ICAR-Central Plantation Crops Research Institute, Regional Station, Vittal, Karnataka, India*

Anitha Karun, *ICAR-Central Plantation Crops Research Institute, Kasaragod, Kerala, India*

Antoinette Sankar, *Cocoa Research Centre, The University of the West Indies, St. Augustine, Trinidad, Trinidad and Tobago, West Indies*

Aparna V., *ICAR-Central Plantation Crops Research Institute, Kasaragod, Kerala, India*

Arunachalam V., *ICAR-Central Coastal Agricultural Research Institute, Goa, India*

Arvind K. Yadav, *ICAR-NRC on Plant Biotechnology, New Delhi, India*

Basavaraju L., *Hindustan Unilever Research Centre, Bengaluru, Karnataka, India*

Bhat A.I., *ICAR-Indian Institute of Spices Research, Kozhikode, Kerala, IndiA*

Bindu Roy C., *Rubber Research Institute of India, Kottayam, Kerala, India*

Bini K., *Rubber Research Institute of India, Kottayam, Kerala, India*

Chowdappa P., *ICAR-Central Plantation Crops Research Institute, Kasaragod, Kerala, India*

Christine Dubreuil-Tranchant, *Institut de Recherche pour le Développement (IRD), centre de Montpellier, BP64501 Montpellier Cedex5, France*

David Gopaulchan, *Cocoa Research Centre, The University of the West Indies, St. Augustine, Trinidad, Trinidad and Tobago, West Indies*

Dhurendra Singh, *ICAR-Central Institute for Arid Horticulture, Bikaner, Rajasthan, India*

Dominique Crouzillat, *Nestlé R&D, 101 Av. G. Eiffel, ND d'Oé BP 4976 F-37097 Tours Cedex 5, France*

Elena Popova, *Gosling Research Institute for Plant Preservation, University of Guelph, Ontario, Canada*

Emmanuel Couturon, *Institut de Recherche pour le Développement (IRD), centre de Montpellier, BP64501 Montpellier Cedex5, France*

Geetha S.P., *Arya Vaidya Sala, Kottakkal, Kerala, India*

Ginny Antony, *Central University of Kerala, Padannakad, Kerala, India*

Guttapadu S., *Hindustan Unilever Research Centre, Bengaluru, Karnataka, India*

Haeng-Hoon Kim, *Sunchon National University, Suncheon, Jeonnam, Korea*

Hemanta Saikia, *Tocklai Tea Research Institute, Jorhat, Assam, India*

Jayanthi Madhavan, *ICAR-Indian Agricultural Research Institute, New Delhi, India*

Jayashree R., *Rubber Research Institute of India, Kottayam, Kerala, India*

Jeena Devasia, *Central Coffee Research Institute, Chikamagalur, Karnataka, India*

Kala R. G., *Rubber Research Institute of India, Kottayam, Kerala, India*

Kalyana Babu B., *ICAR-Indian Institute of Oil Palm Research, Pedavegi, Andhra Pradesh, India*

Karun Gurjar, *ICAR-Central Institute for Arid Horticulture, Bikaner, Rajasthan, India*

Krishna Prakash, *ICAR-Central Plantation Crops Research Institute, Kasaragod, Kerala, India*

Kumari Jayasree P., *Rubber Research Institute of India, Kottayam, Kerala, India*

Lambert A. Motilal, *Cocoa Research Centre, The University of the West Indies, St. Augustine, Trinidad, Trinidad and Tobago, West Indies*

Lestari P., *Balai Besar Penelitian dan Pengembangan Bioteknologi dan Sumberdaya Genetik Pertanian, Bogor, Indonesia*

Madhurjya Gogoi, *Tocklai Tea Research Institute, Jorhat, Assam, India*

Malik S. K., *ICAR-National Bureau of Plant Genetic Resources, Pusa Campus, New Delhi, India*

Manimekalai R., *ICAR-Sugarcane Breeding Institute, Coimbatore, Tamil Nadu, India*

Maskromo I., *Bogor Agricultural University, Bogor, Indonesia; Indonesian Palm Research Institute, Indonesian Agency for Agricultural Research and Development, Manado, Indonesia*

Mathur R.K., *ICAR-Indian Institute of Oil Palm Research, Pedavegi, Andhra Pradesh, India*

Michel Rigoreau, *Nestlé R&D, 101 Av. G. Eiffel, ND d'Oé BP 4976 F-37097 Tours Cedex 5, France*

Minoo Divakaran, *Providence Women's College, Kozhikode, Kerala, India*

Mukundan U., *Ramniranjan Jhunjhunwala College, Mumbai, Maharashtra, India*

Muralikrishna K.S., *ICAR-Central Plantation Crops Research Institute, Kasaragod, Kerala, India*

Murugesan P., *ICAR-Indian Institute of Oil Palm Research, Research Centre, Palode, Thiruvananthapuram, Kerala, India*

Nagaraja N. R., *ICAR-Central Plantation Crops Research Institute, Regional Station, Vittal, Karnataka, India*

Neema M., *ICAR-Central Plantation Crops Research Institute, Kasaragod, Kerala, India*

Nirmal Babu K., *ICAR-Indian Institute of Spices Research, Kozhikode, Kerala, India*

Pathmanathan Umaharan, *Cocoa Research Centre, The University of the West Indies, St. Augustine, Trinidad, Trinidad and Tobago, West Indies*

Patil G. B., *AnandAgricultural University, Anand, Gujarat, India*

Patil J., *Ramniranjan Jhunjhunwala College, Mumbai, Maharashtra, India*

Perera L., *Genetics Coconut Research Institute, Lunuwila, Sri Lanka*

Perla Hamon, *Institut de Recherche pour le Développement (IRD), centre de Montpellier, BP64501 Montpellier Cedex5, France*

Peter K.V., *Kerala Agricultural University, Thrissur, Kerala, India*

Pinaki Acharyya, *ICAR-Central Institute for Arid Horticulture, Bikaner, Rajasthan, India*

Prakash N. S., *Central Coffee Research Institute, Chikamagalur, Karnataka, India*

Pranab Kumar Mandal, *ICAR-National Research Centre for Plant Biotechnology, New Delhi, India*

Prasanta K. Dash, *ICAR-NRC on Plant Biotechnology, New Delhi, India*

Prasath D., *ICAR-Indian Institute of Spices Research, Kozhikode, Kerala, India*

Praveen K. Saxena, *Gosling Research Institute for Plant Preservation, University of Guelph, Ontario, Canada*

Pritom Choudhury, *Tocklai Tea Research Institute, Jorhat, Assam, India*

Raghuramulu Y., *Central Coffee Research Institute, Chikamagalur, Karnataka, India*

Rajesh M. K., *ICAR-Central Plantation Crops Research Institute, Kasaragod, Kerala, India*

Rajesh M.K., *ICAR-Central Plantation Crops Research Institute, Kasaragod, Kerala.*

Ramesh K. Aggarwal, *Centre for Cellular & Molecular Biology, Hyderabad, Telengana, India*

Ramesh S.V., *ICAR-Central Plantation Crops Research Institute, Kasaragod, Kerala.*

Ravindran P.N., *ICAR-Indian Institute of Spices Research, Kozhikode, Kerala, India*

Rebijith K.B., *University of Cambridge, Cambridge, UK*

Rekha Chaudhury, *ICAR-National Bureau of Plant Genetic Resources, Pusa Campus, New Delhi, India*

Rekha K., *Rubber Research Institute of India, Kottayam, Kerala, India*

Rhitu Rai, *ICAR-NRC on Plant Biotechnology, New Delhi, India*

Romain Guyot, *Institut de Recherche pour le Développement (IRD), centre de Montpellier, BP64501 Montpellier Cedex5, France*

Sagarika Das, *Tocklai Tea Research Institute, Jorhat, Assam, India*

Saji K.V., *ICAR-Indian Institute of Spices Research, Kozhikode, Kerala, India*

Sajini K.K., *ICAR-Central Plantation Crops Research Institute, Kasaragod, Kerala, India*

Sangeeta Borchetia, *Tocklai Tea Research Institute, Jorhat, Assam, India*

Serge Hamon, *Institut de Recherche pour le Développement (IRD), centre de Montpellier, BP64501 Montpellier Cedex5, France*

Sinkar V., *Hindustan Unilever Research Centre, Bengaluru, Karnataka, India*

Sivalingam P.N., *ICAR-National Institute of Biotic Stress Management, Raipur, Chhattisgarh, India*

Sobha S., *Rubber Research Institute of India, Kottayam, Kerala, India*

Sreenath H. L., *Central Coffee Research Institute, Coffee Board, Manasagangothri, Mysore, Karnataka, India*

Subhash N., *Anand Agricultural University, Anand, Gujarat, India*

Sudarsono S., *Bogor Agricultural University, Bogor, Indonesia*

SunilKumar K., *ICAR-Indian Institute of Oil Palm Research, Research Centre, Palode, Thiruvananthapuram, Kerala, India*

Tai Y.L., *State Key Lab of Tea Plant Biology and Utilization, Anhui Agricultural University, Hefei, China*

Tanoy Bandyopadhyay, *Tocklai Tea Research Institute, Jorhat, Assam, India*

Thakurdas Saha, *Rubber Research Institute of India, Kottayam, Kerala, India*

Thulaseedharan A., *Rubber Research Institute of India, Kottayam, Kerala, India*

Tony Grace, *Kansas State University, Manhattan, Kansas, USA*

Valérie Poncet, *Institut de Recherche pour le Développement (IRD), centre de Montpellier, BP64501 Montpellier Cedex5, France*

Wan X.C., *State Key Lab of Tea Plant Biology and Utilization, Anhui Agricultural University, Hefei, China*

Wei C.L., *State Key Lab of Tea Plant Biology and Utilization, Anhui Agricultural University, Hefei, China*

Yamuna G., *ICAR-Indian Institute of Spices Research, Kozhikode, Kerala, India*

Yang H., *State Key Lab of Tea Plant Biology and Utilization, Anhui Agricultural University, Hefei, China*

Regeneration Systems

Chapter 1

Rubber

☆ *P. Kumari Jayasree*

1. Introduction

Para rubber tree (*Hevea brasiliensis* Muell. Arg.), popularly known as rubber, is native to Amazonian river basin and belongs to the botanical family *Euphorbiaceae* and the genus *Hevea*. Of the ten species of *Hevea* (Dijkman, 1951), Brazilian rubber is the only cultivated tree species for the commercial production of natural rubber due to factors like high rubber content, yield of acceptable quality and convenience of harvesting (Wycherley, 1992). In *Hevea,* natural rubber is synthesied in the highly specilaized laticifer cells, which are located adjacent to the phloem and is obtained by the processing of latex collected by wounding the bark of the tree. Chemically, natural rubber is a high molecular weight isoprenoid polymer (*cis*-1,4-polyisoprene) composed of 320 to 35,000 molecules and is one of the most industrial commodity due to its high performance characteristics. Today, natural rubber has vast applications in day to day life from the very basic personal articles to automobile industry. Many industries make use of this material for the manufacture of more than 50,000 rubber based products.

From its centre of origin in Brazil, rubber tree was domesticated as a plantation crop. Presently, rubber is commercially grown in the tropical regions of Asia, Africa and South America (Varghese *et al.,* 2000). Over the past twenty years, *Hevea* breeders have focussed their attention on evolving high yielding clones. However, the inherent nature of the crop including long breeding cycle with extended juvenile phase, high heterozygosity, seasonal flowering with low fruit set etc, makes breeding of *Hevea* more difficult and time consuming. In addition, the unidirectional selection practiced over the years has further narrowed the genetic base resulting in a slow down in genetic improvement (Varghese *et al.,* 2000). Biotechnological approaches ranging from tissue culture to molecular biology has contributed many novel methods to overcome the barriers of conventional breeding and to genetically modify rubber plants for further genetic improvement.

2. Tissue Culture Techniques

2.1. Micropropagation

For a long time, the accepted practice of clonal propagation was through bud grafting, where buds of selected clones were grafted into cross pollinated seedlings. However, in most of areas under rubber cultivation, tree to tree variation was largely seen among budded plants (Clombe, 1975) which had a profound influence on both growth and yield. The yield variation within the clones could be eliminated by the use of uniform self rooted plants and thus the production of tissue culture rubber assumes importance. Micropropagation *via* shoot tip or through somatic embryogenesis (SE) has been the method of choice for multiplication of self rooted plants.

During 1970s, attempts for *in vitro* multiplication were based on seedling explants. Paranjothy and Ghandimathi (1975, 1976) initially cultured shoot apices of seedlings and induced roots. Since 1980, at CIRAD (France), Carron and Enjalric (1982) achieved shoot sprouting from stem nodes of greenhouse plants. Thereafter, Gunatilleke and Samaranayake (1988), Carron *et al.* (1989) and Te-Chato and Muangkaewngam (1992) used aseptic seedling shoot tips and to successfully produce plants plants. However, propagation of clonal material was limited due to the highly recalcitrant nature of mature trees and failure in inducing a tap root (Carron and Enjalric, 1983) and presence of bacterial and fungal contaminations (Seneviratne, 1991). Due to the difficulty of *in vitro* rooting, a micrografting technique was later adopted by Perrin *et al.* (1994)) who succeeded in grafting of shoot apices from mature trees into *in vitro* grown seedlings. Explant juvenility was another factor affecting micropropagation and Seneviratne and Flegmann (1996) induced multiple shoots and rooting from nodal explants of juvenile origin. Seneviratne and Wijeskara (1996). showed that polyvinylpyrrolidone (PVP) aided reduction of oxidation problems in *in vitro* stem explants from mature material and axillary bud development could be achieved with supplementation of thidiazuron. Buds of nodal explants taken from dormant branches exhibited better *in vitro* growth response (Seneviratne and Wijeskara, 1997; Lardet *et al.*, 1998). Mendanha *et al.* (1998) also developed plants from axillary buds by culturing on MS medium with growth hormones.

Attempts to develop a propagation protocol at Rubber Research Institute of India was based on shoot tip explants excised from mature trees, which however, failed to root (Sinha *et al.*, 1985; Sobhana *et al.*, 1986). Asokan *et al.* (1988) successfully induced rooting in clonal shoot tips on medium containing 1.5 - 3.0 mg/l indole-acetic acid (IAA) and 0.5-1.5 mg/l kinetin and after hardening, rooted plants were established in the field. Conditions were also optimized for micrografting and 30 days old rootstock and 21 days old scion had considerably enhanced the success rate to 85 per cent (Kala *et al.*, 2002). Efficiency of three fungicides and antibiotics was assessed on microbial contamination and nystatin at 25 mg/l along with streptomycin at 10 mg/l controlled contamination and also favoured explant proliferation (Kala *et al.*, 2004). Shoot tip derived plants were developed for four clones *viz.*, RRII 105, RRII 208, PB 311, RRIM 600 and were field planted. The incidence of tapping panel dryness (TPD) in these four clones was compared after 10 years of tapping and it was observed that TPD incidence was more or less similar in both tissue culture and bud grafted trees (Thulaseedharan *et al.*, 2006).

In recent years, Brazil, Nigeria and Thailand have embarked on research on *in vitro* propagation of *Hevea*. Hui *et al.* (2009) made attempts with stem of mature tree, Reyan 7-33-97, and observed that stem segments in different stages of growth have different contamination levels and therefore different sterilization techniques were needed. Thereafter, a Nigerian team identified a suitable medium for inducing tap root from zygotic embryos with MS basal salts containing 0.075 mg/l benzyl adenine (BA) and 0.001 mg/l naphthalene acetic acid (NAA) (Ighere Dickson *et al.*, 2011). Attempts made on *in vitro* shoot multiplication by Thailand group investigated the effect of peptone and silver nitrate on shoot formation of *ex vitro* raised seedlings. A multiplication rate of five shoots per explant was achieved in medium with 1.0 mg/l silver nitrate while, addition of peptone (0-2 per cent) did not induce multiple shoots (Sirisom and Te-Chato, 2012). Recently, Malaysian Rubber Board experimented with axillary buds of mature budded stumps of clone RRIM 2020. When lateral meristem was cultured, 93 per cent shoot formation was noticed and both callus as well as rooted embryoids was obtained. Meanwhile, with shoot apical meristem, 79 per cent shoot growth was reported, however, in both cases, subsequent elongation and induction of roots appeared to be retarded (Nor Mayati and Jamnah, 2014).

2.2. Somatic Embryogenesis

Most of the high yielding clones are relatively susceptible to tapping panel dryness and other diseases and hence clones with improved resistance to TPD and diseases are very significant. In the scenario of global warming, the development of location specific clones with increased yield and tolerance to abiotic stresses are the need of the day. Additionally, the wider application of rubber as molecular farming factories and engineering with latex biosynthesis gene is also very promising. To exploit the full potential of genetic engineering for the aforesaid achievements, development of highly efficient, reliable *in vitro* plant regeneration system *via* SE is very essential.

2.2.1. Work on Somatic Embryogesis in Laboratories Abroad

The earliest attempt on callus induction was made by Bouychou (1953)., with the aim of using callus for the study of laticiferous system. Later, Chua (1966) initiated callus from plumule tissue of seedlings and Wilson and Street (1975) induced callus from stem explants. Paranjothy (1974) first achieved somatic embryo differentiation from anther wall derived calli, but failed to achieve plant regeneration. Paranjothy and Rohani (1978) subsequently succeeded in shoot development from somatic embryos. Since then, plant development has been achieved from anther wall and plantlets were successfully established in soil (Wang *et al.*, 1980, 1984). Wan *et al.* (1982) revealed clonal differences in embryogenesis from anther callus. Later, Wang and Chen (1995) developed plantlets from stamen cultures through somatic embryogenesis. Wang *et al.* (1998). optimized temperature conditions for callus induction, embryo induction and plant regeneration as 26°C, 24-25°C and 26-27°C respectively and a plant regeneration frequency of upto 40.5 per cent could be achieved.

At the Agricultural Research Centre for International Development (CIRAD) (France), researchers experimented with inner integument tissue of immature fruits and anther for somatic embryogenesis. Carron (1981) induced somatic embryos and plantlets from integument tissue of clone PB 260 while only callus and embryos were obtained from anthers (Carron and Enjarlic, 1985). According to Michaux-Ferriere and Carron (1989)., frequent culturing in callus induction medium reduced the formation of embryogenic callus. EI Hadrami *et. al.* (1989) and EI Hadrami and Auzac (1992), studied the effect of polyamines on SE using polyamine biosynthesis inhibitors. Auboiron *et al.* (1990). investigated the importance of culture vessel and observed that release of CO_2 and ethylene adversely affected embryo induction. EI Hadrami *et al.* (1991) concluded that an excess of auxin and cytokinin and its prolonged exposure in the medium suppressed embryogenic capacity of callus. Etienne *et al.* (1991a) observed a higher relative water content along with lower concentrations of 2,4-dichlorophenoxyacetic acid (2,4-D), BA and abscisic acid (ABA) favoured embryogenic competence of callus. Etienne *et al.* (1991b) also investigated the role of mineral and carbohydrate nutrition on somatic embryogenesis. Michaux-Ferriere *et al.* (1992) confirmed a two way mode of ontogeny *viz.*, unicellular and multicellular origin of SE in rubber. Etienne *et al.* (1993 a, b) observed that a slow desiccation with 120 g/l sucrose and 0.3 mg/l ABA enhanced germination ability and conversion into plantlets and on estimation of endogenous ABA content, it was observed that embryogenic calli accumulated low levels of ABA whereas higher levels were seen in non-embryogenic callus. Montoro *et al.* (1993, 1995) reported the interaction of growth regulators, sucrose and calcium on friability of callus and further induction of somatic embryogenesis. Veisseire *et al.* (1994) a, b) enhanced callus proliferation through a liquid phase and inclusion of ABA combined with BA or adenine in liquid medium stimulated embryo development. Cailloux *et al.* (1996) established a long term embryogenic line by the use of recurrent embryogenesis and the cultures were maintained upto three years on hormone free medium. Etienne *et al.* (1997a) improved somatic embryo induction to 2-fold by the enrichment of $CaCl_2$ (9 mM) in the medium and Etienne *et al.* (1997b) used a pulsed air immersion technique for embryo enhancement. Later, Linossier *et al.* (1997) used polyethylene glycol for obtaining more number of torpedo shaped embryos. Lardet *et al.* (1999) biochemically compared the conversion ability of somatic embryos with zygotic embryos and proved that total mass of starch and protein reserves available in somatic embryos is much smaller. The influence of various carbohydrates also affected SE and among the sugars, maltose combined with sucrose significantly promoted somatic embryo production (Blanc *et al.*, 1999, 2002). During the last decade, Martre *et al.* (2001) attempted temporary immersion technique in 'Automated Temporary Immersion Apparatus' and concluded that the immersed stage induced a substantial oxidative stress. Lardet *et al.* (2008a) developed a secondary SE system and the number of subcultures needed for establishing callus lines was shortened, however, success rates were low and unpredictable. Lardet *et al.* (2008b) further investigated the ontogenetic and physiological aging of explant source on embryogenic capacities and came to the findings that embryogenic potential was much higher in somatic plant derived explant than that of explant excised from mature trees.

At the Rubber Research Institute of Sri Lanka, attempts made on plant development through SE from anther as well as integument tissues resulted in good callus induction; however embryo induction could not be achieved (Nayanakantha and Seneviratne, 2007). Presently, several team from other rubber producing countries have initiated their research on somatic embryogenesis. Hua *et al.* (2010) developed a secondary SE procedure for propagation based on *in vitro* studies at Rubber Research Institute of China. Secondary SE was found to be improved when MS medium was supplemented with 6 mM and 9 mM $CaCl_2$ and also when embryo fragments were placed close to the base of primary embryos. By this procedure, plant regeneration frequencies upto 85 per cent and 75 per cent was achieved. Simultaneously, Zhou *et al.* (2010) established a system from *in vitro* root explants of anther plants. Callus induced on MS medium containing 1.0 mg/l kinetin combined with 0.2 mg/l BA, was differentiated into embryos on medium with 3.0 mg/l kinetin, 1.0 mg/l BA, 0.2 mg/l NAA and 0.05 mg/l gibberellic acid (GA_3). Later, Tan *et al.* (2011). correlated the frequency of laticifer cells with SE ability in five genotypes. Clones PR 107, RRIM 600, Reyna 7-33-97 and Reyan 8-79, possessing higher frequency of laticifer cells, exhibited lower embryogenic ability compared to Haiken 2. Recently, Srichuay *et al.* (2014) induced callus and embryogenesis from anther and the diploid nature of plants was confirmed in 100 per cent of the regenerated plantlets with somatic origin.

2.2.2. Work on Somatic Embryogenesis at RRI, India

Till date in RRII, majority of work on SE has been carried out on the Indian clone RRII 105. At RRII, SE was first described from inner integument tissue explants (Asokan *et al.*, 1992). Maximum callus was induced on B5 medium containing 2.0 mg/l 2,4-D and embryogenesis was achieved on medium with a growth regulator regime of 0.5 mg/l NAA, 1.0 -2.0 mg/l BA and 0.4-0.8 mg/l IAA. Thereafter, Kumari Jayasree *et al.* (1999). developed a system with immature anthers. Modified MS medium supplemented with 2.0 mg/l 2,4-D and 0.5 mg/l kinetin induced maximum callus while 0.7 mg/l kinetin combined with 0.2 mg/l NAA was found to be optimum for embryo induction. Regenerated plantlets have been established in field and diploid nature of plants was proved. Simultaneously, Sushamakumari *et al.* (1999a) induced multiple shoots from germinated somatic embryos of inflorescence derived callus and the cytokinins BA and TDZ at a range of 5.0 mg/l and 0.4 mg/l and shoot primordial wounding of somatic embryos stimulated multiplication rate. Percentage of rooting from BA supplemented medium was rapid and higher (90 per cent) than that of TDZ medium (50 per cent). Sushamakumari *et al.* (2000) discussed the role of sucrose on SE and higher sucrose was found to influence both callus and embryo induction; however, a low level sucrose was essential for plant recovery. Later, cultural parameters affecting somatic embryo induction and germination were studied in detail by Kumari Jayasree *et al.* (2001a). Explants pre-treated for 10 days in liquid callus induction medium reduced duration of callus induction.

Polyamines showed no significant improvement however, slight enhancement on embryo induction was noted with 0.5 mg/l spermidine. Embryo induction medium enriched with 200 mg/l glutamine and 400 mg/l casein hydrolysate was found to improve the efficiency of embryo induction. For induction as well as proliferation of callus and for acquisition of embryogenic competence, *Hevea* needed a continuous dark incubation, while embryo induction and development occurred either in dark or light conditions (Kumari Jayasree *et al.*, 2001). Experimentation with GA_3 revealed that lower levels (2.0 mg/l) increased induction and germination of somatic embryos and subsequent recovery of full plantlets (Kumari Jayasree and Thulaseedharan, 2001b). A biochemical assay with six isozymes was performed for the characterization of embryogenic callus and a marked difference in the expression of isozymes has been observed between embryogenic and non-embryogenic callus (Asokan *et al.*, 2001). A repetitive embryogenesis system was developed from integument derived primary somatic embryos and a combination of 0.5 mg/l NAA, 0.5 mg/l IAA, 2.0 mg/l kinetin along with 5 per cent sucrose induced secondary embryos. Inclusion of 4.0 mg/l 2,4-D in medium also promoted secondary embryos (Asokan *et al.*, 2002). A long term embryogenesis system was attempted and embryogenic callus bearing small embryos was found to be the most viable source for long term establishment and cultures retained embryogenic competence for over three years (Kumari Jayasree and Thulaseedharan, 2004). Effects of cytokinins on germination and plant regeneration were studied with BA, zeatin, Kinetin and TDZ and of the four cytokinins, TDZ at 0.25 mg/l was found to be superior with a regeneration frequency of 82 per cent. Initial growth of plants in soil with respect to height, shoot length and number of leaves was found to be higher with TDZ than plants derived from BA and ZEA treatment (Figures 1.1a–f) (Kumari Jayasree and Thulaseedharan, 2005). Kumari Jayasree *et al.* (2009) investigated the anatomical variations associated with somatic plants during acclimatization. Scanning electron microscopic studies on leaves of healthy plants confirmed the presence of epicuticular wax which was found to be increased during the course of hardening. In acclimatized plant, pattern of wax deposition was almost identical with that in field grown control plants. After acclimatisation, stomatal frequency of healthy plants were slightly higher or more or less equal to that of field grown plants though stomatal aperture size did not increase. Leaves of weak *in vitro* plantlets showed less epicuticular wax accumulation. Vascular continuity and distribution of latex vessels were well established in somatic plants. Kumari Jayasree *et al.* (2010) assessed the genetic stability of immature anther derived somatic plants using RAPD markers. The amplified products of mother tree and all the plants regenerated through somatic embryogenesis were monomorphic, while polymorphism was detected among monoclonal seedlings (Figure 1.1g: lanes 1-marker, 2- mother tree, 3-12 somatic plants, 13-22 seedlings).

Kumari Jayasree *et al.* (2012a) attempted characterisation of embryogenic callus through histochemical localization of storage reserves. Histochemical examinations revealed the accumulation of considerable amount of storage starch, lipid and protein in embryogenic callus at later phase, whereas low levels of major storage reserves were detected in non-embryogenic callus. Kumari Jayasree *et al.* (2012b) also compared changes in protein profile during the sequential stages of SE and

Figures 1.1a-f: Plantlet Regeneration from Immature Anthers and (g) RAPD Analysis with Primer OPB20.

found more proteins were accumulated at embryo induction stage followed by embryogenic callus. Protein profiles of somatic embryos at three developmental stages displayed uniform banding pattern, however, the intensity of bands slightly decreased as the somatic embryos advanced from globular to cotyledon stage and this could be the reason for low conversion ability of somatic embryos to plants and further establishment.

From early 2000's, exploring the possibility of vegetative tissues, such as *ex vitro* and *in vitro* leaf explants as well as *in vitro* root explants, for SE induction gained momentum. Callus was induced from leaf explants of bud-grafted plants on modified MS medium supplemented with 2,4-D (1.2 mg/l), NAA (0.2 mg/l) and BA (1.0 mg/l). Upon subculture, embryogenesis could be achieved in medium containing BA (2.0 mg/l), NAA (0.2 mg/l) and GA_3 (1.0 mg/l) and the somatic embryos could be further regenerated into plantlets in hormone free ½ MS medium (Kala *et al.*, 2007). In an earlier report, callus was induced from *Hevea* leaves; however, no further growth was achieved (Carron and Enjalric, 1982). Mendanha *et al.* (1998) produced abundant callus from leaf explants, but the callus could not be regenerated further by subsequent subculturing. Later at RRII, secondary embryos were successfully induced from the hypocotyl region of leaf derived primary embryos. Embryo maturation (40 per cent) was achieved in WPM medium with BA (0.5 mg/l), kinetin (0.5 mg/l), GA_3 (0.8 mg/l) and 25 per cent of embryos were regenerated into normal plants (Kala *et al.*, 2008). Studies with different sources of leaves proved that *in vitro* derived leaves gave maximum embryogenic competence in shortest time followed by budded plants, while with mature tree derived leaves, no embryogenic potential was observed. Combined effect of $(CaNO_3)_2 \cdot 4H_2O$ (1.5 mM), sucrose (80 g/l) and phytagel (0.5 per cent) also improved the efficiency of embryogenic callus initiation (Kala *et al.*, 2009). Relationship of explant juvenility on *in vitro* culture responses was investigated. A partial (0.5 Kb) and a full length (0.8 Kb) chlorophyll A/B binding protein (*Cab*) gene was characterized from genomic and cDNA respectively. Expression analysis with RT-PCR revealed higher expression of *Cab* in seedlings and juvenile somatic plants than mature budgrafted clonal materials and field grown trees (Kala *et al.*, 2012). Very recently, Kala *et al.* (2015) improved embryo vigour by a three day slow desiccation on PEG containing medium which in turn enhanced embryo quality as well as germination frequency to 87 per cent. Embryos desiccated in medium with IAA/GA_3 and additionally supplemented with 100 mg/l phloroglucinol enhanced lateral root induction in 40 per cent of embryos.

Of late, *in vitro* derived leaf and root explants have also been proven to respond for embryogenesis. Mature leaf of somatic plant induced callus on modified MS medium containing 2,4-D (1.0 mg/l), NAA (0.4 mg/l) and kinetin (0.7 mg/l). Embryogenic callus at a frequency of 71 per cent was produced which further differentiated into embryos on WPM medium supplemented with BA (1.0 mg/l), GA_3 (1.0 mg/l) and NAA (0.2 mg/l). Thirty percent of mature embryos have been converted into plants and the survival rate was enhanced by preconditioning the plants in medium with reduced basal salts and sucrose (Sushamakumari *et al.*, 2010). Most recently, plants were developed from roots of somatic plants. Modified MS medium fortified with 0.4 mg/l 2,4-D and 0.2 mg/l kinetin induced callus that further differentiated into embryos on medium containing 2.0 mg/l BA and 1.5 mg/l GA_3. Plantlet regeneration frequency of 60 per cent could be achieved on modified WPM medium supplemented with 2.0 mg/l BA and 1.0 mg/l GA_3 and the plantlets could be successfully hardened (Sushamakumari *et al.*, 2014).

2.2.3. Limitations of Somatic Embryogenesis

Intensive research conducted at laboratories worldwide on various aspects of SE have led to a very rapid progress in refining the technique, however, due to intrinsic limitations success has been limited. Some of the problems frequently encountered with SE are i) production of embryogenic callus, which was tedious (Engelmann *et al.*, 1997) and difficult process (Kumari Jayasree *et al.*, 2012a) and ii) low rates of germination and plant conversion (Cailloux *et al.*, 1996; Sushamakumari *et al.*, 1999a). The lack or deficiency in major storage reserves of *Hevea* somatic embryos was suspected to inhibit embryo conversion into plantlets (Lardet *et al.*, 1999; Kumari Jayasree *et al.*, 2012b). Another prominent factor playing a key role in the successful SE is the nature of callus. Different callus lines with varying morphogenic potentials were induced from the same genotype, even from the same explants. Differential display analysis and molecular markers for the early diagnosis of callus with embryogenic potential were identified (Charbit *et al.*, 2004). More recently, at RRII, attempts were made on the timely identification of embryogenic callus leading to the characterization of a potent marker, SE receptor kinase (*SERK*) gene (personal communication) and the cloned partial sequence with 0.6 Kb size was deposited in Genbank (KJ451561). Age and early browning in proliferating callus also led to a loss in their embryogenic capacity (Carron *et al.*, 1992; Kumari Jayasree *et al.*, 2001a) and such browning was associated with the accumulation of phenols and ethylene (Michaux- Ferriere and Carron, 1989; Auboiron *et al.*, 1990). A recent study by Piyatrakul *et al.* (2012) corroborated the aforesaid findings of ethylene production and provided insights into the mechanism behind the embryogenic potential of callus and concluded that ethylene production and signalling genes played an important role during *Hevea* somatic embryogenesis. Finally, hardening and acclimatization is also a hindrance to large scale SE-based propagation.

2.2.4. Establishment in Soil and Field Performance

In *Hevea*, which has a life span of 30-35 years, even a small increament in yield per tap will be a great achievement and an increased growth and vigour has already been reported for plants propagated through tissue culture (Carron *et al.*, 1995). A study on root architecture of tissue culture plants derived by SE (clones PB 260 and PR107), microcuttings (clones IRCA 111, PB 235, RRIM 600) revealed that tap root and lateral root system of *in vitro* plants is similar to that of seedling root (Carron *et al.*, 2000). Somatic plants of seven clones were propagated and field planted and data showed that girth at 50 cm was 109.1-135.2 per cent and the average dry rubber yield/tap/year was 129.9- 146.3 per cent more than the donor clone during the first four years of tapping (Chen *et al.*, 2001). Field performance of IRCA 18 showed a gain of 20 per cent in production (Carron *et al.*, 2003) and clone PB 260 recorded a production gain of 16 per cent at 2nd year of tapping. Later, field trials of three clones *viz.*, somatic plant PR 107 and micropropagated plants IRCA 18 and RRIM 600, showed a gain in trunk volume from 9.93 per cent to 16.83 per cent and in dry rubber production/tree of 3.5 per cent to 32.35 per cent (Dibi *et al.*, 2010). So far, SE has been achieved from many clones and assessment of their field performance is being undertaken in different countries. In general, the plants exhibited better growth and latex yield than budgrafted plants. However, long term embryogenesis derived plants of six clones showed many abnormalities which leds to low vigour with trunk girth only 50- 80 per cent that of control plants after five to eight years in the field (Montoro *et al.*, 2012).

2.3. Suspension Culture

Cell suspension system offers a great potential including faster growth rate, direct contact with the medium nutrients, high dispensability and uniformity and good experimental reproducibility. Among tissue culture techniques, suspension culture is a very ideal tool for studies of primary and secondary metabolism, protoplast isolation, large scale propagation, germplasm storage, establishment of bioreactors and finally as target tissue for genetic transformation (Ziyun *et al.*, 2009). Wilson and Street (1975) and Wilson *et al.* (1976) have subcultured stem-derived callus to liquid medium and returning it to solid medium resulted in the production of homogenous, rapidly growing callus. Paranjothy and Ghandimathi (1976) also initiated suspension cultures with anther callus. Many researchers have used cell suspensions as starting material for their experiments *viz*, for studying the metabolism of ethephon (Audley and Wilson, 1978), protoplast isolation (Rohani an Paranjothy, 1980; Haris Ndarussamin and Dodd, 1993; Sushamakumari *et al.*, 1999b), cryopreservation and SE (Veisseire *et al.*, 1993, 1994a); however, a protocol on system establishment is very limited. More recently, Ziyun *et al.* (2009) established cell suspension culture on Modified MS medium containing 2,4-D (0.2 mg/l), NAA (2.0 mg/l), Kinetin (1.0 mg/l) along with 0.2 g/l asparagine, 0.4 g/l casein hydrolysate and coconut water (5 per cent) and under this condition, majority of cells were small in size, rich in cytoplasm and less vacuolated and cultures were multiplied and continued good growth for more than one year.

2.4. Protoplast Technology

Protoplast culture, and their fusion to obtain somatic hybrids, is a potent tool for the induction of genetic variability. Somatic hybridization or cybridization mediated by protoplast may be beneficial for rubber, since agronomically important traits are cytoplasmically controlled (Sushamakumari *et al.*, 2000b). Protoplasts are also a very convenient material for gene transfer thereby gene of desired traits can be incorporated (Thulaseedharan *et al.*, 2006). In rubber, though few attempts were made in the past on protoplast culture, regeneration of plantlets from isolated protoplasts were not successful till 2000. Rohani and Paranjothy (1980) tested leaves of various ages and attempts on pith and cell suspension culture yielded viable protoplast and cell wall regeneration was also evident. Wilson and Power (1989) isolated protoplasts from stem tissues and achieved cell division and Haris Ndarussamin and Dodd (1993) used suspension culture of anther callus for protoplast isolation. Later, Cazaux and d' Auzac (1994) used tobacco nurse culture and observed that viability and division of protoplasts could be maintained and microcalli were obtained for the first time. Cazaux and d' Auzac (1995) reported that *Hevea* stem protoplasts were found to be recalcitrant to division and increased ethylene production decreased protoplast viability. Sushamakumari *et al.* (1999b) at RRII observed that among various tissues *viz.*, leaf, young stem, friable callus, fine suspension cells of inflorescence derived callus and cell suspension was the best source for high yield of viable protoplasts. Preculture of protoplasts in liquid medium did not show mitotic division, but when the protoplast was plated in semi-solid medium, microcalli formation was achieved within one day of culture. Sushamakumari *et al.* (2000b) reported a yield of 3.1×10^7- 3.2×10^7 protoplasts g^{-1} f.

wt from suspension culture. Sustained division and growth was obtained with *Lolium multiflorum* nurse cells and microcallus was induced within two months of culture. Upon further subculturing in MS based medium, 40 per cent of callus differentiated into embryos and later germinated into plantlets. In recent years, Das and Dey (2009) proved young leaves are potential source of protoplasts as 95 per cent healthy and viable protoplast were released from young leaves and the viability was ascertained using FDA test. Most recently, Sushamakumari *et al.* (2012) isolated viable protoplasts from intact pollen grains, which on culturing in medium with 0.8 mg/l 2,4-D and 0.5 mg/l BA in the presence of *Hevea* nurse culture, led to the formation of microcolonies. Dai *et al.* (2014) regenerated plantlets from protoplast derived callus and by this procedure a highest yield of 3.6 x10^7 protoplasts/ packed cell volume was obtained from anther calli cell suspension culture. Enzyme composition, age of cell suspension, medium composition, nurse culture and plating densities were identified as key factors influencing protoplast culture.

2.5. Haploid Production

Anther and pollen culture or culture of unpollinated ovules has generated considerable interest in haploid production for crop improvement. The haploids have offered great potentials for easy production of homogygous doubled haploids and to facilitate the genomic mapping. With the view of producing haploid plants, Satchuthananthavale and Irugalbandra (1972) first cultured anther and later, Satchuthananthavale (1973), Ghandimathi and Paranjothy (1975) obtained callus and noticed a continued growth from anther and pollen grains after 4-5 weeks culture. Since then, Chen *et al.* (1978) and Chen *et al.* (1981, 1982) reported production of pollen plantlets and Chen (1984) could successfully establish the plantlets in the soil. In most of the haploid studies, mature floral buds at 3-4 mm in length were used and thus induction of both somatic and haploid callus was common, however, somatic callus proliferation was suppressed by the incorporation of 7-8 per cent sucrose and coconut water in the medium. Cytological studies revealed many aneuploidy cells in the root tips along with haploid cell; however, no diploid cell was seen (Carron *et al.*, 1989). A simultaneous effort on haploid production was continued by gynogenesis and Guo *et al.* (1982) developed plantlets by the culture of unpollinated ovules, however, the ploidy level of the regenerants was not reported. Shijie *et al.* (1990) produced plantlets from 13 clones of *Hevea* through anther culture which were field planted. Attempts at RRII on haploid production were initiated in 1990's and Asokan *et al.* (1992). regenerated plantlets from mature anther which were successfully established in the field; however, the ploidy of regenerants was not confirmed. Das *et al.* (1994)) cultured anther of several clones on medium supplemented with 6 per cent sucrose and induced callus. Later, Jayashree *et al.* (2005a) attempted haploid induction from isolated microspores and reported microcalli formation for the clone RRII 105. Pretreatment of late uninucleate to early binucleate stage anther with 0.3 M mannitol at 33°C as well as a static liquid culture promoted the division of microspores. When ovary co-culture was included in callus induction medium, the rate of cell proliferation was increased (Jayashree *et al.*, 2005a) When the microspores were pre-treated in 100 mg/l colchicine for 96 hrs, maximum callus was induced. Also higher sucrose level (180 g/l) in the presence of growth regulator kinetin and zeatin (3.0 mg/l) favoured callus growth (Jayashree *et al.*, 2005b) Very recently, *in vitro* culture of unpollinated ovlues of clones GT1, PB 217, PB 260 and PB 280 were attempted with the objective of haploidization. Callus and embryos were induced from clones PB 217, PB 260 and PB 280, however, the ploidy level of the embryos were not determined (Kouassi *et al.*, 2008).

2.6. Cryopreservation

In rubber, production of embryogenic callus was cumbersome and required more number of subcultures (Engelmann *et al.*, 1997), longer time coupled with low frequency of callus production (Lardet *et al.*, 2007). Additionally, long term maintenance of callus by regular subculturing leads to degeneration of embryogenic competence (Blanc *et al.*, 2006) Nowadays, embryogenic callus based genetic manipulation was found to be a more viable route for *Hevea* transgenic technology (Blanc *et al.*, 2006; Lardet *et al.*, 2011; Kumari Jayasree *et al.*, 2015). To overcome all these limitations, possibility of cryopreservation is to be explored. Limited research has been carried out on cryopreservation of rubber. Preliminary study on *Hevea* cryopreservation was reported by Normah *et al.* (1986) with zygotic embryos. When excised embryogenic axes were desiccated for 2-3 hrs, followed by stepwise cooling or direct immersion into liquid nitrogen and rapid thawing, embryogenic axes could withstand at a survival rate of 69-71 per cent. Later, efforts were concentrated with embryogenic callus and Veisseire *et al.* (1993) successfully froze embryogenic cell suspensions of one commercial clone and discussed the pregrowth and preculture conditions on the survival of callus. Subsequently, Engelmann *et al.* (1997) attempted cryopreservation of embryogenic callus of two clones, either using a classical freezing or a simplified freezing processes. Under optimal conditions, higher survival and rapid regrowth of callus was observed. Control and cryopreserved callus revealed no significant difference in somatic embryo induction efficiency following both freezing protocols. Lardet *et al.* (2007) developed a cryopreservation technique for embryogenic callus and effects of $CaCl_2$ concentration was thoroughly studied; reduction of $CaCl_2$ in preculture medium from 9 mM to 1 mM or removal of $CaCl_2$ before cryopreservation promoted regrowth of callus. Also, it was found that low level of $CaCl_2$ (1 mM) was found to be optimum for embryogenesis from cryopreserved callus. Zhou *et al.* (2012) reported a simple vitrification procedure by which the callus was successfully cryopreserved in liquid nitrogen and subsequently regenerated into plantlets. By this procedure, the viability of callus after cryopreservation was maximum when precultured for three days and loading time for 20 min and dehydration time for 40 min was found to be optimum for regrowth of callus. A survival of 71.1 per cent viability was obtained from cryopreserved callus. At RRII, initial attempts to develop a suitable cryogenic method with embryogenic callus are underway (personal communication).

2.7. *In vitro* Zygotic Embryo Culture

In vitro embryo culture has emerged as an important tool for crop improvement where interspecific and intergeneric crosses could be developed and hybrid embryos could be rescued. Moreover, introduction of polyembryony, *in vitro* fertilization (IVF) technology and further rescuing of embryos could greatly increase the propagation of true-to-type seedling.

2.7.1. *In vitro* Fertilization and Embryo Rescue

In rubber, one of the earliest applications of *in vitro* culture technique was in embryology, when Muzik (1956) experimented mature zygotic embryo culture and found that mature embryos developed rapidly, while immature embryos failed. Thereafter, Paranjothy and Ghandimathi (1976) used mature embryogenic axes, which usually fail to grow unless a portion of the cotyledon is attached. Toruan and Suryatmana (1977) used decotyledonized embryos and regenerated seedlings in MS medium which were planted in soil. Paranjothy *et al.* (1978) developed seedlings by culturing freshly fallen seeds and after three weeks maintenance in culture, seedlings were established in soil. At RRII, an attempt on IVF resulted, at the first time, in successful fertilization and embryo formation. *In vitro* ovular pollination and flowers, one day before anthesis, was identified as the suitable method and optimum stage for *in vitro* pollination. MS medium containing complex organic supplements promoted ovule/embryo development and fertilized ovules were grown upto three months and embryos upto heart shaped stage. Anatomical studies with 90 day old *in vitro* raised ovule revealed resemblance to a mature seed with a characteristics mottling (Figure 1.2a) in the surface (Rekha *et al.*, 2002). Later, studies were directed to develop a protocol for rescuing of immature embryos. In this study, half *ovulo* embryo culture technique was applied where the ovules were cut into two halves and inoculated by touching the micropylar end in medium. Embryos could be rescued from five week old immature fruits and the embryo recovery was enhanced to 42 per cent and the plantlets could be field planted (Figures 1.2b-e) (Rekha *et al.*, 2010).

2.7.2. Induction of Zygotic Polyembryony

Self-rooted plants with genetic homogeneity could minimize the influence of rootstock and hence the induction of polyembryony, where more than one embryo could be induced from one seed, with zygotic origin is an ideal method for uniform rootstock production. Based on previous observations, and recent works initiated at RRII for rescuing of embryos, open pollinated fruits with different maturity were experimented for polyembryony induction. Fruits with age of 8-10 weeks and half ovulo embryo culture were found to be most suited for inducing polyembryony. Nistch medium supplemented with a combination of GA_3 (2.0 mg/l), Kinetin (3.0 mg/l) and zeatin (0.3 mg/l) along with very high sucrose (100 g/l) enhanced the number of embryos from a single zygote. Embryos were successfully regenerated and due to the better rooting capacity, plants showed a higher survival rate during hardening and established in the field (Figure 1.3) (Rekha *et al.*, 2015). The RAPD profiles of all seedlings were different from their maternal parent and moreover, a similar allelic segregation pattern for all plants proved that all plants were derived from a single zygote. Studies with epigenetic markers confirmed the absence of genetic variability and the lack of variation in the *Msp*1 digests in all primer combination revealed the existence of no methylation variation across the multiple seedlings and all the data supported that the polyembryony derived seedlings exhibited genetic and epigenetic stability (Rekha *et al.*, 2015).

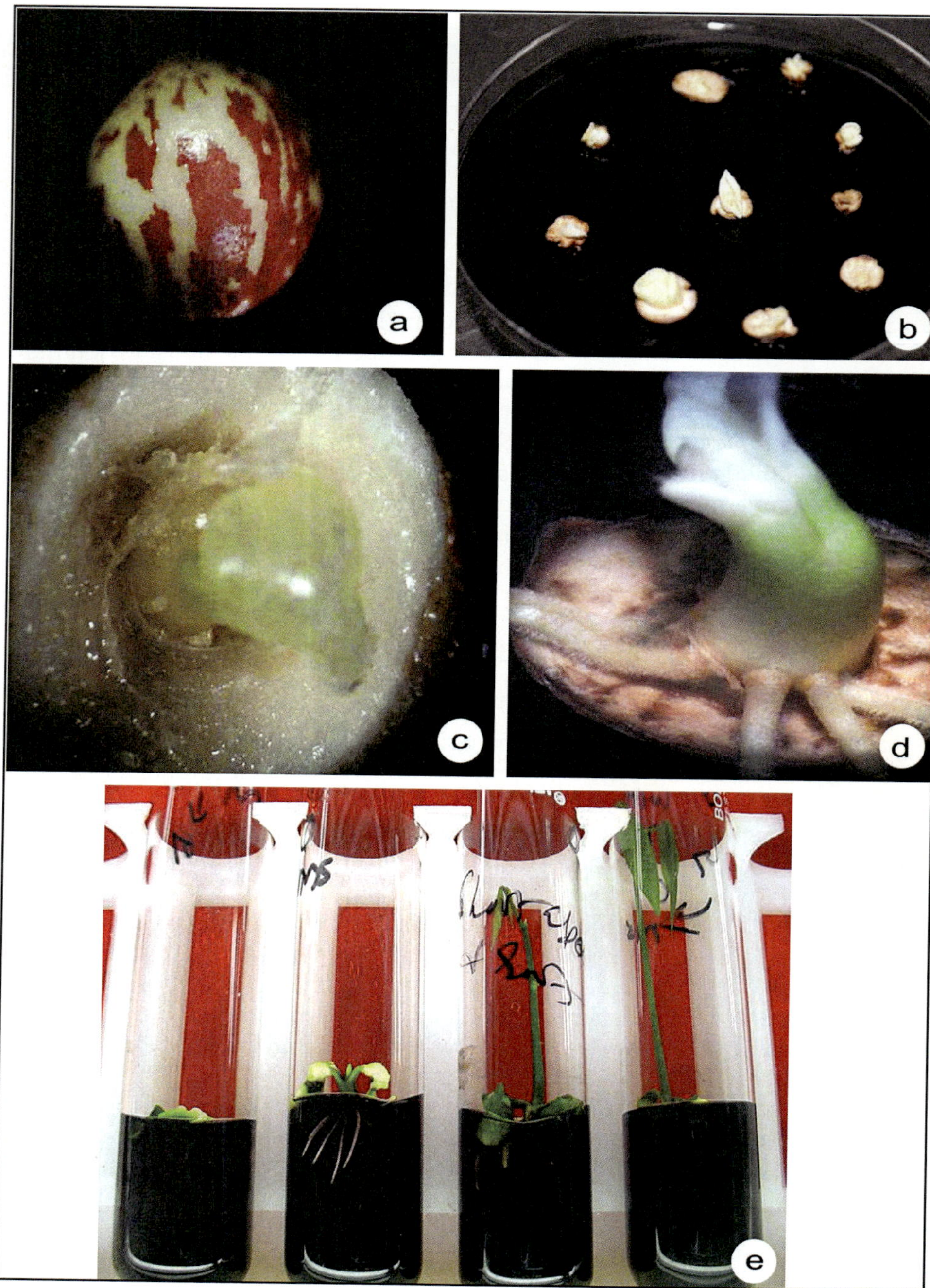

Figure 1.2: Different Stages of Plant Regeneration via Embryo Rescue: (a) Seed development, (b) Ovule culture, (c) Developing embryo, (d) Root differentiation and (e) Plantlet regeneration.

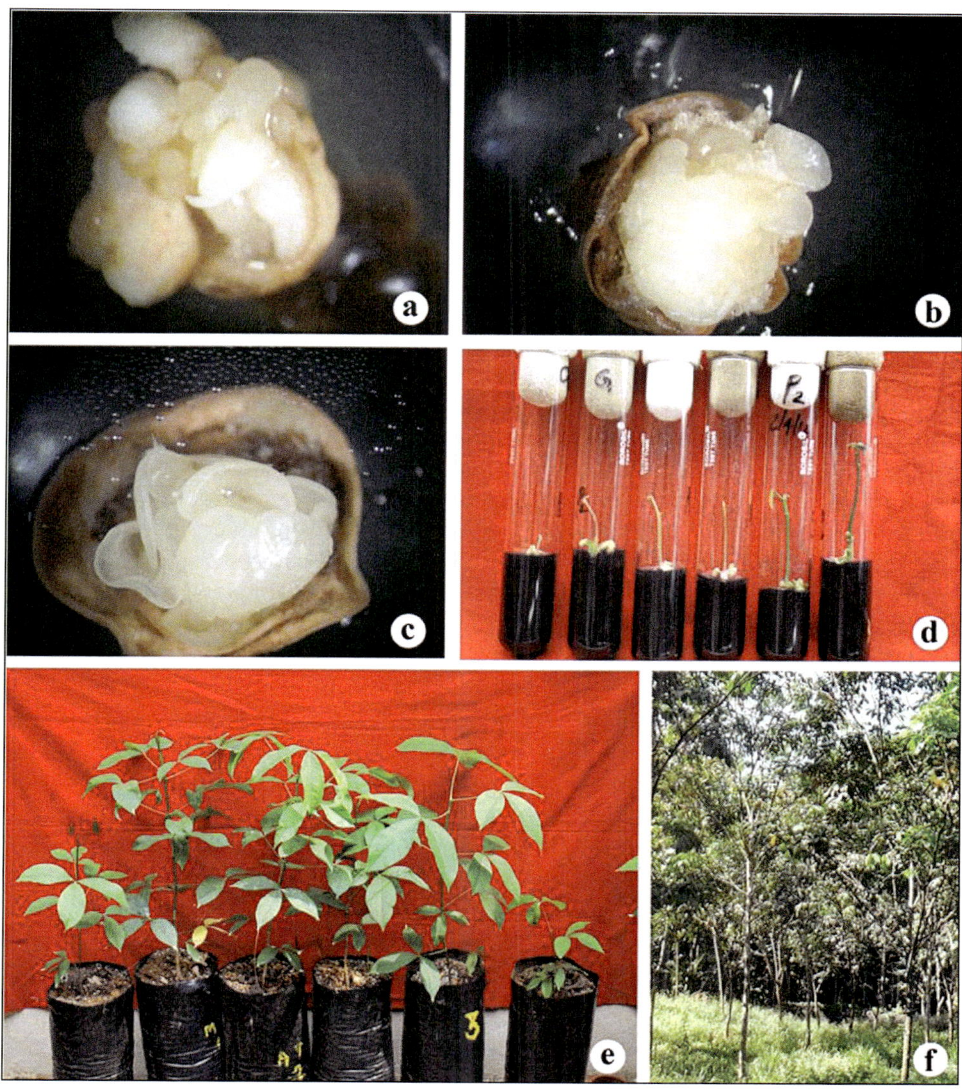

Figure 1.3: Induction of Polyembrony and Plantlet Regeneration.

3. Conclusion

From early 1970's till date, much progress has been made on tissue culture of rubber, which is evident from the publication of a large number of research papers. *In vitro* multiplication through micropropagation would be highly advantageous for obtaining uniform own rooted plants. Extensive studies across the world proved the recalcitrant nature of the crop and therefore *in vitro* rooting of mature explants limits the utility of the technique at a larger scale. A steady increase in research activites was noted on the induction of SE from various explant sources for a number of genotypes. Many factors affecting SE is being controlled to some extent, however, protocol for regeneration at commercial scale is still lacking. Limitations of SE were well described by several researches and studies on molecular mechanism are underway for the characterization of genes directly associated with somatic embryogenesis. Suspension culture of *Hevea* callus was established and a high yield of viable protoplast was isolated and microcalli successfully induced, but research impetus is needed in the field of plant regeneration. Attempts on haploids has generated a considerable interest, however, intensive efforts are required on plantlet development and further ploidy confirmation of regenerants. Long term preservation of embryogenic callus, either for propagation or ensuring the constant supply of target material of transformation, is yet another approach and promising results were achieved with cryopreservation. *In vitro* fertilization has been successfully performed in *Hevea* and the advances in IVF and embryo rescue method are opening the scope of genetic diversity. Recent studies on the induction of polyembryony and molecular analysis proved the scope for rootstock production with the entire characteristic feature of seedlings. This technique could also complement transgenic technology particularly for the production of drought tolerant rootstock which otherwise are difficult to produce conventionally. In spite of all these achievements, more investigations are needed to successfully integrate *in vitro* approaches with conventional breeding progrmammes for rubber genetic improvement.

Acknowlegement

The author thanks Dr. James Jacob, Director of Research, RRII, for permission to publish the chapter and for helpful suggestions. Thanks are also due to Dr. Rekha K. for providing the images presented in Figures 1.2 and 1.3 and all colleagues of Biotechnology Division for kind co-operation.

References

Asokan, M. P., Sobhana, P., Sushamakumari, S. and Sethuraj, M.R. (1988).. Tissue culture propagation of rubber (*Hevea brasiliensis* Wild ex Adr. De Juss. Muell. Arg.) clone GT1. *Indian Journal of Natural Rubber Research*, **1:** 10-12.

Asokan, M. P., Kumari Jayasree, P. and Sushamakumari, S. (1992). Plant regeneration by somatic mebryogenesis of rubber tree (*Hevea brasiliensis*). Paper presented at the International Natural Rubber Conference, Bangalore, India, pp. 49.

Asokan, M. P., Kumari Jayasree, P., Sushamakumari, S. and Sobhana, P. (1992). Plant regeneration from anther culture of rubber tree (*Hevea brasiliensis*). Paper presented at the International Natural Rubber Conference, Bangalore, India. pp. 51.

Asokan, M. P., Narasimhan, K., Kumari Jayasree, P., Thulaseedharan, A., Sushamakumari, S., Kala, R.G. and Jayashree, R. (2001). Isozyme markers for distinguishing embryogenic and non-embryogenic callus during somatic embryogenesis in *Hevea brasiliensis* (Muell.) *Arg. Journal of Tree Sciences*, **20:** 78-85.

Asokan, M. P., Kumari Jayasree, P., Thomas, V., Sushamakumari, S., Kala, R.G., Kala, R.G., Jayashree, R., Sethuraj, M.R. and Thulaseedharan, A. (2002). Influence of 2,4-D and sucrose on repetitive embryogenesis in rubber (*Hevea brasiliensis* Muell. Arg. cv.GT1). *Journal of Tree Sciences*, **21:** 18-26.

Auboiron, E., Carron, M.P. and Michaux-Ferriere, N. (1990). Influence of atmospheric gases particularly ethylene on somatic embryogenesis of *Hevea brasiliensis. Plant Cell Tissue Organ Culture*, **21:** 31-37.

Audley, B.G. and Wilson, H.T. (1978).. Metabolism of 2-chloroethylphosphonic acid (ethephon) in suspension cultures of *Hevea brasiliensis. Journal of Experimental Botany*, **29:** 50-53.

Blanc, G., Michaux-Ferriere, N., Teisson, C., Lardet, L. and Carron, M.P. (1999). Effects of carbohydrate addition on the induction of somatic embryogenesis in *Hevea brasiliensis. Plant Cell Tissue Organ Culture*, **59:** 103-112.

Blanc, G., Lardet, L., Martin, A., Jacob, J.L. and Carron, M.P. (2002). Differential carbohydrate metabolism conducts morphogenesis in embryogenic callus of *Hevea brasiliensis* (Muell. Arg.). *Journal of Experimental Botany*, **53:** 1453-1462.

Blanc, G., Baptiste, C., Oliver, G., Martin, F. and Montoro, P. (2006). Efficient *Agrobacterium-tumefaciens* mediated transformation of embryogenic calli and regeneration of *Hevea brasiliensis* Muell. *Arg. Plants. Plant Cell Reports*, **24(2):** 724-733.

Bouychou, J.G. (1953).. La culture *in vitro* des tissuess d' *Hevea. Archives of Rubber Cultivation*, **30:** 50-53.

Cailloux, F., Guerrier, J., Linossier, L. and Coudret, A. (1996). Long term somatic embryogenesis and maturation of somatic embryos in *Hevea brasiliensis. Plant Science*, **120:** 185-196.

Carron, M.P. (1981). Germination *in vitro* d' embryons immatures d' hevea. *Caoutchoucs et Plastiques*, **612:** 93.

Carron, M.P. and Enjalric, E. (1982).. Studies on vegetative micropagation of *Hevea brasiliensis* by somatic embryogenesis and *in vitro* on microcutting. In: Fujiwara, A. (Ed.), *Plant Tissue Culture*. Maruzen, Tokyo, Japan. pp. 751-752.

Carron, M.P. and Enjalric, E. (1983).. Prospects for the mocropropagation of *Hevea brasiliensis. Revue Generale des Caoutchoucs et Plastiques*, **628:** 65-68.

Carron, M.P. and Enjalric, E. (1985). Somatic embryogenesis from inner integument of the seed of *Hevea brasiliensis* (Muell. Arg.). Comptes Rendus de L Academic des Sciences, Paris, Series III, **300:** 653- 658.

Carron, M.P., Enjalric, E., Lardet, L. and Deschamps, A. (1989). Rubber (*Hevea brasiliensis* Muell. Arg.). In: Bajaj, Y.P.S. (Ed.), *Biotechnology in Agriculture and Forestry II* (*Vol. 5*). Springer, Berlin. pp. 222- 245.

Carron, M.P., d Auzac, J., Etienne, H., El-Hadrami, I., Houst, F., Michaux- Ferriere, N. and Montoro, P. (1992). Biochemical and histological features of somatic embryogenesis in *Hevea brasiliensis. Indian Journal of Natural Rubber Research*, **5(1 and 2):** 7-17.

Carron, M.P., Etienne, H., Michaux-Ferriere, N. and Montoro, P. (1995). Somatic embryogenesis in rubber tree (*Hevea brasiliensis* Muell. Arg.). In: Bajaj, Y.P.S. (Ed.), *Biotechnology in Agriculture and Forestry* (*Vol. 30*). Springer, Berlin. pp. 353- 369.

Carron, M.P., le Roux, Y., Tison, J., Dea, B.C., Caussanel, V., Clair, J. and Keli, J. (2000). Compared root architectures in seedlings and *in vitro* plantlets of *Hevea brasiliensis* in the initial years of growth in the field. *Plant and Soil*, **223:** 73-85.

Carron, M.P., Lardet, L., Leconte, A., Boko, C., Dea, G.B. and Keli, J. (2003). Field growth and rubber yield of *Hevea brasiliensis* (Muell. Arg.) from budded versus *in vitro* micropropagated plants from clone IRCA 18. *Acta Horticulturae*, 616: 283- 293.

Cazaux, E. and d' Auzac, J. (1994).). Microcallus formation from *Hevea brasiliensis* protoplast isolated from embryogenic callus. *Plant Cell Reports*, **13:** 272-276.

Cazaux, E. and d' Auzac, J. (1995). Explanation for the lack of division of protoplast from stems of rubber tree (*Hevea brasiliensis*). *Plant Cell Tissue Organ Culture*, **41:** 211- 219.

Charbit, E., Legavre, T., Lardet, L., Bourgeois, E., Michaux-Ferriere, N. and Carron, M.P. (2004). Identification of differentially expressed cDNA sequences and histochemical characteristics of *Hevea brasiliensis* calli in relation to their embryogenic and regenerative capacities. *Plant Cell Reports*, **22:** 539-548.

Chen, C., Chen, F., Chein, C., Wang, X., Chang, S., Hu, H., Ho, Y. and Li, U.T. (1979). A process of obtaining pollen plants of *Hevea brasiliensis* Muell. Arg. Scientia Sinica, **XXII:** 81-90.

Chen, Z., Qian, C., Qin, M., Wang, C., Suo, C., Xiao, Y. and Xu, X. (1981). Relationship between somatic cells and microspores in the process of anther culture of *Hevea brasiliensis* Muell. *Arg. Acta Botanica*, **1:** 1-7.

Chen, Z., Qian, C., Qin, M., Xu, X. and Xiao, Y. (1982). Recent advances in anther culture of *Hevea brasiliensis* (Muell. Arg.). *Theoretical and Applied Genetics*, **62:** 103-138.

Chen, Z. (1984). Rubber (*Hevea*). In: Sharp, W.R., Evans, DA., Ammirato, P.V. and Yamada Y. (Eds.), *Handbook of Plant Cell Culture* (Vol. 2). MacMillan Publishers, New York. pp. 546-571.

Chen, X., Wang, Z., Wu, H. and Zhang, Z. (2001). Selection of optimum planting material of *Hevea brasiliensis:* self rooting juvenile clone. In: Sainte-Beuve, *J. (Ed.), IRRDB Natural Rubber Conference, Montepiller*, France, CIRAD.

Chua, S.E. (1966) Studies on tissue culture of *Hevea brasiliensis*: Role of osmotic concentration, carbohydrate and pH value in induction of callus growth in plumule tissue from rubber seedling. *Journal of Rubber Research Institute of Malaya*, **19(5):** 272-276.

Clombe, J.C. (1975).. Demonstration of intraclonal variability in young graft trees. *Revue Generale des Caoutchoucs et Plastiques*, **52:** 91-94.

Dai, X.M., Li, Z., Hua, Y.W., Huang, T.D., Sun, A.H., Zhou, Q.N. and Huang, H.S. (2014). Plant regeneration from protoplast culture of Reyan 8-79 *Hevea brasiliensis* (Muell. Arg.). *Journal of Southern Agriculture*, **45 (12):** 2040-2045.

Das, K., Sinha, R.R., Potty, S.N. and Sethuraj, M.R. (1994).). Embryogenesis from anther derived callus of *Hevea brasiliensis* (Muell. Arg.). *Indian Journal of Hill Farming*, **7:** 90-95.

Das, K. and Dey, S. K. (2009). Isolation of protoplast from leaf mesophyll cells of *Hevea brasiliensis. Natural Rubber Research*, **22(1 and 2):** 93-98.

Dibi, K., Boko, C., Obouayeba, S., Gnagna, M., Dea, G.B., Carron, M.P. and Anno, A.P. (2010). Field growth and rubber yield of *in vitro* micropropagated plants of clones PR 107, IRCA 18 and RRIM 600 of *Hevea brasiliensis* (Muell.Arg.). *Agriculture and Biology Journal of North America*, **1(6):** 1291-1298.

Dijkman, M.J. (1951).. *Hevea*: Thirty years of research in in the Far East. University of Miami Press, Florida. pp. 5-7.

EI Hadrami, I., Michaux- Ferriere, N., Carron, M.P. and d' Auzac, J. (1989). Polyamines a possible limiting factor in somatic embryogenesis of *Hevea brasiliensis. Comptes Rendus de L Academic des Sciences*, **308:** 205-211.

EI Hadrami, I., Carron, M.P. and d' Auzac, J. (1991).. Influence of exogenous hormone on somatic embryogenesis of *Hevea brasiliensis. Annals of Botany*, **67:** 511-15.

EI Hadrami, I. and d' Auzac, J. (1992). Effects of polyamine biosynthetic inhibitors on somatic embryogenesis and celluar polyamines in *Hevea brasiliensis. Journal of Plant Physiology*, **140:** 33-36.

Engelmann, F., Lartaud, M., Chabrillange, N., Carron, M.P. and Etienne, H. (1997). Cryopreservation of embryogenic callus of two commercial clones of *Hevea brasiliensis. Cryo Letters*, **18:** 107-116.

Etienne, H., Berger, A. and Carron, M.P. (1991a). Water status of callus from *Hevea brasiliensis* during somatic embryogenesis. *Physiologia Plantarum*, **82:** 213-218.

Etienne, H., Montoro, P. and Carron, M.P. (1991b) Incidence des parameters hydriques sure le development des cals d' *Hevea brasiliensis en culture in vitro. Annales des Sciences Forestieres*, **48:** 253-265.

Etienne, H., Montoro, P., Michaux – Ferriere, N. and Carron, M.P. (1993a). Effects of desiccation, medium osmolarity and abscisic acid on the maturation of *Hevea brasiliensis* somatic embryos. *Journal of Experimental Botany*, **44:** 1613-19.

Etienne, H., Sott, B., Montoro, P., Miginiac, E. and Carron, M.P. (1993b) Relations between exogenous growth regulators and endogenous indole-acetic acid and abscisic acid with expression of somatic embryogenesis in *Hevea brasiliensis* Muell. *Arg. Plant Science*, **88:** 91-96.

Etienne, H., Lartaud, M., Carron, M.P. and Michaux-Ferriere, N. (1997a). Use of calcium to optimize long term proliferation of friable embryogenic calluses and plant regeneration in *Hevea brasiliensis* (Muell. Arg.). *Journal of Experimental Botany*, **48:** 129-37.

Etienne, H., Lartaud, M., Michaux – Ferriere, N., Carron, M., Berthouly, M. and Teisson, C. (1997b). Improvement of somatic embryogenesis in *Hevea brasiliensis* (Muell. Arg.) using the temporary immersion technique. *In Vitro Cellular Developmental Biology*- Plant, **33(2)**: 81-87.

Ghandimathi, H. and Paranjothy, K. (1975).. Anther culture attempts at induction of haploidy in *Hevea* and other plants. *In: Proccedings of National Plant Tissue Culture, Kuala Lumpur,* Malaysia. pp. 32-35.

Guo, G., Jia, X. and Chen, L. (1982).. Induction of plantlets from ovules *in vitro* of *Hevea brasiliensis. Hereditas,* **4(1):** 27-28.

Gunatilleke, I.D. and Samaranayake, G. (1988). Shoot tip culture as a method of micropropagation of *Hevea. Journal of Rubber Research Institute of Sri Lanka,* **68:** 33-44.

Haris Ndarussamin, A. and Dodd, W.A. (1993). Isolation of rubber tree *Hevea brasiliensis* (Muell. Arg.) protoplasts from callus and cell suspensions. *Menara-Perkebunan,* **61:** 25-31.

Hua, Y.W., Huang, T.D. and Huang, H.S. (2010). Micropropagation of self rooting juvenile clones by secondary somatic embryogenesis in *Hevea brasiliensis. Plant Breeding,* **129:** 202-207.

Hui, Z., Ming, P., Xu, W., HuiCa, Z. and Xiong Ting, C. (2009). Micropropagation of rubber tree (*Hevea brasiliensis*) by employing mature stem as explants. *Genomics and Applied Biology,* **28(6):** 1169-1176.

Ighere Dickson, A., Okere, A., Elizabeth, J., Mary, O., Olatunde, F. and Abiodun, S. (2011). *In vitro* culture of *Hevea brasiliensis* (rubber tree). *Journal of Plant Breeding and Crop Science,* **3(9):** 185-189.

Jayashree, R., Rekha, K., Sushamakumari, S. and Thulaseedharan, A. (2005a). Establishment of callus cultures from isolated microspores of *Hevea brasiliensis. Paper presented at ICAR National Symposium on Biotechnological Interventions for Improvement of Horticultural Crops, Issues and Strategies,* Trichur, India, pp. 385-390.

Jayashree, R., Rekha, K., Kumari Jayasree, P., Sanju, R. and Thulaseedharan, A. (2005b). Effect of colchicine on callus induction from isolated microspores in *Hevea brasiliensis. Plant Cell Biotechnology and Molecular Biology,* **8 (1 and 2):** 89-92.

Kala, R.G., Asokan, M.P., Kumari Jayasree, P., Sobha, S., Jayashree, R., Rekha, K. and Thulaseedharan, A. (2002). Optimization of conditions for *in vitro* micrografting in rubber (*Hevea brasiliensis*). *Indian Journal of Natural Rubber Research,* **15(2):** 165-171.

Kala, R.G., Leda, P. and Thulaseedharan, A. (2004). Effect of fungicides and antibiotics to control microbial contamination in *Hevea cultures. Plant Cell Biotechnology and Molecular Biology,* **5:** 51-58.

Kala, R.G., Kumari Jayasree, P., Sushamakumari, S., Jayashree, R., Rekha, K., Sobha, S. and Thulaseedharan, A. (2007). *In vitro* regeneration of *Hevea brasiliensis* from leaf explants. In: Kesavachandran R. *et al.* (Eds.), *Recent Trends in Horticultural Biotechnology,* New India Publishing Agencies, New Delhi, India. pp. 223-228.

Kala, R.G., Kuruvilla, L., Kumari Jayasree, P., Sushamakumari, S., Jayashree, R., Rekha, K., Sobha S. and Thulaseedharan, A. (2008). Secondary somatic embryogenesis and plant regeneration from leaf derived somatic embryos of *Hevea brasiliensis. Journal of Plantation Crops,* **36(3):** 218-222.

Kala, R.G., Gimisha, G.C., Kumari Jayasree, P., Sushamakumari, S., Sobha, S., Jayashree, R., Rekha, K., and. Thulaseedharan, A. (2009). Somatic embryogenesis in leaf cultures of *Hevea brasiliensis*: Effect of explant source plant. *Natural Rubber Research,* **22 (1 and 2):** 117-126.

Kala, R.G., Supriya, R., Sunie, A.M., Anju, V., Kumari Jayasree, P., Sobha, S. and Thulaseedharan, A. (2012). Chlorophyll A/B binding protein gene expression in juvenile and mature leaf explants and its relationship with *in vitro* culture response in *Hevea brasiliensis. Paper presented at International Rubber Conference,* Kovalam, Kerala, India, p. 90.

Kala, R.G, Tisha, L.T., Sobha, S., Jayashree, R., Leda, P. and Thulaseedharan, A. (2015). Somatic embryo germination in *Hevea brasiliensis.* Effect of embryo dessiccation, *phytohormones and phloroglucinol. Rubber Science,* **28(1):** 52-61.

Kouassi, K., Koffi, K.E. Gnagne, Y.M. N'nan, O., Coulibaly, Y. and Sangare, A. (2008). Production of *Hevea brasiliensis* embryos from *in vitro* culture of unpollinated ovules. *Biotechnology,* **7(4):** 793-797.

Kumari Jayasree, P., Asokan, M.P., Sobha, S., Sankariammal, L., Rekha, K., Kala, R.G., Jayashree, R. and Thulaseedharan, A. (1999). Somatic embryogenesis and plant regeneration from immature anthers of *Hevea brasiliensis* (Muell. Arg.). *Current Science,* **76:** 1242-1245.

Kumari Jayasree, P., Thomas, V., Saraswathyamma, C.K. and Thulaseedharan, A. (2001). Optimization of parameters affecting somatic embryogenesis in *Hevea brasiliensis. Indian Journal of Natural Rubber Research,* **14:** 20-29.

Kumari Jayasree, P. and Thulaseedharan, A. (2001). Gibberllic acid related embryo induction and germination in *Hevea brasiliensis. Indian Journal of Natural Rubber Research,* **14:** 106-111.

Kumari Jayasree, P. and Thulaseedharan, A. (2004). Initiation and maintenance of long term somatic embryogenesis in *Hevea brasiliensis. Paper presented at IRRDB Biotechnology Workshop,* Kuala Lumpur, Malaysia, p. 56.

Kumari Jayasree, P. and Thulaseedharan, A. (2005). *In vitro* germination of *Hevea* somatic embryos: Effect of cytokinins. *Plant Cell Biotechnology and Molecular Biology,* **6:** 61-64.

Kumari Jayasree, P., Thomas, V. and Thulaseedharan, A. (2009). Anatomical changes during acclimatization in somatic embryogenesis derived plants of *Hevea brasiliensis. Natural Rubber Research,* **22 (1 and 2):** 156-162.

Kumari Jayasree, P., Sajeevan, R.S., Supriya, R., Rekha, K., Kala, R.G., Jayashree, R. and Thulaseedharan, A. (2010). Genetic analysis of somatic embryo derived plants of *Hevea brasiliensis* (clone RRII 105). *Natural Rubber Research,* **23 (1 and 2):** 20-27.

Kumari Jayasree, P., Reghu, C.P., Kala, R.G. and Thulaseedharan, A. (2012a). Histochemical changes in embryogenic and non-embryogenic calli *of Hevea brasiliensis*. *Natural Rubber Research*, **25 (1):** 86-90.

Kumari Jayasree, P., Sajeevan, R.S. and Thulaseedharan, A. (2012b). Changes in protein profile during different developmental stages of somatic embryogenesis in *Hevea brasiliensis. Rubber Science*, **25(2):** 183-188.

Kumari Jayasree, P., Divya, S., Supriya, R. and Thulaseedharan, A. (2015). *Agrobacterium*-mediated transformation of *Hevea brasiliensis* with apple cDNA encoding sorbitol-6-phosphate dehydrogenase. *Rubber Science*, **28(1):** 31-39.

Lardet, L., Aguilar, M.E., Michaux-Ferriere, N. and Berthouly, M. (1998). Effect of strictly plant related factors on the response of *Hevea brasiliensis* and *Theobroma cacao* nodal explants cultured *in vitro. In Vitro Cellular Developmental Biology-Plant*, **34:** 34-40.

Lardet, L., Piombo, G., Orioi, F., Deschamps, A. and Carron, M.P. (1999). Relations between biochemical characteristics and conversion ability in *Hevea brasiliensis* zygotic and somatic embryos. *Canadian Journal of Botany*, **77:** 1168-1177.

Lardet, L., Martin, F., Dessailly, F., Carron, M.P. and Montoro, P. (2007). Effect of exogenous calcium on post-thaw growth recovery and subsequent plant regeneration of cryopreserved embryogenic calli of *Hevea brasiliensis* (Muell. Arg.). *Plant Cell Reports*, **26:** 559-569.

Lardet, L., Dessailly, F., Carron, M.P., Rio, M.A., Michaux-Ferriere, N. and Montoro, P. (2008a). Secondary somatic embryogenesis in *Hevea brasiliensis* (Muell. Arg.): An alternative process for long term somatic embryogenesis. *Journal of Rubber Research*, **12(4):** 215-228.

Lardet, L., Dessailly, F., Carron, M.,P, Montoro, P. and Monteuuis, O. (2008b). Influences of aging and cloning methods on the capacity for somatic embryogenesis of a mature *Hevea brasiliensis* genotype. *Tree Physiology*, **29:** 291-298.

Lardet, L., Leclercq, J., Benistan, E., Dessailly, F., Oliver, G., Martin, F. and Montoro, P. (2011). Variation in GUS activitiy in vegetatively propagated *Hevea brasiliensis* transgenic plants. *Plant Cell Reports*, **30:** 1847-1856.

Linossier, L., Veisseire, P., Cailloux, F. and Coudret, A. (1997). Effect of abscisic acid and high concentration of PEG on *Hevea brasiliensis* somatic embryos development. *Plant Science*, **124:** 183-91.

Martre, P., Lacan, D., Just, D. and Teisson, C. (2001). Physiological effects of temporary immersion on *Hevea* brasiliensis callus. *Plant Cell Tissue Organ Culture*, **67:** 25-35.

Mendanha, A.B.I., de Almeida Torres, R.A. and de Barros Freire, A. (1998). Micropropagation of rubber tree (*Hevea brasiliensis* Muell. Arg.). *Genetics and Molecular Biology*, **21:** 395-398.

Michaux-Ferriere, N. and Carron, M.P. (1989). Histology of early somatic embryogenesis in *Hevea brasiliensis:* The importance of the timimg of subculturing. *Plant Cell tissue Organ Culture*, **33:** 331-338.

Michaux-Ferriere, N., Grout, H. and Carron, M.P. (1992). Origin and ontogenesis of somatic embryos in *Hevea brasiliensis* (*Euphorbiaceae*). *American Journal of Botany*, **79(2):** 174- 180.

Montoro, P., Etienne, H., Michaux-Ferriere, N. and Carron, M.P. (1993). Callus friability and somatic embryogenesis in *Hevea brasiliensis. Plant Cell Tissue Organ Culture*, **33:** 331-38.

Montoro, P., Etienne, H. and Carron, M.P. (1995). Effect of calcium on callus friability and somatic embryogenesis in *Hevea brasiliensis* (Muell. Arg.): Relations with callus mineral nutrition, nitrogen metabolism and water parameters. *Journal of Experimental Botany*, **46:** 255-61.

Montoro, P., Carron, M.P., Granet, F., Lardet, L., Leclercq, J., Dessailly, F, Martin, F., Gaurel, S., Uche, E., Rio, M. and Oliver, G. (2012). Development of new varietal types based on rejuvenation by somatic embryogenesis and propagation by conventional budding or microcutting in *Hevea brasiliensis*. In: Geelan, D. (Ed.), *Proceedings of Seventh International Symposium on In Vitro Cuture and Horticultural Breeding*. Ghent, Belgium. pp. 2-25

Muzik, T.J. (1956). Studies on the development of of the embryo and seed of *Hevea brasiliensis* in culture. *Lloydia*, **19:** 86-91.

Nayanakantha, N.M.C. and Seneviratne, P. (2007). Tissue culture of rubber: Past, present and future prospects. *Ceylon Journal of Sciences*, **36(2):** 116-125.Mok DWS.

Nor Mayati, C.H. and Jamnah, A.R. (2014). Induction of shoots and roots from vegetative tissue culture of *Hevea brasiliensis* RRIM 2020. *Journal of Tropical Plant Physiology*, **6:** 1-9.

Normah, M.N., Chin, H.F. and Hor, Y.L. (1986). Desiccation and cryopreservation of embryogenic axes of *Hevea brasiliensis Muell. Arg. Pertanika*, **9:** 299-303.

Paranjothy, K. (1974). Induced root and embryoid differentiation in *Hevea* tissue culture. *Presented at 3rd International Congress Plant Tissue Cell Culture*, University, Leicester, Abstract 67.

Paranjothy, K. and Ghandimathi, H. (1975). Morphogenesis in callus cultures *Hevea brasiliensis* Muell. *Arg. Proceedings of National Plant Tissue Culture Symposium*, Kuala Lumpur, pp, 19-25.

Paranjothy, K. and Ghandimathi, H. (1976). Tissue and organ culture of *Hevea*. Proceedings' of International *Rubber* Conference 1975 (Vol 2), Kuala Lumpur, Malaysia, pp, 59-84.

Paranjothy, K. and Rohani, O. (1978). Embryoid and plantlet development from cell culture of *Hevea*. 4th International Congress *Plant Tissue Cell Culture*, University, Calgary Abstract 134.

Paranjothy, K., Lim, T.M., Rohani, O., Ismail, H., Aziz, S. and Kwai, L.S. (1978). Transfer of *Hevea* seed germplasm. *IRRDB Meet Technical Experts*, Kuala Lumpur.

Perrin, Y., Lardet, L., Enjalric, F. asnd Carron, M.P. (1994).) Rajeuissement de clones matures d'(*Hevea brasiliensis*) Muell. *Arg. Par microgreffage in vitro. Canadian Journal of Plant Science*, **74:** 623-630.

Piyatrakul, P., Putranto, R.A., Martin, F., Rio, M.A., Dessailly, F., Leclercq, J., Dufayard, J.F., Lardet, L. and Montoro, P. (2012). Some ethylene biosynthesis and AP2/ERF genes reveal a specific pattern of expression during somatic embryogenesis in *Hevea brasiliensis. BMC Plant Biology*, **12:** 2 44-**5.**

Rekha, K., Jayashree, R., Thomas, V., Kumari Jayasree, P., Kala, R.G., Sobha, S., Sushamakumari, S. and Thulaseedharan, A. (2002). *In vitro* fertilization in *Hevea brasiliensis:* A preliminary investigation. *Proceedings of PLACRYOSYM XV*. pp. 239-245.

Rekha, K., Jayashree, R., Gireesh, T., Sushamakumari, S., Jomini, K.T., Priya, S.A., Kumari Jayasree, P., Sobha, S., Kala, R.G. and Thulaseedharan, A. (2010). Embryo rescue and plant regeneration in *Hevea brasiliensis. Natural Rubber Research*, **23 (1 and 2):** 47-54.

Rekha, K., Thomas, K.U., Sobha, S., Divya, U.K., Saha, T. and Sushamakumari, S. (2015). Genetic and epigenetic uniformity of polyembryony derived multiple seedlings of *Hevea brasiliensis. Protoplasma*, **252:** 783-796.

Rohani, O. and Paranjothy, K. (1980). Isolation of *Hevea* protoplast. *Journal of Rubber Research Institute of Malaysia*, **28(2):** 61-66.

Satchuthananthavale, R. and Irugalbandra, Z.E. (1972). Propagation of callus from *Hevea* anthers. *Quartely Journal of Rubber Research Institute of Ceylon*, **49:** 65-68.

Satchuthananthavale, R. (1973). *Hevea* tissue culture. *Quartely Journal of Rubber Research Institute of Ceylon*, **50:** 91-97.

Seneviratne, P. (1991). Micropropagation of juvenile and mature *Hevea brasiliensis*. PhD Thesis, University of Bath, UK, 278 p.

Seneviratne, P. and Wijeskara, G.A.S. (1996). The problem of phenolic exudates in *in vitro* cultures of mature *Hevea brasiliensis*. *Journal of Plantation Crops*, **24:** 54-62.

Seneviratne, P. and Wijeskara, G.A.S. (1997). Effect of episodic growth pattern on *in vitro* growth of *Hevea brasiliensis*. *Journal of Plantation Crops*, **25:** 52-56.

Seneviratne, P. and Flagmann, A. (1996). The effect of thidiazuron on axillary shoot proliferation of *Hevea brasiliensis in vitro*. *Journal of the Rubber Research Institute of Sri Lanka*, **77:** 1-14.

Shijie, Z., Zhenghua, C. and Xueng, X. (1990). A summary report on anther culture for haploid production of *Hevea brasiliensis*. *In:* Proceedings of IRRDB Symposium on the Breeding of *Hevea brasiliensis*, Kumming, China, pp. 69-78

Sinha, R.R., Sobhana, P. and Sethuraj, M.R. (1985). Axillary buds of some high yielding clones of *Hevea* in culture. First IRRDB *Hevea Tissue Culture Workshop*, Kuala Lumpur, Malaysia, p. 16.

Sirisom, Y. and Te-Chato, S. (2012). The effect of peptone and silver nitrate on *in vitro* shoot formation in *Hevea brasiliensis* Muell. *Arg. Journal of Agricultural Technology*, **8(4):** 1509-16.

Sobhana, P., Sinha, R.R. and Sethuraj, M.R. (1986). Micropropagation of *Hevea brasiliensis*: retrospects and prospects. In: *International Congress of Plant Tissue and Cell Culture*, Minnesota, USA.

Srichuay, W., Kalawong, S., Sirisom, Y. and Te-Chato, S. (2014). Callus induction and somatic embryogenesis from anther cultures of *Hevea brasiliensis* (Muell. Arg.). *Kasetsart Journal of National Science*, **48:** 364-375.

Sushamakumari, S., Rekha, K., Thomas, V., Sobha, S. and Jayashree, R. (1999a). Multiple shoot induction from somatic embryos of *Hevea brasiliensis* (Muell. Arg.). *Indian Journal of Natural Rubber Research*, **12:** 23-28.

Sushamakuamri, S., Sobha, S., Jayashree, R. and Asokan, M.P. (1999b). Evaluation of parameters affecting the isolation and culture of protoplast of *Hevea brasiliensis* (Muell. Arg.). *Current Science*, **77:** 1580-1581.

Sushamakuamri, S., Sobha, S., Rekha, K. and Jayashree, R. (2000a) Influence of growth regulators and sucrose on somatic embryogenesis frm immature inflorescence of *Hevea brasiliensis* (Muell. Arg.). *Indian Journal of Natural Rubber Research*, **13:** 19-29.

Sushamakuamri, S., Asokan, M.P., Antony, P., Lowe, K.C., Davey, M.R. and Power, J.B. (2000b). Plant regeneration from embryogenic suspension derived protoplast of rubber (*Hevea brasiliensis*) *Plant Cell Tissue Organ Culture*, **61:** 81-85.

Sushamakumari, S., Kala, R.G., Sobha, S, Rekha, K., Jayashree, R. and Jayalakshmi, A. (2010) Regeneration and acclimatisation of somatic plants from *in vitro* derived leaf explants of *Hevea brasiliensis. Paper presented at 19th Biennial Symposium on Plantation Crops*, India, p. 74.

Sushamakuamri, S., Sneha Joseph., Sobha, S., Rekha, K., Jayashree, R. and Leda Pavitran. (2012) Effect of nurse culture on inducing division of isolated pollen protoplast of *Hevea brasiliensis. Paper presented at International Rubber Conference*, India, p. 49.

Sushamakuamri, S., Rekha, K., Sobha, S. and Divya, U.K. (2014). Plant regeneration via somatic embryogenesis from root explants in *Hevea brasiliensis. Rubber Science*, **27(1):** 45-53.

Tan, D., Sun, X. and Zhang, J. (2011). Histochemical and immunohistochemical identification of laticifer cells in callus cultures derived from anthers of *Hevea brasiliensis. Plant Cell Reports*, **30:** 1117-1124.

Te-Chato, S. and Muangkaewngam, A. (1992). Tissue culture of rubber 1. *In vitro* micropropagation of rubber. *Songklanakarin Journal of Science and Technology*, **14:** 123-132.

Thulaseedharan, A., Kala, R.G. and Kumari Jayasree, P. (2006). Preliminary report on the incidence of Tapping Panel Dryness in tissue culture plants of *Hevea brasiliensis*. In: James, J., Krishnakumar, R. and Mathew, N.M. (Eds.), *Tapping Panel Dryness of Rubber Trees. RRII, Kottayam*, India. pp. 130-133.

Toruan, N.L. and Suryatmana, N. (1977). Kultur jaringan *Hevea brasiliensis* Muell. *Arg. Menara Perkebunan*, **45:** 17-21.

Varghese, Y.A., Thulaseedharan, A. and Kumari Jayasree, P. (2000). Rubber. In: Parthasarathy, V.A., Bose. T. K., Dekha, P. A., Das, P., Mitr, S. K. and Mohandas, S. (Eds.), *Biotechnology of Horticultural Crops* (Vol. 1). Naya Prakash, Calcutta, India. pp. 630-660.

Veisseire, P., Guerrier, J. and Courdet, A. (1993). Cryopreservation of embryogenic cell suspension of *Hevea brasiliensis. Cryo Letters,* **14:** 295-302.

Veisseire, P., Cailloux, F. and Courdet, A. (1994).a). Effect of conditional media on the somatic embryogenesis of *Hevea brasiliensis. Plant Physiology Biochemistry,* **32:** 571-576.

Veisseire, P., Linossier, L. and Coudret, A. (1994).b). Effect of abscisic acid and cytokinins on the development of somatic embryos in *Hevea brasiliensis. Plant Cell Tissue Organ Culture,* **39:** 219-23.

Wang, Z., Zeng, X., Chen, C., Wu, H., Li, Q., Fan, G. and Lu, W. (1980). Induction of rubber plantlets from anther of *Hevea brasiliensis* Muell. Arg. i*n vitro. Chinese Journal of Tropical Crops,* **1:** 25-26.

Wang, Z., Wu, H., Zeng, X., Chen, C. and Lu, W. (1984). Embryogeny and origin of anther plantlets of *Hevea brasiliensis. Chinese Journal of Tropical Crops,* **5:** 9-13.

Wan, A.R., Ghandimathi, H., Rohani, O. and Paranjothy, K. (1982). Recent developments in tissue culture of *Hevea*. In: Rao, A.N. (Ed.), *Tissue Culture of Economically Important Crops*. COSTED, Singapore. pp. 152-158.

Wang, Z.Y. and Chen, X.T. (1995). Effect of temperature on stamen culture and somatic plant regeneration in rubber. *Acta Agronimics Sincia,* **21:** 723-726.

Wang, Z.Y., Wu, H.D. and Chen, X.T. (1998). Effects of altered temperature on plant regeneration frequencies in stamen culture of rubber trees. J*ournal of Tropical Subtropical Botany,* **6:** 166-168.

Wilson, H.M. and Street, H.E. (1975). The growth, anatomy, and morphogenic potential of callus and cell suspension cultures of *Hevea brasiliensis. Annals of Botany,* **39:** 671-682.

Wilson, H.M., Eisa, M.Z. and Irwin, S.W.B. (1976). The effects of agitated liquid medium *on in vitro* cultures of *Hevea brasiliensis. Physiologia Plantarum,* **36:** 399-402.

Wilson, Z.A., and, Power, J.B. (1989). Elimination of systemic contamination in explants and protoplast cultures of *Hevea brasiliensis* (Muell.Arg.). P*lant Cell Reports,* **7:** 166-168.

Wycherley, P.R. (1992). The genus *Hevea*: Botanical aspects. In: Sethuraj, M.R. and Mathew, N.M. (Eds.), *Natural Rubber: Biology, Cultivation and Technology* (pp. 50-56). Elsevier, Amsterdam.

Zhou, Q.N., Jiang, Z.H., Huang, T.D., Li, Z., Sun, A.H., Dai, X.M. and Li, Z. (2010). Plant regeneration via somatic embryogenesis from root explants of *Hevea brasiliensis. African Journal of Biotechnology,* **9 (48):** 8168-8173.

Zhou, Q.N., Sun, A.H., Li, Z, Hua, Y.W., Jiang Z.H., Huang, T.D., Dai, X.M. and Huang, H.S. (2012). Cryopreservation and plant regeneration of anther callus in *Hevea* by vitrification. *African Journal of Biotechnology*, **11 (28):** 7212-7217.

Ziyun, W., Huang, T.D., Huang, H.S. and Hua, Y.W. (2009). Establishment and characterisation of *Hevea* cell suspension cultures. *Journal of Rubber Research*, **12(1):** 45-57.

Chapter 2

Coconut

☆ *Anitha Karun, K.S. Muralikrishna, M.K. Rajesh and P. Chowdappa*

1. Introduction

Coconut (*Cocos nucifera* L.) is an important perennial palm crop, predominantly cultivated in about 12 million hectares of land in tropical and subtropical coastal lowlands (APCC, 2014). With about 2.1 million hectares of coconut plantations, India is the largest producer of coconut. About 10 million farmers and their families depend exclusively on coconut, while many others in rural and semi-urban locations depend on it marginally for their livelihood (Rethinam, 2006). This palm is primarily grown for its edible oil from dry kernel, but all other parts have one or the other uses which earned the name 'tree of life'. Products from coconut palm possess high nutritional and medicinal value, in addition to numerous industrial applications (Foale, 2003). Coconut production has declined slightly in the last few years owing to several factors. For stabilized yield in the traditional coconut-based farming systems, growing of high-yielding, disease-resistant cultivars has been advocated. Unfortunately, production and supply of homogeneous quality planting material to the farmers has been a major constraint in coconut productivity. Conventional methods can produce only about 3.5 million seedlings annually while there is a demand for 10 million seedlings (Karun *et al.*, 2015). Coconut, with cultivar specific vegetative phase, enters juvenile phase in about 4 to 7 years and has a single apical meristem, without branches or suckers. This restricts its propagation through vegetative means with sole dependence on seed propagation. With no known methods of propagating the coconut palm through vegetative methods, *in vitro* culture is the only option for rapid multiplication of elite, high yielding and disease-resistant germplasm. Successful protocol for clonal propagation has been reported in various palm species such as arecanut (Karun *et al.*, 2004), oil palm (Rabechault *et al.*,

1970), peach palm (Steinmacher *et al.*, 2007) and date palm (Tisserat and Demason, 1980). Regeneration in these palms has been achieved using several explants such as young leaf, immature inflorescence and immature zygotic embryo via both direct and indirect somatic embryogenesis.

Somatic embryogenesis (SE) is a process where somatic cells differentiate into somatic embryos which share morphological similarity with zygotic embryos. It is a kind of asexual embryogenesis considered as an evolved strategy to overcome various environmental and genetical factors (Arnold *et al.*, 2002). Indirect SE has an intermediate callus phase, whereas direct SE does not. SE results in high number of regenerates with a very limited level of somaclonal variation (Ahloowalia, 1991; Henry, 1998). SE is feasible because plants possess cellular totipotency, wherein individual somatic cells possess the capability of regenerating into a whole plant- this makes it an attractive option in comparison to organogenesis as a plant regeneration system.

Research on SE in coconut was initiated four decades ago at Wye College, UK (Eeuwens and Blake, 1977), and later at Office de la Recherche Scientifique et Technique Outre-mer (ORSTOM, France) (Pannetier and Buffard-Morel, 1982). These experiments made use of plant somatic tissues such as young leaves, meristem region of young seedlings, sections from rachillae of young inflorescences, as initial explants to generate embryogenic calli (Branton and Blake, 1983; Gupta *et al.*, 1984). Recent studies related to SE have utilized zygotic tissues apart from somatic tissues such as immature inflorescences and ovaries. Zygotic tissues *viz.*, immature or mature embryos and embryo-derived plumules, were found to be easier to manipulate for achieving SE. In spite of several concerted efforts, reproducible protocol for clonal propagation in coconut has not been achieved. Thus coconut has been considered as one of the most recalcitrant species for *in vitro* culturing. *In vitro* recalcitrance in coconut has been attributed to many factors which include influence of genotype and explant maturity, adsorption of nutrients and hormones by activated charcoal making culture conditions undefined, production of compact calli, less percentage of plantlet regeneration, underperformance of regenerated plantlets and very slow rate of growth during *in vitro* culturing (Fernando *et al.*, 2010). Optimization of culture media, type of explant, plant growth regulators and their concentrations, subculturing periods and other additives have paramount significance in developing a reproducible tissue culture protocol. Studies carried out till date in coconut tissue culture is reviewed briefly below.

2. Explants

The initial starter for tissue culture purpose is termed as explants with meristems generally used for this purpose. For *in vitro* propagation, choice of explants is considered as a key element for successful outcome; hence identification of suitable explants is of primary importance. The potential of various coconut tissues, such as leaves, inflorescence, ovary, anthers and zygotic embryos to undergo callo-genesis, has been tested.

2.1. Leaves

Juvenile leaves are an excellent source of explant in many plants. However, the recalcitrant nature of some plants, especially palms, limits the use of leaves as explants. Very few studies have reported the use of leaves as explants for callusing and SE in coconut. Pannetier and Buffard-Morel (1982) reported asexual embryogenesis from young leaf explants of juvenile coconut palms. Even though callus was induced in leaf explants obtained from mature palms, further development into embryoids was not observed. Plantlets were obtained using immature leaves as explants by Raju *et al.* (1984), but the procedure could not be repeated. Reports indicate that the embryogenic capacity of leaf explants lasts for very short durations which limit their use as explants in clonal propagation studies (Karunaratne *et al.*, 1991). Uncoupling of cell and nuclear size has been reported to interrupt cell co-ordination leading to the recalcitrant nature of leaf explants to *in vitro* culture (Jesty and Francis, 1992). Thus, there has not been much progress using leaves as a source of explants in coconut palm.

2.2. Immature Inflorescence

Immature inflorescence is a potential and promising source of explants to clonally propagate important crop plants since they contain numerous meristematic points. Success depends on the selection of inflorescence of correct maturity stage. Regeneration of plantlets from immature inflorescence explants has been successful in arecanut (Karun *et al.*, 2004), date palm (Fki *et al.*, 2003; Abul-Soad and Mahdi, 2010), peach palm (Steinmacher *et al.*, 2007), oil palm (Teixeira *et al.*, 1994) and juçara palm (Guerra and Handro, 1988). The first report on use of immature inflorescence of coconut as a source of explant dates back to the early 1980s when sections of inflorescence rachillae were observed to proliferate in the medium to form callus, termed as 'calloids' by Branton and Blake (1983).. Explants were cultured in a medium with a range of 2,4-D as auxin source and activated charcoal. Callogenesis of inflorescence tissue depended on its age and the concentration of auxins in the medium. Somatic embryo formation was achieved successfully with a functional bipolar organization and completely differentiated shoot meristem (Verdeil *et al.*, 1994; Sandoval-Cancino *et al.*, 2016). Callus induction in segments of explants depended on basal media used and the size of inflorescence explants. Among the tested media *viz.*, modified Eeuwens Y3 (Eeuwens, 1976), CRI72, anther culture medium and modified Blake medium, callus induction was observed only in CRI72 medium (Vidhanaarachchi and Weerakoon, 1997).

2.3. Unfertilized Ovary

Ovary culture has been used in many crops as a means of clonal propagation since it is a potential tissue for the induction of SE due to the juvenilizing influence of nearby meiotic tissues (Bonga, 1982). Studies by Griffis and Litz (1997) indicated callogenesis and formation of adventitious roots from unfertilized ovary tissues in coconut, but SE could not be induced. Increased efficiency of unfertilized ovaries of coconut for callogenesis was reported by Perera *et al.* (2007), suggesting the possibility of mass production of homogenous planting materials of improved coconut varieties. Consistent callogenesis of about 40 per cent was observed when unfertilized ovaries were cultured in CRI72 medium containing 2,4-D. However plantlet regeneration was limited with a total of 83 plantlets being produced from 32 cultured ovaries (Perera *et al.*, 2007).

2.4. Anthers and Microspores

Anthers and microspores are used as basal explants to produce haploid plantlets or double haploids. Production of double haploids through anther and microspore culture has considerable potential for shortening the breeding cycle in coconut. Inflorescence growth and development in coconut is complex and requires careful observations for choosing anthers and microspores as an explant source. The developmental stage of inflorescence is very important at which collections are made. The stages of coconut inflorescence at which anthers and micropsores are to be samples have been standardized and well defined (Perera, 2003). Anthers consist of both diploid and haploid cells making selective cell division impossible- this may lead to the formation of both diploid and haploid plantlets. In coconut, ploidy analysis of anther derived plantlets indicated that half of the regenerated plantlets were haploid and the rest were diploid (Perera *et al.*, 2008). Auxin sources for culturing coconut anthers were standardized indicating 2,4-D in combination with NAA enhanced calli production, but picloram and IAA had repressive effects. Media supplemented with 2-iP and kinetin, as cytokinin sources, yielded good results as compared to zeatin (Perera *et al.*, 2009). Solid Eeuwens Y3 (Eeuwens, 1976) medium was superior when compared to liquid medium for culturing coconut anthers as indicated by the higher percentage of embryo formation and embryo conversion (Perera *et al.*, 2011). As a whole, even though anthers and microspores served as an excellent source of explants for *in vitro* culture in many crops with successful plantlet regeneration, the final conversion to plantlets was low in coconut.

2.5. Immature Zygotic Embryo

The level of maturity of zygotic embryos is a very important factor since differences in induction frequencies are often observed and immature zygotic embryos have been found to possess better potential and competence to respond *in vitro* than their mature counterparts in coconut. Callusing frequency depends on the developmental stage of the embryo. Under the best conditions, callusing was reported to be around 75 per cent in immature zygotic embryo (Diyasena, 1998).

Immature embryos of coconut (6-7 months post-anthesis) cultured in a medium supplemented with 2,4-D, produced calli with 50 per cent of them turning in to globular embryos. Around 22 per cent of germination was observed in these cultures. Age of the embryo was suggested as an important factor determining callus proliferation and subsequent embryogenesis (Karunaratne and Periyapperuma, 1990).

2.6. Plumular Tissues

In coconut, even though immature embryos were found to be responsive, the response from mature embryos could be improved by slicing the embryos (Adkins *et al.*, 1998; Samosir, 1999; Rajesh *et al.*, 2005, 2014) or by selective excision of plumular tissue from zygotic embryo (Figure 2.1) (Chan *et al.*, 1998; Lopez-

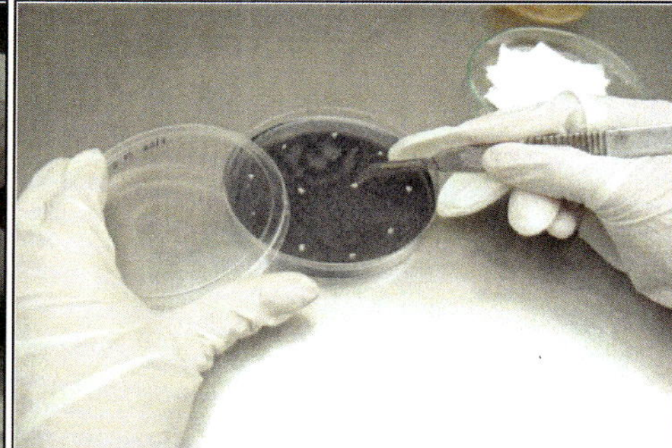

Figure 2.1: Excision of Plumules from Mature Zygotic Embryos of Coconut and Inoculation on to Culture Medium.

Villalobos, 2002; Perez-Nunez *et al.*, 2006). Bhavyashree *et al.* (2016a) compared three methods of isolation of plumule *viz.*, excision of shoot meristem aseptically from *in vitro* germinated embryo after 10-12 days, excision of shoot meristem from *in vitro* germinated embryo subjected to GA3 treatment for five days and excision of shoot meristem from fresh embryo. It was observed that initiation of callus and embryogenic calli was significantly high in plumular explants isolated from fresh embryo. More consistent results in induction of embryogenic calli, formation of somatic embryos and *in vitro* regeneration have been obtained with plumular explants (Chan *et al.*, 1998; Lopez-Villalobos, 2002; Fernando *et al.*, 2004; Perez- Nunez *et al.*, 2006; Saenz *et al.*, 2006; Rajesh *et al.*, 2005, 2014) (Figures 2.2–2.4).

Figure 2.3: Regeneration of Plantlet from Coconut Plumule.

3. Media Composition

3.1. Basal Media

Apart from explant selection, choosing a suitable medium for growing the tissues

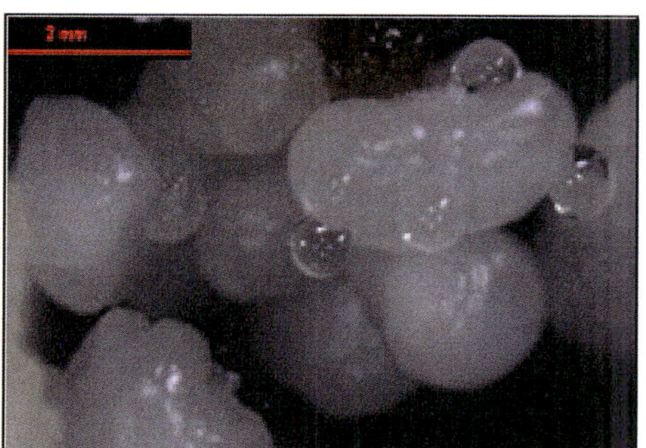

Figure 2.2: Development of Embryogenic Calli from Plumular Explants of Coconut.

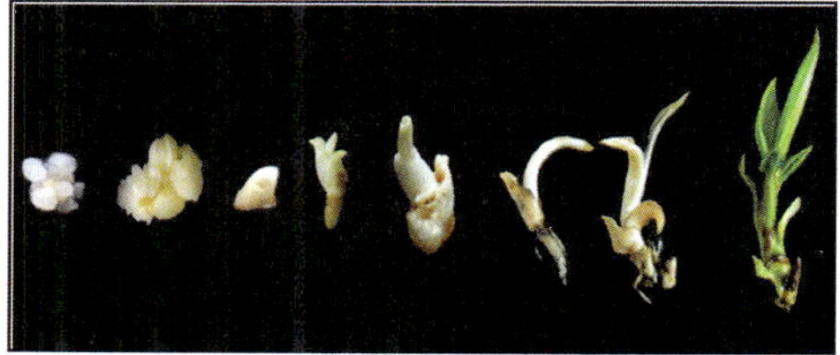

Figure 2.4: Different Growth Stages in Plumule Culture.

is decisive for successful micropropagation. Optimal growth and morphogenesis of tissues may vary for different palms according to their nutritional requirements. Moreover, tissues from different parts of palms may also have different requirements for satisfactory growth. In coconut, Y3 (Eeuwens, 1976) and CRI72 (Karunaratne and Periyapperuma, 1989) media have been used frequently and found to be better than MS (Murashige and Skoog, 1962) and B5 (Gamborg *et al.*, 1968) for callus induction (Branton and Blake, 1983; Bhalla-Sarin *et al.*, 1986).

3.2. Plant Growth Regulators and other Additives

Role of plant growth regulators in culture medium is very important in the induction of callus and further development into somatic embryos. Callus induction medium, in general, requires a strong auxin source and 2,4-D has been the auxin most commonly used in coconut, with the working concentration of 2,4-D differing for different cultivars and explants and also between laboratories (Perez- Nunez *et al.*, 2006; Saenz *et al.*, 2006; Rajesh *et al.*, 2005, 2014; Perera *et al.*, 2009; Bhavyashree *et al.*, 2016 a, b; Sandoval-Cancino *et al.*, 2016). In general, explants were inoculated initially in a higher concentration of 2,4-D and subsequently sub-cultured onto a medium supplemented with lower concentrations of 2,4-D at regular intervals. The uptake of 2,4-D, studied using radio-labeling technology, indicate maximum absorption by the explants was within 24 hours of inoculation (Oropeza and Taylor, 1994). Verdeil *et al.* (2001). reported the significance of 'gametophytic-like' conditions produced by 2, 4-D for the successful transition from the vegetative into the embryogenic state. Fate of 2, 4-D in coconut tissues indicated its conversion into fatty acid analogues and incorporation into triacylglycerol derivatives (Lo´pez-Villalobos *et al.*, 2004). Stable and stored form of 2, 4-D, incorporated into triacylglycerol, may influence SE even when 2, 4-D has been removed from the medium (Nguyen *et al.*, 2015).

Callus initiation has also been achieved using other auxins apart from 2, 4-D. NAA (27 µM), in combination with 2, 4-D (452 µM), has been reported to promote callogenesis in rachillae explants (Gupta *et al.*, 1984). Picloram has been used successfully as an auxin source for callogenesis in various palm species such as arecanut (Karun *et al.*, 2004) and in peach palm (Steinmacher *et al.*, 2007) and has shown promising results in coconut. Optimizing the type of auxin and its concentration for induction of callus and its subsequent multiplication would play a pivotal role in establishing a standard tissue culture protocol in coconut.

Reports have indicated the potential of thidiazuron (TDZ) to enhance callus production in woody plants (Huetteman and Preece, 1993). In coconut, supplementation of the callus induction medium with TDZ has been shown to enhance callus induction and formation of embryogenic calli from plumular explants (Rajesh *et al.*, 2005, 2014; Jayaraj *et al.*, 2014; Bhavyashree *et al.*, 2016a, b) A positive effect of TDZ was observed on coconut ovary culture which enhanced callus induction; however response was dependent on the concentration, with higher concentrations of TDZ (above 18 µM) reducing the frequency of callus formation (Perera *et al.*, 2009). Thus, a combination of auxin and TDZ could enhance callusing in a short period of time.

Perez-Nunez *et al.* (2006) and Chan *et al.* (1998) have reported promotion of SE by incorporating 6-benzylaminopurine (BAP; between 50 to 300 µM) in the medium, which also led to an enhanced number of viable plantlets at the end of the culture phase.

Supplementation of abscisic acid (ABA) in the tissue culture medium has also been reported to enhance induction of SE in coconut from immature embryo explants (Fernan-do, 2001) and plumular tissues (Samosir *et al.*, 1999; Fernando and Gamage, 2000; Fernando *et al.*, 2004). However, the plantlet regeneration was low, around 10 per cent. (Weerakoon, 2004). Adkins *et al.* (1998) and Rajesh *et al.* (2005, 2014) have shown that exogenous supply of polyamines, such as spermine and putrescine, could enhance induction of SE in coconut from plumular explants. Adkins *et al.* (1998) demonstrated the beneficial effects of aminoethoxyvinylglycine (AVG) an ethylene production inhibitor and silver thiosulphate (STS) an ethylene action inhibitor in callus multiplication and SE.

Incorporation of osmotically active agents such as polyethyleneglycol (PEG 3 per cent) along with ABA (45 µM) in tissue culture medium has yielded positive results on formation, maturation and germination of somatic embryos (Samosir *et al.*, 1998). Significant beneficial effect of growth retardant ancymidol (30 µM) was demonstrated by Antonova (2009) on somatic embryo germination using immature inflorescence as explants. SE in coconut has also been reported to be positively influenced by supplementation of polyvinylpolypyrrolidone (PVPP) in the medium (Samosir, 1999).

3.3. Activated Charcoal

Activated charcoal (AC) has been an integral part of the culture medium especially for crops such as coconut. AC adsorbs the unwanted phenols and other growth inhibitory compounds and in turn reduces tissue browning under *in vitro* conditions. However, AC can also adsorb exogenously supplied plant growth regulators (hormones, vitamins) and minerals (Pan and Staden, 1998) which can lead to ambiguities in the precise functional concentrations of these additives in the culture medium. Therefore, AC can be considered a 'necessary evil' as far as coconut is considered. Ebert *et al.* (1993) reported that AC adsorbs auxins and cytokinins to the tune of 99 and 98 per cent, respectively after five days of culture media preparation.

4. Callus Multiplication and Maintenance

In general, coconut explants give rise to highly heterogeneous compact callus, which is not of friable nature (Fernando, 2001). Callus multiplication is important in scaling up protocols for *in vitro* regeneration in coconut and this was achieved by Perez-Nunez *et al.* (2006). Studies have suggested that multiplication of embryogenic callus could be achieved by subdividing the callus and repeated subculturing into media supplemented with lower levels of auxin, which has yielded promising results. Callus maintenance is important as it can provide year round embryogenic calli for further conversion. A recent report indicates that coconut embryogenic callus, obtained from plumular tissues, could be maintained for 21 weeks without compromising on the embryogenic potential (Bhavyashree *et al.*, 2015).

5. Studies of Gene Expression Patterns during Somatic Embryogenesis

Molecular approaches, such as gene expression studies, can help to decipher molecular mechanisms underlying SE in coconut (Perez-Nunez *et al.*, 2009). Efficiency of existing clonal propagation protocol in coconut could be enhanced if a thorough knowledge of molecular events during SE is available. A major breakthrough was achieved with the isolation of a coconut gene (*CnANT*) homologous to the *Arabidopsis AINTEGUMENTA*-like gene, encoding two APETALA2 domains and a linker region (Bandupriya *et al.*, 2013). Analysis of *CnANT* has demonstrated its role in SE as indicated by enhanced expression levels during the callus induction phase (Bandupriya *et al.*, 2013, 2014). The upregulation of *CnANT* gene resulted in increased shoot growth in absence of growth regulators (Bandupriya and Dunwell, 2012). However, *CnANT* did not aid in spontaneous formation of somatic embryos as observed earlier with other *PL/AIL* genes (Bandupriya and Dunwell, 2012; Boutilier *et al.*, 2002; Tsuwamoto *et al.*, 2010). Perez-Nunez *et al.* (2009) isolated *CnCDKA* and *CnSERK* homologs from coconut which are reported to be associated with the induction of SE. A cyclin-dependent kinase, encoded by *CnCDKA*, regulates cell division (Montero-Cortes *et al.*, 2010a) while *CnSERK* encodes a protein receptor (Perez-Nunez *et al.*, 2009), which may be a component of a signaling cascade involved in regulating the rate of SE (Hecht *et al.*, 2001; Schmidt *et al.*, 1997). Similar to *KNOX* class I gene, which is associated exclusively in tissue with meristematic activity, *CnKNOX1* gene was isolated in coconut and its role in increased rate of somatic embryo formation and germination, through the addition of gibberellin during coconut SE, has been demonstrated (Montero-Cortes *et al.*, 2010b).

Rajesh *et al.* (2016) undertook transcriptome analysis of coconut embryogenic calli using Next Generation Sequencing (NGS) which resulted in the identification of 14 genes known to be involved in SE in other plants. Quantitative real-time PCR (qRT-PCR) analyses of these 14 genes were carried in six developmental stages. The results revealed differential gene expression of these 14 genes: *CLV* was upregulated in the initial stage of callogenesis; *GLP*, *GST*, *PKL*, *WUS* and *WRKY* were upregulated during SE, whereas expression of *SERK*, *MAPK*, *AP2*, *SAUR*, *ECP*, *AGP*, *LEA* and *ANT* were higher in the embryogenic callus stage compared to initial culture and somatic embryo stages.

Bhavyashree *et al.* (2016b) carried out comparative studies of gene expression patterns, using RT-qPCR, of eight genes during various *in vitro* developmental stages of calli obtained from shoot meristem explants in WCT (West Coast Tall) and COD (Chowghat Orange Dwarf) cultivars of coconut. Enhanced expression of *PKL*, *SERK* and *WUS* was observed in embryogenic calli compared to non-embryogenic calli. Expression of *GLP*, *ECP* and *GST* was observed to be higher in normal somatic embryos compared to abnormal somatic embryos, whereas expression of *ECP*, *LEAFY*, *GLP* and *WRKY* was higher in normal meristemoids compared to aberrant meristemoids. Higher expression of *SERK*, *PKL* and *WUS* was reported in embryogenic calli of WCT compared to COD. Somatic embryos of COD showed high expression of *GLP* and *GST* compared to WCT, where as in case of *ECP* gene, higher expression was observed in WCT compared to COD. Higher levels of expression *WRKY* and *LEC* were observed in WCT meristemoids compared to COD. These results revealed existence of genotypic differences of cultivars to *in vitro* culture and the use of these genes as markers for coconut SE was also suggested. Despite several studies, understanding of the molecular mechanisms that underlies coconut SE is limited warranting more prioritized research in this area for refining SE protocol.

6. Future Prospects

6.1. Establishment of Cell Suspension Cultures

The well established fact in plant tissue culture is that most plants grow better in liquid than on solid media. Several approaches have been proposed in order to enhance the productivity of *in vitro* culture depending on the final product desired and the species investigated. One such innovative approach is the use of cell suspensions (Figure 2.5) and bioreactors for plant culture. Cell suspensions in specific medium would be ideal for producing large number of somatic embryos and to extract commercially important plant metabolites. Culturing and maintaining of the embryogenic calli obtained from the explants in a liquid medium with appropriate nutrients, auxin source and additives under stable microclimatic condition is referred to as suspension culture. However, several factors such as aeration, agitation, light, temperature would influence the process as suspensions are maintained in flask culture. The embryogenic cells produced in cell suspension culture could be used in bioreactors to enhance SE. Highly recalcitrant nature of coconut to *in vitro* culture necessitates alteration in conventional tissue culture approaches. Some of the factors such as pH, temperature, dissolved oxygen; CO_2 concentrations could play a major role in SE. Keeping several of these factors constant could lead to an effective protocol with enhanced SE. The adaption of bioreactors in plant tissue culture is considered a major milestone since they offer several advantages *viz.*, time saving, labour-saving, relatively easy to scale-up, allow enhanced growth and

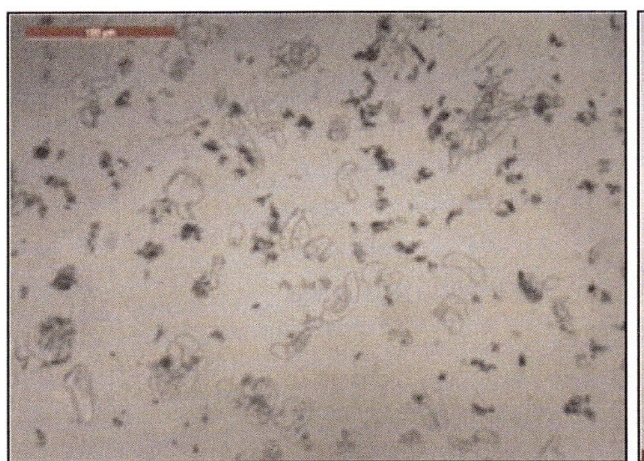

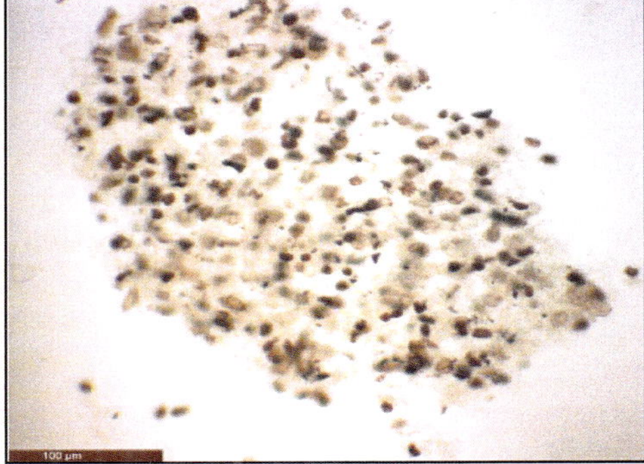

Figure 2.5: Initiation of Cell Suspension from Coconut Embryogenic Calli Derived from Plumular Explants.

multiplication and improved nutrient availability due to the use of liquid medium over traditional tissue culture techniques. Different kinds of bioreactor designs are in use such as aeration-agitation bioreactor, spin filter bioreactor, gaseous phase bioreactor, rotating drum bioreactor and air-driven bioreactor.

Many crop plants have been mass multiplied using bioreactors ever since its inception into plant tissue culture practices (Peak *et al.*, 2001; Ziv, 2005). As a plant production technique, bioreactors are far superior to traditional *in vitro* methods for all the species thus far tested. It is worth noting that with bioreactors, even the difficult-to-propagate woody and tree species can be produced relatively easily at high frequency. For instance, an efficient, somatic embryo-based mass propagation system for the recalcitrant species *Coffea arabica* was developed using a bioreactor (Barry-Etienne *et al.*, 1999). It is evident from several peer reviewed research papers that SE process could be scaled up many folds using bioreactors. A hybrid reactor would be ideal to reduce the *in vitro* culture duration in coconut and also to enhance the rate of SE and conversion of somatic embryos into plantlets.

6.2. Use of Temporary Immersion Systems

A system wherein explants are flooded with nutrient medium containing growth regulators at regular time intervals has been successfully used in scaling up of SE. The system named as temporary immersion systems (TIS) which offer the possibility of automating some culture stages. First of its kind was developed for pineapple (Escalona *et al.*, 1999), and use of similar system is shown to improve the regeneration rate and plantlet quality for other plant species (Etienne and Berthouly, 2002). Since then, the use of TIS in regeneration protocols for plant species has increasing continually (Niemenak *et al.*, 2008; Sankar-Thomas *et al.*, 2009). Culture containers provide additional head space which would improve the efficiency of regenerated plantlets. For coconut, which is classified as one of the toughest crops to clonally propagate, an improved protocol involving temporary immersion system needs to be developed (Figure 2.6). This would be an interesting strategy for up-scaling plantlet regeneration potential in coconut tissues under *in vitro* conditions.

6.3. Transformation Studies

A number of genes have been identified which govern SE. Several transcription factors control the transition from vegetative to embryogenic growth. Among them, members of APETALA2/Ethylene-responsive element binding protein domain family play an important role in promoting embryo development. In coconut, AINTEGUMENTA-like gene (*CnANT*), which codes for two AP2 domains, has been identified. Over-expression of *CnANT* in transgenic *Arabidopsis* yielded regeneration in hormone-free conditions (Bandupriya *et al.*, 2013). Genetic transformation studies could be a tool in developing a viable clonal propagation protocol in coconut. Micro-projectile bombardment method was used initially for inserting *GUS* gene in to embryogenic calli and young leaf tissues of coconut (Samosir, 1999). This was the first reported genetic transformation effort in coconut. *Agrobacterium*-mediated genetic transformation studies in various tissue of coconut such as immature anthers, excised zygotic embryos, plumule-derived embryogenic calli and SE-derived roots

Figure 2.6: Use of Temporary Immersion Systems in Coconut Plumule Culture Experiments.

and leaves were reported by Andrade-Torres *et al.* (2011). Genetic modification in coconut is still a long way away from becoming a reality. This could be useful for the improvement of coconut SE by introducing genes or by over expressing these genes which are known to regulate SE in other plant species.

6. Conclusion

It is evident from literature and from our laboratory experience that progress of coconut tissue culture work is slow. To achieve a commercially viable protocol, it warrants the requirement of explant, media and PGR optimization. SE obtained using plumular explant is promising; however, more efforts are required towards refining and developing a protocol which could be upscaled. Inclusion of automatic systems such as temporary immersion systems and bioreactors would reduce the cost and increase the efficiency of SE. Recent advancement in biotechnological tools will open up avenues to scale-up SE in coconut.

References

Abul-Soad, A.A. and Mahdi, S.M. (2010). Commercial production of tissue culture date palm (*Phoenix dactylifera* L.) by inflorescence technique. *Journal of Genetic Engineering and Biotechnology*, **8**: 39-44.

Adkins, S.W., Samosir, Y.M.S., Ernawati, A., Godwin, I.D. and Drew, R.A. (1998). Control of ethylene and use of polyamines can optimize the conditions for somatic embryogenesis in coconut (*Cocos nucifera* L.) and papaya (*Carica papaya* L.). In: Drew, R. A. (Ed.), *Proceedings of the International Symposium of Biotechnology in Tropical and Subtropical Species*. Australia, Brisbane. pp. 459-466.

Ahloowalia, B.S. (1991). Somatic embryos in monocots: Their genesis and genetic stability. Revue de Cytolgie et de Biologie Vegetales Le Botaniste, **14**: 223-235.

Andrade-Torres, A., Oropeza, C., Sáenz, L., González-Estrada, T., Ramírez-Benítez, J.E., Becerril, K., Chan, J.L. and Rodríguez-Zapata, L.C. (2011). Transient genetic transformation of embryogenic callus of *Cocos nucifera* L. Biologia (Section Cellular and Molecular Biology), **66**: 790-800.

Antonova, I.D. (2009). Somatic embryogenesis for micropropagation of coconut (*Cocos nucifera* L.). PhD Thesis, The University of Queensland, Australia.

APCC (Asian and Pacific Coconut Community) (2014). Statistical Year Book.

Arnold, S.V., Sabala, I., Bozhkov, P., Dyachok, J. and Filonova, L.H. (2002). Developmental pathways of somatic embryogenesis. *Plant Cell Tissue and Organ Culture*, **69**: 233-249.

Bandupriya, H.D.D. and Dunwell, J. (2012). Overexpression of *CnANT*, coconut *BABYBOOM* homologue alters plant growth and morphology in transgenic *Arabidopsis* plants. *Tropical Agricultural Research*, **23**: 249-260.

Bandupriya, H.D.D., Gibbings, J.G. and Dunwell, J.M. (2013). Isolation and characterization of an *AINTEGUMENTA*-like gene in different coconut (*Cocos nucifera* L.) varieties from Sri Lanka. *Tree Genetics and Genomes*, **9**: 813-827.

Bandupriya, H.D.D., Gibbings, J. G. and Dunwell, J.M. (2014). Overexpression of coconut *AINTEGUMENTA*-like gene, *CnANT*, promotes *in vitro* regeneration in transgenic Arabidopsis. *Plant Cell Tissue and Organ Culture*, **116**: 67-79.

Barry-Etienne, D., Bertrand, B., Vasquez, N. and Etienne, H. (1999). Direct sowing of *Coffea arabica* somatic embryos mass-produced in a bioreactor and regeneration of plants. *Plant Cell Reports*, **19(2):** 111-117.

Bhalla-Sarin, N., Baggaa, S.S., Sudhir, K. and Guhar-Mukherjee, S. (1986). Induction and differentiation of callus from embryo of *Cocos nucifera* by IAA conjugates. *Plant Cell Reports*, **5**: 322-324.

Bhavyashree, U., Lakshmi Jayaraj, K., Fayas, T.P., Muralikrishna, K.S., Sajini, K.K., Rajesh, M.K. and Karun, A. (2016a). A comparative study of three different methods of shoot meristem excision for induction of embryogenic calli in coconut. *Research in Biotechnology*, **7**: 50-57.

Bhavyashree, U., Rachana, K.E., Jayaraj, K.L., Rajesh, M.K. and Anitha, K. (2016b). Comparative gene expression profiling during *in vitro* regeneration in two coconut cultivars. *Emirates Journal of Food and Agriculture*, **28(8)**: 572-580.

Bhavyashree, U., Lakshmi Jayaraj, K., Rachana, K.E., Muralikrishna, K.S., Sajini, K.K., Rajesh, M.K. and Karun, A. (2015). Maintenance of embryogenic potential of calli derived from shoot meristem of West Coast Tall cv. of coconut (*Cocos nucifera* L.). *Journal of Plantation Crops*, **43(2):** 105-116.

Bonga, J.M. (1982). Clonal propagation of mature trees: problems and possible solutions. In: Bonga, J.M. (Ed.), *Tissue Culture in Forestry*. Martinus Nijhoff Publ, Dordrecht. pp. 249-271.

Boutilier, K., Offringa, R., Sharma, V.K., Kieft, H., Ouellet, T., Zhang, L.M., Hattori, J., Liu, C.M., van Lammeren, A.A.M., Miki, B.L.A., Custers, J.B.M. and Campagne,

M.M.V. (2002). Ectopic expression of *BABYBOOM* triggers a conversion from vegetative to embryonic growth. *The Plant Cell* **14**: 1737-1749.

Branton, R.L. and Blake, J. (1983). Development of organized structures in callus derived from explants of *Cocos nucifera*. *Annals of Botany*, **52(5)**: 673-678.

Chan, J.L., Saenz, L., Talavera, C. and Hornung, R. (1998). Regeneration of coconut (*Cocos nucifera* L.) from plumule explants through somatic embryogenesis. *Plant Cell Reports*, **17**: 515-521.

Diyasena, R.A.C.N.M. (1998). Development of a charcoal-free protocol for callogenesis and somatic embryogenesis in immature embryo explants of coconut (*Cocos nucifera* L.). M.Sc. thesis, University of Colombo, Sri Lanka.

Ebert, A., Taylor, F.H. and Blake, J. (1993). Changes of 6-benzylaminopurine and 2,4-dichlorophenoxyacetic acid concentrations in plant tissue culture media in the presence of activated charcoal. Plant Cell Tissue and Organ Culture, **33**: 157-162.

Eeuwens, C.J. and Blake, J. (1977). Culture of coconut and date palm tissue with a view to vegetative propagation. *Acta Horticulturae*, **78:** 277-286.

Eeuwens, C.J. (1976). Mineral requirements for growth and callus initiation of tissue explants from mature coconut palms (*Cocos nucifera* L.) and date (*Phoenix dactylifera*) palm cultured *in vitro*. *Physiologia Plantarum*, **36**: 23-28.

Escalona, M., Lorenzo, J.C., Gonzaler, B., Daquinta, M., Gonzaler, J.L., Desjarodin, Y. and Borrota, C.G. (1999). Pineapple (*Ananus Comsus* L. Merr.) micropropagation in temporary immersion systems. *Plant Cell Reports*, **18**: 743-748.

Etienne, H. and Berthouly, M. (2002). Temporary immersion systems in plant micropropagation. *Plant Cell Tissue and Organ Culture*, **69**: 215-231.

Fernando, S.C. and Gamage, C.K. (2000). Abscisic acid induced somatic embryogenesis in immature embryo explants of coconut (*Cocos nucifera* L.). *Plant Science*, **151(2):** 193-198.

Fernando, S.C. (2001). Studies on *in vitro* somatic embryo-genesis of *Cocos nucifera* L. (coconut). PhD thesis, University of Colombo, Sri Lanka.

Fernando, S.C., Vidhanaarachchi, V.R.M., Weerakoon, L.K. and Santha, E.S. (2010). What makes clonal propagation of coconut difficult? *Asia-Pacific Journal of Molecular Biology and Biotechnology*, **18**:163-165.

Fernando, S.C., Weerakoon, L.K., Perera, P.I.P., Bandupriya, H.D.D., Ambagala, I.A., Gamage, C.K.A., Santha, E.S., Gunathilake, T.R. and Perera, L. (2004). Genetic fidel-ity and *ex vitro* performance of tissue-cultured coconut plants. In: *Proceedings of the International Conference on 75th Anniversary of Coconut Research Institute of Sri Lanka. pp*. 47-57.

Fki, L., Masmoudi, R., Drira, N. and Rival, A. (2003). An optimised protocol for plant regeneration from embryogenic suspension cultures of date palm, *Phoenix dactylifera* L., cv. Deglet Nour. *Plant Cell Reports*, **21**: 517-524.

Foale, M. (2003). Coconut in the Human Diet - an Excellent Component. *Coco Info International*, **10(2):** 17-19.

Gamborg, O.L., Miller, R.A. and Ojima, O. (1968). Nutrient requirements of suspension cultures of soybean root cell. *Experimental Cell Research*, **50**: 151-158.

Griffis, J.L. Jr. and Litz, R.E. (1997). Advances in the *in vitro* morphogenesis of several coconut tissues in Florida. In: *Proceedings of International Cashew and Coconut Conference*, Dar-es- Salaam, Tanzania. pp. 349-357.

Guerra, M.P. and Handro, W. (1988). Somatic embryogenesis and plant regeneration in embryo cultures of *Euterpe edulis* Mart. (Palmae). *Plant Cell Reports*, **7**:550–552.

Gupta, P.K., Kendurkar, S.V., Kulkarni, V.M., Shirgurkar, M.V. and Mascarenhas, A.F. (1984). Somatic embryogenesis and plants from zygotic embryos of coconut (*Cocos nucifera* L.) *in vitro*. *Plant Cell Reports*, **3**: 222-225.

Hecht, V., Vielle-Calzada, J.P., Hartog, M.V., Schmidt, E.D.L., Boutilier, K., Grossniklaus, U. and de Vries, S.C. (2001). The *Arabidopsis* Somatic Embryogenesis Receptor Kinase 1 gene is expressed in developing ovules and embryos and enhances embryogenic competence in culture. *Plant Physiology*, **127**: 803-816.

Henry, R.J. (1998). Molecular and biochemical characterization of somaclonal variation. In: Jain, S.M., Brar, D.S. and Ahloowalia, B.S. (Eds.) *Somaclonal Variation and Induced Mutations in Crop Improvement*. Kluwer Academic Publishers, Dordrecht. pp. 485-499.

Huetteman, C.A., and John, E.P. (1993). Thiadiazuron: A potent cytokinin for woody plant tissue culture. *Plant Cell Tissue and Organ Culture*, **33**: 105-119.

Jayaraj, K.L., Bhavyashree, U., Fayas, T.P., Sajini, K.K., Rajesh, M.K. and Karun, A. (2014). Histological studies of cellular differentiation during somatic embryogenesis of coconut plumule-derived calli. *Journal of Plantation Crops*, **43(3)**: 196-203.

Jesty, J.H.F. and Francis, D. (1992). Cellular responses of leaf explants of *Cocos nucifera* L. *in vitro*. *Plant Cell Tissue and Organ Culture*, **28(3)**: 235-244.

Karun, A, Siril, E.A., Radha, E. and Parthasarathy, V.A. (2004). Somatic embryogenesis and plantlet regeneration from leaf and inflorescence explants of arecanut (*Areca catechu* L.). *Current Science*, **86**: 1623-1628.

Karun, A., Rajesh, M.K. and Muralikrishna K.S. (2015). Somatic embryogenesis and bioreactors: Meeting report. *Current Science*, **108(10)**: 1782-1783.

Karunaratne, S. and Periyapperuma, K. (1990). Culture of immature embryos of coconut, *Cocos nucifera* L.: Callus proliferation and somatic embryogenesis. *Journal of the Coconut Research Institute of Sri Lanka*, **8**: 13-22.

Karunaratne, S., Santha, S. and Kovoor, A. (1991). An *in vitro* assay for drought-tolerant coconut germplasm. *Euphytica*, **53(1)**: 25-30.

Karunaratne, S.M. and Periyapperuma, K. (1989). Culture of immature embryos of coconut (*Cocos nucifera*) callus proliferation and somatic embryogenesis. *Plant Science*, **62**: 247-253.

Lopez-Villalobos, A. (2002). Roles of lipids in coconut (*Cocos nucifera* L.) embryogenesis. Thesis (PhD) University of London.

Lopez-Villalobos, A., Hornung, R. and Dodds, P.F. (2004). Hydrophobic metabolites of 2,4-dichlorophenoxyacetic acid (2,4-D) in cultured coconut tissue. *Phytochemistry*, **65**: 2763-2774.

Montero-Cortes, M., Rodriguez-Paredes, F., Burgeff, C., Perez-Nunez, T., Cordova, I., Oropeza, C., Verdeil, J.L. and Sa´enz, L. (2010a). Characterisation of a cyclin-dependent kinase (*CDKA*) gene expressed during somatic embryogenesis of coconut palm. *Plant Cell Tissue and Organ Culture*, **102**: 251-258.

Montero-Cortes, M., Saenz, L., Cordova, I., Quiroz, A., Verdeil, J.L. and Oropeza, C. (2010b). GA3 stimulates the formation and germination of somatic embryos and the expression of a *KNOTTED*-like homeobox gene of *Cocos nucifera* L. *Plant Cell Reports*, **29(9)**: 1049-1059.

Murashige, T. and Skoog, F. (1962). A revised medium for rapid growth and bioassays with tobacco tissue culture. *Physiologia Plantarum*, **15**: 473-479.

Nguyen, Q.T., Bandupriya, H.D.D., Lo´pez-Villalobos, A., Sisunandar, S., Foale, M. and Adkins, S.W. (2015). Tissue culture and associated biotechnological interventions for the improvement of coconut (*Cocos nucifera* L.): A review. *Planta*, **242**: 1059-1076.

Niemenak, N., Saare-Surminski, K., Rohsius, C., Ndoumou, D.O. and Lieberei, R. (2008). Regeneration of somatic embryos in *Theobroma cacao* L. in temporary immersion bioreactor and analyses of free amino acids in different tissues. *Plant Cell Reports*, **27**: 667-676.

Oropeza, C. and Taylor, H.F. (1994).) Uptake of 2, 4-D in coconut (*Cocos nucifera*) explants. In: Lumsden, P.J., Nicholas, J.R. and Davies, W.J. (Eds.), *Physiology, Growth and Development of Plants in Culture.* Kluwer Academic Publishers, Netherlands. pp. 284-288.

Pan, M.J. and Staden, J.V. (1998). The use of charcoal in *in vitro* culture. *Plant Growth Regulation*, **26**: 155-163.

Pannetier, C. and Buffard-Morel, J. (1982). Production of somatic embryos from leaf tissues of coconut, *Cocos nucifera* L. In: Proceedings of the 5th International *Plant Tissue Culture Congress*, Tokyo, Japan.

Peak, K., Hahn, E. and Son, S. (2001). Application of bioreactors for large-scale micropropagation systems of plants. *In Vitro Cellular and Developmental Biology-Plant*, **37**:149-157.

Perera, P.I.P. (2003). Cytological Examination of Microspore Development for Microspore and Anther Culture of Coconut (*Cocos nucifera* L.) cv Sri Lanka Tall. *Cocos*, **15**: 53-59.

Perera, P.I.P., Hocher, V., Verdeil, J.L., Bandupriya, H.D.D., Yakandawala, D.M.D. and Weerakoon, L.K. (2008). Androgenic potential in coconut (*Cocos nucifera* L). *Plant Cell Tissue and Organ Culture* **92**: 293-302.

Perera, P.I.P., Hocher, V., Verdeil, J.L., Doulbeau, Yakanda-wala, D.M.D. and Weerakoon, L.K. 2007. Unfertilized ovary: A novel explant for coconut (*Cocos nucifera* L.) somatic embryogenesis. *Plant Cell Report*, **26**: 21-28.

Perera, P.I.P., Kularatene, J.D.J.S. and Weerakoon, L.K. (2011). Effect of gelling agent and selective sub-culturing on hyperhydricity in anther-derived coconut embryos. *CORD*, **27 (2):** 26-37.

Perera, P.I.P., Yakandawala, D.M.D., Hocher, V., Verdeil, J.L. and Weerakoon, L. K. (2009). Effect of growth regulators on microspore embryogenesis in coconut anthers. Plant Cell Tissue Organ Culture, **96**: 171-180.

Pérez Núñez, M.T., Chan, J.L., Sáenz, L., González, T., Verdeil, J.L. and Oropeza, C. (2006). Improved somatic embryogenesis from *Cocos nucifera* (L.) plumule explants cultured *in vitro. In Vitro Cellular and Developmental Biology-Plant*, **42**: 37-43.

Perez-Nunez, M. T., Souza, R., Saenz, L., Chan, J.L., Gonzalez, T., Zuniga, J.J. and Oropeza, C. (2009). Detection of a *SERK*-like gene in coconut *in vitro* cultures and analysis of its expression during the formation of embryogenic callus and somatic embryos. *Plant Cell Reports*, **28**: 11–19.

Rabechault, H., Ahée, J. and Guenin, G. (1970). Colonies cellulaires et forms embryoides obtenues *in vitro* a partir de cultures d'embryons de Palmier a huile (*Elaeis guineensis* Jacq. var. dura Becc.). *Comptes Rendus de l'Académie des Sciences*, **270:** 3067–3070.

Rajesh, M.K., Radha, E., Sajini, K.K. and Karun, A. (2014). Polyamine-induced somatic embryogenesis and plantlet regeneration *in vitro* from plumular explants of dwarf cultivars of coconut (*Cocos nucifera*). *Indian Journal of Agricultural Sciences*, **84** (4): 527-30.

Rajesh, M.K., Radha, E., Sajini, K.K., Karun, A. and Parthasarathy, V.A. (2005). Plant regeneration through organogenesis and somatic embryogenesis from plumular explant of coconut. *Journal of Plantation Crops*, **33(1)**: 9-17.

Rajesh, M.K., Fayas, T.P., Naganeeswaran, S., Rachana, K.E., Bhavyashree, U., Sajini, K.K. and Karun, A. (2016). *De novo* assembly and characterization of global transcriptome of coconut palm (*Cocos nucifera* L.) embryogenic calli using Illumina paired-end sequencing. *Protoplasma*, **253(3)**: 913-28.

Raju, C.A. (1984). Effect of methyl bromide fumigation on the fungi associated with seed coconuts. *Philippine Journal of Coconut Studies*, **9:** 1-2.

Rethinam, P. (2006). Asian and Pacific Coconut Community activities, achievements and future outlook. In: Adkins, S.W., Foale, M. and Samosir, Y.M.S. (Eds.), *Coconut Revival: New Possibilities for the 'Tree of Life'. ACIAR Proceedings. pp.* 15-21.

Saénz, L., Azpeitia, A., Chuc-Armendariz, B., Chan, J.L., Verdeil, J.L., Hocher, V. and Oropeza, C. (2006). Morphological and histological changes during somatic embryo formation from coconut plumule explants. *In Vitro Cellular and Developmental Biology-Plant*, **42**: 19–25.

Samosir, Y.M.S. (1999). Optimization of somatic embryogenesis in coconut (*Cocos nucifera* L.). Ph.D Thesis, The University of Queensland, Australia.

Samosir, Y.M.S., Godwin, I.D. and Adkins, S.W. (1998). An improved protocol for somatic embryogenesis in coconut (*Cocos nucifera* L.). In: Drew, R.A. (Ed.), *Proceedings of the International Symposium of Biotechnology in Tropical and Subtropical Species*. Australia, Brisbane, pp. 467-475.

Samosir, Y.M.S., Godwin, I.D. and Adkins, S.W. (1999). The use of osmotically active agents and abscisic acid can optimise the maturation of coconut somatic embryos. *Current Plant Science and Biotechnology in Agriculture*, **35**: 341-354.

Sandoval-Cancino, G., Sáenz, L., Chan, J.L. and Oropeza, C. (2016). Improved formation of embryogenic callus from coconut immature inflorescence explants. *In Vitro Cellular and Developmental Biology-Plant*, **52**: 367-378.

Sankar-Thomas, Y.D. (2009). *In vitro* culture of *Camptotheca acuminata* (Decaisne) in Temporary Immersion System (TIS): Growth, development and production of secondary metabolites. *PhD Thesis. Biology Department*. University of Hamburg Hamburg. 136 p.

Schmidt, E.D.L., Guzzo, F., Toonen, M.A.J. and deVries, S.C. (1997). A leucine rich repeat containing receptor-like kinase marks somatic plant cells competent to form embryos. *Development*, **124**: 2049-2062.

Steinmacher, D.A., Clement, C.R. and Guerra, M.P. (2007). Somatic embryogenesis from immature peach palm inflorescence explants: Towards development of an efficient protocol. *Plant Cell Tissue and Organ Culture*, **89**: 15–22.

Teixeira, J.B., Söndahl, M.R. and Kirby, *E.G.* (1994). Somatic embryogenesis from immature inflorescences of oil palm. *Plant Cell Reports*, **13**: 247–250.

Tisserat, B. and Demason, D.A. (1980). A histological study of development of adventive embryos in organ cultures of *Phoenix dactylifera* L. *Annals of Botany*, **46**: 465–472.

Tsuwamoto, R., Yokoi, S. and Takahata, Y. (2010). *Arabidopsis* EMBRYOMAKER encoding an AP2 domain transcription factor plays a key role in developmental change from vegetative to embryonic phase. *Plant Molecular Biology*, **73**: 481-492.

Verdeil, J.L., Hocher, V., Huet, C., Grosdemange, F., Escoute, J., Ferrie're, N. and Nicole, M. (2001). Ultrastructural changes in coconut calli associated with the acquisition of embryogenic competence. *Annals of Botany*, **88**: 9-18.

Verdeil, J.L., Huet, C., Grosdemange, F. and Buffard-Morel, J. (1994).) Plant regeneration from inflorescence of coconut (*Cocos nucifera* L.): Evidence for somatic embryogenesis. *Plant Cell Reports*, **13**: 218-221.

Vidhanaarachchi, V.R.M. and Weerakoon, L.K. (1997). Callus induction and direct shoot formation in *in vitro* cultured immature inflorescence tissues of coconut. *Cocos*, **12:** 39-43.

Weerakoon, L.K. (2004). Coconut tissue and embryo culture research in Sri Lanka: Current developments and future challenges. In: Peiris, T. S. G. and Ranasinghe, C. S. (Eds.), *Proceedings of the International Conference of the Coconut Research Institute of Sri Lanka*, Lunuwila. pp. 41-61.

Ziv, M. (2005). Simple bioreactors for mass propagation of plants. *Plant Cell Tissue and Organ Culture*, **81**: 277–285.

Chapter 3

Oil Palm

☆ *Jayanthi Madhavan and Pranab Kumar Mandal*

1. Introduction

Oil palm (*Elaeis guineensis* Jacq.) accounts for 30 per cent of world oil production and is the largest vegetable oil traded (Mielke, 2013). Production of palm oil from a unit area is 7-10 times more than that of any other oil seed crop. Palm oil is mainly used as food (90 per cent) and as oleochemical substitutes for mineral oil (10 per cent) (Corley and Tinker, 2003). Palm oil is the cheapest vegetable oil available, and the current interest in biodiesel has led to a high demand and all-time high prices for the commodity (Yusof, 2007). High-yielding oil palm varieties developed by breeding programmes can produce over 20 tonnes of fresh fruit bunches (FFB)/ha/yr under ideal management, which is equivalent to 5 tonnes oil/ha/year (excluding the palm kernel oil) (FAO, 2002). Oil palm breeding is hampered by the long breeding cycle, cross pollinated nature and the propagation method *i.e.* via seed. Oil palm is an exceptional example where a quantum leap in yield improvement was obtained through a single gene. It is reported that the "first wave" in yield improvement of up to 30 per cent was by switching from the *dura* to the *tenera* (D x P) planting materials. The approach of enhancing competitiveness through expansion of land for cultivation has limited options. A viable alternative is to increase productivity per unit land area, leading to the increase in FFB yield and oil extraction rate (OER). It has been reported that the best experimental plot produced oil yield of 8.6 t/ha/yr (Corley *et al.*, 1986), selected progenies producing upto 12.2 t/ha/yr (Rajanaidu *et al.*, 1990), individual palm yielding 15 t/ha/yr and the maximum theoretical oil yield of 18.2 t/ha/yr (Corley, 1998). Kushairi *et al.* (2006). reported that cloning such palms would provide high-yielding planting materials for the industry, which would be the "second wave" in yield improvement in the oil palm.

It is impossible to propagate oil palm by conventional vegetative propagation since oil palm lacks any axillary branching or axillary meristem. Due to this, propagation by *in vitro* methods has gained significant importance. Oil palm is a truly allogamous species and commercial plants are hybrids called as *teneras* obtained from crosses of *duras* and *pisiferas*. These plantations of hybrids from controlled crosses consist of genetically dissimilar individuals. Thus the identical reproduction of elite germplasm requires the use of *in vitro* cloning. As compared to seed production, tissue culture of oil palm offers several advantages. It allows rapid multiplication of uniform planting materials with desired characteristics. This enables improvement of planting materials using existing individuals which have all or most of the desired qualities such as good oil yield and composition, slow vertical growth and disease resistance. Additionally, it also opens new avenues for producing novel planting materials via genetic engineering, because tissue culture provide means for regeneration of tissues transformed with genes for traits of interest. Oil palm tissue culture is employed both as a means for producing good *tenera* palms for commercial planting and to multiply good parents (both *dura* and *pisifera*) for seed production. It is also practiced to expedite the exploitation of progenies from interspecific *E. oleifera* X *E. guineensis* crosses.

2. History and Current Status of Oil Palm Tissue Culture

The first attempt at tissue culture of oil palm was made by Startisky (1970), who cultured the leaf primordia and other tissues, by destructive sampling, on a modified Miller medium. This was followed by the first successful reports by Jones (1974) and Rabechault and Martin (1976) by using oil palm leaf tissues. The first clonal plants were planted in Malaysia in 1977 (Corley and Tinker, 2003). The early success of plantlet production in the 1970s (Jones, 1974; Rabéchault and Martin,1976) inspired many oil palm organizations to exploit the *in vitro* propagation technique. In 1986, the problem of clonal mantled fruits, floral and vegetative abnormalities of clonal palms, planted at United Plantations Berhad, Malaysia, was brought to the attention of the scientific community during the *Colloquium on Breeding and Selection for Clonal Oil Palm* in 1985, organized by the International Society for Oil Palm Breeders (ISOPB). This caused a major furore among oil palm tissue culturists. Later, Corley *et al.* (1986) published the first report on abnormality in clonal oil palm. The mantling problem is very severe and involves the inflorescence where the carpels develop into a thick fleshy tissue surrounding the fruit like a mantle over it. These fruits contain little or no oil. In the following two decades, as more information and understanding of the tissue culture process and the problems arising from it accumulated, there was renewed interest to proceed with large-scale propagation of oil palm clones to increase productivity. In countries where the research on oil palm tissue culture has advanced, higher yields have been obtained from oil palm clonal materials. If the yield performance of clones is compared, it was reportedly superior to those of commercial D x P seedlings (Khaw and Ng, 1997). The clones yielded at least 25 per cent higher than seedlings (Rohani *et al.*, 2000; Tan *et al.*, 2003). In Sarawak, Malaysia, the cumulative oil yields of 6-7 year-old clones exceeded that of the D x P by 10 per cent to 37 per cent (Simon and Koh, 2005). Similarly, in Sabah,

Malaysia for 9-10 year-old palms, the clones maintained an advantage over D x P cumulative oil yield by 28 per cent to 55 per cent. Felda (Felda brochure) and UPB reported that the oil yield increment of clones was more than 20 per cent. Generally, the average cumulative oil yield produced in clones gave an advantage over D x P by 44 per cent or 48 t/ha (Simon and Koh, 2005).

3. World Scenario of Oil Palm

There are four different planting materials produced in oil palm. The most common are the seeds obtained from tenera palms *i.e.* the cross between duras and pisiferas (D X P). The other three are cloned materials, mono/bi clonal seeds and seeds from interspecific crosses. The details of the production of these are mentioned below:

3.1. D X P Seeds

Indonesia, which started producing D X P seeds since 1995, has been the largest producer of such seeds followed by Malaysia (Kushairi *et al.*, 2006). Other major oil palm D x P seed producing countries in Asia are Papua New Guinea, Thailand and India. Thailand has become a significant seed producer in the past five years. In Central-South America, Costa Rica is a major oil palm seed producer, producing 30 million D x P seeds annually. Centre de Coopération Internationale en Recherche Agronomique pour le Développement (CIRAD), a French based organization, has developed a network in Central-South America in seed production. It is expected that a significant amount of CIRAD seed production will be in Latin America. The major oil palm seed producers in Africa are CIRAD and its partners in Côte d'Ivoire (CNRA), Cameroon (IRAD, SOCFINCO), Benin (INRAB), Ghana (OPRI), Nigeria (NIFOR) and Democratic Republic of Congo (Unipalma). An estimated 25 million oil palm seeds per annum are produced in Africa (Kushairi *et al.*, 2006).

3.2. Clonal Material

Currently, there are several countries producing oil palm through tissue culture. However the three countries which are major producers of clonal materials are Malaysia, Indonesia and Costa Rica. In Malaysia, there are 12 tissue culture laboratories, including MPOB (Malaysian Palm Oil Board), producing clonal oil palm ramets. However, the production of ramets is far below the amount needed by the industry. It is reported that the tissue culture industry is expanding their existing tissue culture facilities, while more new ones are being established. About three million oil palm tissue culture plantlets are produced annually worldwide (Kushairi *et al.*, 2006). In Malaysia, the current production of clonal oil palm planting materials is two and half million ramets annually from 11 commercial tissue culture laboratories. The annual requirement worldwide is around 40 million ramets. Most tissue culture labs, such as Advanced Agriecological Research (AAR) and Felda, are poised to produce about one million ramets per year. Productions by these companies are expected to increase to at least two million each over the next 3-5 years. The mean oil yield of AAR clones is 7.5 t/ha/yr compared to 6.5 t/ha/yr that of its D x P hybrid seeds.

3.3. Clonal Seeds

Another alternative to developing high yielding planting materials is through clonal seed production (Veerappan *et al.*, 2000; Soh, 2005). United Plantations Berhad (UPB) in Malaysia is the pioneer company in the world to produce biclonal seeds. The estimated annual production by UPB is one million D x P biclonal seeds. In this method of seed production, the *dura* and/or *pisifera* palms are cloned as parents. The clonal parents are crossed to produce seeds the same way as that in conventional D x P seed production. The selection of parents for cloning is largely based on specific combining ability (SCA). Biclonal seeds are produced when both the parents are cloned, while semi-clonal seeds are those with either of the parent is a clone. It is preferable to clone the maternal parent (*dura*) and use normal (non-clonal) progeny-tested *pisiferas* to produce the D x P clonal seeds. This will ensure a large number of female *dura* parents are available for use in seed production. It is sufficient to have a small number of *pisifera* parents as the pollen source.

3.4. Interspecific Hybrid Seeds

It is estimated that 2.5 million *E. oleifera* x *E. guineensis* (OG) interspecific hybrid seeds are produced worldwide. The production is localized in South America. Interspecific hybrids are relatively tolerant to spear rot and the palms are short and compact with more liquid oil compared to the D x P intraspecific hybrids. The OG interspecific hybrids based on Taisha (Ecuador) *E. oleifera* is expected to produce oil palm yields close to that of the D x P.

4. Tissue Culture Techniques

Oil palm is a monocotyledonous plant, and vegetative propagation is made possible only *via* tissue culture. Oil palm tissue culture is time consuming as the whole cycle including callusing and embryogenesis processes takes around 2-5 years. The basic method in micropropagation of oil palm involves culturing the meristematic tissues in a callus induction media followed by sub culture to induce embryogenic callus and subsequent production of embryoids. Figure 3.1 depicts the callus induction obtained from different explants. The most reported pathway of regeneration in oil palm is via somatic embryogenesis (SE). For this, the first step is the induction of callus that is further induced to form embryoids. Once the embryoids are formed they are transferred to the proliferation media. After the sufficient multiplication has taken place, it is allowed for shoot development and rooting and then transferred for hardening. The different process of SE obtained with oil palm explants is given in Figure 3.2. Field transfer follows this. The whole procedure is under controlled light and atmosphere (Hartley, 1988).

5. Factors that Affect Clonal Propagation

5-.1. Explant

The first and foremost step in oil palm tissue culture is the selection of suitable explants from the mother plant or the ortet. If clonal propagation of elite palm is desired, then explants should be taken from the mature palm. Several explant sources have been used to establish tissue cultures of oil palm and they are mature

Figure 3.1: Callus Induction from Different Explants: (a) Spear leaf, (b) Roots, (c) Inflorescence and (d) Zygotic embryos.

embryos, immature embryos, seedling leaves, roots, spear leaves, inflorescences and immature inflorescences. It appears that the choice of explants depends largely on the resources (in terms of breeding and planting materials) and the objectives (whether for the production of elite individuals or the production of clonal seeds) of the individual organization. Zygotic embryos have several distinct advantages: they have a high response to culture induction; they are generally free of endophytes and pathogens; and minimal damage to the mother plant is required for explant harvest. For this reason, zygotic embryos have been the most common explant used for SE studies (Teixeira *et al.*, 1993; Rajesh *et al.*, 2003). Zygotic embryo maturity affects culture response: immature zygotic embryos produced more embryogenic tissue than mature zygotic embryos (Teixeira *et al.*, 1993). Intact zygotic embryos

Figure 3.2: Different Stages of Somatic Emrbyogenesis and Plantlet Regeneration: (a) Embryogenic callus, (b) Formation of somatic embryos, (c) Germination of somatic embryos and (d) Plantlet obtained form somatic embryo.

may be cultured, or individual tissues can be isolated and used as explants. Direct embryogenesis has also been reported from oil palm cotyledonary nodes (Jayanthi *et al.*, 2011). The main problem of using shoot meristems is that each palm will have only one shoot apical meristem and their harvest is destructive to the source palm. Full recovery is therefore necessary before sampling is undertaken. Sampling from spear leaf is depicted in Figure 3.3. The apical meristems are harvested by first removing the outer leaves to expose the apical meristems and then excising them. These need to be surface sterilised for preventing contamination. Inflorescence explants also seems to be a very promising material for culturing. The advantage of using immature oil palm inflorescences is that they are protected by sheaths that prevent contamination and it precludes the need for sterilisation (Teixeira *et al*,. 1994; Jayanthi *et al.*, 2015). But inflorescence tissues materials may not be readily available in all age group of palms. Excising of young leaf tissue is advantageous as it is

Figure 3.3a-d: Sampling of Explants from Spear Leaf.

completely enclosed by the leaf bases of older leaves and therefore does not require severe disinfection (Hartley, 1988). Further, young leaf tissues are abundant and represent true to type nature of mother plant. However, they have less embryogenic potential than shoot tip explants and there are few reports of leaf-based SE in oil palm. Root tissue can be easily obtained but there is a danger of taking in roots from an adjoining palm. To avoid this, soil can be mounted up around the palm base and root growing into it from the required palm can be taken; however, this requires vigorous disinfection (Wooi, 1984). A detailed review of the explant used, media details and response are provided in Table 3.1.

5.2. Genotype

It is reported that genotype has a large influence on *in vitro* response. The basis for such differences is not explained properly due to the complex dynamics of epigenetics and molecular regulation during callus formation and SE (Us-Camas *et al.*, 2014; Elhiti *et al.*, 2013). Two varieties of interspecific *Elaeis oleifera* × *E. guineensis* hybrids showed a varied response for embryogenic callus formation on media supplied with activated charcoal and 2,4-dichlorophenoxyacetic acid (2,4-D); one

Table 3.1: Details of Tissue Culture Experiments Reported in Oil Palm

Sl.No.	*Explant*	*Media*	*Response*	*Reference*
1.	Meristem tissue	Modified Miller's media + Kinetin+NAA	Leaves: roots and callus nodules	Startisky (1970).
2.	Seedling tissues	Liquid medium Heller's salts+ 2,4-D+KIN +NAA	Nodular cell colonies	Rabechault *et al.* (1970).
3.	Stem apex and leaf bases	Liquid mediaum Knop. Heller or MS salts+ 2,4-D+KIN or BA +NAA or IAA	Callus, roots shoots but no plantlets	Rabechault *et al.* (1972).
4.	Seedling roots, shoot apex from germinated seedlings	MS salts and growth regulators	Callus and embryoids	Jones (1974).
5.	Leaf tissues	MS media +2,4-D	Callus, embryoids and rooted shoots	RabechaultandMartin(1976).
6.	Embryos	MS basal devoid of KI + 2,4-D	Callus	Smith and Thomas (1973).
7.	Roots from mature palms	MS salts + growth regulators	Callus, embryoids and plantlets	Wooi *et al.* (1981)
8.	Young leaf sections from nursery and mature palms	Modified MS + growth regulators	Callus embryoids and plantlets	Hanower and Pannetier (1982).
9.	Roots and young leaves from aseptically grown seedlings	MS + 2,4-D or NAA	Callus, embryoids, shoots and rooted shoots	Paranjothy and Othman (1982).
10.	Leaves and roots from pisifera seedlings	MS + NAA or 2,4-D	Callus, embryois and shoots	Nwanko and Krikorian (1983).
11.	Young leaves of 6 month old seedlings	MS + 2,4-D	Callus and embryoids and rooted shoots	Thomas and Rao (1985).
12.	Seedling leaves	MS + 24-D	Callus. Somatic embryos and plantlet regeneration	Raju *et al.* (1989).
13.	Nodular calli from embryogenic immature leaf	Modified MS, Nitch's micro, Morel and Wetmore vitamins	Embryogenic suspension culture with embryoids and shoots	De-Touchet *et al.* (1991).
14.	Immature zygotic embryos	Modified Y3 media with 2,4-D +PVP+cysteine	Callus, embryogenic tissues and plantlet development	Texeira *et al.* (1993).
15.	Immature inflorescence of pisifera	Modified MS + 2, 4-D	Embryogenic calli an somatic embryos and plantlets	Texeira *et al.* (1994).)

Contd...

Table 3.1–*Contd...*

Sl.No.	*Explant*	*Media*	*Response*	*Reference*
16.	18 month old *dura* and 6 month old tenera seedlings	1/2MS +2,4-D	Embryogenic calli, shoots and plantlets	Karun and Sajini (1996).
17.	Mature leaf	Y 3 media with NAA	Embryogenic calli and plantlets	Sogeke (1996).
18.	Mature embryos	MS + 2,4-D	Callus and somatic embryos	Kanchanapoom and Domyoas (1999).
19.	Mature embryos	Modified MS media with 2,4-D and 2,iP and Blaydes medium with polyamines	Callus, somatic embryosm secondary somatic embryos and shoot meristemoids	Rajesh *et al.* (2003).
20.	Primary callus from leaves	MS + dicamba+ ascorbic acid	Secondary SE and plantlet regeneration	Techato and Hilae (2007).
21.	Mature embryos	MS + dicamba+ ascorbic acid	Embryogenic calli and haustorium embryos and shoots and plantlets	Chehmalee and Te-chato (2008).
22.	Embryogenic calli	DF liquid medium +2,4-D + Kinetin	Germination of somatic embryos and plantlets	Sumaryono (2008).
23.	Zygotic embryos	MS medium + picloram	Embryogenic calli and plantlet formation	Balzon *et al.* (2013).
24.	Explants from regenerated plants (Recloning)	N6 + 2,4-D	Embryogenic calli and plantlet formation	Thuzar *et al.* (2012).
25.	Mature zygotic embryos	MS + picloram	Embryogenic calli and plantlet formation	Silva *et al.* (2012).
26.	Inflorescence	MS + picloram	Embryogenic calli	Guedes *et al.* (2011).
27.	Zygotic embryos	MS	Embryogenic calli and plantlet formation	Jayanthi *et al.* (2011).
28.	Zygotic embryos	N6 + 2,4-D	Embrogenic calli and plantlet formation	Thuzar *et al.* (2011).

Contd...

Table 3.1–*Contd...*

Sl.No.	*Explant*	*Media*	*Response*	*Reference*
29.	Thin cell layer	MS + 2,4-D and picloram	Embrogenic calli and plantlet formation	Periera *et al.* (2010).
30.	Embryogenic calli (temporary immersion system)	MS medium + amino acids	Multiplicaiton of embryogenic calli	Marbun *et al.* (2015).
31.	Inflorescence	MS + 2,4-D + picloram	Embrogenic calli and plantlet formation	Jayanthi *et al.* (2015).

showed highest rate of embryogenic tissue formation on media with lowest 2,4-D concentration whereas the other variety displayed the best response at the highest 2,4-D concentration (Alves *et al.*, 2011).

5.3. Media Components and Additives

In the absence of specific or standard callogenic medium composition suitable for all the genotypes, most laboratories formulate media composition that optimally initiates callus and subsequently embryos. The basic media reported in most cases is the Murashige and Skoog media (MS) (Murashige and Skoog, 1962) with several modifications. Eeuwens media (Eeuwens, 1976) has also been reported in the oil palm tissue culture. Despite several reports regarding *in vitro* regeneration in oil palm, the method does not seem to be common and simple. Specific explants require specific media. For example, different working groups (Teixeira *et al.*, 1993, 1994, Jayanthi *et al.*, 2011, 2015) have reported a different media for regeneration from zygotic embryos and inflorescence. Texeira *et al.* (1993) reported the use of Y3 media with additives for zygotic embryos and MS media with additives for inflorescences. In palms, somatic embryo protocols also have several stages like induction, embryo differentiation, culture multiplication, maturation, and conversion. Each stage uses media comprised of basic media components, including basal salts, vitamins, and carbohydrates, plant growth regulators, and other additives, such as activated charcoal (AC). Several optimization studies have compared various basal media formulations, though none have been shown to be universally superior. Moreover, there are a large number of variables involved in the culture process and different laboratories obtain different results from superficially similar treatments. Sometimes, even in the same laboratory, results are not always reproducible (Corley and Tinker, 2003). Hence it is always essential to develop a simple media which works out for the particular laboratory depending on several factors. Reducing the MS basal concentration to half has also been tested in oil palm. Hilae and Te-chato (2005) reported that full strength MS was better than half strength MS medium. Auxin type and concentrations have been compared in several studies. 2, 4-D and NAA are the most common auxins reported for callus induction but several other auxins have also been reported. Use of picloram has been reported from several palms including oil palm. Picloram is reported to be a better auxin than 2,4-D and improved regeneration in case of arecanut (Karun *et al.*, 2004), peach (Steinmacher *et al.*, 2007b), oil palm (Jayanthi *et al.*, 2015), and açaí palm (Scherwinski-Pereira *et al.*, 2012). Jayanthi *et al.* (2015). have also reported that 2,4-D was preferable to napthalene acetic acid (NAA) for SE induction in oil palm cultures. A common finding was that NAA was capable of inducing SE but high concentrations caused culture oxidation and death. Cytokinins like BAP has also been reported from several palms and they improve SE during embryo differentiation. Activated charcoal is the main antioxidant that is used for absorption of phenolic compounds in palm tissue culture and its use has been reported in all palms (Valverde *et al.*, 1987; Karun *et al.*, 2004; Huong *et al.*, 1999; Steinmacher *et al.*, 2007 a,b c; Luis and Pereira, 2014). However the use of activated charcoal should be coupled with the use of high concentration of auxins since most of these auxins are also absorbed by activated charcoal. It has been reported that use of 100-140 mg/l of 2,4-D in the presence activated charcoal will effectively supply only 2-2.8 mg/l of the auxin and the remaining will be absorbed by activated charcoal. Additives like adenine sulfate in MS media are added for callus induction from immature leaves (de-Touchet *et al.*, 1991; Aberlenc-Bertossi *et al.*, 1999). ABA is reported to help in embryo differentiation in case of immature zygotic embryos (Teixeira *et al.*, 1993). Use of polyamines like putrescine, spermine and spermidine, have been reported to improve SE in oil palm (Rajesh *et al.*, 2003) and date palm (Hegazy and Aboshama, 2010). Most of the calli grow slowly, but suitable media can induce a faster growing callus; this can be subcultured with regular transfer until embryoids develop. Once embryogenic calli are developed, they are transferred to a medium which induces proliferation. Shoots appear and are removed as they grow on the proliferating material, which can then be recirculated to produce more shoots. The latter are grown in a rooting medium and, after hardening, can be planted in soil or compost (Hartley, 1988). Once a good root system has been obtained in the laboratory, a ramet can be transferred to a prenursery or to a 'ramet house' and then to a pre nursery. It is important to maintain high humidity around the young leaves for at least four weeks and to avoid temperatures above 32°C, but nevertheless, to ensure a light transmission of at least 17 per cent (Turner, 1984). Use of suspension cultures for multiplication of embryogenic calli have been reported to improve the multiplication rate in oil palm (De Touchet *et al.*, 1991; Teixeira *et al.*, 1995; Kramut and Te-Chato, 2010). Suspension culture is the preferred method when commercialization is intended. Temporary immersion systems in which cultures are exposed to liquid media for fixed time are also advantageous than solid cultures because they facilitate greater gas exchange and avoids hyperhydricity (Etienne and Berthouly, 2002).

6. Challenges in Oil Palm Tissue Culture

The tissue culture process is costly and labour intensive and the entire process requires very specialized infrastructure to ensure a clean and controlled environment, ample laboratory space to house the cultures and most importantly skilled workers. The other challenges in tissue culture of oil palm are mentioned below:

6.1. Availability of Ortets

One of the most important requirements for a successful venture in the production of oil palm clonal materials is the availability of high-quality ortets, arising from breeding and selection programmes. A breeding programme demands large areas for ortets selection. Furthermore, a large-scale propagation of oil palm clones requires a large number of elite ortets. Based on an ortet selection in progeny trials, some 2 per cent to 11 per cent of palms are selected as ortets. Currently, palms selected as ortets are coincidental, *i.e.* selected from trials with other objectives, such as progeny trials in breeding and selection programmes, and was not initially meant for ortet selection. As such the number of ortets selected would be few compared to total palms planted. Thus, it would be beneficial if 'ortet gardens' were to be developed for ortet selection (Kushairi *et al.*, 2006). This could be carried out by selecting the cream of high yielding *dura* and *pisifera* palms to create high yielding progenies. Unlike D x P seed production, where majority of the parents are selected based on general combining ability (GCA), parents in crosses to create progenies for ortet gardens should be based on specific combining ability (SCA). The same parents are repeatedly used to produce the desired number of D x P (*tenera*) progenies for ortet gardens. The other source for palms to be planted in ortet gardens is the reclones of the proven clonal palms.

6.2. Slow Growth

Oil palm tissue culture is much more difficult and inefficient as compared to its monocot counterparts because of its slow and poor growth and low regeneration habit. Beside low and asynchronous embryo development, there are marked influence shown by genotype and cultural conditions in the micro propagation process. Several reports describe oil palm callus as slow growing (Smith and Thomas, 1973; Turnham and Northcote, 1982; Jones, 1974), with doubling times varying from 30-40 days on solid medium. Inoculation into liquid medium has resulted in reducing the multiplication time of cultures (Wong *et al.*, 1999). By using such liquid cultures Texeira *et al.* (1995) has obtained high rates of growth for oil palm cells. The entire process of oil palm tissue culture from the explant inoculation to regeneration and field testing takes around eight to ten years (Mutert and Fairhust, 1999).

6.3. Low Embryogenesis Percentage and Embryoid Survival

After so many years of research, formation of callus and somatic embryos remain one of the major bottlenecks in oil palm tissue culture. Most of the commercial laboratories are reporting a mean percentage of embryogenesis at about 6 per cent. This value differs according to the genotype of the palm selected for cloning. It has been reported that embryogenesis rates for *dura* and *pisifera* palms were similar but both were significantly less than tenera palms. (Wooi, 1995). Even with tenera palms some families are reported to be easy for cloning than others (Paranjothy, 1987; Duval *et al.*, 1987). The rate of callogenesis of oil palm explants remains low, at about 19 per cent. It was also reported that the average rate of embryogenesis in leaf derived callus ranged from 3 per cent to 6 per cent (Low *et al.*, 2008). However, in inflorescence derived explants, high rates of callus induction have been reported but a low SE percentage has been observed (Jayanthi *et al.*, 2015).

6.4. Contamination

In a laboratory condition when cultures are infected due to bacterial or fungal contaminants it is possible to manage them by careful transfer techniques. But for large scale production, losses can be alarmingly high. Another source of contamination reported is cross contamination when subculturing process is done. Pest infestation of the culture rooms and cultures will also result in considerable loss. It is mentioned that mite infestation may wipe out production some times. (Wooi, 1995).

6.5. Culture Stability

Cytology studies and fruit characters of plants regenerated through tissue culture in early 1980's suggested that oil palm cultures were stable. But the stability of oil palm cultures were questioned when variation as floral abnormalities were reported in 1986 (Corley *et al.*, 1986). Somaclonal variation is common in oil palm tissue culture since all the reported regeneration protocols in oil palm are via a callus phase. Though somaclonal variation is desired in other agricultural crops, in case of oil palm, emphasis is on true to type multiplication of elite palms and hence variation is not desired.

Variations in length, stiff, erect and drooping nature of palms, stunted palms, palms which remain juvenile for a long time, ramets with white stripes, brown petioles, flat tops etc may be observed in vegetative characters. In inflorescence, terminal inflorescence and truncated leaf syndrome may be observed. But the most devastating and well-studied variation is the one causing mantled flowers, which has a direct effect on the bunch yield. Extensive research has been carried out to find out the factors and the events leading to mantled flowers and bunch failure. Till date, there is no fool proof method to avoid somaclonal variation. However this variation has been brought down to a great extent by suitable modifications of culture procedures. Now a variation upto 5 per cent is acceptable by commercial firms. An average of 5-10 per cent of palms may be affected according to Rival *et al.* (2000). The worst affected palms have 100 per cent mantled and parthenocarpic fruits in every bunch and suffer complete bunch failure. There are also reports that the oil content from normal bunches of partially abnormal palms were same as that from normal bunches of the same clone (Corley, 1993; Maheran *et al.*, 1995).

7. Mantled Flowers: A Somaclonal Variant

Since oil palm is a commercial crop, any factor affecting the yield would be of great concern. During the 1990's, commercial exploitation of tissue culture techniques for mass clonal propagation of elite oil palm planting material was held back by an unacceptably high incidence of bunch failure in some clones. It was in 1986 when it was observed that some of the oil palm clones planted in 1982 produced abnormal flowering. They had high incidence of flowers with mantled characters. This was due to the feminization of the androecium in both male and female flowers with the result that in its most extreme form all the stamen primordia develop as supplementary carpels. The flowers are described as "mantled" (Corley *et al.*, 1986) and are non-functional and there is no fruit development resulting in parthenocarpy and severe bunch failure. Due to this, several hundred hectares of clonal palms were uprooted and destroyed. All laboratories engaged in large-scale clonal propagation of oil palm have reported varying levels of mantled flowering affecting many clones of different genetic origin. Generally the incidence of mantling increases with the time cultures that have been maintained at the embryoid multiplication stage though some clones can be maintained for many years without appreciable levels of mantled flowering. In some less extreme cases, only some female flowers are affected and these developed one or other supplementary carpels. Such palms will produce a reduced amount of fruits and will often revert to completely normal flowering within a few years (Rival *et al.*, 1998b; Eeuwens *et al.*, 2002). However, this kind of abnormality has a direct effect on the oil production. Some clones were more susceptible to abnormal flowering than others (Durand-Gasselin *et al.*, 1999). An average of 5-10 per cent of palms may be affected according to Rival *et al.* (2000). The worst affected palms have 100 per cent mantled and parthenocarpic fruits in every bunch and suffer complete bunch failure. The oil content from normal bunches of partially abnormal palms were same as that from normal bunches of the same clone (Corley, 1993; Maheran *et al.*, 1995). Hence, this led to an upsurge of interest and studies to find out the causes and remedies for this abnormal flowering.

8. Factors Leading to Mantled Flowering

Several factors were thought to be responsible for this mantled flowering. Corley *et al.* (1986) found that frequency of mantling increased in successive years of planting of the clones of the same plant. This suggested that time spent in culturing may be a causal factor. It was believed that proportion of abnormal flowering increased with the number of sub-cultures a clone had undergone. However, Durand-Gasselin *et al.* (1999) found no relation between time in culture and degree of abnormality with some clones still giving normal plants after 11 years. Eeuwens *et al.* (2002) described the results of a 15 year programme of trials on the effects of culture media on the incidence of abnormality and it is clear from their work that the medium on which embryoids are cultured has a large effect. In general, a relatively long transfer interval (8 weeks) and a high auxin/cytokinin ratio resulted in the lowest incidence of mantled flowering with all types of material. Reducing the transfer interval down to 2 or 4 weeks and or using a media with relatively high concentrations of kinetin (0.25mg 1-1) and low concentrations of naphthalene acetic acid (upto 0.1mg/l), resulted in a high incidence of mantled flowering (above 30 per cent). Exclusion of plant growth regulators from the embryoid multiplication medium did not prevent some mantled flowering (12 per cent). Several studies using biochemical and cytogenetic markers did not reveal any basis for the production of abnormal clones (Shah and Ahmed Parveez, 1995). Two types of calli were observed in oil palm tissue culture- one is characteristically compact organized and nodular, called nodular compact callus (NCC), and the other is soft and friable fast growing callus, called fast growing callus (FGC) (Pannetier *et al.*, 1981; Hanower and Hanower, 1984). The nodular calli is stable and produces good clonal fidelity whereas FGC was involved in malformation of oil palm inflorescences (Marmey *et al.*, 1991; Duval *et al.*, 1988). They found low levels of endogenous cytokinin in FCG and suggested that malformation of oil palm inflorescences is a physiological disorder associated with tissue disorganization and abnormally low endogenous cytokinin levels. Investigations of cytokinin content indicated that higher levels were found in nodular callus and zygotic embryos than in friable calli (Besse *et al.*, 1992). Analysis of two cell lines of a single oil palm genotype, which produced clones with only normal flowers or only abnormal flowers respectively, revealed that cytokinin levels were significantly lower in embryoids and shoots of the ramets produced by the line with abnormal flowers. The abnormal inflorescences had higher concentration of biologically active zeatin riboside and dihydrozeatin- a glucoside and less zeatin than the normal inflorescence at a comparable stage of development. In all other cases, however, differences in cytokinin level between clones exceeded those found between abnormal and normal flowers (Jones, 1990; Jones *et al.*, 1995). Hence the theory of endogenous cytokinins did not explain the reason for the formation of abnormal flowers. The flow cytometry method was applied to study the ploidy level in *in vitro* cultures and in seed-derived adult oil palms and detect putative variations in DNA content between NCC, FGC and friable calli (FC). The three types of calli (NCC, FGC and FC) did not differ in their nuclear DNA content and reinforced the hypothesis of an epigenetic origin for somaclonal variation in oil palm (Rival *et al.*, 1997). Michellotti *et al.* (2007) has attributed somaclonal variation in oil palm to a multi-cause phenomenon. The mantled character is epigenetic in nature as demonstrated firstly, by the fact that reversion to a normal floral phenotype may occur in the field, secondly, in that weak non-Mendelian transmission of the abnormality occurs via seeds compared with strong transmission through tissue culture (Rival *et al.*, 1997, 1998b; Matthes *et al.*, 2001; Tregear *et al.*, 2002). Further all the recent molecular studies also point to an epigenetic basis of this phenomenon.

9. Advances in Tissue Culture of Oil Palm

9.1. Biomarkers and Gene Expression Studies

Molecular research for the tissue culture area has been geared towards developing biomarkers to address the issues of tissue culture amenity and abnormality. Part of it entails gene expression studies for both embryogenesis and abnormality. Expression of certain genes has been linked to embryogenic competence. Somatic embryogenesis responsive kinase (*SERK*) is a highly conserved gene involved in SE across many types of plant species. *SERK* transcription was detected in both callus and zygotic embryos of *E. guineensis* × *E. oleifeira* hybrids (Angelo *et al.*, 2013). Jouannic *et al.* (2007) reported the expression of an oil palm homolog of *KNOX*, *EgKNOX1*, in both somatic embryo and embryogenic callus tissues, while no expression was detected in non-embryogenic tissue. Genes activated during SE include the Aux/IAA gene *EgIAA9* isolated from oil palm (Ooi *et al.*, 2012), and the expression of both genes decreased in response to increasing concentration of exogenous 2,4-D. These authors suggested that endogenous auxin levels may be greater in palms with high capacity for SE. The same researchers, Ooi *et al.* (2013) reported that the potential for callogenesis in cultured leaf explants of oil palms was significantly correlated to the expression changes of a putative brassinosteroid leucine-rich repeat (LRR) receptor kinase (*EgBrRK*), a putative cytokinin dehydrogenase (*EgCKX*) and a putative response regulator type A gene (*EgRR1*). The larger reduction in the expression of *EgRR1* and another cytokinin responsive gene, *EgCK* REGULATED KINASE, in cultures exhibiting higher callogenesis rates suggested an increase in cytokinin signalling output and cytokinin levels. The use of these markers for the prediction of callogenesis rate in uncultured and one-day cultured leaf explants, was used to provide an early assessment of the callogenesis potential of oil palms. Lin *et al.* (2009) constructed suppression subtractive hybridization libraries to find differentially expressed genes during culture initiation and proliferation, finding almost 2000 expressed sequence tags with differential expression between the two stages of development, suggesting large changes in cellular activity. Biomarkers that are being developed will eventually lead to production of a diagnostic tool for clonal amenity and conformity. To date, some potential biomarkers (gene expression-based) have been identified and have undergone some validation. Biomarkers for embryogenesis may be used for screening stages as early as the explants. However, studies in MPOB indicated that the biomarkers may be specific to different tissue culture laboratories due to differences in genotypes and culturing protocol/media used. To avoid this tissue culture laboratories continue to practice 'basket sampling' to ensure that demand for elite oil palm planting materials can be met. A detailed review on oil palm genes isolated and genetic transformation studies are available for further reading [Mandal and Jayanthi (2011a, 2011b)].

9.2. Improving Efficiency in Tissue Culture Process

Low embryogenesis rates (3 per cent to 6 per cent) remain a stumbling block to large-scale ramet production. Since the new millennium, Malaysian Palm Oil Board (MPOB) has been actively developing innovations that can help improve the efficiency of the tissue culture process. Examples of technologies or methods to simplify parts of the tissue culture process include the "double-layer rooting" technique (Zamzuri, 1999; 2001) and "flameless sterilizer" (Zamzuri, 2002).

9.3. Innovations for Suspension Cultures

Suspension cultures not only add to the efficiency of tissue culture in oil palm but also open up the automation possibilities. Propagation via liquid media enables the increase in embryogenic cultures by several folds. However, as the conventional solid culture system does not permit the regeneration of cultures/ embryoids directly in liquid, the best option is to synergize the use of solid cultures with liquid systems (shake flask and bioreactor). In order to mass produce or for bulking up the production of ramets, the suspension culture system was developed (de Touchet *et al.*, 1991; Teixeira *et al.*, 1995; Wong *et al.*, 1999). The bioreactor is the preferred alternative if rapid large scale proliferation is needed (Tarmizi *et al.*, 2003). In relation to this, the MPOB developed the Fast Transfer Technique in liquid culture system (Tarmizi and Zaiton, 2005), the 2-in-1 MoSlim (MPOB Simple Impeller for liquid culture) with later upgraded to SLIM-FaTT (simple impeller with fast transfer technique) (Tarmizi and Zaiton, 2006a; 2006b) and the MPOB Modified Vessel (MoVess) (Tarmizi *et al.*, 2007). Besides regeneration of embryogenic aggregates from liquid culture system, these cultures could also be encapsulated as artificial seeds.

9.4. Barcoding for Monitoring and Recording

There is a need for proper management of information flow generated from the laboratory to the field as well as integration of information collected from other disciplines *e.g.* breeding and molecular biology. MPOB has developed a tissue culture database system using a relational database management software for computerized audit trail (Zamzuri, 2001b) and further enhanced with bar-coding for monitoring and recording purposes in OPTRACKS (Tarmizi *et al.*, 2003). OPTRACKS has been licensed to three oil palm agencies in Malaysia.

9.5. Molecular Markers to Assess Somaclonal Variant - Mantled Flowering

RAPD and SSR analysis was carried out to find out markers for differentiating normal/abnormal clones (Chowdhury, 1995; Jayanthi *et al.*, 2008, 2013) But several of these studies could only detect polymorphism between lines and they were not able to distinguish clones with defective floral morphology (Shah *et al.*, 1994;

Rival *et al.*, 1998a; Jayanthi *et al.*, 2013). It was found that there was significant difference in DNA Methylation in variant versus normal regenerants. Studies were initiated to find out the levels of DNA methylation by HPLC quantification of 5-methyl deoxycytidine. Shah and Ahmed Parveez (1995) have shown that levels of 5-methylcytosine measured by HPLC were significantly higher in regenerants with abnormalities than in normal regenerated clones. In yet another study with oil palm clones, it was reported that global methylation in leaf DNA of abnormal regenerates is 0.5-2.5 per cent lower than in their normal counterparts (20.8 per cent vs. 22 per cent respectively). This work once again confirmed the co-relation between DNA hypomethylation and mantled somaclonal variation in oil palm (Jaligot *et al.*, 2000, Kubis *et al.*, 2003). A set of 27 oil palm cDNA probes was screened for methylation-sensitive restriction fragment length polymorphisms (RFLPs) using callus genomic DNA digested with the isoschizomeric enzymes *Msp*I and *Hpa*II. Only two probes (CPHO 62 and 63) were found to differentiate reproducibly in two different genotypic backgrounds between nodular compact calli (NCC) and fast-growing calli (FGC), which generate 5 per cent and 100 per cent "mantled" plantlets, respectively (Jaligot *et al.*, 2002). Recently, AFLP analysis was used to compare mother plant and regenerated plants. No polymorphisms was observed when standard AFLP'S were produced with 10 different primer combinations; in contrast, polymorphism could be detected with studies using methylation sensitive AFLP's. Polymorphism appeared as new bands in the regenerated plants suggesting a reduction in methylation. Matthes *et al.* (2001) observed that loss of methylation during oil palm tissue culture is in agreement, however, with Jaligot *et al.* (2000), who described a trend towards hypomethylation in abnormal plant material. Further, they reported that none of the polymorphisms identified was consistently different between normal or abnormal clones in all sets. This suggests that the tissue culture process induces changes in methylation in oil palm regenerants at many sites within the genome although the possibility that certain sequences are pre-disposed than others. Oil Palm cDNA sequences were used as probes in methylation sensitive AFLP-southern blot studies. Two oil palm cDNAs, namely CPH 062 and CPH 063, were found to display a differential methylation pattern between normal and abnormal embryogenic calli (Jaligot *et al.*, 2002). The use of AFLPs with methylation-sensitive enzymes has made possible the identification of some tissue culture induced polymorphisms in oil palm regenerants that were not detected by other methods. Their results suggested that different approaches would be required to identify the causal basis of the mantled fruit abnormality. In yet another study using Methylation Sensitive Amplification Polymorphism (MSAP), the methylation status of CCGG sites was compared in three normal vs. the mantled regenerants (Jaligot *et al.*, 2004). They reported that their methylation sensitive markers could only discriminate between the two phenotypes within the same clonal progeny. Morcillo *et al.* (2006) has described two novel oil palm genes namely *EgM39A* and *EgIAA1*, both of which display increased transcript accumulation in epigenetically abnormal calli and were found to be potential as early markers of clonal conformity. CIRAD had embarked on a major research programme in collaboration with MPOB to identify molecular markers of the *mantled* flowering abnormality found in typically 5-10 per cent of regenerant palms. The *mantled* character was shown to be epigenetic in nature on the basis of ploidy, RAPD and AFLP studies previously carried out in the CIRAD-CP/IRD Montpellier laboratory. Given that the *mantled* abnormality does not appear to involve genomic modifications at the nucleotide sequence level, clonal conformity markers were identified firstly at the mRNA level and secondly, by studying sequence-specific DNA methylation. In order to study gene expression via mRNA accumulation, differential display RT-PCR analysis was performed on three different types of oil palm tissue culture material, namely callus, somatic embryos and shoot apex segments excised from leafy shoots. Differential expression markers have been cloned and their expression patterns rechecked by Northern hybridization using a range of material of different genotypes in order to assess their reliability as clonal conformity markers. Subsequently, markers representing mRNAs that accumulate in a consistently mantled-related fashion have been characterized further by isolating full-length cDNA clones. These studies have enabled the identification and characterization of 15 different marker genes, which were, assessed for their potential as clonal conformity markers (CIRAD, 2007). Thus, experiments designed to decipher reasons for abnormal clones have led to series of studies that revealed molecular regulation of flower development in oil palm. As a first step in elucidating the molecular mechanisms underlying oil palm ower development and the 'mantled' phenotype, 14 MADS box genes from oil palm have been isolated by Alwee *et al.* (2006). cDNAs of 12 MADS box genes, belonging to seven distinct subfamilies, were isolated and characterized. A broad conservation of oral homeotic gene functions between oil palm and *Arabidopsis* was found (Adam *et al.*, 2007a). But interestingly these studies have not shed any light on the sex determination and none of the oil palm MADS box genes described are found to display a sex-dependent expression pattern (Adam *et al.*, 2007b). Rival *et al.* (2008). reported the identification of floral homeotic genes of the "MADS box" transcription factor family which were affected by the chain of events resulting in the "mantled" abnormality. However, epigenetic effects have recently been implicated in the formation of mantled fruits. DNA hypomethylation of a LINE retrotransposon related to rice Karma, in the intron of the homeotic gene DEFICIENS, was common to all mantled clones and is associated with alternative splicing and premature termination. Dense methylation near the Karma splice site (termed the Good Karma epiallele) predicts normal fruit set, whereas hypomethylation (the Bad Karma epiallele) predicts homeotic transformation, parthenocarpy and marked loss of yield. Loss of Karma methylation and of small RNA in tissue culture contributes to the origin of mantled, while restoration in spontaneous revertants accounts for non-Mendelian inheritance. The ability to predict and cull mantling at the plantlet stage will facilitate the introduction of higher performing clones and optimize environmentally sensitive land resources (Abdullah *et al.*, 2015). This was reported by the multinational team from MPOB, Orion Genomics, and Cold Spring Harbor Laboratory. The discovery was enabled by the combination of MPOB's and the oil palm industry's vast collection of highly characterized clonal palms with a solid knowledge of oil palm and tissue culture and Orion's MethylScope® technology, a tool used to precisely map DNA methylation across entire genomes. (http://www.prnewswire.com)

10. Conclusion

However, with all the limitations encountered, the success in tissue culture of oil palm has been proven. Several techniques, like suspension cultures have been standardized, to reduce mantled flowering problem (Ho *et al.*, 2009). For example, Applied Agricultural Resource (AAR), a company located at Malaysia, has fine tuned the entire tissue culture process; they have the capacity to produce 1.5 million ramets annually and they have successfully brought down the mantling to less than 5 per cent. They have achieved a mean callusing rate of 11-20 per cent with 7 per cent embryogenesis. But, they could obtain 100 per cent callusing and 88 per cent embryogenesis by using the recloning technique. They have reported their yield data of clones over five years at five different locations- it has been mentioned that the FFB yield of clones ranged from 20.8 to 27.6 t FFB/ha/yr and oil yield ranged from 6.1 to 7.9 t oil/ha/yr, while the control (D X P) yield was 18.9 to 26.9 FFB/ ha/yr and 5.2 to 7.0 tons oil/ha/yr. Clones planted in commercial fields also have shown 7-23 per cent more FFB and 8-22 per cent more oil than D X P crosses. In Indian context, the immediate prospect is to develop biclonal seeds by developing clones of the best *duras* and *pisiferas* so that these could be preserved for posterity, in addition to solving the issue of space constraints. In fact, bi-clonal seed production programme will be successful even with a regeneration protocol which would give rise to a lesser percentage of true-to-type plantlets. Corley and Startford (1998). suggested that bi-clonal seed is advantageous for new developments rather than in the existing plantation. Hence, considering their views, it is highly relevant in Indian condition to go for bi-clonal seed production as oil palm development and area expansion processes are continuing in the country and the number of parent plants is limited. Considering the constraints of height and the water requirement of the palms, the clonal propagation should be for the multiplication of elite tested palms not only for yield but for drought resistance and dwarfness. The existing protocols need to be refined for larger multiplication rate and commercial success.

References

Abdullah, M. O., Ordway, J. M., Jiang, N. and Martienssen R. A. (2015). Loss of karma transposon methylation underlies the mantled somaclonal variant of oil palm. *Nature*, **525**: 533–537.

Aberlenc-Bertossi, F., Noirot, M. and Duval, Y. (1999). BA enhances the germination of oil palm somatic embryos derived from embryogenic suspension cultures. *Plant Cell Tissue and Organ Culture*, **56**: 53-57.

Adam, H., Jouannic, S., Orieux, Y., Morcillo, F., Richaud, F., Duval, Y. and Tregear, J.W. (2007a). Functional characterization of *MADS* box genes involved in the determination of oil palm ower structure. *Journal of Experimental Botany*, **58**: 1245–1259.

Adam, H., Jouannic, S., Morcillo, F., Verdeil J. L., Duval, Y. and Tregear, J.W. (2007b). Determination of flower structure in *Elaeis guineensis*: Do palms use the same homeotic genes as other species. *Annals of Botany*, **100**: 1–12.

Alves, S. A. O., de Lemos, O. F., dos Filho, B. G. and da Silva, S. (2011). *In vitro* protocol optimization for development of interspecific hybrids of oil palm (*Elaeis oleifera* (H.B.K) Cortés × *Elaeis guineensis* (Jacq.). *Journal of Biotechnology and Biodiversity*, **2**: 1–6.

Alwee, S. S., Van-der Linden, C. G., Van-der Schoot, J., deFolter, S., Angenent G.C., Cheah, S. C. and Smulders M. J. M. (2006). Characterization of oil palm *MADS* box genes in relation to the mantled ower abnormality. *Plant Cell Tissue and Organ Culture*, **85**: 331–344.

Angelo, P. C., Steinmacher, D. A., Lopes, R., da Cunha, R. N. V. and Guerra, M. P. (2013). Histological analysis and transcription profiles on somatic embryogenesis in interspecific hybrids of *Elaeis guineensis* × *E. oleifera*. *Agricultural Sciences*, **4**:1–11.

Balzon, T. A., Luis, Z. A. and Scherwinski-Pereira, J. E. (2013). New approaches to improve the efficiency of somatic embryogenesis in oil palm (*Elaeis guineensis* Jacq.) from mature zygotic embryos. *In Vitro Cellular and Development Biology-Plant*, **49**: 41-50.

Besse, I., Verdeil, J. L., Duval, Y., Sotta, B., Maldiney, R. and Miginiac, E. (1992). Oil palm (*Elaeis guineesis* Jacq.) clonal fidelity, endogenous cytokinin and indoleacetic acid in embryogenic callus culture. *Journal of Experimental Botany*, **43**: 983-989.

Chehmalee, S. and Te-chato, S. (2008). Induction of somatic embryogenesis and plantlet regeneration from cultured zygotic embryo of oil palm. *Journal of Agricultural Technology*, **4**: 137–46.

Chowdhury, M.K.U. (1995). Detection of variation of tissue culturederived normal and abnormal clones of oil palm using RAPD method. In: Rao,V., Henson,I.E. and Rajanaidu, N. (Eds.), *Recent Development in Oil Palm Tissue Culture and Biotechnology. Proceedings 1993 ISOPB International Symposiumon Recent Development in Oil Palm Tissue Culture and Biotechnology. PORIM, Malaysia.* pp.190194.

CIRAD. (2007). http://www.cirad.fr/presentation/programmes/biotrop/reultats/biositecirad/embryo/oiles.htm.

Corley, R.H. V. and Stratford, R. (1998). Biotechnology and oil palm: Opportunities and future impact. In: Corley,R.H.V., Hardon,J.J. and Wood,B.J. (Eds.), *Proceedings of the 1998 International Oil Palm Conference- Oil Palm Research.* Bali, Indonesia. pp. 80-91.

Corley, R.H.V. and Tinker, P.B. (2003). *The Oil Palm*. Blackwell Publishers, USA.

Corley, R.H.V. (1993). Fifteen years experience with oil palm clones- a review of progress. In: Basiron, Y. (Ed.), *Proceedings of International Oil Palm Conference-Agriculture*. PORIM. pp. 69-81.

Corley, R.H.V. (1998). What is the upper limit to oil extraction ratio?.In *Proceedings Oil and Kernel Production in Oil Palm – A Global Perspective*. PORIM. pp. 256-269.

Corley, R.H.V., Lee, C. H. and Wong, C.Y. (1986). Abnormal flower development in oil palm clones. *Planter,* **62**: 233–240.

De-Touchet, B., Duval, Y. and Pannetier, C. (1991). Plant regeneration from embryogenic suspension cultures of oil palm (*Elaeis guineensis* Jacq). *Plant Cell Reports,* **10**: 529-532.

Durand- Gasselin,T., Baudouin, L., Maheran, A.B., Konan, K.E. and Noiret, J.M. (1999). Performance of oil palm clones. In: Rajanaidu,N. and Jalani,B.S. (Eds.), *Proceedings of the Seminar on 'World wide performance of DXP oil palm planting materials, clones and interspecific hybrids'*. PORIM. pp.151-170.

Duval, Y., Duran, T.D., Konan, K. and Pannetier, C. (1988). *In vitro* vegetative propagation of oil palm (*Elaeis guineensis* Jacq.). *Oleagineux,* **43**: 39 47.

Duval, Y., Gasselin T. D., Konan, K. and Pannetier, C. (1987). *In vitro* vegetative micropropagation of oil palm (*Elaeis guineensis* Jacq.) : choice of strategy and results. In: Hassan, A. H., Chew, P. S., Wood. B. J. and Pushparaj, E. (Eds.), International Oil Palm Conference: *Progress and Prospects- Agriculture,* PORIM,Kuala Lumpur, Malaysia. pp. 191-196.

Eeuwens, C. J. (1976). Mineral requirements for growth and callus initiation of tissue explants excised from mature coconut palms (*Cocos nucifera* L.) and culture *in vitro*. *Physioligia Plantarum,* **36**:23-28.

Eeuwens, C.J., Lord, S., Donough,C.R., Rao,V., Vallejo, G. and Nelson, S. (2002). Effects of tissue culture conditions during embryoid multiplication on the incidence of "mantled" flowering in clonally propagated oil palm. *Plant Cell Tissue and Organ Culture,***70**: 311-323.

Elhiti, M., Stasolla, C. and Wang, A. (2013). Molecular regulation of plant somatic embryogenesis. In Vitro Cellular and Development Biology- *Plant,* **49**: 631–642.

Etienne, H. and Berthouly, M. (2002). Temporary immersion systems in plant micropropagation. *Plant Cell Tissue Organ Culture,* **69**: 215–231.

FAO (2002). Small-Scale Palm oil processing in Africa. *FAO Agricultural Services* Bulletin 148 ISSN 1010-1365. Rome, Italy.

Guedes, R. D. S., Silva, T. L. D., Luis, Z. G. and Scherwinski-Pereira, J. E. (2011). Initial requirement for embryogenic callus initiation in thin cell layer explants from immature female oil palm inflorescence. *African Journal of Biotechnology,* **52**:10774-10780.

Hanower, J. and Hanower, P. (1984). Inhibition et stimulation, en culture in vitro,de 'embryogenese des souches issues d'explantsfoliaries de palmier a huile. *Comptes Rendes de l'Academie des Sciences,* **298**:45-48.

Hanower, J. and Pannetier, C. (1982). *In vitro* vegetative propagation of the oil palm (*Elaeis guineensis* Jacq.). In: Fujiwara, A. (Ed.), *Proceedings of the 5th Congress in Plant Tissue and Cell Cultures. The Japanese Association for Plant Tissue Culture.* Tokyo, Japan. pp.745-746.

Hartley, C.W.S. (1988). *The Oil Palm.* Third edition.Longman, London, N.Y.

Hegazy, A. E. and Aboshama, H. M. (2010). An efficient novel pathway discovered in date palm micropropagation. In: IV International Date Palm Conference. *Acta Horticulturae*, **882**:167–176.

Hilae, A. and Te-chato, S. (2005). Effect of carbon sources and strength of MS medium on germination of somatic embryos of oil palm (*Elaeis guineensis* Jacq.). *Songklanarin Journal of Science and Technology*, **27**: 629–635.

Ho, Y. W., Tan, C.C., Soh, A.C., Wong, G., Chong, S.P., Choo, C.N. and Norazura, A. (2009). Biotechnological approaches in producing oil palm planting material - A success story. In: Rethinam, P., Reddy,V. M., Mandal,P. K., Suresh,K. and Prasad, M.V. (Eds.), *Oil Palm for Farmer's Prosperity and Edible oil Security. Proceedings of National Conference on Oil Palm*, SOPOPRAD. pp. 86-93.

Huong, L. T. L., Baicco, M., Hug, B. P., Burno, M., Santilachi, R. and Rosati, P. (1999). Somatic embryogenesis of Canary Island date palm. *Plant Cell Tissue Organ Culture*, **56**: 1–7.

Jaligot, E., Beule,T. and Rival,A. (2002). A methylation-sensitive RFLPs: Characterization of two oil palm markers showing somaclonal variation-associated polymorphism. *Theoretical and Applied Genetics*, **104**: 1263-1269.

Jaligot, E., Beule, T., Baurens, F.C., Billotte, N. and Rival, A. (2004). Search for methylation-sensitive amplification polymorphisms associated with the "mantled" variant phenotype in oil palm (*Elaeis guineensis* Jacq.). *Genome*, **47**:224-228.

Jaligot, E., Rival, A., Beule, T., Dussert, S. and Verdeil, J. L. (2000). Somaclonal variation in oil palm(*Elaeis guineensis* Jacq): DNA methylation hypothesis. *Plant Cell Reports*, **19**:684-690.

Jayanthi, M., Mandal, P. K., Sujatha, G., Jaya Sri, K. S., Srinivasa Rao, G., Sunitha, B. and KochuBabu, M. (2008). Simple sequence repeats and RAPD primers for assessment of genetic uniformity among the field planted clones of oil palm. *Journal of Plantation Crops*, **36(3)**: 235-238.

Jayanthi, M., Murali Mohan, N. and Mandal, P.K. (2011). Direct embryogenesis and plantlet regeneration in oil palm. *Journal of Plant Biochemistry and Biotechnology*, **20(2)**: 249-251.

Jayanthi, M., Sarika, N., Sujatha, G., Mathur, R. K., Rao, C. S. and Mandal, P. K. (2013). Evaluation of SSRs (microsatellites) for detecting genetic variability in oil palm (*Elaeis guineensis*) clone. *Current Horticulture*, **1(1)**: 03–06.

Jayanthi, M.,Susanthi, B., Murali Mohan, N. and Mandal, P. K. (2015). *In vitro* somatic embryogenesis and plantlet regeneration from immature male inflorescence of adult *dura* and *tenera* palms of *Elaeis guineensis* (Jacq). *Springer Plus*, **4**: 256 -264.

Jones, L.H. (1974). Production of clonal oil palm by tissue culture. Oil Palm News, **17**: 1-6.

Jones, L.H. (1990). Endogenous cytokinin in oil palm (*Elaeis guineensis* Jacq.) callus, embryoids and regenerated plants measured by radioimmunoassay. *Plant Cell Tissue and Organ Culture*, **20**: 201209.

Jones, L.H., Hanke, D.E. and Eeuwens, C.J. (1995). An evaluation of the role of cytokinin in the development of abnormal inflorescence in oil palm (*Elaeis guineensis* Jacq.) regenerated from tissue culture. *Journal of Plant Growth Regulation*, **14**: 135142.

Jouannic, S., Collin, M., Vidal, B., Verdeil, J. L. and Tregear, J. W. (2007). A class I *KNOX* gene from the palm species *Elaeis guineensis* (Arecaceae) is associated with meristem function and a distinct mode of leaf dissection. *New Phytology*, **174**: 551–568.

Kanchanapoom, K. and Domyoas, P. (1999). The origin and development of embryoids in oil palm (*Elaeis guineensis* Jacq) embryo culture. *Science Asia*, **25**:195-202.

Karun, A. and Sajini, K.K. (1996). Plantlet regeneration from leaf explants of oil palm seedlings. *Current Science*, **71**: 922-926.

Karun, A., Siril, E.A., Radha, E. and Parthasarathy, V.A. (2004). Somatic embryogenesis and plantlet regeneration from leaf and inflorescence explants of arecanut (*Area catechu* L.).*Current Science*, **86**: 1623-1628.

Khaw, C. H. and Ng, S. K. (1997). Performance of commercial scale clonal oil palm (*Elaeis guineensis* Jacq.) planting in Malaysia. *Paper presented at International Society for Horticultural Science Symposium*. Brisbane, Australia. pp.8

Kramut, P. and Te-Chato, S. (2010). Effect of culture media, plant growth regulators and carbon sources on establishment of somatic embryo in suspension culture of oil palm. *Journal of Agriculture and Technology*, **6**:159–170.

Kubis, K.E., Castilho, A. M. M. F., Vershinin, A.V. and Seymour, J. (2003). Retroelements, transposons and methylation status in the genome of oil palm (*Elaeis guineensis*) and relationship to somaclonal variation. *Plant Molecular Biology*, **52**: 69-79.

Kushairi, A., Tarmizi, A.H., Zamzuri I., Ong-Abdullah M., Rohani, O., Samsul Kamal, R., Ooi S.E., Ravigadevi, S. and Mohd Basri, W. (2006). Current status of oil palm tissue culture in malaysia. In: Kushairi, A., Sambanthamurthi, R., Abdullah, M.O. and Choong, C.K. (Eds.), In: *Proceedings of Clonal and Quality Replanting Material.Malaysian Palm Oil Board*. pp. 3-14.

Lin, H. C., Morcillo, F., Dussert, S., Tranchant-Dubreuil, C., Tregear, J.W. and Tranbarger, T.J. (2009). Transcriptome analysis during somatic embryogenesis of the tropical monocot *Elaeis guineensis*: Evidence for conserved gene functions in early development. *Plant Molecular Biology*, **70**:173–192.

Low, E. L., Alias, H., Boon, S., Shariff, E. M., Tan, C. A.,Ooi, L., Cheah, S., Raha, A., Wan, K. and Singh, R. (2008). Oil palm (*Elaeis guineensis* Jacq.) tissue culture ESTs: Identifying genes associated with callogenesis and embryogenesis. *BMC Plant Biology*, **8**:62-66.

Luis, Z. G. and Pereira, J. E. S. (2014). An improved protocol for somatic embryogenesis and plant regeneration in macaw palm (*Acrocomia aculeata*) from mature zygotic embryos. *Plant Cell Tissue Organ Culture*, **118**: 485-496.

Maheran, A. B., Abu Zarin, O., Aw, K.T. and Chin, C.W. (1995). FELDA's early experiences with vegetative propagation of the oil palm (*Elaeis guineensis* Jacq) In: Jalani, B.S., Rajanaidu, N., Cheah, S.C. and Kushairi, A. (Eds.), *Proceedings of 1993 PORIM International Palm Oil Congress- Agriculture*. PORIM. pp. 99-113.

Mandal, P. K. and Jayanthi, M. (2011a). Molecular marker and Marker assisted selection in oil palm–present status and future strategies. In: Singh, H.P., Parthasarathy, V.A. and Nirmal Babu, K. (Eds.), *Advances in Horticulture Biotechnology-Molecular Markers and Marker Assisted Selection – Fruit Crops, Plantation Crops and Spices,*Westville Publishing House, New Delhi. pp. 197-22.

Mandal, P. K. and Jayanthi, M. (2011b). Gene cloning and transgenics in oil palm. In: Singh, H.P., Parthasarathy, V.A. and Nirmal Babu, K. (Eds.), *Advances in Horticulture Biotechnology – Gene Cloning – Fruit Crops, Plantation Crops and Spices*, Westville Publishing House, New Delhi.

Marbun, C.L.M., Toruan-Mathius, N., Utomo, C.R. and Liwang, T. (2015). Micropropagtion of embryogenic callus of oil palm using temporary immersion system. *Procedia Chemistry*, **14**: 122-129.

Marmey, P., Besse, I. and Verdeil, J.L. (1991). A protein marker found to differentiate two types of callus of the same clones of oil palm (*Elaeis guineensis* Jacq.).*Comptes Rendus de l'Académie des Sciences*, Paris, **313**: 333-338.

Matthes, M., Singh, R., Cheah, S.C. and Karp, A. (2001). Variation in oil palm (*Elaeis guineensis* Jacq.) tissue culture-derived regenerants revealed by AFLPs with methylation-sensitive enzymes. *Theoretical and Applied Genetics*, **102**: 971-979.

Michelotti, V., Giorgetti, L., Ruffini Castiglione, M., Pugliesi, C. and Geri, C. (2007). Studies of meiotic process in oil palm (*Elaeis guineensis* jacq.): A FISH approach. *Proceedings of the 51st Italian Society of Agricultural Genetics, Annual Congress Riva del Garda*, Italy. pp.23-26.

Mielke, T. (2013). Oil world annual 2013.ISTA MielkeGmbh, Hamburg, Germany.

Morcillo, F., Gagneur, C., Adam, H., Richaud, F., Singh, R., Cheah, S.C., Rival, A., Duval, Y. and Tregear, J. W. (2006). Somaclonal variation in micropropagated oil palm: Charaterization of two novel genes with enhanced expression in epigenetically abnormal cell lines and in response to auxin. *Tree Physiology*, **26**: 585–594.

Murashige, T. and Skoog, F. (1962). A revised medium for rapid growth and bioassays with tobacco tissue cultures. *Physiologia Plantarum*, **15**: 473-497.

Mutert, E. and Fairhurst, T.H. (1999). Oil palm clones: Productivity enhancement for the future. *Better Crops International*, **13**:45-47.

Nwankwo, B.A. & Krikorian, A.D. (1983). Morphogenic potential of embryo and seedling-derived callus of *Elaeis guineensis* Jacq. var. *pisifera*. *Annals of Botany*, **51:** 65-76.

Ooi, S.E., Choo, C.N., Ishak, Z. and Ong, A. M. (2012). A candidate auxin-responsive expression marker gene, *EgIAA9*, for somatic embryogenesis in oil palm (*Elaeis guineensis* Jacq.). *Plant Cell Tissue and Organ Culture*, **110**:201–212.

Ooi, S.E., Choo, C.N., Ishak, Z. and Ong, A. M. (2013). Candidate hormone-responsive genes for callogenesis of oil palm (*Elaeis guineensis* Jacq.). *Journal of Oil Palm Research*, **25**: 201-211.

Pannetier, C., Arthuis, P. and Liebooux, D. (1981). Neoformation de juunes plantes d' *Elaeis guineensis*a partir de cals primaries obtenus sur fragments foliaires cultives *in vitro*. *Oleagineux*, **3**: 119-122.

Paranjothy, K. and Othman, R. (1982). *In vitro* propagation of the oil palm (*Elaeis guineensis* Jacq.). In: Fujiwara, A. (Ed.), *Proceedings of the 5th Congress in Plant Tissue and Cell Cultures. The Japanese Association for Plant Tissue Culture*, Tokyo, Japan. pp. 747-748.

Paranjothy, K. (1987). Recent developments in cell and tissue culture of oil palm. *Malaysian Applied Biology*, **16**:119-127.

Rabéchault, H. and Martin, J. P. (1976). Multiplication végétative du palmier à huile (*Elaeis guineensis* Jacq.) à l'aide de cultures de tissus foliaires. *Comptes Rendus de l'Académie des Sciences, Paris*. **238**: 1735-1737.

Rabechault, H., Ahee, J. and Guenin, G. (1970). Colonies cellulares et formese mbryoidesobtenus *in vitro* partir de cultures d'embryons de palmier a huile. *Comptes Rendus de l'Académie des Sciences*, **270**: 3067-3068.

Rabechault, H., Martin, J. P. and Cas, S. (1972). Recherches sur da culture de palmier a huile (*Elaeis guineensis* Jacq). *Oleagineux*, **27**: 531-534.

Rajanaidu,N., Rao, V., Halim, A. H. and Ong, S.H. (1990). Genetic resources: *New developments in oil palm breeding*. *Elaeis*, **1**: 1-10.

Rajesh, M. K., Radha, E., Karun, A. and Parthasarathy,V.A. (2003). Plant regeneration from embryo-derived callus of oil palm- the effect of exogenous polyamines. *Plant Cell Tissue and Organ Culture*, **75**: 41-47.

Raju, C.R., Sajini, K.K., Balachandran, S.M., Saji, K.V., Geetha, K.M., Rajashekharan, P.E., Geetha, L. and Bavappa, K.V.K. (1989). Clonal multiplication of oil palm (*Elaeis guineensis* Jacq). *Journal of Plantation Crops*, **16**: 1720.

Rival, A. (2000). Somatic Embryogenesis in oil palm. In: Jain, S.M., Gupta, P.K. and Newton, R. (Eds.), *Somatic Embryogenesis in Woody Plants*. Kluwer Academic, Dordrecht. pp. 249-289.

Rival,A., Bertrand, L., Beule, T., Combes, M.C., Trouslot, P. and Lashermes, P. (1998a). Suitability of RAPD analysis for the detection of somaclonal variants in oil palm (*Elaeis guineensis* Jacq). *Plant Breeding*, **117:** 7376.

Rival, A., Beule, T., Barre, P., Hamon, S., Duval, Y. and Noirot, M. (1997). Comparative flow cytometric estimation of nuclear DNA content in oil palm (*Elaeis guineensis* Jacq) tissue cultures and seed derived plants. *Plant Cell Reports*, **16**: 884-887.

Rival, A.,Jaligot, E., Beulé, T. and Finnegan, J. (2008). Isolation andexpression analysis of genes encoding *MET*, *CMT*, and *DRM* methyltransferases in oil palm (*Elaeis guineensis* Jacq.) in relation to the 'mantled' somaclonal variation. *Journal of Experimental Botany*, **59**:3271-3281.

Rival, A., Tregar, J., Verdeil, J.L., Richard, F., Beule, I., Hartmann, C., Rode, A. and Duval,Y. (1998b). Molecular search for mRNA and genomic markers of the oil palm "mantled" somaclonal variation. *Acta Horticulturae*, **46**:165-171.

Rohani, O., Sharifah, S. A., Mohd, R.Y., Ong, M., Tarmizi, A. H. and Zamzuri, I. (2000). Tissue culture of oil palm. In: Yusof, B., Jalani, B.S. and Chan, K.W. (Eds.), *Advances in Oil Palm Research*, Malaysian Palm Oil Board. **1**:238-283.

Scherwinski-Pereira, J.E., Guedes, R.S., da Silva, R.A., Fermino, P.C.P. Luis, Z.G. and Freitas, E. (2012). Somatic embryogenesis and plant regeneration in açaí palm (*Euterpe oleraceae*). *Plant Cell Tissue and Organ Culture*, **109**:501–508.

Scherwinski-Pereira, J. E., Guedes, R., Fermino, P., Silva, T. and Costa, F. (2010). Somatic embryogenesis and plant regeneration in oil palm using thin cell layer technique. *In Vitro Cellular and Developmental Biology- Plant*, **46**:378-385.

Shah, F.H. and Ahmed-Parveez, G. K. (1995). DNA variation in abnormal tissue culture regenerants of oil palm (*Elaeis guineensis*). *Asia Pacific Journal of Molecular Biology and Biotechnology*, **3**:49-53.

Shah, F.H., Rasid, O., Simon, A.J. and Dunsdon, A. (1994).) The utility of RAPD markers for the determination of genetic variation in oil palm (*Elaeis guineensis*). *Theoretical and Applied Genetics*, **89**:713-718.

Silva, R. D.C., Luis, Z. G. and Scherwinski-Pereira J. E. (2012). Differential responses to somatic embryogenesis of different genotypes of Brazilian oil palm. *Plant Cell Tissue and Organ Culture*, **111**: 59-67.

Simon, S. and Koh, H. L. (2005). *An Update on Performance of Tissue Cultured Oil Palm Clones in PBP Oil Palms Bhd – East Malaysia*. In: Proceedings 2005 National Seminar on Practices Super High Yielding Plantation. *Malaysian Palm Oil Board*, Kuala Lumpur. pp. 20

Smith, W.K. and Thomas, J.A. (1973). The isolation and *in vitro* cultivation of cells of *Elaeis guineensis*. *Oleagineux*, **28**: 123127.

Sogeke, A. K. (1996). Rapid callus proliferation, somatic embryogenesis and organogenesis of oil palm (*Elaeis guineensis* Jacq.). *Elaeis*, **8**: 92-103.

Soh, A. C. (2005). Super yielding oil palm – Strategic Breeding Plan. In: *Advances in Breeding and Clonal Technologies for Super Yielding Planting Materials, 2005 National Seminar*. Malaysian Palm Oil Board, Kuala Lumpur.

Staritsky, G. (1970). Tissue culture of the oil palm as a tool for its vegetative propagation. *Euphytica*, **19:** 288.

Steinmacher, D.A., Cangahuala-Inocente, G. C., Clement, C. R. and Guerra, M.P. (2007a). Somatic embryogenesis from peach palm zygotic embryos. *In vitro Cellular and Developmental Biology- Plant*, **43**:124–132.

Steinmacher, D.A., Clement, C.R. and Guerra, M.P. (2007b). Somatic embryogenesis from immature peach palm inflorescence explants: Towards development of an efficient protocol. *Plant Cell Tissue and Organ Culture*, **89**:15-22.

Steinmacher, D. A., Krohn, N.G., Dantas, A.C.M., Stefenon, V.M., Clement, C.R. and Guerra, M.P. (2007c) Somatic embryogenesis in peach palm using the thin cell layer technique: Induction, morphohistological aspects and AFLP analysis of somaclonal variation. *Annals of Botany*, **100**: 699–709.

Sumaryono, L., Riyadi, I., Kasi, P. D. and Ginting, G. (2008). Growth and differentiation of embryogenic callus and somatic embryos of oil palm (*Elaeis guineensis* Jacq.) in temporary immersion system. *Indonesian Journal of Agriculture*,**1**: 109-114.

Tan, C. C., Wong, G., Soh, A. C., Hor, T. Y., Chong, S. P. and Gopal, K. (2003). Experiences and lessons from oil palm clonal evaluation trials and commercial test plantings. In: *Proceedings of the 2003 PIPOC International Palm Oil Congress*. MPOB, Bangi. pp. 1093-1119.

Tarmizi, A.H, Bahiyah, Nur, R., Zaiton, R., Samsul Kamal, R. and Zuraida, A.R. (2005). Technologies for Improving Efficiency of Oil Palm Tissue Culture. In:*Proceedings of 2005 Conference on Biotechnology of Plantation Commodities*, 9-11 August 2005, Seri Kembangan, Selangor. pp. 64-85.

Tarmizi, A.H. and Zaiton, R. (2006a). Simple Impeller with Fast Transfer Techniques (SLIM-FaTT) in Liquid Cultrure System. *MPOB Information Series*, TT No. 304.

Tarmizi, A.H. and Zaiton, R. (2006b). Two-in-One MPOB-Simple Impeller (2 in 1 MO--SLIM) in Liquid System. *MPOB Information Series*, TT No. 303.

Tarmizi, A.H., Norjihan, M.A. and Zaiton, R. (2003). Multiplication of oil palm suspension cultures in a bench-top two litre bioreactor. *Journal of Oil Palm Research*, **2**: 44-49.

Tarmizi, A. H., Zamzuri, I. and Hashim, H. (2003). Oil Palm Tissue Culture Tracking System (OPTRACKS): Version 1. *MPOB Information Series*, TT No. pp. 185.

Tarmizi, A. H., Zamzuri, I., Ong-Abdullah, M., Samsul Kamal, R., Ooi, S. E., Rohani, O. and Rajinder, S. (2007). MPOB innovation to improve oil palm cloning. *Seminar on Sourcing of Oil Palm Planting Materials for Local-overseas Joint Ventures and Nursery Management*, 23-24 August 2007, Kuala Lumpur.

Te-chato, S. and Hilae, A. (2007). High-frequency plant regeneration through secondary somatic embryogenesis in oil palm (*Elaeis guineensis* Jacq. *var.* tenera). *Journal of Agricultural Technology*, **2**: 345-357.

Teixeira, J. B., Sondahl, M. R. and Kirby, E.G. (1994).) Somatic embryogenesis from immature inflorescences of oil palm. *Plant Cell Reports*, **13**: 247-250.

Teixeira, J. B., Sondahl, M. R. and Kirby, E.G. (1993). Somatic embryogenesis from immature zygotic embryos of oil palm. *Plant Cell Tissue and Organ Culture*, **34**: 227-233.

Teixeira, J.B., Sondahl, M.R., Nakamura,T. and Kirby, E.G. (1995). Establishment of oil palm cell suspensions and plant regeneration. *Plant Cell Tissue and Organ Culture*, **40**:105-111.

Thomas, V. and Rao, P.S. (1985). *In vitro* propagation of oil palm (*Elaeis guineensis* Jacq. Var. Tenera) through somatic embryogenesis in leaf- derived callus. *Current Science*, **5**:184-185.

Thuzar, M., Vanavichit, A., Tragoonrung, S. and Jantasuriyarat, C. (2012). Recloning of regenerated plants from elite oil palm (*Elaeis guineensis* Jacq.) cv Tenera. *African Journal of Biotechnology*, **82**: 14761-14770.

Thuzar, M., Vanavichit, A., Tragoonrung, S. and Jantasuriyarat, C. (2011). Efficient and rapid plant regeneration of oil palm zygotic embryos cv Tenera through somatic embryogenesis. *Acta Physiolia Plantarum*, **33**: 123-128.

Tregear, J.W., Morcillo, F., Richaud, F., Berger, A., Singh, R., Cheah, S.C., Hartmann, C., Rival, A. and Duval, Y. (2002). Characterization of a defensin gene expressed in oil palm inflorescences: Induction during tissue culture and possible association with epigenetic somaclonal variation events. *Journal of Experimental Botany*, **373**: 1387-1396.

Turner, J. M. (1984). Ramet establishment and survival. Oil Palm Breeders Meeting. Unifield, T.C. Ltd. Mimeograph.

Turnham, J. and Northcote, D. H. (1982). The use of acetyl-CoA carboxylase activity and changes in wall composition as measures of embryogenesis in tissue cultures of oil palm (*Elaeis guineensis*). *Biochemical Journal*, **208**:323-332.

Us-Camas, R., Rivera-Solís, G., Duarte-Aké, F. and De-la-Peña, C. (2014). *In vitro* culture: An epigenetic challenge for plants. *Plant Cell Tissue and Organ Culture*, **118**:187–201.

Valverde, R., Arias, O. and Thorpe, T.A. (1987). Picloram-induced so-matic embryogenesis in pejibaye palm (*Bactris gasipacs* H.B.K.). *Plant Cell Tissue and Organ Culture*, **10**: 149–156.

Veerappan, P., Bilal, M., Nazeeb, M. and Loong, S. G. (2000). Early performance and potential of semiclonal progenies. In: Pushparajah,E. (Ed.), *Proceedings of International Planters Conference on Plantation Tree Crops in the New Millennium*: The Way Ahead. *The Incorporated Society of Planters*, Kuala Lumpur (Vol.I). pp.117- 130.

Wong, G., Chong, S.P., Tan, C.C. and Soh, A.C. (1999). Liquid suspension culture-a potential technique for mass production of oil palm clones: *Proceedings of the 1999 PORIM International Palm Oil Congress on "Emerging technologies and opportunities in the next Millennium (Agriculture)*, 1-6 February 1999, Kuala Lumpur, Malaysia.

Wooi, K.C. (1984). Palm tissue culture. In: Micropropagation of Selected Root Crops, Palms, Citrus and Ornamental Species. *FAO Plant Production Paper*, No. 59, FAO, Rome. pp. 88112.

Wooi, K.C. (1995). Oil palm tissue culture: Current practice and constraints. In: Rao, V., Henson, I.E. and Rajanaidu,N. (Eds.), *Proceedings of 1993 ISOPB International Symposium on Recent Development in Oil Palm Tissue Culture and Biotechnology*. PORIM, Malaysia. pp. 21-32.

Wooi, K.C., Wong, C. Y. and Corley, R. H. V. (1981). Tissue culture of palms – a review. In:Rao A. N. (Ed.),*Tissue Culture of Economically Important Plants.* COSTED and ANBS. pp. 138-144.

Yusof, B. (2007). Palm oil production through sustainable plantations. *European Journal of Lipid Science and Technology,* **109**:289 – 295.

Zamzuri, I. (1998). Efficient rooting of oil palm *in vitro* plantlets using double-layer technique. *PORIM Bulletin,* **36**: 23-36.

Zamzuri, I. (2001). Double-layer technique in rooting of oil palm *in vitro* plantlets. *MPOB Information Series,* TT No. 99.

Zamzuri, I. (2002). Flameless sterilizer. MPOB Information Series, TT No.139, MPOB.

Chapter 4

Date Palm

☆ *G.B. Patil and N. Subhash*

1. Introduction

Date palm (*Phoenix dactylifera* L.) is one of the oldest cultivated plants (Riad, 2006). About 15 wild species of *Phoenix* native to tropics and subtropics of old world, from Canary Island through Africa and the near East to South East Asia to East Indies have been reported (Shah, 2014), which are interfertile with *Phoenix dactylifera*. In India and Pakistan, *P. dactylifera* was pollinated with *P. sylvestris* which, after human selection, led to evolution of present day date palm cultivars (Al-Khalifah *et al.*, 2013).

Date palm is a dioecious, perennial, monocotyledonous fruit tree that belongs to the family *Arecaceae*. It is one of the most important fruit trees of arid and semi-arid regions of the world due to its tolerance to alkaline/saline soils and resistance to drought, heat, frost and water logging conditions. In India, commercial cultivation of date palm is restricted to Kutchh regions and North West Rajasthan. Date palm is one of the important fruit trees of Kutchh region of Gujarat state, where it is also known as '*Kalpavriksha* of Kutchh'. Dates are produced and marketed at the '*khalal*' (fresh fruit) stage in India because the climatic conditions do not favour full ripening of fruits on the tree to produce '*tamar*' stage of dates (processed fruit).

Being dioecious and cross pollinated in nature, propagation through seeds results in highly heterogeneous and heterozygous populations. The seed-raised progeny has about 50 per cent unproductive male trees and 50 per cent female trees with poor/varying productivity in terms of both yield and quality. Traditionally, date palm is propagated through off-shoots or suckers; however, this propagation method has several limitations which include:

- ☆ Production of off-shoots in limited numbers (10-15) for a certain period in the lifetime of a young palm tree

- Number of offshoots produced by an individual date palm tree is highly variable and also differs from one cultivar to another (Taha *et al.*, 2003; Sudhersan and Aboel Nil, 2004).
- Non-availability of sufficient amount of planting material of local elite genotypes.
- Transmission of disease-causing pathogens and insects along with the offshoot.
- The traditional method of vegetative propagation through offshoot is slow, laborious, time-consuming and expensive (Sudhersan and Aboel Nil, 2004).

These limitations have led to concerted efforts on development of micropropagation techniques for large scale production of true to type plants through *in vitro* techniques in selected superior cultivars.

2. Micropropagation in Date Palm

Tissue culture propagation in date palm have advantages *viz.*, production of rapidly growing large number of true to type, uniform quality disease free plants from fewer high quality genotypes/cultivars, apart from genetic improvement through development of tolerant genotypes to biotic and abiotic stresses employing cell and protoplast culture, *in vitro* mutagenesis, *in vitro* selection of cells and gene transfer (Al-Maarie, 1995).

The earliest efforts on propagation of date palm by tissue culture can be traced back to 1970s. The earliest reports on date palm tissue culture were made by Tisserat (1978) and Reuveni (1979). Subsequently, several reports, dealing with standardizing and refining the technique with respect to various date palm cultivars, adopting different methods, were published. There are mainly two methods of *in vitro* propagation adopted for date palm tissue culture *viz.*, direct organogenesis and somatic embryogenesis (Raj Bhansali, 2010). Explants like zygotic embryos, root segments (Sharma *et al.*, 1980), young leaves (Sharma *et al.*, 1984), shoot apices (Al-Khayri, 2001; Raj Bhansali *et al.*, 1988; Zaid and Tisserat, 1983), immature inflorescence (Bhaskaran and Smith, 1992; Drira and Benbadis, 1985; Loutfi and Chlyah, 1998) and adventitious buds (Bouguedoura *et al.*, 1990) have been used for this purpose.

Murashige and Skoog (1962) medium (MS medium) is most widely used for date palm tissue culture (Raj Bhansali, 2010). The basal medium is supplemented with amino acids (arginine, asparagine, glycine, adenine and glutamine), vitamins (inositol, biotin, pyridoxine, nicotinic acid and thiamine), sucrose, *etc.* Activated charcoal and polyvinylpyrrolidone (PVP) have been reported to reduce the problems associated with phenolic compounds (Raj Bhansali, 2010). Various auxins [2,4-dichlorophenoxy acetic acid (2,4-D), indole acetic acid (IAA), β-naphthoxy acetic acid (BNOA) and naphthalene acetic acid (NAA)] and cytokinins [6-benzyl aminopurine (BA), kinetin and isopentenyl adenine (2-iP)] have been also used in date palm regeneration (Raj Bhansali, 2010).

Tissue culture also has a large potential role to play in date palm germplasm conservation (Johnson, 1996). Long-term storage of tissue cultured material could be achieved by *in vitro* cryopreservation in which cultures of different genotypes could be preserved at ultra low temperatures (-196°C) and as slow growth cultures.

3. Approaches for Tissue Culture Multiplication

Date palm is a highly difficult crop to work with, with respect to tissue culture, like other members of *Arecaceae*, especially coconut. The success of date palm cloning by tissue culture techniques, based on organogenesis and somatic embryogenesis, has been investigated by many workers. Organogenesis *via* shoot tip multiplication, with or without the intervention of callus phase, gave only very limited number of shoots (15-20), required longer periods of culture, had lower multiplication rate and showed strong influence of the variety/genotype, which made it expensive (Poulain *et al.*, 1979; Beauchesne, 1983). An alternative method for multiplication is through somatic embryogenesis (SE), which theoretically gives tremendous multiplication rate. Moreover, shoot tips from offshoot could be used as initial explants without destructive sampling of the mother palm (Tisserat, 1979, 1982).

4. Somatic Embryogenesis

Protocols for *in vitro* SE in date palm have been developed by several researchers (El Hadrami *et al.*, 1995; Raj Bhansali *et al.*, 1988; Sharma *et al.*, 1986; Sudhersan *et al.*, 1993). The explants were incubated in complete darkness for 3–6 months in culture rooms for the production of embryogenic callus. The phenolics interference and consequent browning, could be overcome by the use of activated charcoal, PVP, cysteine-HCl, ascorbic acid and citric acid (Dass *et al.*, 1989; Raj Bhansali, 1988; Raj Bhansali and Kaul, 1991; Zaid and Tisserat, 1983). Frequent subculturing is also adopted to overcome browning problems. Regeneration of somatic embryos occurs when the callus is subcultured on suitable media, often hormone-free. Suspension cultures of date palm friable callus for rapid somatic embryogenesis have been established by many laboratories (Bhaskaran and Smith, 1992; Fki *et al.*, 2003; Othmani *et al.*, 2009; Sharma *et al.*, 1986). Hundreds of somatic embryos could be developed from suspension cultures within a short time. Refinement of protocols for direct organogenesis in date palm has been accomplished by several researchers (Sudhersan *et al.*, 1993). However, the rate of multiplication has been reported to be less as compared to somatic organogenesis (Raj Bhansali, 2010). The explants enlarge in culture on establishment medium within 2–3 months and then regenerate into adventitious buds. These further develop into shoots within another six months. The shoots develop into plantlets on transfer into rooting medium.

In India, among the efforts made by various laboratories, research at Plant Tissue Culture Laboratory of Anand Agricultural University, Anand, Gujarat, has yielded a commercially viable protocol after successful demonstration of micropropagation (mass multiplication through somatic embryogenesis) in Indian date palm genotypes (Figure 4.1).

Figure 4.1: Various Stages of Somatic Embryogenesis Based Technology Developed at AAU, Anand.

4.1. Explant Selection and Surface Sterilization

To induce embryogenic calli, various explants have been used, which include zygotic embryos (Ammar and Benbadis, 1977; Reynolds and Murashige, 1979), shoot tips (Veramendi and Navarro, 1997), lateral buds (Bouguedoura *et al.*, 1990; Drira, 1983), leaves (Bhaskaran and Smith, 1992; Fki *et al.*, 2003) or inflorescence tissues (Bhaskaran and Smith, 1992; Drira and Benbadis, 1985; Fki *et al.*, 2003).

The starting material collected from shoot tip of healthy offshoots from mother trees, having selected qualities, has been used for effective callus induction. Due to easy availability and non-destructive excision, use of immature inflorescences has been reported as source of explants in date palms by many researchers (El-Kosary *et al.*, 2007; Masmoudi *et al.*, 2007; Feki and Drira, 2007; Abul-Soad, 2007; Tisserat, 1979). When an immature inflorescence is used as a primary explant, only calli, which originated from the proliferation of floral tissues, showed embryogenic competency. Also, only female inflorescence explants responded well to culturing and the use of male flowers as explants failed to induce embryogenesis (Al-Khairy, 2007) or yielded a success rate of less than 5 per cent (Zaid *et al.*, 2007).

4.2. Callus Induction and Callus Multiplication

MS medium has been commonly used for date palm tissue culture. The auxin 2,4-D has been reported as the most popular auxin for callogenesis (Fki *et al.*, 2003). Among cytokinins, picloram (0.2–0.5 mg/L) was found best for induction of callus that generally resulted in formation of non-embryogenic calli or abnormal somatic embryos (Fki, 2005). Like other palms, date palm also shows slow callogenesis process which may require 4–8 months for its initiation. Use of charcoal throughout the embryogenesis was found effective in controlling the massive release of phenolic compounds. Liquid media for plant regeneration has also been widely used for callus induction and multiplication (Bhaskaran and Smith, 1992; Daguin and Letouze, 1988; Fki *et al.*, 2003; Sharma *et al.*, 1986).

4.3. Induction and Multiplication of Somatic Embryos

Use of solid medium, with partial or complete removal of auxins, led to the formation of embryogenic calli and multiplication of somatic embryos. Increased quantity of undifferentiated mass of cells is very important for the faster production of large number of somatic embryos as the origin of every embryo is from a single cell. These cells, once obtained in sufficient quantities, should be shifted to a vessel containing a different medium for the conversion of somatic cells to different stages of embryo formation and later to a medium for embryo development.

4.4. Maturation of Somatic Embryos

In the final stage of development, the somatic embryos are elliptical in shape. The somatic embryos could be matured uniformly by exposing the embryos to embryo maturation medium mostly consisted of higher concentration of sucrose and ABA (Fki, 2005). The promotive effect of ABA is mainly exerted during the development of the cotyledon due to accumulation of storage proteins and prevention of precocious germination.

4.5. Germination of Somatic Embryos

Fully matured somatic embryos, on transferring to modified hormone-free medium, will germinate to form shoot and root simultaneously. Al-Khayri (2003). showed the positive effect of IBA (0.2–0.4 mg/L) on germination rates of somatic embryos which were produced on solid medium. Media containing an additional source of inorganic phosphate (170 mg/L sodium dihydrogen phosphate+100 mg/L potassium dihydrogen phosphate) resulted in faster germination of somatic embryos (Sharon and Shander, 1998).

4.7. Hardening

Pre-hardening of *in vitro* rooted plantlets was found to be essential for maximum survival of plants during primary hardening which may lead to better adaptation of these plantlets to *ex vitro* hardening conditions. This step is essential for maximum survival of plants during primary hardening. Primary hardening by gradually changing the conditions like reduced humidity, increased temperature, increased illumination (light) and substrate (cocopeat : soilite : sand, 1:1:1) require three to four months period. Well developed plants can be transferred to net house for secondary hardening which may take a period of six months, at the end of which the plants would develop compound leaves and become suitable for planting in the open field. Secondary hardened date palm plants with at least one compound leaf have been proved to be most suitable for 100 percent field establishment (Unpublished data).

Taha and Hassan (2014). reported hardening of the plantlets in pots containing a mixture of peat: perlite (2:1) utilizing polyethylene sheets for tunneling in a greenhouse maintained at 27±2°C, under sunlight and 80-90 per cent relative humidity.

Table 4.1: The Schedule of Mass Multiplication of Tissue Culture Raised Date Palm Plantlets Using Somatic Embryogenesis Developed at Anand Agricultural University, Anand

Sl.No.	*Stage of Protocol*	*Approx. Duration Required*
1.	Collection of suckers for tissue culture and establishment of aseptic cultures	1 to 2 months
2.	Initiation of callusing	6 to 8 months
3.	Callus multiplication	6 to 8 months
4.	Induction of somatic embryogenesis	3 to 4 months
5.	Maturation of somatic embryos	3 to 4 months
6.	Germination of somatic embryos	2 to 3 months
7.	*In vitro* hardening	2 to 3 months
8.	Primary hardening	4 to 6 months
9.	Secondary hardening	6 to 8 months

5. Direct Organogenesis

Organogenesis provides date palm buds that eventually produce plantlets without passing through the callus stage. Relatively few plantlets can be produced with this procedure in longer period as compared to embryogenesis. However, since plantlets are produced directly from tissues of mother plant without passing through callus stage, they are typically identical to mother plant (Aaouine, 2000).

Beauchesne and Rhiss (1978) established organogenesis as an alternative to somatic embryogenesis for date palm tissue culture. Many explants like leaves (Sharma *et al.*, 1980), inflorescence tissues (Tisserat and De Mason, 1980), roots (Smith, 1975) and zygotic embryos (Reuveni, 1979; Zaid and Tisserat, 1983) have been experimented for this purpose. Success in direct organogenesis has also been reported by some researchers using axillary branching of shoot tips (Tisserat, 1984). Al-Maarie and Al-Ghamid (1997) and Al-Khateeb *et al.* (2002). were successful in enhancing adventitious bud formation on shoot tips.

Method of organogenesis consisted of following steps *viz.*, vegetative buds initiation, and shoot bud multiplication, elongation of regenerated shoot, rooting and hardening. Initiation is the most crucial phase of organogenesis which is directly related to the various problems encountered during *in vitro* propagation. Efficiency of organogenesis based protocol is low due tc less numbers of explants responsive to *in vitro* culture, a long initiation phase, low rates of multiplication and strong influence of genotype and medium used. Despite avoiding a callus phase, genetic variation can occur as a result of low subculture frequency or excessive use of growth regulators to maximize shoot formation (Abahmane, 2015).

6. Problems Encountered during Micropropagation of Date Palm

6.1. Browning of Cultures

Tissue browning during initiation of the asceptic cultures is the most common phenomenon in almost all woody tree species (Block and Lankes, 1996). Polyphenols released in the cultures are oxidized by polyphenol oxidases leading to the formation of quinones which are highly toxic to cultured tissues (Abahmane, 2015). The most effective strategies for controlling browning include use of activated charcoal, storage of cultures in dark conditions and frequent and regular transfer of cultures.

6.2. Contamination

Endophytic bacterial contamination is observed in newly initiated cultures through offshoots for upto 4-5 months. These contaminants were identified as the genus *Bacillus* (Leary *et al.*, 1986). Use of antibiotics during initiation phase, collection of offshoots during winter season, sterilizing the dissection tools and use of young offshoots have been reported for effective control of contamination (Benjama *et al.*, 2001). Identification of contamination creating bacteria followed by its culturing (individually) followed by antibiotic sensitivity tests will be useful in controlling bacteria during initiation. However, incorporation of antibiotics in the medium is not recommended as it delays the time for callus initiation.

6.3. Vitrification/Hyperhydricity

Hyperhydricity due to greater accumulation of water inside tissue is a common physiological disorder in date palm somatic embryogenesis and organogenesis (Mazri and Meziani, 2015). It occurs due to many factors such as use of liquid media, use of high concentrations of plant hormones and/or ammonium, presence of high humidity levels and gases, particularly, ethylene inside culture tubes. Among different methods to control vitrification, increasing agar concentration, use of container covers that allow proper release of gases, reduction of hormonal and ammonium concentrations and the use of semi solid instead of liquid media were reported be most effective (Al Khateeb, 2006).

6.4. Precocious Rooting of Shoot Buds

In organogenesis, early rooting of the shoot buds resulted in the diversion of nutrients to root formation rather than shoot formation thereby reducing the multiplication. Rooting in shoot buds could be reduced by lowering the concentration of auxins particularly NAA (Al Khateeb *et al.*, 2006).

7. Importance of Genetic Fidelity Testing in Date Palm

Genetic uniformity of the tissue culture plants is the most important criteria for commercial production and adoption of technology by the farmers. Different types of somaclonal variations has been reported in tissue culture, which include abnormal leaves with wide leaflet, slow growth, variegated leaves and no flowering and fruit setting (McCubbin *et al.*, 2000) and abnormal multicarpel flowers and fruits with 6-7 carpels (Al-Khalifah *et al.*, 2008). Somaclonal variations occur due to genetic and epigenetic changes in response to stresses due to long transfer cycles, more number of multiplication, higher concentration of hormones, mainly 2,4-D, and faulty procedures of inoculation. Stringent quality control practices under commercial set up through certification and early detection of the variants using molecular marker techniques, as followed in other crops, is the need of hour for date palm micropropagation.

Application of Random Amplified Polymorphic DNA (RAPD) and Inter-Simple Sequence repeat (ISSR) markers have been successfully demonstrated to detect the genetic similarities or dissimilarities in regenerated plants in fidelity testing (Kumar *et al.*, 2010). Micropropagation protocol based on somatic embryogenesis was found to be efficient with clear, distinct, monomorphic and reproducible amplicons across 27 micropropagated plants based on 13 RAPD and 12 ISSR markers (Figure 4.2).

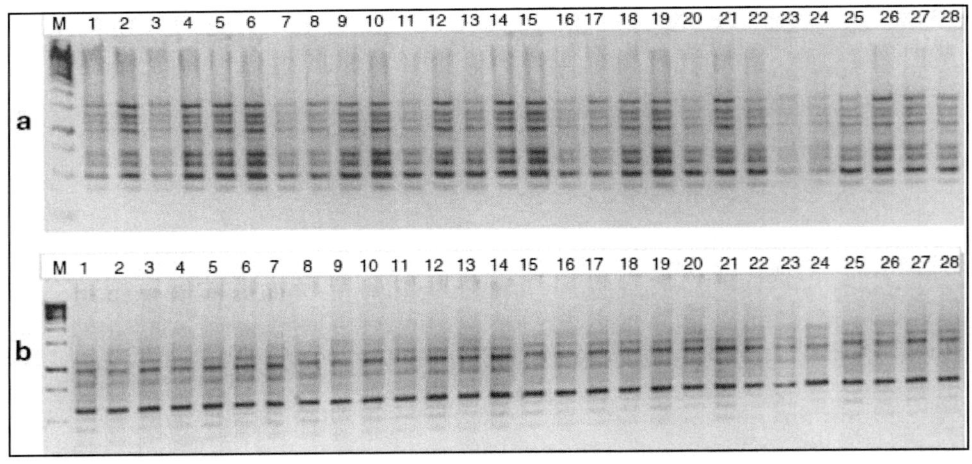

Figure 4.2: PCR Amplification Products Obtained with (a) RAPD primer OPE-15 and (b) ISSR primer UBC-835. Lane M, represents one Kb ladder; lane 1, represents mother plant; lanes 2-28 represent tissue cultured raised plants.

8. Conclusion

Date palms exhibit slower response under *in vitro* conditions, which is mainly influenced by genotype, physiology of explants and media composition. SE-based micropropagation is the most desired method owing to feasibility to scale up on a large scale. However, clonal fidelity testing of the regenerated plants is an important step for assuring quality and true to type nature of tissue culture plantlets. Direct organogenesis-based micro-propagation protocols are limited by low plantlets production and higher costs. Problems of browning, vitrification and contamination can be resolved by efficient management and modification of media constituents.

References

Aaouine, M. (2000). Production of date palm vitro-plants: The Moroccan experience. *Proceedings of Date Palm International Symposium. Windhoek,* Namibia, pp. 46–52.

Abahmane, L. (2015). Date palm micropropagation via organogenesis. In: Al-Khayri, J.M., Jain, S.M. and Johnson, D.V. (Eds.), *Date Palm Genetic Resources and Utilization* (Vol. 2: Asia and Europe). *Springer Science,* Dordrecht, Netherlands. pp. 69-90.

Abul-Soad, A. A. (2007). Inflorescence tissue culture utilization for date palm (*Phoenix dactylifera* L.) micropropagation. In: Books of Abstracts of 4th Symposium on *Date Palm in Saudi Arabia. King Faisal University,* Al-Hassa, Saudi Arabia. p. 144.

Al- Maarie, K.W. (1995). Propagation of date palm (*Phoenix dactylifera* L.) by plant tissue culture technique. *Altandeed Altaswerie Corp.* (Dabs), Damuscus, Syria.

Al-Khalifah, N.S., Askari, E., and Shanavaskhan, A. E. (2013). Date palm tissue culture and genetical identification of cultivars grown in Saudi Arabia. National Center for Agriculture Technologies, King Abdulaziz City for Science and Technology, Kingdom of Saudi Arabia.

Al-Khateeb, A. A., Abdalla, G.R., Alt-Dinar, H.M., Abdalhameed, A. and Abugulia, K.A. (2002). Auxin: Cytokinin interactions in the *in vitro* micropropagation of date palm (*Phoenix dactylifera* 'L.') Egypt. *Journal of Applied Sciences,* **17**(10): 409-415.

Al-Khateeb, A.A., Aljaber, A.M.S. and Aljabr, A.M.H. (2006). Date palm in Kingdom of Saudi Arabia. *The National Date Palm Research Center, Ministry of Agriculture,* Kingdom of Saudi Arabia, pp. 138.

Al-Khayri, J.M. (2001). Optimization of biotin and thiamine requirements for somatic embryogenesis of date palm (*Phoenix dactylifera* L.). *In Vitro Cellular and Developmental Biology-Plant,* **37**: 453–456.

Al-Khayri, J. M. (2003). *In vitro* germination of somatic embryos in date palm: effect of auxin concentration and strength of MS salts. *Current Science,* **84**: 101–104.

Al-Khayri, J.M. (2007). Micropropagation of date palm *Phoenix dactylifera* L. In: Jain, S. M. and Haggman, H. (Eds.), *Protocols for Micropropagation of Woody Trees and Fruits.* Springer, Berlin. pp. 509-526.

Al-Maari, K.W. and Al-Ghamdi, A.S. (1997). Micro-propagation of five date palm cultivars through *in vitro* axillary buds proliferation. *DUJ for Agricultural Sciences,* **13**: 55-71.

Ammar, S. and Benbadis, A. (1977). Multiplication végétative du palmier dattier (*Phoenix dactylifera* L.) par la culture de tissus de jeunes plantes issues de semis. *Comptes Rendus Biologies,* **284**: 1789–1791.

Beauchesne, G. and Rhiss, A. (1978). Micropropagation of selected palms. Proceedings of Symposium of Plant Tissue Culture, FAO, Vienna.

Beauchesne, G. (1983). Vegetative propagation of date palm (*Phoenix dactylifera* L.) by *in vitro* culture. In: *Proceedings of the First Symposium on Date Palm. King Faisal University*, Saudi Arabia, pp. 698–700.

Benjama, A., Cherkaoui, B. and Al-Maii, S. (2001). Origin and detection of *Bacillus* contaminating date palm vitro-culture and importance of manipulations conditions. *Al-Awamia*, **104**: 73-74.

Bhaskaran, S. and Smith, R.H. (1992). Somatic embryogenesis from shoot tip and immature inflorescence of *Phoenix dactylifera* cv Barhee. *Plant Cell Reports,* **12**: 22–25.

Block, R. and Lankes, C. (1996). Measures to prevent tissue browning of explants of the apple rootstock M9 during *in vitro* establishment. *Gartenbauwissenschaft*, **61**: 11-17.

Bouguedoura, N., Michaus-Ferriere, N. and Bompard, J.L. (1990). Comportment *in vitro* de bourgeons axillaries de type indéterminé du palmier dattier (*Phoenix dactylifera*). *Canadian Journal of Botany*, **68**: 2004–2009.

Daguin, F. and Letouze, R. (1988). Régénération du palmier dattier (*Phoenix dactylifera* L.) par embryogenèse somatique: amélioration de l'efficacité par passage en milieu liquide agité. *Fruits*, **43**: 191–194.

Dass, H.C., Kaul, R.K., Joshi, S.P. and Raj Bhansali, R. (1989). *In vitro* propagation of *Phoenix dactylifera* L. *Current Science*, **58**: 22–24.

Drira, N. and Benbadis, A. (1985). Multiplication vegetative du palmier dattier (*Phoenix dactylifera*) par reversion on culture in vitro debauches florales de pieds femelles. *Journal of Plant Physiology*, **119**: 227–235.

Drira, N. (1983). Multiplication végétative du palmier dattier (*Phoenix dactylifera* L.) par la culture in vitro de bourgeons axillaires et de feuilles qui en dérivent. *Comptes Rendus Biologies*, **296**: 1077–1082.

El Hadrami, I. (1995). L'embryogenèse somatique chez *Phoenix dactylifera* L. quelques facteurs limitants et marqueurs biochimiques. *Thesis Doctorat d'Etat, Faculté des Sciences Semlalia*,Université Cadi Ayyad, Marrakech.

El-Kosary, Bakr. E. I., Haseeb, G. M. and Mtwaly, A. A. (2007). Micropropagation of selected date palm trees using inflorescence explants. In: Book of abstracts of 4th Symposium on Date Palm in Saudi Arabia. King Faisal University, Al-Hassa, Saudi Arabia. p. 141.

Feki, L. and Drira, N. (2007). Development of an efficient plant regeneration system for date palm. In: *Book of abstracts of 4th Symposium on Date Palm in Saudi Arabia. King Faisal University*, Al-Hassa, Saudi Arabia. p. 477.

Fki, L. (2005). Application des suspensions cellulaires embryogenes au clonage et àl'amélioration *in vitro* du Palmier dattier. Thèse de doctorat, Faculté des Sciences de Sfax-Tunisie.

Fki, L., Masmoudi, R., Drira, N. and Rival, A. (2003). An optimised protocol for plant regeneration from embryogenic suspension cultures of date palm (*Phoenix dactylifera* L.) cv. Deglet Nour. *Plant Cell Reports*, **21**: 517–524.

Johnson, D.V. (1996). Palms: Their Conservation and Sustained Utilization. Status Survey And Conservation Action Plan. *IUCN, Gland, Switzerland and Cambridge.*

Kumar, N., Modi, A. R., Singh, A. S., Gajera, B. B., Patel, A. R., Patel, M. P. and Subhash, N. (2010). Assessment of genetic fidelity of micropropagated date palm (Phoenix dactylifera L.) plants by RAPD and ISSR markers assay. *Physiology and Molecular Biology of Plants*, **16(2)**, 207-213.

Leary, J. V., Nelson, N., Tisserat, B. and Allingham, E. A. (1986). Isolation of pathogenic *Bacillus circulans* from callus cultures and healthy offshoots of date palm (*Phoenix dactylifera* L.). *Applied and Environmental Microbiology*, **52(5)**: 1173-1176.

Loutfi, K. and Chlyah, H. (1998). Vegetative multiplication of date palm from the *in vitro* cultured inflorescence: effect of some growth regulator combinations and organogenetic potential of various cultivars. *Agronomie*, **18**: 573–580.

Masmoudi, F., Chaari, B., Mezio, B. and Drira, N. (2007). Date Palm floral initials and perspectives in tissue culture cloning and improvement. In: *Book of abstracts of 4th Symposium on Date Palm in Saudi Arabia*. King Faisal University, Al-Hassa, Saudi Arabia. p. 142.

Mazri, M. A., Meziani, R. (2015). Micropropagation of Date Palm: A Review. *Cell and Develpmental Biology*, **4**: 160.

McCubbin, M. J., Van Staden, J., and Zaid, A. (2000). A southern African survey conducted for off-types on date palms produced using somatic embryogenesis. In: *Procedings of date palm international symposium, Windhoek, Namibia.*

Murashige, T., and Skoog, F. (1962). A revised medium for rapid growth and bioassay with tobacco tissue cultures. *Physiologia Plantarum*, **15**: 473–497.

Othmani, A., Bayoudh, C. and Drira, N. (2009). Somatic embryogenesis and plant regeneration in date palm *Phoenix dactylifera* L., cv. "Boufeggous" is significantly improved by fine chopping and partial desiccation of embryogenic callus. *Plant Cell Tissue and Organ Culture*, **97**: 71–79.

Poulain, C., Rhiss, A. and Beauchesne, G. (1978). Multiplication vegetative en culture *in vitro* du palmierdattier (*Phoenix dactylifera L.*). *Comptes Rendus de l'Academie d'Agriculture de France*, **11**: 1151–1154.

Raj Bhansali, R. and Kaul, R.K. (1991). Into future – date through tissue culture. *Indian Journal of Horticulture*, **36**: 6–10.

Raj Bhansali, R. (2010). Date palm cultivation in the changing scenario of Indian arid zones:challenges and prospects. In: Ramawat, KG, (Ed.), *Desert Plants*. Springer Verlag, Berlin, pp. 423–459.

Raj Bhansali, R., Kaul, R.K. and Dass, H.C. (1988). Mass cloning of date palm plantlets through repetitive somatic embryogenesis. *Journal of Plant Anatomy and Morphology*, **5**: 73–79.

Reuveni, O. (1978).Embryogenesis and plantlets growth of date palm (*Phoenix dactylifera* L.) derived from callus tissues. *Plant Physiology*, **63**: 63-138.

Reynolds, J. and Murashige, T. (1978).Asexual embryogenesis in callus cultures of palms. *In Vitro*, **15**: 383–387.

Riad M. (2006). The date palm sector in Egypt. CIHEAM – *Options Mediterraneanes*, 45–53.

Shah, J. J. (2014). Date palm cultivation in India: An overview of activities. Emirates *Journal of Food and Agriculture*, **26**(11): 987-999.

Sharma, D.R., Dawra, S. and Chowdhuri, J. B. (1984). Somatic embryogenesis and plant regeneration in date palm (*Phoenix dactylifera*) cv Khadrawy through tissue culture. *Indian Journal of Experimental Biology*, **22**: 763–766.

Sharma, D. R., Deepak, S. and Chowdhuri, J.B. (1986). Regeneration of plantlets from somatic tissues of date palm (*Phoenix dactylifera*). *Indian Journal of Experimental Biology*, **24**: 763–766.

Sharma, D.R., Kumari, R. and Chowdhuri, J. B. (1980). *In vitro* culture of female date palm (*Phoenix dactylifera*) tissue. *Euphytica*, **29**: 169–174.

Sharon, M. and Shankar, C. (1998). Somatic embryogenesis and plant regeneration from leaf primordial of *Phoenix dactylifera* cv Yakubi. *Indian Journal of Experimental Biology*, **36**: 526–529.

Smith, S.N. (1975). Vegetative propagation of date palm by root tip culture. *Bulletin D'Agron Saharienne*, **1**: 67.

Sudhersan, C. and AboEl-Nil, M. (2004). Axillary shoot production in micropropagated date palm. *Current Science*, **86**: 771–773.

Sudhersan, C., El-Nil, A. and Al-Baiz, A. (1993). Occurrence of direct somatic embryogenesis on the sword leaf of *in vitro* plantlets of *Phoenix dactylifera* cultivar Barhee. *Current Science*, **65**: 887–888.

Taha, H.S., Bakheet, S.A. and El-Bahr (2003). Alternative approach for micropropagation of the date palm c.v. Zaghlool. *Arab Journal of Biotechnology*, **6**: 103-112.

Taha, R.A. and Hassan, M.M. (2014). Using low levels of seawater to enhance growth and development of Date Palm embryogenic cultures. *Asian Journal of Agricultural Sciences*, **6(2)**: 69-74.

Tisserat, B. and De Mason, D. (1980). A histological study of the development of adventive embryos in organ cultures of *Phoenix dactylifera* L. *Annals of Botany*, **46**: 465–472.

Tisserat, B. (1978).Propagation of date palm (*Phoenix dactylifera* L) *in vitro*. *Journal of Experimental Botany*, **30**: 1275-1283.

Tisserat, B. (1982). Factors involved in the production of plantlets from date palm callus cultures. *Euphytica* **31**: 201–214.

Tisserat, B. (1984). Propagation of date palm by shoot tip cultures. *Hort Science,* **19**: 230-231.

Veramendi, J. and Navarro, L. (1997). Influence of explant sources of adult date palm (*Phoenix dactylifera* L.) on embryogenic callus formation. *Journal of Horticultural Sciences,* **72**: 655–671.

Zaid, A. and Tisserat, B. (1983). *In vitro* shoot tip differentiation in *Phoenix dactylifera* L. Date Palm Journal, **2**: 163–182.

Zaid, A., Al-Kaabi, A.A. and El Korchi, B. (2007). Large scale *in vitro* propagation of rare and unique male date palm (*Phoenix dactylifera* L.). In: *Proceedings of 3rd International Date Palm Conference,* Abu Dhabi.

Chapter 5

Arecanut

☆ *Anitha Karun, Krishna Prakash, M.K. Rajesh and P. Chowdappa*

1. Introduction

Arecanut (*Areca catechu* L.) is commonly known as betel nut or 'Supari'. The importance of this nut is due to its use for masticatory purposes and is grown widely in most of the tropical Pacific, Asia, and parts of East Africa. The genus *Areca* belongs to the family Arecaceae (2n = 32) under the tribe Areceae are native to South and South-East Asia and the Pacific islands and is reported to contain about 76 species, of which *A. catechu* is the only cultivated species, the nuts of which are chewed as a mild stimulant (Bavappa *et al.*, 1982). Apart from its popularity as a masticatory nut, indigenous communities traditionally also use it in religious and social functions and it finds use as an ingredient in traditional medicines. The pharmaceutical importance of arecanut is due to the presence of an alkaloid, arecoline. Tannins extracted from tender arecanut are considered to be an excellent source of natural dye, tanning agent and adhesive (Bavappa *et al.*, 1982).

Arecanut palm is very widely cultivated plant in countries like India, Bangladesh, China, Indonesia, Vietnam, the Philippines, Thailand, Malaysia, Sri Lanka and Japan. India ranks first in both area (49 per cent) and production (50 per cent) of arecanut and at the same time, is the largest consumer also (FAOSTAT, 2013). In India, arecanut is cultivated in an area of 453 thousand hectares with an annual production of 6.22 lakh tonnes (NHB, 2013-14). Its cultivation is localized in southern and north-eastern states of India with the major states cultivating this crop being Karnataka, Kerala, Assam, Tamil Nadu, Meghalaya and West Bengal.

Arecanut palm is a slender, unbranched, single-trunked monoecious palm that can grow up to 100 ft, commonly growing in hot, humid tropical regions. Seed is the only propagule of arecanut as is the case of many other palm species. This, together with the long juvenile phase, heterozygous and outbreeding nature of the crop, makes the populations highly heterogeneous. Requirement of large area for experimentation further limit the scope of genetic improvement programmes in arecanut. Since the demand for quality planting material exceeds the supply and capacity, the additional requirements is being met from the unselected local cultivars. As compared to seed production, tissue culture of areca palm offers several advantages; mainly, it allows rapid multiplication of elite genotypes with desired characteristics.

Palm trees have been considered recalcitrant in tissue culture, although successful regeneration of palms through *in vitro* clonal propagation have been achieved using several explants such as roots, young leaves, shoot tips, immature inflorescence and zygotic embryos, mainly through direct and indirect somatic embryogenesis (Wang *et al.*, 2006). Somatic embryogenesis is the process by which embryo forms and develops from bipolar structures from somatic cells that parallel the developmental path of zygotic embryos. Indirect somatic embryogenesis has an intermediate callus phase whereas direct somatic embryogenesis does not have these. Successful protocol for clonal propagation via somatic embryogenesis has been reported for several palms such as African oil palm (*Elaeis guineensis*; Rabechault *et al.*, 1970), date palm (*Phoenix dactylifera*; Tisserat and Demason, 1980), peach palm (*Bactris gasipaes*; Valverde *et al.*, 1987), coconut (*Cocos nucifera*; Verdeil *et al.*, 1994), bottle palm (*Hyophorbela genicaulis;* Sarasan *et al.*, 2002), arecanut (Karun *et al.*, 2004) and interspecific crosses *Elaeis guineensis* × *E. oleifera* (Angelo *et al.*, 2013).

2. Arecanut Tissue Culture

Reports of arecanut tissue culture are rather limited. Arecanut tissue culture work was initiated for the first time by Ganapathi *et al.* (1997); *in vitro* germination studies of excised mature embryos of arecanut was carried out. Mathew and Philip (2000) first reported the protocol for *in vitro* propagation *via* direct adventitious shoot bud differentiation from cotyledon explants. A repeatable protocol for arecanut somatic embryogenesis and plantlet regeneration from leaf and immature inflorescence explants was developed by Karun *et al.* (2004) and Radha *et al.* (2006) at ICAR-Central Plantation Crops Research Institute, Kasaragod. Optimization of culture media, type of explant, plant growth regulators and their concentrations, subculturing periods, other additives, have paramount significance in developing a reproducible tissue culture protocol.

2.1. Explant

The initial step in tissue culture is the selection of suitable explants from the mother plants. Several explants sources have been tested for the establishment of tissue culture in arecanut. Ganapathi *et al.* (1997) studied in *vitro* germination of excised mature zygotic embryos of arecanut. Mathew and Philip (2000) first reported the protocol for *in vitro* propagation via direct adventitious shoot bud differentiation from cotyledonary explants. A protocol for *in vitro* germination of zygotic embryos, which facilitates rapid development of plantlets, was reported by Karun *et al.* (2002). A technique for vegetative propagation through *in vitro* shoot bud regeneration from shoot tip derived callus from four-week old sprouts of arecanut was described by Wang *et al.* (2002). Plant regeneration through somatic embryogenesis from zygotic embryo-derived callus was first reported by Wang *et al.* (2003). A repeatable protocol for arecanut somatic embryogenesis and plantlet regeneration from leaf and immature inflorescence (size 10–25 cm) explants was developed by Karun *et al.* (2004). Plantlet regeneration of arecanut through somatic embryogenesis using root, leaf and stem derived callus was reported by Wang *et al.* (2006). Radha *et al.* (2006) reported plantlet regeneration *via* direct and indirect somatic embryogenesis from immature inflorescence (length size 18.5–39cm) explants of arecanut palm. This occurrence of direct somatic embryogenesis is the first report in arecanut palm tissue culture.

2.2. Culture Method and Media

The basic technique in micropropagation of arecanut involves culturing the meristematic tissues in a callus induction media followed by subculture to induce embryogenic callus and subsequent production of embryoids. The preferred pathway of regeneration is indirect somatic embryogenesis through callus phase. For this, the first step is the induction of callus and the callus is induced to form somatic embryos. Role of basal medium and plant hormones in the culture medium is very important in induction of callus and further development into somatic embryos and plantlets.

In vitro propagation of arecanut *via* direct adventitious shoot bud differentiation from cotyledon explants on Murashige and Skoog (MS), White's, Branton and Blake's (BB) medium was reported by Mathew and Philip (2000). Results obtained showed combinations of auxins and cytokinins with activated charcoal, 2,4-D and high levels of phosphate in BB medium were critical for the differentiation of additional shoots from the cotyledon. The darkening effect of activated charcoal induced rooting in shoot cultures. Synergistic action of abscisic acid and auxins in the rooting medium enhanced the frequency of rooting.

Plantlet formation through shoot formation from shoot tip derived callus of arecanut was reported by Wang *et al.* (2002). Greenish soft callus was formed on shoot tip explants within four weeks, when cultured on MS basal medium supplemented with BA and TDZ (best result at 0.2 mg l^{-1} each). Most of calli proliferated during subculture on the same medium for callus induction, and 50–60 per cent of them formed shoots. About 90 per cent of shoots formed roots on medium containing 0.1 mg l^{-1} NAA after four weeks in culture. Plants have also been obtained through somatic embryogenesis from zygotic embryo derived callus of arecanut (Wang *et al.*, 2003). When segments of zygotic embryos were cultured on MS medium supplemented with dicamba (2, 4, 6 and 8 mg dm^{-3}) for 7 to 8 weeks in darkness, wounded regions of explants formed callus like structure. After 7- 8 weeks, callus showed pale yellow, compact, nodular structures which converted into somatic embryos within 2 to 4 months. These somatic embryos developed into plantlets after 10 weeks of culturing on hormone-free basal medium. After subculturing every month for three months, the plantlets were ready for transfer for acclimatization in the green house, 24 per cent survival rate was reported.

A viable protocol for somatic embryogenesis and plantlet regeneration from immature inflorescence explants of arecanut was reported for the first time at CPCRI, Kasaragod (Karun *et al.*, 2004). The protocol was standardized with leaf explants excised from one-year-old seedlings and later modified for immature inflorescence (spadix length 10–15 cm) sampled from adult palms of different arecanut varieties *viz.* Mangala, Sumangala and Mohitnagar (Figure 5.1). The basal medium used was MS medium and picloram as most suitable callogenic agent. Serial transfer of explants from higher to lower auxin concentration (from 200μM – 100μM – 50μM –10μM – 5μM to hormone free medium) at 30–35 days interval was essential for sustained growth of callus and somatic embryo induction. Hormone free MS medium was used for somatic embryo formation. Germination of somatic embryo was achieved in half strength MS medium supplemented with cytokinins; 20 μM 6-benzylaminopurine (BA) was found to be the best. For rapid germination of somatic embryos, MS liquid medium supplemented with 5 μM BAP was used. Subsequent plantlet development was achieved in MS medium supplemented with 10 mg l^{-1} BA, 5 mg l^{-1} indole-3-butyric acid (IBA) and 0.5 mg l^{-1} naphthaleneacetic acid (NAA). Plantlets with 2-4 leaves and fairly good root system were transferred to potting mixture, consisting of sterilised sand and soil in the ratio of 5:1 and kept inside poly house for hardening for 12-18 months.

Figure 5.1: Arecanut Clonal Propagation through Somatic Embryogenesis: (A) Inflorescence explants, (B) Callusing and formation of somatic embryos, (C) Germination of somatic embryos, (D) and (E) Plantlet development.

This protocol for mass multiplication of elite arecanut palms has been successfully applied for propagation of field resistant Yellow Leaf Disease (YLD) palms (Karun *et al.*, 2005). Direct and indirect somatic embryogenesis from inflorescence explants of arecanut was reported by Radha *et al.* (2006). However the number of direct somatic embryos formed in Eeuwens Y3 medium was very few and the plantlet development was completed within 9-12 months. Five cytokinins *viz.*, thidiazuron (TDZ), BA, kinetin, N6-(2-isopentenyl) adenine (2iP) and zeatin were tested for their effect on regeneration. Amongst these, TDZ at a concentration of 2 mg l^{-1} gave maximum shoot length, number of leaves and root growth and was found to be the most efficient cytokinin for maturation and conversion of direct somatic embryos of arecanut into complete plantlets.

Wang *et al.* (2006) reported somatic embryogenesis and plantlet regeneration using arecanut root, leaf and stem parts as explants. They were able to induce and maintain embryogenic calli on MS basal medium contained full-strength macro- and micro-elements of MS salts supplemented with 2, 4-dichlorophenoxyacetic acid (2, 4-D) or 3,6-dichloro-2-methoxybenzoic acid (dicamba) at concentrations of 2, 4, 6 and 8 mg dm^{-3} in darkness. Somatic embryos were formed in the presence of 2-4 mg dm^{-3} dicamba on primary callus and during subculture on 2-8 mg dm^{-3} 2, 4-D or 2-4 mg dm^{-3} dicamba-containing media. Earlier, Wang *et al.* (2003) also reported formation of somatic embryos from zygotic embryo-derived callus after 2-3 passages of subculture (at an interval of 8 weeks) on 2-8 mg dm^{-3} dicamba-containing medium. Plantlet conversion from embryos was successfully achieved on growth regulator-free medium kept under a 16 h photoperiod condition.

Assessing the genetic fidelity of tissue culture derived plantlets is very much important for establishing the genetic uniformity in perennial crops, as these crops will remain in the field for a long time. Among the different molecular markers available, RAPD markers are preferred due to their cost effectiveness, technical simplicity and non-requirement of sequence information of template DNA. To evaluate clonal fidelity of somatic embryogenesis derived plantlets from Yellow Leaf Disease (YLD) resistant arecanut palms, Random amplified polymorphic DNA (RAPD) markers were used by Karun *et al.* (2008). Pair wise genetic similarities generated by Jaccard's coefficient between each mother palm and its progenies (eight plantlets/palm) were showing high similarity (99 per cent in one case and 98 per cent in another).The low level of genetic variability using RAPD markers in plantlet derived from direct somatic embryogenesis of inflorescence culture shown by plantlets of direct somatic embryogenesis can be exploited for large-scale multiplication of elite arecanut palms with desirable qualities.

3. Future Aspects

Several approaches have come afore in order to enhance the productivity of *in vitro* culture depending on the final product desired and the species investigated. One such innovative approach is the use of cell suspensions and bioreactors for plant culture. Cell suspension culture is the method of culturing and maintaining the embryogenic calli obtained from the explants in a liquid medium with appropriate nutrients under stable microclimatic condition.The technique enables the production of individualized embryos in synchronous growth with both root and shoots; compared to routine process with polyembryonic cultures and compulsory rooting treatment (De Touchet *et al.*, 1991). The embryogenic cells produced in cell suspension culture could be used in bioreactors to enhance somatic embryogenesis.

Zouine *et al.* (2005) demonstrated a procedure for the rapid development of a high number of somatic embryos from embryogenic suspension culture in date palm. This method might be efficient for mass propagation of arecanut. The adaption of bioreactors in the plant tissue culture is considered as a major milestone since they offer several advantages *viz.*, short time and labour-saving, relatively easy to scale-up, allow enhanced growth and multiplication and improved nutrient availability due to the use of liquid medium over traditional tissue culture techniques. A system named as temporary immersion systems (TIS), wherein explants are flooded with nutrient medium containing growth regulators at regular time intervals, has been successfully used in scaling up of somatic embryogenesis. This system offers the possibility of automating some culture stages. This would be an interesting strategy for up-scaling plantlet regeneration potential in arecanut tissues under *in vitro* conditions.

References

Angelo, P.C., Steinmacher, D.A., Lopes, R., da Cunha, R.N.V. and Guerra, M.P. (2013). Histological analysis and transcription profiles on somatic embryogenesis in interspecific hybrids of *Elaeis guineensis* × *E. oleifera*. *Agricultural Sciences*, **4**:1–11.

Bavappa, K.V.A., Nair, M.K. and Premkumar, T. (1982). The Areca palm (*Areca catechu* Linn.). *Central Plantation Crops Research Institute*, Kasaragod, Kerala, India.

De Touchet, B., Duval, Y. and Pannetier, C. (1991). Plant regeneration from embryogenic suspension cultures of oil palm (*Elaeis guineensis* Jacq.). *Plant Cell Reports*, **10**:529–532.

Ganapathi, T.R., Suprasanna, P., Bapat, V. and Rao, P.S. (1997). In vitro culture of embryos of arecanut (Areca catechu L.). *Fruits*, **52**:313–316.

Karun, A., Manimekalai, R., Priya, S. and Radha, E. (2005). Genetic fidelity of tissue derived plantlets and histology of somatic embryogenesis in arecanut (*Areca catechu* L.). *Poster paper presented at International Conference on Plant Genomics and Biotechnology*: Challenges and Opportunities, *Indira Gandhi Agricultural University*, Raipur, India.

Karun, A., Radha, E., Vijayan, P.S., George, J., Rajesh, M.K. and Ananda, K.S. (2008). Assessment of genetic fidelity of arecanut plantlets derived through direct somatic embryogenesis by RAPD markers. *Journal of Plantation Crops*, **36**(3):228–231.

Karun, A., Siril, E.A., Radha, E. and Parthasarathy, V.A. (2004). Somatic embryogenesis and plantlet regeneration from leaf and inflorescence explants of arecanut (*Areca catechu* L.). *Current Science*, **86**:1623–1628.

Karun, A., Siril, E.A., Radha, E. and Parthasarathy, V.A. (2002). *In vitro* embryo retrieval technique for arecanut (*Areca catechu* L.). *Proceedings of PLACROSYM XV. pp.* 209 – 214.

Mathew, M. and Philip, V.J. (2000). *In vitro* adventitious shoot formation from embryos of *Areca catechu* L. *Phytomorphology*, **50**:221–227.

Rabechault, H., Ahée, J. and Guenin, G. (1970). Colonies cellulaires et forms embryoides obtenues in vitro a partir de cultures d'embryons de Palmier a huile (*Elaeis guineensis* Jacq. var. dura Becc.). *Comptes Rendus de l'Académie des Sciences* **270**:3067–3070.

Radha, E., Karun, A., Ananda, K.S. and Kumaran, P.M. (2006). Plantlet regeneration via direct and indirect somatic embryogenesis from inflorescence explants of arecanut palms. *Journal of Plantation Crops,* **34**(3):50–54.

Sarasan, V., Ramsay, M. and Roberts, A. (2002). *In vitro* germination and induction of direct somatic embryogenesis in "Bottle Palm" [*Hyophorbe lagenicaulis* (L. Bailey) H.E. Moore], a critically endangered Mauritian palm. *Plant Cell Reports,* **20**:1107–1111.

Tisserat, B. and Demason, D.A. (1980). A histological study of development of adventive embryos in organ cultures of *Phoenix dactylifera* L. *Annals of Botany* **46**:465–472.

Valverde, R., Arias, O. and Thorpe, T.A. (1987). Picloram-induced somatic embryo genesis in pejibaye palm (*Bactris gasipaes* HBK). *Plant Cell Tissue Organ Culture,* **10**:149–156.

Verdeil, J.L., Huet, C., Grosdemange, R. and Buffard-Morel, J. (1994).) Plant regeneration from cultured immature inflorescences of coconut (*Cocos nucifera* L.): evidence for somatic embryogenesis. *Plant Cell Reports,* **13**:218–221.

Wang, H.C., Chen, J.T. and Chang, W.C. (2006). Somatic embryogenesis and plant regeneration from leaf, root and stem-derived callus of *Areca catechu* L. *Biologia Plantarum,* **50**:279–282.

Wang, H.C., Chen, J.T., Wu, S.P., Lin, M.C. and Chang, W.C. (2002). Plant regeneration through somatic embryogenesis from zygotic embryo-derived callus of *Areca cathecu* L. (Arecaceae). *In Vitro Cellular and Developmental Biology - Plant* **39**:34–36.

Wang, H.C., Chen, J.T., Wu, S.P., Lin, M.C. and Chang, W.C. (2003). Plant regeneration through shoot formation from callus of *Areca cathecu* L. *Plant Cell Tissue Organ Culture,* **75**:95–98.

Zouine, J., El Bellaj, M., Meddich, A., Verdeil, J.L. and El Hadrami, I. (2005). Proliferation and germination of somatic embryo from embryogenic suspension cultures in *Phoenix dactylifera. Plant Cell Tissue Organ Culture,* **82**:83–89.

Chapter 6

Spices

☆ *K. Nirmal Babu, Minoo Divakaran, S.P. Geetha, P.N. Ravindran and K.V. Peter*

1. Introduction

Spices and herbs, the group of aromatic plants which placed India on the global map, are now cultivated throughout the world and widely used as fresh or dried plant parts, to flavour food, confectionery and also in medicine and perfumery. It has been estimated that these crops are grown on an area of 8 million ha globally contributing to 31.6 million tons of spices annually and India's share to world spices production is 52 per cent in volume and 58.6 per cent in value (Spices Board of India, 2014). Black pepper, cardamom, ginger, turmeric, vanilla, cinnamon, clove, nutmeg and tamarind constitute the major spices, while coriander, cumin, fennel and fenugreek are important seed spices and saffron, lavender, thyme, oregano, celery, anise and sage are important herbal spices.

The productivity of many of these crops is low due to the lack of high yielding, pest and disease resistant varieties. Advent of biotechnology has advanced the knowledge of fundamental botany and found wide applications in plant propagation, production of secondary metabolites, production of pathogen-free plants, plant genetic manipulations and conservation of plant genetic resources *etc.* and the past few years have witnessed a quantum jump in utilization of biotechnological tools to achieve the above through commercial propagation, development of novel varieties and marker assisted breeding.

2. Black Pepper

2.1. Micropropagation and Plant Regeneration

Availability of quality planting material is the prerequisite for adoption of large scale production of spices. High rate of multiplication coupled with the additional advantage of obtaining disease-free planting materials makes micropropagation a viable alternative to conventional propagation (Nirmal Babu and Minoo, 2003). Direct shoot proliferation from pre-existing meristems of apical or axillary buds or from the meristems, formed *de novo* from explants such as leaf segments and stem segments without the mediation of callus, are utilized. Black pepper, *Piper nigrum* L., is native to India and is the most important spice in the world. Conserving the genetic diversity and development of *Phytophthora* foot rot resistant lines is an immediate priority for all breeding programmes. Reports on micropropagation of black pepper using various explants are available (Ravindran *et al.*, 1996; Nirmal Babu *et al.*, 1997, 2012). Multiple shoots can be induced using BA in the culture medium (MS or SH medium) either alone or in combination with auxins. Endogenous contamination severely hampers establishment of black pepper cultures. A commercially viable protocol for large scale *in vitro* multiplication of black pepper overcoming these problems was reported by Nazeem *et al.* (2004). Protocols were standardized for micropropagation of other endangered and medicinally important species of *Piper* like *P. longum, P. chaba, P. betle, P. barberi* and *P. colubrinum* (Nirmal Babu *et al.*, 1997, 2012). Joseph *et al.* (1996) and Yamuna (2007) reported somatic embryogenesis from zygotic embryos, while Nair and Gupta (2003, 2006) reported cyclic somatic embryogenesis from the maternal tissue, which has tremendous potential for automated micropropagation (Figure 6.1). Nirmal Babu *et al.* (2005) reported somatic embryogenesis from mature leaf tissues; such systems are useful in transgenic experiments. Plant regeneration has also been reported in other *Piper* species like *P. longum, P. betle, P. chaba, P. attenuatum* and *P.colubrinum* through direct and indirect organogenesis (Nirmal Babu *et al.*, 2012), Attempts on induction of variability on somaclones for tolerance to *Phytophthora* foot rot resistance (Shylaja *et al.*, 1996) resulted in identification of tolerant somaclones through *in vitro* selection of calli as well as somaclones using crude culture filtrate and toxic metabolite isolated from *Phytophthora capsici*.

Figure 6.1: Somatic Embryogenesis in Black Pepper.

2.2. Protoplast Culture

Removal of the cell wall (a major barrier to distant hybridizations) to release 'protoplasts' makes the protoplast technology suitable for genetic transformation by introduction of trans-gene DNA and somatic hybridization by protoplast fusion of species or subspecies resistant to traditional cross breeding, or isolation of sub-cellular organelles *etc.* Reliable procedures are available for isolation, culture and fusion of protoplasts from a range of spices. Successful isolation and culture of protoplasts were reported in leaf tissues in black pepper (Shaji *et al.*, 1998) and these protoplasts could be successfully developed up to microcalli stage.

3. Cardamom

3.1. Micropropagation and Plant Regeneration

Cardamom, the 'Queen of Spices' is also native to India. The productivity of cardamom is hampered by various diseases of viral etiology. Utilization of virus-free planting material is an important input into disease management strategy. Cardamom is one of the first crops where commercialization of micropropagation has been achieved. Efficient *in vitro* methods for rapid clonal propagation of cardamom are available (Nadgauda *et al.*, 1983; Nirmal Babu *et al.*, 1997, 2011b; Nirmal Babu and Minoo, 2003). Successful regeneration of plantlets from callus of seedling explants and anthers of cardamom has been reported (Nirmal Babu, 1997, Nirmal Babu *et al.*, 2011b) (Figure 6.2). Manohari *et al.* (2008) also reported an efficient protocol for the induction of somatic embryogenesis and plant regeneration in small cardamom. Identification of a few 'Katte' disease tolerant somaclones was reported by Nirmal Babu *et al.* (2003a)..

Figure 6.2: Micropropagation in Cardamom from Seedling Explants.

3.2. Protoplast Culture

Propotplasts were successfully isolated and cultured from cell suspensions and leaf tissues in cardamom (Geetha *et al.*, 2000) and these protoplasts could be developed up to microcalli stage.

4. Ginger

4.1. Micropropagation and Plant Regeneration

Ginger is the third most important spice that originated in South Asia. There is no seed set in ginger leading to limited variability and this hampers crop improvement programmes. Rhizome rot caused by *Pythium aphanidermatum* and bacterial wilt caused by *Ralstonia solanacearum* are the major diseases affecting ginger. Diseases of ginger are often spread through infected seed rhizomes. The tissue culture will help in the production of pathogen-free planting material. Clonal multiplication of ginger from vegetative buds (Nadgauda *et al.*, 1980; Nirmal Babu *et al.*, 1997, 1998; Sharma and Singh, 1997), optimization of media composition for micropropagation of ginger (Nirmal Babu *et al.*, 2011a), regeneration of plantlets *via* callus phase from leaf, vegetative bud, ovary and anther explants (Kacker *et al.*, 1993; Nirmal Babu,

1997; Nirmal Babu *et al.*, 2005) and plant regeneration from ginger anther has been reported (Samsudeen *et al.*, 2000) (Figure 6.3). This system was used for inducing somaclonal variability in ginger where lack of seed set hampers conventional breeding. A few promising high yielding rhizome rot tolerant somaclones also have been identified (Nirmal Babu *et al.*, 1996a; Sumathi, 2007).

Figure 6.3: Plant Regeneration in Ginger from Callus Cultures.

In nature, ginger fails to set fruit. However, by supplying required nutrients to young flowers and by *in vitro* pollination 'fruit'development and subsequently plants could be recovered (Nirmal Babu *et al.*, 1992a; 1992b, Samsudeen *et al.*, 2000). *In vitro* pollination was attempted by Nazeem *et al.* (1996) to overcome the pre-fertilization barriers like spiny stigma, long style and coiling of pollen tube *etc.* and successful seed set was obtained. Induction of tetraploids ginger through *in vitro* colchicine treatment and tetraploid somaclone with extra bold rhizomes have also been reported (Adaniya and Shirai, 2001; Smith *et al.*, 2004; Nirmal Babu *et al.*, 1996b, 2005; Wang *et al.*, 2010)

4.2. Protoplast Culture

Successful isolation and culture of protoplasts have been reported from cell suspensions and leaf tissues in ginger (Nirmal Babu, 1997; Geetha *et al.*, 2000) and these protoplasts could be successfully developed up to microcalli stage. Somatic hybridization of ginger through chemical fusion (PEG mediated) and its regeneration was reported by Guan *et al.* (2010). RAPD technique was used for the identification of hybrids and flow cytometry analysis revealed the diploid nature of all regenerated progenies.

5. Turmeric

5.1. Micropropagation and Plant Regeneration

Turmeric of commerce is the dried rhizomes of *Curcuma longa* L. which belongs to the family *Zingiberaceae*. India is the major producer and exporter of this spice. Curcumin is the important colouring material from turmeric and development of varieties with high recovery of curcumin is the need of the hour. Successful micropropagation of turmeric has been reported (Nadgauda *et al.*, 1978; Nirmal Babu *et al.*, 1997; Sunitibala *et al.*, 2001; Salvi *et al.*, 2002; Panda *et al.*, 2007; Ghosh *et al.*, 2013). This technique is used for production of disease-free planting material. Organogenesis and plantlet formation was achieved *via* callus cultures of turmeric (Nirmal Babu *et al.*, 1997; Sunitibala *et al.*, 2001; Salvi *et al.*, 2000) (Figure 6.4).

Variants with high curcumin content were isolated from tissue cultured plantlets (Nadgauda *et al.*, 1982). Root rot disease tolerant clones of turmeric cv. Suguna were isolated using continuous *in vitro* selection technique against pure culture filtrate of *Pythium graminicolum* (Gayatri *et al.*, 2005).

Figure 6.4: *In vitro* Multiplication in Turmeric.

Renjith *et al.* (2001) reported *in vitro* pollination and hybridization between two short duration types VK-70 and VK-76 and reported seed set and seed development. This reduces the breeding time and helps in recombination breeding which so far was not attempted in turmeric. Protocols for micropropagation of many economically and medicinally important Zingiberaceous species like *Amomum subulatum* (large cardamom), *Curcuma aromatica* (kasturi turmeric), *C. amada* (mango ginger), *C. zedoaria, Kaempferia galanga, K. rotunda* and *Alpinia* spp. have been developed (Chang and Criley, 1993; Geetha *et al.*, 1997; Chan and Thong, 2004; Chithra *et al.*, 2005; Raju *et al.*, 2005. Rahman *et al.* (2004) reported efficient plant regeneration through somatic embryogenesis from leaf base-derived callus of *Kaempferia galanga* L.

5.2. Protoplast Culture

Successful isolation and culture of protoplasts were reported from cell suspensions and leaf tissues in turmeric (Geetha *et al.*, 2000) and these protoplasts could be successfully developed up to microcalli stage.

6. Vanilla

6.1. Micropropagation and Plant Regeneration

Vanilla planifolia, native to Mexico and Central America and now cultivated in other parts of the tropics, is the source of natural vanillin. Micropropagation of vanilla using apical meristem was standardized for large scale multiplication of disease free and genetically stable plants (Kononowicz and Janick, 1984; Minoo, 2002; Minoo *et al.*, 2006; Minoo and Nirmal Babu, 2009). Successful plant regeneration from shoot and seed derived callus was reported in vanilla (Nirmal Babu *et al.*, 1997; Minoo, 2002).

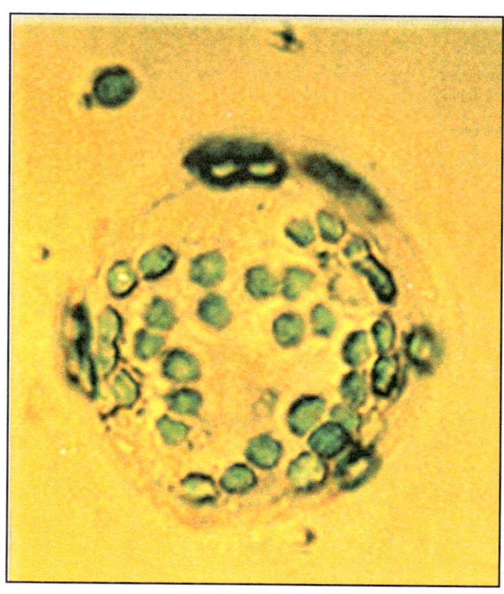

Figure 6.5: Protoplast Isolation in Vanilla.

6.2. Protoplast Culture

Minoo *et al.* (2008) reported the isolation of viable protoplasts in *Vanilla* (Figure 6.5) species, *i.e. V. planifolia* and *V. andamanica* and could successfully induce PEG mediated protoplast fusion between *V. planifolia* and *V. andamanica.* This protoplast fusion technology can be useful in generation of somatic hybrids and for gene transfer of useful traits to *V. planifolia* especially the natural seed set and disease tolerance observed in *V. andamanica.*

7. Tree Spices

7.1. Micropropagation and Plant Regeneration

Cinnamon, clove, nutmeg, curry leaf, pomegranate, tamarind, allspice and garcinia are some of the important tree spices. In these perennial tree crops, identification and clonal multiplication of high-yielding 'elite' genotypes becomes a priority due to long pre-bearing period. Micropropagation of cinnamon, Chinese cassia and camphor was reported from seedlings and mature tree explants (Mini *et al.*, 1997; Nirmal Babu *et al.*, 1997; Huang *et al.*, 1998). Multiple shoots were induced from shoot tips and nodal segments of *Cinnamomum camphora* on Woody Plant Medium (WPM) (Huang *et al.*, 1998) and from a cotyledonary node on MS medium (Azad *et al.*, 2005). Successful micropropagation of Chinese cassia was reported by Inomoto and Kitani (1989) using nodal explants from seedlings on MS medium. Micropropagation protocols for *C. camphora* were developed by Nirmal Babu *et al.* (2003b).

In vitro multiple shoot induction was worked out in *Garcinia indica* by Kulkarni and Deodhar (2002). Murashige and Skoog's medium supplemented with BAP gave optimal response in different genotypes investigated. Micropropagation of three species of *Garcinia* was reported by Huang *et al.* (2000), Malik *et al.* (2005) and Mohan *et al.* (2012). *In vitro* shoot initiation from explants of field grown trees of nutmeg was reported by Mallika *et al.* (1997). Micropropagation of clove from seedling explants have been reported (Mathew and Hariharan 1990; Superman and Blake 1990). MS medium supplemented with IBA or activated charcoal induced root formation. However, there are no reports on successful micropropagation of clove from mature shoot explants.

Reports on micropropagation of curry leaf, pomegranate, camboge and tamarind are also available (Mascarenhas *et al.*, 1987; Hazarika *et al.*, 1995; Rao *et al.*, 1997; Bhuyan *et al.*, 1997; Mathew *et al.*, 1999; Nirmal Babu *et al.*, 2000; Mehta *et al.*, 2000). High-frequency direct shoot proliferation was induced in intact seedlings of *M. koenigii* (Bhuyan, *et al.*, 1997). Shoot proliferation was also reported from different explants like nodal cuttings (Nirmal Babu *et al.*, 2000) and leaves (Mathew and Prasad, 2007). Efficient micropropagation protocols for pomegranate have been reported by Bin and Jiang (2003), El-Agamy *et al.* (2009) and Patil *et al.* (2011). The plantlets grown on WPM were found to be significantly better in average survival, plantlet height and average leaves number per shoot when compared to MS and NN media (El-Agamy *et al.*, 2009; Kaji *et al.*, 2013). *In vitro* regeneration and high frequency regeneration of tamarind was achieved in different media compositions

(Hussain *et al.*, 2004; Pattepur *et al.*, 2010). Thidiazuron can play a major role to induce germination in tamarind seedlings (Mehta *et al.*, 2004). Reports on successful callus induction and plant regeneration in nutmeg, cinnamon, camphor, pomegranate and curry leaf *etc.* are available (Bhansali, 1990; Iyer *et al.*, 2000; Kong *et al.*, 2009; Shi *et al.*, 2009, 2010; Paul *et al.*, 2011).

8. Seed and Herbal Spices

8.1. Micropropagation and Plant Regeneration

Seed spices and herbs constitute a large group of widely different aromatic plants which are used as spices, culinary herbs, medicinal herbs and those which are used in aroma therapy. Micropropagation protocols for many seed and herbal spices are available. They include coriander, anise, thyme, peppermint, spearmint, celery, lavender, savory, ocimum, oregano, basil, sage, fennel, parsley, dill and garlic, saffron, *Eryngium foetidum* and capsicum (Bhojwani, 1980; Cellarova, 1992; Furmanowa and Olszowska, 1992; Panizza and Tognoni, 1992; Patnaik and Chand, 1996; Vandemoortele *et al.*, 1996; Sajina *et al.*, 1997a; Ochoa-Alejo and Ramirez-Malagon, 2001; Gupta and Bhargava, 2001; Sharma *et al.*, 2004; Aflatuni *et al.*, 2005; Karaoglu *et al.*, 2006; Majourhat *et al.*, 2006; Minas, 2009; Song *et al.*, 2009; Ascough *et al.*, 2009; Falk *et al.*, 2009; Kothari *et al.*, 2010; Fadel *et al.*, 2010; Irikova *et al.*, 2011; Samantaray *et al.*, 2012; Nhung and Quynh, 2012; Kara and Baydar, 2012; Navroski *et al.*, 2012; Zeybek *et al.*, 2012; Rodeva *et al.*, 2013; Santoro *et al.*, 2013; Dixit and Chaudhary, 2013; Keller and Senula, 2013). Clonal propagation of chemically uniform fennel plants through somatic embryoids was reported by Miura *et al.* (1987). Shoot regeneration protocols for fenugreek, cumin, coriander has also been reported (Nirmal Babu *et al.*, 1997; Tawfik and Noga, 2001; Ebrahimie *et al.*, 2003; Aasim *et al.*, 2009). Jakhar *et al.* (2003) reported *in vitro* flowering and seed formation in cumin.

Somatic embryogenesis has been established in saffron, garlic and chilli (Blazquez *et al.*, 2003; Sheibani *et al.*, 2006; Munyon *et al.*, 1989). An efficient protocol for adventitious shoot formation in fennel was developed by investigating the effect of plant growth regulators effect by Jakhar and Choudhary (2012). Profuse callus differentiation was observed when medium was supplemented with 1.0 mg/l BAP followed by 1.0 mg/l BAP + 0.5mg/l IBA. The shoot morphogenesis was observed in callus proliferated from the shoot apex explants incubated at 1.0 mg/l BAP + 0.5 mg/l IBA, upon subculture on same levels of plant growth regulator. In order to create variability, organogenesis followed by mutagenesis has also been identified as a potential *in vitro* technique. In this process, stock organogenetic callus are treated with physical or chemical mutagens. The studies have shown positive indications to isolate promising mutants in cumin (Parashar *et al.*, 2014). Similar type of efforts can be made for creation of variability in cumin for resistance to alternaria blight, root rot in fenugreek, and for many other biotic stresses.

In vitro methods of screening could prove highly useful in screening a large germplasm collections or cell lines for resistance to prevalent fungal diseases and tolerance to drought and salt stress. The reports are available on *in vitro* selection for salt tolerance in fenugreek on media containing 0.025-1.5 per cent NaCl (Settu *et al.*,

1997), and drought tolerant cell lines cultured on media containing 0.25-1.50 per cent PEG in coriander (Stephen and Jayabalan 2000) through tissue culture. Selection of somaclonal variants resistant to *Septoria apilicola* by callus culturing in the presence of fungal culture filtrate in celery (Evenor *et al.*, 1994), fusarium yellow resistant celery line, a somaclonal variant (Lacy *et al.*, 1996) and resistance to *Alternaria* blight in cumin (Shukla *et al.*, 1997b) have been reported by different workers.

8.2. Protoplast Culture

Shekhawat and Galston (1983) reported isolation, culture and shoot regeneration from mesophyll protoplasts of fenugreek. Successful regeneration of whole plants from tissue cultured shoot primordial of garlic was reported by Ayabe *et al.* (1995).

9. Development of Synthetic Seeds

Synthetic seeds or artificial seeds are defined as artificially encapsulated somatic embryos, shoot buds, cell aggregates, or any other tissue which can be used for sowing as a seed and those possess the ability to grow into a plant under *in vitro* or

Figure 6.6: Synthetis Seed Production in (a) Pepper, (b) Cardamom and (c) Vanilla.

ex vitro conditions. Artificial or synthetic seeds can be an ideal system for low cost plant movement, propagation, conservation and exchange of germplasm. Synthetic seeds were developed by encapsulation of *in vitro* developed small shoot buds in 3- 5 per cent calcium alginate in black pepper (Figure 6.6a), and cardamom (Figure 6.6b), somatic embryos and *in vitro* regenerated shoot buds in ginger and turmeric, *in vitro* regenerated shoot buds and protocorms in vanilla (Figure 6.6c) (Sharma *et al.*, 1994; Sajina *et al.*, 1997b), somatic embryos in cinnamon and curry leaf and nodal segments in pomegranate leaf (Sundararaj *et al.*, 2010; Minoo, 2002; Gayatri *et al.*, 2005; Naik and Chand, 2006; Nikhil and Shukla, 2013). Synseeds seeds have also been reported in cumin (Tawfik and Noga, 2001, 2002), coriander (Kim *et al.*, 1996a, b; Stephen and Jayabalan, 2000), fennel (Sajina *et al.*, 1997b), celery (Pratap, 1992), dill (Ratnamba and Chopra, 1974; Sehgal, 1978) and nigella (Hamid *et al.*, 2004) and regeneration of these plants have been successfully obtained.

10. Microrhizome

Microrhizome technology is useful for developing disease free planting material, and hence is an ideal source of planting material suitable for germplasm exchange, transportation and conservation. *In vitro* induction of microrhizomes in ginger (Figure 6.7) was reported by many workers (Bhat *et al.*, 1994; Sharma and Singh, 1995; Nirmal Babu, 1997; Nirmal Babu *et al.*, 2003a, 2005; Tyagi *et al.*, 2007; Sumathi, 2007; Zheng *et al.*, 2008). The microrhizome derived plants have more tillers but were shorter. They gave fresh rhizome yield ranging from 100-800 g per plant with an estimated yield of 10 kg per 3 m^2 bed. Many reports are available on *in vitro* micro-rhizome formation in turmeric (Figure 6.8) (Nirmal Babu *et al.*, 2003a; Cousins and Alderberg, 2008; Ravindran *et al.*, 2005). Low sucrose has been reported to decrease the size of microrhizome but optimium microrhizome production at 6-9 per cent sucrose was also reported. Sucrose (6-9 per cent) was most effective in rhizome formation (Sumathi *et al.*, 2014).

Figure 6.7: Microrhizomes in Ginger.

Figure 6.8: Stages in Microrhizome Induction and *in vitro* Rhizome Formation in Turmeric.

11. Production of Secondary Metabolites

Biotechnology could be harnessed to exploit the potential of spices for bio-production of useful plant metabolites. The use of tissue culture for the biosynthesis of secondary metabolites particularly in plants of pharmaceutical significance holds an interesting alternative to control production of plant constituents. This technique is all the more relevant in recent years due to the ruthless exploitation of plants leading to reduced availability. *In vitro* proliferation of nutmeg mace and synthesis of flavour components in culture was reported by Nirmal Babu *et al.* (1992a). Since mace is the source of anti-carcinogenic compound myristicin, this technique with improvement could be used for production of myristicin. Most of the reports in saffron were on the *in vitro* proliferation of stigma and *in vitro* synthesis of colour components and metabolites. Proliferation of stigma of saffron *in vitro* and chemical analysis of metabolites produced through tissue cultures of *Crocus sativus* has been reported (Sano and Himeno, 1987; Himeno *et al.*, 1988; Sarma *et al.*, 1991).

Plant cells cultured *in vitro* produce wide range of primary and secondary metabolites of economic value. Production of phytochemicals from plant cell cultures has been presently used for pharmaceutical products. Production of flavour components and secondary metabolites *in vitro* using immobilized cells is an ideal system for spices crops. Ahmad *et al.* (2013) concluded that regenerated tissues of *P. nigrum* are a good source of biologically active metabolites for antimicrobial activities and callus culture presented itself as a good source for such activities. Production of saffron and capsaicin was reported using such system (Ravishankar *et al.*, 1993, 1995; Johnson *et al.*, 1996; Venkataraman and Ravishankar, 1997). Johnson *et al.* (1996) reported biotransformation of ferulic acid vanillamine to capsacin and vanillin in immobilized cell cultures of *Capsicum frutescens*. Reports on the *in vitro* synthesis of crocin, picrocrocin and safranel from saffron stigma (Himeno and Sano, 1995) and colour components from cells derived from pistils (Hori *et al.*, 1988) are available for further scaling up. Callus and cell cultures were established in nutmeg, clove, camphor, ginger, lavender, mint, thyme, celery *etc.* Cell immobilization techniques have been standardized in ginger, sage, anise and lavender (Ilahi and Jabeen, 1992). Production of essential oils from cell cultures (Ernst, 1989) and accumulation of essential oils by *Agrobacterium tumefaciens* transformed shoot cultures of *Pimpinella anisum* has been reported (Salem and Charlwood, 1995). Regulation of the shikimate pathway in suspension culture cells of parsley (Conn and McCue, 1994) and production of anethole from cell cultures of *Foeniculum vulgare* (Hunault and Manoir, 1992) was reported. Growth and production of monoterpene by transformed shoot cultures of *Mentha citrata* and *Mentha piperata* in flasks and fermentors was reported by Hilton *et al.* (1995). Production of rosmarinic acid in suspension cultures of *Salvia officinalis* has been discussed by Hippolyte *et al.* (1992). Reports on production of phenolic flavour compounds using cultured cells and tissues of vanilla are also available (Dorenburg and Knorr, 1996). *In vitro* production of petroselinic acid was reported from cell suspension cultures of coriander (Kim *et al.*, 1996a). Kintzios *et al.* (2004). reported scaling up of micropropagation of *Ocimum basilicum* L. in an airlift bioreactor and accumulation of rosmarinic acid. Though the feasibility of *in vitro* production of spice principles has been demonstrated, methodology for scaling up and reproducibility need to be developed before it can reach commercial levels. Once standardized this technology has tremendous potential in industrial production of important compounds like capsaicin, vanillin, crocin, picrocrocin, safranol, myristicin, anethole, menthol and curcumin.

12. Conclusion

Biotechnology thus offers solutions to the threats posed by climate change and loss of biodiversity, by non-convetional approaches for large scale cultivation and development of climate resilient crops. It has proved to be a key tool to achieve sustainable agriculture and agri based industry, through improvement of food production in terms of quantity, quality and safety, while preserving the environment. Significant progress has been made in the field of biotechnology for micropropagation, conservation and management of genetic resources, disease and pest mangement and molecular characterization. Identifying markers linked to important agronomic characters will help in marker assisted selection to shorten breeding time. Application of recombinant DNA technology for production of resistant types to biotic and abiotic stress has to go a long way before they can be effectively used in spices improvement. Though programmes have been initiated in many laboratories for *in vitro* secondary metabolite production these techniques are to be refined and scaled up for possible industrial production of the products. Owing to their commercial potential, intensification and application of biotechnology in spices is important and indispensable in the coming decade.

References

Aasim, M., Khawar, K.M., Sancak, C. and Ozcan, S. (2009). *In vitro* shoot regeneration of Fenugreek (*Trigonella foenumgraceum* L.). *American-Eurasian Journal of Sustainable Agriculture*, **3**: 135-138.

Adaniya, S. and Shirai, D. (2001). *In vitro* induction of tetraploid ginger (*Zingiber Officinale* Roscoe) and its pollen fertility and germinability. *Journal of Horticulture*, **88**: 277-287.

Aflatuni, A., Uusitalo, J., Ek, S. and Hohtola, A. (2005). Variation in the amount of yield and in the extract composition between conventionally produced and micropropagated peppermint and spearmint. *Journal of Essential Oil Research*, **17(1)**: 66-70.

Ahmad, N., Abbasi, B.H. and Fazal, H. (2013). Effect of different *in vitro* culture extracts of black pepper (*Piper nigrum* L.) on toxic metabolites - producing strains. *Toxicology and Industrial Health*, **32(3)**: 500-506.

Ascough, G.D., Erwin, J.E. and van Staden, J. (2009). Micropropagation of Iridaceae - a review. *Plant Cell Tissue and Organ Culture*, **97(1)**: 1-19.

Ayabe, M., Taniguchi, K. and Sumi, S.I. (1995). Regeneration of whole plants from protoplasts isolated from tissue-cultured shoot primordia of garlic (*Allium sativum* L.). *Plant Cell Reports*, **15(1-2)**: 17-21.

Azad, M.A.K., Yokota, S., Ishiguri, F., Yahara, S. and Yoshizawa, N. (2005). Large-scale clonal propagation of *Cinnamomum camphora* (L.) Nees and Eberm. Bulletin of the Utsunomiya University Forests.

Bhansali, R.R. (1990). Somatic embryogenesis and regeneration of in plantlets in pomegranate. *Annals of Botany*, **66(3)**: 249-253.

Bhat, S.R., Chandel, K.S.P. and Kacker, A. (1994).) *In vitro* induction of rhizome in ginger *Zingiber officinale* Rosc. *Indian Journal of Experimental Biology*, **32 (5)**: 340–344.

Bhojwani, S.S. (1980). *In vitro* propagation of garlic by shoot proliferation. *Scientia Horticulturae*, **13**: 47-52.

Bhuyan, A.K., Pattnaik, S. Chand, P.K. (1997). Micropropagation of curry leaf tree (*Murraya koenigii* (L.) Spreng by axillary proliferation using intact seedlings. *Plant Cell Reports*, **16**: 779-782.

Bin, Z. and Jiang, L. (2003). Study on micropropagation technology of pomegranate pyaman in Xinjiang. *Journal of Xinjiang Agricultural University*, **2**: 34-39.

Blazquez, S., Piqueras, A., Serna, M.D., Casas, J.L. and Fernández, J.A. (2003). Somatic embryogenesis in saffron: Optimisation through temporary immersion and polyamine metabolism. *ISHS Acta Horticulturae* **650**: 269-276.

Cellarova, E. (1992). Micropropagation of *Mentha* L. In: Bajaj, Y.P.S (Ed.), *Biotechnology in Agriculture and Forestry* (*Vol. 19*). *High Tech and Micropropagation III*. Springer, Heidelberg. *pp*. 262-275.

Chan, L.K. and Thong, W.H. (2004). *In vitro* propagation of *Zingiberaceae* species with medicinal properties. *Journal of Plant Biotechnology*, **6(3):** 181-188.

Chang, B. K. W. and Criley, A. (1993). Clonal propagation of pink ginger *in vitro*. *Horticultural Science*, **28**: 1203.

Chithra, M., Martin, K.P., Sunandakumari, C. and Madhusoodanan, P.V. (2005). Protocol for rapid propagation, and to overcome delayed rhizome formation in field established *in vitro* derived plantlets of *Kaempferia galanga* L. *Scientia Horticulturae*, **104(1)**: 113-120.

Conn, E.E. and McCue, K.F. (1994).) Regulation of the shikimate pathway in suspension cultured cells of parsley (*Petroselinum crispum* L.). In: Ryu, D.D.Y. and Furusaki, S (Ed.), *Advances in Plant Biotechnology*. Elsevier Science. Netherlands. pp. 95-102

Cousins, M.M. and Adelberg, J.W. (2008). Short-term and long-term time course studies of turmeric (*Curcuma longa* L.) microrhizome development *in vitro*. *Plant Cell Tissue and Organ Culture*, **93(3)**: 283-293.

Dixit, V. and Chaudhary, B.R. (2013). *Allium sativum*: Four step approach to efficient micropropagation. *International Journal of Innovative Biological Research*, **2(1)**: 6-14.

Dornenburg, H. and Knorr, D. (1996). Production of phenolic flavour compounds with cultured cells and tissues of vanilla species. *Food Biotechnology*, **10(1)**: 75–92.

Ebrahimie, E., Habashi, A.A., Ghareyazie, B., Ghannadha, M. and Mohammadie, M. (2003). A rapid and efficient method for regeneration of plantlets from embryo explants of cumin (*Cuminum cyminum*). *Plant Cell Tissue and Organ Culture*, **75(1)**: 19-25.

El-Agamy, S.Z., Mostafa, R.A., Shaaban, M.M. and El-Mahdy, M.T. (2009). *In vitro* propagation of Manfalouty and Nab El-gamal pomegranate cultivars. *Research Journal of Agriculture and Biological Sciences*, **5**: 1169-1175.

Ernst, D. (1989). *Pimpinella anisum* L. (Anise): Cell culture, somatic embryogenesis and production of anise oil. In: Bajaj, Y.P.S. (Ed.), *Biotechnology in Agriculture and Forestry* (Vol. 7). Medicinal and Aromatic Crops II. Springer. Verlag Berlin. pp. 381-397.

Evenor, D., Pressman, E., Ben Yephet, Y. and Rappaport, L. (1994).) Somaclonal variation in celery and selection by coculturing toward resistance to *Septoria apiicola*. *Plant Cell Tissue and Organ Culture*, **39:** 203-210.

Fadel, D., Kintzios, S., Economou, A.S., Moschopoulou, G. and Constantinidou, H.I.A. (2010). Effect of different strength of medium on organogenesis, phenolic accumulation and antioxidant activity of spearmint (*Mentha spicata*). *The Open Horticulture Journal*, **3**: 31-35.

Falk, L., Biswas, K., Boeckelmann, A., Lane, A. and Mahmoud, S.S. (2009). An efficient method for the micropropagation of lavenders: Regeneration of a unique mutant. *Journal of Essential Oil Research*, **21(3)**: 225-228.

Furmanowa, M. and Olszowska, O. (1992). Micropropagation of Thyme (*Thymus vulgaris* L.) In: Bajaj, Y. P. S. (Ed.), *Biotechnology in Agriculture and Forestry* (Vol.19). High-Tech and Micropropagation III, Springer. Verlag, Heildelberg. pp. 230-242.

Gayatri, M.C., Roopadarshini, V., Kavyashree, R. and Kumar, C.S. (2005). Encapsulation and regeneration of aseptic shoot buds of turmeric (*Curcuma longa* L.). *Plant Cell Biotechnology and Molecular Biology*, **6(3/4)**: 89-94.

Geetha, S.P., Manjula, C., John, C.Z., Minoo, D., Nirmal Babu, K. and Ravindran, P.N. (1997). Micropropagation of *Kaempferia* spp. (*Kaempferia galanga* L and *K. rotunda* L.). *Journal of Spices Aromatic Crops*, **6(2)**: 129–135.

Geetha, S.P., Nirmal Babu, K., Rema, J., Ravindran, P.N. and Peter, K.V. (2000). Isolation of protoplasts from cardamom (*Elettaria cardamomum* Maton.) and ginger (*Zingiber officinale* Rosc.). *Journal of Spices and Aromatic Crops*, **9(1)**: 23-30.

Ghosh, A., Chatterjee, P. and Ghosh, P. (2013). A protocol for rapid propagation of genetically true to type Indian turmeric (*Curcuma longa* L.) through *in vitro* culture technique. *Advances in Applied Science Research*, **4(3)**: 39-45.

Guan, Q., Guo, Y., Wei, Y., Meng, F. and Zhang, Z. (2010). Regeneration of somatic hybrids of ginger via chemical protoplast fusion. *Plant Cell Tissue and Organ Culture*, **102**: 279-284.

Gupta, D. and Bhargava, S. (2001). Thidiazuron induced regeneration in *Cuminum cyminum* L. *Journal of Plant Biochemistry and Biotechnology*, **10(1)**: 61-62.

Hamid, E., El-Olemy, M.M. and Al-Said, M.S. (2004). Enhancement of somatic embryogenesis and production of developmentally arrested embryos in *Nigella sativa* L. *Horticultural Science*, **39(2)**: 321-323.

Hazarika, B.N., Nagaraju, V. and Parthasarathy, V.A. (1995). Micropropagation of *Murraya koenigii* Spreng. *Annals of Plant Physiology*, **9(2)**: 149-151.

Hilton, M.G., Jay, A., Rhodes, M.J.C. and Wilson, P.D.G. (1995). Growth and monoterpene production by transformEd shoot cultures of *Mentha citrata* and *Mentha piperita* in flasks and fermenters. *Applied Microbiology and Biotechnology*, **43(3)**: 452–459.

Himeno, H. and Sano, K. (1995). Synthesis of crocin, picrocrocin and safranal by saffron stigma like structures proliferated *in vitro*. *Agricultural and Biological Chemistry*, **51(9)**: 2395-2400.

Himeno, H., Matsushima, H. and Sano, K. (1988). Scanning electron microscopic study on the *in vitro* organogenesis of saffron stigma and style like structures. *Plant Science*, **58**: 93-101.

Hippolyte, I., Marin, B., Baccou, J.C. and Jonard, R. (1992). Growth and rosmarinic acid production in cell suspension cultures of *Salvia officinalis* L. *Plant Cell Reports*, **11**: 109-112.

Hori, H., Enomoto, K. and Nakaya, H. (1988). Induction of callus from pistils of *Crocus sativus* L. and production of colour components in the callus. *Plant Tissue Culture Letters*, **5**: 72-77.

Huang, L.C., Huang, B.L. and Murashige, T. (1998). A micropropagation protocol for *Cinnamomum camphora*. *In Vitro Cellular and Developmental Biology–Plant*, **34**: 141-146.

Huang, L.C., Huang, B.L., Wang, C.H., Kuo, C.I. and Murashige, T. (2000). Developing an improved *in vitro* propagation system for slow-growing species using *Garcinia mangostana* L. (Mangosteen). *In Vitro Cellular and Developmental Biology–Plant*, **36(6)**: 501-504.

Hunault, G. and Du Manoir, J. (1992). Micropropagation of fennel. In: Bajaj, Y.P.S. (Ed.), *Biotechnology in Agriculture and Forestry* (*Vol.19*), High-Tech and Micropropagation III, Springer-Verlag, Heidelberg. pp. 199-216.

Hussain, T. M., Chandrasekhar, T., Arifullah, M. and Gopal, G.R. (2004). Effect of benzyladenine and thidiazuron on *in vitro* shoot formation from cotyledonary nodes of *Tamarindus indica* Linn. *Propagation of Ornamental Plants*, **4**: 47-52.

Ilahi, I. and Jabeen, M. (1992). Tissue culture studies for micropropagation and extraction of essential oils from *Zingiber officinale* Rosc. *Pakistan Journal of Botany*, **24(1)**: 54-59.

Inomoto, Y. and Kitani, Y. (1989). *In vitro* propagation of *Cinnamomum cassia*. *Plant Tissue Culture Letters*, **6**: 25-27.

Irikova, T., Grozeva, S. and Rodeva, V. (2011). Anther culture in pepper (*Capsicum annuum* L.) *in vitro*. *Acta Physiologiae Plantarum*, **33(5)**: 1559-1570.

Iyer, R.I., Jayaraman, G., Gopinath, P.M. and Sita, G.L. (2000). Direct somatic embryogenesis in zygotic embryos of nutmeg (*Myristica fragrans* Houtt.). *Tropical Agriculture*, **77(2)**: 98-105.

Jakhar, M.L. and Choudhary, M.R. (2012). Regeneration of *in vitro* plantlets through organogenesis in Fennel (*Foeniculum Vulgare* Mill.). *Journal of Plant Science and Research*, **28(2)**: 203.

Jakhar, M.L., Dhayal, M.S. and Rathore, V.S. (2003). *In vitro* flowering and seed formation in cumin (*Cuminum cyminum* L.). In: *Proceedings of National Seminar on New Perspectives in Spices, Medicinal and Aromatic Plants. Indian Society for Spices, Calicut. pp.* 86.

Johnson, T.S., Ravishanker, G.A. and Venkataraman, L.V. (1996). Biotransformation of ferulic acid vanillamine to capsacin and vanillin in immobilised cell cultures of *Capsicum frutescens. Plant Cell Tissue and Organ Culture*, **44(2)**: 117- 123.

Joseph, B., Joseph, D. and Philip, V.J. (1996). Plant regeneration from somatic embryos in black pepper. *Plant Cell, Tissue and Organ Culture*, **47(1)**: 87-90.

Kacker, A., Bhat, S.R., Chandel, K.P.S. and Malik, S.K. (1993). Plant regeneration *via* somatic embryogenesis in ginger. *Plant Cell Tissue and Organ Culture*, **32(3)**: 289-292.

Kaji, B.V., Ershadi, A. and Tohidfar, M. (2013). *In vitro* propagation of pomegranate (*Punica granatum* l.) Cv.'Males Yazdi'. *Albanian Journal of Agricultural Sciences*. pp. 12.

Kara, N. and Baydar, H. (2012). Effects of different explant sources on micropropagation in Lavender (*Lavandula* sp.). *Journal of Essential Oil Bearing Plants*, **15(2)**: 250-255.

Karaoglu, C., Çocu, S., Ipek, A., Parmaksýz, I., Uranbey, S., Sarýhan, E., Arslan, N., Kaya, M. D., Sancak, C., Özcan, S., Gürbüz, B. (2006). *In vitro* micropropagation of saffron. ISHS Acta Horticulturae 739: 223–227. *II International Symposium on Saffron Biology and Technology*. pp. 28–30.

Keller, E.R.J. and Senula, A. (2013). Micropropagation and cryopreservation of garlic (*Allium sativum* L.). *Methods in Molecular Biology*, **99**: 4353-368.

Kim, S.W., Park, M.K., Bae, K.S., Rhee, M.S. and Liu, J.R. (1996a). Production of petroselinic acid from cell suspension cultures of *Coriandrum sativum. Phytochemistry*, **42(6)**: 1581-1583.

Kim, S., Park, M., Liu, J.R., Kim, S.W. and Park, K.M. (1996b). High frequency plant regeneration via somatic embryogenesis in cell suspension cultures coriander (*Coriandrum sativum* L.). *Plant Cell Reports*, **15**: 751-753.

Kintzios, S., Kollias, H., Straitouris, E. and Makri, O. (2004). Scale-up micropropagation of sweet basil (*Ocimum basilicum* L.) in an airlift bioreactor and accumulation of rosmarinic acid. *Biotechnology Letters*, **26(6)**: 521–523.

Kong, L., Dai, D., Shang, M., Li, K. and Zhang, C.X. (2009). Thidiazuron-induced somatic embryos, their multiplication, maturation, and conversion in *Cinnamomum pauciflorum* Nees (Lauraceae). *New Forests*, **38(2)**: 131-142.

Kononowicz, H. and Janick, J. (1984). *In vitro* propagation of *Vanilla planifolia. Horticultural Science*, **19**: 58-59.

Kothari, S. L., Joshi, A., Kachhwaha, S. and Ochoa-Alejo, N. (2010). Chilli peppers - A review on tissue culture and transgenesis. *Biotechnological Advances*, **28(1)**: 35-48.

Kulkarni, M. and Deodhar, M. (2002). *In vitro* regeneration and hydroxycitric acid production in tissue culture of *Garcinia indica* Choisy. *Indian Journal of Biotechnology*, **1**: 301–304.

Lacy, M.L., Grumet, R., Toh, D.F., Krebs, S.L., Cortright, B.D. and Hudgins, E. (1996). MSU-SHK5: A somaclonally derived *Fusarium* yellows resistant celery line. *Horticultural Science*, **31(2)**: 289-290.

Majourhat, K., Martínez-Gómez, P., Piqueras, A. and Fernández, J. A. (2006). Enhanced plantlet regeneration from cultured meristems in sprouting buds of saffron corms. ISHS Acta Horticulturae, 739: *II International Symposium on Saffron Biology and Technology*.

Malik, S.K., Chaudhury, R. and Kalia, R.K. (2005). Rapid *in vitro* multiplication and conservation of *Garcinia indica*: A tropical medicinal tree species. *Scientia Horticulturae*, **106(4)**: 539-553.

Mallika, V.K., Rekha, K., Marymol, M., Manjula, M. and Vikraman Nair, R. (1997). *In vitro* shoot initiation from explants of field grown trees of nutmeg (*Myristica fragrans* Houtt.). In: Edison, S., Ramana, K. V., Sasikumar, B., Nirmal Babu, K. and Santhosh, J. E. (Ed.), *Biotechnology of Spices, Medicinal and Aromatic Crops, Indian Society for Spices*. pp. 29-34.

Manohari, C., Backiyarani, S., Jebasingh, T., Somanath, A. and Usha, R. (2008). Efficient plant regeneration in small cardamom (*Elettaria cardamomum* Maton.) through somatic embryogenesis. *Indian Journal of Biotechnology*, **7**: 407–409.

Mascarenhas, A.F., Nair, S., Kulkarni, V.M., Agrawal, D.C., Khuspe, S.S. and Mehta, U.J. (1987). Tamarind. In: Bonga, J.M. and Durzan, D.J. (Eds.). *Cell and Tissue Culture in Forestry*, (Vol. 3), Martinus Nijhoff, Dordecht. pp. 316-330.

Mathew, D. and Prasad, M.C. (2007). Multiple shoot and plant regeneration from immature leaflets of *in vitro* origin in curryleaf (*Murraya koenigii* Spreng). *Indian Journal of Plant Physiology*, **12(1)**: 18-22.

Mathew, K.M., Rao, Y.S., Kumar, K.P., Sallykutty, J., Lakshmanan, R. and Madhusoodanan, K.J. (1999). Micropropagation of curry leaf (*Murraya koenigii* L.). *Journal of Spices Aromatic Crops*, **8(1)**: 77-79.

Mathew, M.K. and Hariharan, M. (1990). *In vitro* multiple shoot formation in *Syzygium aromaticum*. *Annals of Botany*, **65**: 277279.

Mehta, U. J., Barreto, S.M. and Hazra, S. (2004). Effect of thidiazuron in germinating tamarind seedlings. *In Vitro Cellular and Developmental Biology–Plant*, **40(3)**: 279-283.

Mehta, U.J., Krishnamurthy, K.V. and Hazra, S. (2000). Regeneration of plants via adventitious bud formation from mature zygotic embryo axis of tamarind (*Tamarindus indica* L.). *Current Science*, **78(10)**: 1231–1234.

Minas, G. J. (2009). Peppermint (*Mentha piperita*) sanitation and mass micropropagation *in vitro*. In: *International Symposium on Medicinal and Aromatic Plants*, **853**: 77-82.

Mini, P.M., John, C.Z., Samsudeen, K., Rema, J., Nirmal Babu, K. and Ravindran, P.N. (1997). Micropropagation of *Cinnamomum verum* (Bercht and Presl.). In: Edison, S., Ramana, K.V., Sasikumar, B., Nirmal Babu, K. and Santhosh, J.E. (Ed.), *Biotechnology of Spices, Medicinal and Aromatic Crops,* Indian Society for Spices. pp. 35-38.

Minoo, D. and Nirmal Babu, K. (2009). Micropropagation and *in vitro* conservation of vanilla (*Vanilla planifolia* Andrews). In: Jain S.M. and Saxena, P.K. (Eds.), *Springer Protocols, Methods in Molecular Biology 547, Protocols for In Vitro Cultures and Secondary Metabolite Analysis of Aromatic and Medicinal Plants, The Humana Press*, Springer, USA. pp. 129-138.

Minoo, D. (2002). SeEdling and somaclonal variation and their characterization in Vanilla. Ph.D Thesis, Calicut University, Kerala, India.

Minoo, D., Geetha, S.P., Nirmal Babu, K. and Peter, K.V. (2008). Isolation and fusion of protoplasts in *Vanilla* species. *Current Science*, **94(1)**: 115-120.

Minoo, D., Nirmal Babu, K., Ravindran, P.N. and Peter, K.V. (2006). Interspecific hybridization in vanilla and molecular characterization of hybrids and selfed progenies using RAPD and AFLP markers. *Scientia Horticulturae*, **108(4)**: 414-422.

Miura, Y., Fukui, H. and Tabata, M. (1987). Clonal propagation of chemically uniform fennel plants through somatic embryoids. *Planta Medica*, **53(1)**: 92-94.

Mohan, S., Parthasarathy, U. and Babu, K.N. (2012). *In vitro* and in vivo adventitious bud differentiation from mature seeds of three *Garcinia* spp. *Indian Journal of Natural Products and Resources*, **3(1)**: 65-72.

Munyon, I.P., Hubstenberger, J.F. and Phillips, G.C. (1989). Origin of plantlets and callus obtain Ed from chilli pepper anther cultures. *In Vitro Cellular and Developmental Biology-Plant*, **25(3)**: 293-296.

Nadgauda, R.S., Khuspe, S.S. and Mascarenhas, A.F. (1982). Isolation of high curcumin varieties of turmeric from tissue culture. In: R.D Iyer (Ed.), *Proceedings of Annual Symposium on Plantation Crops*. CPCRI, Kasaragod. pp. 143-144.

Nadgauda, R.S., Kulkarni, D.B., Mascarenhas, A.F. and Jaganathan, V. (1980). Development of plantlets from tissue cultures of ginger. In: *ProceEding annual symposium on plantation crops*. CPCRI, Kasaragod. pp. 143-147.

Nadgauda, R.S., Mascarenhas, A.F. Madhusoodanan, K.J. (1983). Clonal multiplication of cardamom (*Elettaria cardamomum* Maton.) by tissue culture. *Journal of Plantation Crops*, **11**: 60–64.

Nadgauda, R.S., Mascarenhas, A.F., Hendre, R.R. and Jaganathan, V. (1978). Rapid clonal multiplication of turmeric *Curcuma longa* L. plants by tissue culture. *Indian Journal of Experimental Biology*, **16**: 120-122.

Naik, S.K. and Chand, P.K. (2006). Nutrient-alginate encapsulation of *in vitro* nodal segments of pomegranate (*Punica granatum* L.) for germplasm distribution and exchange. *Scientia Horticulturae*, **108(3)**: 247-252.

Nair, R.R. and Dutta Gupta, S. (2003). Somatic embryogenesis in black pepper (*Piper nigrum* L.): 1. Direct somatic embryogenesis from the tissues of germinating seeds and ontogeny of somatic embryos. *Journal of Horticultural Science and Biotechnology*, **78**: 416–421.

Nair, R. R. and Dutta Gupta, S. (2006). High frequency plant regeneration through cyclic secondary somatic embryogenesis in black pepper (*Piper nigrum* L.). *Plant Cell Reports*, **24**: 699–707.

Navroski, M.C., Waldow, D.A., Pereira, M.D. and Pereira, A.D. (2012). Callus formation *in vitro* and internodal stem apices in savory. *Agroambiente On-line,* **6(3)**: 228-234.

Nazeem, P.A., Augustin, M., Rathy, K., Sreekumar, P.K., Rekha, C.R., Shaju, K. V., Peter, K.V., Girija, D. and Kesavachandran, R. (2004). A viable protocol for large scale *in vitro* multiplication of black pepper (*P. nigrum* L.). *Journal of Plantation Crops,* **32**: 163-168.

Nazeem, P.A., Joseph, L., Rani, T.G., Valsala, P.A., Philip, S. and Nair, G.S. (1996). Tissue culture system for *in vitro* pollination and regeneration of plantlets from *in vitro* raised seeds of ginger - *Zingiber officinale* rosc. *International Symposium on Medicinal and Aromatic Plants*. ISHS, Acta Horticulturae. pp. 426.

Nhung, H.N. and Quynh, N.T. (2012). A study on growth ability of *Thymus vulgaris* L. under impact of chemical and physical factors of culture medium. *Journal of Biology*, **34(3SE)**: 234-241.

Nikhil, A. and Shukla, S. (2013). Production of artificial seeds from nodal region of sweet neem (*Murraya koenigii*). *Journal of Advance Pharmaceutical Research and Bioscience*, **1(2)**: 71-74.

Nirmal Babu, K. and Minoo, D. (2003). Commercial Micropropagation of Spices. In: Chandra, R. and Misra, M. (Ed.), *Micropropagation of Horticultural Crops, International Book Distributing Company*, Lucknow. pp. 345.

Nirmal Babu, K. (1997). *In vitro* studies in *Zingiber officinale* Rosc. PhD Thesis, Calicut University, Kerala, India

Nirmal Babu, K., Anu, A., Remasree, A.B. and Praveen, K. (2000). Micropropagation of curry leaf tree *Murraya koenigii* (L.) Spreng. *Plant Cell Tissue and Organ Culture,* **61(3)**: 199–203.

Nirmal Babu, K., Minoo, D. and Parthasarathy, V.A. (2011a). Ginger. In: Singh, H. P., Parthasarathy, V. A. and Nirmal Babu, K. (Eds.), *Advances in Horticulture Biotechnology* (Vol. 1). Regeneration Systems – *Fruit Crops, Plantation Crops and Spices*. Westville Publishing House, New Delhi. pp. 421-442.

Nirmal Babu, K., Minoo, D., Geetha, S.P., Samsudeen, K., Rema, J., Ravindran, P.N. and Peter, K.V. (1998). Plant biotechnology - it's role in improvement of spices. *Indian Journal of Agricultural Sciences*, **68(8)**: 533-547.

Nirmal Babu, K., Nair, R.R., Saji, K.V. and Parthasarathy, V.A. (2012). Biotechnology. In: Singh, H.P., Parthasarathy, V.A., Srinivasan, V. and Saji, K.V. (Eds.), *Piperaceae*. Westville Publishing House, New Delhi. pp. 57-81.

Nirmal Babu, K., Ravindran, P.N. and Peter, K.V. (1997). *Protocols for micropropagation of spices and aromatic crops*. Indian Institute of Spices Research, Calicut, Kerala. p. 35.

Nirmal Babu, K., Sajina, A., Minoo, D., John, C.Z., Mini, P.M., Tushar, K.V. and Ravindran, P.N. (2003a). Micropropagation of camphor tree (*Cinnamomum camphora*). *Plant Cell Tissue and Organ Culture,* **74(2)**: 179-183.

Nirmal Babu, K., Sajina, A., Minoo, D., John, C.Z., Mini, P.M., Tushar, K.V., Rema, J. and Ravindran, P.N. (2003b). Micropropagation of camphor tree (*Cinnamomum camphora*). *Plant Cell Tissue and Organ Culture,* **74(2)**: 179–183.

Nirmal Babu, K., Samsudeen, K. and Ravindran, P.N. (1996a). Biotechnological approaches for crop improvement n ginger, *Zingiber officinale* Rosc. In: Ravishanker, G. A. and Venkataraman, L. V. (Ed.), *Recent Advances in Biotechnological Applications of Plant Tissue and Cell Culture*. IBH Publishing Co, New Delhi. pp. 321-332.

Nirmal Babu, K., Samsudeen, K., Minoo, D., Geetha, S.P. and Ravindran, P.N. (2005). Tissue culture and Biotechnology of Ginger. In: Ravindran, P.N. and Nirmal Babu, K. (Ed.), *Ginger – The genus Zingiber*. CRC Press, Boca Raton, USA. pp. 181- 210.

Nirmal Babu, K., Samsudeen, K. and Ravindran, P.N. (1992b). Direct regeneration of plantlets from immature inflorescence of ginger (*Zingiber officinale* Rosc.) by tissue culture. *Journal of Spices Aromatic Crops,* **1**: 4348.

Nirmal Babu, K., Samsudeen, K. and Ravindran, P.N. (1996b). Biotechnological approaches for crop improvement in ginger, *Zingiber officinale* Rosc. In: Ravishanker, G.A. and Venkataraman, L.V. (Ed.), *Recent Advances in Biotechnological Applications on Plant Tissue and Cell Culture*. Oxford IBH Publishing Co. New Delhi. pp. 321–332.

Nirmal Babu, K., Samsudeen. K. and Ratnambal, M.J. (1992a). *In vitro* plant regeneration from leaf derivEd callus in ginger, *Zingiber officinale* Rosc. *Plant Cell Tissue and Organ Culture,* **29**: 7174.

Nirmal Babu, K., Usha Rani, T. R. and Parthasarathy, V.A. (2011b). Cardamom. In: Singh, H.P., Parthasarathy, V.A. and Nirmal, K. (Ed.), *Advances in Horticulture* Biotechnology (Vol. 3). Molecular Markers and Marker Assisted Selection - Fruit Crops, *Plantation Crops and Spices*, Westville Publishing House, New Delhi. pp. 261-268.

Ochoa-Alejo, N. and Ramirez-Malagon, R. (2001). *In vitro* chili pepper biotechnology. *In Vitro Cellular and Developmental Biology - Plant,* **37(6)**: 701-729.

Panda, M.K., Mohanty, S., Subudhi, E., Acharya, L. and Nayak,S. (2007). Assessment of genetic stability of micropropagated plants of *Curcuma longa* L. by cytophotometry and RAPD analysis. *International Journal of Integrative Biology,* **1(3)**: 189–195.

Panizza, M. and Tognoni, F. (1992). Micropropagation of lavender (*Lavandula officinalis* Chaix X *Lavandula latifolia* villars cv. Grosso). In: Bajaj, Y.P.S. (Ed.),

Biotechnology in Agriculture and Forestry (Vol.19). *High-Tech and Micropropagation III*, Springer-Verlag, Heidelberg. pp. 295-305.

Parashar, M., Jakhar, M.L., and Malik, C.P. (2014). A review on biotechnology, genetic diversity in cumin (*Cuminum cyminum*). *International Journal of Life Science and Pharma Research*, **4(4)**: L17-L34.

Patil, V.M., Dhande, G.A., Thigale, D.M. and Rajput, J.C. (2011). Micropropagation of pomegranate (*Punica granatum* L.) 'Bhagava'cultivar from nodal explant. *African Journal of Biotechnology*, **10(79)**: 18130-18136.

Patnaik, S. and Chand, P.K. (1996). *In vitro* propagation of medicinal herbs *Ocimum americanum* L. syn. *O. canum* Sims (hoary basil) and *Ocimum sanctum* L. (holy basil). *Plant Cell Reports*, **15(11)**: 846-851.

Pattepur, S.V., Mokashi, A.N. and Ajjappalavara, P.S. (2010). Effect of cytokinins and auxin on shoot proliferation of corylEdonary nodes derived from axenic seedling of tamarind (*Tamarindus indica* L.). *Asian Journal of Horticulture*, **5(1)**: 185-188.

Paul, S., Dam, A., Bhattacharyya, A. and Bandyopadhyay, T.K. (2011). An efficient regeneration system via direct and indirect somatic embryogenesis for the mEdicinal tree *Murraya koenigii*. *Plant Cell Tissue and Organ Culture*, **105(2)**: 271-283.

Pratap, K.P. (1992). Artificial seeds: Vatika from the Seed and Plant People. Spring Issue, **1**: 27-30.

Rahman, M.M., Amin, M.N., Ahamed, T., Ali, M.R., Habib, A. (2004). Efficient plant regeneration through somatic embryogenesis from leaf base-derivEd callus of *Kaempferia galangal* L. *Asian Journal of Plant Sciences*, **3(6)**: 675- 678.

Raju, B., Anita, D. and Kalita, M.C. (2005). *In vitro* clonal propagation of *Curcuma caesia* Roxb and *Curcuma zEdoaria* Rosc from rhizome bud explants. *Journal of Plant Biochemistry and Biotechnology*, **14(1)**: 61-63.

Rao, Y.S., Mary, M.K., Pradip Kumar, K., Salykutty, J., Laxmanan, R., Madhusoodhanan, K.J. and Potty, S.N. (1997). Tissue culture studies on tree spices. In: Edison, S., Ramana, K.V., Sasikumar, B., Nirmal Babu, K. and Santhosh, J. E. (Eds.), *Biotechnology of Spices, Medicinal and Aromatic Crops*. Indian Society for Spices. pp. 39-44.

Ratnamba, S.P. and Chopra, R.N. (1974). *In vitro* induction of embryoids from hypocotyls and cotyledons of *Anethum graveolens* seedlings. *Zeitschrift für Pflanzenphysiologie*, **73**: 452-455.

Ravindran, P. N., Nirmal Babu, K., Peter, K.V., Abraham, Z. and Tyagi, R.K. (2005). Spices. In: B.S. Dhillon., R.K. Tyagi., S, Saxena. and G.J. Randhawa. (Eds.) *Plant Genetic Resources: Horticultural Crops*. Narosa Publishing House, New Delhi. pp. 190- 227.

Ravindran, P.N., Peter, K.V., Nirmal Babu, K., Rema, J., Samsudeen, K., Minoo, D., Geetha, S.P., Sajina, A., Mini, P.M., Manjula, C. and John, C.Z. (1996). Biotechnological approaches in spice crops - present scenario and future prospects. In: Das, M. R. and Mundayoor, S. (Eds.), *Biotechnology for Development. State Committee on Science, Technology and Environment*, Kerala. pp. 175-197.

Ravishankar, G.A., Sudhakar, J.T. and Venkataraman, L.V. (1993). Biotechnological approach of *in vitro* production of capsaicin. In: *Proceedings of the National Seminar on Post Harvest Technology of Spices*. Trivandrum. pp. 75-82.

Renjith, D., Valsala, P.A. and Nybe, E.V. (2001). Response of turmeric (*Curcuma domestica* Val.) to *in vivo* and *in vitro* pollination. *Journal of Spices and Aromatic Crops*, **10(2)**: 135-139.

Rodeva, V., Gudeva, L.K., Grozeva, S. and Trajkova, F. (2013). Obtaining haploids in anther culture of pepper *Capsicum annuum* L. and their inclusion in the breeding process. *Yearbook-Faculty of Agriculture*, **7(1)**: pp-7.

Sajina, A., Geetha, S.P., Minoo, D., Rema, J., Nirmal Babu, K., Sadanandan, A. K., Ravindran, P. N. (1997a) Micropropagation of some important herbal spices. In: Edison, S., Ramana, K.V., Sasikumar, B., Nirmal Babu, K. and Santhosh, J. E. (Eds.), *Biotechnology of Spices, Medicinal and Aromatic Plants*. Indian Society for Spices, Calicut, India. pp. 79- 86.

Sajina, A., Minoo, D., Geetha, P., Samsudeen, K., Rema, J., Nirmal Babu, K., Ravindran, P. N., Peter, K.V. (1997b) Production of synthetic seeds in few spice crops. In: Edison, S., Ramana, K.V., Sasikumar, B., Nirmal Babu, K. Santhosh, J.E. (Eds.) *Biotechnology of Spices, Medicinal and Aromatic Plants*. Indian Society for Spices, Calicut, India, pp 65-69.

Salem, K.M.S.A. and Charlwood, B.V. (1995). Accumulation of essential oils by *Agrobacterium tumefaciens* transformed shoot cultures of *Pimpinella anisum*. *Plant Cell Tissue and Organ Culture*, **40(3)**: 209-215.

Salvi, N.D., George, L. and Eapen, S. (2000). Direct regeneration of shoots from immature inflorescence cultures of turmeric. *Plant Cell Tissue and Organ Culture*, **62(3)**: 235–238.

Salvi, N.D., George, L. and Eapen, S. (2002). Micropropagation and field evaluation of micropropagated plants of turmeric. *Plant Cell, Tissue and Organ Culture*, **68(2)**: 143-151.

Samantaray, A., Sial, P. and Kar, M. (2012). Micropropagation and biochemical analysis of spear mint (*Mentha spicata*). *Indian Journal of Innovations and Developments*, **1(7)**: 489-493.

Samsudeen, K., Nirmal Babu, K., Minoo, D. and Ravindran, P.N. (2000). Plant regeneration from anther derived callus cultures of ginger (*Zingiber officinale* Rosc.). *Journal of Horticultural Science and Biotechnology*, **75(4)**: 447–450.

Sano, K. and Himeno, H. (1987). *In vitro* proliferation of saffron (*Crocus sativus* L.) stigma. *Plant Cell Tissue and Organ Culture*, **11**: 159-166.

Santoro, M.V., Nievas, F., Zygadlo, J., Giordano, W. and Banchio, E. (2013). Effects of growth regulators on biomass and the production of secondary metabolites in peppermint (*Mentha piperita*) micropropagatEd *in vitro*. *American Journal of Plant Sciences*, **4**: 49-55.

Sarma, K.S., Sharada, K., Maesato, K., Hara, T. and Sonoda, Y. (1991). Chemical and sensory analysis of saffron produced through tissue cultures of *Crocus sativus*. *Plant Cell Tissue and Organ Culture*, **26**: 11-16.

Sehgal, C. B. (1978). Differentiation of shoot buds and embryoids from inflorescence of *Anethum graveolens* in cultures. *Phytomorphology*, **28**: 291-297.

Settu, A., Ranjitha Kumari, B. D. and Jeya Mary, R. (1997). *In vitro* selection for salt tolerance in *Trigonella foenumgraceum* using callus and shoot tip cultures. In: Edison, S., Ramana, K.V., Sasikumar, B., Nirmal Babu, K. and Santhosh, J.E. (Ed.), *Biotechnology of Spices, Medicinal and Aromatic Plants*. Indian Society for Spices, Calicut. pp. 119-121.

Shaji, P., Anandaraj, M. and Sharma, Y.R. (1998). Comparative study of protoplast isolation and development in *Piper nigrum* (black pepper) and *P. colubrinum*. In: Mathew, N.M. and Jacob, C.K. (Eds.), *Developments in Plantation Crops Research*. Allied Publishers, New Delhi. pp. 51-53.

Sharma, R.K., Wakhlu, A.K. and Boleria, M. (2004). Micropropagation of *Anethum graveolens* L. through axillary shoot proliferation. *Journal of Plant Biochemistry and Biotechnology*, **13(2)**: 157-159.

Sharma, T.R. and Singh, B.M. (1995). *In vitro* micro rhizome production in *Zingiber officinale* Rosc. *Plant Cell Reports*, **15(3/4)**: 274-277.

Sharma, T.R. and Singh, B.M. (1997). High frequency *in vitro* multiplication of disease free *Zingiber officinale* Rosc. *Plant Cell Reports*, **17(1)**: 68-73.

Sharma, T.R., Singh, B.M. and Chauhan, R.S. (1994).) Production of encapsulated buds of *Zingiber officinale* Rosc. *Plant Cell Reports*, **13**: 300-302.

Sheibani, M., Nemati, S.H., Davarinejad, G.H., Azghandi, A.V. and Habashi, A. A. (2006). Induction of somatic embryogenesis in saffron using thidiazuron. In: *II International Symposium on Saffron Biology and Technology, ISHS Acta Horticulturae*. pp. 739.

Shekhawat, N.S. and Galston, A.W. (1983). Mesophyll protoplasts of fenugreek (*Trigonella foenumgraecum*): Isolation, culture and shoot regeneration. *Plant Cell Reports*, **2(3)**: 119-121.

Shi, X., Dai, X., Liu, G. and Bao, M. (2009). Enhancement of somatic embryogenesis in camphor tree (*Cinnamomum camphora* L.): Osmotic stress and other factors affecting somatic embryo formation on hormone-free medium. *Trees*, **23(5)**: 1033-1042.

Shi, X., Dai, X., Liu, G., Zhang, J., Ning, G. and Bao, M. (2010). Cyclic secondary somatic embryogenesis and efficient plant regeneration in camphor tree (*Cinnamomum camphora* L.). In *Vitro Cellular and Developmental Biology-Plant*, **46(2)**: 117-125.

Shukla, M.R., Subhash, N., Patel, D.R. and Patel, S.A. (1997b) *In vitro* selection for resistance to *Alternaria* blight in cumin (*Cuminum cyminum*). In: Edison,

S., Ramana, K.V., Sasikumar, B., Nirmal Babu, K. and Santhosh, J.E. (Eds.), *Biotecnology of Spices, Medicinal and Aromatic Plants*. Indian Society for Spices, Calicut. pp. 126-128.

Shylaja, M.R., Nair, S.G., Nazism, P.A., Millikan, V.K., and Mathew, M.K. (1996). *In vitro* screening of black pepper for tolerance to *Phytophthora capsici*. *Journal of Plantation Crops*, **24**: 171-178.

Smith, M.K., Hamill, S.D., Gogel, B.J. and Severn-Ellis, A.A. (2004). Ginger (*Zingiber officinale*) autotetraploid with improved processing quality produced by an *in vitro* colchicines treatment. *Australian Journal of Experimental Agriculture*, **44**: 1065-1072.

Song JuYeon, S., Sivanesan, I., Chul Geon, A. and Byoung Ryong, J. (2009). Micropropagation of paprika (*Capsicum annuum*) and its subsequent performance in greenhouse cultivation. *Korean Journal of Horticultural Science and Technology*, **27(2)**: 293–298.

Spices Board of India (2014). http://www.indianspices.com/.

Stephen, R. and Jayabalan, N. (2000). Artificial seed production in coriander (*Coriandrum sativum* L.). *Plant Tissue Cult*, **10(1)**: 45-49.

Sumathi, V. (2007). *Studies on Somaclonal Variation in Zingiberaceous Crops*. PhD Thesis, University of Calicut, Kerala, India.

Sumathi, V., Geetha, S.P., Minoo, D., Smitha, S., Soji Joseph. and Nirmal Babu, K. (2014). Induction and field performance of *in vitro* microrhizomes of turmeric (*Curcuma longa* L) for production of disease free planting material. *International Journal of Innovative Horticulture*, **3(2)**: 142-150.

Sundararaj, S.G., Agrawal, A. and Tyagi, R.K. (2010). Encapsulation for *in vitro* short-term storage and exchange of ginger (*Zingiber officinale* Rosc.) germplasm. *Scientia Horticulturae*, **125(4)**: 761-766.

Sunitibala, H., Damayanti, M. and Sharma, G.J. (2001). *In vitro* propagation and rhizome formation in *Curcuma Ionga* Linn. CYTOBIOS-CAMBRIDGE. pp. 71-82.

Superman, U. and Blake, J. (1990). Studies on tissue culture of clove tree plant. *Indonesian Journal of Crop Science*, **5**: 67-75.

Tawfik, A.A. and Noga, G. (2001). Adventitious shoot proliferation from hypocotyl and internodal stem explants of cumin. *Plant Cell Tissue and Organ Culture*, **66(2)**: 141-147.

Tawfik, A.A. and Noga, G. (2002). Differentiation of somatic embryos in suspension cultures and plant regeneration of cumin (*Cuminum cyminum* L.). *Journal of Applied Botany*, **76**: 144-149.

Tyagi, R.K., Agrawal, A., Mahalakshmi, C., Hussain, Z. and Tyagi, H. (2007). Low-cost media for *in vitro* conservation of turmeric (*Curcuma longa* L.) and genetic stability assessment using RAPD markers. In *Vitro Cellular and Developmental Biology–Plant*, **43**: 51-58.

Vandemoortele, J.L., Billard, J.P., Boucaud, J. and Gaspar, T. (1996). Micropropagation of parsley through axillary shoots proliferation. *Plant Cell, Tissue and Organ Culture*, **44(1)**: 25-31.

Venkataraman, L.V. and Ravishanker, G.A. (1997). Biotechnological approaches for production of saffron and capsaicin - a perspective. In: Edison, S., Ramana, K.V., Sasikumar, B., Nirmal Babu, K. and Santhosh, J.E (Eds.), *Biotechnology of Spices, Medicinal and Aromatic Plants. Indian Society for Spices,* Calicut, India. pp. 156-165.

Wang, Z.M., Niu, Y. and Song, M., Tang, Q.L. (2010). Tetraploid of *Zingiber officinale* Roscoe. *in vitro* inducement and its morphology analysis. *China Vegetables*, **4**: 13.

Yamuna, G. (2007). Studies on cryopreservation of spice genetic resources. PhD Thesis, Calicut University, Kerala, India.

Zeybek, E., Sertaç Önde, S. and Kaya, Z. (2012). Improved *in vitro* micropropagation method with adventitious corms and roots for endangered saffron. *Central European Journal of Biology*, **7(1)**: 138-145.

Zheng, Y., Liu, Y., Ma, M. and Xu, K. (2008). Increasing *in vitro* microrhizome production of ginger (*Zingiber officinale* Roscoe). *Acta Physiologiae Plantarum,* **30(4)**: 513-519.

Chapter 7

Coffee

☆ *H.L. Sreenath*

1. Introduction

Coffee is one of the most important beverages in the world with over 2.25 billion cups consumed every day. For decades, coffee has remained the most important agricultural export commodity in the world, second only to oil. Coffee is prepared from the roasted and ground beans of coffee plants (*Coffea* sps). Although the *Coffea* genus includes more than 124 species, commercial coffee production relies mainly on two related species. *C. arabica* L., popularly known as arabica coffee, and *C. canephora* Pierre, called as robusta coffee, which contribute 65 per cent and 35 per cent to the total global coffee production, respectively (International Coffee Organization: http://www.ico.org). Both are woody perennial species with useful life span of 3-4 decades. Coffee is cultivated in more than 11 million hectares (ha) in tropical and sub-tropical regions of Africa, Asia and the Americas. Small stakeholders, with less than 5 ha, account for about 70 per cent of world coffee production, and more than 80 million people depend on the crop for their income.

Coffee farmers are facing several challenges that need to be overcome to increase the productivity for sustainable cultivation. Tissue culture techniques can offer some unique solutions for improving the productivity of coffee farms. Cultivation of high yielding hybrids or superior clones is a major approach for productivity improvement in both the species. Cultivation of F_1 hybrids in self pollinating *C. arabica* has shown significant increase in yield due to heterosis. However, production of these F_1 hybrids involves the tedious process of manual/hand pollination. Rapid clonal propagation of these *C. arabica* F_1 hybrids is desirable to meet the large scale demand of planting material in time. *C. canephora* is cross pollinated and hence seedling progenies tend to be heterogeneous with many undesirable plants. Thus production of true-to-type plants through vegetative/clonal propagation is desirable for introducing new genotypes into production in both the species. Clonal propagation of heterozygous elite plants and the use of multiple clone lines at one time will assure preservation of heterozygosity and plasticity in coffee plantations. Micropropagation is a way of clonal propagation through tissue cultures (stem cuttings, axillary buds, meristem clusters, leaf cuttings *etc.*) grown under aseptic and controlled environment. Micropropagation is desirable in both the species of coffee for rapid and large scale clonal multiplication of elite genotypes to rise true to type plantations.

Often variant plants are produced in tissue culture. This variation is called somaclonal variation. Somaclonal variation is the expression of the naturally occurring variability in plant cells or the result of *in vitro* induced variability of cells following plant regeneration (Larkin and Scowcroft, 1981; Evans and Sharp, 1986). Genetic diversity is rather limited, particularly in *C. arabica* cultivars, due to its narrow genetic base. Often desirable traits needed for breeding improved cultivars are lacking in arabica selections. Somaclonal variation can be helpful in this regard by enhancing genetic variation. Somaclonal variation is undesirable when the goal is large-scale clonal propagation of true to type elite lines. On the other hand, somaclonal variation is useful to create genetic diversity, thus expanding the germplasm pool with novel traits for crop improvement. Somaclonal variation is an excellent method for shortening breeding programs, since it can provide access to genetic variability within existing cultivars (Evans and Sharp, 1986). Somaclones carry few genetic alterations and so the overall genetic integrity of the original commercial cultivar is preserved. Somaclonal variation has contributed to the release of improved varieties in some plant species (Hammerschlag, 1992). Genetic transformation offers new tool for improving plant varieties for important agronomic traits by allowing direct transfer of genes into plants. This technique can be used well-known cultivars in one or two characters. Reliable tissue culture regeneration is essential for successful genetic transformation. Review of literature outlined here brings out that coffee tissue culture is being applied successfully for meeting the opposing objectives of propagating true to type elite plants on large scale and obtaining useful somaclonal variants. Also, progress in plant regeneration through somatic embryogenesis (SE) in *C. arabica* and *C. canephora* has enabled genetic transformation of both the species.

2. Micropropagation through Somatic Embryogenesis

2.1. Historical Background

Several reviews on coffee tissue culture have been published (Sondahl and Loh, 1988; Sondahl and Baumann, 2001; Dublin, 1991; Sondahl and Lauritis; 1992; Etienne, 2005; Sreenath and Naidu, 1999). Plant regeneration through SE is successful in both arabica and robusta coffee. SE is the ability to induce large quantities of somatic embryos. Subsequent germination of these non-sexual embryos and recovery of normal coffee plants are techniques of paramount importance for multiple applications in coffee improvement programs (Sondahl and Baumann, 2001). Propagation through SE is a type of vegetative (asexual) reproduction exploiting plant cell totipotency. SE can be integrated into the breeding schemes of both arabica and robusta coffee. In robusta, it allows the rapid propagation of selected clones. In arabica, its main application will be for F_1 hybrid production, thereby avoiding manual hybrid seed production. SE also opens the door for other biotechnological methods that require regeneration protocols such as cryopreservation of selected heterozygous clones, obtaining somaclonal variants, somatic hybridization and genetic transformation.

Pioneering work on successful regeneration from coffee tissue culture includes robusta shoot cultures (Staritsky, 1970), high frequency embryogenesis from mature leaf explants of arabica (Sondahl and Sharp, 1977), production of somatic embryos from leaves of a robusta hybrids in auxin free medium (Dublin, 1981) and SE from young leaves of arabica (Yasuda *et al.*, 1985) *etc.* A liquid culture protocol for a highly synchronized somatic embryo production was published by Neuenschwander and Baumann (1992). Large numbers of robusta somatic embryos were produced in 3-litre bioreactor cultures by Zamarripa *et al.* (1991) followed by the work of Noriega and Sondahl (1993) with arabica embryos using a 5-litre bioreactor system. Protocol for *in vitro* propagation through SE was developed for the recently released Kenyan superior commercial arabica variety 'Batian' to meet the demand for planting material (Lubabali *et al.*, 2014). Using a unique apparatus for a temporary immersion culture, a protocol for the development of coffee plantlets was reported by Berthouly *et al.* (1995a). Solid and liquid media for coffee SE have provided the key for a series of applications for coffee improvement programs like micropropagation and somaclonal variation. These achievements have also helped in successful isolation of protoplasts and gene transfer programs.

2.2. Characterizing Somatic Embryogenesis

It is essential to characterize and understand various factors controlling the process of SE in coffee for its successful application. Acuna (1993) presented data for the production of embryogenic tissue (ET) in two selected genotypes and two culture media. Scanning electron microscopic (SEM) studies on coffee embryogenic tissues and early stages of embryo differentiation have been previously reported by Sondahl *et al.* (1978) and Nakamura *et al.* (1992). SEM study was made by Tahara *et al.* (1995) using three types of coffee calli (arabica) - one embryogenic callus (EC) and two non embryogenic calli (yellow callus, NYC; white callus, NWC). EC was composed of yellow, spherical cytoplasm-rich cells, uniform in size, NWC displayed elongated swollen translucent cells; NYC had cells similar in appearance to EC cells, but more dispersed. EC in the presence of 2, 4-D was incapable of regenerating somatic embryos; however, in a 2, 4-D-free medium, globular stage embryos arose after 2 to 3 weeks. The authors reported that EC preserved its embryogenic potential when maintained on 2, 4-D (10 μM) medium for six years.

The effect of asparagine on coffee somatic embryo induction medium was demonstrated by Nishibata *et al.* (1995). The investigators showed that addition of asparagine (10 μM) to the regeneration medium promoted embryogenesis, while the addition of glutamine, glutamate or aspartate strongly inhibited SE. Moreover, the addition of asparagine (10 μM) to 2, 4-D growth medium was able to induce somatic embryos and inhibits further cell proliferation.

The effect of plant growth regulators on SE of leaf cultures of *C. canephora* was reported by Hatanaka *et al.* (1995). It was demonstrated that a cytokinin (5 µM) was essential for the formation of somatic embryos in robusta leaf cultures and that 2-iP was the most effective cytokinin source. The authors also reported that auxins (IAA, IBA, NAA, 4-FA, 2, 4-D) inhibited SE. Culture conditions for induction of SE in arabica and robusta tissues have been reported by Yasuda *et al.* (1995). Using young leaf explants, both species produced somatic embryos on A3 cytokinin-only medium (5 µM 2-iP or BA), genotypic differences were observed in culture reactions.

The induction of SE was tested using young leaf explants of ten F_1 hybrids obtained from commercial arabica cultivars and wild genotypes from Ethiopia (Etienne *et al.*, 1997). Embryogenic cells were produced after six months on solid cultures, multiplied in Petri dishes and transferred to 125-ml Erlenmeyer flasks to establish embryogenic cell suspensions at 100 rpm and 27°C with subculture intervals of 10 weeks. Young somatic embryos were transferred to RITA® vessels under periodic immersion technique for embryo germination and plantlet development. Plantlets with one pair of leaves and a tap root were obtained after 3 to 4 months of cultivation in RITA vessels. A genotypic differential response to SE was observed among the F_1 hybrids. In case of a high-embryogenic material, up to 9000 plantlets were obtained per RITA vessel, but in the case of a low-embryogenic hybrid, only 750 to 1000 plantlets were obtained per vessel.

A critical study on 'direct or low' SE induction from arabica leaf explants was presented by Loyola-Vargas *et al.* (1999). Using soft leaves from *in vitro* plantlets on Yasuda *et al.*'s (1985) medium, somatic embryos were observed directly from mesophyll cells of the explants after 21 days. No embryogenic tissue (friable calli, embryogenic calli) were observed in these cultures. The authors studied the effect of nitrogen on coffee SE and suggested use of 4 to 9 mM to obtain maximum embryos. The optimum ratio of nitrogen sources has been worked out as 1 $N0_3$: 2 NH_4 for maximum response.

2.3. Genetic Control of Somatic Embryogenesis

Somatic embryogenesis and vegetative cutting capacity have been reported to be under distinct genetic control in *Coffea canephora* (Priyono *et al.*, 2010). The authors observed diversity for SE ability observed not only among two groups of *C. canephora* (Congolese and Guinean), but also within these different genetic groups. The study thus showed that under given experimental conditions SE ability is genotype dependent. Further, the study detected eight QTLs determining SE. Six positive QTLs for SE ability were localized on one single chromosome region of the consensus genetic map. Two negative QTLs for SE ability (frequency of micro calli without somatic embryo) were detected on another linkage group (Priyono *et al.*, 2010).

2.4. Molecular Mechanism of Somatic Embryogenesis

Recently, insights into the molecular mechanism regulating SE in coffee have been obtained by profiling the epigenetic changes and extracellular proteome. The embryogenic development of *C. canephora* was shown to involve crosstalk between

DNA methylation and histone modifications during the earliest embryogenic stages of SE (Nic-Can *et al.*, 2013). The authors found that low levels of DNA methylation, histone H3 lysine 9 dimethylation (H3K9me2) and H3K27me3 change according to embryo development. Moreover, the expression of *LEAFY COTYLEDON1* (*LEC1*) and *BABY BOOM1* (*BBM1*) were only observed after SE induction, whereas the expression of *WUSCHEL-RELATED HOMEOBOX4* (*WOX4*) decreased its expression during embryo maturation. It was found that 5-azacytidine strongly inhibited the embryogenic response by decreasing both DNA methylation and gene expression of *LEC1* and *BBM1*. Through chromatin immunoprecipitation (ChIP) assays, *WOX4* was found to be regulated by the repressive mark H3K9me2, while *LEC1* and *BBM1* were epigenetically regulated by H3K27me3. It was concluded that epigenetic regulation plays an important role during somatic embryogenic development, and a molecular mechanism underlying SE was proposed (Nic-Can *et al.*, 2013).

Suspension cultures of both *Coffea canephora* and *Coffea arabica* were used to study the population of proteins secreted into the media (Mukul-Lopez *et al.*, 2012). Two types of cultures were used, one for the propagation of suspension cultures (non-embryogenic) and another for the induction of SE (embryogenic). The evaluated days were 14 and 42 for non-embryogenic condition and 21, 42, and 98 for the embryogenic condition. An embryogenic system was established in the *C. arabica* species, obtaining 4,000 embryos per litre. The proteins secreted into the culture media were analyzed, both under non-embryogenic and SE induction conditions. In *C. canephora* medium, 173 proteins were found after 14 days of culture under non-embryogenic conditions. In *C. arabica*, 523 proteins were found after 14 days under non-embryogenic conditions. Under embryogenic conditions, 379, 409 and 175 proteins were found after 21, 42 and 98 days, respectively. Some proteins were secreted exclusively under embryogenic conditions and others proteins under non-embryogenic conditions (Mukul-Lopez *et al.*, 2012).

2.5. Genetic Fidelity of Plants Propagated through Somatic Embryos

For large scale practical use, the SE technology needs to be validated by confirming the true-to-type status of the produced plants in the field. To evaluate the stability of coffee somatic embryos produced via solid and liquid media, an experiment field was established using *C. arabica* cv Bourbon LC line B. After four years under field conditions and at the second crop, very few differences could be seen among the coffee plants. This field test plot demonstrated that coffee plants derived from somatic embryos could be used for micropropagation (Sondahl and Baumann, 2001).

Five elite robusta plants were selected on the basis of their agronomic traits for micropropagation through SE (Ducos *et al.*, 1999). These clones multiplied through SE were field tested in five coffee producing countries (4000 plants/location), namely the Philippines, Thailand, Mexico, Nigeria and Brazil. Based on visual inspection of 8000 plants under field conditions in the Philippines, all micropropagated robusta plants showed normal flowers and fruits two years after planting.

A total of 20,000 plants from F_1 hybrids of arabica derived through somatic embryos using the RITA temporary periodic immersion technique

(Berthouly *et. al.*, 1995a) were used to establish test field plots in four Central American countries (Etienne *et al.*, 1999). The objective was to evaluate the performance of the embryo-derived plants under distinct farming conditions. It is reported that among 4000 *in vitro* plants under field and nursery conditions evaluated, no somaclonal variation was observed (Etienne *et al.*, 1999).

In Uganda, a project for large scale propagation and distribution of six selected robusta clones to farmers was launched. Propagation was to be accomplished by the cutting process and by *in vitro* methods (Berthouly *et al.*, 1995a). The micropropagation effort utilized the periodic immersion technique (Berthouly *et al.*, 1995a) for cloning via axillary bud development (microcuttings) and SE. A total of 2000 RITA vessels were installed and the expectation was to produce 600000 plants/year from microcuttings and 2.0 to 2.5 million plants/year via SE (Berthouly *et al.*, 1995b).

PROMECAFE successfully implemented a project for technical and economic validation of the SE process for the massive propagation of F_1 hybrids of arabica (Zamarripa and Petiard, 2004). Similarly, INIFAP in association with Nestle started producing 1.5 million robusta plantlets from different selected clones to be distributed to growers in Veracruz State. The results obtained from this project could contribute to the first real commercial application of coffee SE (Zamarripa and Petiard, 2004).

2.6. Comparison of Micropropagated Plants with Seedlings

Plants of *Coffea arabica* derived *via* SE, namely, somaclones, were evaluated with *C. arabica* seedlings grown in the nursery (Menendez-Yuffa *et al.*, 2010). Somaclones of *C. arabica* cvs. Caturra and Costa Rica 95 (Catimor) were smaller and less vigorous than seedlings of the same cultivar at the time of their transfer to the nursery. Following an initial slow growth for a period of 10 weeks, somaclones began to grow faster than seedlings until both groups of plants were equal in size at 21 weeks (entire duration of growth in the nursery). Comparisons of aerial and root systems of 30-cm long somaclones and seedlings of two cultivars revealed that plants of somaclones were more vigorous than seedlings, based on the higher number of leaves, larger leaf area, and greater dry weight of aerial organs. For cv. Caturra, the root dry weight of somaclones was significantly greater than that of seedlings and was attributable to the large diameter roots. Analysis of 176,000 F_1 hybrid somaclones revealed that these exhibited more heterogeneous growth than did the seedlings derived from zygotic embryos; moreover, there was a genotype effect. Almost 9–20 per cent of somaclones required an additional 3–4 months of growth in the nursery, and 8–12 per cent clones culled for other undesirable horticultural attributes. Only 0.10–0.23 per cent of somaclones displayed variant phenotypes. The observed somaclone vigour in the nursery was carried over to field performance as these plants were more precocious than seedlings and yielded coffee beans one year earlier than seedlings (Menendez-Yuffa *et al.*, 2010).

2.7. Scaling-up of Coffee Micropropagation to Commercial Scale

Of all the possible micropropagation techniques, vegetative propagation by SE is by far the most promising one for the rapid, large-scale dissemination of elite individuals. Research on coffee SE began at the end of the 1970s and continued in various institutes. This led to better understanding and control of the system after overcoming several constraints. The major constraints that had to be overcome included genotypic effect, particularly for obtaining embryogenic tissues, or are related to the quality of regenerated somatic embryos and their conversion to plantlets, the incidence of somaclonal variation and, more generally, a lack of reproducibility and efficiency at certain stages of the process, leading to prohibitively high production costs. After four decades of intensive research and development efforts, coffee SE has reached industrial scale production. Thus, coffee is one of the few plants where in micropropagation through SE has been scaled up to commercial production level. The success could be mainly attributed to the efforts of Centre de Coopération Internationale en Recherche Agronomique pour le Développement (CIRAD, France) in collaboration with private players like ECOM for *Coffea arabica* F_1 hybrids and by Nestle for *C. canephora* clones (Figures 7.1 and 7.2). In both instances, regeneration was achieved though SE using temporary immersion bioreactor technology and subsequently, direct sowing of green somatic embryos in nursery. CIRAD and Nestle have evolved slightly different temporary immersion systems for their coffee materials. In both cases, it is a three stage multiplication system. In the first stage, callus is induced from leaf explants on semisolid medium. In the second stage, embryogenic cell suspensions are proliferated and torpedo stage somatic embryos were induced in liquid medium. In the third stage, somatic embryos are matured and pre-germinated by periodic brief immersion of the embryos in liquid medium. Annual production of micropropagated plants of coffee has reached one to several million by the efforts of the two groups.

Successful technology transfer for scaling up micropropagation of *Coffea arabica* to commercial scale through SE was described by Etienne *et al.* (2010, 2012). Research on coffee SE began at the end of the 1970s at various institutes, including CIRAD. Between 1995 and 2001, CIRAD moved the technique forward from a research laboratory scale to a technique enabling industrial dissemination of extremely promising *Coffea arabica* F_1 hybrids. Over that period, two technological innovations made technology transfer economically feasible: mass production of somatic embryos in temporary immersion bioreactors (Figure 7.1) and the possibility of sowing them directly in the nursery. At the same time, reassuring data were obtained on the genetic conformity of regenerated plants (somaclonal variation frequency < 3 per cent). In 2002, in partnership with the ECOM group, CIRAD decided to transfer the SE method on an industrial scale to Central America so that four arabica hybrid clones, that were selected for agroforestry based farming systems, could be disseminated throughout that part of the world. After overcoming various difficulties faced at different stages, successful technology transfer occurred in 2010. This was one of the first examples of SE technology applied at a commercial scale (Etienne *et al.*, 2010, 2012).

A pilot process implemented at Nestle R and D Centre-Tours for mass propagation of selected robusta clones was described by Ducos *et al.* (2011). A batch takes 4 to 6 months to complete and consists of three phases. The development of torpedo embryos is achieved using Erlenmeyer flasks. The pre-germination

Figure 7.1: RITA Vessels Based Temporary Immersion Bioreactor System Developed at CIRAD, France and Adopted for Commercial Scale Micropropagation of F_1 Hybrids of Arabica. Top: Overview of the system; Bottom: Close up view of a few RITA vessels (Etienne *et al.*, 2012).

is conducted in a 10-L temporary immersion bioreactor made of glass or flexible disposable bags (Figure 7.2). The latter type, the so-called 'Box-in-Bags', insures a higher light transmittance to the biomass due to its horizontal design. It allows a higher torpedo-to-cotyledonary stage conversion rate. Prior to shipment, the pre-germinated embryos are maintained under storage by spreading them out in layers

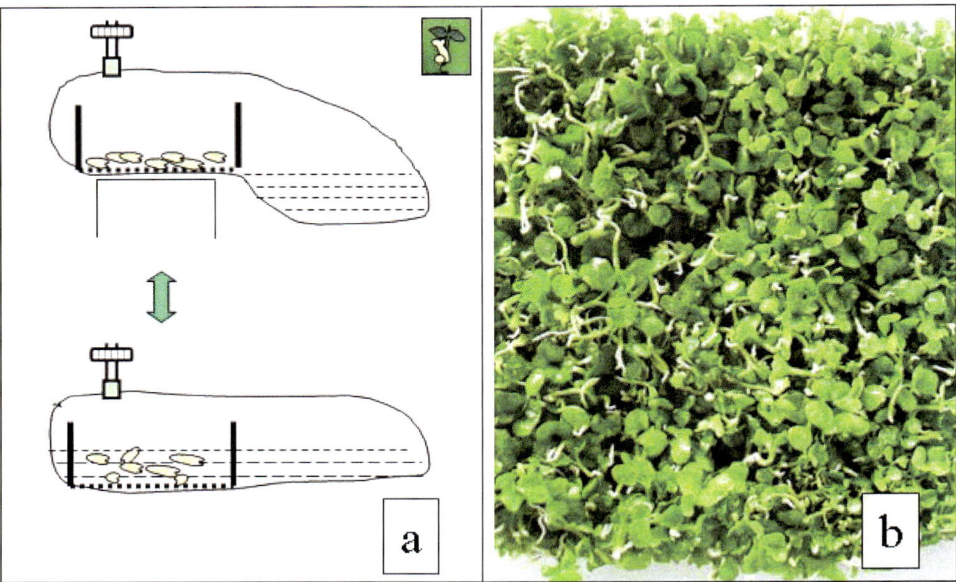

Figure 7.2: Flexible Plastic Based Disposable Temporary Immersion Bioreactors Developed by Nestle R and D Centre for Micropropagation of Robusta Coffee Clones. Top: Overview of a culture room with disposable bioreactors; Bottom left (a): Diagram of a disposable bioreactor; right (b): Pregerminated somatic embryos at the end of the phase (Ducos *et al.*, 2011).

onto coconut fibres in *ex vitro* conditions. These storage conditions preserve their ability to develop plantlets for at least two months and induce their hardening. The production capacity of this pilot process was around 4.0 million pre-germinated embryos per year, able to regenerate plantlets with a frequency of 70-76 per cent. Embryos produced during this implementation step were sent to different coffee producing countries, mainly to Thailand. In the local greenhouses, the embryos were directly transplanted into trays containing commercial peat or coconut fibres. The usage of the "micro-environment" method, combined with media releasing CO_2, was well adapted for the large-scale acclimatization of very small *vitro*-plantlets in tropical greenhouses. By June 2010, 2.9 million somatic seedlings were produced, out of which 1.8 million were already planted in the field (Ducos *et al.*, 2011).

3. Recovering Useful Somaclonal Variants

3.1. Somaclonal Variation

Somaclonal variation (SV) is the expression of the naturally occurring variability in plant cells or the result of *in vitro* induced variability of cells following plant regeneration (Larkin and Scowcroft, 1981; Evans and Sharp, 1986). Most of the spontaneous variability from *in vitro* plants is associated with chromosome alterations such as breakage, translocation, deletions, aneuploidy, polyploidy and somatic crossing over. In addition, SV can also have a single gene origin for example a point mutation, alteration in gene, copy number, activation of transposon elements and epigenetic changes like variation in DNA methylation. SV is undesirable when the goal is large-scale clonal propagation of uniform elite line. On the other hand, SV is useful to create genetic diversity, expanding the germplasm pool with novel traits for crop improvement. SV is an excellent method for shortening breeding programs, since it can provide access to genetic variability within existing cultivars (Evans and Sharp, 1986). Somaclones carry few genetic alterations and so the overall genetic integrity of the original commercial cultivar is preserved. Somaclonal variation can contribute to the release of improved plant varieties (Hammerschlag, 1992; Karp, 1995). A wide array of agronomically important coffee genotypes is used for SV study (Sondahl and Bragin, 1991; Sondahl and Lauritis, 1992). Tissue culture was initiated from mature leaf explants following Sondahl and Sharp (1977) protocol and donor plants were maintained in a green house collection. Plantlets were recovered from both the 'low frequency pathway' (LFSE) and 'high frequency pathway' (HFSE). A total of 14948 *in vitro*-derived plants, representing nine different coffee genotypes, were established in the field in Brazil. Screening was done at the Ro regeneration during the first and second crops. The most interesting variant forms were studied in the next generation by establishing progeny fields. The overall variability found in this *in vitro*-derived coffee population was 10 per cent, but variability was highly genotype-dependent, for instance: 30.6 per cent for Yellow Bourbon and only 3.3 per cent for Red Catuai. The most common mutation was for fruit color (42.35 per cent yellow to red) followed by change in plant stature (3.8 per cent tall to short). Based on 7722 *in vitro* plants evaluated, the frequency of variability was similar for plants originated from HFSE (or indirect embryogenesis; 12 per cent) and LFSE (or direct embryogenesis; 10.4 per cent) (Sondahl and Baumann, 2001). Many interesting variants were selected from this programme and their progenies were studied in subsequent generations. The most interesting mutations were carried forward by standard breeding methods aiming for the release of new varieties in the future. Emphasis was made in selecting superior cup qualities associated with desirable agronomic traits. Three main breeding populations have been derived from this tissue culture program: Laurina somaclones, Icatu somaclones and Aramosa somaclones. Some characteristics of segregating individuals of the Icatu and Aramosa populations are reported by Sondahl *et al.* (1997). Other small populations were also studied, such as short-stature mutants of Mundo Novo and Yellow Bourbon. Another interesting population was based on one Margogype mutant plant that was derived from Yellow Catuai leaf cultures. Second and third generations of this somaclone showed segregation for the typical Margogype phenotype; normal Catuai and an intermediate phenotype with short stature but very large beans (Sondahl and Baumann, 2001).

Effect of the genotype and the age of embryogenic cell suspensions on SV in five F_1 hybrids of *C. arabica* were determined by Etienne and Bertrand (2003). Among the seven phenotypic variants characterized, the Dwarf, Angustifolia and multi-stem variants were the most frequent among the regenerated plants (1.4, 4.8 and 2.9 per cent, respectively). The frequency of variants increased exponentially with the age of the embryogenic suspension. For all genotypes, somaclonal variation was low (1.3 per cent) in plants produced from embryogenic callus or 3-month-old cell suspensions and increased in frequency with increasing suspension age (6, 10 and 25 per cent in plants produced from cell suspensions aged 6, 9 and 12 months, respectively). Large differences in SV among genotypes were found only in plants produced from 12-month-old cell suspensions. For two genotypes, the oldest suspensions produced a majority of somaclonal variants (80–90 per cent), whereas SV ranged between 8 and 18 per cent in the other genotypes. Cell suspension age and genotype also affected the type of variant produced. The severity of somaclonal variations increased with cell suspension age. For all genotypes combined, the Angustifolia variant was the most common. The other somaclonal variations were specific to certain genotypes or distributed randomly among the genotypes (Etienne and Bertrand, 2003). Flow cytometry was applied to verify ploidy instability in embryogenic cell aggregates of *C. arabica*, through successive subcultures (Clarindo *et al.*, 2012). Flow cytometry detected the occurrence of non-true-to-type aggregates in all samples collected after approximately 4 months in liquid medium. These aggregates showed octaploid and/or aneuploid cells, with DNA ploidy level being corroborated by chromosome counting. Considering the results, it is recommended to limit the subcultures to less than four months for true-to-type mass propagation of *C. arabica* cell aggregate suspensions. As flow cytometry was able to detect SV at an early stage of tissue culture in *C. arabica*, it might be a useful tool for quality control in the micropropagation of the species.

High genetic and epigenetic stability is reported in *Coffea arabica* plants derived from embryogenic suspensions and secondary embryogenesis as revealed by AFLP, MSAP and the phenotypic variation analysis (Bobadilla Landey *et al.*, 2013). In view of the need for true-to-type large-scale propagation of *C. arabica* hybrids, suspension protocols based on low 2,4-D concentrations and short proliferation periods have been developed. The effects of embryogenic suspensions and secondary embryogenesis, used as proliferation systems, on the genetic conformity of SE-derived plants (emblings) were assessed in two hybrids. When applied over a six month period, both systems ensured very low somaclonal variation rates, as observed through massive phenotypic observations in field plots (0.74 per cent from 200 000 plants). Molecular AFLP and MSAP analyses performed on 145 three year-old emblings showed that polymorphism between mother plants and emblings was extremely low, *i.e.* ranges of 0–0.003 per cent and 0.07–0.18 per cent respectively, with no significant difference between the proliferation systems for the two hybrids. Chromosome counting showed that seven of the 11 variant emblings analyzed were characterized by the loss of 1–3 chromosomes. This work showed that both embryogenic suspensions and secondary embryogenesis are reliable for true-to-type propagation of elite material. The main change in most of the rare phenotypic variants was aneuploidy, indicating that mitotic aberrations play a major role in somaclonal variation in coffee (Bobadilla Landey *et al.*, 2013).

Long-term cell cultures were used in coffee to study the cytological, genetic and epigenetic changes occurring during cell culture ageing with the objective of identifying the mechanisms associated with SV (Bobadilla Landey *et al.*, 2015). Phenotyping AFLP, MSAP, SSAP molecular markers were done to identify somaclonal variations. The results showed that cell culture ageing is highly mutagenic in coffee and chromosomal rearrangements are directly linked to SV. Conversely, the analysis of methylation and transposable elements changes did not reveal any relation between the epigenetic patterns and SV (Bobadilla Landey *et al.*, 2015).

3.2. Somaclonal Variety in Coffee

Sondahl and Baumann (2001) have outlined the development of a new coffee variety through SV. Laurina is a natural mutant from Red Bourbon, plants having small leaves, thin lateral branches, short stature and elongated fruits and beans. Laurina plants have 50 per cent reduced caffeine content. Among more than 800 *in vitro*-derived plants of Laurina, 15 elite plants were selected at the Ro generation in June 1991. These selected plants were clearly more vigorous than sister plants and donor controls as demonstrated by greater leaf area, lateral branches, plant height, plant diameter and superior yield. Seeds of these selected plants were taken to establish a separate experimental field (4 ha in size) to evaluate the performance of each somaclone line. Growth pattern, yield and caffeine content were monitored for each of the 15 lines (total of 360 plants per line) during the first five years under field conditions (three successive crops). It was observed that the caffeine content was stable and equal to the donor plants, the growth pattern was stable for all lines (no segregation) and the yield from the top five lines was twice as high as for the control plants. Yield evaluation continued up to the sixth harvest (1999 crop) in the experimental field thus confirming the initial selection of the top five lines for the superior yield and reduced biannual cycle. A third generation of selected lines was established in a semi-commercial plot design of 25 ha in size. Seeds from the top five high-yielding lines were scaled under the name 'Bourbon LC'. At the time the first round of selection of elite Laurina somaclones was completed, filings for patent protection were made on the discovery. A utility patent was awarded for a coffee variety (Sondahl *et al.*, 1995) and it is also the first case of the release of a coffee variety derived from natural variability isolated from somatic embryo cultures. Bourbon LC is the only naturally reduced (50 per cent) caffeine variety being produced in commercial quantities (Sondahl and Baumann, 2001).

4. Genetic Transformation

4.1. Genetic Transformation Studies

By allowing direct transfer of genes into plants, genetic transformation offers new tool for improving plant varieties for important agronomic traits. This technique can be used on well-known cultivars to improve them in one or two characters. Progress in plant regeneration through SE system in *C. arabica* and *C. canephora* has enabled genetic transformation of both the species with several marker genes as well as genes of agronomic importance and cup quality. However, just like regeneration through SE, coffee genetic transformation is still a tedious process. The whole process of genetic coffee transformation from primary inoculated explant to plantlet transfer to the green house usually takes 12-20 months. Several reviews are published on coffee genetic transformation from time to time (Sreenath, 2003; Ribas *et al.*, 2006; Etienne *et al.*, 2008; Mishra and Slater, 2012).

Genetic transformation of coffee plants has been reported by several research groups (Spiral and Petiard, 1993; Sugiyama *et al.*, 1995; Van Boxtel *et al.*, 1995; Hatanaka *et al.*, 1999; Leroy *et al.*, 2000; Ogita *et al.*, 2005). Two techniques were tested; direct transformation through particle bombardment (biolistics) (Van Boxtel *et al.*, 1995) and *Agrobacterium*-mediated transformation. Genetic transformation with *A. rhizogenes* was first tested by Spiral and Petiard (1993). It was successful with 10-40 per cent of regenerated roots showing *GUS* expression in histochemical assay and confirmation of vector DNA sequence integration (Spiral and Petiard, 1993; Spiral *et al.*, 2000). However, when transferred to green house, disturbed phenotypes (crinkled leaves, short internodes) were observed in regenerated plantlets due to expression of various hairy root genes (Spiral *et al.*, 2000). Subsequently, interest was shifted from *A. rhizogenes*-mediated transformation to *A. tumefaciens*-mediated one, in order to obtain a normal phenotype in regenerated plantlets and transgenic plantlets were successfully obtained. Two selection systems, hygromycin (Hatanaka *et al.*, 1999) and the herbicide chlorsulfuron (Spiral *et al.*, 2000), proved efficient for selecting transformed embryogenic callus.

The first transgenic coffee plants expressing the *B. thuringiensis cry1Ac* gene conferring insect resistance to leaf miner *Perileucoptera coffeela* were obtained by Leroy *et al.* (2000). Using *A. tumefaciens*-mediated transformation, Spiral *et al.* (2000) successfully transferred the *cry1Ac* gene into *C. canephora* and *C. arabica* genotypes. Efficiency of transformation varied depending on the genotype tested, the arabica genotypes being less amenable to embryo regeneration. Molecular characterization of transformed plants showed that 69 per cent of them carried a unique copy of T-DNA and Cry1Ac protein expression in leaves was obtained for 18 of 23 plantlets tested (Leroy *et al.*, 2000). Three different levels of resistance could be measured with some highly resistant plants, slightly susceptible and fully susceptible plants (Leroy *et al.*, 2000). Ogita *et al.* (2005). reported construction of transgenic coffee plants (*C. canephora*) in which expression of the gene encoding theobromine synthase (*CaMXMT1*) was repressed by RNA interference (RNAi). The caffeine content of these plants was reduced by up to 70 per cent indicating that it should be feasible to produce coffee beans that are genetically deficient in caffeine.

Private and public organizations have worked on coffee transformation with the goal of introducing genes of practical interest into coffee varieties. Such genes might increase resistance to diseases and pests like leaf rust, coffee berry borer, coffee stem borer and control fruit ripening or the caffeine and sugar content. However, the high cost of tests for regulatory approval and possible consumer acceptance and environmental issues seem to have slowed down the companies in developing genetically modified coffee plants.

4.2. Coffee Tissue Culture and Genetic Transformation Studies in India

In India, research is being conducted at Central Coffee Research Institute (CCRI) to develop and evaluate tissue culture technology for both arabica and robusta varieties. The research revealed that SE is the only feasible method for micropropagation in both the species. Propagation through SE involves several steps (Figure 7.3) and each step is influenced by several factors. To optimize the entire process, several factors were investigated. Plant regeneration through SE is achieved in more than 20 Indian selections of arabica and robusta species (Raghuramulu *et al.*, 1987; Babu *et al.*, 1993; Jayashree *et al.*, 1995; Muniswamy and Sreenath, 1995b; Chaithra and Sreenath, 2010). Genotypic differences were found in respect to callus induction, SE and plant regeneration in arabica plants (Naidu *et al.*, 1999). Encapsulation techniques were investigated to develop artificial seeds from embryos (Muniswamy and Sreenath, 1995a). Hardening protocols were developed for small scale (Muniswamy *et al.*, 1994) and medium scale plant production and trial plots of tissue cultured plants were established in different agroclimatic zones (Sreenath, 1998). After conducting detailed studies, protocols were developed to multiply several arabica and robusta genotypes by using leaf explants. Selected genotypes of three improved selections *viz.*, Cauvery, Sln 9 and CxR were used for large scale field evaluation of micropropagated plants against seedling progeny. Plant regeneration was achieved from the apical bud and nodal explants (Ganesh and Sreenath, 1997, 2008) (Figure 7.4), but rate of multiplication was very less and not suitable for large scale propagation. Effect of TIBA and BAP were tested on integument cultures of *C. canephora* (Babu *et al.*, 1997). For the first time in any plant species, plant regeneration was achieved from integument tissues of *C. canephora* cv CxR cultivar (Sreenath *et al.*, 1995) and in *C. arabica* (Jayashree *et al.*, 2006). Plant regeneration was also achieved successfully in anther culture (Muniswamy and Sreenath, 2000), embryo culture (Muniswamy *et al.*, 1993; Muniswamy and Sreenath, 1997; Sreenath *et al.*, 1992) and endosperm culture (Muniswamy and Sreenath, 2001). Culture of zygotic embryos under slow growth conditions was found suitable for germplasm preservation in *C. arabica* (Naidu and Sreenath, 1999). Research was conducted to develop and optimize genetic transformation technique in *C. arabica*

Figure 7.3: Stages of Coffee Micropropagation through Somatic Embryogenesis. Top left: Callus induced on leaf explant; Top right: Clumps of somatic embryos; Middle left: Germinating somatic embryos; Middle right: Well grown plantlets ready for hardening; Bottom left: Plantlets hardened in net pots; Bottom right: Plants in poly bags ready for field transfer.

Figure 7.4: Coffee Multiplication through Axillary Shoot Proliferation. Top left: Nodal explant; Top right; Axillary bud break; Bottom left: Elongation of axillary shoots; Bottom right: Multiplication of axillary shoots.

and *C. canephora* by investigating various factors (Naveen *et al.*, 2002; Mishra *et al.*, 2002; Sreenath and Naveen, 2005; Mishra and Sreenath, 2005; Mishra *et al.*, 2009). Embryogenic calli have been used successfully to get transgenic coffee plants with heterologous chitinase gene in *C.arabica* cv Cauvery (Sreenath and Veluthambi, 2009) and osmotin in C. *canephora* cv. CxR (Sreenath *et al.*, unpublished) genes through *Agrobacterium tumefaciens*. Microcalli were regenerated successfully from the protoplasts liberated from embryogenic calli in *C. arabica* cv Cauvery.

Occurrence of deleterious somaclonal variation is the most important concern in propagation through SE. Mainly to address this issue and to evaluate the performance of the Tissue culture (TC) plants, field trials were conducted. Around 16000 TC plants and 4000 seedlings of respective genotypes were planted in the TC trial plots in 44 locations, mostly comprising growers' fields in the three traditional coffee growing states in South India. In arabica, Cauvery (S4348 and S4350), Sln6 (Swarnagiri clones), Sln9 and a few F_1 hybrids (S2800, S2790, and S2794) were evaluated. These arabica cultivars were selected mainly for their resistance to leaf rust disease. In robusta, some selected genotypes of CxR cultivar were used. TC trial plots were evaluated for a few years against the respective seedling progenies. Overall, the performance of the TC plants was normal and variants were negligible, indicating that incidence of somaclonal variation is not a concern in the technology. Among arabicas, Cauvery and the F_1 hybrid S2800 performed well. The field performance of *C. arabica* selection S2800 against seedling progeny is documented (Muniswamy *et al.*, 2015) (Figure 7.5). The micropropagated plants of arabica selections maintained their leaf rust resistance in the field. In robusta, CxR genotype TL1 performed well (Figure 7.5). However, genetic variability and novel genome organizations was observed in SE-derived plants of *C. arabica* (Rani *et al.*, 2000). Regenerated plants from leaf explants of elite selection *C. arabica* cv. Cauvery (S4347) obtained through high-frequency SE were used. To assess the genetic integrity of the nuclear, mitochondrial and chloroplast genomes among the hardened regenerants, multiple DNA markers (RFLP, RAPD, and ISSR) were employed. Although the nuclear and mitochondrial genomes of the mother plant and five ramets derived from the mother ortet were similar in organization, this was not so in the somatic embryo-derived plants where both nuclear and mitochondrial genomes changed in different, characteristic ways and produced novel genome organizations. A total of 480 genetic loci were scored in 27 somatic embryo-derived plants and a control. Among these, 44 loci were polymorphic. A relatively low level of polymorphism (4.36 per cent) was found in the nuclear genome, while polymorphism in the mitochondrial genome (41 per cent) was much higher. No polymorphism was detected in the chloroplast genome. The polymorphism in the mitochondrial genome was found in only four plants. Such selective polymorphism was not true for the nuclear genome (Rani *et al.*, 2000).

Thus, laboratory scale micropropagation technology for coffee has been developed using semi solid media and demonstrated in India by CCRI. However, still industrial scale production of coffee TC plants has not happened in India. Temporary immersion bioreactor technology needs to be developed for this purpose. However, it is essential to evaluate the entire critical factors specific to a country before embarking on industrializing the technology for a perennial crop like coffee. In arabica, it is essential to identify highly heterotic F_1 hybrids suitable to derive the benefits of the technology. Another crucial factor to be addressed in arabica is the problem of pests like white stem borer (*Xylotrechus quadripes*). All the arabica selections developed so far in India are susceptible to white stem borer. All micropropagated plants of arabica selections were seriously affected by white stem borer in the field trials. For robustas, the problems of self sterility and inbreeding

Figure 7.5: Somatic Embryo Derived Arabica (Top) and Robusta (Bottom) Coffee Plants in the Fields. Top left: Trial plot of *C. arabica* cv S2800; top right: Single plant of *C. arabica* cv S2800 with ripe fruits; bottom left: Trial plot of *C. canephora* cv CXR TL1; bottom right: Single plant of *C. canephora* cv CXR TL1 with ripe fruits.

depression have to be addressed. From some selections, particularly of hybrid origin some interesting somaclonal variants are obtained. These can serve as useful starting materials for breeding somaclonal varieties.

3. Conclusion

Of all the possible micropropagation techniques, vegetative propagation by SE is by far the most promising one for the rapid, large-scale dissemination of elite individuals. Micropropagation *via* SE is the technology that has the scale to satisfy commercial plantations and to be competitive in cost as compared to other propagation methods. In four and half decades of intensive research, enormous progress is made in the induction and maintenance of embryogenic-competent cell lines, as well as, regeneration of somatic embryos from both arabica and robusta species. Greater control of the SE process, coupled with the development of periodic immersion bioreactor techniques has led to the development of reliable methods for mass coffee propagation. Coffee SE has reached industrial scale with the establishment of a couple of production systems with capability of 2 - 4 million plants per year. Large scale field evaluation of somatic embryo-derived coffee plants in both arabica and robusta has demonstrated high genetic fidelity with minimum somaclonal variation. However, the success of the commercial application of SE depends directly on the production cost and the agronomic and technological value added in the selected hybrids. In spite of very high frequency of true-to-type plants in coffee plants propagated through SE, several types of somaclonal variants are also encountered. In recent years, somaclonal variation in coffee plants is being studied in depth, including analysis at the molecular level. This has led to better understanding and control of the process. Cell culture ageing is found to be the most signifant factor directly linked to somaclonal variation. Apparently by regulating the age of the cell cultures in multiplication stage, somaclonal variation may be regulated. In coffee useful somaclonal variants can be exploited for breeding new varieties. There is already an example of a C. *arabica* somaclonal variety with low caffeine. By allowing direct transfer of genes into plants, genetic transformation offers new tool for improving plant varieties for important agronomic and other useful traits. Progress in plant regeneration through SE system in *C. arabica* and *C. canephora* has enabled genetic transformation of both the species with several marker genes, as well as, genes of agronomic importance and cup quality. Thus, tissue culture in coffee is applied successfully for meeting the opposing objectives of rapid multiplication of true to type plants and recovering useful somaclonal variants. It has also enabled successful genetic transformation of coffee with useful traits.

References

Acuna, M.E.A. (1993). Somatic embryogenesis induced by culture on single media in coffee plants from crosses of *Coffea arabica* by Timor hybrid. In: Proceedings of the 15th ASIC Colloqium (Montpellier). ASIC, Paris, France. pp. 82-88.

Babu, K. H., Shanta, H. M., Ganesh, D. S. and Sreenath, H. L. (1993). Plant regeneration through high frequency somatic embryogenesis from leaf tissue, in two cultivars of coffee. *Journal of Plantation Crops,* **21**: 307-312.

Babu, K. H., Jayashree, G., Naidu, M. M. Ganesh, D. S. and Sreenath, H. L. (1997). Effect of TIBA and BAP on integument cultures of CxR cultivar of robusta coffee (*Coffea canephora*). In: Ravishankar, G. A. and Venkataraman, L. V. (Ed.), *Biotechnological Applications of Plant Tissue and Cell Culture.* Oxford and IBH Publishing Co. Pvt. Ltd., New Delhi. pp. 296-300.

Berthouly, M., Dufour, M., Alvard, D., Carasco, C., Alemanno, L. and Teisson, C. (1995a) Coffee micropropagation in a liquid medium using the temporary immersion technique In: *Proceedings of the 16th ASIC Colloqium (Kyoto) ASIC,* Paris, France. pp. 514-519.

Berthouly, M., Alvard, D., Carasco, C. and Duris, D. (1995b) A technology transfer operation: A commercial *Coffea canephora* micropropagation laboratory in Uganda. In: *Proceedings of the 16th ASIC Colloqium (Kyoto).* ASIC, Paris, France. pp. 743-744.

Bobadilla Landey, R., Cenci, A., Georget, F., Bertrand, B., Camayo, G., Dechamp, E., Simpson, J., Herrera, J. C., Santoni, S., Lashermes, P. and Etienne, H. (2013). High genetic and epigenetic stability in *Coffea arabica* plants derived from embryogenic suspensions and secondary embryogenesis as revealed by AFLP, MSAP and the phenotypic variation rate. *PLoS ONE,* **8(2)**: e56372.

Bobadilla Landey, R., Cenci, A., Guyot, R., Bertrand, B., Georget, F., Dechamp, E., Herrera, J.C., Aribi, J., Lashermes, P. and Etienne, H. (2015). Assessment of genetic and epigenetic changes during cell culture ageing and relations with somaclonal variation in *Coffea arabica*. *Plant Cell Tissue and Organ Culture*, **122**: 517–531.

Chaithra, T. R. and Sreenath, H. L. (2010). Induction of high frequency somatic embryogenesis and plant regeneration from cotyledonary leaf and hypocotyl tissues of aseptic seedlings in arabica coffee (*Coffea arabica* L.) cv. cauvery. *Journal of Plantation Crops*, **38**: 219-222.

Clarindo, W. R., Carvalho, C. R. and Mendonca, M. A. C. (2012). Ploidy instability in long-term *in vitro* cultures of *Coffea arabica* L. monitored by flow cytometry. *Plant Growth Regulation*, **68**: 533-538.

Dublin, P. (1981). Embryogenesis somatique directe sur fragments de feuilles de cafeier Arabusta. *Café Cacao Thé*, **25:** 237-241.

Dublin, P. (1991). Les techniques modernes de reproduction asexuce. Impacts sar l' amerliorationgenetique des cafeiers. In: *Proceeding of the 14th ASIC Colloquium* (San Francisco) ASIC, Paris, France. pp. 365-377.

Ducos, J. P., Gianforcaro, M., Florin, B., Petiard, V. and Deshayes, A. (1999). Canephora (robusta) clones: *In vitro* somatic embryogenesis. In: Proceedings of the 18th *ASIC Colloquium* (Helsinki) ASIC Paris, France. pp. 295-301.

Ducos, J. P., Gibault, E. and Lambot, C. (2011). Coffee propagation by somatic embryogenesis at Nestle R and D Center-Tours. In: Proceedings of the IUFRO Working Party 2.09.02 Conference 2011. Somatic Embryogenesis of Forest Trees Conference on "*Advances in Somatic Embryogenesis of Trees and its Application for the Future Forests and Plantations*", August 19-21, 2010, Suwon, Republic of Korea. pp. 72- 77.

Evans, D. A. and Sharp, W. R. (1986). Somaclonal variation in agriculture. *Biotechnology*, **4:** 428-532.

Etienne, H. (2005). Protocol of somatic embryogenesis: coffee (*Coffea arabica* L. and *C. canephora* P.). In: Jain, S.M and Gupta, P.K (Eds.), *Protocols for Somatic Embryogenesis in Woody Plants*. Series: Forestry Sciences, Vol. 77. Springer SBS, Dordrecht. pp. 167–179.

Etienne, H., Barry-Etienne, D., Vasquez, N. and Berthouly, M. (1999). Aportes de la biotechnologia mejoramiento genetico del café; el ejemplo de la multiplication porembrigenesissomatica de hibridos F1 en America Central. In: Bertrand, B. and Rapidel, B. (Eds.), *Desafios de Caficulturaen Centroamerica.* IICA, San Jose. pp. 457-493.

Etienne, H. and Bertrand, B. (2003). Somaclonal variation in *Coffea arabica*: effects of genotype and embryogenic cell suspension age on frequency and phenotype of variants. *Tree Physiology*, **23**: 419–426.

Etienne, H., Bertrand, B., Anthony, F., Cote, F. and Berthouly, M. (1997). L'embryogenese somatique un outil pour l'amelioration genetique du cafeier.

In: *Proceedings of the 17th ASIC Colloquium* (Nairobi) ASIC, Paris, France. pp. 457-465.

Etienne, H., Bertrand, B., Ribas, A., Lashermes, P., Malo, E., Montagnon, C., Alpizar, E., Bobadilla, R., Simpson, J., Dechamp, E., Jourdan, I. and Georget, F. (2010). Current applications of coffee (*Coffea arabica*) somatic embryogenesis for industrial propagation of elite heterozygous materials in Central America and Mexico. In: *Proceedings of the IUFRO Working Party 2.09.02 Conference 2011.* Somatic Embryogenesis of Forest Trees Conference on *''Advances in Somatic Embryogenesis of Trees and its Application for the Future Forests and Plantations''* August 19-21, 2010, Suwon, Republic of Korea. pp. 63-71.

Etienne, H., Bertrand, B., Montagnon. C., Bobadilla Landey, R., Dechamp, E., Jourdan, I., Alpizar, E., Malo, E., Georget, F. (2012). Un exemple de transfert de technologie réussidans le domaine de la micropropagation: la multiplication de Coffea arabica par embryogenèse somatique, CahAgric **21**: 115-24.

Etienne, H., Lashermes, P., Menedez-Yuffa, A., Guglielmo-Croquer, Z. D., Alpizar, E. and Sreenath, H. L. (2008). Coffee. In: Kole, C. and Hall, T.C. (Eds.), *Compendium of Transgenic Crop Plants Volume 8: Transgenic Plantation Crops, Ornamentals and Turf Grasses*. Blackwell Publishing Ltd., USA. pp. 57-84.

Ganesh, D. S. and Sreenath, H. L. (1997). Shoot regeneration from apical bud and node cultures in four clones of robusta coffee (*Coffea canephora*). In: Ravishankar, G. A. and Venkataramanan, L. V. (Eds.), *Biotechnological Applications of Plant Tissue and Cell Culture*. Oxford and IBH, New Delhi. pp. 301-306.

Ganesh, D. S. and Sreenath, H. L. (2008). Micropropagation of *Coffea arabica* using apical buds of mature field grown plants. *Journal of Plantation Crops*, **36**: 1-7.

Hammerschlag, F. A. (1992). Somaclonal variation. In: Hammerschlag, F. A. and Litz, R. E. (Eds.), *Biotechnology of Perennial Fruit Crops*. CAB International, UK. pp. 35-55.

Hatanaka, T., Azuma T., Uchida, N. and Yasuda, T. (1995). Effect of plant hormones on somatic embryogenesis of *Coffea canephora*. In: *Proceedings of the 16th ASIC Colloquium (Kyoto) ASIC*, Paris, France. pp. 790-797.

Hatanaka, T., Choi, Y. E., Kusano, T. and Sano, H. (1999). Transgenic plants of *Coffea* canephora from embryogenic callus via *Agrobacterium tumefaciens*-mediated transformation. *Plant Cell Reports*, **19:** 106-110.

International Coffee Organization (ICO): http://www.ico.org.

Jayashree, G., Ganesh, D. S. and Sreenath, H. L. (1995). Effect of kinetin on germination of somatic embryos in CxR cultivar of coffee. *Journal of Coffee Research*, **25:** 102-105.

Jayashree, G., Sreenath H. L. and Ganesh. D. S. (2006). High frequency somatic embryogenesis and plant regeneration from integument tissues of *Coffea arabica* L. *Journal of Plantation Crops*, **34**: 7-10.

Karp, A. (1995). Somaclonal variation as a tool for crop improvement. *Euphytica*, **85**: 295 - 302.

Larkin, P. J., and Scowcroft, W. R. (1981). Somaclonal variation- a novel source of variability from cell cultures for plant improvement. *Theoretical and Applied Genetics*, **60**: 197-214.

Leroy, T., Henry, A. M., Royer, M., Altosaar, I., Frutos, R., Duris, D. and Philippe, R. (2000). Genetically modified coffee plants expressing the *Bacillus thuringiensis cry1Ac* gene for resistance to leaf miner. *Plant Cell Reports*, **19**: 382-385.

Loyola-Vargas, V. M., Fuentes-Cerda, C. F. J., Monforte-bonzales, M., Mondez-Zeel, M., Rojas-Herrera, R. and Mijangos-Cortes, J. (1999). Coffee tissue culture as a new model for the study of somaclonal variation. In: *Proceedings of the 18th ASIC Colloquium (Helsinki) ASIC*, Paris, France. pp. 302-307.

Lubabali A. H., Alakonya A. E., Gichuru E. K., Kahia J. W. and Mayoli R. N. (2014). *In vitro* propagation of the new disease resistant *Coffea arabica* variety Batian. *African Journal of Biotechnology*, **13**: 2414-2419.

Menendez-Yuffa A., Barry-Etienne D., Bertrand B., Georget F. and Etienne H. (2010). A comparative analysis of the development and quality of nursery plants derived from somatic embryogenesis and from seedlings for large-scale propagation of coffee (*Coffea arabica* L.). *Plant Cell Tissue and Organ Culture*, **102**: 297–307.

Mishra, M. K. and Slater, A. (2012). Recent advances in the genetic transformation of coffee. *Biotechnology Research International*, Article ID 580857.

Mishra, M. K. and Sreenath, H. L. (2005). High-efficiency *Agrobacterium*-mediated transformation of coffee (*Coffea canephora* Pierre ex. Frohner) using hypocotyl explants. In: *20th International Conference on Coffee Science*, Bangalore, India, 11-15 October 2004. ASIC, Paris. pp. 792-796.

Mishra, M. K., Sreenath, H. L., McCormac, A. C., Devi, S., Elliott, M. C. and Slater, A. (2009). Two critical factors: *Agrobacterium* strain and antibiotics selection regime improve the production of transgenic coffee plants. In: *Proceedings of 22nd International Conference on Coffee Science*, ASIC 2008, Campinas, SP, Brazil, 14-19 September, 2008. ASIC, Paris. pp. 843-850.

Mishra, M. K., Sreenath, H. L. and Srinivasan, C. S. (2002). *Agrobacterium*-mediated transformation of coffee: an assessment of factors affecting gene transfer efficiency. In: Sreedharan, K., Kumar, P.K.V., Jayarama and Chulaki B.M. (Eds.), *Proceedings of the 15th Plantation Crops Symposium* (*Placrosym XV*), Mysore, India, December 2002. Central Coffee Research Institute, Coffee Research Station. pp. 251–255.

Mukul-Lopez, H.G., De-la-Pena, C., Galaz-Avalos, R.M. and Loyola-Vargas V.M. (2012). Evaluation of the extracellular proteome profile during the somatic embryogenesis process of *Coffea spp. Journal of Mexican Chemical Society*, **56**: 72-79.

Muniswamy, B. Naidu, M.M. and Sreenath, H.L. (1993). Somatic embryogenesis and plant regeneration from cultured immature zygotic embryos of *Coffea canephora* Pierre. *Journal of Plantation Crops*, **21**: 346-350.

Muniswamy, B. Naidu, M.M. and Sreenath, H.L. (1994).) A simple procedure for hardening of *in-vitro* raised plantlets of *Coffea canephora*. *Journal of Coffee Research*, **24:** 49-53.

Muniswamy, B. and Sreenath, H.L. (1995a) Standardization of encapuslation techniques using zygotic embryos of *Coffea arabica*. *Journal of Coffee Research*, **25:** 24-29.

Muniswamy, B. and Sreenath, H.L. (1995b) High frequency somatic embryogeneis from cultured leaf explants of *Coffea canephora* on a single medium. *Journal of Coffee Research*, **25:** 98-101.

Muniswamy, B. and Sreenath, H.L. (1997). Genotype specificity of direct somatic embryogenesis from cultured embryos in coffee. In: Ravishankar, G.A. and Venkataramanan, L.V. (Eds.), *Biotechnological Applications of Plant Tissue and Cell Culture*. Oxford and IBH, New Delhi. pp. 204-209.

Muniswamy, B. and Sreenath, H. L. (2000). Plant regeneration through somatic embryogenesis from anther cultures of CxR cultivar of coffee. *Journal of Plantation Crops*, **28:** 61-67.

Muniswamy, B. and Sreenath, H. L. (2001). Embryogenesis and plant regeneration from endosperm tissues of *Coffea arabica* L. *Journal of Plantation Crops*, **29:** 18-21.

Muniswamy, B., Sreenath, H. L., Samuel, S. D. and Jayarama (2015). *In vitro* multiplication of *Coffea arabica* F_1 hybrid (S 2800) and its performance in the field. *Journal of Plantation Crops*, **43**: 225-230.

Naidu, M. M., Ganesh, D. S., Jayashree, G. and Sreenath, H. L. (1999). Effect of genotype on somatic embryogenesis and plant regeneration in arabica coffee. In: Kavi Kishor, P.B. (Ed.), *Plant Tissue Culture and Biotechnology- Emerging Trends*. Universities Press, Hyderabad, India. pp. 90-95.

Naidu, M. M. and Sreenath, H. L. (1999). In *vitro* culture of coffee (*Coffee arabica* L.) zygotic embryos for germplasm preservation. *Plant Cell Tissue and Organ Culture*, **55:** 227-230.

Nakamura, T., Taniguchi, T. and Maeda, E. (1992). Studies on somatic embryogenesis of coffee by scanning electron microscope. *Japanese Journal of Crop Science*, **61:** 476-486.

Naveen, K. S., Sreenath, H. L., Sreedevi, G., Veluthambi, K., Naidu, R., Sreedharan, K., Vinod Kumar, P. K. and Chulaki, B. M. (2002). Transgenic coffee (*Coffea arabica*) plants with marker genes through *Agrobacterium tumefaciens*-mediated transformation. In: Proceedings of the 15th Plantation Crops Symposium Placrosym XV, Mysore, India. *Central Coffee Research Institute, Coffee Research Station*. pp. 219-225.

Neuenschwander, B. and Baumann, T.W. (1992). A novel type of somatic embryogenesis in *Coffea arabica*. *Plant Cell Reports*, **10:** 608-612.

Nic-Can, G. I., Lopez-Torres, A., Barredo-Pool, F., Wrobel, K., Loyola-Vargas, V. M. (2013). New insights into somatic embryogenesis: Leafy Cotyledon1, Baby

Boom1 and Wuschel-related Homeobox4 are epigenetically regulated in *Coffea canephora*. *PLoS ONE*, **8**: e72160.

Nishibata, T., Azuma, T., Uchida, N., Yasuda, T. and Yamaguchi, T. (1995). Amino acids on somatic embryogenesis in *Coffea arabica*. In: *Proceedings of the 16th ASIC Colloquium (Kyoto) ASIC*, Paris, France. pp. 839-844.

Noriega, C. and Sondahl, M. R. (1993). Arabica coffee micropropagation through somatic embryogenesis via bioreactors. In: *Proceedings of the 15th ASIC Colloquium (Montpellier) ASIC*, Paris, France. pp. 73-81.

Ogita, S., Uefuji, H., Morimoto, M. and Sano, H. (2005). Metabolic engineering of caffeine production. *Plant Biotechnology*, **22**: 461-468.

Priyono, F. B., Rigoreau, M., Ducos, J-P., Sumirat, U., Mawardi S., Lambot, C., Broun, P., Petiard, V., Wahyudi, T. and Crouzillat, D. (2010). Somatic embryogenesis and vegetative cutting capacity are under distinct genetic control in *Coffea canephora* Pierre. *Plant Cell Reports*, **29**: 343–357.

Raghuramulu, Y., Purushotham, K., Sreenivasan, M. S. and Ramaiah, P. K. (1987). *In vitro* regeneration of coffee plantlets in India. *Journal of Coffee Research*, **17**: 57-64.

Rani, V., Singh, K. P., Shiran B., Nandy, S, Goel, S., Devarumath, R. M., Sreenath, H. L. and Raina S. N. (2000). Evidence for new nuclear and mitochondrial genome organizations among high-frequency somatic embryogenesis-derived plants of allotetraploid *Coffea arabica* L. (Rubiaceae). *Plant Cell Reports*, **19**: 1013–1020.

Ribas, A. F., Pereira, L. F. P. and Vieira, L. G. E. (2006). Genetic transformation of coffee. *Brazilian Journal of Plant Physiology*, **18**: 83–94.

Sondahl, M. R. and Baumann, T. W. (2001). Agronomy II: Developmental and Cell Biology. In: Clarke, R. J. and Vitzthum, O.G. (Eds.), *Coffee - Recent Developments*. Blackwell Science, Oxford. pp. 202-223.

Sondahl, M. R. and Bragin, A. (1991). Somaclonal variation as a breeding tool for coffee improvement. In: *Proceedings of the 14th ASIC Colloqium (San Francisco) ASIC*, Paris, France. pp. 701-710.

Sondahl, M. R. and Lauritis, J. A. (1992). Coffee. In: Hammerschlag, F. A. and Litz, R. E. (Eds.), Biotechnology *of Perennial Fruit Crops*. CAB International, UK. pp. 401-420.

Sondahl, M. R. and Loh, H. T. (1988). Coffee biotechnology. In: Clarke, R.J. and Macrae, R (Eds.), *Coffee Vol.4: Agronomy*. Elsevier Applied Science, London. pp. 235-262.

Sondahl, M. R., Petracco, M. and Zambolin, L. (1997). Breeding for qualitative traits in arabica coffee. In: *Proceedings of the 17th ASIC Colloquium (Nairobi) ASIC*, Paris, France. pp. 447-456.

Sondahl, M. R., Romig, W. R. and Bragin, A. (1995). *Induction and selection of somaclonal variation in coffee. US Patent No. 5,436,395*. Washington, DC: U.S. Patent and Trademark Office.

Sondahl, M. R., Salisbury, J. L. and Sharp, W. R. (1978).SEM characterization of embryogenic tissue and globular embryos during high frequency somatic embryogenesis in coffee callus cells. *Zeitschrift für Pflanzenphysiologie*, **94:** 185-188.

Sondahl, M. R. and Sharp, W. R. (1977). High frequency induction of somatic embryos in coffee callus cells. *Zeitschrift für Pflanzenphysiologie*, **81:** 395-408.

Spiral, J., Leroy, T., Paillard, M. and Petiard, V. (2000). Transgenic coffee (*Coffea* species). In: Bajaj, Y. P. S. (Eds.), *Biotechnology in Agriculture and Forestry*, Vol. 44 Springer-Verlag, Berlin, Heidelberg. pp. 55-76.

Spiral, J. and Petiard, V. (1993). Developpement D'une Methode de transformation appliquee a differenteses peus de cafeieret regeneration de plantules transgeniques. In: *Proceedings of the 15th ASIC Colloquium (Montpellier) ASIC*, Paris, France. pp. 106-114.

Sreenath, H. L. (1998). Micropropagation of superior coffee genotypes. *Indian Coffee*, **62**: 17-19.

Sreenath, H. L. (2003). Genetic transformation and regeneration of coffee. In: Singh, R. P. and Jaiwal, P.K. (Eds.), *Plant Genetic Engineering Vol 3. Improvement of Commercial Plants-1.* Sci Tech Publishing LLC, USA. pp. 131-141.

Sreenath, H. L., Muniswamy, B., Naidu, M. M., Dharmaraj, P. S. and Ramaiah, P. K. (1992). Embryo culture of three interspecific crosses in coffee. *Journal of Plantation Crops*, **20**: 243-247.

Sreenath, H. L. and Naveen, K. S. (2005). Survey of endogenous β-glucuronidase (GUS) activity in coffee tissues and development of an assay for specific elimination of this activity in transgenic coffee tissues. In: *Proceedings of 20th International Conference on Coffee Science*, Bangalore, India, 11-15 October 2004. ASIC, Paris. pp. 656-663.

Sreenath, H. L. and Naidu, R. (1999). Potential, progress and future thrust areas of coffee biotechnology research in India. In: *Proceedings of the 18th ASIC Colloquium (Helsinki) ASIC*, Paris. pp. 281-294.

Sreenath, H. L., Shantha, H. M., Babu, K. H. and Naidu, M. M. (1995). Somatic embryogenesis from integument (perisperm) cultures of coffee. *Plant Cell Reports*, **14:** 670-673.

Sreenath, H. L. and Veluthambi, K. (2009). Transgenic arabica coffee (*Coffea arabica* L cv. Cauvery) plants with rice chitinase (*Chi*11) gene. *Paper presented in the 6th Solanaceae Genome Workshop held between 8th and 13th November 2009 at Le Meridien*, New Delhi, India. p. 268.

Startitsky, G. (1970). Embryoid formation in callus cultures of coffee. *Acta Botanica Neerlandica*, **19:** 509-514.

Sugiyama, M., Matsuoka, C. and Takagi, T. (1995). Transformation of coffee with *Agrobacterium rhizogenes*. In: *Proceedings of the 16th ASIC Colloquium (Kyoto) ASIC*, Paris, France. pp. 853-859.

Tahara, M., Naknishi, T., Yasuda, T. and Yamaguchi, T. (1995). Histological and biological aspects in somatic embryogenesis of *Coffea arabica*. In: Proceedings of the 16th ASIC Colloquium (Kyoto) ASIC, Paris, France. pp. 860-867.

Van Boxtel, J., Berthouly, M., Carasco, C., Dufour, M. and Eskes, A. (1995). Transient expression of β-glucuronidase following biolistic delivery of foreign DNA into coffee tissues. *Plant Cell Reports*, **14**: pp.748-752.

Yasuda, T., Fujii, Y. and Yamaguchi, T. (1985). Embryogenic callus induction from *Coffea arabica* leaf explants by benzyl adenine. *Plant Cell Physiology*, **26:** 595-597.

Yasuda, T., Tahara, M., Hatanaka, T., Nishibata, T. and Yamaguchi, T. (1995). Clonal propagation through somatic embryogenesis. In: *Proceedings of the 16th ASIC Colloquium (Kyoto) ASIC*, Paris, France. pp. 537-541.

Zamarripa, C. A., Ducos, J. P., Tessereau, H., Bollon, H., Eskes. A. B. and Petiard, V. (1991). *Developpement d'un procede de multiplication en masse du cafeier par embryogenese somatiqueen milieu liquide*. In: *Proceedings of the 14th ASIC Colloquium (San Francisco) ASIC*, Paris, France. pp. 392-402.

Zamarripa, C. A. and Petiard, V. (2004). Biotechnologies applied to coffee. In: Wintgens, J. N. (Ed.), *Coffee: Growing, Processing, Sustainable Production.* Wiley-VCH, Weinheim. pp. 137-163.

Chapter 8

Tea

☆ *M. Ambatkar, J. Patil, U. Mukundan, V. Sinkar, L. Basavaraju and S. Guttapadu*

1. Introduction

Two out of three people, all over the world, begin their day with a cup of tea. It is, therefore, not surprising that tea features among the world's most consumed beverages, second only to water. There are many different types of tea, and each type has its characteristic flavour, colour and aroma. Different cultures have integrated tea into their social customs. The decoction of tea leaves was used medicinally in China and parts of North-eastern India for several centuries. The journey of tea from being an herbal remedy to becoming a widely consumed daily beverage gained momentum in the 17th century when it was introduced to the western countries, as an export from China. The consumption of tea took roots in India in the 19th century (Jain and Newton, 1990) possibly as a result of British colonisation. Since then, tea has become an inseparable part of daily and social life in India.

Tea, scientifically *Camellia sinensis,* belongs to the family Theacea. The genus *Camellia* consists of about 250 species of evergreen shrubs that are mainly found in East Asia. Many species of the genus *Camellia* are cultivated as ornamental plants, on account of the large, beautiful flowers that they produce. Tea plants, in nature, can grow as tall as 15 m. However, when cultivated commercially, the height of the tea bush is limited to 60–100 cm, by pruning, to facilitate the harvest of tender leaves.

While most of the *Camellia* plants are ornamental, *C. sinensis* is cultivated for its leaves. The leaves contain a non-protein amino acid, known as 'theanine'. Despite its presence in all *Camellia* spp. in varying amounts, *C. sinensis* produces the highest quantity of this amino acid. Theanine is the primary active compound found in tea which is responsible for stimulating break activity (Nobre *et al.*, 2008), the 'wakeme-up' effect of tea. Theanine also accounts for major part of extract (Harbowy and Balentine, 1997).

This chapter describes the propagation of tea, as a crop, and lays special emphasis on the micropropagation of tea. The different aspects of tea micropropagation and a brief overview of research in this area have been covered. The authors have also included their own findings in the area of culturing of tea callus to increase the production of theanine.

2. Tea Cultivation in the World and in India

Tea is cultivated in at least 30 countries over the world, including China, India, Indonesia, Kenya and Sri Lanka. The relative share of the major tea-producing regions of the world has been shown in Figure 8.1.

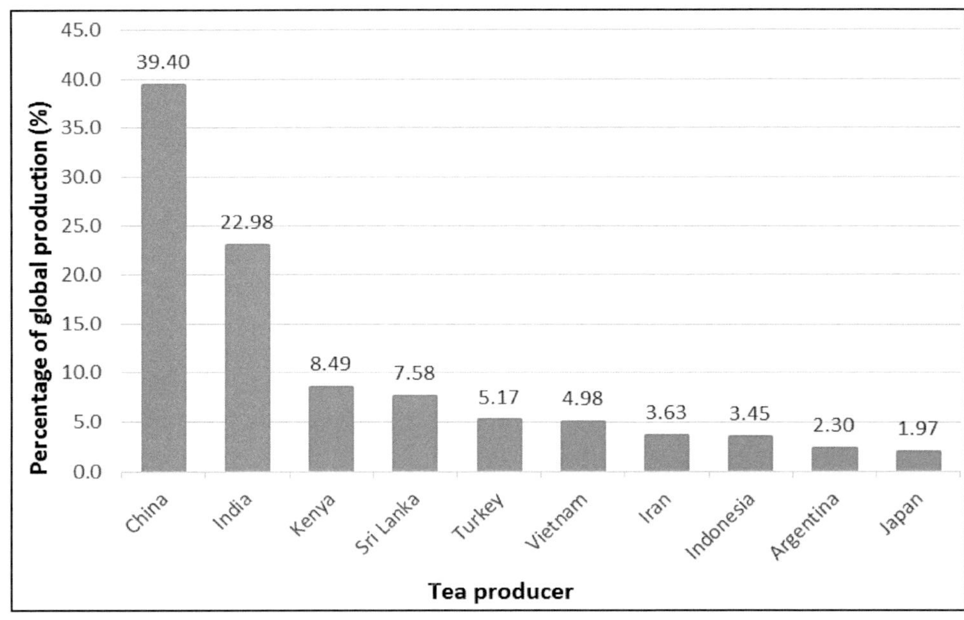

Figure 8.1: Comparison of Tea Production by different Tea-Producing Countries (FAOSTAT Report, 2012).

The cultivation of tea, in India, is widely distributed. Tea plantations are mainly located in Assam in the North-East, on the slopes of the Nilgiri Hills in the South and in Himachal Pradesh and Uttar Pradesh in the North (Jain and Newton, 1990). Each location produces tea with distinct attributes (Table 8.1). The varieties of tea cultivated in India are Assamica, Cambod and Chinary or China type (Mondal *et al.*, 2002, 2004). The primary distinction between the cultivated varieties is the size of the leaves. The comparison of leaf sizes of the three varieties is given in Table 8.2.

Table 8.1: Attributes of Teas grown in different Parts of India

Tea (India)	Attributes
Anamallais	Golden saffron liquor; floral, biscuit-like aroma; brisk and strong flavour
Assam	Deep amber liquor; rich, full-bodied flavour; brisk, strong and malty character
Darjeeling	Golden or amber liquor; delicate, flowery or peachy flavour
Dooars	Clear, black, heavy tea with a good volumetric count
Karnataka	Golden ochre liquor; aromatic; fa rly brisk flavour
Munnar	Golden yellow to orange liquor; strong, brisk and fruity flavour
Nilgiri	Golden yellow liquor; floral aroma; brisk flavour
Terai	Spicy and slightly sweet flavor
Wayanad	Reddish liquor; biscuit-like aroma; mild, full-bodied, brisk flavour

(Compiled from http://www.indiatea.org/tea_growing_regions.php and http://www.teacoffeespiceofindia.com/tea/tea-varieties)

Table 8.2: Comparison of Leaf Size in the Cultivated Tea Varieties

Variety	Relative Leaf Size
Assam	Large
Cambod	Intermediate
China	Small

The production of tea, for daily consumption, is an industry that has generated a large amount of employment and wealth. The tea industry in India involves a large network of individuals working in different capacities: tea producers, retailers, distributors, auctioneers, packers and exporters. This labour-intensive industry is important to an economy like India's because it requires both, skilled and unskilled labour.

3. Health Benefits Associated with Tea

Intensive research on the contents of tea and the effects of tea consumption, especially in the last decade, has shown that tea consumption can provide several health benefits. The demand for tea has increased tremendously over the last decade owing to the numerous health benefits associated with tea consumption. The effect of tea consumption in humans has been studied extensively in relation to cancer and cardiovascular diseases (Yang and Landau, 2000).

3.1. Polyphenols in Tea

The health benefits of tea come from the presence of large amounts of polyphenols. The typical polyphenols present in tea include catechins, such as epigallocatechin-3-gallate (EGCG), epigallocatechin (EGC), epicatechin gallate (ECG) and epicatechin (EC), as well as flavonoids such as theaflavin, theaflavin-3-gallate and theaflavin-32-gallate. It has been demonstrated that the polyphenolic compounds from tea are easily absorbed by the human gastrointestinal tract (Sharma *et al.*, 2007). The catechins from black and green teas have been shown to appear in human plasma and circulate in the lipoprotein fraction after consumption (van het Hof *et al.*, 1999). The concentration of active compounds present in aqueous extracts (teas) prepared from unprocessed and processed tea leaves have been compared in Table 8.3.

Table 8.3: Composition of Tea Prepared from Fresh and Processed Leaves (Powdered tea)

Components in Aqueous Extract	*Fresh Tea Leaves mg/dl*	*Powdered Tea mg/dl*
Caffiene	42	30
Gallic acid	3.6	6.9
Epigallocatechin	2.0	2.3
Epigallocatechin gallate	3.9	4.8
Epicatechin	1.4	2.2
Epicatechin gallate	6.0	3.6
Total catechins	13.3	12.9
Total theaflavins	6.0	2.5
Tea solids	525	467

Adapted from Duffy *et al.*, 2001.

The best-known property of tea polyphenols is their antioxidant action. Tea polyphenols are capable of scavenging reactive oxygen species (ROS) and reactive nitrogen species (RNS). They can chelate metal ions that are needed for the action of peroxidative enzymes. As a result, tea polyphenols can prevent the damage caused to lipid membranes, proteins and nucleic acids in cells (Yang and Landau, 2000). A study conducted on patients with Coronary Artery Disease (CAD) showed that the regular consumption of tea reversed endothelial vasomotor dysfunction. This effect has been shown to be on account of the antioxidants present in tea (Duffy *et al.*, 2001). A standard solution of ascorbic acid was used as the positive control to study the reversal of endothelial vasomotor dysfunction due to antioxidants (Gokce *et al.*, 1999). Another study that investigated the correlation between consumption of green tea and the occurrence as well as progress of cancer in human beings showed a strong negative correlation between the two parameters, especially in women (Imai *et al.*, 1997).

3.2. L-theanine: An Important Constituent of Tea

Tea contains well over 700 active compounds (Harbowy and Balentine, 1997). One of the compounds unique to tea is the non-protein amino acid L-theanine (γ-glutamylethylamide). This amino acid comprises approximately 50 per cent of the free amino acid content of tea (Mukai *et al.*, 1992) and contributes between 1–2 per cent of the dry weight of the leaves. L-theanine is considered among the main components responsible for the relaxing effect of tea (Nobre *et al.*, 2008) as well as the characteristic 'umami' flavour of green tea (Juneja *et al.*, 1992, Mukui *et al.*,

1992). L-theanine has received attention from various research groups because of its ability to facilitate relaxation, to reduce anxiety and stress, to enhance learning ability and memory and to suppress high blood pressure.

Several studies have shown that theanine has a stimulatory effect on the brain. L-theanine has a sedative, relaxing effect on the central nervous system. This amino acid crosses the blood–brain barrier (Nobre *et al.*, 2008). The L-leucine transporter helps theanine to cross the blood–brain barrier. In the brain, L-theanine can increase the levels of the neurotransmitters serotonin and dopamine (Yokogoshi *et al.*, 1998). It has been shown that large doses (as high as 5 g/kg of body weight) of L-theanine may be consumed without serious adverse effects. Tea research has been directed towards increasing the L-theanine content of tea. The consumption of tea has been shown to cause the generation of α brain waves in the occipital and parietal regions of the brain surface within 40 minutes of consumption. The generation of α -brain waves is indicative of a relaxed state without the induction of drowsiness.

4. Susceptibility of Tea as a Crop

The demand for tea is clearly on the rise, all over the world. All tea-producing countries compete with each other to capture the largest market share. Until the year 2009, as seen in Figure 8.1, India was the second-largest tea producing country. Approximately 28 per cent of the world's tea was produced in India. The volume of Indian teas in the market, however, is slowly being lost to some smaller countries like Kenya and Sri Lanka. The high cost of Indian teas in the global market is responsible for the India's shrinking market share.

Tea is susceptible to a variety of biotic and abiotic stresses, including drought, global warming and climate change and diseases like tea-blister blight, thread black rot and root knot. Given below in Table 8.4 is a list of tea diseases and their causative pathogens. Tea research has long focused on the development of new varieties of tea that are more resilient towards abiotic stresses like drought and more resistant towards biotic stresses caused by pathogens.

Table 8.4: Common Diseases Affecting Tea

Disease	*Pathogen*	*Organ Affected*	*Major Symptom*
Tea blister blight	*Exobasidium vexans*	Leaf	Pale yellow-green spots on leaf
Brown blight	*Guignardia camelliae*	Leaf	Small water-soaked brown spots
Bird's eye spot	*Cercospora theae*	Leaf	Numerous irregular leaf spots
Red rust	*Cephaleuros parasiticus*	Stem	Red-Orange patches on stem
Horse hair blight	*Marasmius crinisequi*	Stem	Black fungal threads attached to stem by brown discs
Root knot	*Meliodogyne incognita*	Root	Galls and knots on roots
Brown Root Rot	*Phellinus noxius*	Root	Discoloured wood just below bark, stunting of plant growth

5. Plant Characteristics Affecting Productivity

The selection of tea plants for the cultivation of tea, on a large scale, is an interesting process. Tea bushes, in general, are well trained so that the top surface of the bush is either flat or dome shaped. The season of plucking tender tea leaves is known as 'flush'. The points on the tea bush where tender leaves grow are called 'flushing points'. The flat or dome shape of the top surface facilitates the plucking of tea leaves. These shapes also allow an equal incidence of sunlight on all the leaves without any mutual shadowing. The flat, well-trained top surface of the tea bush is called the 'plucking table'. The selection of high yielding plants is necessary to ensure a steady supply of tea. Traditional methods of plant selection are based on certain morphological characteristics that are indicative of high productivity.

Tea bushes with a larger size and large area are considered capable of high yield. Tea plants that have a large number of actively growing points, or flushing points, are likely to produce high yields. A fascinating relation has been established between the leaf angles (angles that the leaves make with the shoot axis) and productivity. It has been shown that tea bushes with semi-erect leaves give the highest yield (Table 8.5) in terms of dry weight of plucked leaves over a period of five years (Banerjee, 1992). This phenomenon may be explained by the better availability of sunlight to leaves in plants with semi-erect leaves. The maximum weight of leaves (plucked) is produced by plants that have a leaf angle of approximately 63°.

Table 8.5: Correlation between Leaf Angle and Dry Weight of Plucked Leaves

Mean Leaf Angle (°)	*Average (±SE) Plucking Weight (kg) per Plant*	*Mean Leaf Angle (°)*	*Average (±SE) Plucking Weight (kg) per Plant*
40	6.41±2.13	47	6.32±3.12
50	8.45±2.12	56	8.43±3.45
58	8.42±2.75	63	8.70±2.31
68	8.23±3.01	70	8.60±2.93
78	6.89±2.85	80	7.20±2.65
85	6.75±2.73		

Adapted from Banerjee (1992).

6. Need to Increase Tea Productivity

Many of the tea plantations in different parts of the world are at least 50 years old and their productivity is slowly decreasing, as a result of age (Tahardi and Shu, 1992). Tea is a woody plant with a slow growth rate (Patil *et al.*, 2013). Tea is a natural out breeder and, as a result, the seedlings produced display a high degree of heterogeneity. The heterogeneity is seen in terms of yield, quality, leaf characteristics and growth habit (Bezbaruah, 1971). The conventional method of propagation of tea is through seeds. One of the means of overcoming the problem of heterogeneity is through vegetative propagation of a high-yielding, elite tea plant. A cutting with a single internode is used for vegetative propagation. Conventionally propagation of tea is also carried out using cuttings with a single leaf bud. This method, however, is considered inefficient for propagation (Tahardi and Shing, 1992).

The conventional tea breeding method, using seedlings, has contributed to a very large extent to the establishment of tea gardens all over the world. This method has also been used extensively even for improvement of tea as a crop. However, seedling propagation comes with its own set of shortcomings, such as low success rate of pollination, short flowering period, long duration of seed maturation and differences in the flowering time of some clones (Mondal *et al.*, 2004).

Vegetative propagation, using cuttings has been a valuable alternative means of propagation. However, propagation by cuttings also has some limitations such as the unavailability of suitable planting material owing to drought or winter dormancy, poor survival rate due to inadequate rooting of cuttings and season-dependent rooting of cuttings.

7. Micropropagation of Tea

It appears that micropropagation is a potentially useful method to circumvent the limitations of conventional propagation and to produce large numbers of genetically identical tea plants. The slow growth of tea plants has been a concern even in micropropagation. The slow growth rate of the tea plant and the recalcitrant behaviour of the tea seed are also reflected in the slow growth of somatic embryos and callus cultures.

A large number of researchers have tried to enhance the multiplication rate of tea. Thereafter, the thrust of research in tea micropropagation was to improve the hardening percentages and survival rates through the process of hardening (Mondal *et al.*, 2004).

The process of selecting a mother plant for micropropagation is similar to that followed to select a suitable plant for vegetative propagation. The mother plant has to be one that has desirable traits like high yield, pleasant aroma, full-bodied flavour, *etc.* The quality of tea has been found to be affected by the intensity of the green colour and hairiness of the leaf. In this respect, 'quality' refers to liquor colour, strength and the overall aroma of the tea. Morphological markers such as leaf colour and hairiness can be used in the selection of mother plant. The geographical origin of mother plant can also affect the traits observed in it.

In this chapter, we will mainly be considering those instances of tea tissue culture that are performed using Murashige and Skoog (MS) medium. The explants used for tea micropropagation include shoot tips, nodal segments and axillary buds (Mondal *et al.*, 2004); zygotic embryos, mature and immature cotyledons (Iddagoda *et al.*, 1988; Jha and Sen, 1992); epidermal layers of stem segment and stem segments (Kato, 1985) and anthers (Khaliq *et al.*, 2002). Somatic embryos of tea have also been used for the micropropagation of tea. One of the main advantages of using somatic embryogenesis is that adventitious embryos can be developed from explant tissues without going through an intermediate callus phase (Bano *et al.*, 1991). Somatic embryogenesis has alos been mainly used in the production of artificial seeds (Mondal *et al.*, 2004).

In our laboratory, we succeeded in obtaining *in vitro* shoots from apical and nodal segments of 3-month-old tea seedlings of UPASI-9, a drought-resistant variety. The leaves produced by the *in vitro* shoots were used to as explants for the initiation of callus (Patil *et al.*, 2013). A list of explants used by different workers and their responses to *in vitro* culture are compiled in Table 8.6.

Table 8.6: Explants from Different Varieties and their Response in Tissue Culture

Tea species/cultivar	*Explant*	*Response*	*Reference*
C. sinensis (L.) O. Kuntzecv.TRI-2025	Shoot tip	Axillary shoot proliferation	Arulpragasam and Latiff (1986).
C. sinensis (L.) O. Kuntze T-78	Shoot tips and cotyledonary nodes	Axillary bud proliferation and rooting	Jha and Sen (1992).
C. sinensis TV-1	Terminal bud and nodal stem of field plant	Shoot proliferation and rooting	Agarwal *et al.* (1992).
C. sinensis (L.)	Microshoot	Rooting	Jain *et al.* (1993).
C. sinensis (L.) O. Kuntze	Nodal segment	Axillary shoot multiplication and rooting	Bag *et al.* (1997).

(Adapted from Mondal *et al.*, 2004)

8. Browning of Explants during Tea Tissue Culture

One of the main problems that limit the success of the micropropagation of tea is the browning of explants. Ironically, polyphenols, which confer health benefits on tea, are responsible for significant losses during micropropagation. Plants, that are rich in phenolic compounds, faced the problem of browning during micropropagation. Explants often show blackening or browning shortly after being isolated. Blackening and browning leads to inhibited growth and eventually to the death of the explants. When the explants are sterilised, cut and inoculated on the growth medium, phenolic compounds present in the plant are exuded from the cut surfaces of the explants. The oxidation and polymerization of phenolic compounds is manifested as tissue browning. Tea is naturally rich in polyphenols like tannins. It is highly prone to inhibition of growth of explants in tissue culture. If the explants are excised from older tissue, they are more prone to browning and blackening (George and Sherrington, 1984) primarily because the polyphenol content of older tissue is higher than that of younger tissue.

Tissue blackening or browning is mainly due to the action of copper-containing oxidative enzymes, like polyphenoloxidases and tyrosinase (Lerch, 1981), which are released when tissue wounding occurs. The polyphenolic compounds that are formed on oxidation are phytotoxic and cause the irreversible inhibition of the growth of explants. The medium as well as explants the turn brown (Çörduk and Aki, 2011) as a result of the oxidation of polyphenols.

It is possible to reduce the extent of browning using various chemicals including antioxidants, like citric acid (Sarwar, 1985), ascorbic acid (Iddagoda *et al.*, 1988), and polyvinylpyrrolidone or PVP as well as activated charcoal (Çörduk and Aki, 2011). Given below are some commonly used methods to overcome browning of tea explants:

8.1. Leaching or Dispersal

Soaking the explants in distilled water for 24 hours before sterilisation and inoculation on medium can be used to control browning. In some cases, the explants are initially inoculated in liquid medium to enable the dispersion of polyphenols into the medium. The explants are subsequently transferred to solid medium.

8.2. Frequent Subculturing

Moving the explants to new medium after short intervals is one of the most widely used methods to control browning. This method has been used to control browning in plants like *Sideritis trojana* (Çörduk and Aki, 2011).

8.3. Modifying Redox Potential

The tendency of a substance to get oxidized or reduced depends on its redox potential. Soaking the explants in ascorbic acid and citric acid exposes them to reducing agents and low pH (to prevent the action of oxidative enzymes). Other reducing agents include L-cysteine hydrochloride, dithiothreitol, glutathione and mercaptoethanol. Explants may be immersed in sterile solutions of the antioxidants immediately after excision. Those antioxidants that are not heat-labile (and will endure autoclaving) may be added to the medium during medium preparation.

8.4. Darkness

Light activates the biosynthesis of phenols and their enzymatic oxidation. Tissue browning may be prevented if the newly explanted tissues are kept in dark for up to 14 days and subsequently transferred to low light intensity (500–1000 lux).

8.5. Chelating Agents

Ethylenediaminetetraacetic acid (EDTA) has been used to chelate metals like copper and iron, which are essential for the action oxidative enzymes like polyphenol oxidase or can increase the rate of the oxidation, respectively.

8.6. Absorption with Activated Charcoal

The superior absorptive ability of activated charcoal is often utilized in tissue culture to absorb inhibitory compounds. It has been used as an additive in growth media. The use of activated charcoal has been reported to prevent the build-up of polyphenols and improving explant growth (Reuveni and Lilien-Kipnis, 1974; Reynolds and Murashige, 1979). On the other hand, activated charcoal has also been reported to absorb growth regulators like Ferric EDTA (Fe-EDTA) and result in diminished growth of cultured plantlets (Snir, 1981).

8.7. Absorption by Polyvinylpyrolidone (PVP)

PVP is a polyamide that is routinely used in combination with β-mercaptoethanol for the extraction of proteins and nucleic acids. Phenols are absorbed by PVP through hydrogen bonding and thus PVP prevents their oxidation, thereby reducing the occurrence of browning (George and Sherrington, 1984). Explants that were treated with PVP showed better results than those treated with 5 per cent H_2O_2 or 0.005 per cent ascorbic acid (Gupta *et al.*, 1981). Other techniques to reduce browning include the modification of medium composition and plant growth regulators (PGRs) (Çörduk and Aki, 2011).

It is possible to assess how effective these anti-browning chemicals and techniques are in reducing the incidence of browning. A browning scale is used to quantify the extent of browning. It is also possible to use the technique of colorimetry to measure the extent of browning (Lunadei *et al.*, 2011) and to assess the effectiveness of different procedures used to reduce browning. The five-point browning scale has been outlined below in (Table 8.7).

Table 8.7: Browning Scale Described by Lunadei *et al.* (2011).

Browning Level	*Reading (Fraction of surface area showing browning)*
0	None
1	<10 per cent
2	10–25 per cent
3	25–50 per cent
4	>50 per cent

The browning index can be calculated using the formula given by Yang *et al.* (2010).:

Equation 1

Browning index =

$$\sum\left[\frac{\text{Browning level} \times \text{Number of explants with that browning}}{\text{Total number of explants}}\right]$$

7. Plant Growth Regulators for Tissue Culture of Tea

Plant growth regulators (PGRs) or plant hormones are chemical compounds that profoundly influence growth and differentiation in plants. The role of PGRs in growth medium is non-nutritional; they show their effect on plant tissues even at very low concentrations (George and Sherrington, 1984).

Tea does not respond easily when cultured *in vitro*. The use of PGRs to obtain callus formation, shoot proliferation, root initiation or somatic embryogenesis in explants is common. The use of auxins, like α-naphthalene acetic acid (NAA), indole-3 acetic acid (IAA), 2,4-dichlorophenoxy acetic acid (2,4-D) and indole-butyric acid (IBA), and cytokinins like 6-benzyladenine (BA), N6-furfuryladenine (Kinetin) and thidiazuron (TDZ) is most common.

Kato *et al.* (1985) have reported the use of BA (10 mg/L) on explants such as the epidermal layer of stem segments, intact stem segments and stem segment without epidermal layer for indirect organogenesis through callus formation. The auxins 2, 4-D and α-NAA are effectively induce callus, though they are ineffective for the induction, growth and development of tea shoots (Nakamura, 1988). Synthetic auxins like picloram and 2,4,5-trichlorophenoxyacetic (2,4,5-T) have been used in elongation of tea shoots (Arulpragasam and Latiff, 1986; Iddagoda *et al.*, 1988; Jain *et al.*, 1993). IAA (0.1–2.0 mg/L) and kinetin (0.21–8.0 mg/L) have been shown to be necessary for the induction and multiplication of axillary shoots (Sarwar, 1985).

The use of TDZ in tissue culture of tea showed an interesting effect upon comparing the effects of TDZ and BA on shoot bud proliferation (Mondal *et al.*, 2004). Extremely low concentrations of TDZ alone (1 pM–100 nM) resulted in the induction of shoot bud proliferation. Low concentrations of TDZ were effective in inducing shoot multiplication and maintaining high rates of shoot proliferation on hormone-free medium for over 20 subsequent subcultures. On the contrary, higher concentrations of BA (1–10 μM) and its continuous presence is required to to initiate and sustain shoot proliferation. TDZ can induce callus in all tea explants at higher concentrations (5, 10 and 15 μM) in combination with 2,4-D or IBA at 5, 10 and 15 μM. The best combination to induce callus in tea explants is 5 μM TDZ and 10 μM NAA. The application of TDZ is required only initially and in very small quantities. Hence, TDZ is quite a useful cytokinin in tea micropropagation, despite its high cost.

In our experiments with *in vitro* shoot regeneration using explants derived from UPASI-9, we obtained response with MS media containing BA, Kin and TDZ. These three cytokinins were tested at 0.1 (low), 0.5 (medium) and 1.0 (high) mg/L. As outlined as in Table 8.8., the effect of the three tested cytokinins showed that 0.5 mg/L was the optimum concentration. All the explants showed the development of one shoot per explant. Therefore, the effectiveness of the PGRs in shoot induction was compared using the response time of explants, number of leaves produced and time taken to produce leaves *in vitro*. The detailed results are reported in Table

Table 8.8: Comparison of Responses of Apical and Nodal Segments of *C. sinensis* to different Concentrations of Cytokinins

PGR	*0.1 mg L^{-1}*	*0.5 mg L^{-1}*	*1.0 mg L^{-1}*
TDZ	No shoot induction	Callus formation (12 weeks)	Blackening of explant
Kin	No shoot induction	Shoot induction (4 weeks)	Stunted shoot growth
BA	No shoot induction	Shoot induction (6 weeks)	Stunted shoot growth

Adapted from Patil *et al.*, 2013.

Table 8.9: Comparison of Shoot Proliferation Response with different PGRs

PGR (mg L^{-1})	*Mean no. of Leaves*	*Response Time (Weeks)*	*Time for Leaf Growth (Weeks)*
0.5 BA	2.3±0.943	6	17
0.5 Kin	7.3±0.942	4	16
0.5 TDZ	Callus formation	12	12

Adapted from Patil *et al.*, 2013.

8.9. The explants treated with kinetin responded relatively earlier (at 4 weeks) and showed leaf growth 16 weeks after inoculation.

9. Callus Cultures in Tea

A combination of NAA and BA, an auxin and a cytokinin respectively, has been reported to induce callus in some cases and the formation of 4–5 shoots per explant over a period of 8–12 weeks (Phukan and Mitra, 1984; Bag *et al.*, 1997). In our experiments, nodal explants from field-grown saplings and leaves formed *in vitro* were used for the purpose of callus induction. In accordance with the findings of Nakamura (1988), the auxins, 2, 4-D and NAA, were effective to induce callus. The following PGRs and their combinations (Table 8.10) were used to supplement MS medium for callus induction. The most effective combination of PGRs, in the medium, for the induction of callus was 0.5 mg/L each of NAA and 2, 4-D. The experiments with callus cultures of tea, in our laboratory, were primarily to increase the accumulation of L-theanine by the callus cultures.

Table 8.10: List of PGRs Used for Callus Induction

PGR	*Concentration (mg/L)*	*Response induced*	*Explant*
TDZ	0.5	Callus	Leaf
BA	0.5; 1.0	Callus	Leaf
BA	2.0	Shoot	Nodal explant
Kin	0.5	Shoot induction	Nodal explant
Kin	1.0	Shoot induction	Nodal explant
2,4-D	1.0	Callus induction followed root initiation	*In vitro* leaf
2,4-D	0.5; 1.5; 2.0	Callus induction	*In vitro* leaf
NAA	0.5; 1.0	No response	*In vitro* leaf
IBA	0.5; 1.0	No response	*In vitro* leaf
IAA	0.5; 1.0	No response	*In vitro* leaf
BA + IBA	4 + 2	No response	Nodal explant
2,4-D + NAA	0.5 + 1.51.5 + 0.5	No response	*In vitro* leaf
2,4-D + NAA	0.5 + 0.5	Best callus induction	*In vitro* leaf
NAA + IBA	0.5 + 0.5	No response	*In vitro* leaf
NAA + IAA	0.5 + 2.85	No response	*In vitro* leaf

The roots of the tea plant synthesize the largest amount of L-theanine. L-theanine synthesised in the roots is then transported to different parts of the plant, mainly to the leaves. The biosynthetic pathways involved in the synthesis of L-theanine have been studied in great detail (Harbowy and Balentine, 1997). Theanine synthetase is the enzyme responsible for the synthesis of L-theanine. Two amino acids, namely L-alanine and L-glutamic acid are involved in the biosynthesis of L-theanine. These amino acids are used indirectly and directly in the synthesis of L-theanine, respectively. L-alanine serves as the precursor of a compound called ethylamine (Et-NH_2). The conversion of L-alanine to Et-NH_2 is catalysed by the enzyme L-alanine carboxylase. The skeleton of L-glutamic acid directly is used to form that of L-theanine. Preliminary experiments, wherein L-glutamic acid was added to the medium, showed that its addition does not significantly increase the L-theanine content of the callus (data not shown).

9.1. Increasing the Theanine Content of Callus

Our experiments were designed with the aim of increasing the theanine content of the callus, derived from the leaf, by providing the callus with precursors of L-theanine. Accordingly, leaf explants of approximately 1 cm^2 were inoculated on modified MS medium supplemented with 0.5 mg/L NAA + 0.5 mg/L 2,4-D, (optimum medium) and gelled with 8 per cent (w/v) agar (HiMedia, India) to induce callus growth. The callus was allowed to grow on this medium for 60 days (8–9 weeks) before utilizing it for further experiments. Cultures were maintained under a photoperiod regime of 16/8 hours light/dark using 3000 lux cool white fluorescent tubes and the temperature was maintained at 25±2 °C.

9.2. Effect of Precursor on Theanine Formation

The optimum medium was supplemented with different concentrations of Et-NH_2in the form of Ethylamine.HCl (Et-NH_2·HCl, Sigma-Aldrich Pvt. Ltd., Germany). The pH of the medium was adjusted to pH 5.7±0.05 prior to autoclaving. In each concentration, 0.10 g of callus was inoculated and observed for 30 days.

9.3. Estimation of Theanine Content

Callus samples of 0.1 g were subjected to 24 hour extraction in 1 ml of 50 per cent (v/v) ethanol. The samples, along with standard L-theanine (1.0 mg/ml) and L-glutamic acid (1 mg/ml), were then analysed using HPTLC with 70 per cent (v/v) n-propanol as the mobile phase. Ninhydrin in acetone (2 per cent w/v) was used for visualization and derivatization. Plates were observed under visible light. Theanine content in the samples was estimated densitometrically by comparing with the standard of known concentration. The Rf of theanine in the n-propanol mobile phase is 0.48. The maximum accumulation of L-theanine in the calli was observed on the 30th day after inoculation. Increasing the culture duration further up to 60 days resulted in a decrease in the theanine content of the calli (Figure 8.2).

A comparison was made between the L-theanine contents of the calli grown on plain MS medium and optimum medium. When Et-NH_2.HCl was used as supplement, the calli grown on modified MS medium accumulated lower amounts of L-theanine than those grown on modified MS supplemented with 0.5 mg/L each of NAA and 2,4,-D (optimum medium) (Figure 8.3).

The roots are known to be the primary site of L-theanine synthesis in the tea plant. The auxins NAA and 2,4-D promoted the formation of roots in the calli. The induction of rhizogenesis was observed in some calli if the calli were maintained on medium supplemented with NAA and 2, 4-D for more than 90 days.

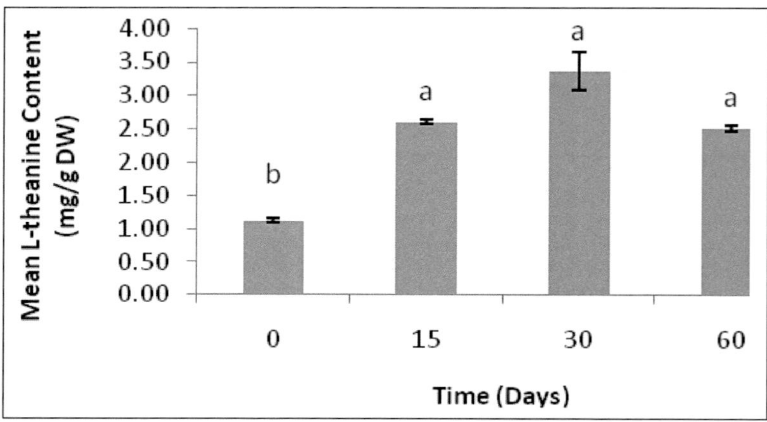

Figure 8.2: Theanine Accumulation over Time in *C. sinensis* Callus grown on Optimum Media.

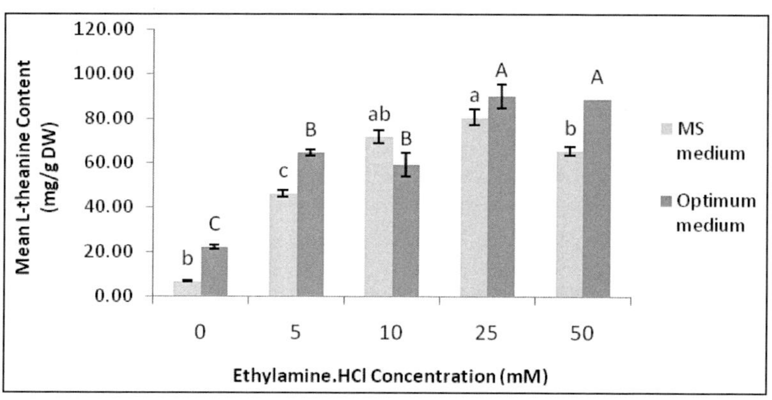

Figure 8.3: L-theanine Content in *C. sinensis* Calli Grown for 30 Days in Media Supplemented with Varying Concentrations of Ethylamine.HCl

In the L-theanine biosynthesis pathway (Figure 8.4), Et-NH_2 is the immediate precursor of L-theanine (Harbowy and Balentine, 1997). Et-NH_2 is also believed to be a limiting factor for theanine synthesis (Matsuuda and Kakuda, 1990). Hence adding Et-NH_2 as a supplement to the growth medium was expected to increase the theanine content of the calli.

Our findings showed that on day 30 of the culture cycle, the accumulation of theanine was highest in calli supplied with 25 mM Et-NH_2.HCl (Figure 8.5 and Table 8.11). Supplementation of growth medium with 25 mM Et-NH_2.HCl resulted in an average theanine content of 90.20 mg/g dry weight of callus on medium and 80.72 mg/g dry weight of callus on MS medium.

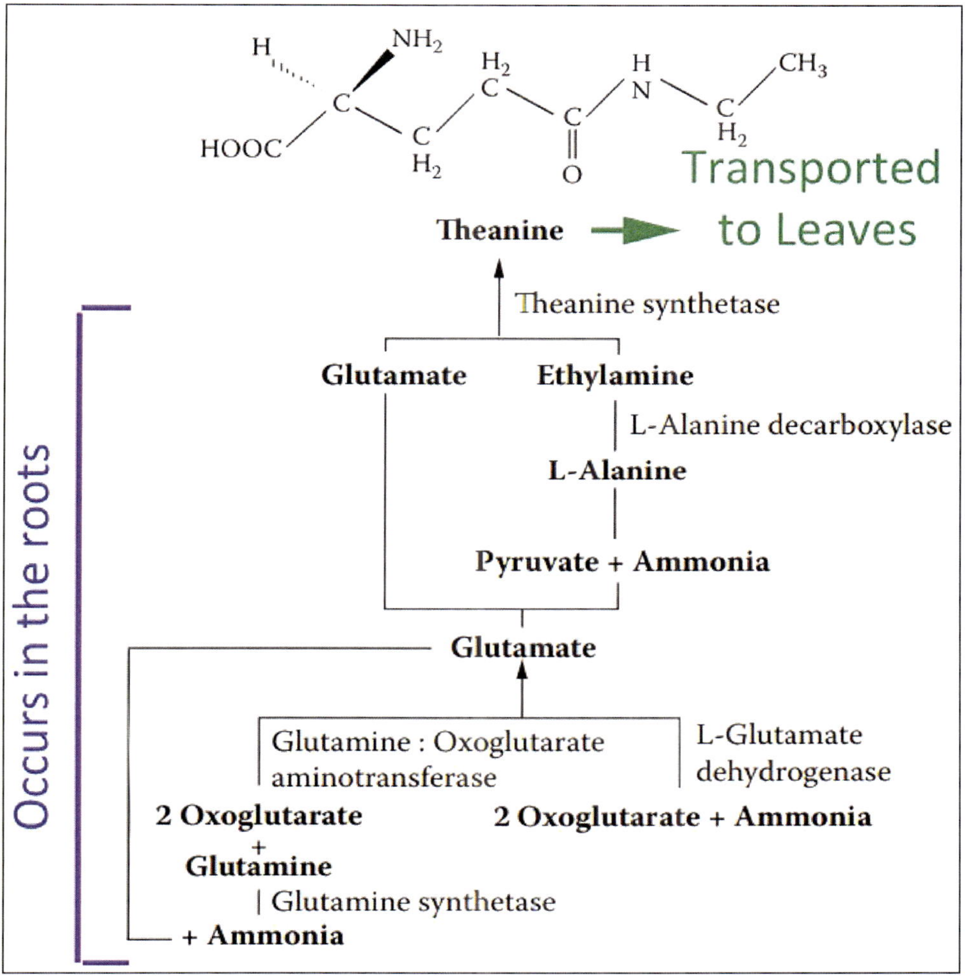

H NH$_2$ H$_2$ H CH$_3$
HOOC–C–C(H$_2$)–C(H$_2$)–C(=O)–N(H)–C(H$_2$)–CH$_3$

Theanine → Transported to Leaves

Theanine synthetase

Glutamate **Ethylamine**

L-Alanine decarboxylase

L-Alanine

Pyruvate + Ammonia

Glutamate

Glutamine : Oxoglutarate aminotransferase

L-Glutamate dehydrogenase

2 Oxoglutarate + Glutamine

2 Oxoglutarate + Ammonia

Glutamine synthetase

+ Ammonia

Occurs in the roots

Figure 8.4: Biosynthesis of L-theanine.

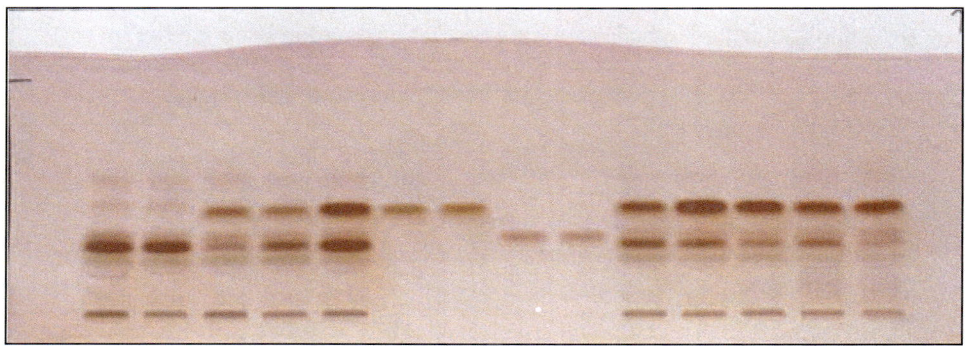

Figure 8.5: HPTLC Analysis of Tea Callus Grown with and without Et-NH$_2$.HCl (Courtesy: HPTLC Application Laboratory, ANCHROM Pvt. Ltd).

Table 8.11: List of Samples Loaded on TLC plate (Figure 8.5)

Lane (Left to right)	*Sample*
Lane 1 and 2	Control (0 mM Et-NH_2.HCl)
Lane 3 and 4	5 mM Et-NH_2.HCl
Lane 5 and 10	10 mM Et-NH_2.HCl
Lane 6 and 7	L-Theanine standard (1 mg/ml)
Lane 8 and 9	L-Glutamate (1 mg/ml)
Lane 11 and 12	25 mM Et-NH_2.HCl
Lane 13 and 14	50 mM Et-NH_2.HCl

These results were similar to those of an earlier investigation of the effect of Et-NH_2 on theanine content in callus cultures of tea (Matsuuda and Kakuda, 1990). The combination of the two precursors was also added to the optimum growth medium as a supplement. Supplying ethylamine and alanine in the growth medium was expected to enhance theanine synthesis since ethylamine would no longer be a limiting factor. The concentration of Et-NH_2.HCl was maintained at 25 mM while the concentration of L-alanine was varied as 5, 10 and 25 mM. The calli grown with a combination of 25 mM each of Et-NH_2.HCl and L-alanine showed the highest L-theanine content (29.93 mg/g dry weight) at 30 days. This content of L-theanine is significantly higher when compared to (data not shown) that of calli grown with only L-alanine (3.85 mg/g) or that of calli grown with only Et-NH_2.HCl (23.14 mg/g). Clearly, when both precursors are present together, they exhibit a synergistic effect, leading to enhanced synthesis of L-theanine.

The productivity of theanine defined as the product of theanine content (mg/g dry weight of callus) and increase in biomass (g) of callus was calculated for each of the different treatments over 30 days (Table 8.12). The productivity of calli treated with a combination of 25 mM Et-NH_2.HCl and 25 mM L-alanine (0.476 mg) was significantly higher than that of the calli treated with 25 mM Et-NH_2.HCl (0.245 mg) and control (0.046 mg). Hence supplementing the growth medium of the *C. sinensis* calli with a combination of 25 mM Et-NH_2.HCl and 25 mM L-alanine significantly improves the theanine content of the calli over the culture duration of 30 days.

Table 8.12: Effect of Supplements on Mean Theanine Productivity in Callus over 30 Days

Supplement	*Mean Theanine Productivity (mg)±SE*
Control	0.046 ±0.001
25 mM EtNH_2.HCl	0.245 ±0.024
25 mM EtNH_2.HCl + 5 mM L-alanine	0.353 ±0.015
25 mM EtNH_2.HCl + 10 mM L-alanine	0.318 ±0.011
25 mM EtNH_2.HCl + 25 mM L-alanine	0.476 ±0.008

L-theanine is a reserve of raw material for the biosynthesis of catechin and catechin-like polyphenols (Harbowy and Balentine, 1997). In our experiments, we found a moderately strong negative correlation between the content of L-theanine and polyphenols (measured as catechin equivalents) in the callus between days 15–60. Hence, it is likely that increasing the culture duration of the callus beyond 30 days, causes the accumulated theanine to be channelled into polyphenol synthesis, thereby decreasing the theanine content of the callus while simultaneously increasing polyphenol content (Figure 8.6).

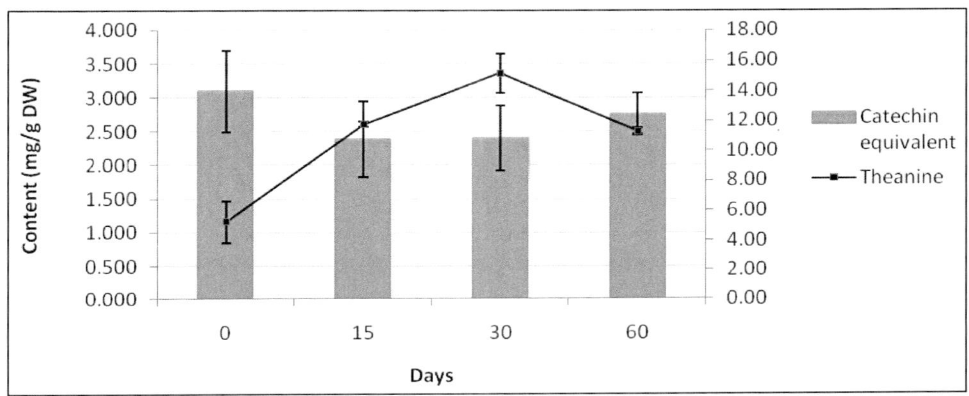

Figure 8.6: Comparison of Polyphenols and Theanine Accumulation in *C. sinensis* Callus grown in Optimum Medium over 60 Days.

10. Hairy Roots for Tea

A plant can be induced to produce hairy roots when it is infected by *Agrobacterium rhizogenes*. This bacterium is a Gram negative soil bacterium which is capable of inducing hairy roots in a variety of plants, which include mostly dicotyledonous plants, though some monocots and gymnosperms are also infected by it (Murugesan *et al.*, 2010). The induction of hairy roots is due to the integration of the root inducing (Ri) plasmid T-DNA into the genome of the plant and its subsequent expression (Villalobos-Amador *et al.*, 2002).

Several species of bacteria are able to cause hairy root formation in plants. However, *Agrobacterium rhizogenes* is preferred because they have been proven to be highly efficient at transformation (Park *et al.*, 2009). *A. rhizogenes* naturally transform host plants into chemical producers of their chemical food, *i.e.* opines. The induction of hairy roots in tea using *A. rhizogenes* was performed for the first time in tea plants by Zehra *et al.* (1996). In their experiments, Zehra and co-workers used 35-day-old *in vitro* leaves as explants, and *A. rhizogenes* strain used was A_4. The leaf discs were co-cultivated with the bacterial culture for 48 h. The leaf discs were cleansed of excess, adhering bacteria and inoculated on MS medium. Hairy roots were observed after 35 days.

Later, Konwar *et al.* (1998) induced hairy roots in, 4–6 months old tea shoots by the co-cultivation method. Thereafter, the explants were grown on MS medium supplemented with 5 mg/L IBA. IBA was used in the medium to stimulate the growth of hairy roots. Matsumoto and Fukui (1997) have reported the use of acetosyringone (100–500 μM) to improve the efficiency of transformation in tea. They succeeded in obtaining transformed tea callus. The callus was declared to be transformed after polymerase chain reaction (PCR) analysis.

Despite the tremendous potential of hairy root cultures in a plant like tea, this technique has not been widely used on a commercial scale. The recalcitrance of tea to genetic transformation is the major hindrance in the large-scale application of this technology to tea.

11. Hardening

The hardening stage is the most crucial stage in tea micropropagation because the risk of loss is highest at this stage. The transfer of tea plantlets from culture vessels to the greenhouse and finally to the fields is a gradual process, that takes at least s months. Stringent control over the microclimate is necessary in the initial stages of transfer. Arulpragasam *et al.* (1988) transferred rooted plantlets of tea, 5–8 cm in length, to small pots containing fumigated soil. These pots were kept in humid chamber for at least 10 days before transfer to a relatively lower humidity set up. These plants were then gradually acclimatized to field conditions. The lack of a tap root system makes the tissue culture raised plantlets very susceptible to drought.

The intensity of light to which the plantlets are exposed is also gradually increased. Plantlets are slowly moved out of low light and high humidity conditions through stages where the humidity decreases and light intensity increases until the plantlet can withstand field conditions (Mondal *et al.*, 2004).

Tissue culture raised plantlets that have been grown in aseptic conditions are very susceptible to microbial infection. Often, these plantlets do not survive the transfer to field soil because of microbial shock. Pandey *et al.* (2000) significantly reduced plantlet mortality by the use of cultures of *Bacillus subtilis* and *Pseudomonas corrugata*. These cultures were inoculated into the soil in which plantlets were hardened before being transferred to the field. The tea plantlets that had been inoculated before field transfer were able to survive the transfer to field soil.

12. Conclusion

Increasing the production of tea is, undoubtedly, a priority for researchers. An adverse impact of low productivity is the loss of market share in the global tea market on account of the high cost of Indian tea. Tissue culture is a promising means to propagate elite, high yielding varieties in large numbers, in a relatively short period. Continuous research and improvement in tissue culture techniques have been able to reduce the losses suffered at different stages. The tremendous potential of callus cultures and hairy root cultures has yet to be utilised fully on the commercial scale. Perhaps the cost of mass producing callus and hairy root cultures of tea plants is not economically feasible, at present. The overall cost of micropropagated tea plants is significantly higher than that of vegetatively propagated tea plants. The challenge that lies before tea research in India is the development of innovative, sustainable and inexpensive methods that will reduce the cost of micropropagation. The reduced cost of micropropagation needs to be coupled with developing superior quality varieties of tea. Providing high quality teas at competitive prices will help India regain, and perhaps expand, its share in the global tea market.

Acknowledgement

The authors Mugdha Ambatkar, Jayesh Patil and Usha Mukundan would like to thank Hindustan Unilever Limited for the financial assistance provided for the research project titled 'Evaluation of root cultures of *Camellia sinensis* for theanine production'.

References

Agarwal, B., Singh, U. and Banerjee, M. (1992). *In vitro* clonal propagation of tea (*Camellia sinensis* (L.) O. Kuntze). *Plant Cell Tissue and Organ Culture*, **30**: 1-5.

Arulpragasam, P.V. and Latiff, R. (1986). Studies on the tissue culture of tea (*Camellia sinensis* (L.) O. Kuntze). 1. Development of a culture method for the multiplication of shoots. *Sri Lanka Journal of Tea Science* **55**: 44-47.

Arulpragasam, P.V., Latiff, R. and Seneviratne, P. (1988). Studies on the tissue culture of tea (*Camellia sinensis* (L.) O. Kuntze). 3. Regeneration of plants from cotyledon callus. *Sri Lanka Journal of Tea Science* **57**: 20-23.

Bag, N., Palni, L.M.S. and Nandi, S.K. (1997). Mass propagation of tea using tissue culture methods. *Physiology and Molecular Biology of Plants*, **3**: 99-103.

Banerjee, B. (1992). Selection and Breeding of Tea. In: Wilson, K.C. and Clifford, M.N. (Eds.), *Tea: Cultivation to Consumption* (1st Ed).Wisconsin-Madison: Chapman and Hall. pp. 53-86.

Bano, Z., Rajaratnam, S. and Mohanty, B. D. (1991). Somatic embryogenesis in cotyledon culture of tea (*Thea sinensis* L.) *Journal of Horticultural Sciences*, **66**: 465-470.

Bezbaruah, H.P. (1971) Cytological investigation in the family theacea-1. *Chromosome numbers in some Camellia species and allied genera. Carylogia*, **24**: 421-426.

Biosynthesis of L-theanine, retrieved on January 21, 2016 from http://www.kyarazen.com/secrets-tea-seasons-aging/biosynthesis.

Chen, T. and Chen, S. (1982). Diseases of tea and their control in the People's Republic of China. *Plant Disease*, **66**: 961-965.

Çörduk, N. and Aki, C. (2011). Inhibition of browning problem during micropropagation of *Sideritis trojana* Bornm., an endemic medicinal herb of Turkey. *Romanian Biological Letters*, **16**: 6760-6765.

Duffy, S.J., Keaney, J.F. Jr, Holbrook, M., Gokce, N., Swerdloff, P.L., Frei, B. and Vita, J.A. (2001). Short- and long-term black tea consumption reverses endothelial dysfunction in patients with coronary artery disease. *Circulation*, **140**:151-156.

Food and Agriculture Organization of the United Nations (2012). *FAO Statistics Yearbook.*

George, E.F. and Sherrington, P.D. (1984). *Plant Propagation by Tissue Culture Handbook and Directory of Commercial Laboratories*. Eastern Press, Reading, Berks, Great Britain.

Gokce, N., Keaney, J.F. Jr, Frei, B., Holbrook, M., Olesiak, M., Zachariah, B.J. Leeuwenburgh, C., Heinecke J.W. and Vita, J.A. (1999). Long-term ascorbic acid administration reverses endothelial vasomotor dysfunction in patients with coronary artery disease. *Circulation*, **99**: 3234 -3240.

Gupta, P.K., Mascarenhas, A. F. and Jagannathan, V. (1981). Tissue culture of forest trees: Clonal multiplication of mature trees of *E. citriodora* Hook by tissue culture. *Plant Science Letters*, **20:** 195-200.

Harbowy, M.E. and Balentine, D.A. (1997). Tea chemistry. *Critical Reviews in Plant Sciences*, **16:** 415-480.

Iddagoda, N., Kataeva, N.N. and Butenko, R.G. (1988). *In vitro* clonal micropropagation of tea (*Camellia sinensis* L.) 1. Defining the optimum condition for culturing by means of a mathematical design technique. *Indian Journal of Plant Physiology*, **31**: 1-10.

Imai, K., Suga, K., and Nakachi, K. (1997). Cancer-preventive effects of drinking green tea among a Japanese population. *Preventive Medicine*, **26**: 769-775.

India Brand Equity Foundation. Tea varieties: Retrieved December 28, 2015 from http://www.teacoffeespiceofindia.com/tea/tea-varieties.

Indian Tea Association. Tea growing regions: Retrieved December 28, 2015 from http://www.indiatea.org/tea_growing_regions.php.

Jain, S.M. and Newton, R.J. (1990). Prospects of biotechnology for tea improvement. *Proceedings of the Indian National Science Academy*, **B56**: 441-448.

Jain, S.M., Das, S.C. and Barman, T.S. (1993). Enhancement of root induction from *in vitro* regenerated shoots of tea (*Camellia sinensis* L.). *Proceedings of the Indian National Science Academy*, **59**: 623-628.

Jha, T.B. and Sen, S.K. (1992). Micropropagation of an elite Darjeeling tea clone. *Plant Cell Reports* **11**: 101-104.

Kato, M. (1985). Regeneration of plantlets from tea stem callus. *Japanese Journal of Breeding*, **35**: 317-322.

Khaliq, A., Rashid, H. and Quraishi, A. (2002). Tissue culture studies of tea (*Camellia sinensis* L.). *Pakistan Journal of Agricultural Research*, **17**: 297-301.

Konwar, B.K., Das, S.C., Bordoloi B.J. and Dutta R.K. (1998). Hairy root development in tea through *Agrobacterium rhizogenes*-mediated genetic transformation. Two and a Bud, **45**: 19-20.

Lerch, K. (1981). *Metal ions in Biological Systems-Vol. 13*. Marcel Dekker, New York. pp. 143-186.

Lunadei, L., Diezma B., Lléo L. and Galléguillos P. (2011). Enzymatic browning in fresh–cut apple slices by different kinds of imaging algorithms. *VI Congreso Ibérico de AgroIngeniería*, Universidade de Évora, Portugal.

Matsumoto, S. and Fukui, M. (1997). Effect of acetosyringone application on *Agrobacterium*- mediated gene transfer in tea plant (*Camellia sinensis*). Bulletin of the National Research Institute of Vegetables, *Ornamental Plants and Tea Japan* **14**: 9-15.

Mondal, T. K. (2002). Camellia biotechnology: A bibliographic search. *International Journal of Tea Science*, **2 and 3**: 28-37.

Mondal, T. K., Bhattacharya, A., Laxmikumaran, M. and Ahuja, P.S. (2004). Recent advances of tea (*Camellia sinensis*) biotechnology. *Plant Cell Tissue and Organ Culture*, **76**: 195-254.

Mukui, T., Horie, H. and Goto T. (1992). Differences in free amino acids and total nitrogen contents among various prices of green tea. *Tea Research Journal*, **76**: 45-50.

Murugesan, S., Manoharan, C., Vijaykumar, R. and Panneerselvam, A. (2010). Isolation and characterization of *Agrobacterium rhizogenes* from the root nodules of some leguminous plants. *International Journal of Microbiological Research*, **1**: 92-96.

Nakamura, Y. (1998). Effects of the kinds of auxins on callus induction and root differentiation from stem segment culture of *Camellia sinensis* (L.) O. Kuntze. *Tea Research Journal*, **68**: 1-7.

Nobre, A.C., Rao, A. and Owen, G.N. (2008). L-theanine, a natural constituent in tea, and its effect on mental state. *Asia Pacific Journal of Clinical Nutrition*, **17**:167-168.

Pandey, A., Palni, L.M.S. and Bag, N. (2000). Biological hardening of tissue culture raised tea plants through rhizosphere bacteria. *Biotechnology Letters*, **22**: 1087–1091.

Park, S.U., Kim, Y.K. and Lee, S.Y. (2009). Establishment of hairy root culture of *Rubia akane* Nakai for Alizarin and Purpurin production. *Scientific Research and Essay*, **4**: 94-97.

Patil, J., Ambatkar, M., Mukundan, U., Guttapadu, S. and Sinkar, V. (2013). Tea tissue culture: *In vitro* shoot regeneration and callus induction in tea. *International Journal of Tea Science*, **9**: 3-5.

Phukan, M.K. and Mitra, G.C. (1984). Regeneration of tea shoots from nodal explants in tissue culture. *Current Science*, **53**: 874-876.

Reuveni, O. and Lilien-Kipnis, H. (1974). Pamphlet 145, 3-39, *Volcani Institute of Agricultural Research*, Bet Dagan Israel.

Reynolds, J.F. and Murashige, T. (1978).Asexual embryogenesis in callus cultures of palms. *In Vitro*, **15**: 383-387.

Sarwar, M. (1985). Callus formation from explanted organs of tea (*Camellia sinensis* L.). *Journal of Tea Science*, **54**: 18-22.

Sharma, V.K., Bhattacharya, A., Kumar A. and Sharma, H.K. (2007). Health benefits of tea consumption. *Tropical Journal of Pharmaceutical Research*, **6**: 785-792.

Snir, I. (1981). Micropropagation of red raspberry. *Scientia Horticulturae*, **14**: 139-143.

Tahardi, J.S. and Shu, W. (1992). Commercialization of clonal micropropagation of superior tea genotypes using tissue culture technology USAID/CDR Network meeting on tea crop biotech Costa Rica Takeda Y Differences in caffeine and tannin contents between tea cultivars and application to tea breeding. Japan *Agricultural Research Quarterly* **28**: 117.

van het Hof, K.H., Wiseman, S.A., Yang, C.S. and Tijburg, L.B.M. (1999). Plasma and lipoprotein levels of tea catechins following repeated tea consumption. *Proceedings of the Society of Experimental Biology and Medicine*, **220**: 203-209.

Villalobos-Amador, E., Rodríguez-Hernández, G. and Pérez-Molphe-Balch, E. (2002). Organogenesis and *Agrobacterium rhizogenes*-induced rooting in *Pinus maximartinezii* Rzedowsky and *P. pinceana* Gordon. *Plant Cell Reports*, **20**: 779-785.

Yang, C.S. and Landau, J.M. (2000). Effects of tea consumption on nutrition and health. *The Journal of Nutrition*, **130:** 2409-2412.

Yang, H., Zhou, C., Wu, F. and Cheng, J. (2010). Effect of nitric oxide on browning and lignification of peeled bamboo shoots. *Postharvest Biology and Technology*, **57**: 72-76.

Zehra, M., Banerjee, S., Mathur, A.K. and Kukreja, A.K. (1996). Induction of hairy roots in tea (*Camellia sinensis* L.) using *Agrobacterium rhizogenes*. *Current Science*, **70**: 84–86

Chapter 9

Cocoa

☆ *Anitha Karun, V. Aparna, K.S. Muralikrishna and M.K. Rajesh*

1. Introduction

Cocoa (*Theobroma cacao* L., 2n=20) is one of the important commercial plantation crops mainly grown for its seeds or beans, which are an important raw materials for chocolate and beverage industries (Cooper *et al.*, 2008). It belongs to the family Malvaceae, categorized under genus '*Theobroma*'. The Latin name '*Theobroma*' literally means 'Food of gods'. This crop has played an important role in many ancient South American cultures (Hurst *et al.*, 2002).

The genus '*Theobroma*' contains 22 species among which *Theobroma cacao* L. is widely cultivated. *Theobroma grandiflorum* L. is the other closely related species of cocoa, which is also the source for a variety of chocolate known as *cupulate* or *cupuacu* (Venturieri, 2011). The tropical plant is a native of Amazon region of South America (Bartley, 2005; Cheesman, 1944); later its cultivation spread to the countries in Asia and Africa (Bartley, 2005; Zhang and Motilal, 2016). The main growing areas of the crop are situated approximately within 20° North and South latitude of the equator. Cocoa needs a temperature of 21-32°C and well distributed rainfall of 100-250 cm for its optimal growth. It grows only below 1000 m of elevation, ideally below 300 m from the mean sea level. Even though majority of the species are found in its native place of South America, half of the world's supply of cocoa comes from the two East African countries Cote D'Ivoire and Ghana, which contribute to 42.4 per cent and 17.4 per cent of world's supply, respectively (International Cocoa Organization, 2015).

Criollo, Forastero, Trinitario and Nacional are the four major types of cocoa cultivated around the world (Clement, 2010). The Criollo type produces very high quality cocoa beans, cultivated mainly in its native land South and Central

America but yields are fairly low. Well known cultivars of Criollo type are Chuao, Porcelana, Puerto Cabello and Carupano (Pohlan and Perez, 2008). Forastero is the other important cultivar contributing to nearly 80 per cent of the world's cocoa production. It is cultivated mostly in Africa and some parts of Central and South America. Its growth is vigorous and yields are higher than other types (Pohlan and Perez, 2008). A number of Forestero varieties are cultivated in different parts of the world and some of them are Carenero Superior, Caracas Natural, Rio Caribe and Forastero Amenolado (http://www.cacaoweb.net/cacao-tree.html). Trinitario is a crossbreed between the Forastero and Criollo. It has inherited qualities of aroma from Criollo type and disease resistance, productivity from Forastero type. This hybrid type is mainly cultivated in Central America, South America and Asia (Motilal *et al.*, 2010). Nacional is cultivated in South America and western parts of the Andes. This type is mostly prone to pest and diseases, but it has an excellent aroma (Solorzano *et al.*, 2012).

Cocoa is highly heterozygous crop because of its self-compatible nature (Pound, 1932; Knight and Rogers, 1953, 1955; Cope, 1958, 1959, 1962). In recent years, genetic advancements have been made in cocoa through different breeding approaches. In order to meet the growing demands of the cocoa farmers for true-to type planting material, there is a need for an efficient propagation system which should be cost-effective and produce a large number of true-to-type elite plants. *In vitro* multiplication tools are better choice over conventional propagation methods to meet the above demands. Plant regeneration *via* somatic embryogenesis provides an alternative approach for clonal propagation of cocoa (Li *et al.*, 1998), since the plants are derived from the genetically identical cells of donor parents. Centre de Coopération Internationale en Recherche Agronomique pour le Développement (CIRAD), Montpellier, France and Pennsylvania State University, USA, have developed viable protocols for *in vitro* multiplication through somatic embryogenesis (SE) in cocoa.

Early attempts for *in vitro* regeneration of cocoa through SE started during 1970s (Esan, 1977, 1992; Pence *et al.*, 1979, 1980) with immature embryo as an explant material. Later, a number of successful studies have been reported in cocoa with different explants, media and plant hormone combinations (Esan, 1977; Pence *et al.*, 1979; Adu-Ampomah *et al.*, 1988; Santos and Machado, 1989; Tan and Furtek, 2002). Axillary bud culture and SE are two basic approaches for clonal multiplication of cocoa.

2. *In vitro* Propagation of Cocoa

2.1. Collection and Sterilization of Cocoa Explant Material

Sterilization of explant material is the primary step in cocoa micropropagation since the explant material is highly prone to various types of contamination. Various sterilization methods have been used in cocoa to prevent the contamination caused by fungi, bacteria, epiphytes and other micro-organisms (Hall and Collin, 1975; Duhem *et al.*, 1988). More than 90 per cent contamination was observed in cocoa explant material collected from the field regardless of different procedures used for surface sterilization. Reduced contamination was reported with sodium hypochlorite, calcium hypochlorite and its commercial formulations like Domestos (Passy and Jones, 1983; Esan, 1985a), Esan (1985b), Legrand and Mississo (1986) and Mallika *et al.* (1992) have suggested regular fungicidal sprays to the mother plants before collecting the explant material. Bavistin (0.2 per cent) and Dithane M-45 (0.3 per cent) sprays were recommended for the mother plants at an interval of three days before collecting explants for sterilization (Mallika *et al.*, 1992). Collection of explant material from the fungicide treated plants and surface sterilization with 70 per cent ethanol, 30 per cent calcium hypochlorite (15 minutes) followed by thorough washing with distilled water, help protect explants from contamination (Esan, 1985a). Surface sterilization of explants with 2 per cent orthodifolatan, followed by 15 minutes calcium hypochlorite rinsing and distilled water wash, was recommended by Legrand and Mississo (1986). Inoculation of surface sterilized material into a media containing the anti-microbial agents helped to minimise contamination chances in cocoa (Yidana *et al.*, 1987). However, Duhem *et al.* (1988) observed tissue necrosis with the addition of anti-microbial agents to the growing media.

Pods and buds can be sterilized by 0.2 per cent warm laundry detergent solution followed by 70 per cent alcohol (2-3 minutes), 10-20 per cent mild commercial bleach solution along with few drops of surfactant added to it and agitated thoroughly. In all sterilisation methods used, distilled water wash is essential at the end (Hall and Collin, 1975). In case of anthers, rinsing should be prolonged to last for 3-5 minutes (Esan, 1992). Pods can be sterilised by alcohol flaming or boiling saline water dip before excision of mature or immature embryos from the pods under aseptic condition (Esan, 1992). Excised embryos can be inoculated into a media without any further sterilization and 100 per cent contamination free cultures were obtained with this method (Esan, 1982). However, Novak *et al.* (1986) recommended further sterilization of excised embryos with NaOCl (5.25 per cent) along with surfactant for 40 minutes followed by sterile water rinsing. Embryos can also be treated with citric acid (50 mg/L) and ascorbic acid (40 mg/L) to check the explant browning as well as contamination (Novak *et al.*, 1986).

2.2. Inoculation and *in vitro* Multiplication of Explants

Starting from the pioneering work of Evans (1951)., there are many reports (Archibald, 1954; Esan, 1977; Townsley, 1974; Hall and Collin, 1975; Prior, 1977; Pence *et al.*, 1979; Novak *et al.*, 1986; Bhavyashree, 2009) of callus induction from almost all types of explant material, *viz.*, stem, bark, nodal cuttings, petiole, leaf, flower, shoot apex, embryos, anthers, floral parts, in various types of media and conditions (Figure 9.1). Some of the workers (Orchard *et al.*, 1979; Passey and Jones, 1983; Flynn *et al.*, 1990) have used shoot tip material of seedlings as an explant. Stem and nodal regions of *in vitro* raised seedlings have also been used as explant material by Esan (1985a), Legrand and Mississo (1986) and Figueira *et al.* (1990). Mallika *et al.* (1992) used explant material collected from budded plants of elite cocoa cultivars maintained under the controlled glass house conditions. Legrand and Mississo (1986) and Flynn *et al.* (1990) observed a positive correlation between the explant size and *in vitro* growth response. A single node stem cutting, with length ranging from 2-3 cm having maximum length towards the lower internode with subtending leaf above, is an ideal explant for cocoa tissue culture (Mallika *et

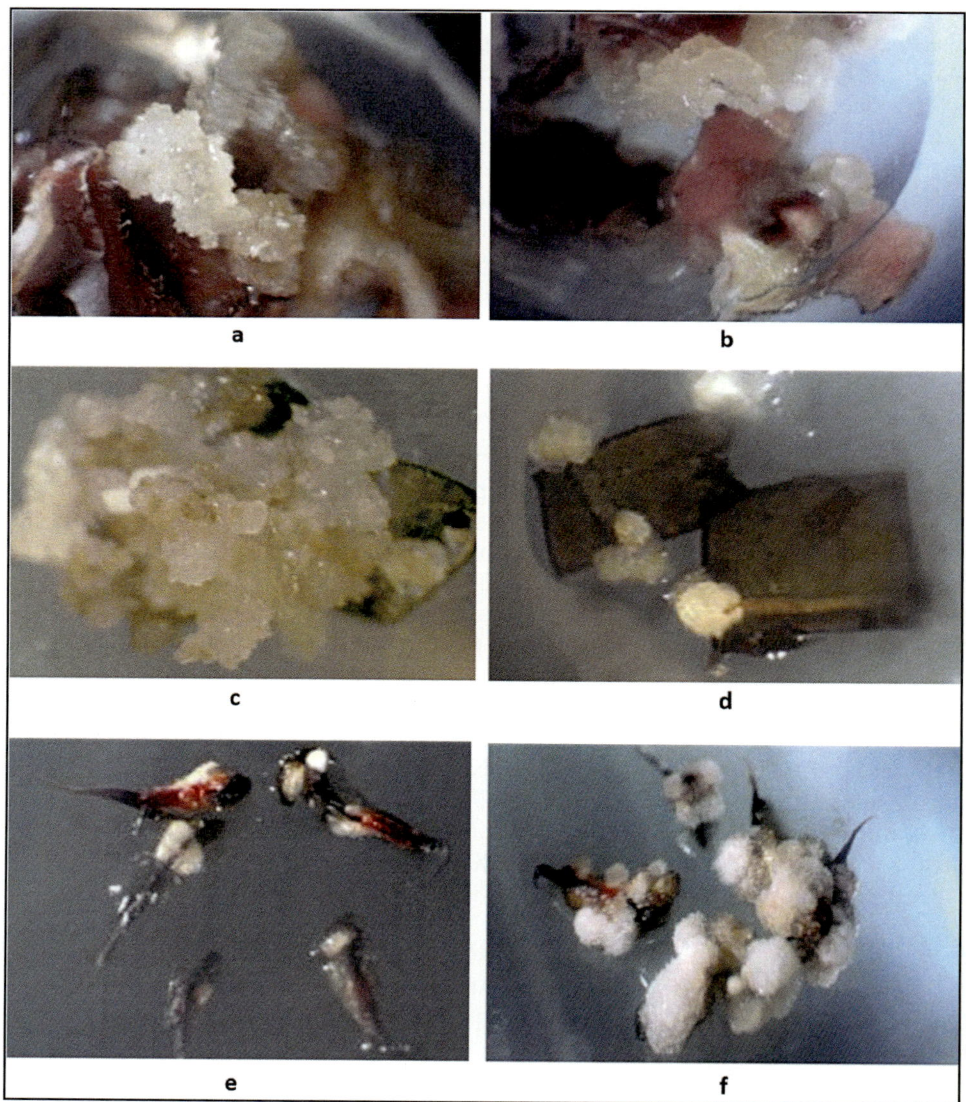

Figure 9.1: Callus Initiation from Different Explants of Cocoa. Callogenesis in cotyledonary explants (a, b), in tender leaves (c, d) and staminodal (e, f) explants.

al., 1992). Explant collection during leaf flushing stage may show some positive effects on *in vitro* grown cultures (Mallika *et al.*, 1992). Many researchers have observed improved bud initiation from pre-existing meristems. Maximova *et al.* (2003) also tried micropropagation of cocoa from somatic embryo derived plants Salzar *et al.* (2006) studied plant regeneration *via* SE technique. Different protocols were standardized in cocoa in order to use staminodes, petal, anther and immature cotyledon as explant materials. Tan *et al.* (1998) reported root and shoot formation from staminodes of young flower buds. Bhavyashree (2009) studied the callus initiation and SE from the leaf, nodal cuttings, cotyledon and staminodal explants of cocoa using MS and DKW basal media. Among different explants used, better embryogenic callus initiation was observed from the leaf explant and more browning was reported with DKW media compared to MS basal media. Reports suggest that callus induction from the cocoa is possible in almost all types of basal media with or without any plant growth regulator. But organogenesis from the callus has been found to be difficult in cocoa. Esan (1985a) observed callus initiation from plumular, radical and hypocotyl regions of cultured embryo axis. Among these three types of callus induced, plumular callus gave rise to shoots and roots.

2.3. Somatic Embryogenesis

Somatic embryogenesis offers scope for *in vitro* seedling production by direct embryogenesis (Esan, 1977; Pence *et al.*, 1979; Elhag *et al.*, 1987; Ad-Ampomah *et al.*, 1988; Santos and Machado, 1989) as well as indirect embryogenesis from embryogenic competent callus (Kononowicz and Kononowicz, 1984). There are many reports on direct embryogenesis from zygotic embryos (Kononowicz and Kononowicz., 1984; Kononowicz and Janick, 1984a, b; Wen *et al.*, 1984). But plantlets obtained from zygotic embryos were not true to type. When somatic embryos are induced from the vegetative tissues or other than zygotic tissue of non-sexual origin, this process allows the mass multiplication of elite and uniform plantlets.

A protocol was developed for indirect SE of cocoa by Tahardi and Mardiana (1995). Embryogenic calli was initiated from the immature flower buds, which was followed by development of somatic embryoids from nodular calli. After two subcultures in a hormone free medium, embryo maturation and germination were achieved. A detailed study was conducted for cocoa flower bud tissues with respect to its physiological age, floral explant types, genotypes, phytohormones and media compositions in order to improve SE (Tan and Furtek, 2002). Salzar *et al.* (2006) reported SE from the Venezuelan cocoa cultivars. Li *et al.* (1998) observed SE and plant regeneration from floral tissues of a number of cocoa cultivars. Effect of different carbon sources (glucose, sucrose, fructose, maltose and sorbitol) on cocoa SE was studied by Guiltinan *et al.* (2002). Among the five carbon sources tested, glucose, fructose and sucrose were found to show positive effect on somatic embryo production, while no somatic embryo production was observed in medium supplemented with maltose or sorbitol. Maximova *et al.* (2002) developed the secondary SE technique in cocoa using primary somatic embryos developed from the cotyledon explants. Primary somatic embryos collected from two different locations were primarily cultured on callus multiplication medium containing 2.4 μM 2,4-D along with 1.4 μM BA. Cultures were further transferred to hormone free embryo development medium for induction of secondary SE.

Scanning electron microscopic studies have been conducted to monitor different developmental stages of SE from immature embryo explant in cocoa (Santos and Machado, 1989). A typical cocoa embryo is characterized by an elongated axis with two well-developed cotyledons (Santos and Machado, 1989). MS basal medium along with auxin component is sufficient for induction of embryoids in cocoa.

Kononowicz and Kononowicz (1984) observed highest number of embryoids with MS medium, containing 2,4-D alone. Improved mitotic index of callus was observed with increased concentrations of 2,4-D (Kononowicz and Kononowicz, 1984). Callus transition towards embryogenesis was accompanied by increased DNA replication and RNA synthesis (Knonowicz and Janick, 1984b). Promotive effect of embryogenesis was observed with the addition of casein hydrolysate and coconut water to the media containing auxin component. Induced somatic embryos were cultured in a basal media containing cytokinin and auxins; this improved maturation and germination rate of somatic embryos. Absicic acid also aided maturation of embryoids (Alemano *et al.*, 1996). Proper germination was reported in embryoids of more than 4 mm size. Sub culturing process could be continued until one or more leaflets appeared from germinated embryos (Alemanno *et al.*, 1996). An efficient *in vitro* clonal propagation method for cocoa was developed by Guiltinan *et al.* (2002), who have also reported the occurrence of primary and secondary SE processes in cocoa.

2.4. Organogenesis

Woody Plant Medium (WPM) was superior over half-MS medium for long term maintenance of cultures without tissue necrosis and abscission (Flynn *et al.*, 1990; Mallika *et al.*, 1992). MS liquid medium was found to be beneficial for shoot growth and elongation (Adu-Ampomah *et al.*, 1987). Addition of ethylene inhibitors (5 ppm $AgNO_3$ or 0.5 mg/L $CoCl_2$) was suggested to overcome the problems of excess callus production at the cut ends of explants, as well as for subsequent improvement in shoot regeneration (Mallika *et al.*, 1992). Addition of additional amino acids and anti-oxidants to basal media helped sustained growth of *in vitro* regenerated shoots in cocoa (Flynn *et al.*, 1990). Addition of vitamins and casein hydrolysate to the basal media improved the proliferation of axillary shoots from cotyledonary node explant. Adu-Ampomah *et al.* (1987) observed plantlet regeneration from shoot tips of *in vitro* grown seedlings. Addition of cytokinin to the media helped axillary shoot induction, shoot elongation and leaf formation from the cotyledon as well as root explants (Figu *et al.*, 1990, 1991). But prolonged treatment with high levels of cytokinins (2-iP) led to shoot tip necrosis instead of its sustained growth (Figueira *et al.*, 1990, 1991).

Increased levels of illumination and CO_2 promoted *in vitro* shoot development in cocoa (Figueira *et al.*, 1991; Figueira and Janick, 1990). Shoot elongation and production of normal leaves from the detached axillary shoots were observed when the explants were exposed to higher levels of CO_2 (20,000 ppm) and 150-200 μM $s^{-1}m^{-2}$ of PPFD (Photosynthetic Photon Flux Density) (Figueira *et al.*, 1991). Shoot elongation as well as leaf development was enhanced when concentration of CO_2 was increased from the ambient to 24,000 ppm (Figueira and Janick, 1990). The same developmental changes were not true under dark, implying that photosynthesis stimulation is an important deciding factor for the performance of *in vitro* cultures. Maintenance of higher CO_2 may also act as an ethylene inhibitor, in addition to its role in stomatal regulation and cellular pH maintenance. Removal of cotyledons from embryos also improved plantlet recovery under dark conditions (Mallika *et al.*, 1992).

Shoots are normally kept for one month in a rooting medium containing auxin. *In vitro* rooting was obtained in cocoa using plant hormones *viz.*, IBA, IAA and phloroglucinol (Passey and Jones, 1983). Sometimes, the roots formed from *in vitro* multiplied shoos were thick, stubby and unbranched resembling tubers and the plantlets failed to survive in hardening stage (Mallika *et al.*, 1992). Anatomical studies revealed lack of vascular continuity between shoot and roots at the collar region. Rooting was normally induced in shoots developed from different explant material by pulsing treatment with various concentrations of IBA. Thin and healthy roots were observed on shoots when auxin treated material was further transferred to hormone free charcoal medium. The rooted plants could be field planted after gradual hardening process (Mallika *et al.*, 1992).

2.5. Induction of Multiple Shoots

Adu-Ampomah *et al.* (1988) developed a protocol for the production of cocoa plantlets using shoot tip culture. Multiple shoot production from the axillary bud culture will be an ideal technique for obtaining large number of plantlets from a single explant. When nodal explants were grown on the medium containing thidiazuron, axillary buds were proliferated and shoot development progressed (Figueira *et al.*, 1990, 1991). Mallika *et al.* (1996) induced the multiple shoots from nodal segments on WPM media with additional supplements.

2.6. Problems Associated with Cocoa Micropropagation

Plantlet production is possible through *in vitro* multiplication in various genotypes of cocoa. Still cocoa is considered as a recalcitrant species because of certain factors, which hinder multiplication rate *in vitro* and limit the elite planting material production through tissue culture methods. Presence of number of phenolic compounds and their oxidation hinders the callus multiplication and SE in cocoa. Naturally, cocoa contains large amounts of polyphenolics and their oxidation could be one of the limiting factors preventing proper tissue multiplication and maintenance (Griffiths, 1958; Kim and Keeney, 1983). In non-embryogenic calli, production of phenolic compounds has been found to be higher than embryogenic calli (Alemonno *et al.*, 1996). Profuse callusing at the cut ends of explant material after bud break is found to arrest shoot growth. Heavy callusing from the cocoa explant material may be due to the presence of higher concentrations of endogenous plant hormones *i.e.* auxin and cytokinins (Mallika, 1992). Callus appearance has been cited as another interfering factor for the success of cocoa *in vitro* cultures (Passey and Jones, 1983; Dublin, 1984; Legrand and Mississo, 1986; Mallika *et al.*, 1992) Khalid *et al.* (1991) observed the inhibitory action of ethylene on tissue organogenesis from callus. Browning, excessive production of slimy exudates, callus over growth, poor organogenesis from the callus, slow multiplication rate and bud dormancy are some of the reported problems during the *in vitro* multiplication of cocoa (Hall and Collin, 1975; Legrand and Mississo, 1986).

2.7. Hardening, Acclimatization and Field Transplanting of Rooted Plantlets

Hardening and transplanting procedures for cocoa for tissue cultured plantlets have been standardized by many workers. After root initiation, plantlets were transferred to a potting mixture and protected from desiccation by covering with polythene bags. Air circulation to the plants could be maintained by making small holes on the polythene bag. After 3-4 months of acclimatization, the plants could be successfully transferred to the bigger pots with standard potting mixture (Mallika *et al.,* 1996). Growth rate was initially slower, which improved gradually. Growth behaviour and morphological appearance of plantlets were comparable after six months of initial establishment. Tissue cultured plantlets in the field was found to be more vigorous than *ex vitro* derived plants (Mallika *et al.,* 1996). Flowering and fruiting of the tissue culture derived plants were on par with the budded or grafted plants. Pollen fertility and viability was reported to be more in budded and seedling derived plants compared to micro propagated ones. However, this did not influence the cocoa pod yield. Variations were observed in case of pod morphology, bean number and bean weight (Mallika *et al.,* 1996).

2.8. Anther Culture

Esan (1977) and Prior (1977) made the earliest attempts for standardizing cocoa anther culture technique. Esan (1982) reported the production of haploid and diploid types of callus from cocoa anther cultures. Prior (1977) observed induction of callus from anther somatic tissues and also demonstrated its use as a nurse tissue for the growth and reproduction of a cocoa fungal pathogen *Oncobasidium theobromae*. To date, no androgenic cocoa plants production has been reported under *in vitro* conditions, although many of the haploid plants have been found in nature (Esan, 1992). Callus induction and root development was observed after 60 and 90-120 days of anthers inoculation, on MS and double strength white medium supplemented with 20 gl^{-1} sucrose,100 mgl^{-1} inositol, 0.2 mgl^{-1} kinetin and 5 $mg1^{-1}$ NAA. With the addition of casein hydrolysate or coconut milk extract, callus development was improved, but rooting was either delayed or even prevented (Esan, 1992). Mallika *et al.* (1992) also could induce callus from anthers; however they could not obtain plant regeneration from anther callus. MS medium supplemented with 2 mg 1^{-1} 2-iP, 0.1 mg 1^{-1} NAA and 126 mg 1^{-1} phloroglucinol was found to be most favourable combination for root development in anther callus induced proembryoids (Sunil, 1992).

A detailed study on cocoa anther culture and influence of different factors for successful plantlet regeneration was conducted by Sunil (1992), who reported that the development stage of anther, minimum temperature in the field of donor plants, type and strength of basal medium, type and concentration of auxins and cytokinins, carbohydrate source, sucrose level, physical environment and gamma and UV rays were main factors that influence anther callus induction (Sunil, 1992). Anther callus multiplication was influenced by type of basal medium, type and concentration of growth regulators, carbohydrate source, sucrose level, presence of amino acids, organic supplements, ethylene releasing and inhibiting chemicals, adenine and its derivatives, unusual regulants, gibberellins and growth inhibitors, light and gamma rays. The factors influencing indirect embryogenesis were found to be stage of anther development and light. Hybrid genotypes responded more favourably to callus induction, callus multiplication, callus rhizogenesis than traditional cultivars and they were also found to be stable with respect to seasonal changes (Sunil, 1992).

Sunil (1992) first reported plantlet recovery from cocoa anthers via indirect embryogenesis. Anthers at tetrad stage were subjected to a two-stage culture procedure involving incubation and subculture (after 4 weeks) on modified H3 basal medium supplemented with 1 mgl^{-1} NAA and 0.1 mgl^{-1} 2-iP for 50 days (stage 1) and subsequent weekly transfer to ½ MS basal media supplemented with 1mgl^{-1} 2-iP and 3 mgl^{-1} GA_3 (stage II). Proembryoids were obtained in stage 1 medium *via* callus within 45 days of culture, only when anthers derived from Criollo, Trinitario and hybrid (H2) were used. Serial subculturing of embryoids in stage II media led to formation of shootlets and rootlets. With four sub-cultures in stage II in a span of one month, the embryoids could be germinated into plantlets of size 2.5 cm with two leaflets and one rootlet (Sunil, 1992).

2.9. Protoplast Isolation and Culture

Thompson *et al.* (1987) used protoplast isolation and fusion techniques for cocoa regeneration. When excised protoplasts from the tissues were grown in dark, cell wall regeneration and embryoids formation was observed further. Studies were also conducted on factors influencing the protoplast isolation from young cocoa leaves at the early F2 stage. Chantrapradist (1999) isolated protoplasts from rapidly growing cocoa cell suspensions using 2 per cent (w/v) driselase, 0.5 M sorbitol, 1 mM MES (2-N-morpholino ethanesulfonic acid) and 10 mM $CaCl_2$ $2H_2O$ at pH 5. After 3 hours of cell incubation with the enzyme solution, protoplasts were separated with an average yield of 4.5×10^5 per gram fresh weight. Best results were obtained when protoplasts were grown in MS basal medium supplemented with 2.3 µM 2, 4-D under dark conditions. Cell wall formation and cell division were observed after 8-10 days of culture. Continuous cell division leads to a formation of cell colonies and small microcalli within four weeks.

2.10. Embryo Rescue

Embryo culture helps to overcome the problems of seed abortion and improves the germination ability of difficult to germinate types. Poor germination of cocoa seeds may be due to toxic or lethal factors of cotyledonary tissue (Ibanez, 1964). Despite high heterozygosity in cocoa, disease resistant genotypes among the cultivated types are limited in cocoa. Distant hybridization techniques are normally used for improving the disease resistance, agronomical traits or other qualitative traits of beans in cocoa (Kononowicz and Janick, 1984c). An *in vitro* embryo culture technique helps to maintain the embryos resulting from distant crosses. Kononowicz and Janik (1984b) successfully excised 100 days old embryo, which could be maintained *in vitro* up to its maturity using embryo culture technique. Palma and Villalobos (1989) successfully obtained 16-87 per cent of plantlets along with 2 per cent of haploid types from two different crosses by embryo rescue technique.

2.11. Suspension Cultures for Somatic Embryogenesis and Secondary Metabolite Production in Cocoa

Hall and Collin (1975) first initiated cocoa suspension cultures from seedling tissues. Jalal and Collin (1978).and Tsai and Kinsella (1981) formulated a synthetic media for maintaining cocoa callus and suspension cultures. MS basal media supplemented with 0.5 mgl^{-1} 2, 4-D and 0.1 mgl^{-1} kinetin was found to be the most successful medium for growing cocoa suspension (Tsai and Kinsella, 1981). Wen *et al.* (1984) established cell suspension using cocoa bean callus tissues and investigated lipid composition in liquid cell suspension in comparison with cocoa beans at different maturities. Proliferations of asexual embryos were observed from immature zygotic embryos when they are cultured on semi-solid or liquid medium (Wen *et al.*, 1984). Niemenak *et al.* (2008) used temporary immersion bioreactor system (TIS) for multiplication of cocoa somatic embryos; better results were obtained with TIS system compared to solid media. TIS also improved the formation of somatic embryos as well as their development regarding their conversion to torpedo shaped forms. High speed multiplication was obtained in cocoa cell suspension upon enzymatic treatment of explant material in liquid suspensions. (Rojas *et al.*, 2012).

Cocoa suspension cultures have been mainly initiated in order to synthesize flavour (purine alkaloids, theobromine, caffeine and theophylline) and butter components (triglycerides, cis-palmito-oleostearin) of cocoa. Townsley (1974) produced a chocolate aromatic product from mature suspension cell of cocoa. But, suspension cultures were found to have low polyphenolic content in comparison with callus and explant tissue (Jalal and Collin, 1997). No purine alkaloids were identified in suspension cultures of cocoa (Jalal and Collin, 1979). When purine precursors (7-methylxanthosine and methionine) were artificially added to callus, theobromine was synthesized; it shows the activity of a part of the purine biosynthetic pathway (Jalal and Collin, 1979). Later, Gurney *et al.* (1992) tested and observed low amount of purine alkaloids production and accumulation in suspensions in comparison with the callus cultures. Leathers and Scragg (1989) observed the effect of different temperatures on suspension growth, lipid content and fatty acid composition of cocoa cell suspension cultures. The optimal temperature for growth of cell suspension was found to be 30°C. Lipid and fatty acid biosynthesis were maximal at temperatures ranging from 15-20°C. The lipid composition of cocoa butter is different from the lipids which were identified in cocoa suspensions (Tsai and Kinsella, 1982). During development of the embryo, fatty acid composition becomes more saturated as cocoa butter is synthesized. An increased sucrose concentration to a medium containing somatic embryos induces synthesis of triglycerides, which constitute cocoa butter (Pence *et al.*, 1981).

2.12. Gene Expression Studies Related to Somatic Embryogenesis

Somatic embryogenesis is an efficient propagation system for rapid and mass multiplication of cocoa. One of the major bottleneck in cocoa SE is that the efficiency of somatic embryo production is highly genotypic dependent. Further, only a lower percentage of plantlet regeneration is achieved from somatic embryos due to improper cotyledon development in cocoa. Studies were conducted to understand the SE process in cocoa by examining the genes related to the process of. Maximova *et al.* (2014) studied the gene expression patterns during differential developmental stages of cocoa somatic and zygotic embryogenesis. The expression of 28,752 genes was determined at four developmental time points during zygotic embryos and two time points during cocoa somatic embryogenesis. During zygotic embryogenesis, 10,288 differentially expressed genes were enriched for functions related to responses to abiotic and biotic stimulus, metabolic and cellular processes. In total, 10,175 genes were differentially expressed in zygotic and SE. Many TF genes, related to ethylene metabolism and response, were more strongly expressed in somatic embryogenesis as compared to zygotic embryogenesis. Genes related to fatty acid metabolism, flavonoid biosynthesis and seed storage functions were also found to be differentially expressed between two stages. The insights gained from the differential gene expression patterns might enable designing of more efficient protocols for cocoa SE.

An orthologue of the *Arabidopsis* Leafy Cotyledon-2 gene (*AtLEC2*) was characterized in *Theobroma cocoa* (*TcLEC2*) and its expression studies were conducted in cocoa cultures (Zhang *et al.*, 2014). The expression pattern of *TcLEC2* was reported to be higher in embryogenic than non-embryogenic calli. Transient overexpression of *TcLEC2* in immature zygotic embryos of cocoa causes a change in the gene expression profiles and fatty acid composition. The overexpression of *TcLEC2* in cocoa explants was found to improve the frequency of regeneration of stably transformed somatic embryos. Another important gene BABY BOOM (*BBM*) was characterized (*TcBBM*) and tested in cocoa (Florez *et al.*, 2015). *TcBBM* gene expression was observed in entire embryo development stages. Expression level of *TcBBM* was reported to be high in SE compared to zygotic embryogenesis. *TcBBM* over-expression alone in cocoa led to the formation of embryogenic structures without addition of any exogenous plant growth regulators. Only moderate enhancements in embryogenic potential was observed with transient ectopic expression of *TcBBM*. Constitutive overexpression of this gene greatly increased SE proliferation but also appeared to inhibit subsequent development of embryo development and regeneration. *TcBBM* and *TcLEC2* could potentially be used as a biomarker for the improvement of the SE process and screen for elite varieties in cacao germplasm (Zhang *et al.*, 2014; Florez *et al.*, 2015).

2.13. Micrografting

In this method, *in vitro* raised shoots are grafted on to root stock of *in vitro* or *ex vitro* origin. This technique helps to save time as well as resources in the micropropagation of cocoa. Aguilar *et al.* (1992) attempted a micrografting technique in cocoa using somatic embryos and young cocoa seedlings as a scion and rootstock material. Best results were observed on simple culture medium with three months old rootstock and somatic embryos without cotyledons. Nearly 10 month time period was required for complete plant regeration after a successful graft attempt was done. Bindu (1997) tried micrografting using *in vitro* raised shoots from nodal segments as a scion material. Axenic seedlings cultured on half MS liquid medium devoid of sucrose was found to be best as rootstock when *in vitro* raised shoots used as a scion material. Side grafting was the most ideal for micrografting procedure among the different grafting techniques available. Success was highest when scions with two or more hardened leaves were grafted 4 cm below the cotyledons in 4-5 weeks old axenic seedlings with a few hardened leaves. Anatomical studies revealed that the graft union was complete in about a month. Successful grafts could be obtained when scion material had one or two hardened leaves. Grafting on *ex vitro* root stock material was more successful and exhibited the rapid and extensive elongation of shoots. Field performance of micro grafted plants was also found to be satisfactory (Bindu and Mallika, 2008).

3. Future Prospects

Plantlets have been raised from the cocoa vegetative or floral plant parts collected from the field, seeds, mature and immature zygotic embryos *etc.* Variations in plant genotype, physiological maturity of explants and seasonal variations are some of the problems hindering the development of a viable protocol. Though SE appears feasible, protocol for recovery of plantlet has still not been perfectly standardized. A 'perfect' protocol for the clonal multiplication and cryopreservation techniques need to be standardized for the multiplication as well as maintenance of elite cocoa gemrplasm. Since cocoa whole genome sequencing has already been completed, the data can be used for selection and modification of different genes which are involved in SE process through different transformation approaches. The prospects for making use of the various biotechnological approaches in cocoa hybridization programme also seem to be very bright.

References

Adu-Ampomah, Y., Novak, F.J., Afza, R. and Van Durren, M. (1987). Embryoid and plant production from cultured cocoa explant. *Proceedings of Tenth International Cocoa Research Conference,* Santo Domingo. pp.129-136.

Adu-Ampomah, Y., Novak, F.J., Afza, R., Van Durren, M. and Perea- Dollos, M. (1988). Initiation and growth of somatic embryos of cocoa (*Theobroma cocoa* L.). *Café Cacao The'*, **32:** 187-200.

Aguilar, M.E., Villalobos, V.M. and Vasquez, N. (1992). Production of cocoa plants (*Theobroma cacao* L.) via micrografting of somatic embryos. In *Vitro Cellular and Developmental Biology- Plant*, **28**: 15-19.

Alemanno, L., Berthouly, M. and Ferrière, M.N. (1996). Histology of somatic embryogenesis from floral tissues in *Theobroma cacao* L. *Plant Cell Tissue and Organ Culture*, **46**: 187-194.

Archibald, J.F. (1954). Culture *in vitro* cambial tissues of cocoa. *Nature*, **173**: 351-352.

Bartley, B.G.D. (2005). The genetic diversity of cacao and its utilization. In: Lockwood, R. (Ed.), *Experimental Agriculture, Volume.42, Issue.1*, CABI Publishing, Wallingford, U.K. p. 341.

Bhavyashree, U. (2009). *In vitro* multiplication of cocoa (*Theobroma cacao* L.). M.Sc. Thesis, Department of Biosciences, Mangalore University, Karnataka, India.

Bindu, M.R. and Mallika, V.K. (2008). Micrografting in cocoa (*Theobroma cacao* L.). *International Journal of Agriculture Sciences*, **4**: 601-603.

Bindu, M.R. (1997). Standardization of *in vitro* techniques for rooting, hardening and micrografting in cocoa (*Theobroma cacao* L.). Ph.D thesis, Faculty of Agriculture, Kerala Agricultural University, Kerala, India.

Chantrapradist, C. and Kanchanapoom, K. (1999). Cell suspension, isolation and culture of cacao (*Theobroma cocoa* L.) protoplasts. *Songklanakarin Journal of Science and Technology*, **21**: 1-8.

Cheesman, E. (1944). Notes on the nomenclature, classiûcation and possible relationships of cocoa populations. *Tropical Agriculture*, **21**: 145-146.

Clement, C.R., Cristo-Arau ́jo, M., d'Eeckenbrugge, G.C., Pereira, A.A., and Rodrigues, D.P. (2010). *Origin and domestication of native Amazonian crops. Diversity*, **2**: 72-106.

Cooper, K.A., Donovan, J.L., Waterhouse, A.L and Williamson, G. (2008). Cocoa and health: A decade of research. *British Journal of Nutrition*, **99**: 1-11.

Cope, F.W. (1962). Mechanism of self incompatibility in *Theobroma cocoa* L. *Heredity*, **17**: 157-182.

Cope, F.W. (1958). Incompatibility in *Theobroma cocoa* L. *Nature*, **181**: 279.

Cope, F.W. (1959). Nuclear fusion and non-fusion in *Theobroma cocoa* L. *Nature*, **183**: 1540.

Dublin, P. (1984). Techniques de reproduction végétative in vitro etamélioration génétique chezles caféiers cultivés. *Café Cacao Thé*, **28**: 231-244.

Duhem, K., Le Mercier, N. and Boxus, P. (1988). Difficulties in the establishment of axenic *in vitro* cultures of field collected coffee and cacao germplasm. *Acta Horticulture*, **225**: 67-76.

Elhag, H.M., Whipkey, A. and Janik, J. (1987). Induction of somatic embryogenesis from callus in cocoa (*Theobroma cacao* L.) in response to carbon source and concentration. *Revista Theobroma*, **17**: 153-162.

Esan, E.B. (1977). Tissue culture studies on cacao (*Theobroma cacao* L.): A supplementation of current research. *Proceedings of the 5th International Cacao Research Conference*, Ibadan, Nigeria. pp. 153-162.

Esan, E.B. (1982). Shoot regeneration from a callus derived from embryo axes cultures of *Theobroma cacao* L. *in vitro*. *Turrialba*, **32**: 359-364.

Esan, E.B. (1985a). Axillary bud transformation into asexual embryos from seedling node cultures of cocoa. *Nigeria Agricultural Journal*, **20**: 13-22.

Esan, E.B. (1985b). Clonal multiplication of *Theobroma cocoa* L. zygotic embryos *in vitro* asexual embryogenesis. *Nigeria Agricultural Journal*, **20**: 23-38.

Esan, E.B. (1992). Micropropagation of Cocoa (*Theobroma cacao* L.). In: Y.P.S. Bajaj (Ed.), *Biotechnology in Agriculture and Forestry, Vol. 18, High-Tech and Micropropagation II*. Springer-Verlag, Berlin Heidelberg. pp. 96-122.

Evans, H. (1951). Investigations on propagation of cocoa. *Tropical Agriculture*, **28**: 147-203.

Figueira, A., Whipkey, A. and Janick, J. (1991). Increased CO_2 and light promote *in vitro* shoot growth and development of *Theobroma cacao* L. *Journal of the American Society of Horticultural Science*, **116**: 585-589.

Figueira, A., Whipkey, A., and Janick, J. (1990). Micropropagation of *Theobroma cocoa L. a recalcitrant species in high CO_2. Horticultural Science 87th Annual Meet, ASHS*, Tueson, Arizona. pp. 608.

Florez, S.L., Erwin R.L., Maximova S.N., Guiltinan M.J. and Curtis W. (2015). Enhanced somatic embryogenesis in *Theobroma cacao* L. using the homologous BABY BOOM transcription factor. *BMC Plant Biology*, **15:** 2-12.

Flynn, W.P., Glicenstein, L.J. and Fritz, P.J. (1990). *Theobroma cacao* L.: An axillary bud *in vitro* propagation procedure. *Plant Cell, Tissue and Organ Culture*, **20**: 111-117.

Griffiths, L.A. (1958). Phenolic acids and flavonoids of *Theobroma cacao* L.: Separation and identification by paper chromatography. *Biochemical Journal*, **70**: 120-125.

Guiltinan, M., Maximova, S.N., Traore, A., Miller, C., Robinson, M., Antunez de Mayolo, G., Barreto, E., Young, A. and Pishak, S. (2002). Integrated systems for the propagation of cacao: Tissue culture, bentwood gardens and rooted cuttings. In: *Proceedings of the 13th International Cocoa Research Conference, Workshop on the New Technologies and Cocoa Breeding*, Kota Kinabalu, Sabah, Malaysia. Cocoa Producers Alliance, Lagos, Nigeria. pp. 323-330.

Gurney, K.A., Evans, L.V. and Robinson, D.S. (1992). Purine alkaloid production and accumulation in cocoa callus and suspension cultures. *Journal of Experimental Botany*, **43**: 769-775.

Hall, I.R.H. and Collin, H.A. (1975). Initiation and growth of tissue cultures of *Theobroma cacao* L. *Annals of Botany*, **39**: 555-570.

Hurst, W.J., Tarka, S.M., Powis, T.G., Valdez, F. and Hester, T.R. (2002). Cacao usage by the earliest Maya civilization. *Nature*, **418**: 289-90.

Ibanez, M.L. (1964). Role of the cotyledon in sensitivity to cold of cacao seed. *Nature*, **201:** 414-415.

International Cocoa Organization (2015/16). *Quarterly Bulletin of Cocoa Statistics*, Vol. XLII, No. 1, Cocoa.

Jalal, M.A.F and Collin, H.A. (1978). Secondary metabolism in tissue culture cultures of *Theobroma cacao* L. *New Phytologist*, **83**: 343-349.

Khalid, M., Charaibi, B., Latche, A., Roustan, J.P. and Fallot, J. (1991). Stimulation of shoot regeneration from cotyledons of *Helianthus annus* by ethylene inhibitors, silver nitrate and cobalt. *Plant Cell Reports*, **10**: 204-207.

Kim, H. and Keeney, P. (1983). *Polyphenols-tannins in Cocoa Beans*. 37th PMCA Production Conference, Lancaster, PA. pp. 60-63.

Knight, R. and Rogers, H.H. (1953). Sterility in *Theobroma* cocoa L. Nature, **172**:164.

Knight, R. and Rogers, H.H. (1955). Incompatibility in *Theobroma* cocoa *L. Heredity*, **9**: 69-77

Kononowicz, A.K. and Janick, J. (1984b). The influence of carbon source on the growth and development of somatic embryos of *Theobroma cocoa* L. *Physiologia Plantarum,* **61**: 155-162.

Kononowicz, A.K and Janick, J. (1984). *In vitro* development of zygotic embryos of *Theobroma cocoa* L. *Journal of American Society of Horticultural Sciences,* **109**: 266-269.

Kononowiczh, J. and Janick, J. (1984). Response of embryogenic callus of *Theobroma cacao* L. to gibberellic acid and inhibitors of gibberellic acid synthesis. *Journal of Plant Physiology,* **113**: 359- 366.

Kononowicz, H. and Kononowicz, A.K. (1984). Asexual embryogenesis via callus of *Theobroma cacao* L. *Journal of Plant Physiology,* **113**: 347-358.

Leathers, R.R. and Scragg, A.H. (1989). The effect of different temperatures on the growth, lipid content and fatty acid composition of *Theobroma cacao* L. cell suspension cultures. *Plant Science,* **2**: 217-222.

Legrand, B. and Mississo, E. (1986). Influence de la taille des explantats et des régulateurs de croissance sur le développement des tissus de *Theobroma cacao* L., var. Amelonado cultivés *in vitro*. Café Cacao The', **30**: 239-246.

Li, Z., Traore, A., Maximova, S. and Guiltinan, M.J. (1998). Somatic embryogenesis and plant regeneration from floral explants of cacao (*Theobroma cacao* L.) using thidiazuron. In *Vitro Cellular and Developmental Biology-Plant,* **34**: 293-299.

Mallika, V.K., Mathew, J.P., Chacko, D.,Vijaykumar, N.K. and Nair, R.V. (1996). Induction of multiple shoots in nodal explants of fully grown trees of cocoa (*Theobroma cocoa* L.). *Journal of Plantation Crops,* **24**: 503-510.

Mallika, V.K., Shankar, A., Sindhu, L., Rekha, M.C., Vijaykumar, N.K. and Nair R.V. (1992). Plantlet regeneration *in vitro* from nodal segments of cocoa (*Theobroma cocoa* L.) and field planting. *Journal of Plantation Crops,* **20**: 114-122.

Maximova, S.N., Alemanno, L., Young, A., Feffiere, N., Traore, A. and Guiltinan, M.J. (2002). Efficiency, genotypic variability, and cellular origin of primary and secondary somatic embryogenesis of *Theobroma cacao* L. In *Vitro Cellular and Developmental Biology-Plant,* **38**: 252-259.

Maximova, S.N., Florez, S., Shen, X., Niemenak, N., Zhang, Y., Curtis, W. and Guiltinan, M.J. (2014). Genome wide analysis reveals divergent patterns of gene expression during zygotic and somatic embryo maturation of *Theobroma cacao* L., the chocolate tree. *BMC Plant Biology,* **14**: 1-17.

Maximova, S.N., Miller, C.R., Antúnez de Mayolo, G., Pishak, S., Young, A. and Guiltinan, M.J. (2003). Stable transformation of *Theobroma cacao* L. and influence of matrix attachment regions on GFP expression. *Plant Cell Reports,* **21**: 872-883.

Motilal, L., Zhang, D., Umaharan, P., Mischke, S., Mooleedhar, V. and Meinhardt, L.W. (2010). The relic Criollo cacao in Belize-genetic diversity and relationship with Trinitario and other cacao clones held in the International Cocoa Genebank, Trinidad. *Plant Genetic Resources: Characterization and Utilization,* **8**: 106-115.

Niemenak, N., Saare-Surminski, K., Rohsius, C., Ndoumou, D.O. and Lieberei, R. (2008). Regeneration of somatic embryos in *Theobroma cacao* L. in temporary immersion bioreactor and analyses of free amino acids in different tissues. *Plant Cell Reports*, **27**: 667–676.

Novak, F.J., Donini, B. and Owusu, G. (1986). Somatic embryogenesis and *in vitro* plant development of cocoa (*Theobroma cacao* L.). *Proceedings on International symposium on nuclear techniques and in vitro culture for plant improvement. IAEA*, Vienna, Austria. pp. 443-449.

Orchard, J.E., Collin, H.A. and Hardwick, K. (1978).Culture of shoot apices of *Theobroma cacao* L. *Physiologia Plantarum*, **47**: 207-210.

Palma,T. and Villalobos, V.M. (1989). Rescate *in vitro* de embriones provenientes de semillas aplanadas de cacao (*Theobroma cacao* L.). *Turrialba*, **39**: 525-529.

Passey, A.J. and Jones, O.P. (1983). Shoot proliferation and rooting *in vitro* of *Theobroma cacao* L. type Amelonado. *The Journal of Horticultural Science and Biotechnology*, **58**: 589-592.

Pence, V.C., Hasegawa, P.M. and Janick, J. (1978).Asexual embryogenesis in *Theobroma cacao* L. *Journal of American Society of Horticultural Sciences*,**104**: 145-148.

Pence, V.C., Hasegawa, P.M. and Janick, J. (1980). Initiation and development of asexual embryos of *Theobroma cacao* L. *in vitro*. *Journal of Plant Physiology*, **98**: 1-14.

Pence, V.C., Hasegawa, P.M. and Janick, J. (1981). Sucrose-mediated regulation of fatty acid composition in asexual embryos of *Theobroma cacao* L. *Physiologia Plantarum*, **3**: 378-384.

Pohlan, H.A.J. and Perez, V.D. (2008). Soil Plant Growth and Crop Production. Vol-III, Growth and Production of Cacao. *Encyclopaedia of Life Supporting Systems*. pp. 1-32.

Pound, F.J. (1932). Studies on Fruitfulness in Cocoa -II, *First Annual Report on Cocoa Research Trinidad*, pp. 26-28.

Prior, C. (1977). Growth of *Oncobasdium theobroma* Talbot and Keane in dual culture with the callus tissues of *Theobroma cocoa* L. *Journal of General Microbiology*, **99**: 219-222.

Rojas, L.F., Florez, C., Atehortua, L., Gil, A. and Londono, J. (2012). Application of a new method to obtain cell biomass of *Theobroma cacao* L. seeds for metabolites production in flasks and Bioreactor. *Acta Horticulture*, **961**: 435-440.

Salzar, V.R., Betancourt, C., Mata, J and Garcia, F. (2006). Somatic embryogenesis in Venezulan cocoa cultivars. *Agronomia-Tropical-Maracay*, **56**: 61-74.

Santos, A.V.P and Machado, R.D. (1989). A scanning electron microscopy study of *Theobroma cacao* L. somatic embryogenesis. *Annals of Botany*, **64**: 293-296.

Solorzano, R.G.L., Fouet, O., Lemainque, A., Pavek, S., Boccara, M., Argout, X., Amore, F., Courtois, B., Risterucci, A.M. and Lanaud, C. (2012). Insight into the wild origin, migration and domestication history of the fine flavour Nacional *Theobroma cocoa* L. variety from Ecuador. *PLoS ONE*, **7**: 1-11.

Sunil, K.P. (1992). Anther culture in cocoa (*Theobroma cacao* L.). M.Sc. Thesis. Vellanikkara, *Department of Agricultural Botany, College of Horticulture*, Kerala, India.

Tahardi, J.S. and Mardiana, N. (1995). Cocoa regeneration via somatic embryogenesis. Menara Perkebunan, **52**: 174- 178.

Tan, C.L. and Furtek, D.B. (2002). Development of *in vitro* regeneration system for *Theobroma cacao* L. from mature tissues. *Plant Science*, **164**: 407-412.

Tan, C.L., Davey, M.R., Mulligan, B.J., Power, J.B., Lowe, K.C., Jaafar, H., Yeang, Y.H. and Lee, M.T. (1998). Somatic embryogenesis in cocoa as a basis for genetic transformation. In: Davey, M.R., Alden, P.G., Lowe, R.C. and Power, J.B. (Eds.), *Tree Biotechnology- Towards the Millennium*. Nottingham University Press, Nottingham, UK. pp.149-154.

Thompson, W., Collins, H.A., Issac, S. and Hardwick, K. (1987). Isolation of protoplasts from cocoa (*Theobroma cacao* L.) leaves. *Café Cacao, The'*, **31**: 115-120.

Townsley, P.M. (1974). Chocolate aroma from plant cells. *Canadian Institute of Food Science and Technology Journal*, **7**: *76-78.*

Tsai, C.H and Kinsella, J.E. (1982). Tissue culture of cocoa bean (*Theobroma cacao* L.): Incorporation of fatty acids into lipids of cultured cells. *Lipids*, 17: 848-852.

Tsai, C.H. and Kinsella, J.E. (1981). Initiation and growth of callus and cell suspensions of *Theobroma cacao* L. *Annals of Botany*, **48**: 549-558.

Venturieri, G.A. (2011). Flowering levels, harvest season and yields of cupuassu (*Theobroma grandiflorum*). *Acta Amezonica*, **41**: 143-152.

Wen, M.C., Germen, B., Kinsella, J.E. (1984). Cocoa bean cell and embryo culture. *Journal of the American Oil Chemists Society*, **61**: 1720-1724.

Yidana, J.A., Withers, L.A. and Ivins, J.D. (1987). Development of single method for collection and propagating cocoa germplasm *in vitro*. *Acta Horticulture*, **212**: 95-98.

Zhang, D. and Motilal, L. (2016). Origin, dispersal and current global distribution of cacao genetic diversity In: Baily, L.W. and Meinhardt, L.W. (Eds.), *Cocoa Diseases*. Springer International Publishing, Switzerland. pp 1-31.

Zhang, Y., Clemens, A., Maximova S.N. and Guiltinan, M.J. (2014). The *Theobroma cacao* B3 domain transcription factor TcLEC2 plays a dual role in control of embryo development and maturation. *BMC Plant Biology*, **14**: 2-16.

Molecular Markers and Marker Assisted Selection

Chapter 10

Rubber

☆ *Thakurdas Saha, C. Bindu Roy and K. Bini*

1. Introduction

Rubber (*Hevea brasiliensis*; 2n=36) is a perennial, cross pollinated tree crop belonging to the family Euphorbiaceae. *H. brasiliensis* is the only cultivated species and contributes to 95 per cent of the world natural rubber production. The breeding of rubber tree faces many constraints due to its perennial nature, long breeding cycles, extended juvenile period and the low genetic diversity exhibited by the cultivated clones due to the unidirectional selection for yield characters. In India, *Hevea* breeding was initiated in the year 1954 in Rubber Research Institute of India (RRII), Kottayam, when the first set of hybrids were evolved including the most popular cultivated clone RRII 105. But the conventional method of breeding, utilizing morphological markers, has its own limitations, since they are very less in number and are also highly influenced by environment.

DNA-based genetic markers have revolutionized the ability to follow chromosome segments and have led to new opportunities such as map-based cloning and marker assisted selection in plant breeding. These markers play a significant role in plant breeding efforts from the detection of polymorphism, thereby assessing the genetic diversity, to the construction of genetic linkage maps, which are essential for marker assisted selection, QTL analysis and finally map-based cloning of important agronomic traits. DNA markers are also valuable tools for genotyping, population studies and genetic resource management.

Contrary to molecular markers, morphological or phenotypic characters are often strongly influenced by the environment. In order to increase the efficiency and precision of crop improvement programmes in rubber, the molecular marker techniques are now widely used as described.

2. Molecular Markers

Molecular markers are the specific DNA fragments that can be identified at specific locations of the genome and are based on naturally occurring polymorphisms in the DNA sequence, *i.e.*, base pair deletions, substitutions or additions (Gupta *et al.*, 1996). They are considered as landmarks on chromosomes, which are essential in finding out where the genes are placed in a genetic map. In fact, the development of marker systems was initiated with the mutations at the loci controlling plant morphology (Stadler, 1929). But the morphological markers provided little information on many complex genetic factors and led to the development of molecular markers. Isozymes were the first developed markers (Markert and Moller, 1959) before the advent of DNA-based molecular markers. The advent of molecular marker technology has revolutionized the entire scenario of biological sciences and widened the existing knowledge on polymorphism between individuals – 'from morphological to molecular level'.

The concept of using variations at DNA level as genetic markers started with Restriction Fragment Length Polymorphism (RFLP) and its first documentation came from viruses (Grodzicker *et al.*, 1974) followed by a subsequent elegant demonstration made in the human α-globin gene cluster (Jefferys, 1979). The recent PCR-based approach, gel free visualization of PCR products and automation at various steps are the boons to the molecular marker approaches adopted for genome mapping and genetic diversity analysis in any organism.

Different types of DNA markers used in molecular breeding include mainly Random Amplified Polymorphic DNA (RAPD), Amplified Fragment Length Polymorphism (AFLP), microsatellites/simple sequence repeats (SSRs) and Single Nucleotide Polymorphisms (SNPs). In *Hevea*, molecular markers like RAPD, AFLP (Lespinasse *et al.*, 2000a), SSR (Feng *et al.*, 2009, 2010; Roy *et al.*, 2004,2012a; Saha *et al.*, 2005, 2012), SNP (Pootakham *et al.*, 2011) have been developed and used in various genetic analysis.

2.1. Random Amplified Polymorphic DNA (RAPD) Markers

RAPD markers, first described by Williams *et al.* (1990), offered the highest potential for generating large number of markers with ease (Russell *et al.*, 1997). RAPD markers have been shown to be very useful for DNA fingerprinting, detection of genetic polymorphism, varietal identification, evaluation of gene flow between species (Arnold *et al.*, 1991), detection of gene introgression in various plant species (Orozco Castillo *et al.*, 1994) and mapping (Grattapaglia and Sederoff, 1994).

In *Hevea brasiliensis*, RAPD markers were initially developed for varietal identification and genetic diversity analysis. The applicability of RAPD markers for genetic analysis in *H. brasiliensis* was evaluated in a set of 24 clones in breeding pool of RRII (Varghese *et al.*, 1997). Maximum genetic distance was displayed by the clone RRIC 100 (a hybrid clone developed in Sri Lanka) and it was used as parent in hybridization programmes in India, resulting in highly heterotic hybrids (Licy, 1997). The genetic relationships among 37 cultivated clones of *H. brasiliensis* was established by Venkatachalam *et al.* (2002) using this marker system. Mathew *et al.* (2005) conducted phylogenetic relationship studies in three species of *Hevea*

(*H. brasiliensis, H. benthamiana* and *H. spruceana*) using RAPD markers along with other marker systems such as chloroplast DNA PCR-RFLP and heterologous chloroplast microsatellites. A dwarf genome specific RAPD marker was identified, cloned and sequenced from rubber by Venkatachalam *et al.* (2004). A RAPD marker was developed which showed partial homology to proline-specific permease gene (Venkatachalam *et al.*, 2006). Differentially expressed thymidine kinase gene related to tapping panel dryness syndrome in the rubber tree was also identified by RAPD markers (Venkatachalam *et al.*, 2006). In rubber, a number of RAPD markers have been used to identify clones (Nurhaimi *et al.*, 1998) and to identify markers related to diseases (Toruan-Mathius *et al.*, 2002). Shoucai *et al.*, (1994) identified mildew resistance genes using this technique. A putative RAPD marker linked to abnormal leaf fall resistance locus in rubber was identified by Saha *et al.* (2015), which was used to profile Wickham clones and wild accessions of rubber for establishing association with *Phytophthora* tolerance. RAPD marker system was used in phylogenetic analysis of wild *Hevea* accessions originating from Acre, Rondonia and Matto Grosso provinces of Brazil (Figure 10.1) and three sub-populations were clearly differentiated based on their geographical locations. The genetic diversity of early introduced clones of *H. brasiliensis* in Southern Thailand was studied by Nakkanong *et al.* (2008) using RAPD markers. RAPD technique was employed in the evaluation of genetic variability of *Hevea* population developed through the hybridization of PB 260 (maternal clone) with five Amazonian genotypes. The structure and genetic diversity of the IRRDB 1981 germplasm collection was studied using six selected RAPD primers by Lam *et al.* (2009) and the UPGMA clustering conformed to the geographical regions of the collections. The phylogenetic relationships among 45 rubber clones was analyzed using 12 RAPD primers and based on the genetic distance obtained, an attempt was made to select the parent trees for further crop improvement programmes (Oktavia and Kuswanhadi, 2011). The genetic similarity among various rubber clones from different Brazilian regions was identified with 19 random primers (Bicalho *et al.*, 2008). Other than the genetic diversity analysis,

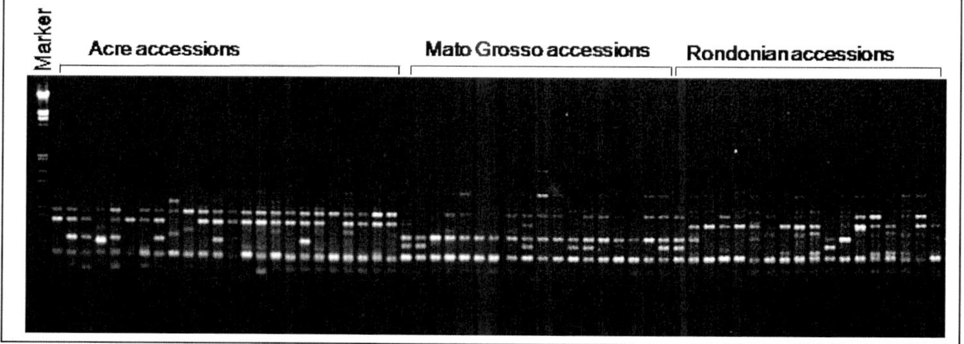

Figure 10.1: Gel Photograph Showing RAPD Profiles of Wild *Hevea* Accessions Originating from Acre, Rondonia and Mato Grosso provinces of Brazil. Acre accessions showed distinct RAPD profiles, differentiating them from other two populations.

RAPD markers were used to construct a genetic linkage map in *Hevea* in an F1 population along with other markers.

2.2. Amplified Fragment Length Polymorphism (AFLP) Markers

This multi-locus marker technology was developed by Vos *et al.* (1995). This was based on the detection of genomic restriction fragments by PCR amplification. AFLP is considered to be information rich due to its ability to analyze a large number of polymorphic loci simultaneously with a single primer combination on a single gel as compared to RAPDs and RFLPs (Powell *et al.*, 1996). They are more reliable and allow establishment of a saturated genetic linkage map. But AFLPs are non-locus specific, dominant markers that can exhibit only two states in a species (presence versus absence of a band) (Seguin *et al.*, 2001).

In *H. brasiliensis*, there are only a few reports regarding the assay of AFLP markers. AFLP markers were utilized by Lespinasse *et al.* (2000a) for the construction of a genetic linkage map in *Hevea* in the progeny derived from an interspecific cross. Roy *et al.* (2012b) used AFLP markers for assessment of genetic diversity among cultivated clones from different South East Asian rubber growing countries. They reported that geographical origin of the clones did not reveal any correlation with

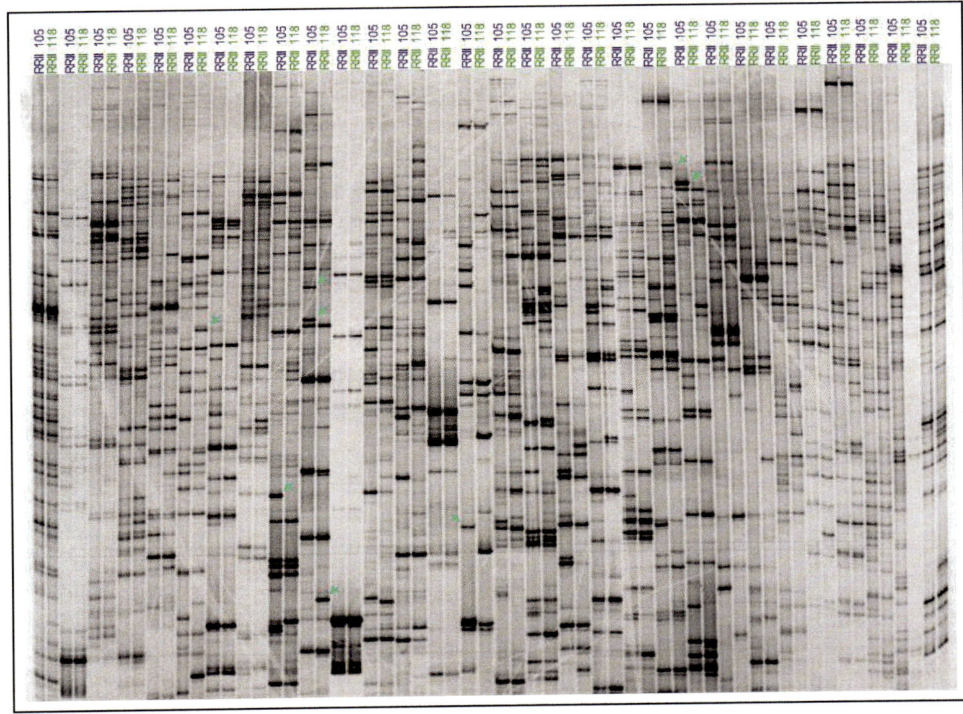

Figure 10.2: AFLP Marker Development for *Hevea* Genome Mapping. AFLP profiles of parents (RRII 105 and RRII 118) of a mapping population with different primer combinations. Low levels of polymorphisms were detected between RRII 105 and RRII 118. Some of the polymorphic bands are marked.

their clustering pattern except for the Sri Lankan clones. In this study, a simple method was demonstrated to identify the potential primer combinations for AFLP analysis using mini PAGE followed by silver staining. AFLP markers were successfully used in linkage map construction along with other DNA markers by Bini (2013). (Figure 10.2).

2.3. Microsatellite/SSR Markers

Microsatellite DNA or simple sequence repeats (SSRs), a relatively new class of DNA markers, are highly informative genetic markers. The existence of microsatellite sequences in plants and algae was first reported in 1986 (Tautz *et al.*, 1986). The term 'microsatellites' was coined by Litt and Luty (1989). These are abundant, dispersed throughout the genome and show higher levels of polymorphism than other genetic markers (Schlotterer and Tautz, 1992). These features, coupled with their ease of detection, have made them useful molecular markers. Their potential for multiplexing and their inheritance in a co-dominant manner are additional advantages (Morgante and Olivieri, 1993; Thomas and Scott, 1993). Simple sequence length polymorphisms are based on the difference in the number of the DNA repeat units (Cho *et al.*, 2000). One of the first reports describing the presence of microsatellites in plant genomes was in the forest trees (Conditt and Hubbell, 1991). The SSR markers are of interest to geneticists and breeders and have been successfully used to infer about genetics, pedigree, phylogeny and/or identity of various traits and/or germplasm accessions.

In *H. brasiliensis* much importance has been given in the recent years for the development of microsatellite markers. The first report of DNA fingerprints in *H. brasiliensis* using heterologous minisatellite probes from humans came from Besse *et al.* (1993). Low *et al.* (1996), for the first time, detected microsatellites in the *Hevea* genome through the database search of some *Hevea* gene sequences. A systematic effort towards developing microsatellite markers was made by the authors' laboratory through screening of *Hevea* genomic library (Roy *et al.*, 2004, 2012a; Bini, 2013). Microsatellites markers were used to identify 27 *H. brasiliensis* clones in our laboratory (Saha *et al.*, 2005). The polymorphisms observed among 40 popular *Hevea* clones with SSR markers (Figure 10.3) could be used for developing markers for screening various traits in *Hevea* breeding programmes.

Along with RAPD markers, four microsatellite pairs (*hmac4, hmac5, hmct1 and hmct5)*, reported by Saha *et al.* (2005), were used for genetic diversity analysis in 53 early introduced clones of *Hevea* collected from different areas of Southern Thailand by Nakkanong *et al.* (2008). Dinucleotide (CT)n repeats detected in *Mn-SOD* gene had been used as SSR markers for genetic relationship studies by Lespinasse *et al.* (2000a) and Lekawipat *et al.* (2003). Lespinasse *et al.* (2000b) described the construction of a genetic linkage map of rubber tree using SSR markers along with different molecular markers. The polymorphic microsatellite loci isolated and characterized from an enriched genomic library of *H. brasiliensis* was highly useful in understanding genetic diversity and gene flow among *Hevea* species (Souza *et al.*, 2009). Two hundred and ninety six new polymorphic microsatellite markers were introduced by Le Guen *et al.* (2011a) for genomic studies in *H. brasiliensis* through screening of an enriched

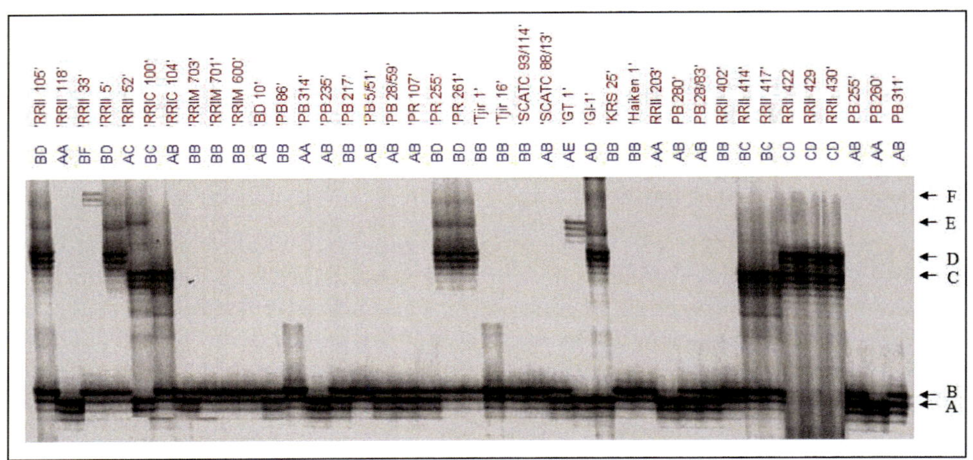

Figure 10.3: Characterization of Popular Clones of *H. brasiliensis* using Genomic SSR Markers: Autoradiogram showing allelic diversity among 40 *H. brasiliensis* clones at the microsatellite locus *hmac4*. Six (A-F) alleles are detected at that locus.

genomic library. Similarly, 100 microsatellite markers from *Hevea* genome was reported by Roy *et al.* (2012a). Dinucleotide repeats (AG)n were detected at the 3′-UTR of mRNA of HMG-CoA reductase encoded by the gene *HMGR* (Figure 10.4A), which is a key enzyme involved in latex biosynthesis in rubber (Saha *et al.*, 2005). SSR polymorphism at this locus (Figure 10.4B) was successfully used for studying the allelic diversity in wild accessions of rubber (Saha *et al.*, 2007).

Cross-species amplification of the markers, developed for *H. brasiliensis*, was also found successful in the wild *Hevea* species *H. guianensis, H. rigidifolia, H. nitida, H. pauciflora, H. benthamiana and H. camargoana* (Saha *et al.*, 2005; Souza *et al.*, 2009). Thus it was reported that a high degree of sequence homology existed in the microsatellite flanking regions of these species. Garcia *et al.* (2011) identified microsatellite markers from GenBank sequences and utilized 15 markers for the genetic diversity analysis of *Hevea* clones. Microsatellite markers were used in diversity analysis and their cross species amplification was tested in six *H. brasiliensis* species by Mantello *et al.* (2012).

Molecular approach is one of the methods for the development of SSR markers. Here genomic libraries are constructed (with or without enrichment for SSRs), screened, candidate clones are sequenced and SSR motifs are identified either manually or using computer programs. Microsatellite enrichment has been developed to increase the proportion of clones in a given library containing the microsatellite motif of interest. Several strategies for microsatellite enrichment have been reported (Kijas *et al.*, 1994; Edwards *et al.*, 1996; Fischer and Bachman, 1998; Koblizkova *et al.*, 1998; Hamilton *et al.*, 1999; Paetkau, 1999; Phan *et al.*, 2000; Zane, et al., 2002). SSRs, once developed, are extremely valuable, though their development is time consuming, laborious and expensive. The construction of a microsatellite-enriched library in *H. brasiliensis* was reported by Atan *et al.* (1996). Development of

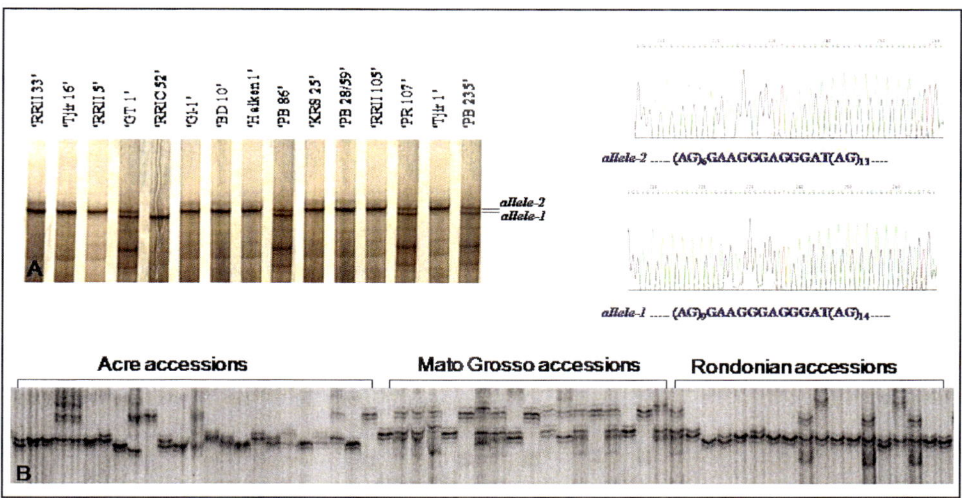

Figure 10.4: Allelic Variation at the Locus *HMGR* in Cultivated Clones of *H. brasiliensis*. Two Alleles are identified, sequencing of these alleles showed two polymorphic AG repeats flanking a 12 bp conserved region (A). Wide allelic diversity (nine alleles) was noticed at this locus among wild *Hevea* accessions originating from Acre, Rondonia and Mato Grosso provinces of Brazil (B).

trinucleotide microsatellite markers from an enriched genomic library of rubber (*H. brasiliensis*) was reported by Bini *et al.* (2008) and used in linkage map construction.

The computational or bioinformatics approaches take advantage of the available sequences such as those in the public databases and by scanning through them for identifying the ones that contain SSRs. They supplement the molecular approaches by identifying SSR repeats in candidate sequences derived from the libraries. Data mining of microsatellites from ESTs makes use of this approach and has proven effective for generating markers for fingerprinting, genetic mapping and comparative mapping among species (Varshney *et al.*, 2005). Developing these markers is less costly and time effective, and may provide abundant information. Sequences from many genomes are continuously made freely available in the public databases and mining of these sources using computational approaches permits rapid and economical marker development.

Expressed sequence tags (ESTs) are ideal candidates for mining SSRs not only because of their availability in large numbers, but also due to the fact that they represent expressed genes. EST-SSR markers were developed first in rubber trees by Feng *et al.* (2009). Multivariate techniques and microsatellite markers were used for genetic divergence estimation in rubber by Gouvea *et al.* (2010). Genetic diversity analysis of wild and cultivated clones of *Hevea* was estimated by Perseguini *et al.* (2012) using EST-derived SSR markers and the cross species amplification of these markers were also estimated successfully. It was found that since rubber tree is highly heterozygous due to its out crossing nature, transcriptome sequencing is an attractive alternative to whole-genome sequencing which focuses on the coding regions of the genome which could be used for SSR mining. Earlier, transcriptome studies of *H. brasiliensis* focused mainly on latex in order to gain insight into the rubber biosynthesis pathways (Han *et al.*, 2000; Ko *et al.*, 2003; Chow *et al.*, 2007). The transcriptome from rubber tree bark was sequenced with Illumina paired-end sequencing and 106 EST-SSR markers were developed by Li *et al.* (2012). In the studies of Triwitayakorn *et al.* (2011), the transcriptome from the vegetative shoot apex was sequenced and synthesised 323 EST derived SSR primer pairs which were amplifiable in *H. brasiliensis* clones. From these, a selected number of 47 SSR markers were utilized for the genetic similarity analysis of 20 *H. brasiliensis* clones. In authors' laboratory, transcriptome data generated from both *Corynespora* pathogen challenged and control leaf samples of RRII 105 (susceptible clone) and GT1 (resistant clone) was used in large scale SSR mining for functional marker generation (Roy and Saha, 2012) (Figure 10.5).

Achievements in the use of microsatellites have been made in areas like selection and diagnostics in segregating populations, genome selection during gene introgression (in back cross programmes), genome mapping, gene tagging, cultivar identification, germplasm characterization and in estimation of genetic relatedness (Gupta *et al.*, 1996). Microsatellites markers are found throughout both the transcribed and non-transcribed regions of a genome (Varshney *et al.*, 2005). Their role in gene regulation and genome evolution has also been discussed widely (Aishwarya and Sharma, 2007). They are considered as valuable tools for phylogenetic, fingerprinting and molecular breeding studies (Sharma *et al.*, 2008).

2.4. Single Nucleotide Polymorphism (SNP)

SNPs represent the most common variations across a genome (Gupta *et al.*, 2001; Kwok *et al.*, 2001). Interest in single nucleotide polymorphisms (SNPs), stimulated in

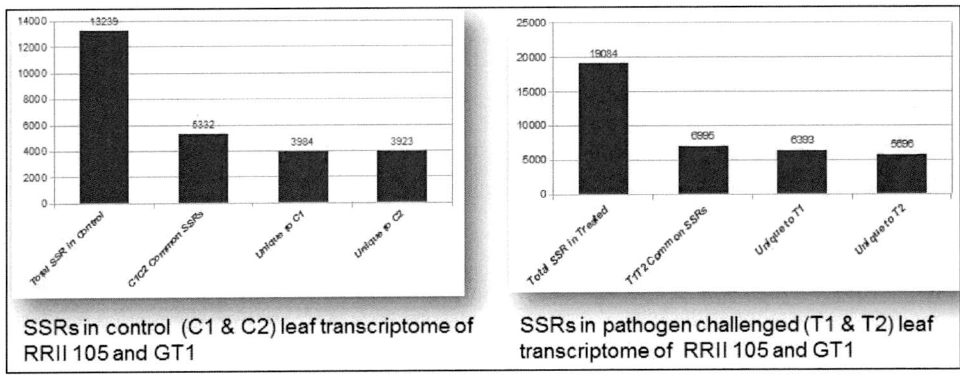

SSRs in control (C1 & C2) leaf transcriptome of RRII 105 and GT1

SSRs in pathogen challenged (T1 & T2) leaf transcriptome of RRII 105 and GT1

Figure 10.5: Large Number of Simple Sequence Repeats (SSRs) Containing Transcripts were Identified from Transcriptome Data of Control (C1 and C2) and Pathogen Challenged (T1 and T2) RRII 105 and GT1. Abundance of common and unique SSRs bearing transcripts are shown in bar diagrams.

part, by the progress of human SNP discovery, is rapidly increasing. SNPs are stable and the relative fidelity of their inheritance is higher than that of the other marker systems like SSRs and AFLPs (Semagn *et al.*, 2006). SNPs occur at a frequency of about one SNP in 1000 nucleotides in human genome (Wang *et al.*, 1998). SNPs are consequences of either transition or transversion events. SNP fall into several classes depending on (i) their precise location in the genome and (ii) impact of their location within coding or regulatory regions onto the encoded proteins or phenotype. Given that majority of SNPs are located in noncoding DNA, they are called noncoding SNPs. SNP that reside on exons and corresponding cDNA are called coding SNPs or exonic SNPs. Exonic SNPs, that do not change the composition of the encoded domains or proteins, are called synonymous SNPs (Weising *et al.*, 2005). SNPs, when synonymous, can change the structure and stability of the messenger RNA, whereas non-synonymous can change the structure and function of the protein and consequently affect the amount of protein produced, as it may be a deleterious mutation (Guimaraes and Costa, 2002). SNPs contribute directly to a phenotype or can be associated with a phenotype as a result of linkage disequilibrium (Risch and Merikangas, 1996). Direct analysis of genetic variation at the DNA level has made SNPs attractive as genetic markers (Bhattramakki *et al.*, 2002; Rafalski, 2002). In plants, SNP seems to be more abundant than those in human genome as evidenced by the existence of one SNP per 20 bp in wheat (Picoult *et al.*, 1999) and one SNP per 70 bp in maize genome (Bhattramakki *et al.*, 2000). SNPs, including insertion/deletions (InDels), can provide a rich source of useful molecular markers in genetic analysis for many crop species (Ronaghi *et al.*, 1996).

Rapid advances in the high-throughput next-generation sequencing (NGS) technologies have facilitated extensive SNP discovery projects in several plant species (Barbazuk *et al.*, 2007). Once discovered, SNPs can be converted into genetic markers that can be assayed in a high-throughput manner (Gut, 2004; Kwok, 2001). SNP markers are very useful both for marker-assisted selection and for gene isolation when found in proximity of the coding sequences. Since SNPs are abundant, they can be used as genetic markers in many applications such as cultivar identification, construction of genetic maps, assessment of genetic diversity or marker-assisted breeding (Flint-Garcia *et al.*, 2005; Chagne *et al.*, 2008; Wu *et al.*, 2008).

2.4.1. SNP Identification Methods

The broad approaches for SNP discovery are resequencing and *in silico* SNP mining. There are other techniques also like denaturing/temperature gradient gel electrophoresis (D/TGGE) (Myers *et al.*, 1988) and single stranded conformational polymorphism (SSCP) (Orita *et al.*, 1989), which have been employed to detect polymorphisms in forest (Plomion *et al.*, 1999) and horticultural trees (Etienne *et al.*, 2002).

Direct amplicon sequencing, first reported by Wang *et al.* (1998), was one of the simplest forms of SNP discovery. In cloned amplicon sequencing, heterozygous SNPs may be unambiguously determined in primary analysis and thereby determining the haplotype structure. In *H. brasiliensis,* there are only limited reports regarding the identification and utilization of SNP markers (Pootakham *et al.*, 2011; Mantello *et al.*, 2014). In authors' laboratory, an effort was made to identify SNPs mainly in the latex biosynthesis genes (Figure 10.6) along with other genes involved in complex biosynthetic pathways *viz.*, ubiquitin precursor, latex patatin homolog, ethylene inducible protein, transcript associated with tapping panel dryness, latex plastidic aldolase, thioredoxin, glutathione peroxidase and hevamin B (Bini *et al.*, 2010). Out of the 12 genes, SNPs could be identified only in five genes *viz.*, geranylgeranyl diphosphate synthase, farnesyl diphosphate synthase, mevalonate kinase, ubiquitin precursor and latex patatin homolog.

Large scale SNP identification could be initiated with transcriptome sequencing using next generation sequencing (NGS) platform. Pootakham *et al.* (2011). identified 5883 biallelic SNPs by transcriptome sequencing and 50 of them were validated in *Hevea* clones. Mantello *et al.* (2014) also performed RNA sequencing (RNA-seq) of *H. brasiliensis* bark tissue on the Illumina GAIIx platform and reported large number of SSR and SNPs. Finally, the transcripts/sequences belonging to the mevalonate (MVA) and 2-C-methyl-D-erythritol 4-phosphate (MEP) pathways, involved in rubber biosynthesis were selected for SNP validation. A total of 78 SNPs were validated in 36 genotypes of *H. brasiliensis*.

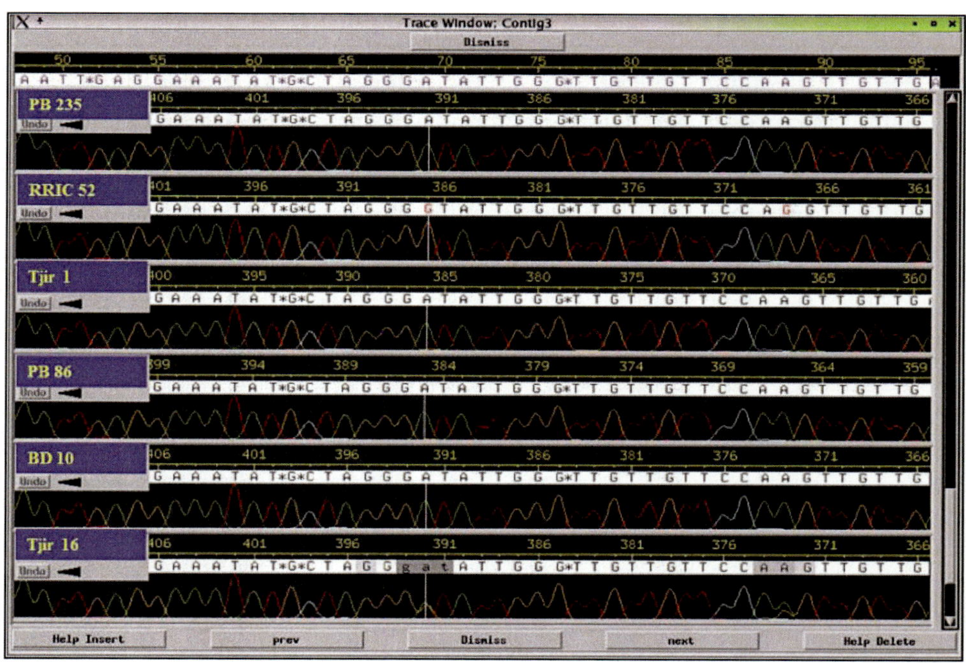

Figure 10.6: Screenshot of the 'Consed' Program Showing the Alignment of the Partial Chromatograms of the Locus Geranylgeranyl Diphosphate Synthase from *Hevea* Clones Showing SNP at the Nucleotide Position 69. Heterozygosity could clearly be detected in Tjir 16.

2.4.2. SNP Genotyping

High-throughput, high density SNP genotyping has become an essential tool for QTL mapping, association genetics, gene discovery *etc.* and applied in several crops (Zhu *et al.*, 2001; Rafalski, 2002). In *Eucalyptus,* the Golden Gate Genotyping Technology developed by Illumina (Ganal *et al.*, 2009) was used for the detection of SNPs (Grattapaglia *et al.*, 2011). Recently, next generation sequencing (NGS) has been coupled with genome complexity reduction techniques and barcoding to identify and genotype a set of common SNPs in a mapping population. This strategy is referred to as genotyping-by-sequencing (GBS) (Elshire *et al.*, 2011). GBS utilizes restriction endonucleases to digest the genome into fragments, which are subsequently sequenced on high-throughput platforms. Pootakham *et al.* (2015). applied a GBS technique to simultaneously discover and genotype SNP markers in two rubber tree populations 'P' (BPM24 x RRIM600) and 'C' (BPM24 x RRIC110). The female parent of both crosses, BPM24, is a descendent of a GT1 x AVROS1734 cross. A total of 21,353 single nucleotide substitutions were identified, 55 per cent of which represented transition events.

Large scale genotyping in minimally equipped laboratories requires a low cost technology. The restriction based techniques was the earliest method used for the detection of single nucleotide polymorphism (Botstein *et al.*, 1980). The cleaved amplification polymorphic site or polymerase chain reaction-restriction fragment length polymorphism (CAPS or PCR-RFLP) and the derived CAPS (dCAPS) (Michaels and Amisino, 1998; Neff *et al.*, 1998) were widely applied (Iwaki *et al.*, 2002; Yanagisawa *et al.*, 2003; Yamanaka *et al.*, 2004). Another alternative is the allele-specific PCR amplification (AS-PCR) also called PCR allele-specific amplification (PASA) (Sommer *et al.*, 1992) or amplification refractory mutation system (ARMS) (Newton et al., 1989). The allele specific PCR amplification follows a simple procedure with a common reverse primer and two allele specific forward primers in normal conditions and allows the discrimination of alleles by gel electrophoresis (Dutton and Sommer, 1991). Detection of amplicons generated through PASA can be effectively done through other techniques also such as fluorescence detection, real-time fluorescence detection, and sequencer (Germer and Higuchi, 1999; Ishiguro *et al.*, 2005; Hansson and Kawabe, 2005; Wu *et al.*, 2005; Hinten *et al.*, 2007). AS-PCR techniques have been adopted in the author's laboratory for genotyping but with minor modifications (Figure 10.7A). Usually two outer primers are used and one allele specific primer at a time to amplify the locus along with the allele. An alternative strategy was also developed where one common reverse primer with an allele specific primer and a pair of primers for amplification of actin or chloroplast gene as PCR control in a reaction were successfully used. PCR control is necessary to discriminate between the failed reaction and the absence of specific alleles in the genotype (Figure 10.7B) (Saha *et al.*, unpublished).

High resolution melt (HRM) analysis, that measures the dissociation of double stranded DNA at high temperature, offers a high-throughput technique for fast detection of variation in nucleic acid sequences. This technique is being used for SNP genotyping in several plant species (Wu *et al.*, 2008; Han *et al.*, 2012). This method was successfully adopted in rubber for SNP genotyping of the mapping population.

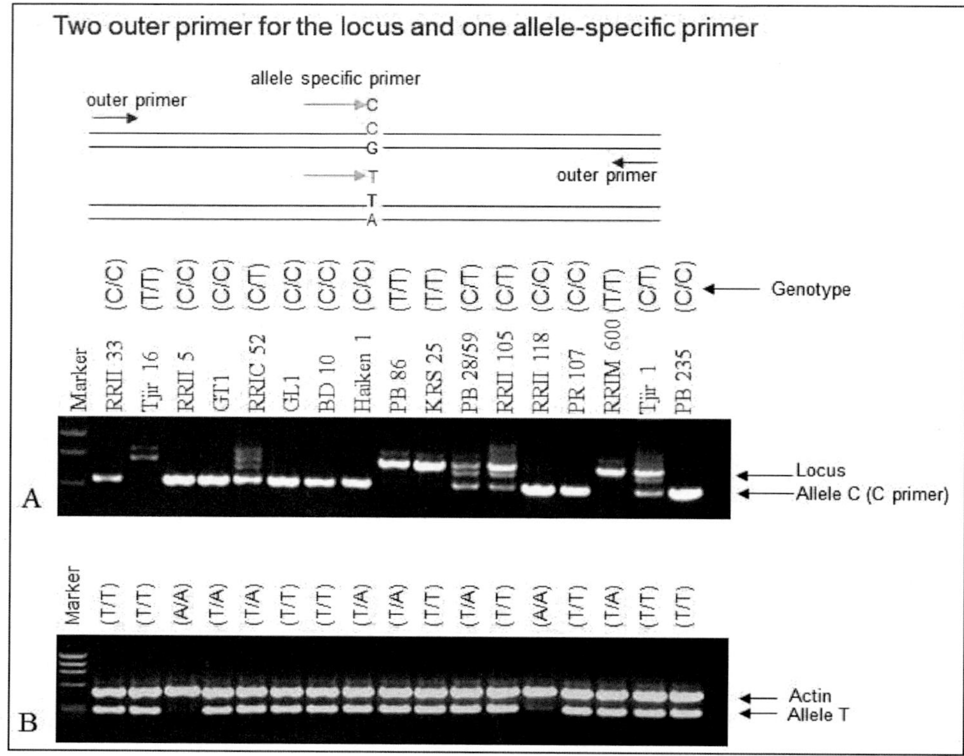

Figure 10.7: Allele-specific Amplification of SNPs for Genotyping in Rubber Clones. Three primers, two locus-specific outer primers and one allele-specific primer (allele T) were used in single PCR reaction for SNP genotyping (A). Actin was successfully used as PCR control (amplified in all genotype) along with one reverse and allele specific (allele T) primers for genotyping(B). Genotyping data obtained through these experiments were validated through Sanger sequencing of the PCR products.

However, single genotyping method is not ideally suited for all applications and a number of good genotyping methods are available to meet the needs of many study designs.

2.5. Genetic Linkage Map and QTL Analysis

Genetic linkage mapping is a method by which location of genes/markers on a chromosome are determined based on recombination frequencies observed in pedigrees or progeny population (Clark and Wall, 1996). The concept of linkage between genetic markers in heredity was first proposed a century ago by Morgan (1911). The first genetic linkage map of X-chromosome of *Drosophila* was published by Sturtevant in 1913 and the first partial genetic map of maize was published by Emerson *et al.* (1935).

The importance of developing a high-density linkage map of *Hevea* genome had already been realized by the researchers from natural rubber growing countries. In *H. brasiliensis*, the number of linkage groups for a saturated linkage map should be equal to the haploid chromosome number of 18. The first linkage map of *Hevea* was constructed in 2000 by the researchers from CIRAD, France using mainly RFLP and AFLP markers (Lespinasse *et al.*, 2000a), which was later used to identify the QTL variants conferring resistance to the South American leaf blight disease (Lespinasse et al., 2000b). Following the construction of first linkage map, there were several attempts in generating linkage maps in different rubber growing countries using different cross combinations and various types of DNA markers. Feng *et al.* (2010) constructed a genetic linkage map in *H. brasiliensis* using microsatellite markers. The genetic linkage map consisted of 91 marker loci in 18 linkage groups and covered 1937.06 cM with an average genetic distance of 21.29 cM between adjacent markers. Souza *et al.* (2011) developed a genetic linkage map based on a mapping population derived from a controlled cross between the cultivars PB217 and PR255. Two hundred twenty five microsatellite markers were genotyped and the map was distributed in 23 linkage groups and 2,471.2 cM in length. EST-SSR developed from transcriptome sequencing together with published SSR markers were used to construct linkage map using a population of 81 individuals derived from hybridization between RRIM600 and RRII105 by Triwitayakorn *et al.* (2011). The map consisted of only 97 loci distributed on 23 linkage groups. The total map distance covered 842.9 cM with a mean interval of 11.9 cM and the average loci per linkage group were approximately four loci. Shearman *et al.* (2015) reported a high resolution linkage map using a total of 12,326 SNPs from 4,244 contigs generated through transcriptome sequencing. A mapping population consisted of 149 offspring from a cross between RRIM 600 and RRII 105 were used to perform target sequence enrichment and high throughput sequencing to genotype progenies along with their parents. There were 18 linkage groups and total map length was 4,160 cM, which appears to be high.

Majority of the research work on QTL analysis for disease resistance in rubber tree was restricted to South American leaf blight (SALB) caused by the fungus *Microcyclus ulei* in South American rubber growing countries including Brazil, the primary centre of origin of rubber. Resistance QTLs to SALB were identified both in controlled conditions (Lespinasse *et al.*, 2000b) and under natural infestation (Le Guen *et al.*, 2003). SALB resistance were mapped using 195 F1 progeny individuals derived from the cross between a susceptible cultivated clone, PB260, and a resistant clone, R038, derived from interspecific hybridization by Lespinasse *et al.* (2000b). The reaction type (RT) and the lesion diameter (LD) were measured on immature leaves after artificial inoculation of the fungus. Eight QTLs for resistance were identified on the R038 map. Only one QTL was detected on the PB260 map. Le Guen *et al.* (2003) investigated genetic resistance components of the *H. benthamiana* RO 38 cultivar to *Microcyclus ulei* causing South American Leaf Blight (SALB) disease using a population of 192 progenies from the same cross combination planted in a field trial in French Guiana in order to evaluate the resistance parameters under real infestation conditions. One major QTL located on linkage group g13 was detected on the RO 38 map, responsible for 36 to 89 per cent of the phenotypic variance of resistance. This resistance QTL corresponded to one that had previously been detected under controlled conditions of infestation and called it M13-1bn. Le Guen *et al.* (2007) detected the presence of eight resistance quantitative trait loci (QTLs), all of them were inherited from the interspecific parent. Among these QTLs, only one contributed to the partial resistance against a highly pathogenic isolate. Another report about the components of genetic resistance of the *H. brasiliensis* cultivar MDF 180 against SALB through QTL mapping was published by Le Guen *et al.* (2011b). Genetic maps were constructed mainly using microsatellites and AFLP markers. A major resistance QTL, denoted as M15md was found in the resistant parent only. A qualitative gene responsible for the resistance against isolates from French Guiana was also identified.

Other than disease resistance, QTL mapping was performed by Souza *et al.* (2013). to understand the genetic architecture of the traits related to growth of rubber tree in two different conditions (winter and summer) using a mapping population (270 individuals) derived from a cross between genotypes PB217 and PR255. Traits evaluated were height and girth growth. The genetic map contained 284 markers assigned to 23 linkage groups with a total length of 2688.8 cM. A total of 18 QTLs for growth traits during the summer and winter seasons were detected. For height, QTLs detected during the summer season were different from the ones detected during winter season. This type of difference was also observed for girth. This study is particularly important when the objective of a breeding program is to obtain phenotypes that are adapted to sub-optimal regions.

In authors' laboratory at RRII, a linkage map of rubber was constructed using a segregating progeny population comprising of 60 individuals, obtained from a cross between two popular cultivated *Hevea brasiliensis* clones: RRII 105 and RRII 118 (Bini, 2013). A total of 227 markers comprising of 96 RAPD markers, 79 AFLP markers, 47 SSR markers and five SNP based markers were utilized for the construction of a genetic linkage map. Marker segregation analysis was performed using the software JoinMap 3.0. Marker groups were determined using a minimum LOD score of 3.0 and a recombination frequency of 0.4. Twenty-one linkage groups were identified for RRII 105 and 20 groups for RRII 118. The total genetic distance covered was 685.84 cM in RRII 105 and 523.33 cM in RRII 118. Eighty-seven markers were unlinked in RRII 105 out of 173 loci analyzed and 79 markers remained unlinked out of 151 markers analyzed in RRII 118. Three linkage groups harboring bridge markers common to both the parents were used to merge the respective groups.

However, the linkage maps created in this study are preliminary and QTL mapping is not possible even after saturation of the existing map as the entire progeny population is not planted in the main field and hence proper phenotyping is not possible. Moreover, this parental combination showed low level of polymorphisms with the existing markers generated in the laboratory. Therefore, an interspecific cross between *H. brasiliensis* (clone RRII 105) and *H. benthamiana* (clone F4542) with varying levels of disease resistance and yield potential was made and a progeny population was raised for construction of a linkage map (Jayashree M., unpublished). RRII 105 is a commercially cultivated high yielding clone with moderate level of susceptibility to *P. meadii* causing Phytophthora leaf fall, whereas, F4542 is a low yielder with high level of tolerance to both. *Phytophthora meadii* and *Corynespora cassiicola*. Genotyping of the mapping population comprising the parents and 86 progenies was carried out using co-dominant SNP markers and dominant *silico* DArT markers derived from DArT sequencing (DArTseq™). A linkage map was developed for each parent by combining the relevant *silico* DArT and SNP markers. Using the markers in common to join the two linkage maps, a consensus map was created which covered 3709 cM. Finally a high density integrated genetic linkage map was constructed using 24004 markers, which assembled into 18 linkage groups, thus reflecting the haploid chromosome number of *Hevea* ($n = 18$) (Roy *et al.*, unpublished). This high density linkage map can be utilised for QTL mapping for disease resistance and also mapping for latex yield in rubber tree. Therefore, application of genomic approaches facilitated by emerging sequencing technologies may significantly accelerate the breeding program for crop improvement in rubber.

3. Conclusion

The advent of genomics tools and technologies in 21st century have provided unique capabilities for understanding the molecular basis of plant growth, development and key traits towards improving crop productivity. Insights derived from genome sequencing and transcriptomic studies are now applied to genome-wide measurements of components of interest, which definitely has the potential in dissecting the molecular, biochemical, physiological and evolutionary basis of traits. In recent years rubber genome has been sequenced (Rahman et al., 2013; Tang et al., 2016),these genomic data along with transcriptomic resources have greatly facilitated the identification of genome wide SSR motifs and SNPs. Presently ,they are converted to markers for various genetic analyses. High-throughput genotyping platforms are essential for application of large number of markers, which is only possible with the SNP markers, as they are amenable to high-throughput automated analysis. This is the reason why SNPs have become the marker of choice over multi-allelic SSR markers, besides their uniform distribution throughout a genome. Now genome wide SNP genotyping can be performed through the development of SNP chip. New technology platforms are available for medium to ultra-high-throughput genotyping for a variety of studies including development of high density genetic maps for QTL mapping and LD based association studies. The knowledge derived from such efforts will accelerate the breeding program towards development of "smart clones" that can yield better through improved growth and ability to withstand biotic and abiotic stresses.

Acknowledgements

Authors express their gratitude to the Rubber Board of India for providing financial support for all the rese arch programs carried out in the Genome analysis Laboratory of RRII.

References

Aishwarya, V. and Sharma, P.C. (2007). UgMicroSatdb: Database for mining microsatellites from unigenes. *Nucleic Acids Research*, **1**: 1-4.

Arnold, M.L., Buckner, C.M. and Robinson, J.J. (1991). Pollen-mediated introgression and hybrid speciation in Louisiana. *Proceedings of the National Academy of Sciences*, **88**: 1398-1402.

Atan, S., Low, F. C. and Saleh, N. M. (1996). Construction of a microsatellite-enriched library from *Hevea brasiliensis. Journal of Natural Rubber Research*, **11**: 247-255.

Barbazuk, W.B., Emrich, S. J., Chen, H. D., Li, L. and Schnable, P. S. (2007). SNP discovery via 454 transcriptome sequencing. *The Plant Journal*, **51**: 910-918.

Besse, P., Lebrun, P., Seguin, M. and Lanaud, C. (1993). DNA fingerprints in *Hevea brasiliensis* (rubber tree) using human minisatellite probes. *Heredity*, **70**: 237--244.

Bhattramakki, D., Dong, J., Chhabra, A. K. and Hart, G. E. (2000). An integrated SSR and RFLP linkage map of *Sorghum bicolor* (L.) *Moench. Genome*, **43**: 988-1002.

Bicalho, K.C., Oliveira, L.E.M., de Santos, J.B. and Mesquita, A.C. (2008). Similaridade genetica entre clones de seringueira *(Hevea brasiliensis)* por meio de marcadores RAPD. *Ciencia Agrotecnica*, **32**: 1510-1515.

Bini, K. (2013). Development of a genetic linkage map using molecular markers in Para rubber tree (*Hevea brasiliensis*). *Doctoral dissertation*, University of Kerala, Trivandrum.

Bini, K., Ravindran, M. and Saha,T. (2008). Development of trinucleotide microsatellite markers from an enriched genomic library of rubber (*Hevea brasiliensis*). In: National Symposium: From *Chromosome to Genomes: Challenges and Prospects*, University of Kerala, Trivandrum, India.

Bini, K., Ravindran, M., Thomas, K.U., Roy, C.B. and Saha, T. (2010). Single nucleotide polymorphisms in rubber (*Hevea brasiliensis*) and their inheritance in RRII 400 series hybrid clones. In: Plantation Crops Symposium - PLACROSYM *XIX*, RRII, Kottayam.

Botstein, D., White, R.L., Skolnick, M. and Davis, R.W. (1980). Construction of a genetic linkage map in man using restriction fragment length polymorphism. *The American Journal of Human Genetics*, **32**: 314-331.

Chagne, D., Gasic, K., Crowhurst, R.N., Han, Y., Bassett, H.C., Bowatte, D.R., Lawrence, T. J., Rikkerink, E.H.A., Gardiner, S.E. and Korban, S. S. (2008). *Development of a set of SNP markers present in expressed genes of the apple. Genomics*, **92**:353-358.

Cho, Y.G., Ishii, T., Temnykh, S., Chen, X., Lipovich, L., McCouch, S.R., Park, W.D., Ayres, N. and Cartinhour, S. (2000). Diversity of microsatellites derived from genomic libraries and GenBank sequences in rice *(Oryza sativa* L.). *Theoretical and Applied Genetics*, **100**: 713-722.

Chow, K.S., Wan, K.L. and Isa, M.N. (2007). Insights into rubber biosynthesis from transcriptome analysis of *Hevea brasiliensis* latex. *Journal of Experimental Botany*, **58**: 2429-2440.

Clark, M.S. and Wall, W.J. (1996). *Chromosomes*: The Complex Code. New York: Chapman and Hall.

Conditt, R. and Hubbell, S.P. (1991). Abundance and DNA sequence of two-base repeat regions in tropical tree genomes. *Genome*, **34**: 66-71.

Dutton, C. and Sommer, S.S. (1991). Simultaneous detection of multiple single-base alleles at a polymorphic site. *BioTechniques*, **11**:700-702.

Edwards, K.J., Barker, J.H.A., Daly, A., Jones, C. and Karp, A. (1996). Microsatellite libraries enriched for several microsatellite sequences in plants. *BioTechniques*, **20**: 758-760.

Elshire, R.J., Glaubitz, J.C., Sun, Q., Poland, J.A., Kawamoto, K., Buckler, E.S. and Mitchell, S.E. (2011). A Robust, Simple Genotyping-by-Sequencing (GBS) *Approach for High Diversity Species. PLoS ONE*, **6**(5): e19379.

Emerson, R., Beadle, G. and Fraser, A. (1935) A summary of linkage studies in maize. *Memoirs of the Cornell University Agricultural Experiments Station*, **180**: 1-83.

Etienne, C., Rothan, C., Moing, A., Plomion, C., Bodene, C., Svanella-Dumas, L., Cosson, P., Pronier, V., Monet, R. and Dirlewanger, E. (2002). Candidate genes and QTLs for sugar and organic acid contents in peach *[Prunus persica* (L.) Batsch]. *Theoretical and Applied Genetics*, **105**:145-159.

Feng, S.P., Li, W.G., Huang, H.S, Wang, J.Y. and Wu, Y.T. (2009). Development, characterization and cross-species/genera transferability of EST-SSR markers for rubber tree *(Hevea brasiliensis). Molecular Breeding*, **23**: 85-97.

Feng, S.P., Li, W.G., Yu, F., Wang, J.Y. and Wu, Y.T. (2010). Construction of genetic linkage map for rubber tree *(Hevea brasiliensis)* based on SSR markers. *Hereditas*, **8**: 857-863.

Fischer, D. and Bachman, K. (1998). Microsatellite enrichment in organisms with large genome *(Allium cepa* L.) *BioTechniques*, **24**: 796-802.

Flint-Garcia, S. A., Thuillet, A., Yu, J. M., Pressoir, G., Romero, S. M., Mitchell, S. E., Doebley, J., Kresovich, S., Goodman, M. M. and Buckler, E. S. (2005). Maize association population: a high-resolution platform for quantitative trait locus dissection. *The Plant Journal*, **44**: 1054-1064.

Ganal, M. W., Altmann, T. and Roder, M. S. (2009). SNP identification in crop plants. *Current Opinion in Plant Biology*, **12**: 211-217.

Garcia, R. I. A., Gonzalez, S. S. M., Montoya, C. D. and Aristizabal, F. (2011). Identification in silico of SSR markers for genotyping *Hevea* sp. clone gardens in Colombia. *Agronomía Colombiana*, **29**: 359-366.

Germer, S. and Higuchi, R. (1999). Single-tube genotyping without oligonucleotide probes. *Genome Research*, **9**: 72-78.

Gouvea, L. R. L., Rubiano, L. B., Chioratto, A. F., Zucchi, M. I. and Goncalves, P. S. (2010). Genetic divergence of rubber tree estimated by multivariate techniques and microsatellite markers. *Genetics and Molecular Biology*, **33**: 308-318.

Grattapaglia, D. and Sederoff, R. (1994).) Genetic linkage maps of *Eucalyptus grandis* and *Eucalyptus urophylla* using a pseudo-testcross: Mapping strategy and RAPD markers. *Genetics*, **137**: 1121-1137.

Grattapaglia, D., Silva, Jr. O. B., Kirst, M., Lima, B. M., Faria, D. M. and Pappas, Jr. G. (2011). High-throughput SNP genotyping in the highly heterozygous genome of Eucalyptus: assay success, polymorphism and transferability across species. *BMC Plant Biology*, **11**: 65.

Grodzicker, T., Williams, J., Sharp, P. and Sambrook, J. (1974). Physical mapping of temperature sensitive mutations. *Cold Spring Harbor Symposia on Quantitative Biology*, **39**: 439-446.

Guimarães, P.E.M. and Costa, M.C.R. (2002). SNPs: sutis diferenças em um código. *Biotecnologia Ciência and Desenvolvimento*, **26**: 24-27.

Gupta, P. K., Balyan, H. S., Sharma, P. C. and Ramesh, B. (1996). Microsatellites in plants: a new class of molecular markers. *Current Science*, **70**: 45-54.

Gupta, P.K., Roy, J.K. and Prasad, M. (2001). Single nucleotide polymorphisms: a new paradigm for molecular marker technology and DNA polymorphism detection with emphasis on their use in plants. *Current Science*, **80**: 524–535.

Gut, I. G. (2004). An overview of genotyping and single nucleotide polymorphisms (SNP). In: Rapley, R. and Harbron, S. (Eds.), *Molecular Analysis and Genome Discovery*. Wiley Chichester, pp. 43-64.

Hamilton, M. B., Pincus, E. L., Di Fiore, A. and Fleischer, R.C. (1999). Universal linkers and ligation procedures for the construction of genomic DNA libraries enriched for microsatellites. *BioTechniques*, **27**: 500-507.

Han, K.H., Shin, D.H., Yang, J., Kim, I. J., Oh, S. K. and Chow, K. S. (2000). Genes expressed in the latex of *Hevea brasiliensis. Tree Physiology*, **20**: 503-510.

Han, Y., Khu, D. and Monteros, M.J. (2012). High-resolution melting analysis for SNP genotyping and mapping in tetraploid alfalfa (*Medicago sativa* L.). *Molecular Breeding*, **29**: 489-501.

Hansson, B. and Kawabe, A. (2005). A simple method to score single nucleotide polymorphisms based on allele-specific PCR and primer-induced fragment-length variation. *Molecular Ecology Resources*, **5**: 692-696.

Hinten, G.N., Hale, M.C., Gratten, J., Mossman, J.A., Lowder, B.V., Mann, M. K. and Slate, J. (2007). SNP-SCALE: SNP scoring by colour and length exclusion. *Molecular Ecology Resources*, **7**: 377-388.

Ishiguro, A., Kubota, T., Soya, Y., Sasaki, H., Yagyu, O. and Takarada, Y. (2005). High- throughput detection of multiple genetic polymorphisms influencing drug metabolism with mismatch primers in allele-specific polymerase chain reaction. *Analytical Biochemistry*, **337**: 256-261.

Iwaki, K., Nishida, J., Yanagisawa, T., Yoshida, H. and Kato, K. (2002). Genetic analysis of Vrn- B1 for vernalization requirement by using linked dCAPS markers in bread wheat *(Triticum aestivum* L.). *Theoretical and Applied Genetics*, **104**: 571-576.

Jeffreys, A. J. (1978).DNA sequence variants in the G gamma-, A gamma-, delta and beta-globin genes of man. *Cell*, **18**: 1-10.

Kijas, J.H.M., Fowler, J.C.S., Garbett, C.A. and Thomas, M.R. (1994).) Enrichment of microsatellites from citrus genome using biotinylated oligonucleotides sequences bound to streptavidin-coated magnetic particles. *BioTechniques*, **16**: 656--662.

Ko, J. H., Chow, K. S. and Han, K.H. (2003). Transcriptome analysis reveals novel features of the molecular events occurring in the laticifers of *Hevea brasiliensis* (para rubber tree). *Plant Molecular Biology*, **53**: 479-492.

Koblizkova, A., Dolezel, J. and Macas, J. (1998). Subtraction with 3′ modified oligonucleotides eliminates amplification artifacts in DNA libraries enriched for microsatellites. *BioTechniques*, **25**: 32-38.

Kwok, P.Y. (2001). Methods for genotyping single nucleotide polymorphisms. *Annual Review of Genomics and Human Genetics*, **2**: 235-258.

Lam, L.V., Thanh, T., Chi, V.T.Q. and Tuy, L.M. (2009). Genetic Diversity of *Hevea* IRRDB′81 Collection Assessed by RAPD Markers. *Molecular Biotechnology*, **42**: 292-298.

Le Guen, V., Garcia, D., Doare´, F., Mattos, C.R.R. and Condina, V. (2011b). A rubber tree's durable resistance to *Microcyclus ulei* is conferred by a qualitative gene and a major quantitative resistance factor. *Tree Genetics and Genomes*, **7**: 877-889.

Le Guen, V., Garcia, D., Mattos, C. R. R., Doaré, F., Lespinasse, D. and Seguin, M. (2007). Bypassing of a polygenic *Microcyclus ulei* resistance in rubber tree, analyzed by QTL detection. *New Phytologist*, **173**: 335–345.

Le Guen, V., Gay, C., Xiong, T. C., Souza, L. M., Rodier-Goud, M. and Seguin, M. (2011a). Development and characterization of 296 new polymorphic microsatellite markers for rubber tree *(Hevea brasiliensis)*. *Plant Breeding*, **130**: 294-296.

Le Guen, V., Lespinasse, D., Oliver, G., Rodier-Goud, M., Pinard, F. and Seguin, M. (2003). Molecular mapping of genes conferring field resistance to South American Leaf Blight *(Microcyclus ulei)* in rubber tree. *Theoretical and Applied Genetics*, **108**: 160-167.

Lekawipat, N., Teerawatanasuk, K., Rodier- Goud, M., Seguin, M., Vanavichit, A., Toojinda, T. and Tragoonrang, S. (2003). Genetic diversity analysis of wild germplasm and cultivated clones of *Hevea brasiliensis* Muell. Arg. by using microsatellite markers. *Journal of Rubber Research*, **6**: 36-47.

Lespinasse, D., Rodier-Goud, M., Grivet, L., Leconte, A., Legnate, H. and Seguin, M. (2000a) A saturated genetic linkage map of rubber tree *(Hevea* spp.) based on RFLP, AFLP, microsatellite and isozyme markers. *Theoretical and Applied Genetics*, **100**: 127-138.

Lespinasse, D., Grivet, L., Troispoux, V., Rodier-Goud, M., Pinard, F. and Seguin, M. (2000b). Identification of QTLs involved in the resistance to South American leaf blight *(Microcyclus ulei)* in the rubber tree. *Theoretical and Applied Genetics*, **100**: 975-984.

Li, D., Deng, Z., Qin, B., Liu, X. and Men, Z. (2012). *De novo* assembly and characterization of bark transcriptome using Illumina sequencing and development of EST-SSR markers in rubber tree *(Hevea brasiliensis* Muell. Arg.) *BMC Genomics*, **13**: 192.

Licy, J. (1997). Variability correlation and heterosis for yield and yield components in certain hybrid clones of para rubber tree *Hevea brasiliensis* Muell. *Arg. Doctoral dissertation,* University of Kerala, Trivandrum, India.

Litt, M. and Luty, J.A. (1989). A hypervariable microsatellite revealed by in vitro amplification of a dinucleotide repeat within the cardiac muscle actin gene. *The American Journal of Human Genetics*, **44**: 397-401.

Low, F. C., Atan, S., Jaafar, H. and Tan, H. (1996). Recent advances in the development of molecular markers for *Hevea* studies. *Journal of Natural Rubber Research*, **11**: 32-44.

Mantello, C. C., Cardoso-Silva, C. B., da Silva, C. C., de Souza, L. M., Scaloppi, Junior E. J., Gonçalves, P. S., Vicentini, R. and de Souza, A. P. (2014) *De Novo* Assembly and transcriptome analysis of the rubber tree (*Hevea brasiliensis*) and SNP markers development for rubber biosynthesis pathways. *PloS ONE* **9**:e102665. doi: 10.1371/*journal.pone*.0102665.

Mantello, C. C., Suzuki, F. I., de Souza, L. M., Goncalves, P. S. and de Souza, A. P. (2012). Microsatellite marker development for the rubber tree *(Hevea brasiliensis):* characterization and cross-amplification in wild *Hevea* species. *BMC Research Notes*, **5**: 329.

Markert, C.L. and Moller, F. (1959) Multiple forms of enzymes: tissue, ontogenetic, and species specific patterns. *Proceedings of the National Academy of Sciences*, **45**: 753-763.

Mathew, R., Roy, B., Ravindran, M., Nazeer, M.A. and Saha, T. (2005). Phylogenetic relationship of *Hevea brasiliensis* species as revealed through molecular markers. *Natural Rubber Research*, **18**: 14-24.

Michaels, S.D. and Amisino, R.M. (1998). A robust method for the detecting single-nucleotide changes as polymorphic markers by PCR. *Plant Journal*, **14**: 381-385.

Morgan, T.H. (1911) An attempt to analyze the constitution of the chromosomes on the basis of sex-limited inheritance in Drosophila. *Journal of Experimental Zoology*, **11**: 365-413.

Morgante, M. and Olivieri, A.M. (1993). PCR amplified microsatellites as markers in plant genetics. *Plant Journal*, **3**: 175-182.

Myers, R.M., Sheffield, V.C. and Cox. D. R. (1988). Detection of single base changes in DNA: ribonuclease cleavage and denaturing gradient gel electrophoresis: A Practical Approach. IRL. In: Davies, K.E. (Ed.), *Genome Analysis*. Oxford. pp. 95-139.

Nakkanong, K., Nualsri, C. and Sdoodee, S. (2008). Analysis of genetic diversity in early introduced clones of rubber tree *(Hevea brasiliensis)* using RAPD and microsatellite markers. *Songklanakarin Journal of Science and Technology*, **30**: 553-560.

Neff, M. M., Neff, J. D., Chory, J. and Pepper, A. E. (1998). dCAPS a simple technique for the genetic analysis of single nucleotide polymorphisms: experimental applications in Arabidopsis thaliana genetics. *Plant Journal*, **14**: 387-392.

Newton, C. R., Graham, A., Heptinstall, L. E., Powell, S. J., Summers, C., Kalsheker, N, Smith, J. C. and Markham, A. F. (1989). Analysis of any point mutation in DNA. The amplification refractory mutation system (ARMS). Nucleic Acids Research, **17**: 2503–2516.

Nurhaimi, H., Woelan, S. and Darusamin, A. (1998). RAPD analysis of genetic variability in plant rubber *(Hevea brasiliensis* Muell. Arg.) clones. *Menara Perkebunan*, **66**: 9--19.

Oktavia, F. and Kuswanhadi, L. M. (2011). Genetic relationship of Wickham and IRRDB 1981 rubber population based on RAPD markers analysis. *HAYATI Journal of Biosciences*, **18**: 27-32.

Orita, M., Iwahana, H., Kanazawa, H., Hayashi, K. and Sekiya, T. (1989). Detection of polymorphisms of human DNA by gel electrophoresis as single-strand conformation polymorphism. *Proceedings of the National Academy of Sciences*, **86**: 2766-2770.

Orozco-Castillo, C., Chalmers, K. J., Waugh, R. and Powell, W. (1994).) Detection of genetic diversity and selective gene introgression in coffee using RAPD markers. *Theoretical and Applied Genetics*, **87**: 934-940.

Paetkau, D. (1999). Microsatellites obtained using strand extension: An enrichment protocol. *BioTechniques*, **26**: 690-697. Perseguini, J., Romão, L., Briñez, B., Scaloppi, E.J.R., Gonçalves, P. and Benchimol L.L. (2012) Genetic diversity of cultivated accessions and wild species of rubber tree using EST-SSR markers. Pesquisa Agropecuária Brasileira, 47: 1087-1094.

Perseguini, J.M.K.C., Oblessuc, P.R., Rosa, J.R.B.F., Gomes, K.A., Chiorato, A.F., Carbonell, S.A.M. and Benchimol-Reis, L.L. (2016). Genome-Wide association studies of anthracnose and angular leaf spot resistance in common bean *(Phaseolus vulgaris L.)*. PloS ONE, **11(3)**, e0150506.

Phan, J., Reue, K. and Peterfy, M. (2000). MS-IRS: A simple method for isolation of microsatellites. *Bio Techniques*, **28**: 18-20.

Picoult-Newberg, L., Ideker, T. E., Pohl, M. G., Taylor, S. L., Donaldson, M. A., Nickerson, D. A. and Boyce-Jacino, M. (1999). Mining SNPs from EST data bases. *Genome Research*, **9**: 167-174.

Plomion, C., Frigerio, J. M., Ridolfi, M., Pot, D., Pionneau, C., Bodenes, C., Kremer, A., Hurme, P., Savolainen, O., Avila, C., Gallardo, F., Canovas, F. M., David, H., Neutelings, G. and Campbell, M. (1999). *Developing SSCP markers in two Pinus species. Molecular Breeding*, **5**: 21-31.

Pootakham, W., Chanprasert, J., Jomchai, N., Sangsrakru, D., Yoocha, T. and Therawattanasuk, K. (2011). Single nucleotide polymorphism marker development in the rubber tree, *Hevea brasiliensis* (Euphorbiaceae). *American Journal of Botany*, **98**: e337-e338.

Pootakham, W., Ruang-Areerate, P., Jomchai, N., Sonthirod, C., Sangsrakru, D., Yoocha, T., Theerawattanasuk, K., Nirapathpongporn, K., Romruensukharom, P., Tragoonrung, S. and Tangphatsornruang, S. (2015). Construction of a high-density integrated genetic linkage map of rubber tree (*Hevea brasiliensis*) using genotyping-by-sequencing (GBS). *Frontiers in Plant Science*, **6**: 367.

Powell, W., Morgante, M. and Andre, C. (1996). The comparison of RFLP, RAPD, AFLP and SSR (microsatellite) markers for germplasm analysis. *Molecular Breeding*, **2**: 225-238.

Rafalski, A. (2002). Applications of single nucleotide polymorphisms in crop genetics. *Current Opinion in Plant Biology*, **5**: 94-100.

Rahman, A. Y. A., , Usharraj , A.O., Misra, B.B. et al. (2013) Draft genome sequence of the rubber tree Hevea brasiliensis. BMC Genomics, 14: 75. doi: 10.1186/1471-2164-14-75. Risch, N. and Merikangas, K. (1996). The future of genetic studies of complex human disease. *Science*, **273**: 1516-1517.

Ronaghi, M., Uhlen, M. and Nyren, P. A. (1996). A sequencing method based on real time pyrophosphate. *Science*, **281**: 363-365.

Roy, C.B. and Saha, T. (2012). Generation of transcriptome resources in rubber (*Hevea brasiliensis*) in response to *Corynespora cassiicola* causing Corynespora leaf disease for gene discovery and marker identification using NGS platform. *International Rubber Conference IRC* 2012, 29-31 October 2012, Thiruvananthapuram.

Roy, C.B., Nazeer, M.A. and Saha, T. (2004). Identification of simple sequence repeats in rubber *(Hevea brasiliensis). Current Science*, **87**: 807-811.

Roy, C.B., Ravindran, M. and Saha, T. (2012a). One hundred microsatellite markers from *Hevea* genome for characterization of cultivated clones of rubber *(Hevea brasiliensis).* 24th Kerala Science Congress, 29-31, January, *Rubber Research Institute of India*, Kottayam.

Roy, C.B., Ravindran, M. and Saha, T. (2012b) Efficient screening of AFLP primer combinations for evaluating genetic diversity among cultivated rubber (*Hevea brasiliensis*) clones. *Natural Rubber Research*, **25:** 21-30.

Russell, J.R., Fuller, J.D., Macaulay, M., Hatz, B.G., Jahoor, A., Powell, W. and Waugh, R. (1997). Direct comparison of levels of genetic variation among barley accessions detected by RFLPs, AFLPs, SSRs and RAPDs. *Theoretical and Applied Genetics*, **95**: 714-722.

Saha, T. and Priyadarshan, P.M. (2012) Genomics of Hevea rubber. In: Schnell, R.J. and Priyadarshan, P.M. (Eds.), *Genomics of Tree Crops.* Springer, New York. pp. 261-298.

Saha, T., Roy, C.B. and Nazeer, M.A. (2005). Microsatellite variability and its use in the characterization of cultivated clones of *Hevea brasiliensis. Plant breeding*, **124**: 86--92.

Saha, T., Roy, C.B., Ravindran, M., Bini, K. and Nazeer, M.A. (2007). Allelic diversity revealed through SSR polymorphisms at locus encoding HMG- CoA reductase in rubber *(Hevea brasiliensis). Silvae Genetica*, **56**: 58-65.

Saha, T., Roy, C.B., Ravindran, M., Bini, K. and Thomas, K. U. (2015). Towards development of molecular markers for *Phytophthora* tolerance in rubber (*Hevea brasiliensis*). In: *Phytophthora Diseases of Plantation Crops*. Westville Publishers, New Delhi. pp. 144-151.

Schlotterer, C., and Tautz, D. (1992). Slippage synthesis of simple sequence DNA. *Nucleic Acids Research*, **20**: 211-215.

Seguin, M., Gay, C., Xiong, T.C. and Rodier-Goud, M. (2001). Microsatellite markers for genome analysis of rubber tree *(Hevea* spp.). Biotechnology and rubber tree: *Proceedings of IRRDB symposium*. pp. 25-28.

Semagn, K., Bjornstad, A. and Ndjiondjop, M.N. (2006). Principles, requirements and prospects of genetic mapping in plants. *African Journal of Biotechnology*, **5**: 2569-2597.

Sharma, R.K., Gupta, P., Sharma, V., Sood, A., Mohapatra, T. and Ahuja, P.S. (2008). Evaluation of rice and sugarcane SSR markers for phylogenetic and genetic diversity analysis in bamboo. Genome, **51**: 91-103

Shearman, J.R., Sangsrakru, D., Jomchai, N., Ruang-Areerate, P., Sonthirod, C. and Naktang, C. (2015). SNP identification from RNA sequencing and linkage map construction of rubber tree for anchoring the draft genome. *PloS ONE* **10**: e0121961.

Shoucai, C. S., Hansuang, H., Dong Qiong, H., Sheng, L. and Xu e-qin, Z. (1994). Identification of mildew resistant gene from *Hevea* tree by RAPD technique. *Chinese Journal of Tropical Crops*, **15**: 26.

Sommer, S.S., Groszbach, A.R. and Bottema, C.D.K. (1992). PCR amplification of specific alleles (PASA) is a general method for rapidly detecting known single-base changes. *BioTechniques*, **12**: 82-87.

Souza, L. M., Mantello, C. C., Santos, M. O., Goncalves, P. S. and de Souza, A. P. (2009). Microsatellites from rubber tree *(Hevea brasiliensis)* for genetic diversity analysis and cross-amplification in six *Hevea* wild species. *Conservation Genetics Resources*, **1**: 75-79.

Souza, L.M., Gazaffi, R., Mantello, C.C., Silva, C.C. and Garcia, D. (2013). QTL mapping of growth-related traits in a full-sib family of rubber tree (*Hevea brasiliensis*) *evaluated in a sub-tropical climate*. *PloS ONE*, **8(4)**: e61238.

Souza, L.M., Mantello, C.C., Suzuki, F., Gazaffi, R., Garcia, D., Le Guen, V., Garcia, A.A.F. and de Souza, A.P. (2011). Development of a genetic linkage map of rubber tree (*Hevea brasiliensis*) based on microsatellite markers. *BMC Proceedings*, **5**(7): 39.

Stadler, L. J. (1929) Chromosome number and the mutation rate in *Avena* and *Triticum*. *Proceedings of the National Academy of Sciences*, **15**: 876-881.

Sturtevant, A.H. (1913) The linear arrangement of six sex-linked factors in *Drosophila*, as shown by their mode of association. *Journal of Experimental Zoology*, **14**: 43-59.

Tang, C., Yang, M., Fang, Y., et al. (2016) The rubber tree genome reveals new insights into rubber production and species adaptation. Nature Plants, doi:10.1038/

nplants.2016.73. Tautz, D., Trick, M. and Dover, G.A. (1986). Cryptic simplicity in DNA is a major source of genetic variation. *Nature*, **322**: 652-656.

Thomas, M.R. and Scott, N.S. (1993). Microsatellite repeats in grape vine reveal DNA polymorphism when analysed as sequence tagged sites. *Theoretical and Applied Genetics*, **86**: 985-990.

Toruan-Mathius, N., Lalu, Z., Soedarsono and Aswidinnor, H. (2002). Genetic variation of rubber (*Hevea brasiliensis* Muell. Arg.) clones resistance and susceptible to *Corynespora cassiicola* using RAPD and AFLP markers. *Menara Perkebunan*, **70**: 35-48.

Triwitayakorn, K., Chatkulkawin, P., Kanjanawattanawong, S., Sraphet, S., Yoocha, T., Sangsrakru, D., Chanprasert, J., Ngamphiw, C., Jomchai, N., Therawattanasuk, K. and Tangphatsornruang, S. (2011). Transcriptome sequencing of *Hevea brasiliensis* for development of microsatellite markers and construction of a genetic linkage map. *DNA Research*, **18**: 471-482.

Varghese, Y.A., Knaak, C., Sethuraj, M.R. and Ecke, W. (1997). Evaluation of random amplified polymorphic DNA (RAPD) markers in *Hevea brasiliensis. Plant Breeding*, **116**: 47-52.

Varshney, R. K., Graner, A. and Sorrells, M. E. (2005). Genic microsatellite markers in plants: Features and applications. *Trends in Biotechnology*, **23**: 48-55.

Venkatachalam, P., Priya, P., Gireesh, T., Saraswathyamma, C. K. and Thulaseedharan, A. (2006). Molecular cloning and sequencing of a polymorphic band from rubber tree [*Hevea brasiliensis* (Muell.)Arg.]: The nucleotide sequence revealed partial homology with proline specific permease gene sequence. *Current Science*, **90**: 1510-1515.

Venkatachalam, P., Priya, P., Saraswathyamma, C.K. and Thulseedharan, A. (2004). Identification, cloning and sequence analysis of a dwarf genome specific RAPD marker in rubber tree *(Hevea brasiliensis). Plant Cell Reports*, **23**: 327-332.

Venkatachalam, P., Thomas, S., Priya, P., Thanseem, I., Gireesh, T., Saraswathyamma, C. K. and Thulaseedharan, A. (2002). Identification of DNA polymorphism among clones of *Hevea brasiliensis* Muell. Arg. using RAPD analysis. *Indian Journal of Natural Rubber Research*, **15**: 172-181.

Vos, P., Hogers, R., Bleeker, M., Reijans, M., van de Lee, T., Hornes, M., Frijters, A., Pot, J., Peleman, J., Kuiper, M. and Zabeau, M. (1995). AFLP: a new technique for DNA fingerprinting. *Nucleic Acids Research*, **23**: 4407-4414.

Wang, D.G., Fan, J.B., Siao, C.J., Berno, A., Young, P., Sapolsky, R., Ghandour, G., Perkins, N., Winchester, E. and Spencer, J. (1998). Large scale identification, mapping, and genotyping of single-nucleotide polymorphisms in the human genome. *Science*, **280**: 1077-1082.

Weising, K., Nybom, H., Wolf, K. and Kahl, G. (2005). DNA finger printing in plants: *Principles, Methods and Applications, 2nd Edn.* CRC Press, Boca Raton. Taylor and Francis Group. pp. 277.

Williams, J.G.K., Kubelik, A.R., Livak, K.J., Rafalski, J.A. and Tingey, S.V. (1990). DNA polymorphisms amplified by arbitrary primers are useful as genetic markers. *Nucleic Acids Research,* **18**: 6531-6535.

Wu, S.H., Wirthensohn, M., Hunt, P., Gibson, J. and Sedgley, M. (2008). High resolution melting analysis of almond SNPs derived from ESTs. *Theoretical and Applied Genetics*, **118**: 1-14

Wu, W.M., Tsai, H.J., Pang, J.H.S., Wang, H.S., Hong, H.S. and Lee, Y.S. (2005). Touchdown thermo cycling program enables a robust single nucleotide polymorphism typing method based on allele-specific real-time polymerase chain reaction. *Analytical Biochemistry*, **339**: 290-296.

Yamanaka, S., Nakamura, I., Watanabe, K.N. and Sato, Y.I. (2004). Identification of SNPs in the *waxy* gene among glutinous rice cultivars and their evolutionary significance during the domestication process of rice. *Theoretical and Applied Genetics*, **108**: 1200-1204.

Yanagisawa, T., Kiribuchi-Otobe, C., Hirano, H., Suzuki, Y. and Fujita, M. (2003). Detection of single nucleotide polymorphism (SNP) controlling the waxy character in wheat by using a derived cleaved amplified polymorphic sequence (dCAPS) marker. *Theoretical and Applied Genetics*, **107**: 84-88.

Zane, I., Bergelloni, L. and Patarnello, T. (2002). Strategies for microsatellite isolation: a review. *Molecular Ecology*, **11**: 1-16.

Zhu, Y. L., Hyatt, S., Quigley, C. and Song, Q. J. (2001). Single nucleotide polymorphisms (SNPs) in soybean genes cDNAs and random genomic sequence. In: *Plant and Animal Genome IX Conference*, San Diego.

Chapter 11

Coconut

☆ *L. Perera, R. Manimekalai, S. Sudarsono, I. Maskromo and P. Lestari*

1. Introduction

Coconut (*Cocos nucifera* L.) is a member of the monocot family *Arecaceae* (*Palmaceae*) and the subfamily *Cocoideae* that includes 27 genera and 600 species, and is currently the only species of the genus *Cocos*. Coconut possesses a diploid genome with 32 chromosomes (2n=2x=32). Many different varieties and forms of coconut have been described by different workers (Narayana and John, 1949; Gangolly *et al.*, 1957; Menon and Pandalai, 1958 and references there in; Liyanage, 1958) based on morphology, breeding habit and more recently according to their place of origin (Ohler, 1984). There is a great diversity in coconut, for example fruits vary for size, shape and colour and the proportions by weight of fruit components *viz.* husk, shell, endosperm and water (the liquid endosperm). There are also coconut variants having high value, such as: Makapuno, Kopyor and Aromatic coconuts (Maskromo *et al.*, 2015a; Maskromo *et al.*, 2013b, c; Maskromo *et al.*, 2011). Realizing the benefits of coconut as revealed by newest research findings, people are accepting coconut today as a wonder food commodity. With all the health benefits attributed to the coconut, there is a big demand opening in the world market for coconut today and therefore there is a need for increased production and supply of coconut. In some countries (*i.e.* the Philippines, Thailand and Indonesia), breeding for unique and specific characters, such as makapuno and kopyor type solid endosperms and aromatic liquid endosperm, which have higher value have also been initiated (Sudarsono *et al.*, 2015; Novarianto *et al.*, 2014; Sudarsono *et al.*, 2014). Thus collection and evaluation of coconut germplasm and breeding coconut for desirable characters has become a priority in many research programmes in many coconut growing countries.

2. Coconut Germplasm

Germplasm is a collection of genetic resources of the target crops. Availability of diverse coconut germplasm is the pre-requisite for coconut breeding because they are the genetic source used by coconut breeders to develop new cultivars. Coconut germplasm is mainly stored as a living collection of trees in the field (*ex situ*) which is a very resource intensive task. A programme for coconut germplasm collection and conservation has been in place in coconut centers around the world for many years, resulting in entries of 1621 coconut accessions, comprising of at least 1200 tall and 421 dwarf coconuts, deposited in the Coconut Genetic Resources Database (CGRD). Selection of those coconut accession/collections is mainly based on phenotype and ecological distribution. Hence this coconut collection would probably contain redundancies and genetically close accessions. Further, appropriately designed collection strategies have not been in place in coconut for the purpose of identification of exact number of locations and collection of exact number of individual per location to ensure capturing the highest diversity and loosing rare alleles at the lowest probability. For this, within and between population variation and population differentiation statistics have to be developed.

The selection of parents in the coconut breeding in many national and international coconut breeding activities has been arbitrary and based on mostly differences in phenotypes and place of origin. Application of molecular markers could help overcome the above weaknesses and strengthen the germplasm conservation and breeding programmes. Indonesian Palm Research Institute (IPRI), Manado in collaboration with Bogor Agricultural University have validated existence of redundancies and genetically close coconut accessions in the germplasm collections, using molecular markers (Sudarsono *et al.*, 2015, Kumaunang and Maskromo, 2007). Therefore, molecular analysis of germplasm collections is an important task.

3. Molecular Markers

Molecular markers have become an important tool in almost every crop plant to manage germplasm collection and conservation, generate diversity indices, track down origins and thereby accelerate breeding programmes. DNA-based marker techniques are considered more effective as opposed to traditional techniques using morphological and descriptive markers, which are laborious, subjective, time consuming and limited. Attempts to incorporate isozyme techniques in coconut breeding have been reported in the Philippines (Benoit, 1979; Carpio, 1982), India (Geethalakshmi *et al.*, 2004; Parthasarathy *et al.*, 2004), Indonesia (Novarianto. 1988; Novarianto *et al.*, 1988a,b; Novarianto *et al.*, 1993) and in Sri Lanka (Fernando, 1995). However, its use for characterizing coconut populations has been limited. Isozymes are relatively cheap, easy to use, independent of environment, and co-dominant markers; however, its use in coconut breeding is limited because the number of detected polimorphic loci are limited.

Currently, many DNA based marker techniques have been employed in various genetic studies to reveal variations at the DNA level. Out of these, the most popular and widely used DNA marker techniques are Restriction Fragment

Length Polymorphism (RFLP), Randomly Amplified Polymorphic DNA (RAPD) and Amplified Fragment Length Polymorphism (AFLP). Simple Sequence Repeats (SSRs) or microsatellites have also been developed to screen individuals. Microsatellites are tandem repeats of short sequence motifs occurring randomly in the eukaryotic genomes, in which the basic repeat unit is around 1- 6 base pairs. These simple repeats in the nucleotide sequences are mostly located in the centromere and are mostly represent non-coding regions of the genomic DNA. Recently, with the availability of sequencing data from transcriptome and genome sequencing projects, Single Nucleotide Polymorphism (SNP) have been developed and utilized to support coconut breeding programs (Mauro-Herrera *et al.*, 2006). All of molecular markers rely on the existence of polymorphism. Polymorphism is simply variation at the DNA base sequences or at the restriction sites (site where the restriction enzyme cleaves the DNA) due to natural occurrence of base substitution or insertion-deletion (InDel) mutations. Several combinations of PCR-based molecular techniques reveal the existence of high degree resolution of polymorphism (variation) and these form of variations could be used as screening methods to ascertain the identity of the individual plant and its relationship to the population.

4. Molecular Markers Availability and their Use in Coconut

Currently, there are many powerful molecular marker techniques and the choice of the markers depend upon several factors such as the information content of the marker, ease of performance, reproducibility, expense and availability of the experts and the facilities. A variety of molecular markers have been used in coconut but the most popular type of markers currently or exhaustively used in coconut include Restriction Fragment Length Polymorphism (RFLP), Randomly Amplified Polymorphic DNA (RAPD), Amplified Fragment Length Polymorphism (AFLP) and Simple Sequence Repeats (SSR) or microsatellites markers. With many DNA sequencing projects becoming reality for coconut, it would be possible to evaluate nucleotide diversity among gene sequences in coconut and identify the existence of single nucleotide polymorphisms (SNPs). Single nucleotide amplified polymorphism (SNAP) marker may be generated based on the presence of SNPs in the sequences of target gene (Maskromo *et al.*, 2013).

The first DNA marker system reported to be used in coconut was the RAPD (Everard, 1996; Ashburner *et al.*, 1997; Duran *et al.*, 1997; Wadt *et al.*, 1999;). The principle of RAPD analysis is amplification of random DNA segments with single short arbitrary primers resulting in the amplification of several discrete DNA products which can be separated on agarose gels by electrophoresis (Williams *et al.*, 1991; Welsh and McClelland, 1991). This method is fast and simple and requires no prior sequence knowledge or the use of radioactivity and requires only small amounts of DNA for the analysis (Babu *et al.*, 2014). However, the disadvantages of this method are that it is a dominant marker and most probably less reproducible between assays and laboratories (Rafalski *et al.*, 1996; Powell *et al.*, 1996a). However, RAPD markers have contributed to the wealth of information in coconut. Manimekalai and Nagarajan (2006) used RAPD markers to derive inter-relationship among coconut germplasm accessions and reported low genetic similarity among South Pacific and South East Asian coconut accessions. Rajesh *et al.* (2013) have identified a RAPD marker capable of differentiating tall and dwarf coconut palms. Sequence characterized amplified region (SCAR) primers were designed from the unique RAPD amplicons and validated. RAPD markers have also been used to analyze intra-population diversity of Bali Tall, Mapanget Tall, Sawarna Tall and Tenga Tall in Indonesia (Novarianto *et al.*, 2001; Pandin, 2009a, b).

RFLP markers have also been employed in coconut (Lebrun *et al.*, 1999a; Lebrun *et al.*, 1998), though in a limited manner. RFLP (Beckmann and Soller, 1983) was one of the first techniques to be used widely to detect variation at the DNA level. Such markers are expected to be numerous, developmentally stable and co-dominant. The major drawbacks of RFLPs are that the process is very labour intensive, expensive and requires large amount of good quality DNA (Powell et al., 1996b). The use of AFLPs in coconut is also limited but a good coverage of genetic diversity and genetic relationships of coconut of Sri Lanka (Perera *et al.*,1996) and entire geographic range of coconut cultivating area in the world (Perera *et al.*, 1996; Teulat *et al.*, 2000) have been published.

In coconut, use of SSR markers provides wider coverage as its applications have been very diverse ranging from assessing genetic diversity in natural populations to creating high-resolution genetic maps (Perera *et al.*, 1999, 2000, 2001, 2003; Teulat *et al.*, 2000, Rivera *et al.*, 1999; Meerow *et al.*, 2003; Akuba, 2002). SSR markers have also been used to study pollen dispersal and determined the level of self and cross pollination among Dwarf and Tall Coconuts (Larekeng *et al.*, 2015a,b). Although SSRs are one of the most effective candidate marker system for population analysis (Karp, 1999), their development used to be expensive as isolation of DNA fragment carrying SSR sequences and DNA sequencing are needed for each and every plant. Some examples of cross amplification of SSRs between closely related species are however possible and an example of such study has also been published (White and Powell, 1997a). Despite the high cost involved in isolation and development of SSRs, substantial number of coconut SSR primers have been developed and published (Perera *et al.*, 1999, 2000; Rivera *et al.*, 1999; Teulat *et al.*, 2000). A set of 39 coconut specific microsatellite primer pairs have been developed by Perera *et al.* (1999) from a small insert genomic library enriched for CA repeats using genomic DNA from Sri Lankan Tall coconut variety, following the method described by White and Powell (1997b). Moreover, another set of 38 polymorphic microsatellite primer pairs have been isolated by Rivera *et al.* (1999) from a small insert genomic library enriched for several types of microsatellite repeats following the method described by Edwards *et al.* (1996). using genomic DNA from the Philippines coconut variety Tagnanan Tall (primer sequences are available in Teulat *et al.*, 2000).

Despite the differences in molecular marker system, laboratory where the study was conducted, number and type of accessions or varieties used among the investigations, the results obtained (Rohde *et al.*, 1995; Lebrun *et al.*, 1998; Perera *et al.*, 2003; Teulat *et al.*, 2000) generally revealed distribution of genetic diversity of coconut and genetic relatedness in coconut accessions. A high level of genetic diversity of tall coconut was observed in all studies compared to dwarf with dwarfs showing great loss of allelic richness. The distribution of genetic diversity between accessions in talls was also found to be higher than that of dwarf group. These finding have led to proposal of changes to the germplasm collection strategies for dwarf and tall groups (Perera, 1999; Ashburner *et al.*, 1997). The phenetic tree (Perera *et al.*, 2003; Teulat *et al.*, 2000) constructed in these studies have revealed the genetic relationships of coconut in the world: all tall coconuts were divided into two main groups, the first group comprising all the tall varieties from Southeast Asia, the Pacific and the West coast of Panama and all dwarfs in a sub-cluster within the tall cluster. The second group consisted of talls from South Asia, East Africa and West Africa. Interestingly, none of the dwarf coconuts grouped with the second main tall group. These results were very much in agreement with the conclusions of Harries (1978) on the evolution and dissemination of coconuts based on morphological, phenotypical and ethno botanical evidences. According to Harries (1978) naturally evolved coconuts; characterized as 'Niu Kafa' type, predominate in South Asia, West and East Africa, the Caribbean and the Atlantic coast of Central America while coconuts selected under cultivation; characterized as 'Niu Vai' type, predominate in Southeast Asia, some Pacific islands and the West coast of Central America. It is generally accepted that the coconut palm came into existence on the Atlantic coast of Africa, South America and around the Caribbean region for only about 500 years (Child, 1974; Purseglove, 1972) and that there is a great similarity between these coconuts and those coconuts in East Africa, India and Sri Lanka (Harries, 1978). The grouping of Panama Tall (Panama Manarge and Panama Aguadulce varieties; both from the Pacific coast of Panama) with Southeast Asian and Pacific Talls is in agreement with Whitehead's (1976) observation of an eastward movement of coconuts from Southeast Asia to the Pacific region and subsequently from there to the Pacific region to the Pacific coast of America. These results are largely in agreement with the results from ISTR (Inverse Sequence-Tagged Repeats) analysis (Rohde *et al.*, 1995), which grouped Panama Tall with Polynesian varieties/populations of coconuts.

The grouping of all dwarf forms from different geographical regions in a single cluster within the main South Asia and Pacific group and the 'Niu Vai' type of coconuts and loss of allelic richness observed in dwarfs suggest that all dwarfs have a common origin and evolved from the Southeast Asia/Pacific group of talls. 'Niu Vai' type of coconuts in the Southeast Asia/Pacific region, were domesticated there and only later introduced to the other regions. The results of Teulat *et al.* (2000) strongly support a common origin of dwarf varieties. Manimekalai and Nagarajan (2007) used SSR markers to derive genetic relationship among coconut accession belonging to different geographic regions. Dwarf and intermediate accessions showed highest similarity among them. The tall accessions belonging to South East Asia, South Asia and South Pacific were clustered based on their geographical regions, but dwarf and intermediate accessions were clustered separately. Clustering of accessions belonging to Atlantic and America revealed the spread of coconut from Far East to South Pacific.

SSR markers have also been utilized for ensuring the legitimacy of coconut hybrids, which overcomes unreliable morphological traits, like petiole colour, which were utilized earlier as a marker for hybrid identification (Perera, 2010; Rajesh *et al.*, 2012). This technique paves the way for ensuring authenticity and quality of hybrid seedlings of coconut before distribution to the farming community. Furthermore Sudarsono *et al.* (2015) have recently reported an innovative approach of identifying better parental combination for producing better coconut hybrids based on SSR markers. Use of SSR markers to evaluate pollen dispersal in Dwarf and Tall coconut has also been demonstrated (Larekeng *et al.*, 2015a,b). Parentage analysis has been used to evaluate pollen dispersal among Dwarf and Tall Kopyor coconut in two different kopyor coconut production centers in Indonesia. Investigations have been undertaken to elucidate the dispersal of pollen, the rate of self and out-crossing and the distance of pollen travel in both Pati Dwarf and Kalianda Tall Kopyor coconut population using six SSRs and four SNAPs.

A recent trend in identification and development of cost-effective SSRs has been mining publicly available ESTs (expressed sequence tags). Even though the extent of polymorphism revealed by EST-SSRs is low compared to genomic SSRs, they possess many advantages which include easy access, a high level of transferability to related species and probably represent functional genetic variation. Within the species studies conducted at the global level allowed identifying two large genetic groups, known as "Pacific" and "Indo-Atlantic" which can be seen as sub-species. The role of humans in the dissemination of coconut was highlighted by Gunn *et al.* (2011).

Recently Xiao *et al.* (2013) have scanned 57,304 coconut unigenes for di-, tri-, tetra-and hexa-nucleotide repeat sequences and an additional 30 microsatellites have been developed. These markers have been utilized to evaluate coconut genetic diversity in 30 individuals representing accessions from China (12 samples) and Southeast Asia (18 samples). Based on the results, it was inferred that Chinese coconuts did not evolve independently of the Southeast Asian populations. Combining population structural analyses and historic information, they have proposed a possible explanation for coconut dispersal patterns from Southeast Asia to China: sea currents could have carried coconuts into the Hainan province, while human dispersal from Southeast Asia may have brought coconuts to the Yunnan province.

With the recent advances in genomic research, there has been a shift towards utilization of gene-targeted, functional markers in lieu of random DNA markers. One such novel marker system is the start codon targeted polymorphism (SCoT), which was utilized for assessment of genetic relationship and diversity among 23 coconut accessions (10 talls and 13 dwarfs), representing different geographical regions (Rajesh *et al.*, 2015).

Although the general trends in placing coconut accessions in sub-clusters within the phenetic trees are generally consistent and are in accordance with their origin and geographical locations, consistency in placing individual accessions relative to each other within them is questionable. Different phenetic trees based not only on different molecular marker systems, but also on the same molecular marker systems has shown that placing the different individual accessions relative to the other accessions is different. Different marker systems are being applied by different research groups; most often both RAPD and SSR as the major tools. Only very few studies have been carried out in the world to compare the results obtained from different marker systems to see the consistency of the results and these represents recent studies on barley (Russell *et al.*, 1997) and cultivated potato (Milbourne *et al.*, 1997). Inconsistent results with different markers systems demonstrated on barley (Russell *et al.*, 1997), soybean (Powell *et al.*, 1996a), and *Brassica* (Thormann *et al.*, 1994), have led to an alarming situation that each PCR-based marker system needs careful evaluation before being applied with a new crop. To date, large number of coconut collections have been made and planted *ex situ* and further collecting of accessions are in progress in view of conserving more diversity. However, lack of adequate knowledge on the amount and distribution of genetic variation in coconut have hindered the implementation of appropriate conservation strategies (*i.e.* number of populations, number of individuals per population *etc.*). Molecular marker based characterization provide efficient method to quantify the genetic variation and to estimate genetic relationships between populations. Selection of molecular marker/s for assessing plant gene pool and germplasm analysis however has to be made with great care as studies on barley; soybean and *Brassica* have demonstrated inconsistent results with different marker assays. Therefore, comparison of outcome of different molecular marker systems is necessary to determine optimum number of markers and number of individuals per accessions with a particular marker system, for effectively determining the correct and consistent genetic relationships (genetic distances and similarities) among the accessions. This procedue will, when applied with different markers system, determine the cross- comparison of the results between systems and guide scientists in different regions or countries with the specifications (which marker to be used with which number of individuals per accession). The results of this kind of a study will help to develop a suitable molecular marker approach for coconut and for molecular characterization. In view of these objectives, highly diverse coconut varieties (*viz.* talls, dwarf greens, San Ramon and King Coconut), each variety represented by 20 individuals or more, should be studied with 20 or more numbers of markers with RAPD primers being the choice for less resource countries and SSR primers the choice of resource and skillful countries. Similarity matrices should be constructed from data for each marker systems with differing number of primer pairs and differing number of individuals per population. Dendrograms would be generated from these matrices to see the genetic relationships to check any changes in the relationships with respect to changing the number of markers and the number of individuals. The amount of genetic diversity should also be studied with the changing number of primers and the number of individuals. This whole procedure should enable estimation of the appropriate number of markers and the number of individuals per population for a proper DNA assay for each marker system. Same similarity matrices should also be used to study the improvement in correlation between the results obtained from different system with changing number of markers and numbers of individuals thus an appropriate package can be designed with each marker system. Based on the allele frequencies of the sample, exact number of individuals that should be collected from a population to capture the maximum genetic diversity would be estimated for future conservation. The effectiveness of each system should be determined in terms of number of loci revealed by each marker system and the amount of polymorphism detected. At the same time, this approach should also study their outcome with phenotypic diversity of coconut. Once an appropriate marker package is developed, approximately 100 coconut populations from the coconut biodiversity conservation programme should be evaluated and characterized in order to develop core collections, eliminate duplicates and identify priority germplasm for breeding.

In view of an appropriate package of a DNA assay for coconut, a microsatellite kit has been developed to identify coconut cultivars (Baudouin and Lebrun, 2002). Its construction involved as stated by Baudouin and Lebrun (2002); several steps such as gathering more than 600 DNA samples belonging to 113 reference populations in order to represent accurately the coconut genetic diversity worldwide, producing a set of 80 polymorphic microsatellite markers, screening them on a reduced set of coconut genotypes, testing the remaining 14 markers on the 600 coconut genotypes, devising protocols, adapted to the use in producing countries and devising suitable statistical methods, in order to identify the reference population, which is the most likely source of a given DNA sample. However this microsatellite kit is unable to distinguish between some coconut cultivars (personal observation).

Further to the considerations above, how the markers are distributed within the coconut genome and whether the markers are genomic or functional (ESTs) are important considerations as depending on those factors the results seems to vary. Thus a framework map of coconut and an improved coconut marker system is necessary. Studies on genome mapping in coconut are still limited, preliminary and recent. The first genome map for coconut (Rohde *et al.*, 1999) was developed based on ISTR markers. This was further extended with a different mapping population using AFLP, ISTR, RAPD and ISSR markers. Three hundred and eighty two makers have been placed in the map resulting in 16 linkage groups and identifying six QTLs for early germination (Herran *et al.*, 2000). Further, QTL for other traits such as, leaf production, girth and height has also been identified for the same mapping population (Ritter *et al.*, 2000). In addition to this, another mapping population in Ivory Coast resulted in 280 [markers] mapped on 16 linkage groups and QTLs related to nut number, bunch number and fruit component related traits (Lebrun *et al.*, 2001, Baudouin *et al.*, 2006) identified. Bandaranayake and Kearsey (2005) stated in their publication that the size of the mapping population is critical in any genome map. Bandaranayake *et al.* (2005) concluded through a simulation study that between 200 to 400 individuals is the effective size of a mapping population for coconut for a steady map resolution. The coconut linkage maps described in Herran *et al.* (2000) and Lebrun *et al.* (2001) were based on less than 65 individuals. Based on the experience and the information generated, a mapping population comprising more than 250 individuals has been developed in Sri Lanka but genotyping results indicated that only 16 out of 300 microsatellite markers are polymorphic between the parents thus making the F_1 not sufficiently polymorphic to be used for construction of a genetic map. Use of new mapping populations is being focused in Jamaica (unpublished) and in CIRAD, France (Luc Baudouin, unpublished data). A mapping population with 120 individuals has also been generated using F_2 intercross derived from Nias Yellow Dwarf x Kopyor Tall Homozygous coconut to find SSR and SNAP marker loci linked to kopyor mutant character in Indonesia. The mapping populations have been genotyped using 51 SSR and 17 SNAP functional marker loci and they were analyzed using linkage analysis software. Preliminary results of the analysis indicated that two SSR loci (CnCIR226 and CnCIR_J2) and two SNAP loci (SUS#3 and WRKY2#1) are probably linked to kopyor character in coconut (Maskromo, 2015). Once a proper framework map of coconut using all the available markers is developed, the choice of markers for the marker package should be developed based on their distribution in the genome. This should include functional markers such as EST-SSRs and SNAP markers too. Results generated utilizing either genomic markers or only with functional markers may differ depending on the genetic variation in the target areas of the genome.

Considering long-term physiological adaptation of the coconut palms (Peries, 1993), a collection and conservation of coconut representing different eco-geographical regions in Sri Lanka was initiated in 1991 (Perera *et al.*, 1996). A molecular marker study in Sri Lanka (Perera *et al.*, 2001) on 33 such tall coconut accessions using SSR markers revealed that there was no genetic differentiation among these accessions as the estimated variation between population accounted only less than 1 per cent. This has led to a change in the strategic plan of coconut germplasm collection in Sri Lanka; further random collection of coconut accessions discontinued and only biased collection is being carried out based on phenotypic characters. Among 33 accessions of coconut only few accessions were characterised to survive a very long and severe drought. The underlining mechanism of survival of those putative drought tolerant palms could be due to genetic factors contributed by drought tolerant genes/QTLs. Thus, a set of genomic and limited number of functional markers or combination of them DNA molecular marker based assay is seriously required. Can the genomic markers which are located in noncoding regions of the genome are qualified for capturing the functional genetic variation (variation in the cording regions) in plants and thus used to make inferences in a germplasm collection? Can the redundancies identified based on molecular markers be actually redundancies? Can co-collections developed based on molecular marker data guarantee that all the useful genes were made available to the breeders? Can even the limited number of functional markers are qualified for capturing the genetic variations occurred in the cording regions in plants, for example pest and disease resistance which are usually govern by major genes? Coconut molecular biologist needs to address these issues seriously and need to use molecular marker data rationally and cautiously depending on the objective of studies. There are instances that molecular markers cannot differentiate coconut accessions which are phenotypically highly distinguishable for many characters or distinguish by single characters. For example Raja brown dwarf, Tebing Tinggi Dwarf and Malayan Red Dwarf are indistinguishable by 12 genomic microsatellites (Perera, 2001). Similarly the Ambakelle special coconut accession, a drought tolerant selection Sri Lanka tall and its original population; Ambakelle tall accessions are indistinguishable by molecular markers (unpublished data). Availability of sequences of the same gene from different accessions would make it possible to evaluate nucleotide diversity among sequences of the target genes and identification of single nucleotide polymorphisms (SNPs), towards this end. SNAP markers could be generated based on the identified SNP in the target gene sequences. If the SNPs exist in the coding region and they are non-synonymous SNPs, it should be possible to evaluate the association between changes in amino acid residues, possible changes in protein function and the possible changes in phenotypes. As functional marker, SNAP is expected to be better than EST-SSR since SNPs are more frequently found in the genes and most probably are more polymorphic. Use of SNAP markers based on identified SNPs of eight different coconut genes (ABI, SACPD, SUS, and five WRKY genes) with 528 accessions of coconut germplasm collections (Dwarf and Tall coconuts) belonging to IPRI, Manado have been reported (Sudarsono *et al.*, 2015). Results of the analysis confirm the informativeness of the evaluated SNAP markers for analyzing coconut germplasm collections. Subsequently, the same SNAP markers have also been used to analyze diversity of Indonesian Kopyor coconuts collected from different locations in Indonesia. Results of the evaluation further confirmed the informativeness of the develop SNAP markers and their usefullness for cluster analysis of coconut germplasm collections.

Of all the described markers above, breeders who utilize them must be cautious when using information generated only through molecular markers. Combining molecular and phenotype data of the individual accessions and using the combine data to manage the existing coconut germplasm would be more resource efficients. Such combined data can be used to design more efficient hybridization and conservation strategies to develop new and more productive coconut cultivars. Thus evaluation and characterization of all germplasm accessions for their morphological characters and yield and yield related characters, for their responses to biotic and abiotic stresses and for their adaptation to different agro-climatic conditions are of also vital importance.

5. Data Analysis and Interpretation

Analysis and interpretation of molecular marker data is of another vital consideration. Analysis and interpretation of molecular marker data need careful examination of the data, their cross compatibility between assays and among laboratories and its repeatability. Need also arises for the involvement of a conventional breeder for sensible data interpretation. For example the most informative polymorphic marker system and the choice of marker today is the SSRs. (Tautz and Renz, 1984). Their high information content, which is directly related to the effective number of alleles at each locus (multi-allelic), hence, that SSRs are an ideal tool for many genetic applications (Bruford and Wayne, 1993; Queller *et al.*, 1993; Dallas *et al.*, 1995). Moreover, these markers are entirely defined by the sequence of the primer and thus can be easily exchanged between laboratories as primer sequences which facilitate collaboration between research groups. Comparison of SSR diversity levels has revealed higher levels of polymorphism compared to other molecular assay procedures and indicates that SSRs are ideal for germplasm assessment and varietal identification due to their high level of allelic diversity (Powell *et al.*, 1996b; Russell *et al.*, 1997). The observed hyper-variability of SSRs is thought to be due to the unique mechanism by which SSR variation is generated. These slippage events during DNA replication are currently the preferred explanation of the origin of SSR polymorphism and are thought to occur more frequently than the point mutation, insertion and deletion events responsible for generating polymorphism detectable by RFLP, AFLP and RAPD analysis (Tautz and Rentz, 1984; Tautz *et al.*, 1986). SSRs are assumed to follow a stepwise mutation model in comparison to the infinite allele mutation model (Valdes *et al.*, 1993; Di Rienzo *et al.*, 1994). The basic idea of the stepwise model is that mutations create new alleles that differ from their previous state by an increase or decrease of one step in the number of repeats. As empirical evidence suggests that mutational changes are often of one repeat unit (Weber and Wong, 1993), the stepwise mutation model has been 'revisited' (Shriver *et al.*, 1993; Valdes *et al.*, 1993; Di Rienzo *et al.*, 1994). Since the SSR data did not always appear to conform to a stepwise mutation model and coconut SSR data seemed to follow the same trend (*i.e.* certain loci did not exhibit the characteristic symmetrical, unimodal allele distribution) genetic distances should be calculated based on the proportion of shared alleles. In addition, it has previously been shown that this distance metric is most suitable for assessing genetic relationships between recently diverged taxa below the species level (Bowcock *et al.*, 1994; Provan *et al.*, 1999).

Although the AFLPs, in assessing the level and distribution of genetic diversity in coconut is a valuable technique, the dominant nature of the AFLP markers limits the amount of information that can be obtained compared to that of co-dominant marker systems such as RFLPs and SSRs. For example co-dominant markers detect heterozygotes from the homozygotes and allelic diversity within a locus. They also allow the direct study of the changes in allele frequencies under natural and artificial selection. However, the level of resolution achievable with RFLPs is limited and the procedure is expensive and laborious and therefore greatest possibility exists for SSR analysis. Moreover as SSRs are sequence tagged sites (STS's), they can also be used for construction of physical maps by STS mapping. Therefore SSR markers on coconut are not only be of value for diversity studies and DNA fingerprinting but also would particularly be beneficial for future studies which will involve genetic linkage mapping and monitoring of gene transmission through generations with collaborations between laboratories.

6. Coconut Genomics: Deciphering the Information from the Genome Sequence

The progress of next generation sequencing (NGS) technology has opened up avenues to understand plant and other organisms at the genome level. NGS techology has been used to generate genome sequences of a number of orphan crops and develop better understanding for the crops at molecular level. The potential progress in alleviating future problems made by availability of such genome sequences remains to be seen. However, breeders for some orphan crops have started to capitalize the potential of genome sequence availability to support their breeding programs. The determination of the whole genome sequence is essential to develop and exploit high-throughput breeding methods and mine the essential genes.

Coconut has also benefitted by the progress in NGS technology. The NCBI GenBank DNA database (http://www.ncbi.nlm.nih.gov/) is one of the repository keeping up the wealth of DNA sequences of almost all studied organisms, including coconut. Searching the NCBI GenBank nucleotide database (http://www.ncbi.

nlm.nih.gov/nucleotide, accessed: 10 Feb 2016, at 10:25) using key word of "*Cocos nucifera*" results in 212,603 entries and out of those 211,744 entries are nucleotide sequences of either chloroplastic, mitochondrial, or nuclear genomes originated from coconuts. Most of the entries come out from either the EST or the NGS sequencing data of coconut.

Further exploration in the NCBI GenBank SRA database (http://www.ncbi.nlm.nih.gov/sra), where publicly accessible outputs of NGS technology are stored, using key word of "*Cocos nucifera*" (accessed: 10 Feb 2016, at 10:25) results in eight entries of research activities describing the transcriptome studies (RNA Seq) of coconuts. Those coconut RNA Seq data deposited in the NCBI GenBank SRA database include: trascriptome profiling of (1) leaf (SRX534428), (2) embryo (SRX534380), (3) embryogenic callus (SRX472157), (4) leaf of root (wilt) disease susceptible cultivar (SRX437650), (5) leaf of root (wilt) disease resistant cultivar (SRX436961); and (6-8) coconut palm (SRX400192, SRX198908, and SRX518095), The sizes of deposited raw data ranged from 4.9 G bases sequences in the form of 3.3 Gb downloadable data up to 24.1G bases sequences in 16.5 Gb dowloadable data. Those are huge publicly available coconut transcriptome data ready for use and further exploration by scientist working with coconut. Since October 2012, at least four institutions have contributed to the wealth of those coconut transcriptome data, including: Academia Sinica (https://www.sinica.edu.tw/); Central Plantation Crops Research Institute (CPRI, http://cpcri.gov.in/); Coconut Research Institute (CRI, http://www.cri.gov.lk/), and University Putra Malaya (UPM, http://www.upm.edu.my/). In November 2015, Bogor Agricultural University (IPB) has initiated the transcriptome data generation by conducting RNA-Seq for normal coconut and Kopyor mutant coconut from zygotic embryo tissues. These data should be available by the end of 2016.

Transcriptomics generated EST sequences from different tissues, different development, or from cultivar with resistance – susceptible to certain disease, may actually be important to idetify the genes involved in the essential processes in coconut. Understanding pathways of fatty acid and other important biosynthetic precesses in coconuts may also results in better understanding of and the genes involved in the processes. Transcriptome and RNA Seq data generation followed by analysis of differentially expressed (DE) genes from such transcriptome data should be able to answer such questions. Published information from some of those coconut RNA seq studies have illustrated the possibilities of gaining such knowledge (Fan *et al.*, 2013; Huang *et al.*, 2013). Using such extensive transcriptome sequencing data, Fan *et al.* (2013). identified 57,304 unique genes from coconut and Huang *et al.,2013* identified the complete chloroplast genome of coconut.

The initial idea to sequence coconut genome has been around since October 2011 in the form of Coconut Genetic Resources Network (COGENT) research strategies (http://www.cogentnetwork.org/48-coconut-projects/research-ideas/110-sequencing-coconut-genome). As follow up of this, the steering committee of COGENT which met in Kochi, India (July 2012) has decided to set up an international thematic group on coconut genomics under the leadership of India to sequence coconut genome and develop tools for coconut improvement. Unfortunately, the initiated idea of sequencing coconut genome has not yet been realized. The second record of the attempt to sequence coconut genome was announced by Alsaihati *et al.* (2014) of Joint Center for Genomics Research, a cooperation between KACST Riyadh, Saudi Arabia and CAS, China. The attempt was announced in a presentation at Plant and Animal Genome XXII conference in Singapore in 2014 (https://pag.confex.com/pag/xxii/webprogram/Paper10752.html). The report stated that the estimated coconut genome size is 2.6 Gb and it contain 50-70 per cent repeat sequences. The assembly of the raw sequence data generated an estimated 0.97 – 2.47 Gb (37-94.5 per cent) of coconut genome sequence coverage (Alsaihati *et al.*, 2014). It also reported a total gap (per cent of draft size) of 251 – 749 Mb (10-29 per cent of draft size). Since that announcement, no further follow up information is reported by the group.

At the end of 2015, Indonesian Palm Genome Project was initiated as a collaborative project among BBBiogen, IPB, and IPRI under KKP3S Project. Part of the KKP3S Project targets are generating genome sequences of coconut and sugar palm. In the project, the genomes of a Tall (Tenga Tall) and a Dwarf (Pati Kopyor Green Dwarf) coconuts have been sequenced. The coconut genome sequencing result in a 148,109,277,288 total read base (bp) for the Dwarf coconut and 99,306,500,882 bp for the Tall one. The total GC and AT content of the total read data for Dwarf coconut are 38 per cent and 62 per cent while for Tall coconut are 37 per cent and 63 per cent, respectively (Lestari *et al.*, 2016). Assembly and downstream analysis of the raw data are on going and hopefully by 2017, preliminary reference of draft genome of coconut would be available for general publics. The generation of a reference coconut genome should be a landmark for coconut molecular breeding.

All molecular marker, transcriptomic, and genomic studies have generated a large amount of data. Most of them have been deposited in a public databases, TROPgeneDB (Tropgene: http://tropgenedb.cirad.fr/tropgene/JSP/index.jsp) and NCBI GenBank Database (http://www.ncbi.nlm.nih.gov/nucleotide and http://www.ncbi.nlm.nih.gov/sra). Making as much information as possible available to the community of coconut breeders should be continued. It is also emphasized here that necessary infra-structure facility and trained staff should be made available to all coconut growing countries in carrying out their own modern biotechnological research in coconut.

References

Akuba, R.H. (2002). Breeding and population genetic studies on coconut (*Cocos nucifera* L.) composite variety using morphological and microsatellite markers. PhD Thesis. *UPLB*, Philippines.

Alsaihati, B., Liu, W., Lin, Q. and Al-Mssallem, I.S. (2014). Coconut genome *de novo* sequencing. Poster PO46, Genome sequencing and ESTs session, January 13, 2014. Plant and Animal Genome XXII *(International Plant and Animal Genome Conference)*, San Diego, California, USA.

Ashburner, G.R., Thompson, W.K. and Halloran, G.M. (1997). RAPD Analysis of South Pacific Coconut Palm Populations. *Crop Science*, **37**: 992-997.

Babu, K.N., Rajesh, M.K., Samsudeen, K., Minoo, D., Suraby, E.J., Anupama, K. and Ritto Paul (2014). Randomly amplified polymorphic DNA (RAPD) and derived techniques. *Methods in Molecular Biology*, **1115**: 191–209.

Bandaranayake, C.K. and Kearsey, M.J. (2005). Genome mapping, QTL analysis and MAS: Importance, principle, constraints and application in coconut. *International Plant Genetic Resources Newsletter*, **142**: 47-54.

Bandaranayake, C.K., Fernando, W.B.S., Fernando, A. and Herath, N. (2005). DNA fingerprinting to distinguish the coconut type San Ramon. CORD, **21**(2): 30-36

Baudouin, L. and Lebrun, P. (2002). The development of a microsatellite kit and dedicated software for use with coconut. *Burotrop Bulletin*, **17**: 52.

Baudouin, L., Lebrun, P., Konan, J.L., Ritter, E., Berger, A. and Billottee, N. (2006). QTL analysis of fruit components in the progeny of a Rennell Island Tall coconut (*Cocos nucifera* L.) individual. *Theoretical and Applied Genetics*, **112**: 258-268.

Beckmann, J.S. and Soller, M. (1983). Restriction fragment length polymorphism in genetic improvement: methodologies, mapping and costs. *Theoretical and Applied Genetics*, **67**: 35-43.

Benoit, H. (1978).Isozymic variation within and between coconut populations. Paper presented during the 5th session of the FAO Technical Working Party on coconut production, *protection and processing*, Manila.

Bowcock, A.M., Ruiz-Linares, A., Tomfohrde, J., Minch, E., Kidd, J.R. and Cavalli-Sforza, L.L. (1994).) High-resolution of human evolutionary trees with polymorphic microsatellites. *Nature*, **368**: 455-457.

Bruford, M. and Wayne, R. (1993). Microsatellites and their applications to population genetic studies. *Current Opinion in Genetics and Development*, **3**: 939-943.

Carpio, C.B. (1982). Biochemical studies of *Cocos nucifera* L. Kalikasan. Philippines *Journal of Biology*, **11**: 319-338.

Child, R. (1974). Coconuts (2nd Ed). London: Longmans, UK.

Dallas, J.F., Dod, B., Boursot, P., Prager, E.M. and Bonhomme, F. (1995). Population subdivision and gene flow in Danish mice. *Molecular Ecology*, **4**: 311-320.

Dekkers, J.C.M. and Hospital F. (2002). The use of molecular genetics in the improvement of agricultural populations. Nature Reviews Genetics, **3**: 22–32.

Di Rienzo, A., Peterson, A.C., Garza, J.C., Valdes, A.M., Slatkin, M. and Freimer, N.B. (1994).) Mutation processes of simple sequence repeat loci in human populations. *Proceeding of the National Academy of Sciences of the USA*, **91**: 3166-3170.

Duran, Y., Rohde, W., Kullaya, A., Goikoetxea, P. and Ritter, E. (1997). Molecular analysis of East African Tall coconut genotypes by DNA marker technology. *Journal of Genetics and Breeding*, **51:** 279-288.

Edwards, K.J.K., Barker, J.H.A., Daly. A., Jones. C. and Carp, A. (1996). Microsatellite libraries enriched for several microsatellite sequences in plants. *Biotechniques,* **20**: 758-760.

Everard, J.M.D.T. (1996). *Use of molecular markers for breeding of the coconut palm* (*Cocos nucifera* L.). M.Sc Thesis. University of New England, Armidale, Australia.

Fan, H., Xiao, Y., Yang, Y., Xia, W., Mason, A.S., Xia, Z. and Tang, H. (2013). RNA-Seq Analysis of *Cocos nucifera*: Transcriptome sequencing and *De novo* assembly for subsequent functional genomics approaches. *PLoS ONE,* **8(3)**: p.e59997.

Fernando, W.M.U. (1995). Patterns of isozyme variation in *Cocos nucifera* L. *Proceedings of Annual Sessions of the Sri Lanka Association for the Advancement of Science,* **51**: 84-86.

Gangolly, S.R., Satyabalan, K., Pandalai, K.M. (1957) Varieties of coconut. *Indian Coconut Journal,* **10**: 3-28.

Geethalakshmi, P., Niral, V. and Parthasarathy, V.A. (2004). Allozyme variation in population of dwarf coconut cultivars. *Journal of Plantation Crops,* **32** (2): 13-15.

Gunn, B.F., Baudouin, L. and Olsen, K.M. (2011). Independent origins of cultivated coconut (*Cocos nucifera* L.) in the old world tropics. *PLoS ONE,* **6(6)**: p. e21143.

Harries, H.C. (1978). The evolution dissemination and classification of *Cocos nucifera* L. Botanical Review, **44**: 205-317.

Herran, A., Estioko, L., Becker, D., Rodriguez, M.J.B., Rohde, W. and Ritter, E. (2000). Linkage mapping and QTL analysis in coconut (*Cocos nucifera* L.). *Theoretical and Applied Genetics,* **101**: 292–300.

Huang, Y.Y., Matzke, A.J.M. and Matzke, M. (2013). Complete sequence and comparative analysis of the chloroplast genome of coconut palm *Cocos nucifera. PLoS ONE,* **8(8)**: p. e74736.

Karp, A. (1999). The use of polymorphic microsatellites for assessing genetic diversity in coconut. In: Santamaria, J. M. (Ed.), *Current Plant Science and Biotechnology in Agriculture- Current Advances in Coconut Biotechnology.* Kluwer Academic Publishers, Dordrecht. pp. 121-129.

Kumaunang, J. and Maskromo, I. (2007). Keragaangenetik plasma nutfahkelapa Dalam (*Cocos nucifera* L) di kebunpercobaan *Mapangetberdasarkanpenanda DNA SSRs. Bulletin Palma,* **33**: 18-27.

Larekeng, S., Maskromo, I., Purwito, A., Matjik, N. and Sudarsono, S. (2015a). Pollen dispersal and pollination patterns studies in Pati kopyor coconut using molecular markers. *CORD,* **31(1)**: 46-60.

Larekeng, S.H., Maskromo, I., Purwito, A., Matjik, N.A. and Sudarsono, S. (2015b) Pollen dispersal based on SSR analysis prolles Kalianda to Kopyor coconut pollinations. *Bulletin Palma,* 77-92.

Lebrun, P., Baudouin, L., Bourdeix, L., Konan, J.L., Barker, J.H.A., Aldam, C., Herran, A. and Ritter, E. (2001). Construction of a linkage map of the Rennell Island Tall coconut type (*Cocos nucifera* L.) and QTL analysis for yield characters. *Genome*, **44**: 962-970.

Lebrun, P., Grivet, L. and Baudouin L. (1999a). Use of RFLP markers to study the diversity of the coconut palm. In: J. M. Santamaria (Ed.) *Current Plant Science and Biotechnology in Agriculture - Current Advances in Coconut Biotechnology*, Kluwer Academic Publishers, Dordrecht. pp. 73-87.

Lebrun, P., N'cho, Y.P., Seguin, M., Grivet, L. and Baudouin, L. (1998). Genetic diversity in coconut (*Cocos nucifera* L.) revealed by restriction fragment length polymorphism (RFLP) markers. *Euphytica*, **101**: 103-108.

Lestari *et al.* (2016). Pemanfaatan Teknologi Genomika dan Bioteknologi untuk Mendukung Percepatan Pemuliaan Palma Penting di Indonesia. In: KKP3S Project Annual Report. IAARD. *Ministry of Agriculture*, Republic of Indonesia.

Liyanage, D.V. (1958) Varieties and forms of coconut palms grown in Ceylon. *Ceylon Coconut Quarterly*, **9**: 1-10.

Manimekalai, R. and Nagarajan, P. (2006). Inter-relationship among coconut (*Cocos nucifera* L) germplasm accessions using RAPD technique. *Genetic Resource Crop Evolution*, **53**: 1137-1144.

Manimekalai, R. and Nagarajan, P. (2007). Use of simple sequence repeat markers for estimation of genetic diversity in coconut (*Cocos nucifera* L.) germplasm accessions. *Journal of Plant Biochemistry and Biotechnology*, **16**: 29-33.

Maskromo, I., Tenda, T., Tulalo, M., Novarianto, H., Sukma, D., Sukendah, and Sudarsono, N. (2015). Keragaman fenotipe dan genetik tiga varietas kelapa genjah kopyor asal Pati Jawa Tengah. *Jurnal Littri*, **2(1)**: 1-8.

Maskromo, I., Novarianto, H. and Sudarsono, N. (2011). Fenologi pembungaan tiga varietas kelapa genjah kopyor Pati. In: Roedhy, P., Slamet, S., Anas, D., Nurul, K., Dewi, S. and Sintho, W.A. (Eds.), *Prosiding Seminar PERHORTI Kemandirian Produk Hortikultura untuk Memenuhi Pasar Domestik dan Ekspor*, Nov 23-24, 2011. Lembang, Indonesia, Bogor (ID): *Perhimpunan Hortikultura Indonesia*. pp. 1002-1010.

Maskromo, I. (2015). Karakterisasi dan pemanfaatan plasma nutfah melalui pendekatan pemuliaan dan molekuler untuk peningkatan hasil buah kopyor dan kualitas benih kopyor. Ph.D Thesis. *Bogor Agricultural University*, Bogor, Indonesia.

Maskromo, I., Sukendah, N., Novarianto, H. and Sudarsono, N. (2013a). Isolation and characterization of sucrose synthase and abscisic acid insensitive genomic fragments and development of SNAP marker for endosperm abnormality in coconut. *Plant and Animal Genomic (PAG)* - Asia (Abstract).

Maskromo, I., Sudarsono, H., Novarianto. (2013b). Potensi produksi pohon induk kelapa dalam kopyorAsal kalianda, lampung selatan. In: *Prosiding Simposium dan Seminar Bersama Peragi Perhorti Peripi Higi*. pp. 430-436.

Maskromo, I., Sudarsonodan, H. and Novarianto. (2013c). *Productivity of three dwarf Kopyor coconut varieties from Pati*, Central Java, Indonesia. *CORD*, **29(2)**: 19-28.

Mauro-Herrera, M., Meerow, A.W., Borrone, J.W., Kuhn, D.N. and Schnell, R.J. (2006). Ten informative markers developed from WRKY sequences in coconut (Cocos nucifera). *Molecular Ecology Notes*, 6 (3): 904–906.

Meerow, A.W., Wisser, R.J., Brown, J.S., Kuhn, D.N., Schnell, R.J. and Broschat, T.K. (2003). Analysis of genetic diversity and population structure within Florida coconut (*Cocos nucifera* L.) germplasm using microsatellite DNA, with special emphasis on the Fiji Dwarf cultivar. *Theoretical and Applied Genetics*, **106**: 715-726.

Menon, K.P.V. and Pandalai, K.M. (1958). The Coconut-A Monograph. Indian Central Coconut Committee. pp. 86-102.

Menon, K.P.V. and Pandalai, K.M. (1958). The coconut palm, a monograph. *Indian Coconut Committee*, Ernakulum.

Milbourne, D., Meyer, R., Bradshaw, J.E., Baird, E., Bonar, N., Provan, J., Powell, W. and Waugh, R. (1997). Comparison of PCR based marker systems for the analysis of genetic relationships in cultivated potato. *Molecular Breeding*, **3**: 127-1360.

Narayana, G.V. and John, C.M. (1949). Varieties and forms of coconut. *The Madras Agricultural Journal*, **36**: 349-366.

Novarianto, H., Maskromo, I., Dinarti, D. and Sudarsono. (2014). *Production technology for Kopyor coconut seednuts and seedlings in Indonesia. CORD*, **30(2)**: 31-40.

Novarianto, H. (1988). Analisis isozim peroksida sepadabibit kelapa KHINA-1. *Jurnal Penelitian Kelapa*, **2**(2): 74-80.

Novarianto, H., Hartana, A. and Gadrinab, L.U. (1988a). Keanekaragaman pola pita isozim peroksidase pada Koleksi kelapa di kebun percobaan pakuwon sukabumi. *Floribunda*, **1(7)**: 25-28.

Novarianto, H., Hartana, A., Makmur, A. and Gadrinab, L.U. (1988b). Analisis isozim daun kelapa hibrida dantetuanya. *Pemberitaan Puslitbangtri XIII*. **(3-4)**: 53-56.

Novarianto, H., Hartana, A., Rumawas, F., Rifai, M.A., Guharja, E. and Nasoetion, A.H. (1993). Kemiripangenetikantarpopulasikelapa di Indonesia berdasarkanpola pita isozim. *Jurnal Penelitian Kelapa*, **6(2)**: 1-8.

Novarianto, H., Kumaunang, J., Tulalo, M.A., Masniawati, A. and Hartana, A. (2001). Keragaman genetic kelapa Dalam Mapanget nomor 32 hasil penyerbukan sendiri berdasarkan penanda RAPD. *Pemberitaan Puslitbangtri*, **7(2)**: 43-48.

Ohler, J.G. (1984). Coconut, Tree of Life. Plant production and Protection Paper 57, *Food and Agricultural Organisation of the United Nations*, Rome.

Pandin, D.S. (2009a). Keragaman genetic kelapa Dalam Tenga (DTA) dan Dalam Mapanget (DMT) berdasarkan penanda RAPD. *Buletin Palma*, No.16.

Pandin, D.S. (2009b). Keragaman genetic kelapa Dalam Bali (DBI) dan Dalam Sawarna (DSA) berdasarkan penanda RAPD. *Buletin Palma* No.17.

Parthasarathy, V.A., Geethalakshmi, P. and Niral, V. (2004). *Analysis of coconut cultivars and hybrids using isozyme polymorphism. Acta Botanica Croatica*, **63**: 69-74.

Perera, L. (2001). Origin domestication and genetic diversity of coconut: *DNA Information. CORD* XVII, **(1)**: 35-51.

Perera, L., Russel, J.R., Provan, J., McNicol, J.W. and Powel, W. (1996). Evaluating genetic relationships between indigenous coconut (*Cocos nucifera* L.) accessions from Sri Lanka by means of AFLP profiling. *Theoretical and Applied Genetics Journal*, **96**: 545-550.

Perera, L., Russell, J.R., Provan, J. and Powell, W. (2003). Studying genetic relationships among coconut varieties/populations using microsatellite markers. *Euphytica*, **132**: 121-128.

Perera, L., Russell, J.R., Provan, J. and Powell, W. (1999). Identification and characterization of microsatellite loci in coconut (*Cocos nucifera* L.) and the analysis of coconut populations in Sri Lanka. *Molecular Ecology*, **8**: 335-346.

Perera, L., Russell, J.R., Provan, J. and Powell, W. (2000). Use of microsatellite DNA markers to investigate the level of genetic diversity and population genetic structure of coconut (*Cocos nucifera* L.). *Genome*, **43**: 15-21.

Peries, R.R.A. (1993). Improving the efficiency of the low input coconut agroecology through conservation and utilisation of biodiversity. *Proceedings of the 10th IFOAM Conference, Lincoln University*, New Zealand. pp. 122.

Powell, W., Machray, G.C. and Provan, J. (1996a). Polymorphism revealed by simple sequence repeats. *Trends in Plant Science*, **1**: 215-222.

Powell, W., Morgante, M., Andre, C., Hanafey, M., Vogel, M.J., Tingey, S.V. and Rafalski, J.A. (1996b). The comparison of RFLP, RAPD, RFLP and SSR (microsatellite) markers for germplasm analysis. *Molecular Breeding*, **2**: 225-238.

Provan, J., Russell, J.R., Booth, A. and Powell, W. (1999). Polymorphic chloroplast simple sequence repeat primers for systematic and population studies in the genus *Hordeum*. *Molecular Ecology*, **8**: 505-511.

Purseglove, J.W. (1972). Tropical Crops: Monocotyledons. Longmans, London.

Queller, D., Strassmann, J. and Hughes, C. (1993). Microsatellite and kinship. *Trends in Ecology and Evolution*, **8:** 285-288.

Rafalski, J.A., Vogel, M.J., Morgante, M., Powell, W., Andre, C. and Tingey, S.V. (1996). Generating and using DNA markers in plants. In: Birren, B. and Lai,

E. (Eds.), *Non-Mammalian Genome Analysis: A Practical Guide*, Academic Press, New York. pp. 75-134.

Rajesh, M.K., Jerard, B.A., Preethi, P., Regi Jacob Thomas, Fayas, T.P., Rachana, K.E. and Anitha Karun (2013). Development of a RAPD-derived SCAR marker associated with tall-type palm trait in coconut. *Scientia Horticulturae*, **150(4)**: 312-316.

Rajesh, M.K., Thomas, R.J., Rijith, J., Shareefa, M. and Jacob, P.M. (2012). Genetic purity assessment of D x T hybrids in coconut with SSR markers. *Indian Journal of Genetics and Plant Breeding*, **72 (4)**: 472-474.

Rajesh, M.K., Sabana, A.A., Rachana, K.E., Shafeeq Rahman, Jerard, B.A. and Anitha Karun (2015). Genetic relationship and diversity among coconut (*Cocos nucifera* L.) accessions revealed through SCoT analysis. *3Biotech*, **5(6)**: 999-1006.

Ritter, E., Rodriguez, M.J.B., Herran, A., Estioko, L., Becker, D. and Rohde, W. (2000). Analysis of quantitative trait loci (QTL) based on linkage maps in coconut (*Cocos nucifera* L.), In: Arencibia, A. (Ed.), *Plant Genetic Engineering Towards the Third Millennium*. Elsevier Science, Amsterdam, The Netherlands. pp. 42-48.

Rivera, R., Edwards, K.J., Arnold, G.M., Ayad, G., Hodgkin, T. and Karp, A. (1999). Isolation and characterisation of polymorphic microsatellites in *Cocos nucifera* L. *Genome*, **42(4)**: 668-675.

Rohde, W., Becker, D., Kullaya, A., Rodriguez, M.J.B., Herran, A. and Ritter, E. (1999). Analysis of coconut germplasm biodiversity by DNA marker technologies and construction of a first genetic linkage map, In: Oropeza, C., Verdeil, J.L., Ashburner, G.R., Cardega, R. and Santamaria, J.M. (Eds.), *Current Advances in Coconut Biotechnology*. Kluwer Academic Publishers, Dordrecht, The Netherlands. pp. 99-120.

Rohde, W., Kullaya, A., Rodriguez, J. and Ritter, E. (1995). Genome analysis of *Cocos nucifera* L. by PCR amplification of spacer sequences separating a subset of *copia*-like 16RI repetitive elements. *Journal of Genetics and Breeding*, **49**: 179-186.

Russell, J.R., Fuller, J.D., Macaulay, M., Hatz, B.G., Jahoor, A., Powell, W. and Waugh, R. (1997). Direct comparison of levels of genetic variation among barley accessions detected by RFLPs, AFLPs, SSRs and RAPDs. *Theoretical and Applied Genetics*, **95**: 714-722.

Shriver, M.D., Jin, L., Chakraborty, R. and Boerwinkle, E. (1993). VNTR allele frequency distributions under a stepwise mutation model: *A computer simulation approach. Genetics*, **134**: 983–993.

Sudarsono, S., Novarianto, H., Hosang, M.L.A., Dinarti, D., Rahayu, M.R. and Maskromo, I. (2014). Produksi bibit kopyor *true to type* dengan persilangan terkontrol dan peningkatan produksi buah kopyor dengan pollinator lebah madu. *Laporan Akhir Program Hi Link*. Bogor (ID): Institut Pertanian Bogor.

Sudarsono, S., Novarianto, H., Hosang, M.L.A., Dinarti, D., Rahayu, M.R. and Maskromo, I. (2015). Pendekatan genomic dan molekuler untuk pengembangan kultivar unggul kelapa eksotik asal Indonesia, penyediaan bibit dan pengendalian hamanya. LaporanAkhir KKP3N. IPB bekerjasama dengan Badan Penelitiandan Pengembangan Pertanian.

Tautz, D. and Rentz, M. (1984). Simple sequences are ubiquitous repetitive components of eukaryotic genomes. *Nucleic Acids Research,* **12**: 4127-4138.

Tautz, D., Tricj, M. and Dover, G.A. (1986). Cryptic simplicity in DNA is a major source of variation. *Nature,* **322**: 652-656.

Teulat, B., Aldam, C., Trehin, R., Lebrun, P., Barker, J.H.A., Arnold, G.M., Karp, A., Baudouin, L. and Rognon, F. (2000). An analysis of genetic diversity in coconut (*Cocos nucifera*) populations from across the geographic range using sequence-tagged microsatellites (SSRs) and AFLPs. *Theoretical and Applied Genetics,* **100**: 764-771.

Thormann, C.E., Ferreira, M.E., Camargo, L.E.A., Tivang, J.G. and Osborn, T. C. (1994).) Comparison of RFLP and RAPD markers to estimating genetic relationships within and among cruciferous species. *Theoretical and Applied Genetics,* **88(8)**: 973-980.

Valdes, A.M., Slatkin, M. and Freimer, N.B. (1993). Allele frequencies at microsatellite loci: The stepwise mutation model revisited. *Genetics,* **133**: 737-749.

Wadt, L.H.O., Sakiyama, N.S., Pereira, M.G., Tupinamba, E.A., Ribeiro, F.E. and Aragao, M.M. (1999). RAPD markers in the genetic diversity study of the coconut palm. In: Santamaria, J.M. (Ed.), *Current Plant Science and Biotechnology in Agriculture - Current Advances in Coconut Biotechnology,* Kluwer Academic Publishers, Dordrecht. pp. 89-97.

Weber, J.L. and Wong, C. (1993). Mutation of human short tandem repeats. *Human Molecular Genetics,* **2**: 1123-1128.

Welsh, J. and Mc Clelland, M. (1991). Genomic fingerprints produced by PCR with consensus tRNA primers. *Nucleic Acids Research,* **19**: 861-865.

Welsh, J. and McClelland, M. (1990). Fingerprinting genomes using PCR with arbitrary primers. *Nucleic Acids Research,* **18**: 7213-7218.

White, G. and Powell, W. (1997a) Cross-species amplification of SSR loci in the Meliaceae family. *Molecular Ecology,* **6**: 1195-1197.

White, G. and Powell, W. (1997b) Isolation and characterization of microsatellite loci in *Swietenia humilis* (Meliaceae): An endangered tropical hardwood species. *Molecular Ecology,* **6**: 851-860.

Whitehead, R.A. (1976). Coconut. In: Simmonds, N.W. (Ed.), *Evolution of Crop Plants,* Longman, London. pp. 221-225.

Williams, J.G.K., Kubelik, A.R., Livak, K.J., Rafalski, J.A. and Tingey, S.V. (1991). DNA polymorphisms amplified by arbitrary primers are useful as genetic markers. *Nucleic Acids Research,* **18**: 6531-6535.

Xiao,Y., Luo, Y., Yang, Y., Fan, H., Xia, W., Mason, A.S., Zhao, S., Sager, R. and Qiao, F. (2013). Development of microsatellite markers in *Cocos nucifera* and their application in evaluating the level of genetic diversity of *Cocos nucifera*. *Plant Omics Journal*, **6**: 193–200.

Chapter 12

Oil Palm

☆ *K. Sunilkumar, R.K. Mathur, B. Kalyana Babu and P. Murugesan*

1. Introduction

The oil palm (*Elaeis guineensis* Jacq.) is the highest oil yielding crop in the world and produces 4-6 MT of palm oil (mesocarp oil) and 0.4 to 0.6 MT of palm kernel oil per hectare per annum (Basri Wahid *et al.*, 2005). The palm belongs to family Arecaceae and the genus *Elaeis*. There are two important species: *Elaeis guineensis* Jacq. or the African oil palm and the *E. oleifera* (HBK) Cortes. or the American oil palm. *E. guineensis* is the high yielding and commercially cultivated species, whereas the *E. oleifera*, a wild type, possesses some of the important traits such as oil quality, short palm height and disease resistance. Interspecific crosses have been developed for introgressing the desirable traits from *E.oleifera* into *E.guineensis*. With only 5 per cent of the total world vegetable oil acreage, palm oil accounts for 33 per cent of vegetable oil production (Singh *et al.*, 2013b). Realizing that the crop is capable of bridging the shortage of availability of edible oil in India, cultivation of oil palm in the country has got a boost from planners, researchers and farmers. With the increase in area of the crop under diverse environments, it has become essential to develop oil palm varieties with high yield, compactness/dwarfness, high oil extraction ratio and tolerance to low moisture regimes.

Variation in growth and yield results from differences in the genetic makeup (genes) and its interaction with the environment (G x E interaction). The genes carry instructions for the synthesis of enzymes (proteins) which lead to expression of a trait/phenotype. Molecular markers help to identify genetic variation in a population. The markers are useful for confirmation of pedigree or legitimacy of parentage, for assessing genetic diversity or in selection of individual or parents for breeding (marker assisted selection, MAS). Selection based on genes (DNA) is expected to be reliable, efficient and precise compared to conventional phenotypic selection which is based on morphological traits influenced by the growing environment. MAS has gained much importance in breeding programmes of many crops due to its accuracy and saving of time and resources and is especially relevant in perennial/plantation crops like oil palm where the breeding cycle is very long and resources like land are limited. In oil palm, each cycle of phenotypic selection (generation) takes 10-12 years and development of a new hybrid takes more than 30 years. With the help of marker assisted selection and breeding, the duration could be considerably reduced.

2. Types of Molecular Markers

2.1. Biochemical Markers

Initial work on molecular markers was with proteins where some enzymes exist in more than one form (isoenzymes or isozymes) which could be separated and their inheritance studied (Ghesquiere, 1984, 1985; Ataga and Fatokun, 1989; Baudouin, 1992; Choong *et al.*, 1996). Generally protein markers are difficult to work as they differ according to tissue types and even within the same tissue, they may vary with physiological status of the plants; hence it is difficult to obtain comparable samples of different plants. Also, there are relatively few enzymes for which isozymes are easily separable (Corley and Tinker, 2003). But even with these limitations, these markers can be effectively used for specific purposes and selection process for the breeding programme. There are reports on the use of protein markers in oil palm since 1984 (Ghesquiere, 1984), where the enzyme polymorphism was studied using nine enzyme systems. Subsequently, Ghesquiere (1985) reported genetic variability of oil palm germplasm accessions from seven different origins using this procedure. Ataga and Fatokun (1989) carried out study of isozymes from pollen of seven accessions of *E. oleifera* and hybrid between *E. oleifera* and *E. guineensis*.

2.2. DNA-based Markers

The most reliable markers are those based on DNA as they depend on the distinctive structure of the genetic material. DNA markers can be used for different purposes: a) for studying genetic diversity, b) phylogenetic relationships and species interrelationship study, c) assessing genetic fidelity of micropropagated plants and identification of somaclonal variants, d) identification of genetically modified plants, e) generation of molecular linkage maps including QTL mapping and tagging useful genes and f) gene pyramiding and marker assisted selection. Commonly used DNA marker systems include restriction fragment length polymorphism (RFLP), random amplified polymorphic DNA (RAPD), amplified fragment length polymorphism (AFLP) and simple sequence repeat (SSR) or microsatellites. A convenient marker system, or combination of them, can be used for different purposes.

2.3. Genetic Diversity Analysis using Molecular Markers

Exploiting the genetic diversity is essential if oil palm improvement is to be successful in the long run (Adon *et al.*, 1998). Thus, assessment of genetic relationship of different palm species/varieties/populations is extremely important before taking up breeding programme of oil palm (Billotte *et al.*, 2001). Traditionally, morphological descriptors were the only information available for assessment of genetic diversity. Once the molecular markers were developed, they have been widely used for detecting the polymorphism and distinguishing individuals or groups.

Use of RAPD markers for genetic diversity study of oil palm was reported for the first time by Shah *et al.* (1994)). Oil palm germplasm accession collected from Africa (Cameroon, Tanzania, Nigeria and Zaire) was studied using 20 primers and high level of genetic variation was reported among the accessions. Subsequently, a variety of DNA probes have been used to screen a diverse set of oil palm accessions in order to identify RLFP markers which would allow genotype discrimination (Jack *et al.*, 1995). Mayes *et al.* (2000) used RFLP markers (40 probes covering 60 per cent oil palm genome) to assess genetic diversity within 54 palms of a specific oil palm breeding program. In addition to the 54 palms, 10 samples encompassing a broader range of origins were also included to provide a wider frame work for comparative analysis. A total of 157 RFLP bands were scored and the data analyzed revealed that South East Asian Deli *dura* material clustered distinctly from AVROS *pisifera* gene pool.

Purba *et al.* (2000) described a new aspect of genetic diversity of Indonesian oil and its consequences on oil palm breeding programmes utilizing isozymes and AFLP based markers. They have used 48 parents representing four populations in the Indonesian oil palm breeding program and analyzed with five selected AFLP primer pairs and four isozymatic systems. About 158 scorable band levels were generated of which 96 (61 per cent) were polymorphic. The results obtained with AFLP showed that crosses among African sub population may be more interesting than the crosses between the African and Deli population. Maizura *et al.* (2001) characterized 11 oil palm germplasm collections *viz.* Nigeria, Cameroon, Congo DR, Tanzania, Angola, Senegal, Sierra Leone, Guinea, Ghana, Madagascar and Gambia using RFLPs and found that all germplasm collections exhibited a higher level of genetic diversity compared to Deli *dura* breeding population. Kularatne *et al.* (2001) screened a total of 687 accessions belonging to 11 African countries and Deli *dura* using AFLP markers with eight primer combinations. A total of 377 bands where recorded with an average of 47 bands per primer pair, of which 97.6 per cent were polymorphic. Overall genetic diversity of oil palm was the highest in Nigerian materials and gradually decreased towards Senegal, Gambia, Angola, and Tanzania. Cluster analysis separated oil palm into three major groups corresponding to three regions of Africa. The study also confirmed the low genetic diversity of Deli *dura* which could have resulted from several cycles of selection it has undergone and was found to be most closely related to Congo oil palm.

In 2002, Barcelos *et al.* reported genetic diversity, its organization and genetic relationship within oil palm germplasm (both *E. oleifera* and *E. guineensis*) using RFLP and AFLP techniques. In case of RFLP study, 241 *E. oleifera* accessions and 48 *E. guineensis* were analyzed using the same 37 cDNA probes. Further, an AFLP analysis was carried out for a subset of 14 accessions of *E. oleifera* and 22 *E. guineensis* using three pairs of enzymes/primer combinations. In the *E. oleifera*, the strong structure of genetic diversity revealed by RFLP was confirmed by AFLP also. Moretzsohn *et al.* (2002) carried out analysis of genetic variations in *E. oleifera* accessions collected from the Amazon forests using RAPD markers. The results showed that Brazilian *E. oleifera* accessions had moderate levels of genetic diversity as compared to *E. guineensis* accessions. The AMOVA showed that most of the genetic variation was found within populations as expected for an allogamous and perennial species. Maizura *et al.* (2006) undertaken characterization of 359 accession of oil palms (*E. guineensis*) originating from 11 African countries by RFLP method using the standard Deli *dura* as check. In general, all the collections exhibited higher levels of diversity than the standard variety. The standard variety (Deli *dura* lost 36 alleles as compared to the natural populations indicating a reduction in genetic variability). Material from Nigeria showed the highest mean number of alleles per locus (1.9) and percentage of polymorphic loci (67.2 per cent).

In India, Mandal *et al.* (2006) attempted genetic diversity study of five *dura* germplasm accessions using RAPD the results of the study revealed a high degree of DNA polymorphism among different palms and also among different accessions studied. Subsequently, five exotic accessions (Guinea Bissau, Zambia, Tanzania, Cameroon, ASD Costa Rica and Palode *dura*) of oil palm have been analyzed for biochemical and molecular characterization using RAPD (Mandal and Susmita, 2006). Accessions from Guinea Bissau were highly homogenous in comparison to other groups. Accessions from Cameroon and ASD Costa Rica were found to be close to each other.

Billote *et al.* (2001) reported the efficiency of SSR markers in revealing the genetic diversity and population structure of the genus *Elaeis* in accordance with known geographical origin and with measured genetic relationship based on previous molecular studies. Bakoume *et al.* (2015) studied the extent of genetic diversity among 194 oil palm genotypes from 49 populations using 16 SSR markers. The genetic diversity was high with a total of 209 alleles detected accounting for an average of 13.1 alleles per locus and a mean expected heterozygosity of 0.644. Morpho-agronomic traits and microsatellite markers (SSRs) were used to evaluate the phenotypic and genetic diversity of accessions of *E. oleifera* from different countries in South America (Arias *et al.*, 2015). SSR marker analysis revealed high genetic diversity (H_T = 0.797) and the presence of specific alleles for each country of origin for *E. oleifera*. The clustering topology obtained showed four distinguishable *E. oleifera* groups, which matched the geographical distribution, in addition to exhibiting a high degree of genetic differentiation (G_{ST} = 0.512) and low gene flow (Nm =0.238) among countries. The results enabled them to conclude that *E. oleifera* has a specific genetic structure and the accession from each country of origin had contributed to the increase in genetic diversity.

Of late, expressed sequence tags (ESTs) are markers of choice for genetic diversity analysis and mapping studies. EST-SSR markers are valuable because their development cost is inexpensive; they represent transcribed genes and a putative function that often can be inferred by homology search. They have been developed and mapped in several crop species and proved useful for marker assisted selection in oil palm. Ting *et al.* (2010) developed a set of EST-based SSR markers through *in silico* approaches from a large collection of 19243 *Elaeis guineensis* ESTs. The ESTs were assembled to give 10258 unigenes; 629 ESTs were found to contain 722 SSRs with a variety of motifs. Dinucleotide repeats formed the largest group (45.6 per cent) consisting of 66.9 per cent AG/CT, 21.9 per cent AT/AT, 10.9 per cent AC/GT and 0.3 per cent CG/CG motifs. Primer pairs from sequences flanking 289 EST-SSRs were tested to detect polymorphisms in elite breeding parents and their crosses. SSR polymorphisms were found in sequences encoding AP2-like, bZIP, zinc finger, MADS-box, and NAC-like transcription factors in addition to other transcriptional regulatory proteins and several RNA interacting proteins. In another study, 10 novel EST- SSRs were developed and used to evaluate genetic diversity of 76 accessions of oil palm originating from seven countries in Africa along with Deli *dura* population (Singh *et al.*, 2008). The average number of observed and effective alleles was 2.56 and 1.84 respectively. The EST-SSR markers were found to be polymorphic with a mean polymorphic information content of 0.53 and the genetic differentiation (F_{ST}) among the population studied was 0.2492 indicating a high level of genetic divergence. Further, UPGMA analysis revealed a strong association between genetic distance and geographic location of the population studied. EST- SSRs were also found to be useful as a new source of polymorphic markers for phylogenetic analysis, because a high percentage of markers showed transferability across species and palm taxa.

Taeprayoon *et al.* (2015) assessed a total of 121 breeding plants from three different populations in Thailand utilizing 96 SSR markers of which 20 were polymorphic. A total of 109 alleles were scored with polymorphic information content ranging from 0.45 to 0.87 for mEgCIR 580 and mEgCIR787 markers respectively. The number of alleles per locus ranged from 3 to 8 and analysis of molecular variance (AMOVA) revealed variations among population and among individuals within population at 33 per cent and 67 per cent respectively. Hence, the SSR markers used in the study were able to classify the genetic background of major oil palm genetic stocks in Thailand. Utility of microsatellite markers in diversity analysis of oil palm accessions have also been reported by Ajambang *et al.* (2012); Ting *et al.* (2010); Arias *et al.* (2013, 2014) and Zaki *et al.* (2010, 2012).

Singh *et al.* (2013b) reported 1.8 Gb genome sequence of the African oil palm *E. guineensis.* A total of 1.535 Gb of assembled sequence and transcriptome data from 30 tissue types were used to predict at least 34802 genes including oil biosynthesis genes and homologue of *WRINKLED 1* (*WRI 1*) and other transcriptional regulators which are highly expressed the kernel. They also reported the draft sequence of South American oil palm *E. oleifera*, which has the same number of chromosome (2n=32) and produces fertile inter-specific hybrid with *E. guineensis,* but seems to have diverged in the New World.

2.4. Markers for Detection of Somaclonal Variations and Clonal Fidelity Testing

During 1986, it was observed that some of the oil palm clones planted in 1982 did not flower normally and produced inflorescence with 'mantled' flowers (Corley *et al.*, 1986). The mantled flowers were non-functional and there was no fruit development resulting in severe bunch failure. Wider adoption of tissue culture technology or clonal propagation of oil palm was hampered by large scale incidence of bunch failure in some clones. This kind of abnormality was expected to have a direct effect on oil production and hence considerable efforts were made to understand the causes and possible remedies of this abnormality.

Several factors were attributed to the occurrence of mantled flower abnormality. The proportion of abnormal flowering increased with the number of subcultures a clone had under gone (Paranjothy *et al.*, 1995a). Eeuwens *et al.* (2002) reported, from a 15 year trial, that the medium on which embryoids are cultured has a profound effect. Rao and Donough (1990) described the preliminary evidence for a genetic cause for the abnormalities. The availability of a reliable marker associated with the floral abnormality would be an important asset in screening out abnormal cultures and will be useful in identifying the *in vitro* protocol, if any, that induces abnormalities (Mathius and Ginting, 1998).

Various techniques have been employed in checking the clonal materials which included RAPD and RFLP markers (Meunier, 1995; Paranjothy *et al.*, 1995b; Mayes *et al.*, 1996) and markers derived from ESTs (Sharifaha *et al.*, 1999). Singh *et al.* (2007) demonstrated the use of SSR markers in oil palm tissue culture. They reported the development of 12 SSR markers suitable as molecular probes for DNA fingerprinting of oil palm clones for clonal identification, monitoring line uniformity within the clone and detecting the culture mix-up.

Alteration of the methylation status could be one of the major changes induced by *in vitro* culture. It was found that there were significant differences in DNA methylation status in variant vs normal regenerants. Jaligot *et al.* (2002) reported substantial demethylation in severely mantled palms. Shah and Parveez (1995) have shown that levels of 5- methyl cytosine were significantly higher in regenerants the abnormalities than normal ones. Matthes *et al.* (2001) observed loss of methylation during oil palm tissue culture and Jaligot *et al.* (2000a,b) described the trend towards hypomethylation in abnormal plant materials. The use of AFLPs with methylation sensitive enzymes has made possible the identification of some tissue culture induced polymorphisms in oil palm regenerants that could not be detected by other methods. However, methylation sensitive markers could only discriminate between the two phenotypes within the same clonal progenies (Jaligot *et al.*, 2004). Two oil palm cDNAs, namely *CPH062* and *CPH 063*, were found to display a differential methylation pattern between normal and abnormal embryogenic calli (Jaligot *et al.*, 2002). Morcillo *et al.* (2006) reported two novel oil palm genes namely *EgM39A* and *EgIAA1*, both of which displaced increased transcript accumulation in epigenetically abnormal calli and were found to have potential as early markers of clonal conformity. A gene encoding a putative plant defensin (*EGDA1*), has been identified and characterized from differential display studies performed on oil palm tissue cultures bearing or lacking an epigenetic homeotic flowering abnormality (Tregear *et al.*, 2002). At the callus stage of the *in vitro* regeneration procedure, a differential accumulation of *EGAD1* transcripts was observed which was correlated with the presence or absence of the mantled flowering abnormality. Hence, it was presumed that *EGAD1* gene expression may be a marker of epigenetic somaclonal variation events. Homologous of genes involved in many aspects of flower development in oil palm were identified among the EST collection, such as *CONSTANS–like, AGAMOUS–like, AGL2, AGL20, LFY – like, SQUMOSA* and *SQUMOSA binding protein etc.* (Ho *et al.*, 2007). Helene *et al.* (2007a) suggested that MADS box genes probably lie at the heart of many key evolutionary events in plants through the fundamental role which they play in the regulation of reproductive development in general and floral structure in particular. The type of morphological changes suggested that it involves homeotic MADS box genes that regulate the identity of flower whorls. A number of MADS box genes from oil palm inflorescences have been isolated by a MADS box directed mRNA display approach. In India, variation of esterase isozymes in the embryogenic shoots derived from tenera seedling explants of oil palm was reported by Upadhyay and Karun (2000). RAPD and SSR analysis of clonal oil palm from farmer's field was carried out by Jayanthi *et al., (*2008). Out of the ten RAPD primers, three exhibited polymorphism and all the ten SSR primers studied produced monomorphic bands with the DNA of the clones. Chan *et al.* (2014) reported selection of reference gene for accurate quantification of gene expression level in oil palm. Accordingly, PD00380 and PD 00569 were identified as the most stably expressed gene in total sample of tissue culture lines for RT- qPCR.

The oil palm 'mantled' fruit abnormality was widely regarded as an epigenetic phenomenon. Ong-Abdulla *et al.* (2015) identified the *MANTLED* locus using epigenome – wide association studies of the African oil palm (*E. guineensis*). DNA hypo-methylation of a LINE retrotransposon, related to rice *Karma* in the intron of the homeotic gene *DEFICIENS*, was common to all mantled clones and was associated with alternative splicing and premature termination. Dense methylation near the *Karma* site (termed the *Good karma* epiallele) resulted in normal fruit set, where as hypomethylation (*termed* the *bad karma* epiallele) resulted in homeotic transformation, parthenocarpy and marked loss of yield. Loss of *Karma* methylation and of small RNA in tissue culture was found to contribute to the origin of mantled plants; and while restoration in spontaneous revertants accounted for non-Mendelian inheritance. The ability to predict and cull mantling at the plantlet stage will facilitate the introduction of higher performing cross and optimize environmentally sensitive land resources.

2.5. Marker for Shell Thickness and Identification of Hybrids

Oil palm has three fruit forms *dura* (thick shelled), *pisifera* (shell less) and *tenera* (thin shelled), which is a hybrid between *dura* and *pisifera*. The *pisifera* palms are usually female sterile. The *tenera* palms yield far more oil than *dura* and hence constitute the commercial variety in most of the palm oil producing countries. There is much importance given for the identification of the fruit form or variety owing to the time and resources employed in the breeding and hybrid seed production of oil palm. A single character (shell thickness) distinguishes these fruit types/varieties.

AFLP analysis revealed a high level of polymorphism between Columbian *E. oleifera* (female) and Nigerian *E. guineensis* (male) parents used to generate the inter-specific hybrids (Singh and Cheah, 1999). Using only ten primer pair combination,

674 bands were amplified of which 91 showed segregation. All markers detected were scored as dominant and segregation ratio indicated that majority of these markers (80 per cent) were inherited in Mendelian manner. The AFLP technique was also suitable for detecting contaminants and would therefore be useful for assessing the fidelity of controlled crosses. The mapping and identification of *SHELL* genes responsible for different fruit forms by Singh *et al.* (2013a) was a breakthrough on the subject. Accordingly, using homozygosity mapping by sequencing, they found two independent mutations in the DNA binding domain of a homologue of the MADS box gene SEEDSTICK (*STK*, also known as *AGAMOS-like 11*), which controls ovule identity and seed development in *Arabidopsis*. The *SHELL* gene is responsible for the *tenera* phenotype in both cultivated and wild palm from sub–Saharan Africa, and their finding provided a genetic explanation for the single gene hybrid vigor attributed to *SHELL* via heterodimerization. This mutation explained the single most important economic trait in oil palm.

Ritter *et al.* (2016) have developed a molecular marker system composing of three primer pairs and two restriction enzymes that allowed in differentiation of three different *Sh* alleles. The developed marker system has been validated in *dura* and *pisifera* genotypes from different origins which covered the standard gene pool that was currently used by the most of oil palm breeders.

3. Linkage Mapping and QTL Studies

The basic requirement for MAS is identification of appropriate markers associated with the targeted traits. By screening a large population of sibs of many different markers, pair or groups of markers that are linked, and tend to be inherited together, can be identified. Such groups are expected to be on the same chromosome and the closeness of the linkage, calculated statistically, shows the relative position of the markers along the chromosome (Corley and Tinker, 2003). The basis for constructing physical maps in coconut and oil palm was established as a result of the collaborative project entitled "construction and exploitation of high density DNA marker and physical maps in the perennial tropical oil crops coconuts and oil palm; from biotechnology toward marker assisted breeding". The consortium consisted of molecular biologists, breeders and producers in South East Asia (the Philippines, Malaysia and Indonesia) along with European partners. The project had established the basis for construction of physical maps in oil palm as well as coconut; details of maps, markers and alignments as well as QTLs were made available on the project web page (http://www.NEIKER.net/link2palm).

Jack and Mayes (1993) reported a programme on oil palm RFLP linkage map for the identification of markers linked to shell thickness, *Fusarium* wilt resistance, oil quality and yield. The first map published was based on RFLPs (Mayes *et al.*, 1996). A mapping population generated by the selfing of an important breeding material segregated for the shell thickness character (*Sh*) which enabled mapping of the gene conferring this major commercial trait. The nearest RFLP marker (pOPgSP1282) was located 9.8 cM from *Sh* in the mapping population and 6.6 cM in a related (A137/30xE80/29) smaller population of 45 palms. The map was updated by Jack *et al.* (1998), when they reported tagging of two genes, namely shell thickness (*Sh*) and virescence (*Vr*). Two single tree linkage maps were constructed for a maternal *tenera* (*sh+ sh-*) palm and for paternal *pisifera* (*sh- sh-*) palm using pseudo test cross mapping strategy in combination with RAPD markers through the analysis of an F1 *tenera* x *pisifera* progeny. RAPD and bulk segregant analysis were used to identify markers more tightly linked to the *sh+* locus. Two RAPD markers (R11-1282 and T19-1046) were identified to be linked on both sides of the *sh+* locus on linkage group 4. The estimated map distance from *sh+* to R11-1282 and T19-1046 were 17.5 and 23.9 cM respectively (Moretzsohn *et al.*, 2000). Rance *et al.* (2001). carried out development of an oil palm RFLP marker for quantitative trait loci (QTL) mapping. Information on 153 RFLP markers were used in combination with phenotypic data from an F2 population to estimate the position and effects of QTLs for traits including yield of fruit and its components as well as measures of vegetative growth. The mapping population consisted of 84 palms and the marker data were analysed to produce a linkage map consisting of 22 linkage groups. Using both single marker and interval mapping analysis, significant marker associated QTL effects were found for 11 of the 13 traits analysed. The results of interval mapping analysis of fruit weight, petiole cross section and the rachis length, and the ratios of shell: fruit, mesocarp: fruit and kernel: fruit indicated significant QTLs ($p<0.05$) at the genome-wide threshold. It was suggested to employ such QTLs in novel breeding strategies *viz.* marker assisted selection for commercial populations. Chua *et al.* (2001) reported a *dura* map that covered a genetic distance of 555 cM and a *pisifera* map covering a genetic distance of 885 cM. On the three maps generated by the same group, loci for carotene content and clonal abnormalities have been identified. The approach of functional genomics was fruitful in identification of several differentially expressed genes (Singh *et al.*, 2001). Ho *et al.* (2007) generated a total of 14537 ESTs from these libraries, from which 6464 tentative unique contigs (TUCs) and 2129 singletons were obtained. Gene homologues for enzymes involved in oil bio-synthesis, utilization of nitrogen sources, and scavenging of oxygen radicals were uncovered among the oil palm ESTs.

Billotte *et al.* (2005) published a microsatellite-based high density linkage map for oil palm from a cross between two heterozygous parents, a *tenera* palm from the LaMe population and a *dura* from the Deli population. A set of 390 SSR markers were utilized. A dense and genome-wide microsatellite frame work as well as saturating AFLPs allowed the construction of a linkage map consisting of 225 microsatellites, 688 AFLPs and the locus of the *Sh* gene, which controls the presence or absence of a shell in the oil palm fruit. An AFLP marker E-Agg/M-CAA132 was mapped at 4.7 cM from the *Sh* locus. The 944 genetic markers were distributed on 16 linkage groups (LGs) and covered 1743 cM. This linkage map was reported to be the first in oil palm having 16 independent linkage groups corresponding to the palm's 16 homologous chromosome pairs. This high density linkage map represented an important step towards quantitative trait loci analysis and physical mapping in the *E. guineensis* species.

In addition to the direct DNA markers, *in vitro* synthesized protein from mRNA was reported as marker for character tagging. Singh and Cheah (2000) identified a flower specific protein in the oil palm (*E. guineensis*) by two-directional gel electrophoresis. Differences in banding pattern was reported between the inflorescence (flower) and vegetative (leaf and young etiolated seedlings) tissues, indicating differential gene expression. Some changes were also observed in the pattern of *in vitro* translation products for mRNA extracted from different stages of flowering.

Once the linkage map is developed, it could be used to choose a set of markers that covers the entire chromosome with reasonably even distribution. Other populations could be screened for this subset of markers and studied for the linkage with useful characters (Corely and Tinker, 2003). It was suggested that heritable traits of economic importance such as shell thickness, fruit color, height and some fruit and bunch components and fatty acid composition could be tagged with DNA markers (Lim and Rao, 2004). Identification of RFLP markers linked to FFB yield, oil/bunch and its components, and vegetative characteristics was reported by Rance *et al.* (2001). Markers linked to carotene and oleic acid content of the oil was identified by Singh *et al.* (2001).

A QTL analysis designed for a multi-parent population was carried out and tested in oil palm (*Elaeis guineesis*) by Billotte *et al.* (2010). A consensus genetic map was produced using 251 microsatellite loci, the locus of the *Sh* major gene controlling fruit shell presence and an AFLP marker of that gene. A set of 76 QTLs involved in 24 quantitative phenotypic traits was identified. Accordingly, across-family analysis proved to be efficient due to the interconnected families. Further, the possibility of identifying QTL markers for small progeny numbers and for MAS was suggested.

A map was constructed using AFLP, RFLP and SSR markers for an interspecific cross involving a Columbian *E. oleifera* (UP 1026) and a Nigerian *E. guineesis* (T 128) by Singh *et al.* (2009). Interval mapping and multiple-QTL model (MQM) mapping (also known as composite interval mapping, CIM) were used to detect QTLs controlling oil quality (measured in terms of iodine value (IV) and fatty acid composition). At a 5 per cent genome wide significance threshold level, QTLs associated with myristic acid (C14:0), palmitic acid (C16:0), palmitoleic acid (18:0), oleic acid (18:1), and linoleic acid (18:2) content were detected. One genomic region on Group I appeared to influence C14:0, C16:0, 18:0 and 18:1 content. Significant QTLs for C14:0, 16:1, 18:0 and 18:1 content were detected around the same locus on Group 15, thus revealing another major locus influencing fatty acid composition in oil palm.

Ting *et al.* (2013), in an effort to identify QTLs associated with tissue culture amenability of oil palm, reported linkage maps improved with SSR markers for identification of QTLs associated with callogenesis and embryogenesis in oil palm. The SSR markers were mapped on to the earlier reported parental maps based on AFLP and RFLP markers. The new linkage map of ENL48 contained 148 markers (33 AFLPs, 71 RFLPs and 77 SSRs) in 23 linkage groups covering a total map length of 798 cM. The ML161 map contained 240 markers and 24 linkage groups. Using the improved maps, two QTLs associated with tissue culture ability were identified each for callusing and embryogenesis rate.

High density genetic maps of crosses of different genetic backgrounds are indispensable tools for investigating oil palm genetics. They are further useful for comparative mapping analysis to identify markers closely linked to traits of interest.

Ting *et al.* (2014) genotyped SNP and SSR markers on two mapping populations, an intra-specific cross with 87 palms and an interspecific cross with 108 palms. Parental maps with 16 linkage groups were constructed for the three fruit forms of *E. guineensis* (*dura, pisifera* and *tenera*). Map resolution was further increased by integrating the dura and pisifera maps into an intra-specific integrated map with 1331 markers spanning 1867cM. They also reported the first map of a Columbian *E. oleifera* comprising 10 LGs with 65 markers spanning 471 cM. Direct comparison between the parental maps identified 603 transferable markers polymorphic in at least two of the parents. Further analysis revealed a high degree of marker transferability covering 1075 cM between the intra and interspecific integrated maps. The high density SNP and SSR based genetic map have greatly improved marker density and genome coverage in comparison with the first reference map based on AFLP and SSR markers. Singh *et al.* (2009) identified 164 QTLs associated with 21 oil yield component traits. It was also noticed that several QTLs were pleiotropic as suggested by QTL clustering of inter related traits on almost all linkage groups.

Identification of marker for fruit colour in oil palm was reported by Singh *et al.* (2014).. They revealed the identification of the VIRESCENS (*VIR*) gene, which controls fruit exocarp colour and is an indicator of ripeness. VIR is a R2R3-MYB transcription factor with homology to *Lilium LhMYB12* and similarity to *Arabidopsis PRODUCTION OF ANTHOCYANIN PIGMENT1* (*PAP1*). They identified five independent mutant alleles of *VIR* in over 400 accessions from sub-Saharan Africa that account for the dominant negative *virescens* phenotype. Each mutation resulted in premature termination of the carboxy-terminal domain of VIR, resembling McClintock's C1-I allele in maize. The identification of *VIR* will allow selection of the trait at the seed or early nursery stage, 3-6 years before fruits are produced and thus greatly advancing introgression into elite breeding material.

Gan (2014) carried out marker development studies in oil palm for genetic linkage mapping and QTL analysis for use in MAS. The use of AFLP method resulted in identification of 29 primer pairs that yielded 49 putative shell-thickness related polymorphic bands. The use of relatively new Diversity Array Technology "Genotyping-by-Sequencing (DArTSeq) platform through genotyping of two closely related *tenera* self-pollinated F2 populations, generated a total of 11675 DArTSeq polymorphic markers of good quality. These markers were used in the construction of the first reported DArTSeq based high density maps for oil palm. Saturation of the shell thickness (*Sh*) region with all available DArTSeq markers as well as map integration around the *Sh* regions for both the populations resulted in identification of 32 SNPs and DArT markers mapped within a 5cM flanking region of the *Sh* gene. Homology search of the DArTSeq marker sequence tag (64 bp) against the recently published oil palm genome assembly confirmed that 23 out of the 32 (72 per cent) DArTSeq markers were located on the p5_sc00060 scaffold in which the *SHELL* gene was identified.

Pootakham *et al.* (2015) reported genome wide SNP discovery and identification of QTLs associated with agronomic traits in oil palm using genotyping-by-sequencing (GBS). Out of 3417 fully informative SNP markers, they were able to place 1085 on a linkage map, which spanned 1429.6 cM and had an average of one marker in every 1.26 cM. Three QTLs affecting trunk height was detected on linkage group 10, 14 and 15. Further, a single QTL associated with fruit bunch weight was identified on linkage group 3. Accordingly the use of GBS approach proved to be rapid, cost effective and highly reproducible in both the oil palm species. Teh *et al.* (2016) identified three key loci for high mesocarp oil content using genome wide association study approach using SNP markers. They performed GWAS for oil-to-dry-mesocarp content (O/DM) on 2,045 genotyped *tenera* palms using 200K SNPs that were selected based on the short-range linkage disequilibrium distance, which is inherent with long breeding cycles and heterogeneous breeding populations.

4. More Applications of DNA Markers

In addition to the use in breeding for specific traits through MAS, molecular markers are also useful is diagnosis and characterization of diseases, determining legitimacy of genotypes/progenies, protection of IPR *etc.* In most of the agricultural crops, diagnosis of diseases are carried out based on the presence of causal organism, which are precisely detected by DNA markers, mainly PCR-based DNA markers. Molecular diagnosis in oil palm included, basal stem rot (BSR) caused by *Ganoderma* spp. and the spear rot disease caused by *Phytoplasma* that are effectively diagnosed by PCR based markers.

BSR disease is caused by fungus *Ganoderma boninense* which is a major threat to oil palm. Hama-Ali *et al.* (2015) reported the association between BSR disease and SSR markers in oil palm. 58 SSR markers were utilized with three progeny types namely KA4G1, KA4G8 and KA14G8 to perform a comparative molecular mapping for association with BSR. A total of 319 alleles were identified with average of 5.51 alleles per locus. Five markers, mEgCIR0793:180, mEgCIR0894:200, mEgCIR03295:210, mEgCIR3737:146 and mEgCIR3785:299 were found to be associated with *Ganoderma* disease in single progeny analysis. However, in pooled data only two alleles mEgCIR0804:213 and, mEgCIR3292:183 were found to be associated with *Ganoderma* disease. The study confirmed that progeny type KAG1 was tolerant whereas the other two were susceptible progenies and they have suggested use of these markers in future work on BSR resistance breeding in oil palm. Tee *et al.* (2013) examined the transcriptional responses of oil palm roots treated with a causal agent of BSR, *Ganoderma boninense* using a cDNA microarray approach. A total of 61 from 3,748 transcripts examined were found to be significantly up- or down-regulated in oil palm roots infected with *G. boninense* at 3 and 6 weeks post inoculation compared to those from uninfected roots.

Ten monomorphic SSR markers and two half sib families were used for detection of illegitimacy in oil palm as reported by Hama-Ali and Tan (2014). Illegitimate offspring IDs 97 and 180 were detected by four monomorphic loci mEgCIR0425, mEgCIR3769 and mEgCIR3902 in family-1 and family-2. In addition, five loci detected one illegitimate offspring ID180. Mayes *et al.* (1996) reported DNA finger printing of 111 elite breeding palms using a comparatively limited number of highly variable RFLP probes. Using this method all but eight of the palms could be distinguished uniquely.

Nurfahisza *et al.* (2014) reported genetic engineering in oil palm to modify fatty acid composition of oil palm with the aim to increase the production of oleic acid that is in high demand for use in oleochemical industries. Biolistic approach was used for inserting an a antisense palmitoyl ACP- thioesterase sequence in to oil palm tissue to down regulate the activity of palmitoyl ACP- thioesterase, and increase the accumulation of oleic acid at the expense of palmitic acid. Genomic DNA was isolated from transgenic oil palm and PCR amplification of the selected marker gene (*bar*) and gene of interest, antisense palmitoyl ACP- thioesterase was successfully carried out and the presence of transgene was confirmed by DNA sequence. The expression profiles of Δ9 stearoyl–acyl carrier protein desaturase (*SAD1* and *SAD2*) and type 3 metallothionein (*MT3-A* and *MT3-B*) were investigated in seedlings of oil palm (*Elaeis guineensis*) artificially inoculated with the pathogenic fungus *Ganoderma boninense* and the symbiotic fungus *Trichoderma harzianum.*

Resolving Intellectual Property Rights (IPR) issues is another application of DNA markers. The principle is same as determining the legitimacy of a genotype through DNA fingerprints. It can be used for protecting the IPR of advanced planting materials, breeding stocks etc as well in resolving IPR disputes.

5. Conclusion

Oil palm, being a perennial crop, having a long breeding cycle, yield assessment would take a considerably very long time. MAS would be of immense benefit in not only reducing the time for breeding cycle, but also will help in most accurate and precise selections than that of conventional methods. With the use of DNA markers, decisions could be made at the seed/seedling level, which will have a lot of impact on reducing the breeding stock in hand thereby optimizing efficiency of the breeders to a greater extent. The use of shell thickness marker has already been commercialized for identifying *dura, pisifera* and *tenera* genotypes at early stage using vegetative tissue. The major issue involved in oil palm MAS is development of proper mapping population involving crossing of the selected parents, planting and evaluation of segregating population for more than ten years. The MAS will also help in screening germplasm/breeding lines against biotic stresses like *Ganoderma*, basal stem rot, bud rot, and abiotic stress tolerance like water stress, salinity stress etc; it will speed up the way for development of stress tolerant oil palm varieties in near future.

References

Adon, B., Baudouin, L., Durand, G. T. and Kouame, B. (1998). Use of non-selected material for oil palm breeding: Angolan germplasm. *Plantations, Recherche, Développement*, **5(3)**: 201-207.

Ajambang, W., Asmono, D. and Toruan, N. (2012). Microsatellite markers reveal Cameroon's wild oil palm population as a possible solution to broaden the genetic base in the Indonesia-Malaysia oil palm breeding programs. *African Journal of Biotechnology*, **11(69)**: 13244-13249.

Arias, D., González, M., Prada, F., Ayala-Diaz, I., Montoya, C., Daza, E. and Romero, H. M. (2015). Genetic and phenotypic diversity of natural American oil palm (*Elaeis oleifera* (HBK) Cortés) accessions. *Tree Genetics and Genomes,* **11(6)**: 1-13.

Arias, D., Montoya, C. and Romero, H. (2013). Molecular characterization of oil palm (*Elaeis guineensis* Jacq.) materials from Cameroon. *Plant Genetic Resources,* **11(02)**: 140-148.

Arias, D., Ochoa, I., Castro, F. and Romero, H. (2014). Molecular characterization of oil palm (*Elaeis guineensis* Jacq.) of different origins for their utilization in breeding programmes. *Plant Genetic Resources,* **12(03)**: 341-348.

Ataga, C. D. and Fatokun, C. A. (1989). Disc polyacrylamide gel electrophoresis of pollen proteins in the oil palm (*Elaeis*). *Euphytica,* **40(1-2)**: 83-88.

Bakoumé, C., Wickneswari, R., Siju, S., Rajanaidu, N., Kushairi, A. and Billotte, N. (2015). Genetic diversity of the world's largest oil palm (*Elaeis guineensis* Jacq.) field genebank accessions using microsatellite markers. *Genetic Resources and Crop Evolution,* **62(3)**: 349-360.

Barcelos, E., Amblard, P., Berthaud, J. and Seguin, M. (2002). Genetic diversity and relationship in American and African oil palm as revealed by RFLP and AFLP molecular markers. *Pesquisa Agropecuária Brasileira,* **37(8)**: 1105-1114.

Basri Wahid, M., Abdullah, S. N. A. and Henson, I. (2005). Oil palm-achievements and potential. *Plant Production Science,* **8**: 288–297.

Baudouin, L. (1992). Utilization of molecular markers for oil palm improvement, I. protein markers. *Oléagineux (Paris),* **47(12)**: 681-691.

Billotte, N., Jourjon, M. F., Marseillac, N., Berger, A., Flori, A., Asmady, H. and Cheah, S. C. (2010). QTL detection by multi-parent linkage mapping in oil palm (*Elaeis guineensis* Jacq.). *Theoretical and Applied Genetics,* **120(8)**: 1673-1687.

Billotte, N., Marseillac, N., Risterucci, A. M., Adon, B., Brottier, P., Baurens, F. C. and Amblard, P. (2005). Microsatellite-based high density linkage map in oil palm (*Elaeis guineensis* Jacq.). *Theoretical and Applied Genetics,* **110(4)**: 754-765.

Billotte, N., Risterucci, A. M., Barcelos, E., Noyer, J. L., Amblard, P. and Baurens, F. (2001). Development, characterisation, and across-taxa utility of oil palm (*Elaeis guineensis* Jacq.) microsatellite markers. *Genome,* **44(3)**: 413-425.

Chan, P.L., Rose, R.J., Murad, A.M.A., Zainal, Z., Low, E.T.L., Ooi, L.C.L., Ooi, S.E., Yahya, S. and Singh, R. (2014). Evaluation of reference genes for quantitative real-time PCR in oil palm elite planting materials propagated by tissue culture. *PloS ONE,* **9(6)**: e99774.

Choong, C. Y., Shah, F. H., Rajanaidu, N. and Zakri, A. H. (1996). Isoenzyme variation of Zairean oil palm (*Elaeis guineensis* Jacq.) germplasm collection. *Elaeis,* **8(1)**: 45-53.

Chua, K. L., Singh, R. and Cheah, S. C. (2001). Construction of oil palm (*Elaeis guineensis* Jacq.) linkage maps using AFLP markers. In: *Cutting-edge technologies for sustained competitiveness.* Proceedings of the 2001 PIPOC International Palm

Oil Congress, *Agriculture Conference. MPOB*, Kuala Lumpur, Malaysia. pp. 461-467.

Corley, R.H.V., Lee, C. H. and Wong, C.Y. (1986). Abnormal ower development in oil palm clones. *Planter*, **62**: 233–240.

Corley, R.H.V. and Tinker, P.B. (2003). *The Oil Palm. World Agriculture Series*, Blackwell Publishing, Oxford, UK.

Eeuwens, C. J., Lord, S., Donough, C. R., Rao, V., Vallejo, G. and Nelson, S. (2002). Effects of tissue culture conditions during embryoid multiplication on the incidence of'mantled flowering in clonally propagated oil palm. *Plant Cell Tissue and Organ Culture*, **70(3)**: 311-323.

Gan, S. T. (2014). The development and application of molecular markers for linkage mapping and quantitative trait loci analysis of important agronomic traits in oil palm (*Elaeis guineensis* Jacq.). Doctoral dissertation, University of Nottingham.

Ghesquiere, M. (1984). Enzyme polymorphism in oil palm (*Elaeis guineensis* Jacq.). 1. *Genetic control of nine enzyme systems. Oleagineux*, **39(12)**: 561-574.

Ghesquiere, M. (1985). Polymorphisme enzymatique chez le palmier à huile (*Elaeis guineensis* Jacq.) II: *Variabilité et structure génétique de sept origines de palmiers. Oleagineux*, **40(11)**: 529-540.

Hama-Ali, E. O., Panandam, J. M., Tan, S. G., Alwee, S. S. R. S., Sheong, T. J., Ling, H. C. and Peng, H. B. (2015). Association between basal stem rot disease and simple sequence repeat markers in oil palm, *Elaeis guineensis* Jacq. *Euphytica*, **202(2)**: 199-206.

Hama-Ali, E.O. and Tan, S.G. (2014). Using monomorphic microsatellite markers in oil palm (*Elaeis guineensis* Jacq.). *Research and Reviews: Journal of Botanical Sciences*, **3(4):** 1-6.

Helene, A., Stefan, J., Yves, O., Fabienne, M., Jean, L. V., Duval, Y. and Tregear, J. W. (2007a). Functional characterization of MADS box genes involved Determination of flower structure in *Elaeis guineensis*: Do palms use the Same Homeotic genes as other species?. *Annals of Botany*, **100**: 1-12.

Ho, C. L., Kwan, Y. Y., Choi, M. C., Tee, S. S., Ng, W. H., Lim, K. A., Lee, Y.P., Ooi, S. and Tan, S. H. (2007). Analysis and functional annotation of expressed sequence tags (ESTs) from multiple tissues of oil palm (*Elaeis guineensis* Jacq.). *BMC Genomics*, **8(1)**: 1.

Jack, P. I., James, C., Price, Z., Rance, K., Groves, L., Corley, R. H. V., Nelson, S. and Rao, V. (1998). Application of DNA markers in oil palm breeding. In: *1998 International Oil Palm Conference.* Bali (Indonesia).

Jack, P. L. and Mayes, S. (1993). *Use of molecular markers for oil palm breeding II use of DNA markers (RFLPs). Oléagineux* (Paris), **48(1)**: 1-8.

Jack, P. L., Dimitrijevic, T. A. F. and Mayes, S. (1995). Assessment of nuclear, mitochondrial and chloroplast RFLP markers in oil palm (*Elaeis guineensis* Jacq.). *Theoretical and Applied Genetics*, **90(5)**: 643-649.

Jaligot, E., Beulé, T. and Rival, A. (2002). Methylation-sensitive RFLPs: characterisation of two oil palm markers showing somaclonal variation-associated polymorphism. *Theoretical and Applied Genetics,* **104(8)**: 1263-1269.

Jaligot, E., Beulé, T., Baurens, F. C., Billotte, N. and Rival, A. (2004). Search for methylation-sensitive amplification polymorphisms associated with the mantled variant phenotype in oil palm (*Elaeis guineensis* Jacq.). *Genome,* **47(1)**: 224-228.

Jaligot, E., Laroche, A., Beulé, T., Oakeley, E. J., Verdeil, J. L., Tregear, J. and Rival, A. (2000a). Is DNA methylation underlying somaclonal variation in oil palm?. In: *International Symposium on Tropical and Subtropical Fruits*. pp. 47-59.

Jaligot, E., Rival, A., Beule, T., Dussert, S. and Verdeil, J. L. (2000b). Somaclonal variation in oil palm (*Elaeis guineensis* Jacq.): The DNA methylation hypothesis. *Plant Cell Reports,* **19(7)**: 684-690.

Jayanthi, M., Mandal, P. K., Sujatha, G., Sri, K. S. J., Rao, G. S., Sunitha, B., Babu, M.K. and Bhat, M. G. (2008). Simple sequence repeats and RAPD primers for assessment of genetic uniformity among the field planted clones of oil palm. In: *Journal of Plantation Crops,* **36(3)**: 235-238.

Jouannic, S., Argout, X., Lechauve, F., Fizames, C., Borgel, A., Morcillo, F., Aberlenc-Bertossi, F., Duval, Y. and Tregear, J. (2005). *Analysis of expressed sequence tags from oil palm* (*Elaeis guineensis*). *FEBS Letters,* **579(12)**: 2709-2714.

Kularatne, R. S., Shah, F. H. and Rajanaidu, N. (2001). The evaluation of genetic diversity of Deli dura and African oil palm germplasm collection by AFLP technique. *Tropical Agricultural Research,* **13**: 1-12.

Lim, C.C. and Rao, V. (2004). DNA marker technology and private sector oil palm breeding. *Planter,* **80**: 611-628.

Maizura, I., Cheah, S. C. and Rajanaidu, N. (2001). Genetic diversity of oil palm germplasm collections using RFLPs. In: *Cutting-edge technologies for sustained competitiveness:* Proceedings of the 2001 PIPOC International Palm Oil Congress, *Agriculture Conference*. Kuala Lumpur, Malaysia. pp. 526-535.

Maizura, I., Rajanaidu, N., Zakri, A. H. and Cheah, S. C. (2006). Assessment of genetic diversity in oil palm (*Elaeis guineensis* Jacq.) using restriction fragment length polymorphism (RFLP). *Genetic Resources and Crop Evolution,* **53(1)**: 187-195.

Mandal, P. K. and Susmitha, D. (2006). Characterization of oil palm (*Elaeis guineensis* Jacq.) germplasm using biochemical and molecular parameters. *Journal of Plantation Crops,* **34 (3)**: 534-539.

Mandal, P. K., Sujatha, G. and Chandrasekhar, G.V.V.N. (2006). Designing of oil palm (*Elaeis guineensis* Jacq.) microsatellite primers and testing their functionality in different palms. In the National Symposium on Molecular Breeding in Crop Plants. *Indian Institute of Vegetable Research,* Varanasi, Uttar Pradesh. pp. 30-31.

Mathius, N., Ginting, G. (1998). Use of biomolecular techniques in studies of abnormalities in oil palm clones. In: 1998 *International Oil Palm Conference.* Bali (Indonesia).

Matthes, M., Singh, R., Cheah, S. and Karp, A. (2001). Variation in oil palm (*Elaeis guineensis* Jacq.) tissue culture-derived regenerants revealed by AFLPs with methylation-sensitive enzymes. *Theoretical and Applied Genetics*, **102**: 971-979.

Mayes, S., Jack, P. L. and Corley, R. H. V. (2000). The use of molecular markers to investigate the genetic structure of an oil palm breeding programme. *Heredity*, **85(3)**: 288-293.

Mayes, S., James, C. M., Horner, S. F., Jack, P. L. and Corley, R. H. V. (1996). The application of restriction fragment length polymorphism for the genetic fingerprinting of oil palm (*E. guineensis* Jacq.). *Molecular Breeding*, **2(2)**: 175-180.

Meunier, J. (1995). Biotechnologies: What prospects for tropical tree crops?. *Plantations Recherche Développement*, **2(1)**: 7-18.

Morcillo, F., Gagneur, C., Adam, H., Richaud, F., Singh, R., Cheah, S. C., Rival, A., Duval, Y. and Tregear, J. W. (2006). Somaclonal variation in micropropagated oil palm. Characterization of two novel genes with enhanced expression in epigenetically abnormal cell lines and in response to auxin. *Tree Physiology*, **26(5)**: 585-594.

Moretzsohn, M. D. C., Ferreira, M. A., Amaral, Z. P. S., Coelho, P. J. D. A., Grattapaglia, D. and Ferreira, M. E. (2002). Genetic diversity of Brazilian oil palm (*Elaeis oleifera* HBK) *germplasm collected in the Amazon Forest. Euphytica*, **124(1)**: 35-45.

Moretzsohn, M. D. C., Nunes, C. D. M., Ferreira, M. E. and Grattapaglia, D. (2000). RAPD linkage mapping of the shell thickness locus in oil palm (*Elaeis guineensis* Jacq.). *Theoretical and Applied Genetics*, **100(1)**: 63-70.

Nurfahisza, A. R., Rafiqah, M. A., Masani, M.Y.A., Hanin, A.N., Rasid, O.A., Parveez, G. K. A. and Ismail, I. (2014). Molecular analysis of transgenic oil palm to detect the presence of transgenes. *Journal of Oil Palm Research*, **26:** 73-80.

Ong-Abdullah, M., Ordway, J.M., Jiang, N., Ooi, S.E., Kok, S.Y., Sarpan, N., Azimi, N., Hashim, A.T., Ishak, Z., Rosli, S.K. and Malike, F.A. (2015). Loss of Karma transposon methylation underlies the mantled somaclonal variant of oil palm. *Nature*, **525(7570)**: 533-537.

Paranjothy, K., Ong, L. and Sharifah, S.S.A. (1995b). DNA and protein changes in relation to clonal abnormalities in relation to clonal abnormalities. In: Rao, V., Henson, I.E. and Rajanaidu, N. (Ed.), Recent development in oil palm, tissue culture and biotechnology, PORIM Malaysia. pp.86-89.

Paranjothy, K., Othman, R., Than, C. C., Wongm, G. and Andsoh, A.C. (1995a). Incidence of abnormalities in relation to clonal abnormalities. In: Rao, V., Henson, I.E. and Rajanaidu, N. (Ed.), Recent development in oil palm tissue culture and biotechnology, *Proceedings 1993 ISOPB International Symposium On Recent Development in Oil Palm Tissue Culture and Biotechnology*. PORIM, Malaysia. pp. 77-85.

Pootakham, W., Jomchai, N., Ruang-areerate, P., Shearman, J. R., Sonthirod, C., Sangsrakru, D., Tragoonrung, S. and Tangphatsornruang, S. (2015). Genome-wide SNP discovery and identification of QTL associated with agronomic traits in oil palm using genotyping-by-sequencing (GBS). *Genomics*, **105(5)**: 288-295.

Purba, A. R., Noyer, J. L., Baudouin, L., Perrier, X., Hamon, S. and Lagoda, P. J. L. (2000). A new aspect of genetic diversity of Indonesian oil palm (*Elaeis guineensis* Jacq.) revealed by isoenzyme and AFLP markers and its consequences for breeding. *Theoretical and Applied Genetics*, **101(5-6)**: 956-961.

Rance, K. A., Mayes, S., Price, Z., Jack, P. L. and Corley, R. H. V. (2001). Quantitative trait loci for yield components in oil palm (*Elaeis guineensis* Jacq.). *Theoretical and Applied Genetics*, **103(8)**: 1302-1310.

Rao, V. and Donough, C. R. (1990). Preliminary evidence for a genetic cause for the floral abnormalities in some oil palm ramets. *Elaeis*, **2**: 199-207.

Riju, A., Chandrasekar, A. and Arunachalam, V. (2007). Mining for single nucleotide polymorphisms and insertions/deletions in expressed sequence tag libraries of oil palm. *Bioinformation*, **2(4)**: 128-131.

Ritter, E., de Armentia, E. L., Erika, P., Herrero, J., Niggrum, Y. P., Santika, B., Endang, Y., Sarimana, Y., Sembiring, Z., Asmono, D. and Hernandez, M. (2016). Development of a molecular marker system to distinguish shell thickness in oil palm genotypes. *Euphytica*, **207(2)**: 367-376.

Shah, F. H. and Parveez, A. (1995). DNA variation in abnormal tissue culture regenerants of oil palm, *Elaeis guineensis*. *Asia-Pacific Journal of Molecular Biology and Biotechnology*, **3(1)**: 49-53.

Shah, F. H., Rashid, O., Simons, A. J. and Dunsdon, A. (1994). The utility of RAPD markers for the determination of genetic variation in oil palm (*Elaeis guineensis*). *Theoretical and Applied Genetics*, **89(6)**: 713-718.

Sharifah, S. S. A., Singh, R. and Cheah, S.C. (1999). Molecular dissection of the floral clonal abnormality in oil palm. **In:** *Proceedings 1999 PORIM International. Palm Oil Congress*. Palm Oil Research Institute. Kuala Lumpur, Malaysia. pp. 321-329.

Singh, R. and Cheah, S. C. (1999). Analysis of the inheritance of AFLP markers in an interspecific cross of oil palm using the pseudo-testcross strategy. *Journal of Oil Palm Research*, Special Issue: 64-73.

Singh, R. and Cheah, S. C. (2000). Identification of flower specific proteins in the oil palm (*Elaeis guineensis*) by two-dimensional gel electrophoresis. *Journal of Oil Palm Research*, **12(2)**: 46-54.

Singh, R., Cheah, S.C., Madon, M., Ooi, L. C. L. and Rahimah, A. R. (2001). Genomic strategies for enhancing the value of the oil palm. *Cutting-edge technologies for sustained competitiveness:* In: *Proceedings of the 2001 PIPOC International Palm Oil Congress*, Agriculture Conference. Kuala Lumpur, Malaysia. pp. 3-17.

Singh, R., Low, E. T. L., Ooi, L. C. L., Ong-Abdullah, M., Nookiah, R., Ting, N. C. and Nagappan, J. (2014). The oil palm VIRESCENS gene controls fruit colour and encodes a R2R3-MYB. *Nature communications*, doi:10.1038/ncomms5106.

Singh, R., Ong-Abdullah, M., Low, E. T. L., Manaf, M. A. A., Rosli, R., Nookiah, R. and Azizi, N. (2013a). Oil palm genome sequence reveals divergence of interfertile species in old and new worlds. *Nature*, **500(7462)**: 335-339.

Singh, R., Low, E. T. L., Ooi, L. C. L., Ong-Abdullah, M., Ting, N. C., Nagappan, J. and Chan, K. L. (2013b). The oil palm shell gene controls oil yield and encodes a homologue of seedstick. *Nature*, **500(7462)**: 340-344.

Singh, R., Nagappan, J., Tan, S. G., Panandam, J. M. and Cheah, S. C. (2007). Development of simple sequence repeat (SSR) markers for oil palm and their application in genetic mapping and fingerprinting of tissue culture clones. *Asia Pacific Journal of Molecular Biology and Biotechnology*, **15**: 121-131.

Singh, R., Tan, S. G., Panandam, J. M., Rahman, R.A., Ooi, L. C., Low, E. T., Sharma, M., Jansen, J. and Cheah, S. C. (2009). Mapping quantitative trait loci (QTLs) for fatty acid composition in an interspecific cross of oil palm. *BMC Plant Biology*, **9(1)**: 1.

Singh, R., Zaki, N. M., Ting, N. C., Rosli, R., Tan, S. G., Low, E. T. L. and Cheah, S. C. (2008). Exploiting an oil palm EST database for the development of gene-derived SSR markers and their exploitation for assessment of genetic diversity. *Biologia*, **63(2)**:227-235.

Suan Choo, C., Ives, C. L. and Bedford, B. M. (1998). The biotechnology of oil palm. In: *Agricultural Biotechnology in International Development*. pp. 161-170.

Taeprayoon, P., Tanya, P., Lee, S. H. and Srinives, P. (2015). Genetic background of three commercial oil palm breeding populations in Thailand revealed by SSR markers. *Australian Journal of Crop Science*, **9(4)**: 281.

Tee, S. S., Tan, Y. C., Abdullah, F., Ong-Abdullah, M. and Ho, C. L. (2013). Transcriptome of oil palm (*Elaeis guineensis* Jacq.) roots treated with *Ganoderma boninense*. *Tree Genetics and Genomes*, **9(2)**: 377-386.

Teh, C. K., Ong, A.L., Kwong Q.B., Apparow, S., Chew, F.T., Mayes, S., Mohamed, M., Appleton, D. and Kulaveerasingam, H. (2016). Genome-wide association study identifies three key loci for high mesocarp oil content in perennial crop oil palm. *Scientific Reports*, **6**: 19075. doi: 10.1038/srep19075.

Ting, N. C., Jansen, J., Mayes, S., Massawe, F., Sambanthamurthi, R., Ooi, L. C. L. and Ithnin, M. (2014). *High density SNP and SSR-based genetic maps of two independent oil palm hybrids. BMC Genomics*, **15(1)**: 1.

Ting, N. C., Jansen, J., Nagappan, J., Ishak, Z., Chin, C. W., Tan, S. G. and Singh, R. (2013). *Identification of QTLs associated with callogenesis and embryogenesis in oil palm using genetic linkage maps improved with SSR markers. PLoS ONE*, **8(1)**: e53076.

Ting, N. C., Zaki, N. M., Rosli, R., Low, E. T. L., Ithnin, M., Cheah, S. C. and Singh, R. (2010). SSR mining in oil palm EST database: application in oil palm germplasm diversity studies. *Journal of Genetics*, **89(2)**: 135-145.

Tregear, J. W., Morcillo, F., Richaud, F., Berger, A., Singh, R., Cheah, S. C. and Duval, Y. (2002). Characterization of a defensin gene expressed in oil palm inflorescences: induction during tissue culture and possible association with epigenetic somaclonal variation events. *Journal of Experimental Botany*, **53(373)**: 1387-1396.

Upadhyay, A. and Karun, A. (2000). Variation of esterase isozyme in regenerated plantlets of oil palm (*Elaeis guineensis* Jacq.). *Journal of Plantation Crops*, **28**: 82-85.

Zaki, N.M., Ismail, I., Rosli, R., Chin, T.N. and Singh, R. (2010). Development and characterization of *Elaeis oleifera* microsatellite markers. *Sains Malaysiana*, **39(6)**: 909-912.

Zaki, N.M., Singh, R., Rosli, R. and Ismail, I. (2012). *Elaeis oleifera* genomic-SSR markers: Exploitation in oil palm germplasm diversity and cross-amplification in arecaceae. *International Journal of Molecular Sciences*, **13(4)**: 4069-4088.

Chapter 13

Date Palm

☆ *P.N. Sivalingam, Karun Gurjar, Dhurendra Singh and Pinaki Acharyya*

1. Introduction

Date palm (*Phoenix dactylifera* L.; 2n = 2x = 36; Arecaceae) is one of the important fruit crops of arid and semi-arid regions of the world. The palm is one of the oldest cultivated fruit trees, grown for at least 7000 years and is believed to have originated in southern Near East (Zohary and SpiegelRoy, 1975) and Mesopotamia (Wrigley, 1995), where it was domesticated. Currently, date palm is found to be distributed in the Middle East countries of Asia, Africa, parts of Europe and USA. The fruit, the economical part of date palm, is a rich source of sugar, minerals, and vitamins etc (Barreveld, 1993). Date palm is dioecious, highly heterozygous, monocotyledonous plant, with a very slow growth rate and a late reproductive phase. Other species of genus *Phoenix* such as *P. canariensis* (Canary Islands date palm), *P. reclinata* (Senegal date palm) and *P. sylvestris* (Indian sugar palm) are also utilized for various purposes.

Date palm has long been an integral part of desert culture, providing food, building material, shade, fuel and income for the rural people (Al-Najm *et al.*, 2016). Egypt, Iran and Saudi Arabia constitute the top three producing countries, while India is the largest importer of dates (Khierallah, 2015). The total number of known cultivars is approximately 5000; of these, 650 are cultivated in Iraq (Ibrahim, 2008), 340 in Saudi Arabia (Al-Mssallem, 1996) and about 135 in United Arab Emirates (Ghaleb, 2008).

The success of any genetic conservation or breeding programme is dependent on understanding the amount and distribution of the genetic variation already present in the gene pool (Jubrael *et al.*, 2005). Morphological [fruit (shape, weight, colour, skin aspect, consistency and texture), leaves, spines], biochemical characters and cytological information have been utilized to assess the extent of genetic diversity among date palm genotypes. On the basis of morphological characters, origin and fruit quality, 531 cultivars have been distinguished by Al-Baker (1962). In a later study, Al-Baker (1972) also described 627 Iraqi cultivars of date palm based on the above morphological parameters, in addition to fruit traits. Similarly, in another study, 110 cultivars of date palm were also identified from Iraq by Al-Jboory (1971). Husien (2002) and Husien and Graib (2004) described about 36 and 50 cultivars of date palm respectively utilizing morphometric traits. Twelve phenotypic traits have also been utilized to describe 110 date palm cultivars (Al-Saleh and Al-Ansary, 2005).

In India, unique morphological descriptors to differentiate cultivars/varieties of date palm are being developed at ICAR-Central Institute for Arid Horticulture (ICAR-CIAH), Bikaner, Rajasthan under the project 'DUS Descriptors for Date Palm'; these parameters are also being utilized for morphological identification of date palm cultivars. However, identification of date palm cultivar using morphological traits is not possible till fruiting stage in addition to requirement for a large set of phenotypic data (Khierallah, 2015). The major disadvantages of these markers are that they are not stable and their levels of expression are greatly influenced by different environmental conditions and developmental stages of the plant.

Biochemical markers, such as isozymes and peroxidase activity, have also been utilized to characterize date palm cultivars in Morocco and Tunisia (Baaziz, and Saaidi, 1988; Ould Salem Mohamed *et al.*, 2001; Majourhat *et al.*, 2002). Main disadvantages of these markers are low abundance, lesser polymorphism and environmental influences; also, these markers do not precisely reflect the existing polymorphism. Because of the limitations with morphological and biochemical markers, DNA-based markers are being presently utilized in date palm to understand genetic diversity in the germplasm pool, genome stability and somaclonal variations in tissue culture derived date palm plantlets, genetic map construction and in date palm breeding programmes (Khierallah, 2015).

2. Use of DNA-based Markers in Date Palm

2.1. Restriction Fragment Length Polymorphism (RFLP) Markers

Despite its reproducibility and co-dominant nature, application of RFLP markers in date palm cultivar identification and differentiation has been limited as this technique needs good quality genomic DNA and use of radio-labelled probes. RFLP markers were developed and later used to differentiate Tunisian date palm cultivars using homologous and heterologous probes (Corniquel and Mercier, 1994; Trifi *et al.*, 2000). However, the number of genotypes analyzed was restricted to a small set of cultivars, which were widely grown in oases or characterized by fleshy dates, which were not a true representative of the germplasm. A modification of this technique is the restriction of PCR amplified products either from genomic or plastid DNA followed by hybridization (PCR-RFLP). Tunisian date palm accessions were characterised using PCR, with consensus primers in the chloroplast genome region, and analysed by RFLP (ctDNA PCR-RFLP). Using this technique, two haplotypes were identified among 15 genotypes of date palm by *HinfI* restriction site (Hela *et al.,* 2004). Similarly, PCR-RFLP was successfully used to differentiate sex in date palm with 90 percent accuracy (Al-Mahmoud *et al.,* 2012).

2.2. Random Amplified Polymorphic DNA (RAPD) Markers

RAPD marker system has been widely used in date palm for the assessment of genetic diversity among different cultivars, genetic fidelity testing of tissue culture derived plantlets, gender differentiation *etc.* Five Egyptian cultivars of date palm *viz.,* Zghloul, Amhat, Samany, Haiany and Siwiand, were characterized using RAPD markers by Saker and Moursy (1999). Sedra *et al.* (1998) described RAPD markers linked to resistance to bayoud disease in a sub-set of Morocco date palm cultivars. DNA fingerprinting for five Egyptian date palm cultivars (Zaghloul, Amhat, Samany, Haiany and Siwi) using four RAPD primers was carried out by Saker and Moursy (1999). Furthermore, Jubrael *et al.* (2001) used RAPD markers and identified nine Iraqi date palm female cultivars. A genetic diversity study was carried out in Egyptian date palm cultivars by RAPD; four famous female cultivars and four unknown males were analyzed (Soliman *et al.,* 2003). There are other reports which characterized RAPD profiles in date palm cultivars from Saudi Arabia and Gaza strip (Askari *et al.,* 2003; El Kichaouil *et al.,* 2013). Results obtained from RAPD technique have also revealed a narrow genetic base in most of the Saudi date palm cultivars, the cultivars possessing more than 50 per cent similarity (Askari *et al.,* 2003; Al-Khalifah and Askari, 2003; El-Tarras *et al.,* 2007); Barhi (an Iraqi cultivar) exhibited only 34 per cent genetic similarity in a study conducted by Al-Khalifah and Askari (2003). In another study by Al-Moshileh *et al.* (2004), five Saudi date palm cultivars (Barhi, Nabtet Ali, Rothanah, Ajwa, and Sokkari) were subject to DNA fingerprint analysis. In order to evaluate the somaclonal variation before and after long term cryopreservation, Letouze *et al.* (1998) and Saker *et al.* (2000) carried out RAPD fingerprinting of Egyptian date palm varieties and found that they are age-dependent, which was in conformation with a previous observation that duration of callus culture plays an important role in determining the level of somaclonal variation (Gaj and Maluszynski, 1987). Al-Qurainy *et al.* (2002) supported the concept that the RAPD technique can be successful in determining the genetic fidelity of micropropagated date palms. Al-Khalifah and Askari (2005). used RAPD markers to detect somaclonal variation in tissue culture derived plant from four date palm varieties. Since reports have shown that RAPD markers possess serious limitations with respect to cultivar characterization because of low polymorphism, irreproducibility and construction of weak grouping associations (Yang *et al.,* 1996; Benkhalifa, 1999; Sedra *et al.,* 1998; Trifi *et al.,* 2000), a need was felt, therefore, to apply more robust techniques to date palm genetic diversity analysis.

2.3. Amplified Fragment Length Polymorphism (AFLP) Markers

AFLP involves two techniques; digestion by restriction enzyme and PCR amplification. DNA is digested by restriction enzymes followed by ligation of double stranded adaptor to the DNA-cut end and DNA amplification. The technique utilizes inherent variation in restriction enzyme recognition sites and nucleotide sequences among the cultivars. El-Khishin *et al.* (2003) conducted DNA profiling of five Egyptian cultivars using AFLPs and the dendrogram generated based on UPGMA (unweighted pair group method using arithmetic averages) showed two major clusters where Siwi and Hayany being the most genetically similar cultivars. On the second cluster, Amhat and Samany was grouped together, whereas Zaghloul was the most distinct cultivar. Jubrael *et al.* (2005) described that the high levels of inter-varietal AFLP polymorphism detected, using five primer combinations among 18 Iraqi date palm cultivars, might be partly due to the strong out-crossing mechanism in this species. Researchers from Tunisia and Egypt have also tested different sets of AFLP primers and generated banding patterns in date palm (El Assar *et al.*, 2005; Rhouma *et al.*, 2007); the data obtained from these studies suggests that AFLP constituted a very attractive and informative procedure for providing evidence of the genetic diversity among date palm ecotypes. Khierallah *et al.* (2011a). carried out AFLP analysis for 18 Iraqi cultivars and detected polymorphic AFLP fragments and helped in discriminating date palm cultivars. However, for date palm cultivars which were growing in Iraq for several centuries, Jaccard's similarity index and Principle Component Analysis (PCA) based on AFLP data revealed diverse relationships among them and they were clustered independently of their geographic origin, in spite of their phenotypic uniqueness.

2.4. Microsatellites or Simple Sequence Repeats (SSRs) Markers

SSR markers have been widely used for studying genetic diversity in date palm (Billotte *et al.*, 2004). Zehdi *et al.* (2004) used these markers to characterize Tunisian cultivars, while Al-Ruqaishi *et al.* (2008) utilized these primers to screen and analyze the genetic diversity among clonal genotypes of Omani cultivars. Elshibli and Korpelainen (2008) examined genetic diversity in Sudan using SSR primers and Ahmed and Al-Qaradawi (2009) utilized SSR markers to analyze genetic diversity among date palm cultivars. On the other hand, Khierallah *et al.* (2011b) tested genetic diversity of Iraqi date palm using SSR primer pairs and showed highly polymorphic patterns with a great number of alleles (188) distributed in 30 date palm cultivars. The number of alleles per locus in this study (8.54) was higher than 7.6 alleles per locus scored by Zehdi *et al.* (2004). in 46 date palm cultivars cultivated in Tunisia, and those studied by Ahmed and Al-Qaradawi (2009) in 15 cultivars grown in Qatar. The heterozygosity value detected in the Iraqi cultivars indicated the presence of high genetic diversity. Domestication of date palm unknown origin (Wrigley 1995) and the nature of date palm culture, might have played an important role in the composition of date palm genome (Elshibli and Korpelainen, 2008). Thus, new cultivars may appear as a result of the continuous selection carried out by farmers following sexual reproduction. Thus the process of domestication and breeding procedures together might resulted have resulted in a mixed genome of date palm within the same country (Elshibli and Korpelainen, 2008). From the genome shotgun next generation sequenced data, Hamwieh *et al.* (2010) developed over 1000 SSR primer pairs. Elmeer *et al.* (2011). found that these new markers could be an initial step towards making use of such markers for genetic mapping and diversity analysis of date palm. Arabnezhad *et al.* (2011) assessed genetic relationships among Iran date palm genotypes grown in different geographical regions by using newly developed SSR markers and differentiated Iran, Iraq and Africa grown date palms.

Cluster analysis significantly distinguished African cultivars from Iranian and Iraqi ones which signified that domestication of African cultivars has followed a different route than those grown in the Middle-East, an hypothesis which was supported by data from Mantel's test and Bayesian analysis.

Expressed Sequence Tags derived SSRs or genic SSRs are mainly used to map candidate genes. A large number of EST gene sequences in date palm was identified by Al-Dous *et al.* (2011). Zhao *et al.* (2013) used 22889 EST sequences from date palm genome data base and identified 4600 ESTs containing SSRs. A total of 4967 primer pairs were designed as EST-SSR markers based on computational analysis.

2.5. Inter Simple Sequence Repeat (ISSR) Markers

The ISSR marker system entails amplification of DNA segments present between two identical microsatellite repeats oriented in opposite directions. In this technique, single primer of length approximately 20 nucleotides is used. Ten high quality fruit yielding date palm accessions from Tunisia, including those introduced from Iraq and Algeria, were analysed by ISSR. Native accessions had 50 per cent genetic divergence and no significant difference was observed between native and introduced accessions indicating their narrow genetic base (Karim *et al.*, 2010). Fifteen cultivars of date palm, collected from the South and South West of Iran, were subjected to 89 ISSR and 84 RAPD markers. The results showed that the markers were linked to morphological traits and could be utilized as informative markers (Marsafari *et al.*, 2014). Similarly, Hamza *et al.* (2012) reported grouping of cultivars from Tunisia, having similar fruit characters, together with the data obtained from profiling these cultivars using ISSR markers. In Iraq, 17 well known cultivars were analysed by both ISSR and RAPD markers. Jaccard's genetic similarity index ranged from 0.422 between the two males 'Risasy' and 'Ghanami Akhhder' to 0.789 between 'Hilawi' and 'Ghanami Ahmer'. These primers were found efficient in analysing genetic diversity of date palm cultivars in Iraq (Khierallah *et al.*, 2014).

2.6. Inter-primer Binding Site (iPBS) Markers

The inter-primer binding site (iPBS) markers were developed for exploring genetic diversity and relationships in plants (Kalendar *et al.*, 2010) iPBS amplification, is based on the virtually universal presence of a tRNA complement as a reverse transcriptase primer binding site (PBS) in LTR retrotransposons (Kalendar *et al.*, 2010). Al-Najm *et al.* (2016) investigated the genetic relationships of 54 Australian collections and 12 Iraqi genotypes using iPBS technique. From the results obtained, the mean expected heterozygosity indicated a high level of inbreeding among the accessions tested. With a few exceptions, the genetic relationships among all accessions could be based on the geographic origin; specifically Iraqi female cultivars, exotic female cultivars collected in Australia and male accessions also collected in Australia. The results revealed that date palm accessions collected in Australia and Iraq were found to be highly divergent.

2.7. Single Nucleotide Polymorphism (SNP) Markers

Genome sequence of date palm, reported by Al-Dous *et al.* (2011) and Al-Mssallem *et al.* (2013), has accelerated genomic study and marker development for its improvement. SNPs were developed in date palm for the first time by Mathew *et al.* (2014). Sequencing on 70 female cultivar samples from across the date palm-growing regions, including four *Phoenix* species as outgroup generated 13,000-65,000 SNPs in a diverse set of date palm fruit and leaf samples. Using these SNPs, date palm cultivars could be segregated into two main regions of shared genetic background from North Africa and the Arabian Gulf. In another study, mitochondrial and plastid genomes of nine Saudi Arabian cultivars were sequenced. For each species, about 60 million 100 bp paired end reads were generated. Based on plastid and mitochondrial reference genomes, SNPs were identified. The results identified cultivar specific SNPs for eight of the nine cultivars. Filtering all sequencing reads that mapped to both organellar genomes nearly eliminated mitochondrial heteroplasmy but all plastid SNPs remained heteroplasmic. This investigation provides valuable insights into inter-organellar DNA transfer in performing SNP analyses from total genomic DNA. Low levels of sequence variation in plastid and mitochondrial genomes identified were compared to nuclear SNPs for molecular characterization of date palm cultivars (Sabir *et al.*, 2014).

3. Molecular Markers for Sex Determination in Date Palm

The sex of date palm cannot be identified until the palm reaches flowering stage, which takes about 4-7 years. Early identification of sex in date palm can enhance breeding programme and genetic studies. El-Kharbotly *et al.* (1998) used AFLP marker to initiate genetic map for both male and female palms and observed their segregation pattern. They identified the trait 'erect and slanting leaf' to be controlled by single factor and inherit according to Mendelian pattern. Among 32 AFLP primer combinations, primers which produced unique banding patterns in male and female palms were identified. RAPD markers have also been used to differentiate four each of male and female trees from Egypt (Soliman *et al.*, 2003) as well as for identification of male and female cultivars of 'Barhee' and 'Sukkary' (Al-Khalifa *et al.*, 2006). Sexual embryos of date palm were used for identification of sex by isozyme and RAPD markers. High levels of peroxidase activity observed in female offshoots, acid phosphatase and glutamate oxaloacetate enzymes also helped in differentiation of sex (Bekheet *et al.*, 2008). Genomic region date palm linked to gender was identified by Al-Mahmoud *et al.* (2012). Microsatellite markers were identified to differentiate male and female date palm as among the 254 microsatellite loci studied, 22 were linked to male (Elmeer and Mattat, 2012). SCAR markers were developed to differentiate male and female genotypes by Dhawan *et al.* (2013). RAPD primer OPA-02 amplified ~1.0 kb fragment which was present in all male genotypes, absent in all female genotypes. SCAR marker that amplifies 354 bp only in male genotypes has also been developed.

4. Marker Assisted Selection (MAS)

MAS is one of the breeding approaches that combines traditional genetics with the use of molecular tools. As date palm is a perennial and long duration fruit crop, usage of molecular marker in the breeding programme has been limited. AFLP markers have been used to study two F_1 and F_4 backcrossed populations to map traits such as fruit and seed weights, seed volume, total soluble solids and measured morphological characters (El-Kharbotly *et al.*, 1998).

5. Conclusion

Various molecular markers have been utilized mainly to understand genetic diversity and phylogenetic relationships among the date palm cultivars. The genetic diversity studies have mainly been focused on either localized cultivars or introduced ones. It is necessary to develop reference collection or core collection or core set of date palm accessions representing the total date palm collections from different regions of the world. For developing this collection, there is a need for concerted efforts to remove duplicates and ambiguity in names of genotypes of date palm by using valid reproducible molecular markers such as SSRs and SNPs. With the availability of genome sequence of date palm, developing and validation of such markers has become much easier. The core collection, which would be developed, would act as nodal centre of date palm germplasm and could be utilized by all countries for improving date palm, identification of alleles for various biotic and abiotic stress tolerances and traits for fruit quality and yield.

References

Ahmed, T.A. and Al-Qaradawi, A. (2009). Molecular phylogeny of Qatari date palm genotypes using simple sequence repeats markers. *Biotechnology*, **8**: 126–131.

Al-Baker, A.J. (1962). The Iraqi dates varieties. *The Government Press*, Iraq.

Al-Baker, A.J. (1972). The date palm: A review of its past and present status, and the recent advances in its culture, industry and trade. Al-Ani Press, Iraq, (In Arabic).

Al-Dous, E.K., George, B., Al-Mahmoud, M.E., Al-Jaber, M.Y., Wang, H., Salameh, Y.M., Al-Azwani, E.K., Chaluvadi, S., Pontaroli, A.C., DeBarry, J., Arondel, V., Ohlrogge, J., Saie, I.J., Suliman-Elmeer, K., Bennetzen, J.L., Kruegger, R.R. and Malek, J.A. (2011). *De novo* genome sequencing and comparative genomics of date palm (*Phoenix dactylifera*). *Nature Biotechnology*, **29**: 521–527.

Al-Jboory, N.M., Khalaf, Z. and Numan, K. (1971). A description of 110 Iraqi dates palm cultivars. Bulletin 38, General Horticulture Directorate, *Ministry of Agriculture*, Iraq.

Al-Khalifah, N.S. and Askari, E. (2003). Molecular phylogeny of date palm (*Phoenix dactylifera* L.) cultivars from Saudi Arabia by DNA fingerprinting. *Theoretical and Applied Genetics*, **107**: 1266–1270.

Al-Khalifah, N.S. (2005). Micropropagation and DNA fingerprinting of date palm trees of Saudi Arabia. *Association of Agricultural Research Institutions in the Near East and North Africa*, Aarinena Publications Amman, Jordan.

Al-Khalifah, N.S., Khan, F.A., Askari, E. and Hadi, S. (2006). *In vitro* culture and genetic analysis of male and female date palm. *Acta Horticulturae*, **725**: 653–661.

Al-Mahmoud, M.E., Al-Dous, E.K., Al-Azwani, E.K. and Malek, J.A. (2012). DNA-based assays to distinguish date palm (Arecaceae) gender. *American Journal of Botany*, **99**: 7–10.

Al-Moshileh, A.M., Motawei, M.I., Al-Wasel, A. and Abdel-Latif, T. (2004). Identification of some date palm (*Phoenix dactylifera* L.) cultivars in Saudi Arabia using RAPD fingerprints. *Journal of Agricultural and Marine Sciences*, **9**: 1–3.

Al-Mssallem, I.S. (1996). Date palm. *Arabian Global Encyclopedia*, **7**: 182-187.

Al-Mssallem, I.S., Hu, S., Zhang, X., Lin, Q., Liu, W., Tan, J., Yu, X., Liu, J., Pan, L., Zhang, T., Yin, Y., Xin, C., Wu, H., Zhang, G. and Ba Abdullahm, M.M. *et al.* (2013). Genome sequence of the date palm (*Phoenix dactylifera* L.). *Nature Communications*, **4**: 2274.

Al-Najm, A., Luo S., Ahmad N.M. and Trethowan, R. (2016). Molecular variability and genetic relationships of date palm (*Phoenix dactylifera* L.) cultivars based on inter-primer binding site (iPBS) markers. *Australian Journal of Crop Science*, **10(5)**: 732-740.

Al-Qurainy, F., Ai-Saad, F. and Filfilan, S. (2002). Comparative study between four cultivars of date palm (*Phoenix dactylifera* L.) produced from tissue culture and offshoot origins by RAPD technology. *Saudi Journal of Biological Science*, **9**: 3-11.

Al-Ruqaishi, I.A., Davey, M., Alderson, P. and Mayes, S. (2008). Genetic relationships and genotype tracing in date palm (*Phoenix dactylifera* L.) in Oman based on microsatellite markers. *Plant Genetic Resources*, **61**: 70–72.

Al-Saleh A.A. and Al-Ansary N.A. (2005). A pictorial for Iraqi date palm cultivars. Part 1. Ministry of Agriculture, Al-Ezza Press, Iraq, (in Arabic).

Arabnezhad, H., Bahar, M., Mohammadi, H.R. and Latifian M. (2011). Development, characterization and use of microsatellite markers for germplasm analysis in date palm (*Phoenix dactylifera* L.). *Scientia Horticulturae*, **134**: 150–156.

Askari, E., Al-Khalifah, T., Al-Hafidh, Y., Khan, F., Al-Hindi, A. and Okawara, R. (2003). Molecular phylogeny of seven date palm (*Phoenix dactylifera* L.) cultivars by DNA fingerprinting. *Pakistan Journal of Botany*, **35**: 323-330.

Baaziz, M. and Saaidi, M. (1988). Preliminary identification of date palm cultivars by esterase isoenzymes and peroxides activities. *Canadian Journal of Botany*, **66**: 89–93.

Barreveld, W.H. (1993). Date palm products: FAO *Agricultural Services Bulletin No. 10. Rome*, Italy.

Bekheet, S.A., Taha, H., Hanafy, M. and Solliman, M.E. (2008). Morphogenesis of sexual embryos of date palm cultured *in vitro* and early identification of sex type. *Journal of Applied Science Research*, **4**: 345–352.

Benkhalifa, A. (1999). Gestion De La Diversite' Ge' Ne'Tique Du Palmier Dattier En Alge'Rie. Paper presented at the workshop constitution at organisation D'E'Quipesde Recherché Scientifique Dans Les Domainnes De Foresterie Et *Des Arbres Fruitiers*, Marrakech, Morocco.

Billotte, N., Marseillac, N., Brottier, P., Noyer, J.L., Jacquemoud-Collet, J.P., Moreau, C., Couvreur, T., Chevalier, H., Pintaud, C. and Risterucci, A.M. (2004). Nuclear microsatellite marker for date palm (*Phoenix dactylifera* L): Characterization and utility across the genus phoenix and in other palm genera. *Molecular Ecology Notes*, **4**: 256–258.

Corniquel, B. and Mercier, L. (1994).) Date palm (*Phoenix dactylifera* L.) cultivar identification by RFLP and RAPD. *Plant Science*, **101**: 163–172.

Dhawan, C., Kharb, P., Sharma, R., Uppal, S. and Aggarwal K.R. (2013). Development of male-specific SCAR marker in date palm (*Phoenix dactylifera* L.). *Tree Genetics and Genomes*, **9**: 1143–1150.

El Kichaouil, A., Abu Zayed, M.A. and Ayesh, B.M. (2013). Genotyping and identification of six date palm (*Phoenix dactylifera* L.) cultivars of the Gaza Strip by random amplification of polymorphic DNA. *Emirates Journal of Food and Agriculture*, 25 **(11)**: 916-925.

El-Assar, A.M., Krueger, R.R., Devanand, P.S. and Chao, C.C.T. (2005). Genetic analysis of Egyptian date (*Phoenix dactylifera* L.) accessions using AFLP markers. *Genetic Resources and Crop Evolution*, **52**: 601–607.

El-Kharbotly, A., El-Mardi, M., Al-Saadi, N. and Al-Mahruki, Y. (1998). Towards the construction of a genetic map of date palms using the amplified fragment length polymorphism technique (AFLP). In: *Proceedings of the First International Conference on Date Palms*, Al-Ain, United Arab Emirates, pp. 194–207.

El-Khishin, D.A., Adawy, S.S., Hussein, E.H.A. and El-Itriby, H.A. (2003). AFLP fingerprinting of some Egyptian date palm (*Phoenix dactylifera* L.) cultivars. *Arab Journal of Biotechnology*, **6**: 223- 234

Elmeer, K. and Mattat, I. 2012. Marker-assisted sex differentiation in date palm using simple sequence repeats. *Biotechnology*, **2**: 241–247.

Elmeer, K., Sarwath, H., Malek, J., Baum, M. and Hamwieh, A. (2011). New microsatellite markers for assessment of genetic diversity in date palm (*Phoenix dactylifera* L.). 3 *Biotech*, **1**: 91–97.

Elshibi, S. and Korpelainen, H. (2008). Microsatellite markers reveal high genetic diversity in date palm (*Phoenix dactylifera* L.) germplasm from Sudan. *Genetica*, **134**: 251–260

El-Tarras, A., Al-Tawatti, N. and Al-Malki, F. (2007). Genetic fingerprinting of some KSA cultivars using modern biotechnological techniques. *Biotechnology*, **6**: 263–267.

Gaj, M.D., and Maluszynski, M. (1987). Genetic variation in callus culture of *Arabidopsis thaliana* (L.) Heynh. IAEA *Scientific and Technical Publications*, **698**: 147-153.

Ghaleb, H.H.A. (2008). Date palm cultivars atlas of United Arab Emirates. In: Ghaleb, H.H.A. (Ed.), Center for Heritage and History Publications. Al-Ain, UAE.

Haider, N., Nabulsi, I. and MirAli, N. (2012). Phylogenetic relationships among date palm (*Phoenix dactylifera* L.) cultivars in Syria using RAPD and ISSR markers. *Journal of Plant Biology Research*, **1**: 2–24.

Hamwieh, A., Farah, J., Moussally, S., Al-Shamaa, K., Elmeer, K., Khierallah, H., Udupa, S., Lababidi, S., Malek, J.A., Aaouine, M. and Baum, M. (2010). Development of 1000 microsatellite markers across the date palm (*Phoenix dactylifera* L.) genome. *Acta Horticulturae*, **882**: 269–277.

Hamza, H., Benabderrahim, A M., Elbekkay, M., Ferdaous, G., Triki, T. and Ferchichi, A. (2012). Investigation of genetic variation in Tunisian date palm (*Phoenix dactylifera* L.) cultivars using ISSR marker systems and their relation with fruit characteristics. *Turkish Journal of Biology*, **36**: 449–458.

Hela, S., Salwa, Z., Ould Mohamed Salem, A., Abdelmajid, R., Mohamed, M. and Mokhtar, T. (2004). Genetic polymorphism of plastid DNA in Tunisian date-palm germplasm (*Phoenix dactylifera* L.) detected with PCR-RFLP. *Genetic Resources and Crop Evolution*, **51**: 479–487.

Husien, F.A. and Graib, S.H. (2004). A description of 50 Iraqi dates palm cultivars. Ministry of Agriculture, Iraq.

Husien, F.A. (2002). A description of some Iraqi dates palm cultivars. Ministry of Agriculture, Iraq.

Ibrahim, A.O. (2008). Date palm, the tree of life. Arab Centre for the Study of Arid Zones and Dry Lands. Damascus, Syria.

Jubrael, J., Udupa, S. and Baum, M. (2005). Assessment of AFLP-based genetic relationships among date palm (*Phoenix dactylifera* L.) varieties of Iraq. *Journal of the American Society for Horticulture Science*, **130**: 442–447.

Jubrael, J.M.S. (2001). Genetic characterization for some date palm cultivars in Iraq using RAPD markers. *IPA Journal for Agricultural Research*, **11**: 138–148.

Karim, K., Chokri, B., Amel, S., Wafa, H., Richid, H. and Nouredine, D. (2010). Genetic diversity of Tunisian date palm germplasm using ISSR markers. *International Journal of Botany*, **6**: 182-186.

Kalendar, R., Antonius, K., Smýkal, P. and Schulman, A.H. (2010) iPBS:A universal method for DNA fingerprinting and retrotransposon isolation. *Theoretical and Applied Genetics*, 121: 1419–1430.

Khanam, S., Sham, A., Bennetgen, J.L. and Mohammed, A.M.A. (2012). Analysis of molecular marker-based characterization and genetic variation in date palm (*Phoenix dactylifera* L.). *Australian Journal of Crop Science*, **6**: 1236–1244.

Khierallah, H.S.M., Bader, S.M., Baum, M., Hamwieh, A. (2011a). Genetic diversity of Iraqi date palms revealed by microsatellite polymorphism. *Journal of American Society of Horticultural Science*, **136**: 282- 287.

Khierallah, H., Bader, S.M., Baum, M. and Hamwieh, A. (2011b). Assessment of genetic diversity for some Iraqi date palms (*Phoenix dactylifera* L.) using amplified fragment length polymorphisms (AFLP) markers. *African Journal Biotechnology*, **10**: 9570-9576.

Khierallah, H.S.M. (2015). Applications of molecular markers in date palm genome analysis and breeding, *Research Signport, Trivandrum*, Kerala, pp. 47-90.

Khierallah, H.S.M., Al-Sammarraie, S.K.I. and Mohammed, H.I. (2014). Molecular characterization of some Iraqi date palm cultivars using RAPD and ISSR markers. *Journal of Asian Scientific Research*, **4**: 490–503.

Letouze, R., Daguin, F., Satour, P., Hamama, L. and Marrionnet, F. (1998). Somatic embryogenesis and mass micropropagation in date palm (*Phoenix dactylifera* L.) characterization and genetic stability of regenerated plantlets by RAPD markers. In: *Proceedings of the 1st International Conference on Date palms*. Al-Ain UAE, pp. 158-167.

Majourhat, K., Bendiab, K., Medraoui, L. and Baaziz, M. (2002). Diversity of leaf peroxidases in date palm (*Phoenix dactylifera* L.) as in an example of marginal (seedling derived) palm groves. *Scientia Horticulturae*, **95**: 31–38.

Marsafari, M., Mehrabi A.A. and Tahmasebi, Z. (2014). The identification of RAPD and ISSR informative markers with some quality traits of fruit in some of Iranian date palm. *International Journal of Plant, Animal and Environmental Sciences*, **4**: 714-722.

Mathew, L.S., Spannag, M., Al-Malki, M., George, B., Torres, M.F., Al-Dous, E.K., Al-Azwani, E.K., Hussein, E., Mathew, S., Mayer, K.F.X., Mohamoud, Y.A., Suhre, K., and Malek, J.A. (2014). A first genetic map of date palm (*Phoenix dactylifera*) reveals long-range genome structure conservation in the palms. *BMC Genomics*, **15**: 285–295.

Ould Salem Mohamed, A., Trifi, M., Sakka, H., Rhouma, A. and Marrakchi, M. (2001). Genetic inheritance analysis of four enzymes in date palm (*Phoenix dactylifera* L.). *Genetic Resources and Crop Evolution*, **48**: 361–368.

Rhouma, S., Zehdi-Azouzi1, S., Ould, M.S.A., Rhouma, A., Marrakchi, M. and Trifi, M. (2007). Genetic diversity in ecotypes of Tunisian date-palm (*Phoenix dactylifera* L.) assessed by AFLP markers. *The Journal of Horticultural Science and Biotechnology*, **82**: 929–933.

Sabir, J.S.M., Arasappan, D., Bahieldin, A., AboAba, S., Bafeel, S., Zari, T.A., Edris, S., Shokry, A.M., Gadalla, N.O., Ramadan, A.M., Atef, A., AlKordy, M.A., ElDomyati, F.M. and Jansen, R.K. (2014). Whole mitochondrial and plastid genome SNP analysis of nine date palm cultivars reveals plastid heteroplasmy and close phylogenetic relationships among cultivars. *PLoS ONE*, **9**: e94158.

Saker, M.M. and Moursy, H.A. (1999). Molecular characterization of Egyptian date palm cultivars: RAPD Fingerprints. *Arab Journal of Biotechnology*, **2 (1)**: 71-78.

Saker, M.M., Bekheet, S.A., Taha, H.S., Fahmy, A.S. and Moursy, H.A. (2000). Detection of somaclonal variation in tissue culture derived date palm plants using isozyme analysis and RAPD fingerprints. *Biologia Plantarum*, **43**: 347-351.

Sedra, M.H., Lashermes, P. and Trouslot, P. (1998). Identification and genetic diversity analysis of date-palm (*Phoenix dactylifera* L.) varieties from Morocco using RAPD markers. *Euphytica*, **103**: 75–82.

Soliman, S.S., Ali, B.A. and Ahmed, M.M.M. (2003). Genetic comparisons of Egyptian date palm cultivars (*Phoenix dactylifera* L.) by RAPD-PCR. *African Journal of Biotechnology*, **2:** 86–87.

Trifi, M., Rhouma, A. and Marrakchi, M. (2000). Phylogenetic relationships in Tunisian date palm (*Phoenix dactylifera* L.) *germplasm collection using DNA amplification fingerprinting. Agronomie*, **20:** 655–671.

Wrigley, G. (1995). Date palm. In: Smartt, J. and Simmonds, N.W. (Eds.), *Evolution of Crop Plants*, 2nd Edn. Longman Group, Essex, England. pp. 399–403.

Yang, W., Oliveira, A.C., Godwin, I.D., Schertz, K. and Bennetzen, J.L. (1996). Comparison of DNA marker technologies in characterizing plant genome diversity: Variability in Chinese sorghums. *Crop Science*, **36**: 1669-1676.

Zehdi, S., Sakka, H., Rhouma, A., Salem, A.O.M., Marrakchi, M. and Trifi, M. (2004). Analysis of Tunisian date palm using simple sequence repeat primers. *African Journal of Biotechnology*, **3**: 215-219.

Zhao, Y., Williams, R., Prakash, C.S. and He, G. (2013). Identification and characterization of gene-based SSR markers in date palm (*Phoenix dactylifera* L.). *BMC Plant Biology*, **12**: 237-245.

Zohary, D. and Spiegel-Roy, P. (1975). Beginning of fruit growing in the Old World. *Science*, **187**: 319–327.

Chapter 14

Arecanut

☆ *N.R. Nagaraja, K.S. Ananda and M.K. Rajesh*

1. Introduction

The genus *Areca* L. belongs to the sub-tribe Arecinae and tribe Arecae in the family Arecaceae and comprises 45 species (Plant List, 2013), growing in hot and humid tropical regions of the world. Among those, *Areca catechu* L. is the only cultivated species, cultivated mainly in Bangladesh, China, India, Indonesia, Malaysia, Myanmar, the Philippines, Sri Lanka, Thailand and Vietnam. It is believed that the South East Asian region is the centre of origin for *A. catechu* L. (Bavappa *et al.*, 1982). It is a monoecious, unbranched palm, widely used for masticatory purposes, either alone or along with slaked lime, betel leaf (*Piper betle* L.) and tobacco. The nuts also play a major role in many of the social, religious functions, known for its medicinal importance. The nuts are being sold as ripe, dried, cured and processed forms.

In India, the *Areca* palm is commonly cultivated in the plains and foothills of the Western Ghats region of states such as Kerala, Karnataka, Goa and some parts of Maharashtra and Gujarat and in the North Eastern states of Assam, Meghalaya and West Bengal. India stands top in both area and production of arecanut in the world, where it is grown in an area of 446 thousand hectares with production of 609 thousand tones (FAOSTAT, 2013). The country accounts for 57 per cent in area and 53 per cent of the total world production, of which the three states *viz.*, Kerala, Karnataka and Assam contributing 90 per cent of area under cultivation and 95 per cent of the production (Rajagopal and Balasimha, 2004). The crop provides economic security for millions of people and for many sole means of livelihood in the Indian sub-continent, South East Asian countries and also in some of the Pacific islands (Sankaran *et al.*, 2013).

2. Genetic Diversity in Arecanut

A number of cultivars (ecotypes) have been identified from various arecanut growing regions within the country as well as from other parts of the world (Bavappa, 1963). A field gene bank is being maintained at ICAR-Central Plantation Crops Research Institute (CPCRI), Regional Station, Vittal, Karnataka, which is considered to be the largest assemblage of the *Areca* germplasm in the world. A total of 173 accessions have been collected so far (ICAR-CPCRI, 2016), that includes 23 exotic accessions from various South East Asian countries representing three species *viz.*, *Areca catechu* L., *A. concinna* Thw. and *A. triandra* Roxb. The indigenous collections, numbering 150, comprise of collections from Assam, Goa, Gujarat, Karnataka, Kerala, Maharashtra, Meghalaya, Tamil Nadu, West Bengal, and Andaman and Nicobar group of Islands (Ananda, 2006).

Earlier work in arecanut were mainly focused on studies on floral morphology, cytogenetics, biochemistry, plant pathology and plant physiology (Bhat, 1982; Joshi and Reddy, 1982; Bhat, 1985). Evaluation of arecanut germplasm has been attempted using morphological traits and yield criteria (Ananda *et al.*, 2000; Rajesh, 2007). The information gathered regarding the diversity, relationship, phylogeny among arecanut accessions using morphological and biochemical approaches are not reliable as environmental factors greatly influence yield, growth and development of perennial crops. Hence, DNA-based marker systems have been utilized for assessment of genetic diversity in arecanut in recent years

3. Using DNA-based Markers in Arecanut

In arecanut, the most widely used molecular marker has been the Random Amplified Polymorphic DNA (RAPD), since it is cost effective, easy to handle, versatile, and can distinguish slight changes in the polymorphic DNA among the populations. Rajesh *et al.* (2007) carried out optimization of RAPD protocol for arecanut. Purushotham *et al.* (2008) carried out assessment of extent of genetic diversity in 11 arecanut cultivars collected from the Western Ghats regions of India using RAPD technique. Two major clusters were formed of which ten cultivars formed the major cluster ('A') while Mohitnagar alone formed a separate cluster ('B') (Figure 14.1). Two unique amplicons produced by primers OPA-13 and OPA-15 were specific to all the cultivars in cluster 'A'. The cluster 'A' was subdivided into two minor clusters, 'A_1' and 'A_2' comprising of nine cultivars and one cultivar respectively. The sub-cluster 'A_1' consisted of two minor clusters 'A_{1a}' and 'A_{1b}', with Mohitnagar *inter se* forming one group while eight cultivars grouped in to another, segregated into three groups. The Maidhan varieties were closely linked with each other while the exotic Saigon-12 formed a distinct clade in this group. An unique band produced by the primer OPD-05 was specific Cluster 'A_2'. The results of the study revealed that despite their narrow distribution in the Western Ghats, the accessions showed moderate polymorphism.

Genetic fidelity of arecanut plantlets that were derived through direct somatic embryogenesis from the Yellow Leaf Disease (YLD) resistant arecanut palms was carried out by Karun *et al.* (2008) utilizing RAPD markers (Figure 14.2). Eight plantlets derived through direct somatic embryogenesis from the YLD mother

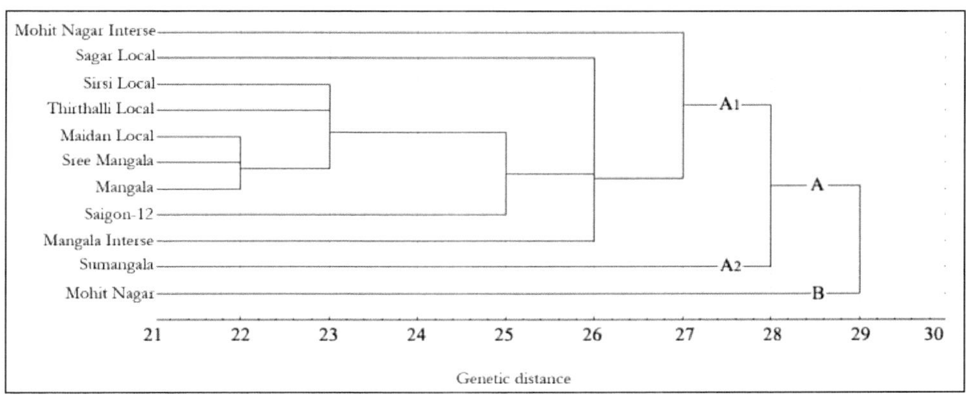

Figure 14.1: Dendrogram Showing RAPD Marker Based Genetic Distance among the 11 Arecanut Cultivars from Western Ghats (Purushotham *et al.*, 2008).

palms were studied. It was concluded that the plantlets derived from direct somatic embryogenesis showed less variation under *in vitro* conditions and hence the inflorescence culture derived plantlets could be employed for mass multiplication for the desirable qualities in the elite palms.

Sankaran *et al.* (2013) assessed the genetic diversity among 10 arecanut accessions from the Andaman and Nicobar Islands using RAPD markers. Among the 30 RAPD primers, 11 were selected to detect polymorphism based on their reliability in pooled DNA. They considered the primers OPF-16, which produced maximum number of bands (49), OPF-8 and OPH-35 (producing 48 bands each), OPF-41 and OPF-9 each produced a minimum number of 39 bands. Their studies showed that the percentage of polymorphism ranged from 50 to 100, with primers OPH-8, OPH-35, OPP-46 and OPF-8 showing 100 per cent polymorphism while OPF-1 showing only 50 per cent. The diversity analysis revealed that the accessions were grouped in two clusters as wild ones segregated from the cultivated nine accessions. The cluster of cultivated ecotypes was further split into two clades with four and five accessions respectively. These results were in conformity with those obtained

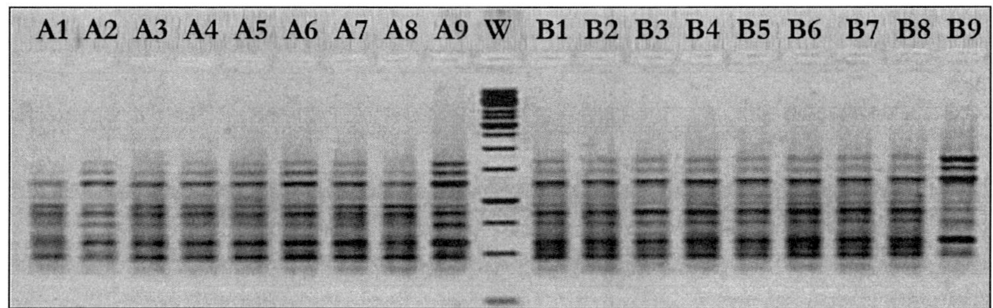

Figure 14.2: RAPD Banding Profile of *in vitro* Propagated Plantlets and Field Grown Mother Palms [A1-A8: Plantlets derived from the mother palm A9; B1-B8: Plantlets derived from the mother palm B9] (Karun *et al.*, 2008).

by Purushotham *et al.* (2008) as huge variability among the wild and cultivated genotypes was observed. It implied that selection of appropriate accessions of arecanut for hybridization or breeding programs is imperative.

Hu *et al.* (2009) reported isolation of nine novel microsatellite loci from *A. catechu* germplasm conserved in Taiwan. When these SSRs were utilized for germplasm evaluation, 5-15 alleles were detected; null alleles were also recorded in five loci. The first large scale studies of estimation of genetic diversity in arecanut were conducted by Bharath *et al.* (2012), who studied the genetic diversity among 60 arecanut accessions utilizing microsatellite markers. Nine microsatellite markers specific to arecanut, isolated earlier by Hu *et al.* (2009), were employed for analysis. The accessions studied were the collections from Konkan, North East region and Andaman and Nicobar Islands of Indian sub-continent, and exotic collections from different arecanut growing regions of South East Asia and Pacific Islands. The results showed that all of the microsatellites showed polymorphism except one. The cluster analysis revealed that they formed two major clusters: Cluster I comprised of exclusively the Konkan collections and Cluster II was formed due to collections from North East and exotic collections. Among the collections from India, Konkan collections formed two distinct clusters.

Bharath *et al.* (2015) carried out RAPD analysis using 14 polymorphic in 60 accessions [43 indigenous (Konkan I and II, Maidhan, North East and Andaman and Nicobar Islands) and 17 exotic germplasm] collected from various parts of South East Asia and Pacific region (British Solomon Islands, China, Fiji, Indonesia, Singapore, Sri Lanka and Vietnam), and conserved in the field gene bank of ICAR-CPCRI, Regional Station, Vittal, Karnataka, India. While analyzing the results obtained from RAPD analysis (Figure 14.3), a maximum of 13 bands were obtained using the primers OPF-6 while only five amplified fragments were obtained for OPAF-19. Shannon's indices showed a great variation with minimum values for OPAF-19 and maximum for OPAF-6 and the minimum gene diversity was recorded for OPAF-6 and maximum for OPM-13. The cluster analysis revealed that the genotypes were segregated into two clusters with Cluster I comprised of all exotic accessions, North East accessions and those from Andaman and Nicobar Islands,

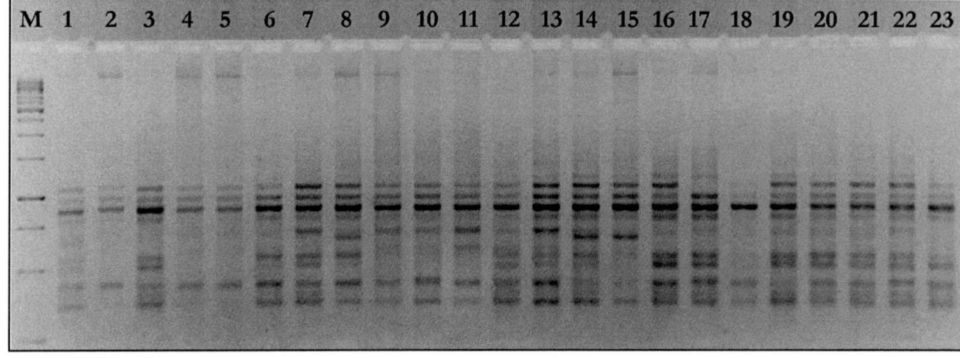

Figure 14.3: RAPD Banding Profiles Generated using Primer OPAF-2 in Arecanut Accessions. M: Standard 1 Kb ladder (M) (Bharath *et al.*, 2015).

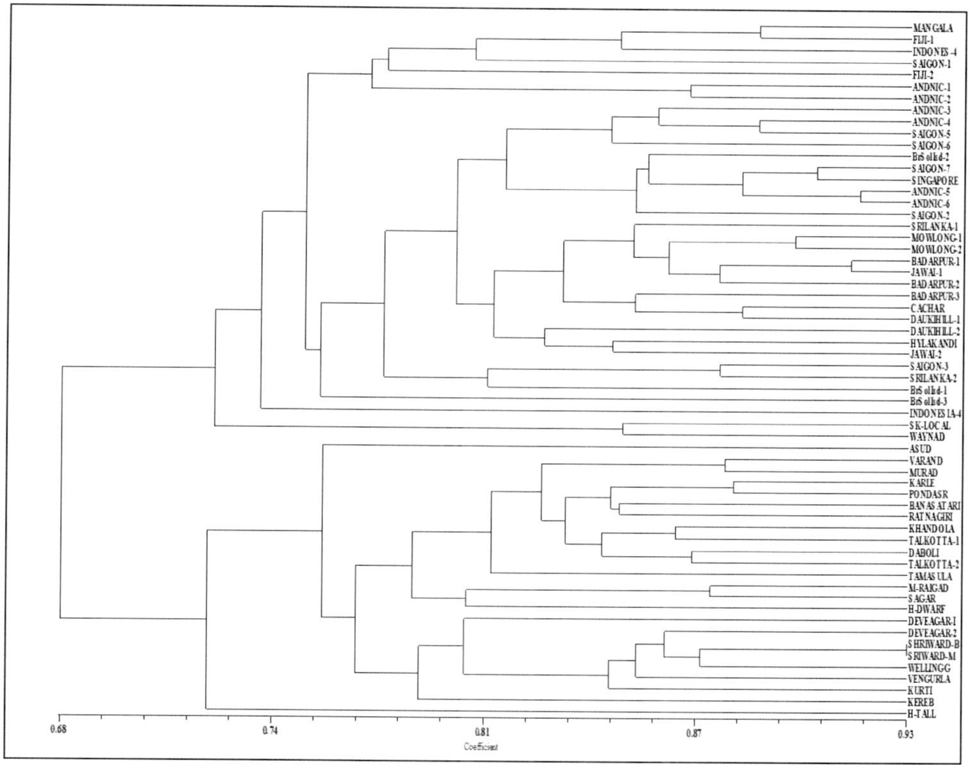

Figure 14.4: A UPGMA Dendrogram Based on RAPD Data for the 60 Arecanut Accessions (Bharath *et al.*, 2015).

while Cluster II comprised of purely the indigenous ones, *viz.*, Konkan I, Konkan II and Maidhan collections. It was observed that collections from North East and Andaman and Nicobar Islands shared similarity with the local South Kanara and Wayanad accessions in the Cluster I, despite their geographical distances. Hirehalli Tall, a Maidhan collection, formed a distinct accession in Cluster II. It was concluded that there existed a high level of genetic diversity among those sixty accessions that were analyzed.

Rajesh *et al.* (2016a) undertook analysis of genetic diversity among six arecanut accessions *viz.*, Mangala, Sumangala, Sreemangala, Mohitnagar, Swarnamangala and a natural dwarf mutant (Hirehalli Dwarf), using SCoT markers to evaluate the applicability of these markers in genetic diversity studies in arecanut. Using 10 SCoT primers, described by Collard and Mackill (2009), 82 band were produced among the accessions, of which 58 (70.73 per cent) were found to be polymorphic. The highest genetic similarity value of 0.89 was found between the Swarnamangala and Mohitnagar and the lowest value of 0.63 was noticed between the Hirehalli Dwarf and Mohitnagar. The dendrogram constructed utilizing the UPGMA grouped the cultivars based on their geographical origins, with Hirehalli Dwarf forming a distinct accession (Figure 14.5).

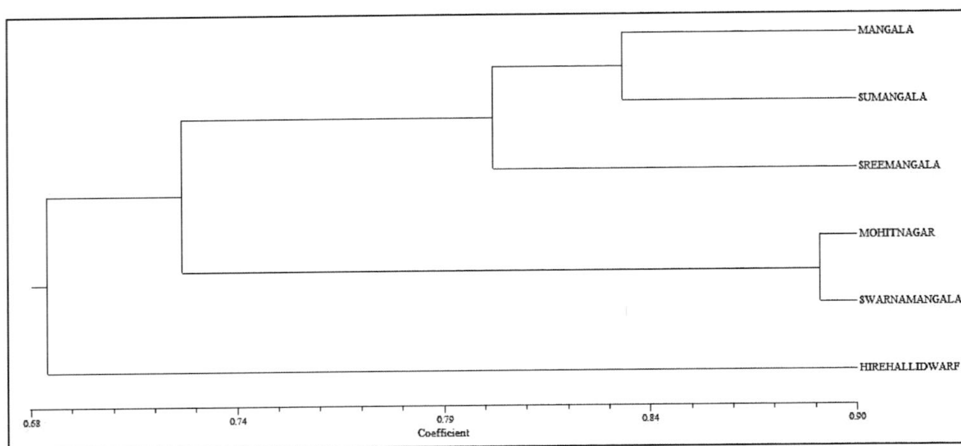

Figure 14.5: Dendrogram Showing the Genetic Relationship among the Six Arecanut Accessions using SCoT Analysis (Rajesh *et al.*, 2016a).

Rajesh *et al.* (2016b) carried out studies utilizing Start Codon Targeted (SCoT) markers to identify molecular marker(s) capable to differentiate tall/dwarf trait in arecanut. Tall cultivars (Mangala, Sumangala, Sreemangala, Mohitnagar, Swarnamangala and Hirehalli Tall) and the natural mutant of arecanut (Hirehalli Dwarf) were screened utilizing 25 SCoT primers described by Collard and Mackill (2009). One of the primers, SCoT11, produced an amplicon of around 1300 bp band specific to all the tall cultivars, which was absent in the dwarf cultivars (Figure 14.6). The DNA fragment was purified, cloned and sequenced. A SCAR marker, capable of distinguishing tall/dwarf trait in arecanut, was also developed and validated, which could ensure supply of genuine hybrid planting material to the farming community.

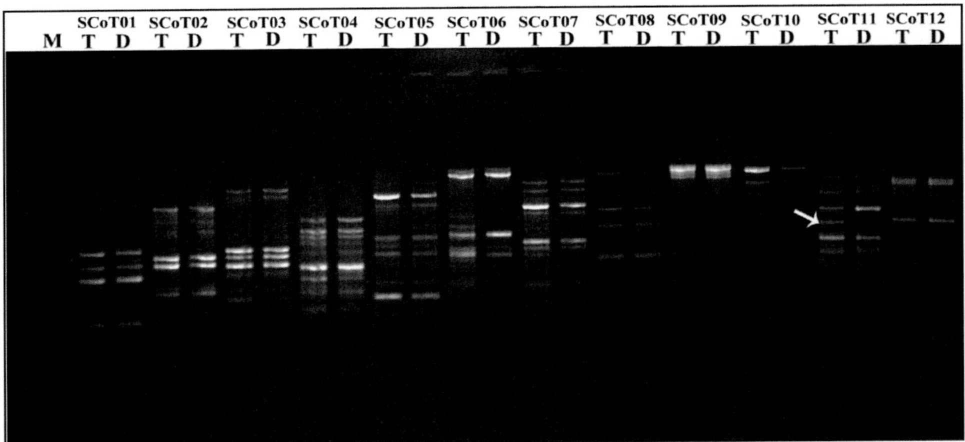

Figure 14.6: Banding pattern of pooled DNA of tall and dwarf palms with the primers SCoT01- SCoT12. Arrowhead represent polymorphic band of SCoT11 primer specific to tall accession. M: High range ladder. T: Tall bulk. D: Dwarf bulk.

4. Conclusion

In perennial crops such as arecanut, the morphological and biochemical methods for characterization of germplasm/evaluation of genetic relationships, yield parameters *etc.* have their own limitations. Hence, advanced biotechnological approaches, such as utilization of DNA-based molecular markers, are to be explored. Despite its demonstrated utility in other crops, only few studies have been carried out in arecanut using molecular markers. As arecanut is an economically important plantation crop, there is a need to characterize the arecanut germplasm using molecular markers. Such studies would aid in selection of desirable parents and clones with desirable traits among arecanut germplasm grown in various agro climatic zones and geographical regions of the world.

References

Ananda, K.S. (2006). *Arecanut Descriptors Part – I. Central Plantation Crops Research Institute*, Kasaragod, Kerala. pp. 172.

Ananda, K.S. and Rajesh, B. (2002). Breeding behaviour in arecanut (*Areca catechu* L.) varieties under South Kanara conditions, In: Manjunath, B. L., Ramesh, R., Faliero, J. R. and Korikanthimanth, V. S. (Ed.), *Proceedings of the National Conference on Coastal Agricultural Research*. ICAR Research Complex for Goa, Old Goa, India. pp. 222-224.

Ananda, K.S., Sane, A. and Choudhary, B.S. (2000). Growth and yield performance of arecanut varieties under Malnad condition. *Journal of Plantation Crops,* **28:** 105-109.

Bavappa, K.V.A., Nair, M.K. and Premkumar, T. (1982). *The Areca Palm* (*Areca catechu* Linn.) *Central Plantation Crops Research Institute*, Kasaragod, Kerala, India.

Bavappa, K.V.A. (1963). Morphological and cytological studies in *Areca catechu* Linn. And *Areca triandra* Roxb. M.Sc. Thesis, University of Madras, p. 63.

Bharath, B.G., Ananda, K.S., Rijith, J., Nagaraja, N.R., Chandran, K.P., Karun, A. and Rajesh, M.K. (2015).. Studies on genetic relationships and diversity in arecanut (*Areca catechu* L.) germplasm utilizing RAPD markers. *Journal of Plantation Crops,* **43(2):** 117-125.

Bharath, B. G., Ananda, K. S., Rajesh, M. K., Nagaraja, N. R. and Rijith, J. (2012). Evaluation of microsatellite markers for genetic diversity analysis among different accessions of arecanut. *22nd Swadeshi Science Congress,* ICAR-CPCRI, Kasaragod, Kerala. p.102.

Bhat, S.K. and Nair, C.P.R. (1985). *Arecanut: Research and Development*. Central Plantation Crops Research Institute, Kasaragod, Kerala. p. 233.

Bhat, S.K. (1982). *25 years of Arecanut Research and Development*. Central Plantation Crops Research Institute, Kasaragod, Kerala. p. 32.

Chawla, H.S. (2009). *Introduction to Plant Biotechnology, 3rd Ed*. Oxford and IBH Publishers, New Delhi, India. p. 718.

Collard B.C.Y. and Mackill D.J. (2009). Start Codon Targeted (SCoT) Polymorphism: A simple, novel DNA marker technique for generating gene-targeted markers in plants. *Plant Molecular Biology Reporter*, **27**: 86-93.

FAOSTAT. 2013. http://faostat.fao.org/

Hu, C.H., Huang, C.C., Hung, K.H., Hsu, T.W. and Chiang, T.Y. (2009). Isolation and characterization of polymorphic microsatellite loci from *Areca catechu* (Arecaceae) using PCR-based isolation of microsatellite arrays. *Molecular Ecology Resources*, **9:** 658-660.

ICAR-CPCRI (2016). Annual Report 2015-16, ICAR- *Central Plantation Crops Research Institute*, Kasaragod-671 124, Kerala, India, p. 9.

Joshi, Y. and Reddy, N.R. (1982).. *Arecanut Bibliography. Central Plantation Crops Research Institute*, Kasaragod, Kerala. p. 116.

Karun, A., Radha, E., Vijayan, P.S., George, J., Rajesh, M.K. and Ananda, K.S. (2008). Assessment of genetic fidelity of arecanut plantlets derived through direct somatic embryogenesis by RAPD markers. *Journal of Plantation Crops*, **36(3):** 228-231.

Purushotham, B., Narayanaswamy, P., Simon, L., Shyamalamma, S. Hegde, M. and Jayapalgowdu, B. (2008). Genetic relationship between cultivars of areca nut (*Areca catechu)* determined by RAPD. *The Asian and Australasian Journal of Plant Sciences and Biotechnology*, **2(1):** 31-35.

Rajagopal, V. and Balasimha, D. (2004). *Arecanut*. Central Plantation Crops Research Institute, Kasaragod, Kerala. p. 306.

Rajesh, B. (2007). Genetic variability for morphological, biochemical and yield traits in arecanut (*Areca catechu* L.) accessions. Ph.D dissertation, *University of Mangalore*, Karnataka, India, p. 202.

Rajesh, M.K., Sabana, A.A., Rachana, K.E., Ananda, K.S. and Anitha Karun (2016a) Potential of start codon targeted (SCoT) markers for assessment of genetic diversity of arecanut (*Areca catechu* L.). *Indian Journal of Horticulture*, **73**(3): 423-426.

Rajesh, M.K., Sabana, A.A., Rachana, K.E., Shafeeq Rahman, Ananda, K.S. and Anitha Karun (2016b) .Development of a SCoT-derived SCAR marker associated with tall type palm trait in arecanut and its utilization in hybrid (dwarf x tall) authentication. *Indian Journal of Genetics and Plant Breeding*, **76**(1): 119-122

Rajesh, M.K., Bharathi, M. and Nagarajan, P. (2007). Optimization of DNA isolation and RAPD technique in arecanut (*Areca catechu* L.). *Agrotrópica*, **19**: 31-34.

Sankaran, M., Chandrasekar, P., Singh, D.R. and Damodaran, V. (2013). Assessment of genetic diversity among arecanut accessions by using RAPD markers. *Indian Journal. of Horticulture,* **70 (3)**: 428-430.

The Plant List (2013). *Version 1.1.* Published on the internet http://www.theplantlist.org/accessed 3rd December 2015.

Chapter 15

Spices

☆ *Minoo Divakaran, K.B. Rebijith, M.K. Rajesh and K. Nirmal Babu*

1. Introduction

The past few decades have witnessed introduction of an array of marker technologies like DNA sequencing and high throughput genotyping methods. Molecular markers have established their position as one of the most powerful and sought-after tools over morphological data, because the bulk of variation at the nucleotide level is not visible at the phenotypic level. Utilization in studies on synteny, genome arrangement within taxa, germplasm analysis, trait mapping, and marker assisted selection, genome mapping *etc.* have led to incorporation of marker technologies to complement and validate phenotypic data.

Molecular tools have been used to help elucidate some aspects of genetic diversity in aromatic species, the genetic relationships between different cultivars and comparisons of molecular marker analysis to the chemical composition of plants. An overview of some important techniques involving molecular markers in spices, understanding what controls flavour and aroma production in different plants and its diverse forms, host-pathogen interactions in deciphering the disease progression pathway and identifying important genes in important genotypes, is presented in this chapter.

The route from genomics to proteomics is not well documented; however, Trindade (2010) has addressed the question of the synthesis of volatiles, with two different approaches first, the biochemical and genetic approach; and second, approaches involving functional genomics and finally, a brief survey of bioinformatics resources. The underlying fact useful for molecular systematics is that different genes accumulate mutations at different rates. This difference depends on how much change a gene can tolerate without losing its function. For example, histone molecules may become non-functional if some of its amino acids are replaced with different ones. On the other hand, ITS of ribosomal RNA can still fold properly even if many of its nucleotides are changed. Thus, ITS can accumulate mutations more rapidly than histones, reflecting the different functional constraints on their gene product. Ribosomal RNA is considered as one of the best target for studying phylogenetic relationship because it is universal and is composed of highly conserved as well as variable domains. Marker-assisted gene pyramiding (MAGP) is one of the most important applications of DNA markers to plant breeding. Gene pyramiding has been proposed and applied to enhance resistance to disease and insects by selecting for two or more genes at a time. For example, such pyramids have been developed in rice against bacterial blight and blast (Huang *et al.*, 1997; Luo *et al.*, 2012)

The advantages of using molecular data is obvious - molecular data are more numerous than fossil records and easier to obtain. There is no sampling bias involved, which helps to correct the gaps in real fossil records. A clearer and robust phylogenetic tree can be constructed with the help of molecular data. When variation in morphological data become insufficient to distinguish two organisms at different levels *viz.*, phyla, class, order, family *etc.*, the analysis of the biomolecules are considered, which are large in number and occur in various forms in species. Therefore, biomolecular markers have become a favourite and sometimes the only reliable information available for researchers to reconstruct evolutionary history.

2. Commonly Utilized Molecular Markers in Spices

Molecular markers are features determined by allelic forms of genes or genetic loci and can be transmitted from one generation to another, thus are being used to tag target genes. Classical markers and DNA markers are used in breeding programmes, however classical markers (including morphological markers, cytological markers and biochemical markers) are not associated with important economic traits always, but DNA markers have developed into many systems based on different polymorphism-detecting techniques or methods (Southern blotting, nuclear acid hybridization, PCR and DNA sequencing), such as RFLP, AFLP, RAPD, SSR, SNP, *etc.*

2.1. Molecular Characterization of Germplasm

2.1.1. Genus Piper

Black pepper, derived from *Piper nigrum* L., is the most important spice of the world. India is the native home of black pepper and is its major producer and exporter. Conserving and fingerprinting the genetic diversity in pepper and development of resistant varieties will increase the production and productivity of black pepper. Use of molecular markers to augment breeding through marker assisted selection is a recent development in black pepper. Most of the studies are confined to molecular profiling of germplasm and indexing genetic fidelity. For the molecular characterization of black pepper germplasm, RAPD, ISSR and SSR markers were used. Pradeep Kumar *et al.* (2001, 2003) characterized 24 black pepper (*Piper nigrum* L.) accessions using RAPD markers and cultivar specific bands were developed except for Panniyur 3. Nirmal Babu *et al.* (2003b). used RAPD markers for studying genetic inter-relationships among fourteen major cultivars and ten released varieties of black pepper and depicted distinct differences between most of the cultivars and varieties. Sreedevi *et al.* (2005) characterized seven high yielding black pepper cultivars using RAPD markers. Nazeem *et al.* (2005) analyzed the variability and relatedness among 49 cultivars/accessions of black pepper using RAPD and AFLP markers. Joy *et al.* (2007) studied the genetic relationships among popular and agronomically important cultivars of black pepper using AFLP analysis. The dendrogram grouped the 49 accessions into three major clusters and four diverse cultivars with only 30 per cent similarity. Menezes *et al.* (2009) developed nine SSR markers from microsatellite enriched library of black pepper. These SSR markers were used to study the genetic diversity of 20 black pepper varieties of germplasm collection in Brazil. Joy *et al.* (2011) developed seven microsatellite markers for black pepper of which four polymorphic markers generated 62 alleles with an average of 15.5 alleles over 4 loci.

The genus *Piper* shows extreme reduction of floral characters which are key factors for phylogenetic analysis. Hence molecular approaches have been used to gain better insights. Sebastian *et al.* (1996) used isoenzyme variation in 11 *Piper* species to study the inter relationships. Gaia *et al.* (2003) studied the genetic diversity in 78 clones of *Piper nigrum* from South America using eight isoenzymes and concluded that most of the clones of *P. nigrum* L. were quite similar and homogeneous. Lebot *et al.* (1991) studied 300 accessions of *P. methysticum* and *P. wichmannii* from Polynesia, Micronesia and Melanesia for isoenzyme variation and concluded that the genetic base of *P. methysticum* was narrow and most of the morphotypes and chemotypes apparently originated through human selection and preservation of somatic mutations in a small number of original clones. He suggested that *P. wichmannii* is the wild progenitor of *P. methysticum*. Molecular data was used to further corroborate taxonomic and morphometric information for better understanding of their origins and species inter-relationships in the genus *Piper*. Jaramillo and Manos (2001) used phylogenetic analysis of sequences of the Internal Transcribed Spacers (ITS) of nuclear ribosomal DNA based on a worldwide sample of the genus *Piper*. Sequences from a 51 species of *Piper* were aligned to yield 257 phylogenetically informative sites. A single un-rooted parsimony network suggested that taxa representing major geographic areas could potentially form three monophyletic groups: Asia, the South Pacific, and the Neo tropics.

Inter- and intra-specific relationships among different species were studied by Chaveerach *et al.* (2002) who demonstrated a closer relation between *P. retrofractum* and *P. kudsura* than between *P. chaba* and *P. retrofractum* using RAPD profiles. Johnson *et al.* (2003) reported that ISSR-PCR along with RAPD was a valuable tool for genetic diversity analysis in *Piper* species. The ISSR markers were also used in identifying selected cultivars and hybrids of black pepper.

Nirmal Babu *et al.* (2003b). studied molecular inter-relationships between 24 Piper species using RAPD profiles. The phylogenetic trees grouped *P. longum*, *P. hapnium* and *P. mullesua* in one group and *P. attenuatum* and *P. argyrophyllum* in another group. *P. pseudonigrum*, *P. nigrum* and *P. galeatum* are clustered together. The study helped in identifying the core collections of holotypes of *Piper* species. Nirmal Babu (2003) studied RAPD and AFLP polymorphism among nine species of *Piper*. The exotic species *P. colubrinum* and *P. arboreum* were found to be highly distinct. *P. argyrophyllum, P. attenuatum, P. bababudani* and *P. nigrum* were found to be closely related. *P. betle, P. chaba,* and *P. longum* L. stood independently. Specific marker probes for each species were developed and were and successfully hybridized with corresponding genomic DNA. Wadt *et al.* (2004), who studied 49 genotypes belonging to three species of *Piper viz., Piper hispidinervum, Piper aduncum,* and *Piper hispidum,* used RAPD markers to prove the demarcation of *P. hispidinervum* and *P. aduncum* as two separate species. Liao *et al.* (2009) reported the isolation and characterization of eleven polymorphic microsatellites loci from an endemic species, *Piper polysyphonum* from China. Sen *et al.* (2010) evaluated genetic diversity of eight *Piper* species *viz., P. nigrum, P. longum, P. betle, P. chaba, P. argyrophyllum, P. trichostachyon, P. galeatum,* and *P. hymenophyllum* using RAPD markers. Jiang and Liu (2011) used RAPD and SRAP (Sequence Related amplified polymorphism) to study genetic diversity among 74 *Piper* spp. in Hainan Island and SRAP technique could efficiently distinguish all *Piper* spp from each other. Patra *et al.* (2011) used RAPD and ISSR markers to analyse the genetic diversity among 15 cultivars of betel vine. Sheeja *et al.* (2013) reported the genetic diversity analysis of 27 *Piper* species using ISSR markers and generated 35 species specific bands for 19 different *Piper* species. The genetic diversity analysis of six different species of *Piper* from North East region of India using RAPD marker was reported by Chowdhury *et al.* (2014). Yoshida *et al.* (2014) developed and characterized nine microsatellite loci for natural populations of *Piper solmsianum*, a potential source of bioactive secondary metabolites. Anupama *et al.* (2015) evaluated six microsatellite markers, developed from black pepper, for cross-species amplification and genetic diversity analysis in 23 *Piper* species from different location in India which included Western Ghats, North East and Exotic regions. The dendrogram generated could discriminate the 23 *Piper* species evaluated into eight clusters; three of the clusters could be further divided into sub clusters. The black pepper-specific microsatellite markers were found to clearly demarcate diversity among Indian and exotic species (Figure 15.1).

2.1.2. Zingiberales

A phylogenetic analysis of the tribe *Zingibereae* (Zingiberaceae) was performed by Ngamriabsakul *et al.* (2003) using nuclear ribosomal DNA (ITS1, 5.8S and ITS2) and chloroplast DNA [*trn*L (UAA) 5′exon to *trn*F (GAA)]; based on the results obtained, it was suggested that the tribe *Zingibereae,* as well as the genus *Curcuma,* are monophyletic. Chase (2004) attempted to have an overview on the phylogeny and relationships in monocots based on analysis of DNA sequence data of seven genes representing all three genomes and reported high bootstrap support to the clades which included Zingiberales. Monocots have been shown in molecular clock studies to be at least 140 million years old, and all major clades and most families date to well before the end of the Cretaceous. Kress *et al.* (2002) studied the phylogeny of the ginger (Zingiberaceae) based on DNA sequences of the nuclear internal transcribed spacer (ITS) and plastid *matK* regions and proposed a new classification of the

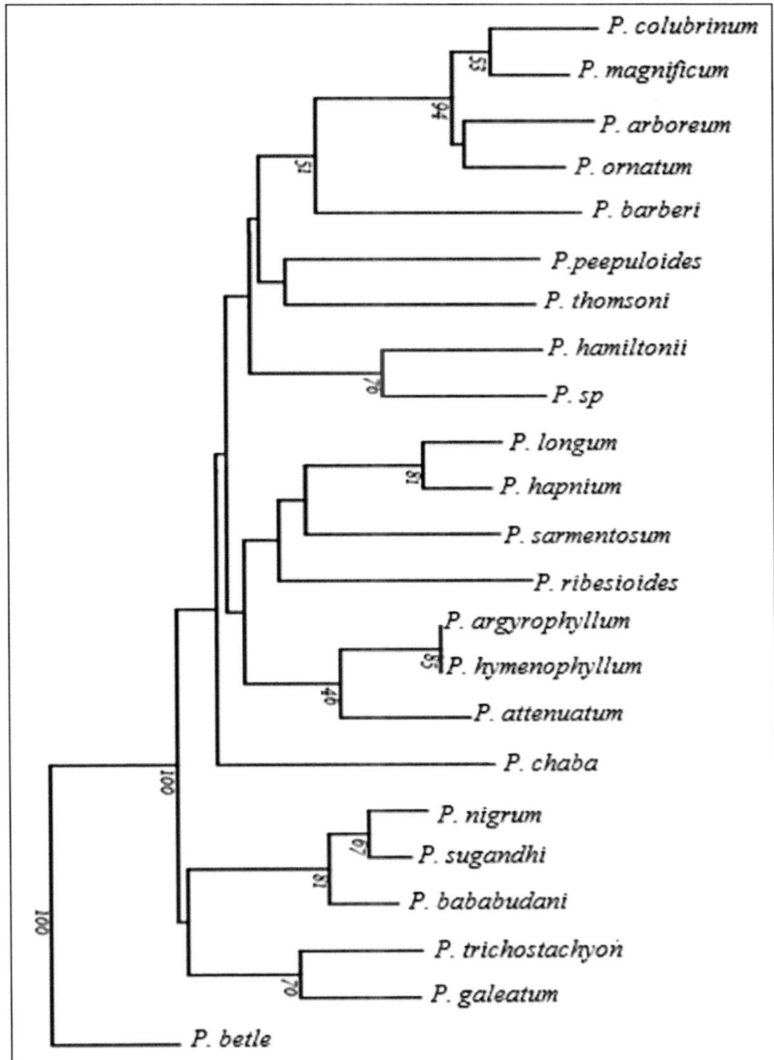

Figure 15.1: Dendogram Based on Six Polymorphic Microsatellite Markers to Study Genetic Relationship among 23 *Piper* Species Constructed by Neighbor Joining Method. Bootstrap values (based on 1000 permutations) are indicated in each node of the phylogram. (Source: Anupama *et al.*, 2015).

Zingiberaceae that recognizes four subfamilies and four tribes: Siphonochiloideae, Tamijioideae, Alpinioideae and Zingiberoideae (Zingibereae, Globbeae).

Jayakumar *et al.* (2005) studied about 11 species representing five major tribes *viz., Amomum, Aframomum, Alpinia, Hedychium* and *Elettaria* and collections of 96 cardamom genotypes of wide genetic variability using for RAPD, PCR-

RFLP and ISSR polymorphism. The collections of cardamom germplasm, which include released varieties, promising lines and local cultivars of cardamom, were characterized using 50 RAPD primers, six ISSR primers and two PCR-RFLP primers to study their relationships and diversity. The study indicated a clear divergence in Kerala and Karnataka collections, the two main regions of cardamom diversity and comparatively less divergence within the population is due to open pollinated seed origin (siblings). The collections and hybrids from Indian Cardamom Research Institute (ICRI), Myladumpara, Kerala, India formed one cluster and they are well separated from rest of the genotypes. The results indicated that controlled breeding, rather than selection from open pollinated progeny, is a preferred approach in cardamom to genrate more genetic variability in germplasm (Nirmal Babu *et al.*, 2005). *De novo* isolation, characterisation and utilization of microsatellite markers for the genetic diversity analysis of small cardamom has been recently reproted by Anu *et al.* (2016).

2.1.2.1 Curcuma Species

Genetic diversity of *Curcuma alismatifolia* Gagnep. Populations, from both cultivated and wild habitats, were studied by Paisooksantivatana *et al.* (2001). Cao *et al.* (2003) used trn K nucleotide sequencing, for identification of six medicinal *Curcuma viz., C. longa, C. phaeocaulis, C. sichuanensis, C. chuanyujin, C. chuanhuangjiang* and *C. chuanezhu* found in Sichuan, China. Hybrid detection and characterization of *Curcuma* spp. from Thailand using sequence characterized amplified (SCAR) DNA markers was done by Anuntalabochai *et al.* (2007). In the study, the tropical cut flower, *C. alismatifolia* variety 'Patumma', was analysed was using 11 RAPD primers. A robust sequence characterized amplified region with 600 bp in length was present in all 'Patumma' varieties and hybrids, which did not amplify in an additional series of 24 distinct *Curcuma* species used as an independent test. The molecular marker developed was useful for cost effective, morphologically independent characterization of *Curcuma* hybrids. Five species of *Curcuma* namely *C. longa, C. zedoaria, C. aromatica, C. amada and C. caesia* were characterized based on the RAPD profiles of DNA isolated from fresh rhizomes by Sreeja (2002). Eleven polymorphic bands were produced in the five species studied using three RAPD markers.

Kress *et al.* (2002) suggested that *Curcuma* is paraphyletic with *Hitchenia, Stahlianthus* and *Smithatris.* Molecular genetic fingerprints of 15 *Curcuma* species were developed using Inter Simple Sequence Repeats (ISSR) and Random Amplified Polymorphic DNA (RAPD) markers to elucidate the genetic diversity/relatedness among the species (Syamkumar, 2008). Siju *et al.* (2010) developed 18 genomic microsatellite markers for turmeric (*Curcuma longa* L.). These markers when used to evaluate 20 turmeric accessions generated 103 alleles with an average of 5.7 alleles per locus. The 18 EST SSR markers developed by Siju *et al.* (2010) in turmeric (*Curcuma longa* L.) revealed 100 per cent cross species transferability among the 13 related species.

Nayak *et al.* (2006) carried out 4C nuclear DNA content and RAPD analysis of seventeen promising cultivars of turmeric (*C. longa*) from India. RAPD analysis clearly showed the genetic variation among the seventeen cultivars using twenty random decamer primers. The inter-cultivar polymorphism ranged from 35.6 per cent to 98.6 per cent among the seventeen cultivars studied and the amplification fragments per primer ranged from four to seventeen with fragment size ranging from 0.4 kb to 3 kb. Genetic diversity analysis of twenty accessions of *C. longa* from different parts of Brazil using RAPD markers produced forty-five polymorphic loci and the dendrogram produced by UPGMA grouping using Jaccard's Index of similarity formed two groups. Among the groups, 44. 4 per cent genetic variability was observed and most part of the variation was found within the groups (Pinheiro *et al.*, 2003). ISSR profiles were generated for seven released varieties of turmeric (Figure 15.2) by Nirmal Babu *et al.* (2016a).

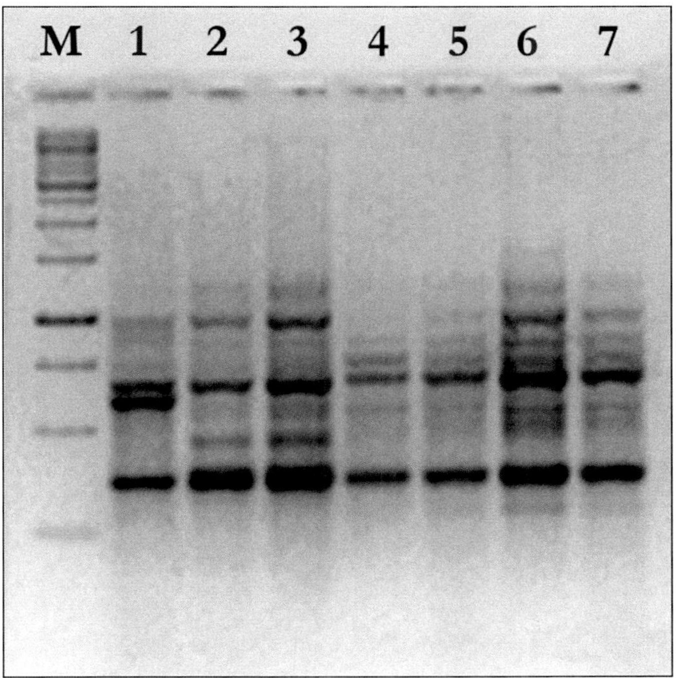

Figure 15.2: ISSR Profiling of Released Varieties of Turmeric using the Primer UBC 834a. Lanes 1–7: Suvarna, Suguna, Sudharsana, Prabha, Prathiba, Alleppey Supreme and Kedaram. M: Molecualr weight ladder.

2.1.2.2. Ginger

Wahyuni *et al.* (2003) studied genetic relationships among ginger accessions based on AFLP markers. Kavitha and Thomas (2008) reported *Zingiber zerumbet* (L) Smith, a wild species related to the cultivated ginger is a potential resistance donor for soft rot disease in ginger caused by *Pythium aphanidermatum* (Edson) Fitzp. They studied the genetic diversity and *P. aphanidermatum* resistance of 74 *Z. zerumbet* accessions belonging to 15 populations. Lee *et al.* (2007) reported isolation and characterization of eight polymorphic microsatellite markers for *Zingiber officinale*

Rosc. (Ginger) from a microsatellite enriched library. These were used to detect a total of 34 alleles across the 20 accessions with an average of 4.3 alleles per locus. The data generated indicated moderate level of genetic diversity among the ginger accessions genotyped with eight markers. ISSR profiles were generated for eight high yielding varieties of ginger (Figure 15.3) by Nirmal Babu *et al.* (2016b).

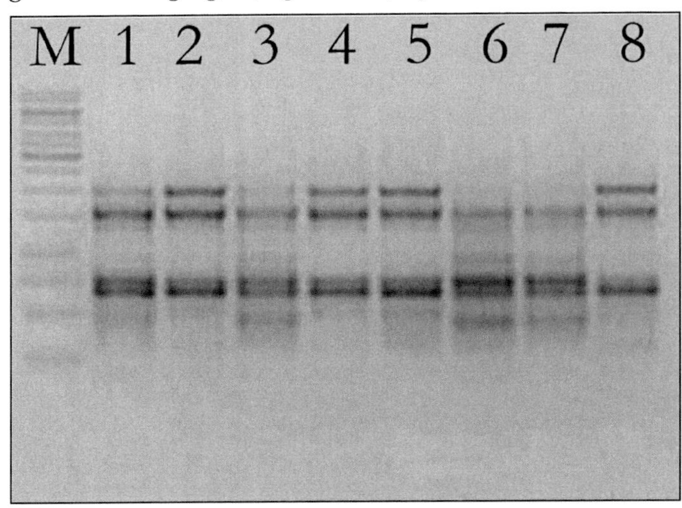

Figure 15.3: ISSR Profiling of High-yielding Varieties of Ginger using Primer UBC 810. Lanes 1-8: Varada, Rejatha, Mahima, Suprabha, Suruchi, Athira, Karthika and OCP 1222. M: Molecualr weight ladder (Source: Nirmal Babu *et al.*, 2016b)

2.1.2.3. Tree Spices

Perennial tree crops face problems like long pre-bearing period, dioecy in nutmeg and narrow genetic base in clove; hence, priority need to be given to characterization of germplasm based on molecular markers.

Genetic identification among cinnamon species was studied analyzing nucleotide sequences of chloroplast DNA from four species (*Cinnamomum cassia*, *C. zeylanicum*, *C. burmannii* and *C. sieboldii*). The two regions studied were the intergenic spacer region between the *trnL* 3'exon and *trnF* exon (*trnL -trnF* IGS) and the *trnL* intron region. We found nucleotide variation at one site in the *trnL-trnF* IGS, and at three sites in the *trnL* intron. With the sequence data from analysis of these regions, the four *Cinnamomum* species used in this study were correctly identified. Furthermore, single-strand conformation polymorphism (SSCP) analysis of PCR products from the *trnL-trnF* IGS and the *trnL* intron resulted in different SSCP band patterns among *C. cassia*, *C. zeylanicum* and *C. burmannii*. Judy (IISR, 2005) conducted a preliminary RAPD study on eleven species of cinnamon *viz. C. verum*, *C. cassia*, *C. malabatrum*, *C. riparium*, *C. macrocarpum*, *C. perottitii*, *C. weightii*, *C. citronella*, *C. tamala*, *C. sulphuratum*, *C. glaucens and C. camphora* using three primers (OPA 1, OPB 5 and OPC 15). The dendrogram indicated that *C. malabatrum* is closely related to C. *verum*. The species like *C. cassia*, *C. weightii* and *C. citronella*, *C. camphora* are grouped together. *Exotic* species like *C. sulphuratum, C. glaucens and C. camphora* were also placed in single group indicating similarities between them.

Molecular fingerprinting of selected elite accessions of *Myristica fragrans* by RAPD, ISSR and rDNA-RFLP markers was done by Sheeja *et al.* (2006). High yielding elite accessions, with high sabinene and low myristicin contents revealed good deal of polymorphism in RAPD but no polymorphism was detected in ISSR. An average genetic distance of 25.5 per cent was observed among the accessions due to the high degree of relatedness among them. Unique bands were identified in some of the rare accessions like A9/4, a very high yielding epicotyl graft with plagiotropic shoots, A9/150, possessing very thick mace and apple shaped bold fruits and A4/22 with unique character of high number of erect shoots for their unequivocal identification. All the accessions showed polymorphism and could be distinguished from each other through RAPD markers and qualified to be included in a core collection.

Garcinia species are distributed widely throughout the old world especially Asia and Africa. *Garcinia* belongs to the family Guttiferae. RAPD polymorphism was used to study species inter relationships between six *Garcinia* species namely *G. indica, G. gummi-gutta, G. cowa, G. mangostana, G.* tinctoria and G. *hombroniana* (Pulla Rao (2003).

In tamarind (*Tamarindus indica* L), it was observed that all the characters varied significantly, except thickness of pod, seed to pod ratio, number of seeds per pod, seed weight per pod, stamen length, filament length, and style length. The variability observed was narrow at molecular level compared to the variability at phenotypic level. Four elite tamarind genotypes were identified which are productive and regular bearing.

2.1.2.4. Vanilla

Continuous clonal propagation has resulted in very little variability for crop improvement programmes in vanilla. An attempt was made to increase the spectrum of variation by interspecific hybridization with *Vanilla aphylla,* an Indian species which is tolerant to *Fusarium*. Interspecific hybrids were successfully produced and morphological characters and molecular profiles revealed the true hybridity of the progenies. Seedling progenies of *V. planifolia,* and interspecific hybrids were evaluated and random amplified polymorphic DNAs (RAPDs) loci were marked (Figure 15.4). The profiles indicate similarity between the parents, selfed progenies and interspecific hybrids and that all the progenies tested were variable when compared to each other, which can be exploited for crop improvement in vanilla (Minoo *et al.*, 2006, 2016) (Figure 15.5).

Use of microsatellites allowed the first molecular based estimation of heterozygosity levels in vanilla, which was not possible with dominant markers like RAPD and AFLP. Bory *et al.* (2008) developed 14 microsatellite primers form di-nucleotide CT/GT enriched genomic library of *Vanilla planifolia*. These markers were monomorphic within cultivated accessions. The 14 SSR markers developed in cultivated Vanilla (*Vanilla planifolia* L.) were transferable to *Vanilla tahitensis* however, 11 loci were found to be polymorphic between these two species.

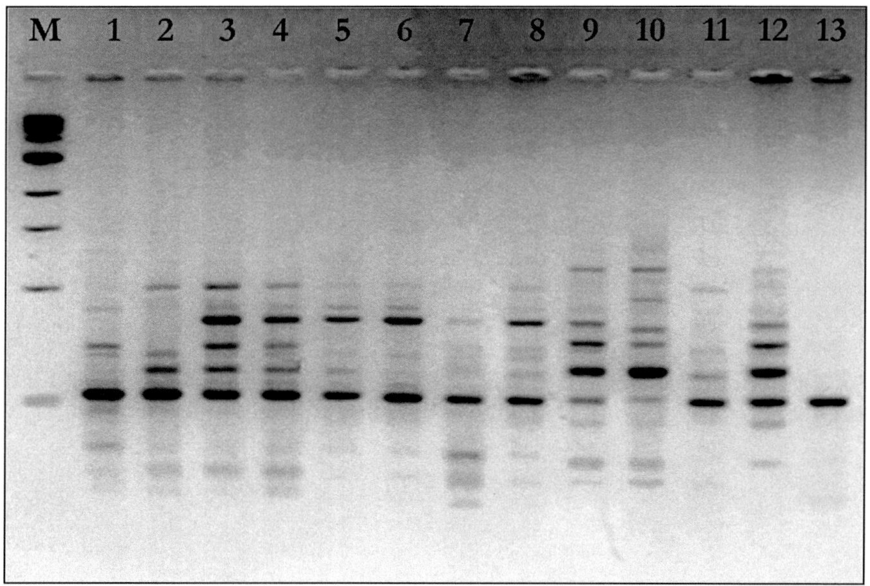

Figure 15.4: RAPD Profiles of Inter Specific Hybrids of Vanilla using OPERON primer OPB20 (Source: Minoo *et al.,* 2016).

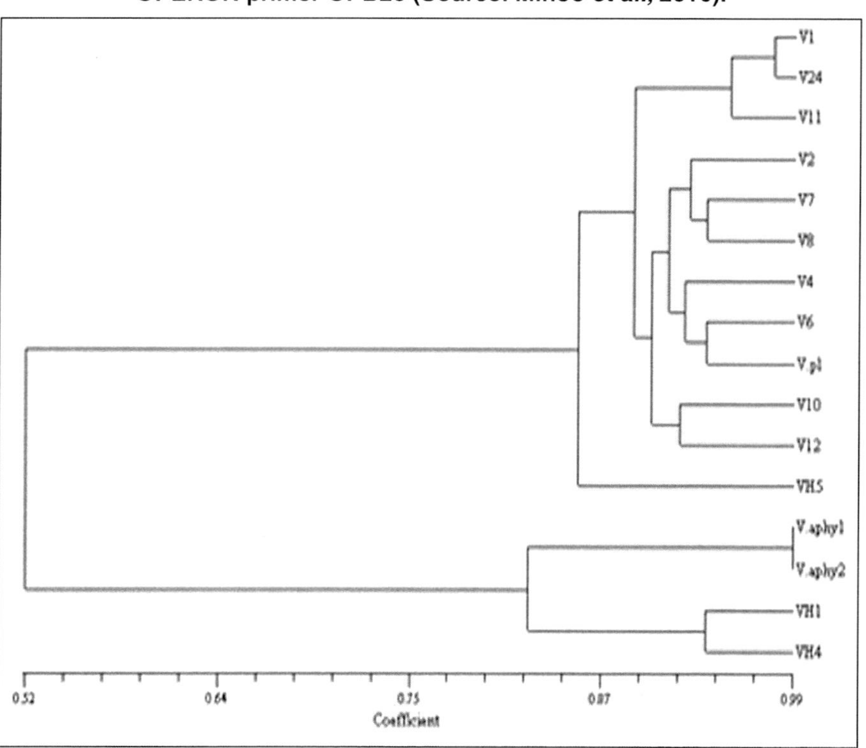

Figure 15.5: Dendrogram Showing Linkage Groups between Selfed Progenies and Interspecific Hybrids based on by RAPD Markers (Source: Minoo *et al.*, 2016).

Sequencing of nuclear genes was used for reconstructing the evolutionary history of Vanilloid orchids (Cameron, 2004). Nuclear (ITS) and plastid (*rbcL* gene) DNA sequences were also used for unraveling the origin of Tahitian Vanilla (Lubinsky *et al.*, 2008). The length polymorphism of neutral caffeic acid O-methyl transferase gene was used to analyse 20 vanilla species and confirmed the strong differentiation of Old World *vs* New World species in the genus (Besse *et al.*, 2009). Cameron (2004) has utilized plastid *psaB* gene sequences for investigating intrafamilial relationships within Orchidaceae.

Minoo *et al.* (2008) have utilized RAPD polymorphism to assess the levels of genetic diversity and interrelationships among different collections of *Vanilla planifolia* Andr. and some related species, which included leafless and leafy types such as *V. tahitensis* J.W.Moore, *V. andamanica*, Rolfe, *V. pilifera* Holtt. and *V. aphylla* Blume (Figure 15.6). Studies revealed the existence of limited variations within collections of *V. planifolia*, indicating its narrow genetic base. Furthermore, *V. tahitensis* was found to be the closest relative of nearest to *V. planifolia*.

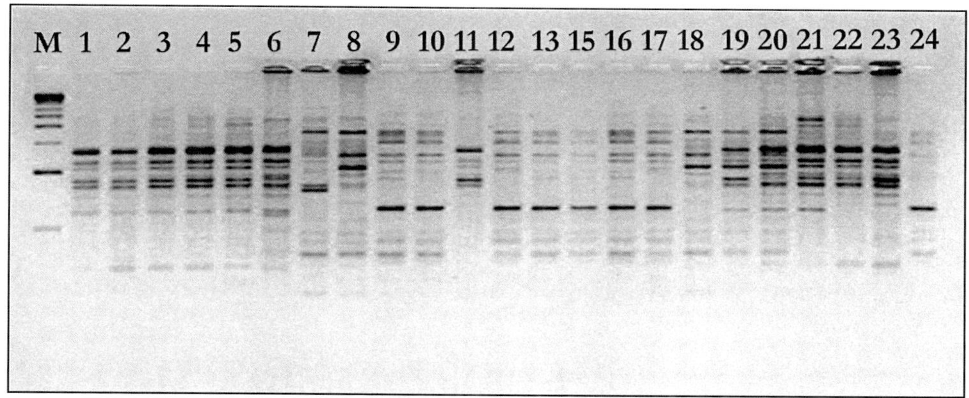

Figure 15.6: RAPD Polymorphism Observed in *Vanilla planifolia, V. tahitensis, V. andamanica, V. pilifera* and *V. aphylla* using OPERON Primer OPB14 (Source: Minoo *et al.*, 2008).

Scientists at the Fujian Agriculture and Forestry University and National Orchid Conservation Centre of China have colloborated to produce the world's first orchid genetic map. The collaboration have yielded and have complete genome sequence of *Vanilla shenzhenica*, in 2015, giving way to the possibility of altering vanilla properties and pathways through transgenic technology in the future.

2.2. Estimation of Genetic Fidelity of Micropropagated Plants using RAPDs

Morphological characters coupled with RAPD profiles have been used by Nirmal Babu *et al.* (2003a). to estimate the genetic fidelity of micropropagated plants pepper where the profiles indicated that the clones were genetically stable and that the micropropagation technology could be used for commercial cloning of black pepper. Shahanas *et al*, (2003) reported intra clonal fidelity of rooted cuttings of cv. Sreekara, derived from bamboo nursery using RAPD. Genetic fidelity of micropropagated *Piper longum* using RAPD profiling was reported by Ajith *et al.* (1997) and occurrence of somaclonal variation was observed.

In turmeric, Nirmal Babu *et al.* (2003a). studied morphological and molecular variations among micropropagated and callus regenerated plants and found variations in both but with higher percentage of variation in callus regenerated somaclones. *In vitro* plants developed through microrhizome exhibited least amount of variations. They inferred that this is due to the accumulated vegetative mutations (mosaic) in turmeric. The genetic fidelity studies of turmeric germplasm conserved in *in vitro* genebank using RAPD profiling showed their genetic integrity (Geetha, 2002; Ravindran *et al.*, 2004).

RAPD profiling, within the replicates of *in vitro* conserved and cryopreserved lines of ginger using operon random (RAPD) primers was studied, but it did not detect any polymorphism between the conserved lines in any of the primers tested, indicating the genetic stability (Geetha, 2002; Peter *et al.*, 2002).

Micropropagated multiple shoots of *Vanilla planifolia* developed from axillary bud explants and established in culture for 10 years, have been used to determine somaclonal variation using random amplified polymorphic DNA (RAPD) and inter-simple sequence repeats markers (ISSR). No difference was observed in banding patterns of any of the samples for a particular primer, indicating the absence of variation among the micropropagated plants, concluding that the micropropagation protocol used for *in vitro* proliferation of vanilla plantlets could be applied for the production of clonal plants over a considerable period of time (Sreedhar *et al.*, 2007).

2.3. Development of Mapping Population and Preliminary Molecular Map

A mapping population of 200 segregating progenies was developed between Subhakara X Panniyur 1 for preparation of preliminary genetic map of black pepper (Nirmal Babu *et al.*, 2003b). RAPD and ISSR profiling of 96 progenies of first mapping population Subhakara X Panniyur 1 was done along with their parents and over 200 polymorphic markers segregating in the population were scored. Preparation of frame work molecular map is in progress using pseudo test cross approach.

2.4. Identification of Markers Linked to Important Characters

2.4.1. Sex Specific Markers

Banerjee *et al.* (1999) studied molecular basis of genotypic differentiation between the male and female plants of dioecious *P. longum* L. and identified two RAPD markers specific to male plants. Philip *et al.* (2000) used RAPD profiles to differentiate three female varieties of *P. longum*. Manoj *et al.* (2004) studied the molecular basis of differentiation between male and female plants of *Piper longum* using RAPD.

Shibu *et al.* (2000) identified sex specific DNA markers for identifying the productive female trees in nutmeg. Ganeshaiah *et al.* (2000) attempted to identify sex specific DNA markers that could potentially be used to determine the sex of

Myristica fragrans seedlings. Out of 60 random primers screened using RAPD-PCR, one primerm OPE II, gave a female sex specific DNA amplification product. Sheeja *et al.* (2006), reported that male seedlings could be characterized by presence of specific bands of 1300 and 1000 bp, during their study

2.4.2. Identification of Hybrids

Isoenzymes were utilized to prove hybrid nature of plants derived from inter specific crosses. Sasikumar *et al.* (1999) used isozyme analysis to characterize inter specific hybrids between *Piper nigrum.* x *P. attenuatum* and *P. nigrum* x *P. barberi* to identify the true hybrids by the presence of hybrid specific as well as male parent specific bands. Johnson *et al.* (2005) used male parent-specific RAPD markers for identification of hybrids in black pepper (*Piper nigrum* L.).

In other spices like *Thymus caespititius*, a low correlation was found between essential oil composition and RAPD analysis; however, molecular data clustered plants according to their geographic origin (Trindade *et al.*, 2008), suggesting that molecular tools should be explored in order to fully understand the inuence of both environmental and genetic factors on volatiles composition. RAPDs have also been used in understanding the genetic relationships and as reliable tools for the discrimination of the two parental taxa and the putative hybrid in natural populations of *Origanum* × *intercedens*, a hybrid between *O. onites* and *O. vulgare*. The results were compared to those obtained with the essential oil composition and morphological characteristics. It was found that DNA fingerprinting and general morphology placed the hybrid closer to *O. onites*, while its essential oil composition showed the hybrid to be more similar to *O. vulgare* (Gounaris *et al.*, 2002).

2.4.3. Markers Linked to Phytophthora Resistance

Nirmal Babu *et al.* (2003b). used for RAPD profiling to identify RAPD markers linked to *Phytophthora* resistance, among eleven lines each of *Phytophthora* susceptible and tolerant cultivars. The study indicated that in general the tolerant lines formed a cluster of their own and a few suscepts were grouped with tolerant lines. A marker at 700 bp region was found to be present in tolerant lines and absent in suscepts. Sheji *et al.* (2006) identified a RAPD marker associated with *Phytophthora* resistance and converted it in to SCAR marker (Figure 15.7). RAPD profiling of disease tolerant and susceptible lines of black pepper was done. A DNA fragment representing a RAPD marker linked to resistance was cloned, sequenced and converted into a SCAR marker. SCAR primers to the resistant allele were developed based upon a deletion region between susceptible and resistant plants. A unique band 360 base pair appeared in all the *Phytophthora* resistant lines but was absent in the susceptible lines, and this marker can be used for screening black pepper germplasm.

2.4.4. Identification of Adulterants

Dhanya *et al.* (2007) reported simple protocol for characterizing and checking adulteration in traded black pepper. Syamkumar *et al.* (2005) reported standardization of protocol for the isolation of amplifiable genomic DNA from dried capsules of traded cardamom. This protocol will help in the PCR-based characterization of different grades of commercially traded cardamom and to identify adulterants if any.

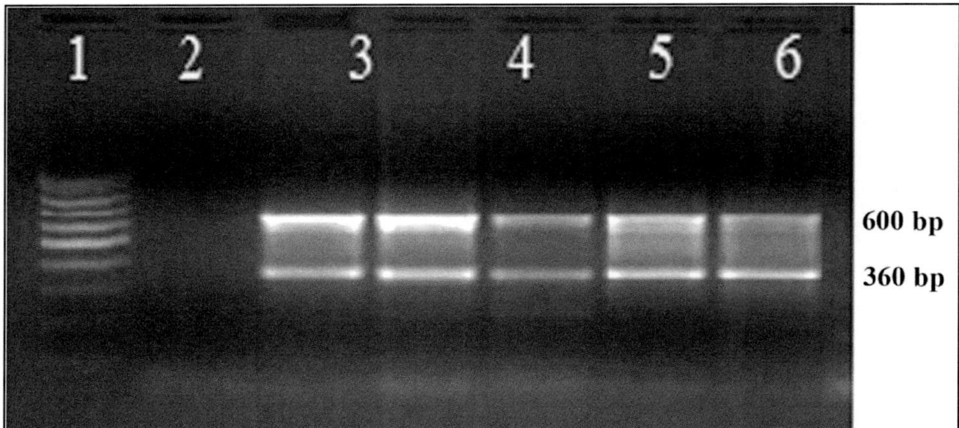

Figure 15.7: Differential Amplification of SCAR Markers among Phytophthora Susceptible and Resistant Lines of Black Pepper (Source: Sheji Chandran *et al.*, 2006).

Detection of extraneous *Curcuma* sp. contamination of powdered samples of turmeric using random amplified polymorphic DNA was reported by Sasikumar *et al.* (2004). The analysis of three market samples of turmeric powder from Kerala, and India, revealed the prevalence of *C. zedoaria* (wild species) powder over the *C. longa* (the common culinary turmeric) powder, although the curcumin levels of the samples were in accordance with the quality standards prescribed for the commodity. The other studies on studies on genetic purity and checking adulteration are those of Cao *et al.* (2001) and Sasaki *et al.* (2002, 2004) who used sequence analysis of Chinese and Japanese *Curcuma* drugs on the 18S rRNA gene and *trnK* gene and the application of amplification-refractory mutation system analysis for their authentication. Application of single nucleotide polymorphism analysis based on species specific nucleotide sequence was developed by Sasaki *et al.* (2004) to identify the plants and drugs derived from *C. longa, C. phaeocaulis, C. zedoaria* and *C. aromatica.* Based on the difference in the nucleotide positions at 177, 645, 724 and a 4 base indel on the *trn*K gene obtained using three different lengths of (26 mer, 30 mer and 34 mer) reverse primers helped to identify the four *Curcuma* species studied. The SNP analysis method developed become a useful method for the identification of botanical origins of *Curcuma* drugs such as 'Ezhu' used in Chinese medicine, which was difficult to identify morphologically and phytochemically.

Xia *et al.* (2005) used molecular (5S-rRNA spacer domains) and chemical fingerprints for quality control and authentication of *Rhizoma Curcumae*, a traditional Chinese medicine used in removing blood stasis and alleviating pain.

Genetic profiling of traded ginger from India and China using 20 RAPD primers and 15 ISSR primers gave consistent amplification pattern. Significant variation was observed between the produces from the two countries (un published, IISR 2008). Jiang *et al.* (2006) used metabolic profiling and phylogenetic analysis for authentication of ginger. They used these tools to investigate the diversity within the ginger species and between ginger and closely related species in the genus

Zingiber. Phylogenetic analysis demonstrated that all *Zingiber officinale* samples from different geographical origins were genetically indistinguishable. In contrast, other *Zingiber* species were significantly divergent, allowing all species to be clearly distinguished using this analysis. In the metabolic profiling analysis, the *Z. officinale* samples derived from different origins showed no qualitative differences in major volatile compounds, although they did show some significant quantitative differences in non-volatile composition, particularly regarding the content of [6]-, [8]-, and [10]-gingerols, the most active anti-inflammatory components in this species. The metabolic profiles of other *Zingiber* species were very different, both qualitatively and quantitatively, when compared to *Z. officinale* and to each other. Comparative DNA sequence/chemotaxonomic phylogenetic trees showed that the chemical characters of the investigated species were able to generate essentially the same phylogenetic relationships as the DNA sequences. This supports the contention that chemical characters can be used effectively to identify relationships between plant species.

2.4.5 Tagging and Isolation of Candidate Genes

2.4.5.1. Black Pepper

Isolation of genes responsible for agronomically important characters, especially for biotic and abiotic stresses is an area of high importance. Johnson *et al.* (2005) reported a method for isolation and reverse transcription of high quality RNA from *Piper* species. They reported amplification, isolation and sequencing of putative β-1,3-glucanase gene from *Piper colubrinum* was achieved using glucanase specific primer. Jebakumar *et al*, (2001) has reported differential induction of various PR proteins like Phenly alanine ammonia lyase (PAL), β-1,3- glucanase in *Phytophthora* tolerant black pepper upon inoculation with *P. capsici*.

In an attempt to isolate and cloning of a cDNA fragments encoding the defense related protein β-1,3-glucanase in black pepper (*P. nigrum* L.) and methyl glutaryl CoA reductase in *Piper colubrinum* were also reported (Girija *et al.*, 2005a, b). A PCR-based suppression subtractive hybridization (SSH) was used by Dicto and Manjusha (2005) to identify *P. colubrinum* resistance genes that are differentially expressed in response to the signaling molecule, salicylic acid (SA). A subtracted library of SA-Induced genes was synthesized and one of the clones showed sequence homology to osmotin, a member of Class-V group of pathogenesis-related (PR) gene family. Northern blot analysis revealed that osmotin is dominantly expressed in SA/ethylene-treated tissue. This indicates that SSH can be used to identify and clone PR genes in *P. colubrinum*. Bioprospecting of novel genes form black pepper was attempted by Sujatha *et al.* (2005). They used heterologus probes to identify the presence of pea lectin genes and tomato protease inhibitor genes in black pepper. Nazeem *et al.* (2008) also reported the role of ß-1, 3 glucanase and related enzymes in the defense mechanism in foot rot tolerant black pepper variety and in resistant genotype *P. colubrinum*. Varma *et al.* (2009) has explained differential induction of chitinase in *Piper colubrinum* in response to inoculation with *P. capsici*. Cloning and sequence characterization of two isoforms of *osmotin*, an antifungal *PR-5* gene homologue, from a salicylic acid-induced subtracted cDNA library was generated in *Piper colubrinum* (Mani and Manjula, 2010).

Piperine is the main bioactive compound among black pepper alkaloids, which performs unique physiological functions. However, the mechanisms of piperine synthesis were poorly understood and studies by Hu *et al.* (2015) were the first to describe the fruit transcriptome of black pepper by sequencing on Illumina HiSeq 2000 platform. A total of 56,281,710 raw reads were obtained and assembled, from which 44,061 unigenes with an average length of 1,345 nt were generated. During functional annotation, 40,537 unigenes were annotated in Gene Ontology categories, Kyoto Encyclopedia of Genes and Genomes pathways, Swiss-Prot database, and Nucleotide Collection (NR/NT) database. In addition, 8,196 simple sequence repeats (SSRs) were detected. In a detailed analysis of the transcriptome, housekeeping genes for quantitative polymerase chain reaction internal control, polymorphic SSRs, and lysine/ornithine metabolism-related genes were identified which could provide useful data for further research on piperine synthesis.

2.4.5.2. Cardamom

Development of structured populations for tagging of Katte and Rhizome rot resistance are in the process of field planting (IISR unpublished 2008). One putative RAPD marker was also identified associated with Katte Resistance (IISR 2003).

2.4.5.3. Ginger

Swetha and Subramanian (2008) reported isolation and molecular analysis of R-gene in resistant *Zingiber officinale* (ginger) varieties against *Fusarium oxysporum.* They observed that the R-gene is present only resistant varieties. These cloned R-genes provide a new resource of molecular markers for marker assisted selection (MAS) and rapid identification of *Fusarium* yellows resistant ginger varieties. Kavitha and Thomas (2006) reported *Zingiber zerumbet,* a close relative of ginger, as a potential donor for soft-rot resistance in Ginger. They employed AFLP markers and mRNA differential display to identify genes whose expression was altered in a soft rot-resistant accession of *Zingiber zerumbet* before and after inoculating it with. *Pythium aphanidermatum,* which is the principal causative agent of soft-rot disease in ginger A few differentially expressed transcript-derived fragments (TDFs) were isolated, cloned and sequenced. Homology searches and functional categorization of some of these clones revealed the presence of defence/stress/signalling group which are homologous to genes known to be actively involved in various pathogenesis-related functions in other plant species. They found *Z. zerumbet* shows adequate variability both at DNA level and in response to *Pythium.* (Kavitha and Thomas, 2006; 2008). Nair *et al.* (2010) identified a member of the pathogenesis-related protein group 5 (*PR5*) gene family in *Z. zerumbet* that is expressed constitutively but upregulated in response to infection by *P. aphanidermatum*. Nair and Thomas (2013) isolated the full length sequence of *ZzR1* resistance gene from *Zingiber zerumbet* with potential for imparting resistance to soft rot in ginger. Isolation of resistance genes from such related genera will help in ginger improvement via transgenic approaches.

2.4.5.4. Turmeric

Joshi *et al.* (2010) have reported isolation and characterization of resistance gene candidates in *Curcuma longa* cv.surama. R gene conferring resistance to

Pythium aphanidermatum was characterized in *Curcuma zedoaria* (Basudeba *et al.*, 2013). Annadurai *et al.* (2013) reported the presence of novel transcripts related to anti-cancer and anti-malarial terpenoides in the transcriptome of *Curcuma longa*.

2.4.5.5. *Vanilla*

Large gene sequence datasets from vanilla pods at different times of development, and representing different tissue types, including the seeds, hairs, placental and mesocarp tissues, were generated using next-generation sequencing technologies, for interrogation of pathways of vanillin and C-lignin biosynthesis in the pod and seed, respectively. The combined 454/Illumina RNA-seq platforms provide both deep sequence coverage and high quality *de novo* transcriptome assembly for this non-model crop species. The annotated sequence data provided a foundation for understanding multiple aspects of the biochemistry and development of the vanilla bean, as exemplified by the identification of candidate genes involved in lignin biosynthesis. Transcriptome data indicated that C-lignin formation in the seed coat involves coordinate expression of monolignol biosynthetic genes with the exception of those encoding the caffeoyl coenzyme A 3-*O*-methyltransferase for conversion of caffeoyl to feruloyl moieties. This database provides a general resource for further studies on this important flavor species (Xiaolan Rao *et al.*, 2014).

Gallage *et al.* (2014) reported that a single hydratase/lyase type enzyme designated vanillin synthase (*Vp*VAN) catalyses direct conversion of ferulic acid and its glucoside into vanillin and its glucoside, respectively. The enzyme shows high sequence similarity to cysteine proteinases and is specific to the substitution pattern at the aromatic ring and does not metabolize caffeic acid and *p*-coumaric acid as demonstrated by coupled transcription/translation assays. *Vp*VAN localizes to the inner part of the vanilla pod and high transcript levels are found in single cells located a few cell layers from the inner epidermis. Transient expression of *VpVAN* in tobacco and stable expression in barley in combination with the action of endogenous alcohol dehydrogenases and UDP-glucosyltransferases result in vanillyl alcohol glucoside formation from endogenous ferulic acid. A gene encoding an enzyme showing 71 per cent sequence identity to *Vp*VAN was identified in another vanillin-producing plant species *Glechoma hederacea* and was also shown to be a vanillin synthase as demonstrated by transient expression in tobacco.

Recently *de novo* biosynthesis of vanillin in *V. planifolia* was found to be catalysed by a single enzyme, vanillin synthase that catalyses ferulic acid and its glucoside to produce vanillin and vanillin glucoside. This promises transgenic plants with high vanillin synthase activity for use as vanillin production sources.

2.4.6 Comparative Genomics for Gene Tagging

2.4.6.1. *Black Pepper*

In perennial polyploids like black pepper, conventional mapping strategies are laborious. Hence genomics approach and using information from other sources, like comparisons with heterologus genomes or genes, could provide the necessary leads for tagging. Candidate genes responsible for pathogenesis can also be identified from sequence information available on *R* genes and information available from

Arabidopsis genome (Aarts *et al.*, 1998). With the advent of next generation sequencing techniques, genome wide analysis of NBS-LRR-encoding genes in *Arabidopsis* (Meyers *et al.*, 1999; Shen *et al.*, 1998) reported that the resistance gene candidates identified by PCR with degenerate oligonucleotide primers map to clusters of resistance genes in lettuce. These form an important approach for isolating R genes using candidate gene approach.

A PCR-based approach for isolating pathogen resistance genes from potato with potential for wide application in plants was reported by Leister *et al.* (1996). The genomic approach was used to isolate R genes in Black pepper. Degenerate primers were designed from the conserved motifs (Ploop and GLPL) of NBS region of known R genes and these were used to amplify similar sequences from genomic DNA of black pepper. A 500 bp fragment was amplified consistently from black pepper variety P 24 using degenerate primers from P LOOP GG – GLPLA3. This fragment was cloned in PCR XL TOPO vector and a library of 48 clones was developed by Nirmal Babu *et al.* (2003b).

Similarly information available on mapping of heterologous loci for example, *Ph-2* locus controlling partial resistance to *Phytophthora infestans* in tomato (Philippe *et al.*, 1998), genetic and physical mapping of molecular markers linked to the *Phytophthora* resistants gene *Rps 1*-k in soybean (Takao *et al.*,1997) can also be used to tag *Phytophthora* resistance in black pepper. Intraspecific comparative genomics to identity avirulence genes from *Phytophthora* was reported by Bos *et al.* (2003). Comparative genomics has already made much headway for solanaceous crops to which *Capsicum* belongs (Tanksley *et al.*, 1988; Livingstone *et al.*, 1999) and where much information on *Phytophthora* resistance is available.

RNA-Seq technology was employed, for the first time (Gordo *et al.*, 2012), to describe the root transcriptome of black pepper. The root transcriptome was sequenced by the NGS SOLiD platform and assembled using the multiple-k method. Blast-2-GO and orthoMCL methods were used to annotate 10338 unigenes. The 4472 predicted proteins showed about 52 per cent homology with the *Arabidopsis* proteome. Two root proteomes identified 615 proteins, which seem to define the plant's root pattern. This dataset of 10,338 unigenes will be important for the biotechnological breeding of black pepper.

2.4.6.2. Ginger

Aswati and Thomas (2006) reported isolation, characterization and expression of resistance gene candidates (RGCs) using degenerate primers based on conserved motifs from the NBS domains of plant resistance (R) genes were used to isolate analogous sequences or resistance gene candidates (RGCs) from cultivated and wild *Zingiber* species. Kavitha and Thomas (2006, 2008) employed mRNA differential display was employed to identify genes whose expression was altered in a soft rot-resistant accession of *Zingiber zerumbet.* A few differentially expressed transcript-derived fragments (TDFs) were isolated, cloned from *Pythium aphanidermatum,* which is the principal causative agent of soft-rot disease in ginger. Sequence and functional categorization of these clones revealed the presence of defence/stress/ signalling group which are homologous to genes known to be actively involved in various pathogenesis-related functions in other plant species. Swetha and Subramanian (2008) reported isolation and molecular analysis of R-gene in resistant *Zingiber officinale* (ginger) varieties against *Fusarium oxysporum*. They observed that the R-gene is present only resistant varieties. Thus the cloned R-genes provide a new resource of molecular markers for marker assisted selection (MAS) and rapid identification of fusarium yellows resistant ginger varieties.

Violaxanthin de-epoxidase (VDE) as the key enzyme of xanthophyll cycle plays an important role in protecting photosynthesis apparatus from the damage of excessive light. Huang *et al.* (2007 reported molecular cloning and characterization of violaxanthin de-epoxidase (VDE) in ginger. A full length (2000 bp) cDNA encoding *violaxanthin deepoxidase* (GVDE) (GenBank accession no. AY876286) was cloned from ginger using RT-PCR and 50, 30 rapid amplification of cDNA ends (RACE). The expression patterns of GVDE in response to light were characterized. GVDE has a 1431 bp open reading frame and the predicted polypeptide contains 476 amino acids with the molecular mass of 53.7 kDa. Northern blot analysis showed that the GVDE was mainly expressed in leaves.

2.4.6.4. Vanilla

Two new Caffeoyl CoA *O*-methyltransferases (OMTs) - caffeoyl CoA OMT-like genes was identified by Widiez *et al.* (2011). by screening a cDNA library from specialized hair cells of pods of the orchid *Vanilla planifolia*. Characterization of the corresponding two enzymes, designated Vp-OMT4 and Vp-OMT5, revealed that both enzymes preferred as a substrate the flavone tricetin *in vitro*, yet their sequences and phylogenetic relationships to other enzymes were distinct from each other. Quantitative analysis of gene expression indicated a dramatic tissue-specific expression pattern for *Vp-OMT4*, which was highly expressed in the hair cells of the developing pod, the likely location of vanillin biosynthesis. Although Vp-OMT4 had a lower activity with the proposed vanillin precursor, 3,4-dihydroxybenzaldehyde, than with tricetin, the tissue specificity of expression suggests it may be a candidate for an enzyme involved in vanillin biosynthesis. In contrast, the *Vp-OMT5* gene was mainly expressed in leaf tissue and only marginally expressed in pod hair cells. Phylogenetic analysis suggests Vp-OMT5 evolved from a cyanobacterial enzyme and it clustered within a clade in which the sequences from eukaryotic species had predicted chloroplast transit peptides. Transient expression of a GFP-fusion in tobacco demonstrated that Vp-OMT5 was localized in the plastids. This is the first flavonoid OMT demonstrated to be targeted to the plastids.

The Vanilla Sustainability Project, an international initiative involving vanilla scientists and researchers (France, Madagascar, Mexico, USA and others), has been actively engaged in the development of genomic resources for *Vanilla planifolia*, with the main objectives of: 1) the development of an annotated reference base for *Vanilla planifolia* using RNA-Seq that functions as a comprehensive reference transcriptome of expressed genes for multiple lines of inquiry including analyses of gene regulation, expression, structure, and number; 2) the analysis of fruit development to understand developmental mechanisms, biosynthetic and metabolic pathways important to fruit quality characteristics (*e.g.*, vanillin content, aroma, yield, non-dehiscence, abiotic stress tolerance) and transcriptional regulation of these important fruit processes; and 3) the analysis of *Vanilla* root responses and root structure to understand the innate defense response to pathogen infection. The factors chosen are critical for survival of the vanilla industry, with the threat of a *Fusarium* pandemic that is destroying vanilla, and the accelerated threat of global warming that has already affected the timing of *Vanilla planifolia* flowering and impacted successful fertilization, thus the database provides a general resource for further studies (Xiaolan Rao *et al.*, 2014).

2.7 Molecular Characterization and Detection of Pathogens

Gosh and Purkauastha (2003) used polyclonal antibodies and antigens of host and pathogen for early diagnosis of rhizome rot disease of ginger caused by *Pythium aphanidermatum* and *Pythium aphanidermatum*, was detected in ginger rhizome after eight weeks of inoculation by agar gel double diffusion and immuoelectrophoretic tests, but only one week after inoculation by indirect ELISA. Kumar and Anandaraj (2006) developed an efficient DNA isolation protocol and PCR based detection of bacterial pathogen in soil. This PCR based method using universal *Ralstonia solanacearum* specific primer offer a rapid method for unambiguous detection of this pathogen at a concentration of 10^3-10^4 cells per gram of soil.

During 2009 and 2010 in Papantla, the greatest vanilla-producing region of Mexico, *Fusarium* was isolated from vanilla roots and stems that showed symptoms of the disease. From 189 isolates 11 morphologically different colonies were selected to verify the species by amplifying and sequencing their ITS regions. The detected species corresponded to *F. proliferatum*, *Fusarium* sp., *F. oxysporum* f. sp. and *vanillae* (which was the most numerous and most pathogenic to vanilla stems and leaves), an undetermined species of *Fusarium* and *F. proliferatum*, which showed no evidence of producing disease symptoms by Adame-Garcia *et al.* (2015).

3. Future Prospects

Uses of molecular markers in the phylogenetic studies of various organisms have become increasingly important in recent times. Availability of fast DNA sequencing techniques along with the development of robust statistical analysis methods provides a new momentum to this field. In this context, utility of different nuclear encoded genes (like 16S rRNA, 5S rRNA, 28S and rRNA) mitochondrial (cytochrome oxidase, mitochondrial 12S, cytochrome b and control region) and few chloroplast encoded genes (like *rbcL*, *matK* and *rpl16*) will reveal important characters of functional value in spices. Though molecular markers are not free entirely of flaws, they can complement the traditional morphology based method for phylogenetic studies. This combinational approach will strengthen the basis of relationships of organisms to a great extent.

Crop specific strategies utilizing biotechnology, thus ensures conservation of resources unaffected by climatic changes, availability of an alternate set of genotypes that could be multiplied in large numbers, engineered with specific genes, and maintained safe from the threats of natural disasters for posterity. Understanding the molecular interactions between the host and pathogens and developing disease diagnostic tools, is important for developing resistant cultivars by biotechnological breeding.

Secondary metabolites impart lot of value to spices and their products; however secondary metabolism is species-specific, which makes the genome sequence of model plants such as *Arabidopsis* of only limited value. Related species can have different secondary metabolite profiles, however, initial metabolic pathways are similar in most plants, and thus homology between genes can be used for strategies to clone genes from other plants. The use of ESTs in combination with functional expression is another approach that could bring important advances in the future. The growing number of plant gene sequences with a known function could result in accumulating data that will result in exponential growth of similar genes identified in other plants. The area of functional genomics faces a major bottleneck due to the unknown biochemical pathways involved, involving assays for the enzymes involved in secondary metabolism. It would also be interesting to examine the molecular processes that bring about the variability in secondary metabolism, at the level of gene regulation, posttranscriptional regulation or protein evolution.

References

Aarts, M.G., Hekkert, B., Holub, E.B., Beynon, J.L., Stiekema, W.J. and Pereira, A. (1998). Identification of R-gene homologous DNA fragments genetically linked to disease resistance loci in *Arabidopsis thaliana. Moleular Plant-Microbe Interactions*, **11**: 251–258.

Adame-García, J., Rodríguez-Guerra, R., Iglesias-Andreu, L.G., Ramos-Prado, J.M. and Luna-Rodríguez, M. (2015). Molecular identification and pathogenic variation of *Fusarium* species isolated from *Vanilla planifolia* in Papantla Mexico. *Botanical Sciences*, **93(3)**: 669-678.

Ajith, A., Parani, M., Rao, C.S., Latha, R. and Balakrishna, P. (1997). Micropropagation and genetic fidelity studies in *Piper longum* L. In: Edison, S., Ramana, K.V., Sasikumar, B., Babu, K.N. and Eapen, S. J (Eds.), *Biotechnology of Spices, Medicinal and Aromatic Plants. Indian Society of Spices*, Calicut, Kerala, India. pp. 94-97.

Annadurai, R.S., Neethiraj, R., Jayakumar, V., Damodaran, A.C., Rao, S.N., Katta, M.A., Gopinathan, S., Sarma, S.P., Senthilkumar, V., Niranjan, V. and Gopinath, A. (2013). *De Novo* transcriptome assembly (NGS) of *Curcuma longa* L. rhizome reveals novel transcripts related to anticancer and antimalarial terpenoids. *PLoS ONE*, **8(2)**: e56217.

Anu, C., Ritto, P., Anupama, K., Senthil kumar, R., Sheeja, T.E., Nirmal Babu, K. and Parthasarathy, V.A. (2016). Isolation and characterization of genomic microsatellite markers for small cardamom (*Elettaria cardamomum* Maton) for utility in genetic diversity analysis. *Physiology and Molecular Biology of Plants*, 1-11.

Anuntalabhochai, S., Sitthiphroma, S., Thongtaksinb, W., Sanguansermsric, M. and Cutlerd, R.W. (2007). Hybrid detection and characterization of *Curcuma spp*. using sequence characterized DNA markers. *Scientia Horticulturae*, **111**: 389 – 393.

Anupama, K., Anu Cyriac, Saji, K.V. and Nirmal Babu, K. (2015). Microsatellite marker based cross species amplification and genetic diversity analysis in the genus *Piper*. *International Journal of Advanced Research*, **3(5)**: 184-191.

Aswati Nair, R. and Thomas, G. (2006). Isolation, characterization and expression studies of resistance gene candidates (RGCs) from *Zingiber* spp. *Theoretical and Applied Genetics*, **116**: 123–134.

Banerjee, N.S., Manoj, P. and Das, M.R. (1999). Male sex-associated RAPD markers in *Piper longum* L. *Currrent Science*, **77**: 693-95.

Basudeba, K., Nanda, S., Nayak, P.K., Nayak, S. and Joshi, R.K. (2013). Molecular characterization and functional analysis of CzR1, a coiled-coil-nucleotide-binding –site-leucine-rich repeat R-gene from *Curcuma zedoria* Loeb. That confers resistance to *Pythium aphanidermatum*. *Physiology and Molecular Plant Pathology*, **83**: 59-68.

Besse, P., Da Silva, D., Bory, S., Noirot, M., and Grisoni, M. (2009). Variation in intron length in caffeic acid o-methyltranferase (COMT) in *Vanilla* species (Orchidaceae). *Plant Science*, **176 (4)**: 452-460.

Bory, S., Da Silva, D., Risterucci, A.M., Grisoni, M., Besse, P. and Duval, M.F. (2008). Development of microsatellite markers in cultivated vanilla: Polymorphism and transferability to other vanilla species. *Scientia Horticulturae*, **115(4)**: 420-425.

Bos, J.I., Armstrong, M., Whisson, S.C., Torto, T.A., Ochwo, M., Birch, P.R. and Kamoun, S. (2003). Intraspecific comparative genomics to identify avirulence genes from *Phytophthora*. *New Phytologist*, **159(1)**: 63-72.

Cameron, K.M. (2004). Utility of plastid *psaB* gene sequences for investigating iontrafamilial relationships within Orchidaceae. *Molecular Phylogenetics and Evolution*, **31 (3)**: 1157-1180.

Cao, H., Komatsu, K., Yao Xue, and Xue Bao. (2003). Molecular identification of six medicinal *Curcuma* plants produced in Sichuan: evidence from plastid *trnK* gene sequences. *Biological and Pharmaceutical Bulletin*, **11**: 871-5.

Cao, H., Sasaki, Y., Fushimi, H., Komatsu, K. and Cao, H. (2001). Molecular analysis of medicinally used Chinese and Japanese *Curcuma* based on 18S rRNA and *trnK* gene sequences. *Biological and Pharmaceutical Bulletin*, **24(12)**: 1389-1394.

Chase, M.W. (2004). Monocot relationships: an overview. *American Journal of Botany*, **91**: 1645-1655.

Chaveerach, R., Kunitake, H., Nuchadomrong, S., Sattayasai, N. and Komatsu, H. (2002). RAPD patterns as a useful tool to differentiate Thai *Piper* from morphologically alike Japanese *Piper*. *Journal of Science Asia*, **28**: 221-225.

Chowdhury, U., Tanti, B., Rethy, P. and Gajurel, P.R. (2014). Analysis of genetic diversity of certain species of Piper using RAPD-based molecular markers. *Applied Biochemistry and Biotechnology*, **174(1)**: 168-173.

Dhanya, K., Jaleel, K., Syamkumar, S. and Sasikumar, B. (2007). Isolation and amplification of genomic DNA from recalcitrant dried berries of Black Pepper

(*Piper nigrum L.*) – a medicinal spice. *Journal of Molecular Biotechnology*, **37**: 165-168.

Dicto, J. and Manjusha, S. (2005). Identification of elicitor-induced *PR5* gene homologue in *Piper colubrinum* Link by suppression subtractive hybridization. *Journal of Current Science*, **88(4)**: 25.

Gaia, J.M.D., Mota, M.G., Da, C., Costa, M.R., Martins, C., Da, S. and Poltronieri, M.C. (2003). Analysis of the genetic diversity in germplasm of *Piper* spp.: a preliminary study for isozyme markers. *Revista de Ciencias Agrarias*, **40**: 9-19.

Gallage, N.J., Hansen, E.H., Kannangara, R., Olsen, C.E., Motawia, M.S., Jørgensen, K., Holme, I., Hebelstrup, K., Grisoni, M. and Møller, B.L. (2014). Vanillin formation from ferulic acid in *Vanilla planifolia* is catalysed by a single enzyme. *Nature Communications*, **5**: 4037.

Ganeshaiah, K.N., Ravishankar, K.V., Anand, L., Shibu, M.P. and Shaanker, U. (2000). Identification of sex-specific DNA markers in the dioecious tree, nutmeg (*Myristica fragrans* Houtt.). *PGR Newsletter*, **121**: 59-61.

Geetha, S.P. (2002). *In vitro technology for genetic conservation of some genera of Zingiberaceae*. Ph.D Thesis, Calicut University.

Girija, D., Beena, P.S and Nazeem, P.A. (2005a). Molecular cloning of a cDNA fragment encoding the defense related protein β-1,3-glucanase in black pepper (*P. nigrum* L.). *Proceedings of the Kerala Science Congress*, January 2005, *KFRI, Peechi*, Kerala. pp. 81-82.

Girija, D., Beena, P.S, Nazeem, P.A. and Puroshothama, M.G. (2005b). Molecular cloning of cDNA fragment encoding hydroxy methyl glutaryl CoA reductase in *Piper colubrinum*. *Proceedings of the National symposium on biotechnological interventions for improvement of horticultural crops*: issues and strategies, Kerala Agricultural University, Thrissur, Kerala. pp. 303-306.

Gordo, S.M., Pinheiro, D.G., Moreira, E.C., Rodrigues, S.M., Poltronieri, M.C., de Lemos, O.F., da Silva, I.T., Ramos, R.T., Silva, A., Schneider, H. and Silva, W.A. (2012). High-throughput sequencing of black pepper root transcriptome. *BMC Plant Biology*, **12(1)**: 1.

Gosh, R. and Purkauastha, R.P. (2003). Molecular diagnosis and induced systemic protection against rhizome rot disease of ginger caused by *Pythium aphanidermatum*. *Journal of Current Science*, **85(12)**: 1782-1786.

Gounaris, Y., Skoula, M., Fournaraki, C., Drakakaki, G. and Makris, A. (2002). Comparison of essential oils and genetic relationship of Origanum × intercedens to its parental taxa in the island of Crete. *Biochemical Systematics and Ecology*, **30(3)**: 249-258.

Hu, L., Hao, C., Fan, R., Wu, B., Tan, L. and Wu, H. (2015). *De Novo* assembly and characterization of fruit transcriptome in black pepper (*Piper nigrum*). *PLoS ONE*, **10(6)**: e0129822.

Huang, J.L., Cheng, L.L. and Zhang, Z.X. (2007). Molecular cloning and characterization of violaxanthin de-epoxidase (VDE) in *Zingiber officinale*. *Plant Science*, **172**: 228–235.

Huang, N., Angeles, E.R., Domingo, J., Magpantay, G., Singh, S., Zhang, G., Kumaravadivel, N., Bennet, J. and Khush, G.S. (1997). Pyramiding of bacterial blight resistance genes in rice: marker-assisted selection using RFLP and PCR. *Theoretical and Applied Genetics*, **95**: 313-320.

IISR. (2003). Annual Report 2002-2003, Indian Institute of Spices Research, Calicut.

IISR. (2005). Abstracts of M.Sc. and Ph.D Dissertations on spice crops, IISR, Calicut.

IISR. (2008). Spices News 2008, Indian Institute of Spices Research, Calicut, Kerala, India.

Jaramillo, M.A. and Manos, P.S. (2001). Phylogeny and patterns of floral diversity in the genus *Piper* (Piperaceae). *American Journal of Botany*, **88**: 706-716.

Jayakumar, V.N., Nirmal Babu, K. and Venugopal, M.N. (2005). Molecular characterization of small cardamom (*Elettaria cardamomum* Maton.) collections and related genera using RAPD, PCR -RFLP and ISSR markers. In: *National Symposium on Biotechnological interventions for improvement of Horticultural Crops*: Issues and Strategies. Kerala Agricultural University, Trissur, Kerala, India. pp.122.

Jebakumar, R.S., Anandaraj, M. and Sarma, Y.R. (2001). Induction of PR-proteins and defense related enzymes in black pepper due to inoculation with *Phytophthora capsici*. *Indian Phytopathology*, **54(1)**: 23-28.

Jiang, H., Xie, Z., Koo, H.J., McLaughlin, S.P., Timmermann, B.N. and Gang, D.R. (2006). *Metabolic profiling and phylogenetic analysis of medicinal Zingiber species*: Tools for authentication of ginger (*Zingiber officinale* Rosc.). ***Phytochemistry*, 67**: 1673–1685.

Jiang, Y. and Liu, J.P. (2011). Evaluation of genetic diversity in Piper spp using RAPD and SRAP markers. *Genetics and Molecular Research*, **10(4)**: 2934-2943.

Johnson, G.K., Ganga, G., Sandeep Varma, R., Sasikumar, B. and Saji, K.V. (2003). Identification of hybrids in black pepper (*Piper nigrum* L.) using male parent-specific RAPD markers. *Current Science*, **88**: 1-2.

Johnson, G.K., Sandeep Varma, R., Ganga, G., Anandaraj, M. and Parthasarathy, V.A. (2005). A method for isolation and reverse transcription of high quality RNA from *Piper* Species *Journal of Spices and Aromatic Crops*, **14**: 1-9.

Joshi, R.K., Mohanty, S., Subudhi, E. and Nayak, S. (2010). Isolation and characterization of resistance gene candidates in turmeric (*Curcuma longa* cv. Surama). *Genetics and Molecular Research*, **9(3)**: 1796-1806.

Joy, N., Abraham, Z. and Soniya, E.V. (2007). A preliminary assessment of genetic relationships among agronomically important cultivars of black pepper. *BMC Genetics*, **8(1)**: 42.

Joy, N., Prasanth, V.P. and Soniya, E.V. (2011). Microsatellite based analysis of genetic diversity of popular black pepper genotypes in South India. *Genetica*, **139(8)**: 1033-43.

Kavitha, P.G and Thomas, G. (2006). *Zingiber Zerumbet*, a potential donor for soft-rot resistance in ginger: *Genetic structure and functional genomics*. In: Extended Abstract, XVIII Kerala Science Congress. pp.169-171.

Kavitha, P.G and Thomas, G. (2008). Defence transcriptome profiling of *Zingiber zerumbet* (L.) smith by mRNA differential display. *Journal of Biosciences*, **33(1)**: 81-90.

Kress, W.J., Prince, L.M. and Williams, K.J. (2002). The phylogeny and a new classification of the gingers (Zingiberaceae): evidence from molecular data. *American Journal of Botany*, **89**: 1682-1696.

Kumar, A. and Anandaraj, M. (2006). Method for isolation of soil DNA and PCR based detection of ginger wilt pathogen, *Ralstonia solanacearum*, *Indian Phytopathology*, **59(2)**: 154-160.

Lebot, V., Aradhya, M.K and Manshardt, R.M. (1991). Geographic survey of genetic variation in kava (*Piper methysticum* Forst.f. and *P. wichmannii* C. DC.). *Pacific-Science*, **45(2)**: 169-185.

Lee, S.Y., Fai, W.K., Zakaria, M., Ibrahim, H., Othman, R.Y., Gwag, J.G., Rao, V.R. and Park, Y.J. (2007). Characterization of polymorphic microsatellite markers, isolated from ginger (*Zingiber officinale* Rosc.). *Molecular Ecology Notes*, **7(6)**: 1009-1011.

Leister, D., Ballvora, A., Salamini, F. and Gebhardt, C. (1996). A PCR-based approach for isolating pathogen resistance genes from potato with potential for wide application in plants. *Nature Genetics*, **14**: 421–428.

Liao, P.C., Gong, X., Shih, H.C. and Chiang, Y.C. (2009). Isolation and characterization of eleven polymorphic microsatellite loci from an endemic species, *Piper polysyphonum* (Piperaceae). *Conservation Genetics*, **10(6)**: 1911-1914.

Livingstone, K.D., Lackney, V.K., Blauth, J.R., van Wijk, R. and Jahn, M.K. (1999). Genoome mapping in *Capsicum* and the evolution of genome structure in the Solanaceae. *Genetics*, **152**: 1183-1202.

Lubinsky, P., Cameron, K.M., Molina, M.C., Wong, M., Lepers-Andrzejewski, S., Gómez-Pompa, A. and Kim, S.C. (2008). Neotropical roots of a Polynesian spice: the hybrid origin of tahitian vanilla, *Vanilla tahitensis* (Orchidaceae). *American Journal of Botany*, **95(8)**: 1-8.

Luo, Y., Sangha, J.S., Wang, S., Li, Z., Yang, J. and Yin, Z (2012). Marker-assisted breeding of *Xa4*, *Xa21* and *Xa27* in the restorer lines of hybrid rice for broad-spectrum and enhanced disease resistance to bacterial blight. *Molecular Breeding*, **30(4)**: 1601-1610.

Mani, T. and Manjula, S. (2010). Cloning and characterization of two osmotin isoforms from *Piper colubrinum*. *Biologia Plantarum*, **54(2)**: 377-380.

Manoj, P., Soniya, E.V., Banerjee, N.S. and Ravichandran, P. (2004). Recent studies on well-known spice, *Piper longum* Linn. *Natural Products Radiance*, **3(4)**: 222-227.

Menezes, I.C., Cidade, F.W., Souza, A.P. and Sampaio, I.C. (2009). Isolation and characterization of microsatellite loci in the black pepper, *Piper nigrum* L. (Piperaceae). *Conservation Genetics Resources,* **1**: 209-212.

Meyers, B.C., Dickerman, A.W., Michelmore, R.W., Sivaramakrishnan, S., Sobral, B.W. Young, N.D. (1999). Plant disease resistance genes encode members of an ancient and diverse protein family within the nucleotide-binding super family. *Plant Journal,* **20**: 317–332.

Minoo Divakaran, Jayakumar, V.N., Veena, S.S., Vimala, J., Basha, A., Saji, K.V., Babu, K.N. and Peter, K.V. (2008). Genetic variations and interrelationships in *Vanilla planifolia* and few related species as expressed by RAPD polymorphism. *Genetic Resources and Crop Evolution,* **55(3)**: 459-70.

Minoo Divakaran, Nirmal Babu, N., and Peter, K.V. (2016). Protocols for biotechnological interventions in improvement of vanilla (Vanilla planifolia Andrews.). In: *Protocols for In Vitro Cultures and Secondary Metabolite Analysis of Aromatic and Medicinal Plants,* Second Edition. pp. 47-63.

Minoo Divakaran, Nirmal Babu, K., Ravindran, P.N. and Peter, K.V. (2006). Interspecific hybridization in vanilla and molecular characterization of hybrids and selfed progenies using RAPD and AFLP markers. *Scientia Horticulturae,* **108(4)**: 414–422.

Nair, R.A. and Thomas, G. (2013). Molecular characterization of *ZzR1* resistance gene from *Zingiber zerumbet* with potential for imparting *Pythium aphanidermatum* resistance in ginger. *Gene,* **516**: 58-65.

Nair, R.A., Kiran, A.G., Sivakumar, K.C. and Thomas, G. (2010). Molecular characterization of an oomycete-responsive PR-5 protein gene from *Zingiber zerumbet*. *Plant Molecular Biology Reporter,* **28**: 128-135.

Nayak, S., Naik, P.K., Acharya, L.K. and Pattnaik, A.K. (2006). Detection and evaluation of genetic variation in 17 promising cultivars of turmeric (*Curcuma longa* L.) using 4C nuclear DNA content and RAPD markers. *Cytologia,* **71(1)**: 49-55.

Nazeem, P.A., Kesavachandran, R., Babu, T.D., Achuthan, C.R., Girija, D. and Peter, K.V. (2005). Assessment of genetic variability in black pepper (*Piper nigrum* L.) varieties through RAPD and AFLP analysis. In: *Proceeding of national symposium on Biotechnological interventions for improvement of horticultural crops:* Issues and Strategies. Thrissur, Kerala. pp. 226-228.

Nazeem, P.A., Achuthan, C.R., Babu, T.D., Parab, G.V., Girija, D., Keshavachandran, R. and Samiyappan, R. (2008). Expression of pathogenesis related proteins in black pepper (*Piper nigrum* L.) in relation to *Phytophthora* foot rot disease. *Journal of Tropical Agriculture,* **46(1-2)**: 45–51.

Ngamriabsakul, C., Newman, M.F. and Cronk, Q.C.B. (2003). The phylogeny of tribe *Zingibereae* (*Zingiberaceae*) based on its (nrDNA) and *trn*l–f (cpDNA) sequences. *Edinburgh Journal of Botany,* **60**: 483-507.

Nirmal Babu (2003). Black pepper biotechnology: A manual. In: Nirmal Babu, K., Nair, R.R. and Krishnamoorthy, B. (Eds.), Indian Instiute of Spices Research, Calicut, India. pp. 16-17.

Nirmal Babu, K., Divakaran, M., Pillai, G.S., Sumathi, V., Praveen, K., Raj, R.P., Akshita, H.J, Ravindran, P.N, and Peter, K.V. (2016a) Protocols for *in vitro* propagation, conservation, synthetic seed production, microrhizome production and molecular profiling in turmeric (*Curcuma longa* L.). In: *Protocols for In Vitro Cultures and Secondary Metabolite Analysis of Aromatic and Medicinal Plants*, Second Edition. pp. 387-401.

Nirmal Babu, K., George, J.K., Anandaraj, M., Venugopal, M.N., and Nair, R.R. (2005). Improvement of selected spices through Biotechnology tools – Black pepper, Cardamom, Ginger, Vanilla. Project report, Department of Biotechnology, Government of India. pp. 111

Nirmal Babu, K., Ravindran, P.N., Sasikumar, B. (2003a). Field evaluation of tissue cultured plants of spices and assessment of their genetic stability using molecular markers. Final Report, Department of Biotechnology, Government of India. pp. 94.

Nirmal Babu, K., Samsudeen, K., Divakaran, M., Pillai, G.S., Sumathi, V., Praveen, K. Ravindran, P.N. and Peter, K.V. (2016b). Protocols for *in vitro* propagation, conservation, synthetic seed production, embryo rescue, microrhizome production, molecular profiling, and genetic transformation in ginger (*Zingiber officinale* Roscoe.). In: *Protocols for In Vitro Cultures and Secondary Metabolite Analysis of Aromatic and Medicinal Plants*, Second Edition. pp. 403-426.

Nirmal Babu, K., Sasikumar, B., Anandaraj, M., George, J.K., Ravindran, P.N., Asha, S. and Praeetha, K. (2003b). Molecular characterization and preparation of molecular maps in black pepper. Final report submitted to National Agricultural Technology Project *(CGP)*, ICAR, New Delhi. p. 50.

Paisooksantivatana, Y., Kako, S. and Seko, H. (2001). Genetic diversity of *Curcuma alismatifolia* Gagnep. (Zingiberaceae) in Thailand as revealed by allozyme polymorphism. *Genetic Resources and Crop Evolution*, **48**: 459-465.

Patra, A.P., Mukherjee, A.K. and Acharya, L. (2011). Comparative study of RAPD and ISSR markers to assess the genetic diversity of betelvine (*Piper betle* L.) in Orissa, India. *American Journal of Biochemistry and Molecular Biology*, **1**: 200-211.

Peter, K.V., Ravindran, P.N., Nirmal Babu, K., Sasikumar, B., Minoo, D., Geetha, S.P. and Rajalakshmi, K. (2002). Establishing *in vitro* conservatory of spices germplasm. ICAR Project report. *Indian Institute of Spices Research,* Calicut, Kerala, India. pp. 131.

Philip, S., Banerjee, N.S. and Das, M.R. (2000). Genetic variation and micropropagation in three varieties of *Piper longum* L. *Current Science*, **78**: 169-173.

Philippe, M., Philippe, T., Jocelyne, O., Henri, L. and Nigel, G. (1998). Genetic mapping of Ph-2, a single locus controlling partial resistance to *Phytophthora infestans*. *Molecular Plant-Microbe Interactions*, **11(4)**: 259-269.

Pinheiro, J.B., Zucchi, M.I., Teles, F.L. and Ázara, N.A. (2003). Diversidade genética molecular em acessos de açafrão utilizando marcadores RAPD. *Acta Scientiarum*, **25**: 195- 199.

Pradeep Kumar, T., Karihaloo, J.L. and Archak, S. (2001). Molecular characterization of *Piper nigrum* cultivars using RAPD markers. *Current Science*, **8** (13): 246-248.

Pradeep Kumar, T., Karihaloo, J.L., Archak, S. and Baldev. A. (2003). Analysis of genetic diversity in *Piper nigrum* L. using RAPD markers. *Genetic Research of Crop Evolution*, **50**: 469-475.

Pulla Rao, V.V. (2003). *Molecular characterization of Garcinia using RAPD polymorphism*. M.Sc Dissertation, Nagarjuna University, Andhra Pradesh, India.

Ravindran, P.N., Nirmal Babu, K., Saji, K.V., Geetha, S.P., Praveen, K. and Yamuna, G. (2004). Conservation of spices genetic resources in *in vitro* gene banks. ICAR Project Report. *Indian Institute of Spices Research*, Calicut, Kerala, India. pp. 81.

Sasaki, Y., Fushimi, H. and Komatsu, K. (2004). Application of single-nucleotide polymorphism analysis of the *trnK* gene to the identification of *Curcuma* plants. *Biological and Pharmaceutical Bulletin*, **27(1)**: 144–146.

Sasaki, Y., Fushimi, H., Cao, H., Cai, S.Q. and Komatsu, K. (2002). Sequence analysis of Chinese and Japanese Curcuma drugs on the 18S rRNA gene and *trnK* gene and the application of amplification-refractory mutation system analysis for their authentication. *Biological and Pharmaceutical Bulletin*, **25(12)**: 1593-9.

Sasikumar, B., Chempakam, B., Johnson, G.K., Remashree, A.B., Devasahayam, S., Dhamayanthi, K.P.M., Ravindran, P.N. and Peter, K.V. (1999). Characterization of two interspecific hybrids of Piper. *Journal of Horticultural Science and Biotechnology*, **74**: 125-131.

Sasikumar, B., Zachariah, T.J., Syamkumar, S. and Remya, R. (2004). PCR based detection of adulteration in the market samples of turmeric powder. *Food Biotechnology*, **18(3)**: 299-306.

Sebastian, A., Sujatha, V.S., Neybe, E.V., Nair, G.S. and Augustine, A. (1996). Isoenzyme variation and species relationship in the genus Piper. *Journal of Tropical Agriculture*, **34(2)**: 85-92.

Sen, S., Skaria, R. and Muneer, P.A. (2010). Genetic diversity analysis in Piper species (Piperaceae) using RAPD markers. *Molecular Biotechnology*, **46(1)**: 72-79.

Shahanas, C.H., Syamkumar, S. and Sasikumar, B. (2003). Analysis of genetic fidelity in rooted cuttings of black pepper. In: Korikanthimath, V.S., John Zacharia, T., Nirmal Babu, K., Susheela Bhai, R. and Kandiannan, K. (Eds.), Proceedings of the *National Seminar on New perspectives in Spices, Medicinal and Aromatic Plants, Indian Society for Spices*, Calicut. pp. 9-14.

Sheeja, T.E., Rajesh, Y., Krishnamoorthy, B. and Parthasarathy, V.A. (2006). DNA polymorphism in clonal and seedling progenies of an elite nutmeg (*Myristica fragrans*) by RAPD. *Journal of Plantation Crops*, **34(3)**: 558-561.

Sheeja, T.E., Uma, G., Sasikumar, B., Saji, K.V. and Rahul, P.R. (2013). Genetic diversity study in *Piper* spp. using inter simple sequence repeat (ISSR) markers. *Journal of Spices and Aromatic Crops,* **22(2)**: 111-119.

Sheji Chandran, Smitha, K.S., Renu Susan George, Ishwara Bhat, A. and Anandaraj, M. (2006). Development of SCAR marker for locating *Phytophthora* resistance in black pepper (*Piper nigrum* L). *Indian Phytopathological Society*, Southern Zone Meeting, Kasargod. pp. 24.

Shen, K.A., Meyers, B.C., Nurul Islam Faridi, M., Chin, D.B., Stelly, D.M. and Michelmore, R.W. (1998). Resistance gene candidates identifed by PCR with degenerate oligonucleotide primers map to clusters of resistance genes in lettuce. *Molecular Plant-Microbe Interactions,* **11**: 815–823.

Shibu, M.P., Ravishankar, K.V., Lalitha, A., Ganeshaiah, K.N. and Shanker, R.U. (2000). Identification of sex specific DNA markers in the dioecious tree, nutmeg (*Myristica fragrans* Houtt.). *Plant Genetic Resources Newsletter,* **121**: 59-61.

Siju, S., Dhanya, K., Syamkumar, S., Sheeja, T.E., Sasikumar, B., Bhat, A.I. and Parthasarathy, V.A. (2010). Development, characterization and utilization of genomic microsatellite markers in turmeric (*Curcuma longa* L.). *Biochemical Systematics and Ecology,* **38(4)**: 641-646.

Sreedevi, M., Syamkumar, S. and Sasikumar, B. (2005). Molecular and morphological characterization of new promising black pepper (*Piper nigrum* L.) lines. *Journal of Spices Aromatic Crops,* **14**: 1-9.

Sreedhar, R.V., Venkatachalam, L. and Bhagyalakshmi, N. (2007). Genetic fidelity of long-term micropropagated shoot cultures of vanilla (*Vanilla planifolia* Andrews) as assessed by molecular markers. *Biotechnology Journal,* **2(8)**: 1007-13.

Sreeja, S.G. (2002). Molecular characterization of *Curcuma* species using RAPD markers. *MSc Thesis, Periyar University,* Tamil Nadu, India. pp. 20.

Sujatha, R., Dash, P.K. and Koundal, K.R. (2005). Identification of plant sources for insect resistance genes using heterologous probes. In: *Proceedings of Seventeenth Kerala Science Congress*. pp. 78-80.

Swetha Priya, R. and Subramanian, R.B. (2008). Isolation and molecular analysis of R-gene in resistant *Zingiber officinale* (ginger) varieties against *Fusarium oxysporum f.* sp. *zingiberi. Bioresource Technology,* **99(11)**: 4540-4543.

Syamkumar, S. (2008). Molecular, biochemical and morphological characterisation of selected *Curcuma* accessions. Unpublished Ph.D Thesis, University of Calicut, Kerala.

Syamkumar, S., Mridula, J. and Sasikumar, B. (2005). Isolation and PCR amplification of genomic DNA from dried capsules of cardamom (*Elettaria cardamom* M.). *Journal of Plant Molecular Biology Reporter,* **23**: 417a-417e.

Takao, K., Shanmukhaswami, S.S., Jinrui, S., Mark, G., Richard, I.B. and Madan, K.B. (1997). High resolution genetic and physical mapping of molecular markers linked to the *Phytophthora* resistants gene *Rps 1*-k in soybean. *Molecular Plant-Microbe Interactions,* **10(9)**: 1035-1044.

Tanksley, S.D., Bernatzky, R., Lapitan, N.L. and Prince, J.P. (1988). Conservation of gene repertoire but not gene order in pepper and tomato. In: *Proceedings of National Academy of Science*. USA. 85. pp. 6419–6423.

Trindade, H. (2010). Molecular biology of aromatic plants and spices. A review. *Flavour and Fragrance Journal*, **25(5)**: 272-281.

Trindade, H., Costa, M.M., Sofia, B.L.A., Pedro, L.G., Figueiredo, A.C. and Barroso, J.G. (2008). Genetic diversity and chemical polymorphism of *Thymus caespititius* from Pico, São Jorge and Terceira islands (Azores). *Biochemical Systematics and Ecology*, **36(10)**: 790-797.

Varma, S.R., Johnson, G.K., Balaji, S. and Parthasarathy, V.A. (2009). Diûerential induction of chitinase in *Piper colubrinum* in response to inoculation with *Phytophthora capsici* the cause of foot rot in black pepper. *Saudi Journal of Biological Sciences*, **16**: 11–16.

Wadt, L., Hde, O., Ehringhaus, C. and Kageyama, P.Y. (2004). Genetic diversity of "*Pimenta Longa*" genotypes (*Piper* spp., Piperaceae) of the Embrapa Acre germplasm collection. *Genetics and Molecular Biology*, **27(1)**: 74-82.

Wahyuni, S., Xu, D.H., Bermawie, N., Tsunematsu, H. and Ban, T. (2003). Genetic relationships among ginger accessions based on AFLP marker. *Journal of Bioteknologi Pertanian*, **8**: 60–68.

Widiez, T., Hartman, T.G., Dudai, N., Yan, Q., Lawton, M., Havkin-Frenkel, D. and Belanger, F.C. (2011). Functional characterization of two new members of the caffeoyl CoA *O*-methyltransferase-like gene family from *Vanilla planifolia* reveals a new class of plastid-localized *O*-methyltransferases. *Plant Molecular Biology*, **76(6)**: 475-488.

Xia, Q., Zhao, K.J., Huang, Z.G., Zang, P., Dong, T.T., Li, S.P. and Tsim, K.W. (2005). Molecular genetics and chemical assessment of *Rhizoma Curcumae* in China. *Journal of Agricultural and Food Chemistry*, **53(15)**: 6019-26.

Xiaolan Rao, Krom, N., Tang, Y., Widiez, T., Havkin-Frenkel, D., Belanger, F.C., Dixon, R.A. and Chen, F.A. (2014). A deep transcriptomic analysis of pod development in the vanilla orchid (*Vanilla planifolia*). *BMC Genomics*, **15(1)**: 1.

Yoshida, N.C., Lima, P.F., Priolli, R.H., Kato, M.J. and Colombo, C.A. (2014). Isolation and characterization of nine polymorphic microsatellite loci in *Piper solmsianum* (Piperaceae). *Applications in Plant Sciences*, **2(4)**.

Chapter 16

Cocoa

☆ *Lambert A. Motilal, Antoinette Sankar, David Gopaulchan and Pathmanathan Umaharan*

1. Introduction

Theobroma cacao L. (chocolate tree; 2n = 2x = 20), is an outbreeding understory tree, commonly referred to as cacao or cocoa, although the latter has been traditionally referred to the processed products of the cacao tree. Cacao was used by the early indigenous peoples of South and Central America more than 3,000 years ago, including the Olmecs, Mayan and later Aztec civilizations in Central America (Young, 1994; Coe and Coe, 1996; Henderson *et al.*, 2007; Powis *et al.*, 2011). The objects of these early exploits, referred to in the literature as "Criollo" cacao, were once taxonomically regarded as a separate species, *T. cacao* ssp. *cacao* (Cuatrecasas, 1964). Although studies have confirmed the Amazon basin as the centre of diversity of cacao and home to the 'Forastero' cacao (Cheesman, 1944; Motamayor et al., 2008; Thomas et al., 2012), there has been little understanding of diversity or phylogenetic relationships between geographical variants or populations. Further, the origin of 'Trinitario' cacao, a hybrid population between Forastero and Criollo and the basis for the fine/flavour cacao industry, has been mired in mystery. The contribution of molecular profiling of cacao has been tremendous, over the past four decades, on reinforcing or refining anthropological studies, resolving taxonomic issues, defining populations, understanding the origin of 'Trinitario' cacao and contributing to the overall understanding of the evolution of cacao.

Cacao has a little understood breeding system enforced by a complex multi-allelic incompatibility system encompassing features of both the sporophytic and gametophytic systems resulting in potential cross-compatible, incompatible and self-compatible reactions depending on the maternal tree and surrounding pollen donors (Knight and Rogers, 1953, 1955; Cope, 1958, 1962; Baker *et al.*, 1997; Ford and

Wilkinson, 2012). Highly inbred landraces also exist in Central and South America, referred to as 'Criollo' or 'Amelonado' or 'Nacional' (Bartley, 2005; Motamayor *et al.*, 2008; Loor Solorzano *et al.*, 2012). With the onset of European colonisation of the Americas, cacao was transported out of the South American mainland into the Caribbean Islands, West Africa, Asia and the Pacific islands (Wood, 1991; Lockwood and End, 1993; Bartley, 2005). Many of these introductions originated from the same geographic area or were descended from seedlings or seeds of few fruits that were transported in oceanic voyages. Plant material conveyed to and established at one site, often served as the secondary focal point for distribution to other distant areas. This, and the fact that a limited number of parents have been used in breeding programmes, has resulted in cultivated cacao possessing a narrow genetic base. The exploitation of genetic diversity in plant breeding has been hindered by poor understanding of the breeding system of cacao, the absence of reliable genetic information on the inheritance of traits and limited genetic markers to support breeding.

Cacao encompasses wild, semi-cultivated and cultivated varieties and the current form of the species is relatively unchanged from the *T. cacao* in the centre of diversity. Due to its recalcitrance and its predominantly outcrossing nature arising from its complex incompatibility system, cacao genetic resources are maintained as living collections in field genebanks (Figure 16.1). Cacao germplasm has been named according to the country, farm, collecting expedition, river system and research station (Lockwood and Gyamfi, 1979; Turnbull and Hadley, 2012). Cacao genetic resources are maintained in over 50 collections worldwide (Motilal and Butler, 2003) with two Universal Collections – the Centro Agronómico Tropical de Investigación y Enseñanza in Costa Rica and the International Cocoa Genebank Trinidad (ICGT) in Trinidad and Tobago. Information on cocoa varieties, or accessions as they are called in the germplasm collections, held in the worldwide collections with accessible records are contained in the International Cocoa Germplasm Database (ICGD) maintained by the University of Reading (Turnbull and Hadley, 2012). Details on accession nomenclature can be found in Lockwood and Gyamfi (1979), Iwaro *et al.* (2003), Bekele *et al.* (2006) and Turnbull and Hadley (2012). In the ICGD, there are about 29,500 accession names of which there are 13,000 synonyms (Turnbull and Hadley, 2012) giving approximately 16,500 putatively unique accessions. Accession nomenclature takes the form of an alphanumeric system where the lettered prefix indicates the accession group and the numeric suffix indicates the fruit, budwood or seed source. Related cacao accessions may therefore be identical to another accession, be within the same family structure as full-sibs or half-sibs, belong to the same accession group or belong to the same population. Cacao accessions may carry the same name but may be genetically distinct from each other, because the same name was unknowingly applied more than once to different germplasm collected in different expeditions; seed-derived descendants from a mother plant were given the same name as the mother plant; or from errors in recording during collection expeditions and reporting the accession name in the scientific literature and germplasm documentation.

Efforts to conserve cacao genetic resources, as repositories for safeguarding the livelihoods of smallholder farmers and cacao businesses, have been hindered by a poor understanding of the diversity, genetic structure and phylogenetic relationships among geographical populations. A more comprehensive understanding of the structure and diversity of cocoa is not only critical to understanding the gaps within the two Universal Collections of cacao germplasm, but is also important to improve the effectiveness and efficiency of collections through establishment of core and minicore collections. Poor fidelity in collections, seed gardens and clonal gardens is a hindrance to the efficiency of breeding programmes and also contributes to the poor quality of planting material supplied to farmers. Molecular markers play an important role in curation, characterisation, augmentation, utilisation and distribution of cacao germplasm. Multilocus molecular profiles can be used for identity analysis, genetic relatedness, genetic diversity, pedigree analysis, phylogenetic assessment and genetic ancestry.

The chapter traces the major historical advances in molecular profiling of cocoa, the current state of the art that has led to refining taxonomical relationships, improving the understanding of the evolution of diversity, its structure and phylogenetic relationships among populations, advancing the conservation and exploitation of genetic resources in breeding, enhancing the deployment of quality planting material in farmer fields, and future possibilities.

2. History of Molecular Profiling in Cocoa

Fruit and seed traits, in conjunction with geographic distribution, were used to classify cocoa into two groups of Criollo and Forastero (van Hall, 1914; Cheesman, 1944; Cuatrecasas, 1964) which were thought to be two separate subspecies *T. cacao* and *T. sphaerocarpum* respectively (Cuatrecasas, 1964). The Forastero group was further partitioned, based on fruit dimension and basal constriction, into Angoleta, Cundeamor, Amelonado and Calabacillo forms (Toxopeus, 1985). However, the limitations and ambiguity of this system have been recognised and a more systematic description of the phenotypic states (Enriquez and Soria, 1966, 1967; Engels *et al.*, 1980; Engels, 1983a,b), as used in most other crops, was instituted using heritable phenotypic features of leaf pubescence and colour; pod shape, features and colour; seed morphology and floral morphology. Phenetic dendrograms have been developed for cacao (Bekele and Bekele, 1996; Aikpokpodion, 2010; Bekele *et al.*, 1994; Bekele *et al.*, 2008a; Bekele *et al.*, 2008b; Maharaj *et al.*, 2011). However, the collection of phenotypic data is time-consuming and burdensome, particularly for reproductive traits in tree crops, phenotypic plasticity, and inconsistent scoring primarily due to improper or insufficient training of data collectors and the influence of modifying factors on trait expression. The use of biochemical and DNA markers, therefore supplanted phenotype characterisation in understanding taxonomic relationships. These molecular markers were heritable, more numerous, less susceptible to environmental vagaries and allowed for a more consistent, reliable, repeatable and reproducible study of genetic variation within any species.

Figure 16.1: Cacao Germplasm, in the International Cocoa Genebank Trinidad, Maintained as Living Trees.

A: Propagated accession with multiple trunks. The genotyped accession (note blue label on central trunk) was sampled. If the trunks are identical to each other, fruits can be used from all trunks for this accession; B: An accession with green fruits which become yellow upon ripening; C: An accession with reddish fruits that will become orange-yellow upon ripening.

Note that the fruit forms in A, B and C are different from each other indicating that these are different accessions. DNA fingerprinting allows for the unique identification of these accessions and provides an estimate of the relatedness of the accessions (Photographs courtesy of Lambert A. Motilal).

2.1. Protein, Allozyme and Isozyme Polymorphism

Protein banding patterns (Kaosiri and Zentmyer, 1980; Erselius and De Vallavieille, 1984; Chowdappa and Chandramohanan, 1995) and isoenzymes (Lanaud and Berthaud, 1985; Atkinson *et al.*, 1986; Lanaud, 1986; Yidana *et al.*, 1987; Oudemans and Coffey, 1991 a,b,c; Oudemans *et al.*, 1994) were the first to be used in molecular profiling in cacao. Allozymes are different forms of the same enzyme that may differ in electrophoretic mobility and can be separated in starch gels, stained and visualised. Following the early work on isozymes studies by Lanaud and Berthaud (1985), Atkinson *et al.* (1986), Lanaud (1986) and Yidana *et al.* (1987), attempted to characterise the genetic diversity held at the ICGT started at the Cocoa Research Centre, then Cocoa Research Unit, in 1984, using the isoperoxidase system although other enzymes including acid phosphatase were also investigated (Yidana and Kennedy, 1984-86). Later, other enzyme systems were incorporated to characterise cacao populations at the Cocoa Research Centre (Warren, 1994). Sounigo *et al.* (2005) reported the classification of 459 cacao accessions from 26 accession groups with isoenzymes and provided evidence for the separation of wild and cultivated types. Warren *et al.* (1995) found that the acid phosphatase and isocitrate dehydrogenase loci were linked to one of the loci in the incompatibility complex of cacao. Although isozymes provided for robust markers capable of being assayed early in the life time of cacao, there were concerns about the number of available markers and their repeatability, particularly across different laboratories.

2.2. Restriction Fragment Length Polymorphism

The use of DNA fingerprinting assays to analyse a plant genome was first reported by Ryskov *et al.* (1988) using restriction fragment length polymorphism (RFLP). The procedure involved digesting the DNA with a restriction enzyme to generate fragments of different sizes, separating the fragments using gel electrophoresis and transferring to a nitrocellulose or nylon membrane. The membrane is then probed with a radioactively labelled DNA fragment and the pattern visualised through autoradiography. Kurt *et al.* (1989) later used synthetic oligonucleotides that targeted simple, repetitive DNA sequences to generate hypervariable DNA fingerprinting profiles in barley and chickpea (*Cicer arietinum*). The technique, however, needed large amounts of DNA. In cacao, RFLPs were used to study the genetic diversity (Laurent *et al.*, 1993, 1994; Figueira *et al.*, 1994; N'Goran *et al.*, 1994; Lerceteau *et al.*, 1997), and later used to create the first genetic maps (Lanaud *et al.*, 1995; Crouzillat *et al.*, 1996). There is considerable support in these studies for the presence of greater diversity in South America and the Upper Amazon Basin, in particular, and the notion of this region as the centre of diversity for cacao. The markers could also be used to separate Criollo from Forastero types.

2.3. Random Amplified Polymorphic DNA

The field of DNA fingerprinting was transformed by the emergence of polymerase chain reaction (PCR)-based technology. Arbitrary primer sequences were used to amplify anonymous segments of genomic DNA to produce polymorphic banding patterns (Williams *et al.*, 1990; Caetano-Anolles and Brant,

1991; Welsh and McClelland, 1990). The assay published by Williams *et al.* (1990). based on the amplification of random segments in the genome with single primers of arbitrary nucleotide sequence called random amplified polymorphic DNA (RAPD), has since been reported in numerous studies. Unlike RFLP, the technique was quick and simple as it did not include blotting or hybridising steps and required only small amounts of DNA. Additionally, there was no requirement for DNA probes or sequence information for primer design as the primers detected polymorphisms in the absence of specific nucleotide sequence information.

The RAPD system was quickly adopted by cacao scientists for fingerprinting purposes with Wilde *et al.* (1992) being the first to report RAPD polymorphism in cacao. These authors assessed 14 primers on 13 genotypes, inclusive of two *Herrania* and one *T. microcarpum* L. individuals. One primer was identified that separated all the individuals. The marker system was adopted by CRC in 1993 and the reproducibility of the system was investigated by Christopher (1993) who used 11 accessions in common to that of Russell *et al.* (1993). Christopher and Sounigo (1995) screened 150 decameric RAPD primers and identified 18 promising primers from which five were selected for identity and cluster analysis. These five primers yielded 17 polymorphic loci, with 1-7 loci per primer and were used to separate 27 of 47 accessions. These authors therefore recommended additional primers (and hence more loci) for identity analysis. Subsequently, with refinement of the technique, RAPDs were used for linkage maps, identity analysis and genetic diversity studies in cacao (Figueira *et al.*, 1994; N'Goran *et al.*, 1994; Lerceteau *et al.*, 1997; Whitkus *et al.*, 1998; Kasran and Subali, 2002; Lanaud *et al.*, 2004; Sounigo *et al.*, 2005; Leal *et al.*, 2008). Analysis of RAPD profiles provided support for unique diversity in southern Mexico (Whitkus *et al.*, 1998), separation of dwarf and vigorous clones (Kasran and Subali, 2002), the distinction of Criollo from Forastero (Figueira *et al.*, 1994; N'Goran *et al.*, 1994; Lerceteau *et al.*, 1997), and the separation of 22 accession groups from each other inclusive of wild and improved cultivars (Sounigo *et al.*, 2005). However, the repeatability and reproducibility of the technique was greatly affected by reaction conditions and comparison of DNA fingerprinting profiles within and among laboratories was severely affected.

2.4. Amplified Fragment Length Polymorphism

Another approach to using PCR in DNA fingerprinting was the development of the amplified fragment length polymorphism (AFLP) assay (Vos *et al.*, 1995). The procedure involved digesting the DNA with a restriction enzyme mixture, ligating oligonucleotide adapters, amplifying the restriction fragments and pre-selective amplification followed by selective-PCR. With this method, a high number of polymorphic bands could be produced and highly reproducible unique DNA fingerprinting patterns generated. Nevertheless, the assay is technically challenging and a larger quantity of DNA than RAPDs is required. The first reported AFLP study in cacao (Perry *et al.*, 1998) reported inter- and intra- varietal differences even among phenotypically similar cacao varieties. AFLPs were mainly used in cocoa to saturate linkage maps (Risterucci *et al.*, 2000; Clément *et al.*, 2003a, b; Quieroz *et al.*, 2003; Lanaud *et al.*, 2004). However, AFLPs have not been adopted for cultivar identification or study of genetic diversity at any of the cocoa genebanks worldwide.

2.5. Inter-simple Sequence Repeat Polymorphism

Inter-simple sequence repeat (ISSR) –PCR, developed by Zietkiewicz *et al.* (1994), is another PCR-based DNA profiling assay. The technique is used to detect the variation in length between two simple sequence repeats (SSRs, microsatellites) in the genome. SSRs are polymorphic loci consisting of multiple tandem repeating sequences usually two to six base pairs in length. The target of an ISSR assay is any DNA segment located within an amplifiable distance between two identical SSR regions oriented in opposite direction. The ISSR markers are mostly inherited as dominant alleles although co-dominance has been reported in maize (Gupta *et al.*, 1994) and citrus (Sankar and Moore, 2001). Although ISSRs were recommended by Charters and Wilkinson (2000)for fingerprinting cacao, its use as a tool for cacao profiling was limited because of the speed of success in the cloning and isolation of microsatellite sequences. A common disadvantage of the RAPD, AFLP and ISSR techniques is the difficulty to determine whether an amplified DNA segment is heterozygous or homozygous at a particular locus. As a result, the marker alleles generated by these assays are dominant which reduces their information content. This limitation was overcome when cloning and sequencing of microsatellite sequences led to SSR markers that were co-dominant in their inheritance.

2.6. Simple Sequence Repeat Polymorphism

Simple Sequence Repeat (SSR) assays, in contrast to the previous markers, provide a combination of locus-specific, co-dominantly inherited bands with high levels of polymorphism. The amplified regions have been termed sequence tagged microsatellite sites (a variant of a sequence tagged site) because the microsatellite is effectively tagged in the genome through the design of a forward and a reverse primer from the unique regions flanking the microsatellite. Microsatellite mutation rate in the genome has been reported to be as high as 10^{-2} per generation (Weber and Wong, 1993; Li *et al.*, 2002), often through alteration of repeat number (Li *et al.*, 2002). Repeat number changes have been theorised to occur through DNA slippage during replication or asymmetric recombination between DNA strands (Tachida and Iizuka, 1992). These polymorphisms have been shown to be co-dominantly inherited, and are therefore valuable for DNA typing and other genetic profiling applications. Another type of marker, SSR-RFLP, combined the power of RFLP with the simplicity of PCR using primers designed around microsatellite-containing regions in the genome. Attempts had been made to evaluate the potential of SSR-RFLP for use in developing country laboratories (P. Umaharan, pers comm). The concept involved the amplification of a fragment, surrounding an SSR locus, long enough to allow subsequent digestion using restriction enzymes, to reveal alleles based on variations in restriction fragment length through gel electrophoresis. However, this method has not been successfully adopted.

The SSR-PCR product size can be predicted based upon the sequence information and used to confirm successful amplification. Sizing of the amplified fragments/alleles can be done manually with size standards or using a high throughput sequencing system for automated sizing. PCR amplification of SSR length polymorphisms in plants was first reported by Akkaya *et al.* (1992) using soybean (*Glycine max*) DNA. Primers that targeted the DNA sequences flanking hypervariable SSR regions in the genome were used in the assay. A limitation of this technique is that species-specific primer pairs are required to produce SSR markers. The development of such primer assays can be time consuming and costly, although assays developed for a species may be transferable across different taxonomic levels depending on the SSR loci and organism involved (Rosetto, 2001; Scott, 2001; Motilal, 2004a).

Cacao-based SSRs were first developed by Lanaud *et al.* (1999). Argout *et al.* (2008) later identified 2252 SSRs from 149,650 ESTs derived from a transcriptome set of 56 cDNA over different cacao organs and environmental conditions. Dimeric to hexameric pure SSRs and compound SSRs were identified with the dimeric and trimeric SSRs accounting for 88 per cent of all SSRs identified and with the poly$(AG)_n$ and poly$(AAG)_n$ motifs being most abundant in the unigenes (Argout *et al.*, 2008).

Variable numbers of SSR loci have been employed in identity and genetic diversity studies. A set of 15 SSR primer pairs (Saunders *et al.*, 2004) were recommended for resolution of identity issues. Cryer *et al.* (2006) later reported on the use of reference genotypes and allelic size standards to unify allele calls. However, ascertainment bias in sample selection affects SSR informativeness (Johnson *et al.*, 2009; Motilal *et al.*, 2009) which may limit the utility of these 15 loci. Errors in SSR typing arise from allele drop out and false alleles and these error rates were locus dependent (Zhang *et al.*, 2006b). Repeat typing was recommended to obtain reliable consensus genotypes (Zhang *et al.*, 2006b).

To date, SSR-PCR has been used for off-types and cacao clone identification (Figueira, 1998; Risterucci *et al.*, 2000, 2001; Motilal, 2004b; Schnell *et al.*, 2004; Cryer *et al.*, 2006; Motilal *et al.*, 2009, 2011; Zhang *et al.*, 2006a,b, 2007, 2009a,b). The combined probability of identity among siblings (PID_{SIB}) is the probability that two sibling individuals drawn at random from a population have identical genotypes (Evett and Weir, 1998; Waits *et al.*, 2001). The PID_{SIB} was recommended to be used in assigning identity match declarations (Zhang *et al.*, 2009a).

Microsatellite polymorphisms have also been used in linkage map generation (Lanaud *et al.*, 2004; Pugh *et al.*, 2004) and for tagging genes for quantitative trait loci towards marker assisted selection (Clement *et al.*, 2003a,b; Brown *et al.*, 2005, 2007; Schnell *et al.*, 2005, 2007a; Lanaud *et al.*, 2009; Marcano *et al.*, 2009). The utility of SSRs in parentage analysis (Schnell *et al.*, 2005) describing genetic diversity and in assessing population ancestry is well known (Lanaud *et al.*, 1999, 2001; Opoku *et al.*, 2007; Motamayor *et al.*, 2008; Marcano *et al.*, 2009; Zhang *et al.*, 2006a,b, 2007, 2009a,b, 2011; Irish *et al.*, 2010; Motilal *et al.*, 2010, 2011, 2012, 2013; Susilo *et al.*, 2011; Trognitz *et al.*, 2011, 2013; Thomas *et al.*, 2012; Dinarti *et al.*, 2015; Santos *et al.*, 2015).

2.7. Single Nucleotide Polymorphism

Single nucleotide polymorphisms (SNPs) are variations at individual nucleotides at specific positions in the genome. SNPs are reportedly the most abundant class of polymorphisms in genomes including the plant genome (Buckler and Thornsberry, 2002; Brumfield *et al.*, 2003). Density estimates have yielded one SNP per 170 bp in the rice genome (Yu *et al.*, 2002), 1900 bp in humans (The International SNP Map Working Group, 2001), 3300 bp in *Arabidopsis thaliana* (The Arabidopsis Genome Initiative 2000), 20,500-24,600 bp in chickpea (Kujur *et al.*, 2015). SNPs can potentially exist at any position throughout the genome, including coding, non-coding and inter-genic regions and hence offers the greatest promise for molecular profiling. Besides its abundance, SNPs present the advantage of being identified without the need for electrophoresis and hence size homoplasy, platform issues and binning problems are avoided. Genotyping error rates are low (less than 1 in 2000; Ranade *et al.*, 2001) and SNP genotypes can be scored with minimal human intervention. Furthermore, similar to SSRs, SNPs are co-dominant but, unlike SSRs, are biallelic rather than multiallelic. However, triallelic states in humans (Hüebner *et al.*, 2007; Hodgkinson and Eyre-Walker, 2010; Jenkins *et al.*, 2014) or multiallelic SNPs (Jenkins *et al.*, 2014) due to copy number variation (MacConaill *et al.*, 2007) have been reported. Multiallelic SNPs have been attributed to sequencing errors (Beissinger *et al.*, 2014). The codominant biallelic state for SNPs will be discussed in this chapter. Three states – homozygous 1, homozygous 2 and the heterozygous condition can therefore be detected by a single biallelic SNP. For example the genotypes AA, GG and AG will be detected using a SNP with biallelic states A/G. Due to this level of detection, SNPs have limited resolving power per locus compared to microsatellites (Glaubitz *et al.*, 2003). However, when the number of sites is factored in, the overall resolving power of SNPs may be comparable to that of other DNA markers. SNPs have therefore emerged as the next generation of molecular markers for species identification and genetic diversity measurements.

SNPs have been used for diverse applications in crops including providing evidence for selection (Beissinger *et al.*, 2014); varietal identification (Ganal *et al.*, 2009); describing population genetic structure (Schmid *et al.*, 2003; Kujur *et al.*, 2015); tagging genes or quantitative trait loci in rice (Konishi *et al.*, 2006), maize (Buckler *et al.*, 2009) and oil palm (Pootakham *et al.*, 2015); generating linkage maps in many crops including *Arabidopsis* (Cho *et al.*, 1999), rice (Nasu *et al.*, 2002; Feltus *et al.*, 2004; Shen *et al.*, 2004) and oil palm (Pootakham *et al.*, 2015); and for genome wide association studies (GWAS) (Bélo *et al.*, 2008; Pajerowska-Mukhtar *et al.*, 2009).

The earliest SNP studies in cacao are those of Borrone *et al.* (2004) and Kuhn *et al.* (2005). These markers in cacao have been derived from single strand conformation polymorphism (Kuhn *et al.*, 2005; Livingstone *et al.*, 2011); expressed sequenced tags (Lima *et al.*, 2009; Allegre *et al.*, 2012) which could come from conserved ortholog set sequences (Kuhn *et al.*, 2012); from microsatellites (Dadzie *et al.*, 2013) using genome walking (Parker *et al.*, 1991) and from RNAseq data (Livingstone *et al.*, 2015). Argout *et al.* (2008) identified 5246 SNPs which were distributed as transitions (A/T - G/C; 54.2 per cent), transversions (32.1 per cent) and indels (13.7 per cent) in their EST database. Subsequently, two cacao genomes were mapped: one a Belizean Criollo cultivar B97-61/B2 (Argout *et al.*, 2011) and another from the Amazonian MATINA 1/6 (Motamayor *et al.*, 2013). These genomes were used to identify numerous SNPs (Allegre *et al.*, 2012). SNPs in cocoa, have been used to generate linkage maps (Allegre *et al.*, 2012; Livingstone *et al.*, 2015), tag genes (Borrone *et al.*, 2004; Lima *et al.*, 2009; Motamayor *et al.*, 2013), determine parentage (Ji *et al.*, 2013; Takrama *et al.*, 2014), evaluate genetic diversity (Ji *et al.*, 2013) and identify hybrids and varieties (Kuhn *et al.*, 2010; Livingstone *et al.*, 2011; Takrama *et al.*, 2012, 2014; Fang *et al.*, 2014).

3. Development of a Global Reference SNP Panel for Resolving Identities in Cacao

Notwithstanding the ongoing cacao SNP studies, neither a logical methodology for selecting SNPs for resolution of identities has been developed nor has there been an attempt to use such an approach to identify a global reference SNP panel for resolution of identities. SNPs were reportedly selected based on their level of polymorphism and to represent all ten chromosomes. Polymorphism estimates, however, have an ascertainment bias as it is affected by the number of samples and the composition of the samples. For instance, even monomorphic SNPs (Kuhn *et al.*, 2010; Livingstone *et al.*, 2011; Allegre *et al.*, 2012; Ji *et al.*, 2013) can be deemed polymorphic depending on the set of cacao plants screened (Livingstone *et al.*, 2011). In this chapter, we present a global reference SNP panel selected based on screening a subset of germplasm from the ICGT for their ability to discriminate the genetic variability for cacao.

The ICGD contains SNP data on 603 accessions from 160 loci (Turnbull and Hadley, 2012) of which 54 bears the "Tc" prefix similar to that of Ji *et al.* (2013) and Takrama *et al.* (2014). These TcSNPs were derived from a set of over 2000 SNPs by CIRAD from Illumina GoldenGate Assays as described by Allegre *et al.* (2012) and prioritised by Michel Boccara at CRC, Trinidad (Michel Boccara, personal communication). Fang *et al.* (2014) employed 44 loci which had the capacity to discriminate amongst the 160 individuals studied but could not resolve the SCA/Ucayali accessions from the MO accessions into their respective population groups of Contamana and Nacional. Ji *et al.* (2013) and Livingstone *et al.* (2015) have suggested that 26 and 30 SNP loci respectively would be sufficient for SNP profiling.

The multilocus profiles of 546 SNPs from 81 cocoa DNA samples were obtained from Illumina GoldenGate Assays as described by Allegre *et al.* (2012). The data was used to obtain estimates of H_e and PIC using PowerMarker (Liu and Muse, 2005), as well as, estimates of simulated power ($simPW_R$) and informativeness for relatedness (I_r) using the program KinInfor v1 (Wang, 2006). Using these outputs and the location of SNPs on the linkage groups (http://cocoagendb.cirad.fr/), different datasets (Table 16.1) were compiled to assess the choice of SNPs on resolution of tree identities.

As was demonstrated for SSR loci, the composition of the panel of polymorphic loci is important and all the members do not need to be the most informative (Motilal *et al.*, 2009). The SNP loci should be selected as a set based on the capacity to discriminate among all individuals. Although, at least 30 loci could give at least 95 per cent resolution, the number of near matches across the different SNP combinations indicate that these combinations would inflate the relatedness of distinct accessions. Relatedness becomes overestimated as the incidence of missing data is increased. If by chance, missing data occurs at the few sites that differentiate the near misses, then even with a reasonably large number (100) of SNPs, distinct but closely related accessions would be deemed equivalent.

Table 16.1: Ascertainment of SNP Selection on Identity Resolution of 81 Cacao Accessions

SNP Selection	*Number of SNPs*	*Resolution(%)*	*Number of Sample Pairs Matching at all but n loci*	*Number of Synonymous Groups*	PID_{SIB}
Random 15	15	90.1	n = 1: 9 n = 2: 33	3 sets: 2 doublets and 1 quadruplet	2.6×10^{-3}
Random 30	30	96.3	n = 1: 6 n = 2: 4	1 triplet	1.5×10^{-5}
Random 45	45	95.1	n = 1: 0 n = 2: 4	1 quadruplet	1.0×10^{-6}
Random 60	60	96.3	n = 1: 3 n = 2: 0	1 triplet	7.6×10^{-10}
Random 75	75	100	n = 1: 4 n = 2: 4	none	1.0×10^{-9}
Random 100	100	96.3	n = 1: 0 n = 2: 3	1 triplet	9.1×10^{-13}
Chromosome 1	92	100	n = 1: 4 n = 2: 2	none	9.9×10^{-12}
Chromosomes 2 and 3	112	97.5	n = 1: 0 n = 2: 3	1 doublet	1.6×10^{-15}
Chromosomes 2 and 8	92	96.3	n = 1: 3 n = 2: 0	1 triplet	2.0×10^{-13}
Chromosomes 4 and 5	112	100	n = 1: 2 n = 2: 1	none	7.0×10^{-16}
Chromosomes 6 and 9	100	97.5	n = 1: 4 n = 2: 5	1 doublet	6.6×10^{-13}
Chromosomes 6, 7, 8, & 10	99	97.5	n = 1: 3 n = 2: 2	1 doublet	1.7×10^{-14}
Ten SNPs distributed per chromosomen = 2: 0	100	95.1	n = 1: 4	1 quadruplet	2.3×10^{-13}
SNPs at every 7.5 cM for each chromosome	100	97.5	n = 1: 0 n = 2: 5	1 doublet	1.6×10^{-13}
[1]Top 100 by H_e	100	97.5	n = 1: 3 n = 2: 3	1 doublet	1.1×10^{-22}
[1]Top 100 by PIC	100	96.3	n = 1: 0 n = 2: 3	1 triplet	2.8×10^{-23}
[2]Top 100 by $simPW_R$	100	100	n = 1: 0 n = 2: 1	None	1.4×10^{-17}
[2]Top 100 by I_r	100	96.3	n = 1: 0 n = 2: 3	1 triplet	2.8×10^{-23}
Select set of 60	60	100	n = 1: 0 n = 2: 0	None	2.1×10^{-13}

1 Choice based on SNPs ranked by expected heterozygosity (H_e), polymorphism information content (PIC) estimated in PowerMarker (Liu and Muse, 2005) from 546 SNPs/81 DNA samples.

2 Choice based on SNPs ranked by simulated power (simPWR) or informativeness for relatedness (I_r) estimated in KinInfor v1 (Wang, 2006) from 546 SNPs/81 DNA samples.

A panel of 192 SNPs (Table 16.2) was compiled based on the above results, on accumulated SNP data at CRC and published information (Ji *et al.*, 2013; Takrama *et al.*, 2014). Eighteen of the 'Select set of 60' loci in Table 1 were retained in the CIRAD-CRC SNP panels but only one of these (TcSNP0886) was common to the set of loci employed by the previously mentioned sources. The complete panel covered 43 of the 54 TcSNP loci in the ICGD database (Turnbull and Hadley, 2012), 56 of the 70 SNP loci employed by Ji *et al.* (2013) and 45 of the 48 SNP loci employed by Fang *et al.* (2014). The complete panel is designed as two complementary 96- member subsets of which Panel A could be used as the first panel. This panel contained 36 loci in common with Fang *et al.* (2014).

Table 16.2: CIRAD-CRC Panel of SNPs for Identity Resolution in Cacao

Panel[1]	*TcSNP*	*Chrom*[2]	*Panel*	*TcSNP*	*Chrom*	*Panel*	*TcSNP*	*Chrom*
A	13	2	A	261	10	A	642	8
A	33	2	**A**	**290**	**3**	A	645	5
A	49	4	A	305	6	A	660	5
A	64	5	**A**	**309**	**6**	**A**	**702**	**10**
A	131	8	A	313	2	***A***	***703***	***3***
A	135	4	***A***	***316***	***2***	A	704	10
A	**139**	**8**	A	326	3	***A***	***736***	***5***
A	143	3	**A**	**329**	**7**	A	749	4
A	**144**	**10**	A	339	9	**A**	**750**	**6**
A	***149***	***5***	**A**	**364**	**9**	A	751	5
A	**150**	**5**	**A**	**372**	**4**	A	791	7
A	**151**	**8**	A	380	2	**A**	**799**	**8**
A	***154***	***1***	A	391	4	A	800	8
A	**189**	**8**	A	397	2	A	814	7
A	192	9	**A**	**429**	**2**	A	835	5
A	**193**	**9**	**A**	**469**	**7**	**A**	**836**	**2**
A	194	1	A	522	5	A	841	5
A	198	4	A	556	6	**A**	**852**	**3**
A	**226**	**9**	**A**	**560**	**10**	**A**	**872**	**4**
A	**230**	**10**	**A**	**591**	**1**	**A**	**878**	**3**
A	***240***	***1***	**A**	**602**	**6**	A	885	2
A	**242**	**9**	**A**	**619**	**6**	**A**	**891**	**2**
A	**899**	**8**	**A**	**1309**	**8**	***B***	***214***	***2***
A	**917**	**10**	A	1350	1	B	256	8
A	929	3	A	1362	8	B	258	7
A	945	3	**A**	**1414**	**9**	B	259	1
A	994	6	**A**	**1484**	**6**	B	280	3
A	**998**	**5**	A	1520	8	B	341	9
A	**999**	**8**	A	1522	2	B	363	3
A	1010	1	A	1527	6	B	421	3
A	1011	1	B	19	3	B	455	9

Contd...

Panel[1]	*TcSNP*	*Chrom*[2]	*Panel*	*TcSNP*	*Chrom*	*Panel*	*TcSNP*	*Chrom*
A	1028	4	**B**	**25**	**9**	B	456	4
A	**1038**	**5**	B	32	4	***B***	***461***	***5***
A	**1060**	**2**	**B**	**75**	**2**	**B**	**480**	**9**
A	**1062**	**3**	***B***	***97***	***2***	B	519	4
A	**1075**	**1**	B	105	8	B	531	1
A	1144	6	B	126	5	**B**	**534**	**1**
A	1165	2	B	141	3	B	546	10
A	1200	6	B	148	3	**B**	**577**	**5**
A	1201	7	B	164	1	B	589	3
A	1216	1	***B***	***173***	***1***	B	606	7
A	**1253**	**9**	**B**	**174**	**4**	B	607	1
A	**1270**	**7**	B	176	3	***B***	***636***	***8***
A	1275	1	B	186	?	B	640	2
B	644	9	B	954	1	B	1237	7
B	669	2	B	964	6	B	1266	6
B	674	10	B	1019	5	B	1293	?
B	**723**	**10**	B	1053	2	B	1302	1
B	737	9	B	1058	1	B	1308	1
B	739	3	**B**	**1063**	**7**	B	1329	1
B	***773***	***4***	B	1074	10	**B**	**1331**	**10**
B	785	1	***B***	***1093***	***4***	***B***	***1349***	***5***
B	**823**	**5**	B	1106	2	B	1401	1
B	833	9	B	1112	5	B	1404	2
B	857	9	B	1136	2	B	1408	5
B	867	9	B	1156	6	B	1416	5
B	871	5	B	1158	5	***B***	***1422***	***2***
B	881	5	**B**	**1159**	**2**	B	1426	8
B	***886***	***4***	***B***	***1160***	***4***	B	1432	1
B	901	6	B	1195	2	**B**	**1453**	**5**
B	***906***	***2***	B	1205	4	B	1457	3
B	933	9	B	1223	5	**B**	**1458**	**1**
B	947	9	B	1228	?	B	1483	8
B	**953**	**4**	B	1229	4	B	1524	2

1. Each panel consists of 96 loci.
2. Chromosome; Bold black font are loci in common with Ji *et al.* (2013). Bold, italic, blue font are loci in the 'Select set of 60' loci from Table 16.1 that was useful in discriminating among accessions. One marker (TcSNP0886) is common to both datasets.
3. SNPs with uncertain chromosome assignment are indicated by ?.

Table 16.3: Number of SNP Markers Used in Varietal Studies in Cacao

Number of SNPs	*Sample size*	*Germplasm Source*	*Genetic Groups Covered*	*Resolution*	*Reference*
6	50	Ghana (CRIG germplasm)	Iquitos, Nanay, Marañón	30 per cent; 8 synonymous groups present	Takrama *et al.* (2012).
13	186	Cameroon, Trinidad	Amelonado, Guiana, Iquitos, Nanay, Nacional	not given; SNPs identified 24.2 per cent offtypes	Livingstone *et al.* (2011).
18	171	USDA-ARS SHRS[1] (Florida, USA)	Seedlings from four mother trees representing three genotypes	Not given; apparently 100 per cent	Livingstone *et al.* (2012).
48 assessed; 44 usable	48	Brazil, Costa Rica, Ecuador, Peru, Trinidad	Amelonado, Criollo, Curaray, Iquitos, Nanay, Marañón	Full; no matching samples	Fang *et al.* (2014).
54 assessed; 53 employed	160	Ghana (CRIG germplasm)	Amelonado, Curaray, Iquitos, Nanay, Marañón	Not given; apparently 100 per cent	Takrama *et al.* (2014).
100 assessed; 70 employed	115	Costa Rica, Honduras, Nicaragua, Trinidad	Amelonado, Criollo, Contamana, Iquitos, Nacional, Nanay,	86.1 per cent; 6 synonymous groups: 5 Criollo sets and 1 Amelonado set	Ji *et al.* (2013).

1 Subtropical Horticulture Research Station.

4. Applications of Molecular Profiling

4.1. Resolving Fidelity Issues

Unambiguous identification of individuals and usage of correctly named accessions are important in managing germplasm collections, seed gardens and clonal gardens, ensuring fidelity during germplasm transfer, parental choice in breeding programmes and reducing the errors associated with experimental analyses especially when dealing with phenotypic data. The problem of mislabelling in crop genebanks is well recognised (van Hintum, 2000; Hurka *et al.*, 2004) and has been shown to be an issue in cocoa germplasm collections (Figueira, 1998; Christopher *et al.*, 1998; Risterucci *et al.*, 2001; Motilal and Butler, 2003; Sounigo *et al.*, 2006a; Motilal *et al.*, 2009, 2011, 2012). Cases where accessions have the same name, but have different genetic profiles, are termed homonymous errors. Cases where accessions have different names but the same genetic profile are termed synonymous errors.

The decreasing cost of molecular marker technologies have led to their routine application for fidelity testing in cocoa germplasm collections, clonal gardens and breeding. A standard set of 15 SSR loci were recommended for cacao identity studies (Saunders *et al.*, 2004). However, full resolution of germplasm accessions was not possible with this set of 15 and a set of nine SSR loci that performed better in separating accessions, was later recommended (Motilal *et al.*, 2009). Similarly two panels of 96 SNPs (192 loci; Section 4.3) are recommended for resolving the identity of cacao accessions. In the past, 48-100 SNPs have been used in cacao for identity studies (Table 16.3). The random match probability (Budowle *et al.*, 2000) and the estimated Amazonian population of 6.81×10^{10} cacao trees (Motilal *et al.*, 2009) was used to obtain a conservative estimate of 1.48×10^{-13} which could represent the upper boundary for identity declaration. Values for PID_{SIB} fall in the range of 10^{-10} (Motilal *et al.*, 2012; 26 SSR loci); 10^{-9} (Takrama *et al.*, 2014; 53 SNPs); 10^{-6} to 10^{-5} (Zhang *et al.*, 2006a,b, 2009a, 2011; 15 SSRs) and 10^{-5} from 26 (Ji *et al.*, 2013) or 44 SNPs (Fang *et al.*, 2014) have been reported. The lower polymorphism of SNP markers relative to SSR loci means that more SNP loci are required than SSR loci to obtain the same PID_{SIB} value. The number of SNP loci suggested to be sufficient (based on the number of, and types of accessions present in the study), to get the same order of magnitude of probability in declaring identity, is nearly twofold that of the number of SSRs previously used among cacao scientists worldwide (Livingstone *et al.*, 2011, 2015; Ji *et al.*, 2013).

The SNPs identified in Table 2 generated a combined PID_{SIB} of 2.9×10^{-31} from a set of 184 loci on 1800 samples inclusive of variable numbers of replicated trees per accession. The suitability of this panel in determining genetic identity and the identification of a best possible minimum set of SNP loci for use in varietal identification has been evaluated using cacao accessions in the ICGT.

4.2. Defining Populations and Phylogenetic Relationships

Molecular fingerprinting techniques such as isozyme electrophoresis, RFLP, RAPD and AFLP and more recently, SSR and SNP assays have allowed for more rapid and accurate investigating of phylogenetic relationships of cacao germplasm using evolutionary models. Sounigo *et al.* (2005), using RAPDs and isozyme electrophoresis techniques, separated germplasm from the ICGT into three major clusters, with one consisting mainly of trees native to Peru, Ecuador and French Guiana; one mainly of all the cultivated Trinitario and one exclusively of trees cultivated in Ecuador. Irish *et al.* (2010) found that accessions generally grouped together according to their geographical origin and traditional genetic background. One cluster contained mainly Trinitario type accessions with varieties mostly from Mexico, Central America, and the Caribbean; while the Amelonado, SIAL and SIC accessions that originated from Brazil, were grouped in another cluster; and the Ecuadorian accessions such as EET and UF which are linked to "Nacional" ancestry, formed another cluster.

Motamayor *et al.* (2008) using 106 SSR markers on samples with provenance from the wild, partitioned cacao genetic diversity into 10 phylogenetic groups– Amelonado, Contamana, Criollo, Curaray, Guiana, Iquitos, Marañón, Nacional, Nanay and Purús. These groups with the exception of Criollo (Mesoamerican origin) are of Amazonian origin. Thomas *et al.* (2012) mapped the 10 genetic groups to distinct geographical locations, with distinct allelic compositions, suggesting that the genetic groups may represent geographically isolated populations. However, as explorations are made in South American countries with endemic cacao, the number of genetic groups presently defined is expected to increase as was recently observed with the discovery of a unique cacao population in Bolivia (Zhang *et al.*, 2011).

The phylogenetic relationships among accessions collected from different expeditions have been further dissected in other studies. Molecular profiles (Sounigo *et al.*, 2002, 2005; Boccara and Zhang, 2006, 2007, 2008; Zhang *et al.*, 2009a; Motilal *et al.*, 2009, 2012, 2013; Loor Solorzano *et al.*, 2012; Boza *et al.*, 2013) and morphological data (Bekele *et al.*, 2005, 2006), have been used to establish the proximity of IMC and NA accessions to each other. This suggests that genetic material collected from the two river systems may have some common ancestry.

The POUND accessions were reportedly collected as budwood from the same area, and supposedly from the same mother trees as the IMC, NA, SCA accessions which were collected as fruits (Pound, 1945; Bartley, 2005). The POUND accessions have been found to be distributed among the SCA, IMC and NA clusters from molecular data (Motamayor *et al.*, 2008; Zhang *et al.*, 2009a; Motilal *et al.*, 2009, 2012; Loor Solorzano *et al.*, 2012) and this was better demonstrated with RAPD data than with isoenzyme data (Sounigo *et al.*, 2005). Using morphological data, the POUND accessions were positioned within the same cluster as NA and PA accessions (Bekele *et al.*, 2006), or dispersed amongst IMC, NA and SCA accessions (Bekele *et al.*, 2005).

Refractario is a group of related accessions that were mass selected for witches' broom disease resistance under natural disease pressure in Ecuador (Pound 1938, 1943; Bartley, 2001, 2005). Phylogenetically, the Refractario accessions cluster separately as a single clade quite distinct from Amelonado, Contamana, Criollo, Guiana, Iquitos, Marañón and Nanay accessions (Motilal *et al.*, 2013). The Refractario germplasm pool has two main subclusters (Zhang *et al.*, 2007; Motilal *et al.*, 2012) and was shown to be genetically distinct from Trinitario accessions based both on molecular profiles (Zhang *et al.*, 2007; Boccara and Zhang, 2007, 2008) and morphological variation (Bekele *et al.*, 2006).

4.3. Corroborating Family Structure

A cacao fruit has variable number of seeds ranging between 15-60 seeds. Each seed can have a potentially different pollen parent which depends on the surrounding trees, the cross-compatibility of the maternal tree and the self-compatibility of the maternal tree. Accessions derived from fruits of a tree are therefore expected to be half-sibs or full sibs with respect to each other. Contrariwise, germplasm collected as budwood from trees in a particular locale, may or may not be related to each other. Furthermore, accessions derived from bulked seed lots or mixed germplasm may have hidden family structure. Since historical records and inventory may be lacking in detail, the use of molecular markers can provide clarity. Knowledge of the family structure is important in planning breeding experiments. Heterosis can be exploited if crosses are between diverse genotypes. Germplasm curators can use family structure in describing the genetic diversity of the collection and obtaining a measure of the redundancy of the collection (Motilal *et al.*, 2013).

For instance, the SCA, IMC, PA and NA accessions were derived from 1, 2, 7-20, and 14-17 mother trees respectively according to collection records (Pound, 1938; Bekele *et al.*, 2005). Molecular evidence has been used to show that the SCA and IMC accessions came from two trees each, the PA accessions from 20 trees, and the NA accessions from 22 mother trees (Zhang *et al.*, 2009a) which is a reasonable fit to the historical literature. The MO accessions were derived from two mother trees (Zhang *et al.*, 2009a) which agreed with the suggestion of Bartley (2005). that fruits were collected from more than one tree.

Additional work in the ICGT with a higher complement of Refractario accessions, but also containing Amelonado, Criollo, GU, NA, IMC, and PA accessions, identified 56 full-sib families and 189 half-sib families each of which contained between 2-17 individuals (Motilal *et al.*, 2013). The Guiana accessions formed a family cluster that could not be linked with the other family units of the 189 half-sib families. High sibship in the Refractario accessions was expected as this germplasm was collected as fruits from about 80 trees that showed disease resistance and these trees were derived from seedlings obtained from disease-free trees (Pound, 1938, 1943). Hybrid types like the ICS and TRD accessions also exhibited family structure and these accession groups contained 12 full-sib families with 2-9 members per family (Johnson *et al.*, 2009). Likewise, up to 12 full-sib families were found in the Indonesian cacao germplasm which were likely derived from Upper Amazon Forastero, Venezuelan germplasm and Criollo germplasm (Susilo *et al.*, 2011).

4.4. Evidence for Anthropogenic Effects

Evidence for anthropogenic effects on cacao genetic diversity can also be gleaned from ancestry and phylogenetic analysis. Thomas *et al.* (2012) provided evidence to support the role of glaciation and human mediation on cocoa population differentiation. The PA accessions, form part of the Marañon group that was collected from the Upper Amazon region, and have been shown to be genetically similar to

Lower Amazon Forastero material from Brazil (Motamayor *et al.*, 2008) Bartley (2005) suggested on the basis of morphological observations, that the PA group probably descended from planting material imported from Brazil. Zhang *et al.* (2009a), using Amelonado accessions from Brazil (Lower Amazon Forastero) showed that these clustered with the PA accessions and merged into the same population group when large groups (K=3) were compiled using the software 'Structure'. These and other data from isozyme and RAPD data (Sounigo *et al.*, 2005) could be taken to support the hypothesis of Bartley (2005). However, the descent of Lower Amazon material from the Upper Amazon via a stopping point at the area where the PA accessions were collected can also be supported by the data. This alternative view is bolstered by the results of Thomas *et al.* (2012) and Sereno *et al.* (2006) who advocated that part of the Lower Amazon cacao population was derived from the Upper Amazon. The Amelonado and the Marañon-Amazon River cluster were also suggested to be derived from the same genepool that led to the differentiation of the Iquitos and Púrus clusters (Thomas *et al.*, 2012). The heterozygosity of descended populations is expected to be less than that of founder populations. Examination of the data of Thomas *et al.* (2012) show that the Marañon related clusters have higher H_e values than those of Guiana or Amelonado clusters. Motilal *et al.* (2013) found that the PA as an accession group had more microsatellite alleles and higher H_e than reference Amelonados from Brasil. The Amelonados in Brasil and the Guiana germplasm may therefore be the result of strong bottlenecks aided by human selection from the same initial pool in the centre of origin and diversity in the Western Upper Amazon that generated the PA accessions. This area includes the Amazon region of Peru and is close to the Brazilian Upper Amazon.

4.5. Evolution of the Trinitario Complex

Trinitario types from different countries have been reported (Toxopeus, 1985; Bartley, 2005) and supposedly arose in several countries as natural hybrids of Forastero and Criollo germplasm (Cheesman, 1944; Toxopeus, 1985; Wood, 1985a,b). Separation of Trinitario types from Honduras and Nicaragua have been reported using SNP data (Ji *et al.*, 2013). However, the Trinitarios from Trinidad appear to cluster with the Trinitarios from Nicaragua according to the data presented by Ji *et al.* (2013). Similarly, the Trinitarios from Trinidad clustered among the Trinitarios from Indonesia, from the figures of Susilo *et al.* (2011) and Dinarti *et al.* (2015). The Trinitarios from Trinidad were shown to have a greater contribution of Forastero ancestry than Criollo ancestry with admixture from Upper Amazon and Lower Amazon according to SSR analyses (Motilal *et al.*, 2010). This was later corroborated by Yang *et al.* (2013), using chloroplast SSRs and SNPs, to establish that three lineages (Upper Amazon Forastero, Lower Amazon Forastero and Criollo) were present in the Trinitario complex in Trinidad.

4.6. New Populations and Landraces

Genetic ancestry in cocoa has been assessed using the software 'Structure' (Pritchard *et al.*, 2000; Hubisz *et al.*, 2009) with SSR and, recently, SNP datasets in a model-based clustering method. Since Motamayor *et al.* (2008), there has been little additional insight on cocoa ancestry although Zhang *et al.* (2011) described a new population from Bolivia. Germplasm collections need to be assessed for genetic structure after identity issues have been clarified. This enables curators to obtain empirical *a priori* proof by genetic information on the representation of the known existing cocoa populations within the germplasm collection. This aspect has been addressed in germplasm collections in Cameroun (Efombagn *et al.*, 2008), Costa Rica (Zhang *et al.*, 2009b), Dominican Republic (Boza *et al.*, 2013), Puerto Rico (Irish *et al.*, 2010), and Trinidad (Zhang *et al.*, 2007, 2009a; Johnson *et al.*, 2009; Motilal *et al.*, 2010, 2012).

Farmer varieties or landraces are also commonly studied for their ancestral contributions. Fine flavour cocoa originates from Criollo (Toxopeus, 1985) and Nacional backgrounds (Loor Solorzano *et al.*, 2012). Flavour sensory attributes that could be influenced by genetic makeup include floral and dry fruit flavour (Ziegleder, 1990; Chanliau and Cros, 1999; Lanaud *et al.*, 2005; Trognitz *et al.*, 2013). Due to the historic movement of germplasm there have been repeated introductions into many countries from few source countries, multiple subsequent transfers between countries within a region, and more recently movement of germplasm from genebanks via plant quarantine centres. These multiple introductions have resulted in diverse ancestral backgrounds and admixed individuals in countries with cultivated but not endemic cacao. The genetic ancestry therefore reflects the historical movement of cacao and can be used to retrace the history where historical records do not exist. The population ancestry of farmer varieties have been reported for Brazil (Santos *et al.*, 2015), Cameroon (Efombagn *et al.*, 2008), Dominican Republic (Boza *et al.*, 2013), Honduras and Nicaragua (Ji *et al.*, 2013), Indonesia (Dinarty *et al.*, 2015), Peru (Zhang *et al.*, 2006a, 2011) and Trinidad (Yang *et al.*, 2013).

In order to derive maximum benefit, studies on cacao ancestry need to incorporate reference profiles, either generated from within the study or obtained from open-access deposited information as in the ICGD. This will allow the proper population ancestries to be identified instead of being unknown contributions as in Trognitz *et al.* (2011, 2013). Although, SSR profiles from different platforms are usually different, the allele calls can be made consistent provided that common accessions were scored between platforms. However, this may not always be possible and the availability of SNP data in the ICGD (Turnbull and Hadley, 2012) is therefore a good alternative especially as the SNP data is expected to be more transferable due to non-reliance on size polymorphisms. Accessions allocated to genetic clusters and with SNP data in the ICGD can therefore be mined to typify their genetic ancestry.

4.7. Core Collection

A core collection is a subset of minimal size with maximum representation of the original germplasm collection. Core collections contain maximal variability in 10-15 per cent of the accessions (Frankel, 1984; Frankel and Brown, 1984; Brown, 1989a,b). Van Hintum *et al.* (2000) reported that most core collections contain 5–20 per cent of the original germplasm collection. Although core collections have been identified in various plant collections such as olive (Belaj *et al.*, 2012), pear (Miranda *et al.*, 2010) and Sea Island cotton (Mei *et al.*, 2012), a core collection for cacao is yet to be adopted. Sounigo *et al.* (2006b) identified a set of 110 accessions in the ICGT based on RAPD and isozyme diversity, disease resistance, and seed traits of importance to the industry. Recently, a core collection of 59 accessions were identified based on the genetic diversity of 24 SSR loci on 414 accessions in the ICGT (Motilal *et al.*, 2013). Santos *et al.* (2015) reported a core collection of 27 plants based on the genetic diversity of 30 SSR loci on 279 varieties from germplasm collection and farmers' fields in Brazil. In the Brazilian core collection, seven plants possessed 70 per cent of the alleles and five of these plants originated from the farmers' fields indicative of the low diversity of the Amelonado accessions in the germplasm collection. However, core collections based on molecular genetic diversity may not adequately represent the agronomic or morphological diversity. Hence, after characterisation of a germplasm collection at the molecular and phenotypic levels, either a composite core collection based on the diversity over both levels can be identified or several core collections can be developed independently from genotypic and phenotypic data.

4.8. Evolutionary Phylogeny

Theobroma was previously placed in the Family Sterculiaceae (Purseglove, 1974), but is now reassigned to Family Malvaceae (Alverson *et al.*, 1999; Bayer *et al.*, 1999). Cuatrecasas (1964) described 22 species of *Theobroma* which are closely allied with the 17 species of *Herrania* Goudot (Whitlock and Baum, 1999). Figueira *et al.* (1994) on the basis of rDNA polymorphism found that all *Herrania* and *Theobroma* species with the exception of *T. cacao* formed a single cluster. Whitlock *et al.* (2001). reported that these two genera formed a single clade based on plastid *ndhF* sequences. Whitlock and Baum (1999), on the basis of sequenced vicilin gene, showed that *Herrania* and *Theobroma* were monophyletic groups but monophyly was more strongly supported for *Herrania* than *Theobroma*. A similar conclusion based on trypsin inhibitor gene sequences was obtained (Silva and Figueira, 2005). Both Whitlock and Baum (1999) and Silva and Figueira (2005) supported the inclusion of *T. mammosum* into the section Glossopetalum rather than as the separate section Andropetalum. In contrast, Borrone *et al.* (2007) obtained much stronger support for the monophyly of *Theobroma* from sequences of five WRKY genes. However, the separation of *Theobroma* into two main clades and the intra-cluster separation of *T. cacao* from its other sister species within the clade was consistently reported (Whitlock and Baum, 1999; Silva and Figueira, 2005; Borrone *et al.*, 2007).

More recently, there has been interest in the DNA barcoding of plants with the aim of identifying and classifying unknown plant species. The barcoding technique involves characterising species based on a DNA segment from a standard and agreed-upon position in the genome. The sequence used is usually short and can be obtained quickly and cheaply. The *trnH-psbA* and ITS spacer regions (Kress *et al.*, 2005) in conjunction with *rbcL* and *matK*, as barcode regions (Fazekas *et al.*, 2012) have been used for species identification in land plants. However, while these regions may distinguish *Theobroma* spp. from each other and from related allies like *Herrania*, *Cola* and *Guazuma*, it has not been established that these barcode regions can reveal intra-species differences. The identification of a set of SNP loci, to generate a unique haplotype per variety, is therefore recommended.

4.9. Genome-Wide Association Studies and Marker Assisted Selection

Extensive and intensive molecular profiling has generated thousands of markers which can be tested for association with a desired trait. The markers so identified have the advantage of being broad-based in application instead of being restricted to a population or populations (Yu and Buckler, 2006). Spurious or false associations are minimised by accounting for population stratification and relatedness (Aranzana *et al.*, 2005; Price *et al.*, 2006; Yu *et al.*, 2006). Association studies have been reviewed and proven useful in understanding the genetic basis for complex traits in many plants (Abdurakhmonov and Abdukarimov, 2008; Zhu *et al.*, 2008; Ingvarsson and Street, 2011; Soto-Cerda and Cloutier, 2012; Gupta *et al.*, 2014). Association mapping is an emergent field in cacao and few studies have been published. Motamayor *et al.* (2013), using association mapping among other techniques, identified a candidate gene from 71 accessions for fruit colour and a single SNP among the 168 that were studied, that affected transcript expressions of the gene.

Markers identified as being strongly associated with a trait may be employed in marker assisted selection programmes. Crouzillat *et al.* (2000b), from simulation studies, demonstrated that, in cacao, the use of molecular markers alone or in combination with phenotypic selection was more effective than phenotypic evaluation only. Reviews on marker-assisted selection have been published (Paterson *et al.*, 1991; Young, 1999; Hospital, 2003; Peleman *et al.*, 2005) and the advantages outlined (Collard *et al.*, 2005). The efficiency of marker assisted versus phenotype assisted selection is higher for traits of low heritability (Collard *et al.*, 2005) but cost effectiveness need to be evaluated for each case (Dreher *et al.*, 2003). Phenotypic evaluations may be time-consuming, difficult or costly (Dreher *et al.*, 2003; Young, 1999; Yu *et al.*, 2000). However, current trends indicate a reducing cost for SNP genotyping which should make marker assisted selection more cost-effective and therefore more favourable than phenotyping. The limited availability of land resources in terms of quantity and tenure issues may also weigh against phenotyping in cacao due to the long vegetative phase and number of years needed to obtain productivity values. Marker assisted selection programmes using quantitative trait loci in cacao have been reported (Schnell *et al.*, 2007a; Schnell *et al.*, 2007b) but those using association mapping are now underway (Motamayor *et al.*, 2010, 2013).

5. Future Prospects

5.1. Methodology

The methodologies for DNA profiling are expected to change as sequencing technologies become more powerful and cheaper, and as researchers develop faster and more efficient bioinformatics tools. On one hand platforms for DNA profiling are becoming more elaborate allowing researchers to rapidly obtain data for 1000's of SNPs across the genome for 1000's of individuals. Advances in DNA hybridisation, florescence microscopy and solid surface DNA capture have allowed for the development of large SNP arrays where 1000's SNPs can be assayed and interpreted in real-time (Ganal *et al.*, 2012, Kim and Misra, 2007). In parallel, there are technologies that seek to improve the efficiency of SNP-based DNA profiling using nanofluidics (Fang *et al.*, 2014, Wang *et al.*, 2009) or other miniaturised PCR volume systems and using robotic platforms, which together seek to reduce the cost of DNA profiling, improve the throughput rate, while reducing the human effort and time spent. Additionally, application of isothermal DNA amplification (Li and Macdonald, 2015, Zhao *et al.*, 2015) has facilitated the development of a field-based rapid method of DNA fingerprinting of organisms. This coupled with access to on-line databases and sophisticated background data processing will allow for sampling, analysis and data processing to occur in the field.

On the other hand, with the improvement in efficiency and the consequent reduction of the cost of next generation DNA sequencing technologies genotyping-by-sequencing has become a method of choice for DNA profiling. This is particularly useful in organisms for which *a priori* genomic information is not available. This approach also allows for sequencing predetermined areas of the genome for genetic variation using various enrichment methods. The technology also allows for the detection of insertions, deletions and microsatellites in addition to SNPs and considerably reduces ascertainment bias compared to array based methods.

With the success of next generation sequencing, and the availability of accurate genome sequence information there are new possibilities for molecular fingerprinting including the use of transposable elements which are known to contribute to a large proportion of genome variation. Polymorphisms in transposable elements can arise from the potential variability in copy number between and within species (Vicient *et al.*, 1999; Pearce *et al.*, 2000; Huang *et al.*, 2008). Sveinsson *et al.* (2013) found that there was active evolution of transposable elements in cacao and suggested that transposon fingerprinting could be used in identifying and characterising cacao accessions.

As deep sequencing and resequencing technologies become cheaper and more extensively used there is likely to be convergence of methodologies of DNA profiling for various purposes, such as genomic selection, genetic diversity, phylogenetics and evolutionary studies.

5.2 Food Forensics

High quality DNA was detected in fermented and dried beans and partially degraded DNA in roasted nibs and chocolates (Crouzillat *et al.*, 2000a). These authors showed that the cacao used to make chocolate could be traced from bean to product using RAPDs, SSRs and sequencing of a fragment of a seed storage protein gene. The adulteration of fermented and dried cocoa beans as a single bean incident could be detected with 44 SNPs (Fang *et al.*, 2014). The establishment of SNP panels that can distinguish fine flavour cocoa from bulk cocoa, as well as discriminate among fine flavour cocoa from different geographic areas would be a useful certification tool. In addition, molecular profiling can be used to support geographical indicators by supporting the uniqueness of cacao genotypes by field, farm or region; establishing the similarity of genotypes to enable bulk fermentation from members of co-operatives; and to brand genetic origin chocolates.

References

Abdurakhmonov, I.Y. and Abdukarimov, A. (2008). Application of association mapping to understanding the genetic diversity of plant germplasm resources. *International Journal of Plant Genomics*, Article ID 574927, doi:10.1155/2008/574927.

Aikpokpodion, P.O. (2010). Variation in agro-morphological characteristics of cacao, *Theobroma cacao* L., in farmers' fields in Nigeria. *New Zealand Journal of Crop and Horticultural Science*, **38**: 157-170.

Akkaya, M.S., Bhagwat, A.A. and Cregan, P.B. (1992). Length polymorphisms of simple sequence repeat DNA in soybean. *Genetics*, **132**: 1131-1139.

Allegre, M., Argout, X., Boccara, M., Fouet, O., Roguet, Y., Bérard, A., Thévenin, J.-M. *et al.* (2012). Discovery and mapping of a new expressed sequence tag-single nucleotide polymorphism and simple sequence repeat panel for large scale genetic studies and breeding of *Theobroma cacao* L. *DNA Research*, **19**: 23-35. doi:10.1093/dnares/dsr039.

Alverson, W.S., Whitlock, B. A., Nyffler, R., Bayer, C. and Baum, D.A. (1999). Phylogeny of the core Malvales: evidence from *ndhF* sequence data. *American Journal of Botany*, **86**: 1474-1486.

Aranzana, M. J., Kim, S., Zhao, K., Bakker, E., Horton, M., Jakob, K., Lister, C., Molitor, J., Shindo, C., Tang, C., Toomajian, C., Traw, B., Zheng, H., Bergelson, J., Dean, C., Marjoram, P. and Nordborg, M. (2005). Genome-wide association mapping in Arabidopsis identifies previously known flowering time and pathogen resistance genes. *PLoS Genetics*, **1**: e60. doi:10.1371/*journal.pgen*.0010060.

Argout, X., Fouet, O., Wincker, P., Gramacho, K., Legavre, T., Sabau, X., Risterucci, A.M., *et al.* (2008). Towards the understanding of the cocoa transcriptome: production and analysis of an exhaustive dataset of ESTs of *Theobroma cacao* L. generated from various tissues and under various conditions. *BMC Genomics*, **9**: 512. doi. 10.1186/1471-2164-9-512.

Argout, X., Salse, J., Aury, J.-M., Guiltinan, M.J., Droc, G., Gouzy, J., Allegre, M., *et al.* (2011). The genome of *Theobroma cacao*. *Nature Genetics*, **43**(2): 101-108.

Atkinson, M., Withers, L.A. and Simpson, M.J.A. (1986). Characterisation of cacao germplasm using isoenzyme markers. *Euphytica*, **35**: 741-750.

Baker, R.P., Hasenstein, K.H. and Zavada, M.S. (1997). Hormonal changes after compatible and incompatible pollination in *Theobroma cacao* L. *HortScience*, **32**: *1231-1997.*

Bartley, B.G.D. (2001). Refractario – an explanation of the meaning of the term and its relationship to the introductions from Ecuador in 1937. *INGENIC Newsletter*, **6**: 10-15.

Bartley, B.G.D. (2005). The genetic diversity of cacao and its utilization. UK: CABI Publishing.

Bayer, C., Fay, M.F., De Bruijn, P.Y., Savolainen, V., Morton, C.M., Kubitzki, K., Alverson, W.S. and Chase, M.W. (1999). Support for an expanded family concept of Malvaceae within a recircumscibed order Malvales: a combined analysis of plastid atpB and rbcL DNA sequences. *Botanical Journal of the Linnaean Society*, **129**: 267-303.

Beissinger, T.M., Hirsch, C.N., Vaillancourt, B., Deshpande, S., Barry, K., Buell, C.R., Kaeppler, S.M., Gianola, D. and de Leon, N. (2014). A genome-wide scan for evidence of selection in a maize population under long-term artificial selection for ear number. *Genetics*, **196**: 829-840.

Bekele, F. and Bekele, I. (1996). A sampling of the phenetic diversity of cacao in the International Cocoa Gene Bank of Trinidad. *Crop Science*, **36**: 57-64.

Bekele, F.L., Bekele, I., Butler, D.R. and Bidaisee, G.G. (2006). Patterns of morphological variation in a sample of cacao (*Theobroma cacao* L.) germplasm from the International Cocoa Genebank, Trinidad. *Genetic Resources and Crop Evolution*, **53**: 933-948.

Bekele, F.L., Bidaisee, G.G., Persad, N. and Bhola, J. (2005). Examining phenotypic relationships among Upper Amazon Forastero clones. In: Annual Report 2004, St. Augustine, Trinidad and Tobago. *Cocoa Research Unit*, The University of the West Indies. pp. 27-42.

Bekele, F.L., Butler, D.R. and Bidaisee, G.G. (2008a). Upper Amazon Forastero cacao (*Theobroma cacao* L.) 1: An assessment of phenotypic relationships in the International Cocoa Genebank, Trinidad. *Tropical Agriculture (Trinidad)*, **85**: 1-15.

Bekele, F.L., Iwaro, A.D., Butler, D.R. and Bidaisee, G.G. (2008b). Upper Amazon Forastero cacao (*Theobroma cacao* L.) 2: An overview of Parinari clones from a breeder's perspective. *Tropical Agriculture (Trinidad)*, **85**: 16-33.

Bekele, F.L., Kennedy, A.J., Mc David, C., Lauckner, F.B. and Bekele, I. (1994).) Numerical taxonomic studies on cacao (*Theobroma cacao* L.) in Trinidad. *Euphytica*, **75**: 231-240.

Belaj, A., Dominguez-García, M.C., Atienza, S.G., Urdíroz, N.M., de la Rosa, R., Satovic, Z., Martín, A. *et al.* (2012). Developing a core collection of olive (*Olea europaea* L.) based on molecular markers (DArTs, SSRs, SNPs) and agronomic traits. *Tree Genetics and Genomes*, **8**: 365-378.

Bélo, A. Zheng, P., Luck, S., Shen, B., Meyer, D.J., Li, B., Tingey, S. and Rafalski, A. (2008). Whole genome scans detects an allelic variant of *fad2* associated with increased oleic acid levels in maize. *Molecular Genetics and Genomics*, **279**(1): 1-10.

Boccara, M. and Zhang, D. (2006). Progress in resolving identity issues among the Parinari accessions held in Trinidad: the contribution of the collaborative USDA/CRU project. In: Annual Report 2005, St. Augustine, Trinidad. *Cocoa Research Unit,* The University of the West Indies. pp. 25-32.

Boccara, M. and Zhang, D. (2007). The contribution of the collaborative USDA/CRU project to resolve identity issues for trees in Marper Farm with missing labels. In: Annual Report 2006, St. Augustine, Trinidad. Cocoa Research Unit, The University of the West Indies. pp. 25-32.

Boccara, M. and Zhang, D. (2008). Identity assessment of Refractario origin cocoa accessions held in Trinidad: the contribution of the collaborative USDA/CRU project. In: Annual Report 2007, St. Augustine, Trinidad. Cocoa Research Unit, The University of the West Indies. pp. 30-36.

Borrone, J. W., Kuhn, D.N. and Schnell, R.J. (2004). Isolation, characterization, and development of WRKY genes as useful genetic markers in *Theobroma cacao*. *Theoretical and Applied Genetics*, **109**: 495-507.

Borrone, J.W., Meerow, A.W., Kuhn, D.N., Whitlock, B.A. and Schnell, R.J. (2007). The potential of the WRKY gene family for phylogenetic reconstruction: An example from the Malvaceae. *Molecular Phylogenetics and Evolution*, **44**: 1141-1154.

Boza, E.J., Irish, B.M., Meerow, A.W., Tondo, C.L., Rodríguez, O.A., Ventura-López, M., Gómez, J.A., Moore, J.M., *et al.* (2013). Genetic diversity, conservation, and utilization of *Theobroma cacao* L.: Genetic resources in the Dominican Republic. *Genetic Resources and Crop Evolution*, **60**: 605-619.

Brown, A.H.D. (1989a). A case for core collections. In: Brown, A.H.D, Frankel, O.H., Marshall, D.R. and Williams, J.T. (Eds.), *The Use of Plant Genetic Resources*. Cambridge University Press, Cambridge. pp. 136–156.

Brown, A.H.D. (1989b). Core collections: A practical approach to genetic resources management. *Genome*, **31**: 818-824.

Brown, J.S., Phillips-Mora, W., Power, E.J., Krol, C., Cervantes-Martinez, C., Motamayor, J.C. and Schnell, R.J. (2007). Mapping QTLs for resistance to frosty pod and black pod diseases and horticultural traits in *Theobroma cacao* L. *Crop Science*, **47**: 1851-1858.

Brown, J.S., Schnell, R.J., Motamayor, J.C., Lopes, U., Kuhn, D.N. and Borrone, J.W. (2005). Resistance gene mapping for witches' broom disease in *Theobroma cacao* L. in a F2 population using SSR markers and candidate genes. *Journal of the American Society for Horticultural Science*, **130**: 366-373.

Brumfield, R.T., Beerli, P., Nickerson, D.A. and Edwards, S.V. (2003). The utility of single nucleotide polymorphisms in inferences of population history. *Trends in Ecology and Evolution*, **18**: 249-256.

Buckler, E.S., Holland, J.B., Bradbury, P.J., Acharya, C.B., Brown, P.J., Browne, C., Ersoz, E., Flint-Garcia, S., *et al.* (2009). The genetic architecture of maize flowering time. *Science*, **325**(5941): 714-718.

Buckler, E.S. and Thornsberry, J. (2002). Plant molecular diversity and applications to genomics. *Current Opinions in Plant Biology*, **5**: 107-111.

Budowle, B., Chakraborty, R., Carmody, G. and Monson, K.L. (2000). Source attribution of a forensic DNA profile. *Forensic Science Communications*, **2**(3): 6 pp. http://www.fbi.gov/hq/lab/fsc/backissu/july2000/source.htm (accessed July 07, 2008).

Caetano-Anolles, G. and Brant, B. (1991). DNA amplification fingerprinting using very short arbitrary oligonucleotide primers. *Nature Biotechnology*, **9**: 553-557.

Chanliau, S. and Cros, E. (1999). Influence du traitement post-récolte et de la torréfaction sue le développement de l'arôme cacao. In: Proceedings of the 12th *International Cocoa Research Conference,* 17-23 November, 1996, Salvador, Bahia, Brazil. COPAL, Malaysia. pp. 959-964.

Charters, M.Y. and Wilkinson, M.J. (2000). The use of self-pollinated progenies as "in-groups" for the genetic characterization of cocoa germplasm. *Theoretical and Applied Genetics,* **100**: 160-166.

Cheesman, E.E. (1944). Notes on the nomenclature, classification and possible relationships of cacao populations. *Tropical Agriculture (Trinidad),* **21**: 144-159.

Cho, R.J., Mindrinos, M., Richards, D.R., Sapolsky, R.J., Anderson, M., Drenkard, E., Dewdney, J., Reuber, T.L., Stammers, M., Federspiel, N., *et al.* (1999). Genome-wide mapping with biallelic markers in *Arabidopsis thaliana. Nature Genetics,* **23**: 203-207.

Chowdappa, P. and Chandramohanan, R. (1995). Electrophoretic protein patterns of three species of *Phytophthora* associated with black pod disease of cocoa (*Theobroma cacao* L.). *Journal of Biosciences,* **20**: 637-644.

Christopher, Y. (1993). The reproducibility of DNA polymorphisms in *Theobroma cacao* L. as exhibited using RAPD analysis. In: Report for 1993. *Cocoa Research Unit,* The University of the West Indies. pp. 26-31.

Christopher, Y., Mooleedhar, V., Bekele, F. and Hosein, F. (1998). Verification of accession in the ICG,T using botanical descriptors and RAPD analysis. In: Report for 1998. *Cocoa Research Unit,* The University of the West Indies. pp. 15-18

Christopher, Y. and Sounigo, O. (1995). The use of RAPD for characterization and genetic assessment of cacao. In: Report for 1995. Cocoa Research Unit, The University of the West Indies. pp. 38-51

Clément, D., Risterucci, A.M., Motamayor, J.C., N'Goran, J. and Lanaud, C. (2003a). Mapping QTL for yield components, vigor and resistance to *Phytophthora palmivora* in *Theobroma cacao* L. Genome, **46**: 204-212.

Clément, D., Risterucci, A.M., Motamayor, J.C., N'Goran, J. and Lanaud, C. (2003b). Mapping quantitative trait loci for bean traits and ovule number in *Theobroma cacao* L. *Genome,* **46**: 103-111.

Coe, S. D. and Coe, M. D. (1996). The true history of chocolate. London: Thames and Hudson, Ltd.

Collard, B.C.Y., Jahufer, M.Z.Z., Brouwer, J.B. and Pang, E.C.K. (2005). An introduction to markers, quantitative trait loci (QTL) mapping and marker-assisted selection for crop improvement: The basic concepts. *Euphytica,* **142**: 169-196.

Cope, F.W. (1958). Incompatibility in *Theobroma cacao. Nature,* **181**: 279.

Cope, F.W. (1962). The mechanism of pollen incompatibility in *Theobroma cacao* L. *Heredity,* **17**: 157-182.

Crouzillat, D., Bellanger, L., Rigoreau, M., Bucheli, P. and Pétiard, V. (2000a). Use of DNA markers for determination of the genetic origin of cacao in fermented or roasted beans and chocolate. In: *Proceedings of the 13th International Cocoa Research Conference*, 9-14 October 2000, Kota Kinabalu, Malaysia. COPAL, Lagos, Nigeria. pp. 71-78.

Crouzillat, D., Menard, B., Mora, A., Phillips, W. and Petiard, V. (2000b). Quantitative trait analysis in *Theobroma cacao* using molecular markers. Yield QTL detection and stability over 15 years. *Euphytica*, **114**: 13-23.

Crouzillat, D., Lerceteau, E., Petiard, V., Morera, J., Rodriguez, H., Walker, D., Phillips, W., Ronning, C., Schnell, R., Osei, J. and Fritz, P. (1996). *Theobroma cacao* L.: A genetic linkage map and quantitative trait loci analysis. *Theoretical and Applied Genetics*, **93**: 205-214.

Cryer, N.C., Fenn, M.G.E., Turnbull, C.J. and Wilkinson, M.J. (2006). Allelic size standards and reference genotypes to unify international cocoa (*Theobroma cacao* L.) microsatellite data. *Genetic Resources and Crop Evolution*, **53**: 1643-1652.

Cuatrecasas, J. (1964). Cacao and its allies. A taxonomic revision of the genus *Theobroma*. In Contributions to the U.S. National Herbarium, **35**(6): 375-614. Washington, DC: Smithsonian Institution.

Dadzie, A.M., Livingstone III, D.S., Opoku, S.Y., Takrama, J., Padi, F., Offei, S.K., Danquah, E. Y., Motamayor, J.C., Schnell, R.J. and Kuhn, D.N. (2013). Conversion of microsatellite markers to single nucleotide polymorphism (SNP) markers for genetic fingerprinting of *Theobroma cacao* L. *Journal of Crop Improvement*, **27**(2): 215-241.

Dinarti, D., Susilo, A.W., Meinhardt, L.W., Ji, K., Motilal, L.A., Mischke S. and Zhang D. (2015). Genetic diversity and parentage in farmer selections of cacao from Southern Sulawesi, Indonesia revealed by microsatellite markers. *Breeding Science*, **65**: 438–446. doi:10.1270/jsbbs.65.438

Dreher, K., Khairallah, M., Ribaut, J. and Morris, M. (2003). Money matters (I): Costs of field and laboratory procedures associated with conventional and marker-assisted maize breeding at CIMMYT. *Molecular Breeding*, **11**: 221-234.

Efombagn, I.B.M., Motamayor, J.C., Sounigo, O., Eskes, A.B., Nyassé, S., Cilas, C., Schnell, R., Manzanares-Dauleux, M.J. and Kolesnikova-Allen, M. (2008). Genetic diversity and structure of farm and GenBank accessions of cacao (*Theobroma cacao* L.) in Cameroon revealed by microsatellite markers. *Tree Genetics and Genomes*, **4**: 821-823.

Engels, J.M.M. (1983a). A systematic description of cacao clones I. The discriminative value of quantitative characteristics. *Euphytica*, **32**: 377-385.

Engels, J.M.M. (1983b). A systematic description of cacao clones II. The discriminative value of qualitative characteristics and the practical compatibility of the discriminative value of quantitative and qualitative descriptors. *Euphytica*, **32**: 387-396.

Engels, J.M.M., Bartley, B.G.D. and Enriquez, G.A. (1980). Cacao descriptors, their states and modus operandi. *Turrialba*, **30**: 209-218.

Enríquez, C.G. and Soria, V.J. (1966) Estudio de la variabilidad de varias características de las mazorcas de cacao (*Theobroma cacao* L.). *Fitotecnica Latinoamericana*, **3**: 99-118.

Enríquez, C.G. and Soria, V.J. (1967) Selección y estudio de los carácteres útiles de la flor para la identificación y descripción de cultivares de cacao. *Cacao (Costa Rica)*, **12**: 8-16.

Erselius, L.J. and De Vallavieille, C. (1984). Variation in protein profiles of *Phytophthora*: comparison of six species. *Transactions of the British Mycological Society*, **83**: 463-472.

Evett, I.W. and Weir, B.S. (1998). Interpreting DNA evidence: statistical genetics for forensic scientists. Sunderland, MA: Sinauer.

Fang, W., Meinhardt, L.W., Mischke, S., Bellato, C.M., Motilal, L. and Zhang, D. (2014). Accurate determination of genetic identity for a single cacao bean, using molecular markers with a nanofluidic system, ensures cocoa authentication. *Journal of Agricultural and Food Chemistry*, **62**: 481-487 doi: 10.1021/jf404402v, Web: 19 Dec 2013

Fazekas, A.J., Kuzmina, M.L., Newmaster, S.G. and Hollingsworth, P.M. (2012). DNA barcoding methods for land plants. In W. J. Kress and D. L. Erickson (eds.), DNA Barcodes: *Methods and Protocols, Methods in Molecular Biology*, **858**, doi. 10.1007/978-1-61779-591-6_11.

Feltus, F.A., Wan, J., Schulze, S.R., Estill, J.C., Jiang, N. and Paterson, A.H. (2004). An SNP resource for rice genetics and breeding based on subspecies indica and japonica genome alignments. *Genome Research*, **14**: 1812-1819.

Figueira, A. (1998). Homonymous genotypes and misidentification in germplasm collections of Brazil and Malaysia. INGENIC Newsletter, **4**: 4-8. http://ingenic.cas.psu.edu/newsletters.htm (accessed July 07, 2008).

Figueira, A., Janick, J., Levy, M. and Goldsbrough, P. (1994).) Reexamining the classification of *Theobroma cacao* L. using molecular markers. *Journal of the American Society for Horticultural Science*, **119**: 1073-1082.

Ford, C.S. and Wilkinson, M.J. (2012). Confocal observations of late-acting self-incompatibility in *Theobroma cacao* L. *Sexual Plant Reproduction*, **25**: 169-183. doi:10.1007/s00497-012-0188-1.

Frankel, O. H. (1984). Genetic perspectives of germplasm conservation. In Genetic manipulation: impact on man and society Part III, Paper No. 15, edited by W. K. Arber, K. Llimensee, W. J. Peacock and P. Starlinger, 161-170. Cambridge: Cambridge University Press.

Frankel, O.H. and Brown, A.H.D. (1984). Plant genetic resources today: a critical appraisal. In Crop Genetic Resources: Conservation and Evaluation, edited by J. H. W. Holden and J. T. Williams, 249–257. London: George Allen and Unwin.

Ganal, M.W., Altmann, T. and Röder, M.S. (2009). SNP identification in crop plants. *Current Opinion in Plant Biology*, **12**: 211-217.

Ganal, M.W., Polley, A., Graner, E.-M., Plieske, J., Wieseke, R., Luerssen, H. and Durstewitz, G. (2012). Large SNP arrays for genotyping in crop plants. *Journal of Biosciences*, **37**: 821-828.

Glaubitz, J.C., Rhodes, O.E. and DeWoody, J.A. (2003). Prospects for inferring pairwise relationships within single nucleotide polymorphisms. *Molecular Ecology*, **12**: 1039-1047.

Gupta, P.K., Kulwal, P.L. and Jaiswal, V. (2014). Association mapping in crop plants: Opportunities and challenges. *Advances in Genetics*, **85**: 109-147.

Henderson, J.S., Joyce, R.A., Hall, G.R., Hurst, W.J. and McGovern, P.E. (2007). Chemical and archaeological evidence for the earliest cacao beverages. *Proceedings of the National Academy of Sciences USA*, **104**: 18937-18940.

Hodgkinson, A. and Eyre-Walker, A. (2010). Human triallelic sites: Evidence for a new mutational mechanism? *Genetics*, **184**: 233-241. doi: 10.1534/*genetics*.109.110510

Hospital, F. (2003). Marker-assisted breeding. In: *Plant Molecular Breeding*, edited by H. J. Newbury, 30-59. Oxford, UK: CRC Press, Blackwell.

Huang, X., Lu, G., Zhao, Q., Liu, X. and Han, B. (2008). Genome-wide analysis of transposon insertion polymorphisms reveals intraspecific variation in cultivated rice. *Plant Physiology*, **148**: 25-40.

Hubisz, M.J., Falush, D., Stephens, M. and Pritchard, J.K. (2009). Inferring weak population structure with the assistance of sample group information. *Molecular Ecology Resources*, **9**: 1322-1332.

Hüebner, C., Petermann, I., Browning, B.L., Shelling, A.N. and Ferguson, L.R. (2007). Triallelic single nucleotide polymorphisms and genotyping error in genetic epidemiology studies: MDR1 (ABCB1) G2677/T/A as an example. *Cancer Epidemiology, Biomarkers and Prevention*, **16**(6): 1185-1192. doi: 10.1158/1055-9965.EPI-06-0759.

Hurka, H., Neuffer, B. and Friesen, N. (2004). Plant genetic resources in botanical gardens. In: Proceedings of the 21st International Symposium on Classical vs. Molecular Breeding of Ornamentals, Part II 25-29 August, 2003, edited by G. Forkmann and S. Michaelis. *Acta Horticulturae*, **651**: 35-44.

Ingvarsson, P.K. and Street, N.R. (2011). Association genetics of complex traits in plants. New Phytologist, **189**: 909–922. doi: 10.1111/j.1469-8137.2010.03593.x.

Irish, B. M., Goenaga, R., Zhang, D., Schnell, R., Brown, J.S. and Motamayor, J.C. (2010). Microsatellite fingerprinting of the USDA-ARS Tropical Agriculture Research Station cacao (*Theobroma cacao* L.) germplasm collection. *Crop Science*, **50**: 656-667.

Iwaro, A.D., Bekele, F.L. and Butler, D.R. (2003). Evaluation and utilisation of cacao (*Theobroma cacao* L.) germplasm at the International Cocoa Genebank, Trinidad. *Euphytica*, **130**: 207-221.

Jenkins, P.A., Mueller, J.W. and Song, Y.S. (2014). General triallelic frequency spectrum under demographic models with variable population size. *Genetics*, **196**: 295-311.

Ji, K., Zhang, D., Motilal, L.A., Boccara, M., Lachenaud, P. and Meinhardt, L.W. (2013). Genetic diversity and parentage in farmer varieties of cacao (*Theobroma cacao* L.) from Honduras and Nicaragua as revealed by single nucleotide polymorphism (SNP) markers. *Genetic Resources and Crop Evolution*, **60** (2): 441-453.

Johnson, E.S., Bekele, F.L., Brown, S.J., Song, Q., Zhang, D., Meinhardt, L.W. and Schnell, R.J. (2009). Population structure and genetic diversity of the Trinitario cacao (*Theobroma cacao* L.) from Trinidad and Tobago. *Crop Science*, **49**: 564-572. doi:10.2135/cropsci2008.03.0128.

Kaosiri, G. and Zentmyer, G.A. (1980). Protein, esterase and peroxidase patterns in the *Phytophthora palmivora* complex from cacao. *Mycologia*, **72**: 998-1000.

Kasran, R. and Subali, S. (2002). Classification of *Theobroma cacao* L. germplasm using random amplified polymorphic DNA. In: *Proceedings of the 13th International Cocoa Research Conference*, 9-14 October 2000, Kota Kinabalu, Malaysia. COPAL, Lagos, Nigeria. pp. 229-235.

Kim, S. and Misra, A. (2007). SNP genotyping: technologies and biomedical applications. *Annual Review of Biomedical Engineering*, **9**: 289-320.

Knight, R. and Rogers, H.H. (1953). Sterility in *Theobroma cacao* L. *Nature*, **172**: 164.

Knight, R. and Rogers, H.H. (1955). Incompatibility in *Theobroma cacao* L., **9**: 69-77.

Konishi, S., Izawa, T., Lin, S.Y., Ebana, K., Fukuta, Y., Sasaki, T. and Yano, M. (2006). An SNP caused loss of seed shattering during rice domestication. *Science*, **312**(5778): 1392-1396.

Kress, W.J., Wurdack, K.J., Zimmer, E.A., Weigt, L.A. and Janzen, D.H. (2005). Use of DNA barcodes to identify flowering plants. *Proceedings of The National Academy of Sciences (USA)*, **102**(23): 8389-8374.

Kuhn, D.N., Borrone, J., Meerow, A.W., Motamayor, J.C., Brown, J.S. and Schnell, R.J. (2005). Single-strand conformation polymorphism analysis of candidate genes for reliable identification of alleles by capillary array electrophoresis. *Electrophoresis*, **26**: 112-125.

Kuhn, D.N., Figueira, A., Lopes, U., Motamayor, J.C., Meerow, A.W., Cariaga, K., Freeman, B., Livingstone III, D.S. and Schnell, R.J. (2010). Evaluating *Theobroma grandiflorum* for comparative genomic studies with *Theobroma cacao*. *Tree Genetics and Genomes*, **6**: 783-792.

Kuhn, D.N., Livingstone III, D., Main, D., Zheng, P., Saski, C., Feltus, F.A., Mockaitis, K., Farmer, A.D., May, G.D., Schnell, R.J. and Motamayor, J.C. (2012). Identification and mapping of conserved ortholog set (COS) II sequences of cacao and their conversion to SNP markers for marker-assisted selection in *Theobroma cacao* and comparative genomics studies. *Tree Genetics and Genomes*, **8**: 97-111.

Kujur, A., Bajaj, D., Upadhyaya, H.D., Das, S., Ranjan, R., Shree, T., Saxena, M.S., Badoni, S., Kumar, V., Tripathi, S., Gowda, C.L.L., Sharma, S., Singh, S., Tyagi, A.K. and Parida, S.K. (2015). Employing genome-wide SNP discovery and genotyping strategy to extrapolate the natural allelic diversity and domestication patterns in chickpea. *Frontiers in Plant Science,* **6** article 162. doi: 10.3389/fpls.2015.00162.

Kumar, L.S. (1999). DNA markers in plant improvement: An overview. *Biotechnology Advances,* **17**: 143-182.

Kurt, W., Weigand, F., Driesel, A.J., Kahl, G., Zischler, H. and Epplen, J.T. (1989). Polymorphic simple GATA/GACA repeats in plant genomes. *Nucleic Acids Research,* **17**: 10128-10128.

Lanaud, C. (1986). Genetic studies of *Theobroma cacao* L. with the help of enzymatic markers. I. Genetic control and linkage of nine enzymatic markers. *Café Cacao Thé,* **30**(4): 259-270.

Lanaud, C. and Berthaud J. (1985). Mise en evidence de nouveaux marqueurs genetiques chez *Theobroma cacao* L. Par les techniques d'electrophese. In: *Proceedings of the 9th International Cocoa Research Conference,* 12-18 February, 1984, Lomé, Togo. COPAL, Lagos Nigeria. pp. 249-253.

Lanaud, C., Boult, E., Clapperton, J., N'Goran, J.N.K., Cros, E., Chapelin, M., Risterucci, A.M., Allaway, D., Gillmore, M., Cattaruzza, A., Fouet, O., Clement, D. and Petithuguenin, P. (2005). Identification of QTLs related to fat content, seed size and sensorial traits in *Theobroma cacao* L. In: Proceedings of the 14th International Cocoa Research Conference, 13-18 October, 2003, *Cocoa Producers' Alliance,* Malaysia. pp. 1120-1126.

Lanaud, C., Clément, D., Flament, M.H., Risterucci, A.M., Kébé, I., Nyassé, S., Sounigo, O., Motilal, L.A., Thévenin, J.M., Paulin, D., Ducamp, M., N'Goran, J.K.A., Fargeas, D., Cilas, C. (2004). Genetic mapping of quantitative trait loci for black pod resistance in cocoa. In: Cilas,C. and Despréaux, D. (Eds.), *Improvement of cocoa tree resistance to Phytophthora diseases.* CIRAD, France. pp. 147-164.

Lanaud, C., Fouet, O., Clement, D., Boccara, M., Risterucci, A. M., Surujdeo-Maharaj, S., Legavre, T. and Argout, X. (2009). A meta-QTL analysis of disease resistance traits of *Theobroma cacao* L. *Molecular Breeding,* **24**: 361-374.

Lanaud, C., Motamayor, J.C. and Risterucci, A.-M. (2001). Implications of new insight into the genetic structure of *Theobroma cacao* L. for breeding strategies. In *Proceedings of the International Workshop on New Technologies for Cocoa Breeding,* 89: 107, Kota Kinabalu, Malaysia: London: Ingenic Press

Lanaud, C., Risterucci, A.M., N'Goran, J.A.K., Clement, D., Flament, M.H., Laurent, V. and Falque, M. (1995). A genetic linkage map of *Theobroma cacao* L. *Theoretical and Applied Genetics,* **91**: 987-993.

Lanaud, C., Risterucci, A-M., Piretti, I., Falque, M., Bouet, A. and Lagoda, P.J.L. (1999). Isolation and characterization of microsatellites in *Theobroma cacao* L. *Molecular Ecology,* **8**: 2141-2143.

Laurent, V., Risterucci, A.M. and Lanaud, C. (1993). Variability for nuclear ribosomal genes within *Theobroma cacao*. *Heredity*, **71**: 96-103.

Laurent, V., Risterucci, A.M. and Lanaud, C. (1994). Genetic diversity in cacao revealed by cDNA probes. *Theoretical and Applied Genetics*, **88**: 193-198.

Leal, J. B., dos Santos, L. M., dos Santos, C. A. P., Pires, J. L., Ahnert, D. and Correa, R. X. (2008). Genetic diversity among farm and germplasm bank accessions of cacao in Bahia, Brazil. Diversidade genetica entre acessos de cacau de fazendas e de banco de germoplasma na Bahia. *Pesquisa Agropecuaria Brasileira*, **43**(7): 851-858.

Lerceteau, E., Robert, T., Pétiard, V. and Crouzillat, D. (1997). Evaluation of the extent of genetic variability among *Theobroma cacao* accessions using RAPD and RFLP markers. *Theoretical and Applied Genetics*, **95**: 10-19.

Li, J. and Macdonald, J. (2015). Advances in isothermal amplification: novel strategies inspired by biological processes. *Biosensors and Bioelectronics*, **64**: 196-211.

Li, Y.-C., Korol, A. B., Fahima, T., Beiles, A. and Nevo, E. (2002). Microsatellites: genomic distribution, putative functions and mutational mechanisms: a review. *Molecular Ecology*, **11**(12): 2453-2465.

Lima, L. S., Gramacho, K. P., Carels, N., Novais, R., Gaiotto, F. A., Lopes, U. V., Gesteira, A. S., Zaidan, H. A., Cascardo, J. C. M., Pires, J. L. and Micheli, F. (2009). Single nucleotide polymorphisms from *Theobroma cacao* expressed sequence tags associated with witches' broom disease in cacao. *Genetics and Molecular Research*, **8**(3): 799-808.

Liu, K. and Muse, S. V. (2005). PowerMarker: An integrated analysis environment for genetic marker analysis. Bioinformatics, **21**: 2128-2129. V 3.25 http://www.powermarker.net (accessed November 07, 2012).

Livingstone III, D.S., Freeman, B., Motamayor, J. C., Schnell, R. J., Royaert, S., Takrama, J., Meerow, A. W. and Kuhn, D.N. (2012). Optimizing of a SNP assay for genotyping *Theobroma cacao* under field conditions. *Molecular Breeding*, 30: 33-52.

Livingstone, D. S., Motamayor, J. C., Schnell, R. J., Cariaga, K., Freeman, B., Meerow, A. W., Brown, J. S. and Kuhn, D. N. (2011). Development of single nucleotide polymorphism markers in *Theobroma cacao* and comparison to simple sequence repeat markers for genotyping of Cameroon clones. *Molecular Breeding*, **27**: 93-106.

Livingstone, D., Royaert, S., Stack, C., Mockaitis, K., May, G., Farmer, A., Saski, C., Schnell, R., Kuhn, D. and Motamayor, J. C. (2015). Making a chocolate chip: development and evaluation of a 6K SNP array for *Theobroma cacao*. *DNA Research*, **22**(4): 279-291.

Lockwood, G. and End, M. (1993). History, technique and future needs for cacao collection. In: Proceedings of International Workshop on Conservation, Characterization and Utilization of Cocoa Genetic Resources in the 21st Century, 13-17 September 1992, Port-of-Spain, Trinidad. *The Cocoa Research Unit*, Trinidad. pp 1-14.

Lockwood, G. and Gyamfi M. M. O. (1978).The CRIG cocoa germplasm collection with notes on the codes used in the breeding programme at Tafo and elsewhere. Technical Bulletin No. 10: Tafo: *Cocoa Research Institute.*

Loor Solorzano, R. G., Fouet, O., Lemainque, A., Pavek, S., Boccara, M., Argout, X., Amores, F., Courtois, B., Risterucci, A. M. and Lanaud, C. (2012). Insight into the wild origin, migration and domestication history of the fine flavour Nacional *Theobroma cacao* L. variety from Ecuador. *PLoS ONE*, **7**(11): e48438. doi:10.1371/*journal.pone*.0048438.

MacConaill, L. E., Aldred, M. A., Lu, X. and LaFramboise, T. (2007). Toward high-throughput SNP genotyping in the presence of inherited copy number variation. *BMC Genomics*, **8**: 211. doi: 10.1186/1471-2164/8/211

Maharaj, K., Maharaj, P., Bekele, F. L., Ramnath, D., Bidaisee, G. G., Bekele, I., Persad, C., Jennings, K. and Sankar, R. (2011). Trinidad selected hybrids: An investigation of the phenotypic and agro-economic traits of 20 selected cacao cultivars. *Tropical Agriculture (Trinidad)*, **88**: 175-185.

Marcano, M., Morales, S., Hoyer, M. T., Courtois, B., Risterucci, A. M., Fouet, O. and Pugh, T. (2009). A genome-wide admixture mapping study for yield factors and morphological traits in a cultivated cocoa (*Theobroma cacao* L.) population. Tree Genetics and Genomes, **5**: 329–337. Mei, Y., Zhou, J., Xu, H. and Zhu, S. (2012). Development of Sea Island cotton (*Gossypium barbadense* L.) core collection using genotypic values. *Australian Journal Crop Science*, **6**: 673-680.

Miranda, C., Urrestarazu, J., Santesteban, L. G., Royo, J. B. and Uribina, V. (2010). Genetic diversity and structure in a collection of ancient Spanish pear cultivars assessed by microsatellite markers. *Journal of the American Society Horticultural Science*, **135**: 428-437.

Motamayor, J., Brown, J., Kuhn, D. and Schnell, R. (2010). Linkage disequilibrium in *Theobroma cacao* L. populations. *Abstract submitted to Plant and Animal Genome Conference*, San Diego, California January 2010. Retrieved 17 December 2012 from http://www.sparc.ars.usda.gov/research/publications/publications.htm?seq_no_115=259537 and pf=1

Motamayor, J. C., Lachneaud, P., da Silva e Mota, J. W., Loor, R., Kuhn, D. N., Brown, J. S. and Schnell, R. J. (2008). *Geographic and genetic population differentiation of the Amazonian chocolate tree* (*Theobroma cacao* L). *PLoS ONE*, **3**(10): e3311. doi:10.1371/journal.pone.0003311.

Motamayor, J. C., Mockaitis, K., Schmutz, J., Haiminen, N., Livingstone III, D., Cornejo, O., Findley, S. D., Zheng, P., *et al.* (2013). The genome sequence of the most widely cultivated cacao type and its use to identify candidate genes regulating pod colour. *Genome Biology*, **14**: r53 doi:10.1186/gb-2013-14-6-r53.

Motilal, L.A. (2004a) Potential of cacao microsatellites amplification across diverse plant taxa. In: Thangadurai, D., Pullaiah, T. and Balatti, P. A. (Eds.), *Genetic Resources and Biotechnology* (*Vol.* 2). Regency Publications, New Delhi. pp. 24-49.

Motilal, L.A. (2004b). Tree identification by simple sequence repeats: a synopsis for 2000-2003. In Annual Report 2003, 13-21. St. Augustine, Trinidad and Tobago: Cocoa Research Unit, The University of the West Indies.

Motilal, L. and Butler, D. (2003). Verification of identities in global cacao germplasm collections. *Genetic Resources and Crop Evolution,* **50**: 799-807.

Motilal, L.A., Zhang, D., Mischke, S., Meinhardt, L.W. and Umaharan, P. (2013). Microsatellite-aided detection of genetic redundancy improves management of the International Cocoa Genebank, Trinidad. *Tree Genetics and Genomes,* **9**: 1395-1411.

Motilal, L.A., Zhang, D., Umaharan, P., Boccara, M., Mischke, S., Sankar, A. and Meinhardt, L. W. (2012). Elucidation of genetic identity and population structure of cacao germplasm within an international cacao genebank. Plant Genetic Resources: Characterization and Utilization, **10**: 232-241.

Motilal, L.A., Zhang, D., Umaharan, P., Mischke, S., Boccara, M. and Pinney, S. (2009). Increasing accuracy and throughput in large-scale microsatellite fingerprinting of cacao field germplasm collections. *Tropical Plant Biology,* **2**: 23-37.

Motilal, L.A., Zhang, D., Umaharan, P., Mischke, S., Mooleedhar, V. and Meinhardt, L.W. (2010). The relic Criollo cacao in Belize – genetic diversity and relationship with Trinitario and other cacao clones held in the International Cocoa Genebank Trinidad. *Plant Genetic Resources: Characterization and Utilization,* **8**(2): 106-115.

Motilal, L.A., Zhang, D., Umaharan, P., Mischke, S., Pinney, S. and Meinhardt, L.W. (2011). Microsatellite fingerprinting in the International Cocoa Genebank Trinidad: accession and plot homogeneity information for germplasm management. *Plant Genetic Resources: Characterization and Utilization,* **9**: 430-438. doi:10.1017//S1479/26211100058X

Nasu, S., Sukuzi, J., Ohta, R., Hasegawa, K., Yui, R., Kitazawa, N., Monna, L. and Minobe, Y. (2002). Search for and analysis of single nucleotide polymorphisms (SNPs) in rice (*Oryza sativa, Oryza rufipogon*) and establishment of SNP markers. *DNA Research,* **9**: 163-171.

N'Goran, J.A.K., Laurent, V., Risterucci, A.M. and Lanaud, C. (1994).) Comparative genetic diversity studies of *Theobroma cacao* L. using RFLP and RAPD markers. *Heredity,* **73**: 589-597.

Opoku, S.Y., Bhattacharjee, R., Kolesnikova-Allen, M., Motamayor, J.C., Schnell, R., Ingelbrecht, I., Enu-Kwesi, L. and Adu-Ampomah, Y. (2007). Genetic diversity in cocoa (*Theobroma cacao* L.) germplasm collection from Ghana. *Journal of Crop Improvement,* **20**: 73-87.

Oudemans, P. and Coffey, M.D. (1991a) Isozyme comparison within and among worldwide sources of three morphologically distinct species of *Phytophthora. Mycological Research,* **95** (1): 19-30.

Oudemans, P. and Coffey, M. D. (1991b) A revised systematics of twelve papillate *Phytophthora* species based on isozyme analysis. *Mycological Research,* **95**(9): 1025-1046.

Oudemans, P. and Coffey, M.D. (1991c) Relationships between *Phytophthora* species: evidence from isozyme analysis. In: Lucas, A., Shattock, R.C., Shaw D.S. and Cooke, L.R. (Eds.), *Phytophthora* (Eds.), Cambridge University Press, United Kingdom. pp. 184-203

Oudemans, P., Forster, H. and Coffey, M.D. (1994).) Evidence for distinct isozyme subgroups within *Phytophthora citricola* and close relationships with *P. capsici* and *P. citrophthora*. *Mycological Research*, **98**: 189–199.

Pajerowska-Mukhtar, K., Stich, B., Achenbach, U., Ballvora, A., Lübeck, J., Strahwald, J., Tacke, E., Hoffbert, H.-R., Ilarionova, E., Bellin,D., Walkemeier, B., Basekow, R., Kersten, B. and Gebhardt, C. (2009). Single nucleotide polymorphisms in the Allene Oxide Synthase 2 gene are associated with field resistance to late blight in populations of tetraploid potato cultivars. *Genetics*, **181**(3): 1115-1127.

Parker, J.D., Rabinovitch, P.S. and Burmer, G.C. (1991). Targeted gene walking polymerase chain reaction. *Nucleic Acids Research*, **19**(11): 3055-3060.

Paterson, A., Tanksley, S. and Sorrels, M.E. (1991). DNA markers in plant improvement. *Advances in Agronomy*, **44**: 39-90.

Pearce, S.R., Knox, M., Ellis, T.H., Flavell, A.J. and Kumar, A. (2000). Pea Ty1-copia group retrotransposons: transpositional activity and use as markers to study genetic diversity in *Pisum*. *Molecular and General Genetics*, **263**: 898-907.

Peleman, J.D., Sørensen, A.P. and van der Voort, J.R. (2005). Breeding by design: exploiting genetic maps and molecular markers through marker-assisted selection. In: Meksem, K. and Kahl, G. (Eds.), *The Handbook of Plant Genome Mapping: Genetic and Physical Mapping*. Weinheim, WILEY-VCH Verlag GmbH and Co. KGaA. pp. 109-129

Perry, M.D., Davey, M.R., Power, J.B., Lowe, K.C., Bligh, F.J., Roach, P.S. and Jones, C. (1998). DNA isolation and AFLP™ genetic fingerprinting of *Theobroma cacao* L. *Plant Molecular Biology Reporter*, **16**(1): 49-59.

Pootakham, W., Jomchai, N., Ruang-areerate, P., Shearman, J. R., Sonthirod, C., Sangsrakru, D., Tragoonrung, S. and Tangphatsornruang, S. (2015). Genome-wide SNP discovery and identification of QTL associated with agronomic traits in oil palm using genotyping-by-sequencing (GBS). *Genomics*, **105**: 288-295.

Pound, F.J. (1938) Cacao and witches' broom disease (*Marasmius perniciosus*) of South America with notes on other species of *Theobroma*. Port-of-Spain: Yuille's Printery.

Pound, F.J. (1943) Cacao and witches' broom disease (*Marasmius perniciosa*). Report on a recent visit to the Amazon territory of Peru, September 1942–February 1943. Port of Spain: Yuille's Printery.

Pound, F.J. (1945) A note on the cocoa population of South America. In: Report and *Proceedings of the 1945 Cocoa Conferenc*e. London, pp. 131-133.

Powis, T.G., Cyphers, A., Gaikwad, N.W., Grivetti, L. and Cheong, K. (2011). Cacao use and the San Lorenzo Olmec. *Proceedings of the National Academy of Sciences USA*, **108**: 8595-8600.

Price, A.L., Patterson, N.J., Plenge, R.M., Weinblatt, M.E., Shadick, N.A. and Reich, R. (2006). Principal components analysis corrects for stratification in genome-wide association studies. *Nature Genetics,* **38**: 904-909.

Pugh, T., Fouet, O., Risterucci, A.M., Brottier, P., Abouladze, M., Deletrez, C., Courtois, B., Clement, D., Larmande, P., N'Goran, J.A.K. and Lanaud, C. (2004). A new cacao linkage map based on codominant markers: development and integration of 201 new microsatellite markers. *Theoretical and Applied Genetics,* **108**: 1151-1161

Purseglove, J.W. (1974). Tropical Crops – Dicotyledons, pp. 543-586. Essex: Longman.

Pritchard, J.K., Stephens, M. and Donnelly, P. (2000). Inference of population structure from multilocus genotype data. *Genetics,* **155**: 945-959.

Quieroz, V.T., Guimarães, C.T., Ahnert, D., Schuster, I., Daher, R. T., Pereira, M.G., Miranda, V.R.M., Loguércio, L.L., Barros, *E.G.* and Moreira, M.A. (2003). Identification of a major QTL in cocoa (*Theobroma cacao* L.) associated with resistance to witches' broom disease. *Plant Breeding,* **122**: 268-272.

Ranade, K., Chang, M.-S., Ting, C.-T., Pei, D., Hsiao C.-F., Olivier, M., Pesich, R., Herbert, J., Chen, Y.-D. I, Dzau, V.J., Curb, D., Olshen, R., Risch, N., Cox, D.R. and Botstein, D. (2001). High-throughput genotyping with single nucleotide polymorphisms. *Genome Research,* **11**: 1262-1268.

Risterucci, A.M., Eskes, B., Fargeas, D., Motamayor, J.C. and Lanaud, C. (2001). Use of microsatellite markers for germplasm identity analysis in cocoa. In: *Proceedings of the Third International Group for Genetic Improvement of Cocoa (INGENIC) International Workshop on the New Technologies and Cocoa Breeding,* 16-17 October, 2000, 25-33. Kota Kinabalu, Malaysia.

Risterucci, A.M., Grivet, L., N'Goran, J.A.K., Pieretti, I., Flament, M.H. and Lanaud C. (2000). A high density linkage map of *Theobroma cacao. Theoretical and Applied Genetics,* **101**: 948-955.

Rossetto, M. (2001). Sourcing of SSR markers from related plant species. In: *Plant genotyping: the DNA fingerprinting of plants,* Henry, R. J. (ed.), CAB International: UK, pp. 211-224.

Russell, J.R., Hosein, F., Johnson, E., Waugh, R. and Powell, W. (1993). Genetic differentiation of cocoa (*Theobroma cacao* L.) populations revealed by RAPD analysis. *Molecular Ecology,* **2**: 89-97.

Ryskov, A.P., Jincharadze, A.G., Prosnyak, M.I., Ivanov, P.L. and Limborska, S.A. (1988). M13 phage DNA as a universal marker for DNA fingerprinting of animals, plants and microorganisms. *FEBS Letters,* **233**: 388-392.

Sankar, A.A. and Moore, G.A. (2001). Evaluation of inter-simple sequence repeat analysis for mapping in *Citrus* and extension of genetic linkage map. *Theoretical and Applied Genetics,* **102**: 206–214.

Santos, E.S.L, Cerqueira-Silva, C B.M., Mori, G.M., Ahnert, D., Mello, D.L.N., Pires, J.L., Corrêa, R.X. and de Souza, A.P. (2015). Genetic structure and molecular diversity of cacao plants established as local varieties for more than two centuries: *the genetic history of cacao plantations in Bahia*, Brazil. PLoS ONE, **10**(12): e0145276. doi:10.1371/journal.pone.0145276

Saunders, J.A., Mischke, S., Leamy, E.A. and Hemeida, A.A. (2004). Selection of international molecular standard for DNA fingerprinting of *Theobroma cacao*. *Theoretical and Applied Genetics*, **110**: 41-47.

Schmid, K.J., Sörensen, T.R., Stracke, R., Törjék, O., Altmann, T., Mitchell-Olds, T. and Weisshaar, B. (2003). Large-scale identification and analysis of genome-wide single-nucleotide polymorphisms for mapping in *Arabidopsis thaliana*. *Genome Research*, **13**: 1250-1257.

Schnell, R.J., Brown, J.S., Olano, C.T., Power, E.J., Krol, C.A., Kuhn, D.N. and Motamayor, J.C. (2004). Frequency of off-type progeny among original ICS 1 x SCA 6 reciprocal families and parental clones used for disease resistance selection in Trinidad. *INGENIC Newsletter* **9**: 34-39.

Schnell, R.J., Brown, J.S., Kuhn, D.N., Cervantes-Martinez, C., Borrone, J.W., Olano, C.T., Motamayor J.C., *et al.* (2007a). Current challenges of tropical tree crop improvement: Integrating genomics into an applied cacao breeding program. *Acta Horticulturae*, **738**: 129-144.

Schnell, R.J., Kuhn, D.N., Brown, J.S., Olano, C.T., Phillips-Mora, W., Amores, F.M. and Motamayor, J.C. (2007b). Development of a marker assisted selection program for cacao. *Phytopathology*, **97**: 1664-1669.

Schnell, R.J., Olano, C.T., Brown, J.S., Meerow, A.W., Cervantes-Martinez, C., Nagai, C. and Motamayor, J.C. (2005). Retrospective determination of the parental population of superior cacao (*Theobroma cacao* L.) seedlings and association of microsatellite alleles with productivity. *Journal of the American Society of Horticultural Science*, **130**: 181-190

Scott, K.D. (2001). Microsatellites derived from ESTs, and their comparison with those derived from other methods. In: *Plant genotyping: the DNA fingerprinting of plants*, Henry, R. J. (ed.), CAB International: UK, pp. 225-237.

Sereno, M.L., Albuquerque, P.S.B., Vencovsky, R. and Figueira, A. (2006). Genetic diversity and natural population structure of cacao (*Theobroma cacao* L.) from the Brazilian Amazon evaluated by microsatellite markers. *Conservation Genetics*, **7**: 13-24.

Shen, Y.J., Jiang, H., Jin, J.P., Zhang, Z.B., Xi, B., He, Y.Y., Wang, G., Wang, C., Qian, L., Li, X., Yu, Q.B., Liu, H.J., Chen, D.H., Gao, J.H., Huang, H., Shi, T.L. and Yang, Z.N. (2004). Development of genome-wide DNA polymorphism database for map-based cloning of rice genes. *Plant Physiology*, **135**: 1198-1205.

Silva, C.R.S. and Figueira, A. (2005). Phylogenetic analysis of *Theobroma* (Sterculiaceae) based on Kunitz-like trypsin inhibitor sequences. *Plant Systematics and Evolution*, **250**: 93-104. doi 10.1007/s00606-004-0223-2.

Soto-Cerda, B.J. and Cloutier, S. (2012). Association mapping in plant genomes, genetic diversity in plants, edited by Mahmut Caliskan, InTech. http://www.

intechopen.com/books/genetic-diversity-in-plants/association-mapping-in-plant-genomes (accessed February 13, 2013)

Sounigo, O., Ramdahin, S. and Christopher, Y. (2002). Etude de la diversite genetique du cacaoyer a l'aide des RAPD. In: *Proceedings of the 13th International Cocoa Research Conference*, 9-14 October 2000, Kota Kinabalu, Malaysia. COPAL, Lagos, Nigeria. pp. 119-125.

Sounigo, O., Risterucci, A-M., Clément, D., Fouet, O. and Lanaud, C. (2006a). Identification of off-types of clones used in the International Clone Trial using DNA analyses. In: *Eskes, A.B. and Efron, Y. Global Approaches to Cocoa Germplasm Utilization and Conservation.* Final report of the CFC/ICCO/IPGRI project on "Cocoa Germplasm Utilization and Conservation: a Global Approach" (1998-2004), edited by, Amsterdam, The Netherlands: CFC/London, UK: ICCO/ Rome, Italy: IPGRI. http://www.bioversityinternational.org/publications/ pubfile.asp?ID_PUB=1172 (accessed July 07, 2008). pp. 82-86.

Sounigo, O., Bekele, F.L., Iwaro, A.D., Thévenin, J-M., Bidaisee, G., Umaharan, R., Sankar, A., Sukha, D., *et al.* (2006b). Description of cocoa clones proposed for the "CFC/ICCO/IPGRI Project Collection. In: Eskes, A.B. and Efron, Y. (Eds.), *Global Approaches to Cocoa Germplasm Utilization and Conservation.* Final report of the CFC/ICCO/IPGRI project on "*Cocoa Germplasm Utilization and Conservation*: a Global Approach" (1998-2004). Amsterdam, The Netherlands: CFC/London, UK: ICCO,/Rome, Italy: IPGRI. pp. 67-81.

Sounigo, O., Umaharan, R., Christopher, Y., Sankar, A. and Ramdahin, S. (2005). Assessing the genetic diversity in the International Cocoa Genebank, Trinidad (ICG,T) using isozyme electrophoresis and RAPD. *Genetic Resources and Crop Evolution*, **52**: 1111-1120.

Susilo, A.W., Zhang, D., Motilal, L.A., Mischke, S. and Meinhardt, L.W. (2011). Assessing genetic diversity in Java fine-flavour cocoa (*Theobroma cacao* L.) germplasm by using simple sequence repeat (SSR) markers. *Tropical Agriculture and Development*, **55**: 84-92.

Sveinsson, S., Gill, N., Kane, N.C. and Cronk, Q. (2013). Transposon fingerprinting using low coverage whole genome shotgun sequencing in cacao (*Theobroma cacao* L.) and related species. *BMC Genomics*, **14**: 502. http://www.biomedcentral.com/1471-2164/14/502

Tachida, H. and Iizuka, M. (1992). Persistence of repeated sequences that evolve by replication slippage. *Genetics*, **131**: 471-478.

Takrama, J., Dadzie, A.M., Opoku, S.Y., Padi, S. K., Adomako, B., Adu-Ampomah, Y., Livingstone III, D.S., Motamayor, J.C., Schnell, R.J. and Kuhn, D.N. (2012). Applying SNP marker technology in the cacao breeding programme in Ghana. *African Crop Science Journal*, **20**: 67-75.

Takrama, J., Kun, Ji., Meinhardt, L., Mischke, S., Opoku, S.Y., Padi, F.K. and Zhang, D. (2014). Verification of genetic identity of introduced cacao germplasm in Ghana using single nucleotide polymorphism (SNP) markers. *African Journal of Biotechnology*, **13**(21): 2127-2136.

The *Arabidopsis* Genome Initiative (2000). Analysis of the genome sequence of the flowering plant *Arabidopsis thaliana*. *Nature*, **408**: 796-815.

The International SNP Map Working Group (2001). A map of human genome sequence variation containing 1.42 million single nucleotide polymorphisms. *Nature*, **409**: 928-933.

Toxopeus, H. (1985). Botany, types and populations. In: Wood, G.A.R. and Lass, R.A. (Eds.), *Cocoa*, 4th Edition. Longman, London. pp. 11-37

Thomas, E., van Zonneveld, M., Loo, J., Hodgkin, T., Galluzzi, G. and van Etten, J. (2012). Present spatial diversity patterns of *Theobroma cacao* L. in the *Neotropics reflect genetic differentiation in Pleistocene refugia followed by human-influenced dispersal*. PLoS ONE, **7** (10): e47676. doi:10.1371/journal.pone.0047676.

Trognitz, B., Cros, E., Assemat, S., Davrieux, F., Forestier-Chiron, N., Ayestas, E., Kuant, A., Scheldeman, X. and Hermann, M. (2013). Diversity of cacao trees in Waslala, Nicaragua: associations between genotype spectra, *product quality and yield potential*. *PLoS ONE*, **8**(1): e54079. doi 10.1371/journal.pone.0054079

Trognitz, B., Scheldeman, X., Hansel-Hohl, K., Kuant, A., Grebe, H. and Hermann, M. (2011). Genetic population structure of cacao plantings within a young production area in Nicaragua. *PLoS ONE*, **6**(1): e16056. doi: 10.1371/journal.pone.0016056

Turnbull, C.J. and Hadley, P. (2012). ICGD 2012: International Cocoa Germplasm Database CD-ROM. CRA Ltd./NYSE Liffe/University of Reading, UK.

Van Hall, C.J.J. (1914) *Cocoa*. Macmillan: London.

Van Hintum, T.J.L. (2000). Duplication within and between germplasm collections. III. A quantitative model. *Genetic Resources and Crop Evolution*, **47**: 507-513.

Van Hintum, Th. J. L., Brown, A.H.D., Spillane, C. and Hodgkin, T. (2000). *Core Collections of Plant Genetic Resources*. IPGRI Technical Bulletin, **3**.

Vicient, C.M., Suoniemi, A., Anamthawat-Jónsson, K., Tanskanen, J., Beharav, A., Nevo, E. and Schulman, A.H. (1999). Retrotransposon BARE-1 and its role in genome evolution in the genus *Hordeum*. *Plant Cell Online*, **11**: 1769-1784.

Vos, P., Hogers, R., Bleeker, M., Reijans, M., Van de Lee, T., Hornes, M., Friters, A., Pot, J., Paleman, J. and Kuiper, M. (1995). AFLP: a new technique for DNA fingerprinting. *Nucleic Acids Research*, **23**: 4407-4414.

Waits, L.P., Luikart, G. and Taberlet, P. (2001). Estimating the probability of identity among genotypes in natural populations: cautions and guidelines. *Molecular Ecology*, **10**: 249-256.

Wang, J. (2006). Informativeness of genetic markers for pairwise relationship and relatedness inference. *Theoretical and Population Biology*, **70**(3): 300-321. doi:10.1016/j.tpb.2005.11.003

Wang, J., Lin, M., Crenshaw, A., Hutchinson, A., Hicks, B., Yeager, M., Berndt, S., Huang, W.-Y., Hayes, R.B. and Chanock, S.J. (2009). High-throughput single nucleotide polymorphism genotyping using nanofluidic Dynamic Arrays. BMC *Genomics*, **10**: 1. doi: 10.1186/1471-2164-10-561.

Warren, J.M. (1994).) Isozyme variation in a number of populations of *Theobroma cacao* obtained through various sampling regimes. *Euphytica*, **72**: 121-126.

Warren, J., Misir, S. and Kalai (1995). Isozyme markers for self-compatibility and yield in *Theobroma cacao* (cacao). *Heredity*, **74**: 354-356.

Weber, J.L. and Wong, C. (1993). Mutation of human short tandem repeats. *Human Molecular Genetics*, **2**: 1123-1128.

Welsh, J. and McClelland, M. (1990). Fingerprinting genomes using PCR with arbitrary primers. *Nucleic Acids Research*, **18**: 7213-7218.

Whitkus, R., de la Cruz, M., Mota-Bravo, L. and Gómez-Pampa, A. (1998). Genetic diversity and relationships of cacao (*Theobroma cacao* L.) in southern Mexico. *Theoretical and Applied Genetics*, **96**: 621-627.

Whitlock, B.A. and Baum, D.A. (1999). Phylogenetic relationships of *Theobroma* and *Herrania* (Sterculiaceae) based on sequences of the nuclear gene *Vicilin*. *Systematic Botany*, **24**: 128-138.

Whitlock, B.A., Bayer, C. and Baum, D.A. (2001). Phylogenetic relationships and floral evolution of the Byttnerioideae ("Sterculiaceae" or Malvaceae s.l.) based on sequences of the chloroplast gene, *ndhF*. *Systematic Botany*, **26**: 420-437.

Wilde, J., Waugh, R. and Powell, W. (1992). Genetic fingerprinting of *Theobroma* clones using randomly amplified polymorphic DNA markers. *Theoretical and Applied Genetics*, **83**: 871-877.

Williams, J.G.K., Kubelik, A.R., Livak, K.J., Rafalski, J.A. and Tingey, S.V. (1990). DNA polymorphisms amplified by arbitrary primers are useful as genetic markers. *Nucleic Acids Research*, **18**: 6531-6535

Wood, G.A.R. (1985a). History and development. In Cocoa, Wood, G.A.R and Lass, R.A. (Eds.) 4th edition, Chapter 1, Longman, London. pp.1-10.

Wood, G.A.R. (1985b). Quality and inspection. In Cocoa, Wood, G.A.R and Lass, R.A. (Eds.) 4th edition, Chapter 14, Longman, London. pp. 505-527.

Wood, G.A.R. (1991). A history of early cocoa introductions. *Cocoa Growers Bulletin*, **44**: 7-12.

Yang, J.Y., Scascitelli, M., Motilal, L.A., Sveinsson, S., Engels, J.M.M., Kane, N., Dempewolf, H., Zhang, D., Maharaj, K. and Cronk, Q.C.B. (2013). Complex origin of Trinitario-type *Theobroma cacao* (Malvaceae) from Trinidad and Tobago revealed using plastid genomics. *Tree Genetics and Genomes*, **9**: 829-840. doi 10.1007/s11295-013-0601-4.

Yidana, J.A. and Kennedy, A.J. (1984-86). Isozyme variation in cocoa. In: Annual Report 1984-86, Cocoa Research Unit: The University of the West Indies. pp. 6-11.

Yidana, J.A., Kennedy, A.J. and Withers, L.A. (1987). Variation in peroxidase isozymes of cocoa (*Theobroma cacao* L.). In: *Proceedings of the 10th International Cocoa Research Conference*, 17-23 May 1987, St. Domingue, Dominican Republic. COPAL, Lagos, Nigeria. pp. 719-722.

Young, A.M. (1994).) The chocolate tree: A natural history of cacao. Washington DC, USA: Smithsonian Nature Books.

Young, N.D. (1999). A cautiously optimistic vision for marker-assisted breeding. *Molecular Breeding*, **5**: 505-510.

Yu, J. and Buckler, E.S. (2006). Genetic association mapping and genome organization of maize. *Current Opinion in Biotechnology*, **17**: 155-160. doi:10.1016/j.copbio.2006.02.003.

Yu, J., Hu, S., Wang, J., Wong, G. K.-S., Li, S., Liu, B., Deng, Y., Dai, L., Zhou, Y., Zhang, X., *et al.* (2002). A draft sequence of the rice genome (*Oryza sativa* L. ssp. *indica*). *Science*, **296**: 79-92. doi 10.1126/science.1068037.

Yu, K., Park, S. and Poysa, V. (2000). Marker-assisted selection of common beans for resistance to common bacterial blight: Efficacy and economics. *Plant Breeding*, **119**: 411-415.

Yu, J., Pressoir, G., Briggs, W.H., Bi, I.V., Yamasaki, M., Doebley, J.F., McMullen, M.D., *et al.* (2006). A unified mixed-model method for association mapping that accounts for multiple levels of relatedness. *Nature Genetics*, **38**: 203-208. doi:10.1038/ng1702

Zhang, D., Arevalo-Gardini, E., Mischke, S., Zúñiga-Cernades, L., Barreto-Chavez, A., Adriazola del Aguila, J. (2006a). Genetic diversity and structure of managed and semi-natural populations of cocoa (*Theobroma cacao*) in the Huallaga and Ucayali Valleys of Peru. *Annals of Botany*, **98**: 647-655.

Zhang, D., Mischke, S., Goenaga, R., Hemeida, A.A. and Saunders, J.A. (2006b). Accuracy and reliability of high-throughput microsatellite genotyping for cacao clone identification. *Crop Science*, **46**: 2084-2092.

Zhang, D., Boccara, M., Motilal, L., Butler, D.R., Umaharan, P., Mischke, S. and Meinhardt, L. (2007). *Microsatellite variation and population structure in the "Refractario" cacao of Ecuador. Conservation Genetics*, **9**(2): 327-337. doi 10.1007/s10592-007-9345-8

Zhang, D., Boccara, M., Motilal, L., Mischke, S., Johnson, E.S., Butler, D.R., Bailey, B. and Meinhardt, L. (2009a). Molecular characterization of an earliest cacao (*Theobroma cacao* L.) collection from Upper Amazon using microsatellite DNA markers. *Tree Genetics and Genomes*, **5**: 595-607. doi:10.1007/s11295-009-0212-2.

Zhang, D., Mischke, S., Johnson, E.S., Phillips-Mora, W. and Meinhardt, L. (2009b). Molecular characterization of an international cacao collection using microsatellite markers. *Tree Genetics and Genomes*, **5**: 1-10.

Zhang, D., Martínez, W.J., Johnson, E.S., Somarriba, E., Phillips-Mora, W., Astorga, C., Mischke, S. and Meinhardt, L.W. (2011). Genetic diversity and spatial structure in a new distinct *Theobroma cacao* L. population in Bolivia. *Genetic Resources and Crop Evolution*, **59**(2): 239-252.

Zhao, Y., Chen, F., Li, Q., Wang, L. and Fan, C. (2015). Isothermal amplification of nucleic acids. *Chemical Reviews*, **115**: 12491-12545.

Zhou, H. and Steffenson, B. (2013). Genome-wide association mapping reveals genetic architecture of durable spot blotch resistance in US barley breeding germplasm. *Molecular Breeding,* **32**: 139-154. doi:10.1007/s11032-013-9858-4.

Zhu, C., Gore, M., Buckler, E. S. and Yu, J. (2008). Status and prospects of association mapping in plants. *The Plant Genome,* **1**: 5-20 doi:10.3835/plantgenome2008.02.0089.

Ziegleder, G. (1990). Linalool contents as characteristics of some flavour grade cocoas. *Zeitschrift für Lebensmittel-Untersuchung und Forschung,* **191**: 306-309.

Zietkiewicz, E., Rafalski, A. and Labuda, D. (1994).) Genome fingerprinting by simple sequence repeat (SSR)-anchored polymerase chain reaction amplification. *Genomics,* **20**: 176-183.

Chapter 17

Coffee

☆ *N.S. Prakash, Jeena Devasia, Ramesh K. Aggarwal and Y. Raghuramulu*

1. Introduction

1.1. Overview of Global Coffee Industry

Coffee, the popular beverage crop, is cultivated in more than 80 counties in both tropical and subtropical regions. An exceedingly admired drink world over, coffee is the second largest traded commodity, next to petroleum products in global market with an annual turn over of about US$70 billion. Coffee is mainly produced

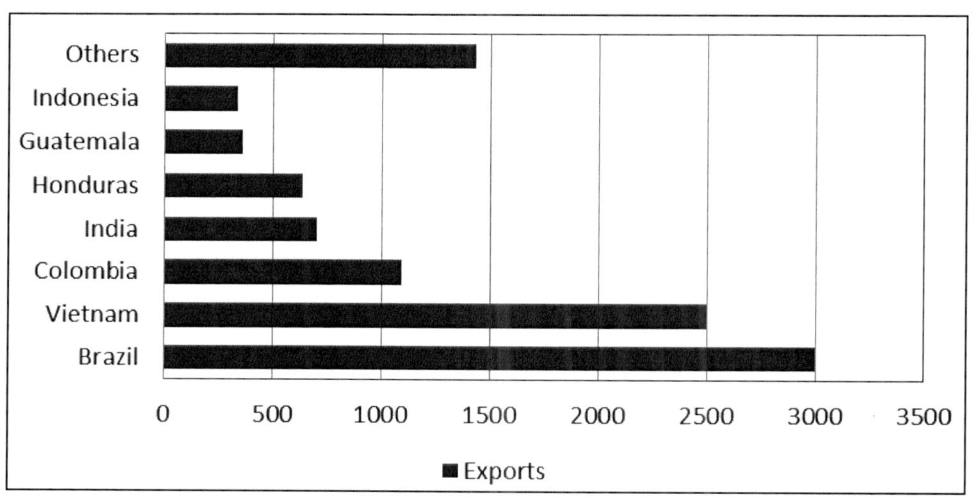

Figure 17.1: Coffee Exports from Leading Coffee Producers (March, 2016) (in '000 60 Kg bags, Data: ICO).

in developing countries while it is largely consumed in developed countries- thus export earnings from coffee form an important source of foreign exchange for producer countries (Figure 17.1). The global coffee production touched 143.2 million bags (60 kg) during 2014 (http://www.ico.org/trade_statistics.asp). The exports are growing at an average rate of about 3 per cent in terms of quantity and 15 per cent in terms of value annually, for the last 10 years period. Global consumption is growing at an average annual rate of 2.5 per cent and only 30 per cent of the produce is consumed within the producing countries. Coffee is predominantly cultivated as a smallholder enterprise, involving around 26 million farmers. Coffee industry also provides employment to an estimated 100 million people in areas of cultivation, processing and marketing.

1.2. Indian Coffee Industry

In India, coffee is one of the important plantation crops grown with a planted area of 0.4 million ha distributed mainly in the Southern states of Karnataka, Kerala, Tamil Nadu, Andhra Pradesh and Odisha (Figure 17.2). India produces around 5.3 million bags (0.31 million MT) contributing to 3.7 percent of world production (ICO Ann Rev, 2012-13) and claims sixth position after Brazil, Vietnam, Columbia, Indonesia and Ethiopiain production. Coffee is predominantly a smallholder enterprise with 99 per cent of coffee farmers possess holdings of less than 10 ha and depends on coffee for their livelihood.

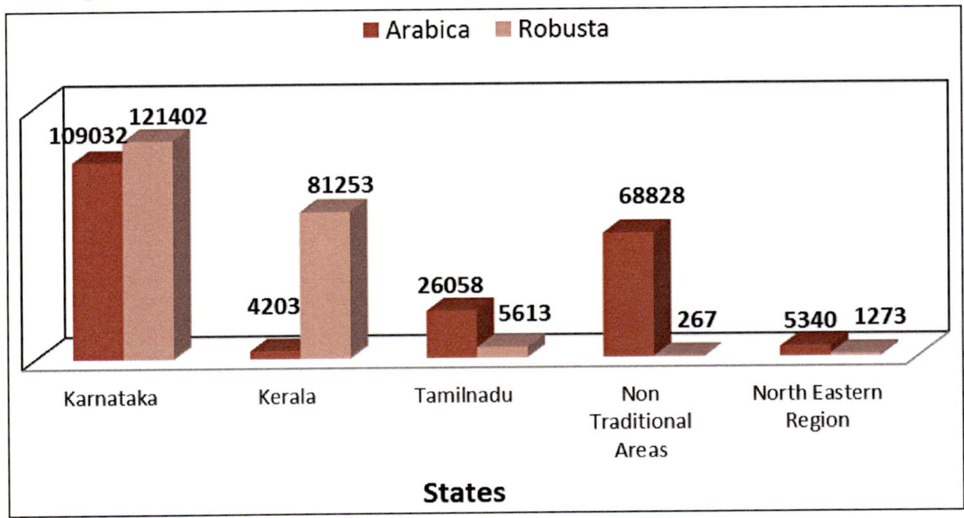

Figure 17.2: Distribution of Coffee Growing Area in India (In hectares) (Source: Database; February 2016).

2. Commercial Coffee Species

2.1. Origin and Distribution

Coffee belongs to the genus *Coffea* of family Rubiaceae representing over 100 species. However, commercial coffee comes from two main species, *Coffea arabica* L., known as arabica coffee, and *C. canephora* Pierre ex A. Froehner, popularly called as robusta coffee, which together accounts for 99 per cent of global coffee production. The third species, *C. liberica* Bull ex Hiern or Liberica coffee, is grown to a smaller extent in East Africa and Asia and accounts for about 1 per cent of global production. Arabica coffee (*Coffea arabica*) has been reported to have originated in the high lands of Abyssinia in South West Ethiopia, where it is grown at altitudes ranging from 1300 to 2000 m above MSL (Lashermes *et al.*, 1996a). Robusta coffee (*Coffea canephora*) has its origin in Central Africa and the species is distributed at altitudes below 1000 m MSL (Berthaud and Charrier, 1988). Apart from adaptability to high and low altitudes, the arabica and robusta coffee types differ in plant morphology, vegetative vigour, ploidy level, breeding behaviour, genetic diversity, yield potential, bean quality and also in genes conditioning resistance for major diseases and pests (Herrera *et al.*, 2011).

The species of the subgenus *Coffea* are distributed in four geographical regions of the inter-tropical African forests: the West and Central Africa regions that consists of the two most important species, *Coffea arabica* and *C. canephora*, and also *C. liberica*; Central Africa with *C. eugenioides*; the East African region that includes *C. salvatrix*, *C. racemosa*, and *C. costatifructa*, among others; and lastly, the Madagascar region (Lashermes *et al.*, 1997). While some species, such as *C. liberica* or *C. canephora*, show a wide distribution, others like *C. arabica* or *C. stenophylla*, are confined to limited areas (Charrier and Berthaud, 1985).

2.2. The Coffee Plant

The phenotype of coffee plants can vary from small perennial bushes to thick, hard, wooden trees, characterized by plagiotropic branches, paired inforescences, hermaphrodite flowers, white or slightly pink corollas, and a long style that sticks out the corolla. The fruit is an indehiscent drupe with two seeds. Each seed exhibits a characteristic deep groove in the ventral portion (Davis *et al.*, 2006). The morphology of coffee plants vary from species to species and some distinct features differentiate the two commercially cultivated species (*C. arabica* and *C.canephora*). Nevertheless, the intra-specific variation is generally low except for certain specific characters. Among the two commercial species, *Coffea arabica* grows into a small tree under natural growth. However, when plant growth is regulated through pruning and handling, the arabica plants attain a dense bushy stature. The second commercial species, *Coffea canephora,* shows robust growth with broad leaves and grows into a bigger bush compared to arabica. Interestingly, in the genus *Coffea*, *C. arabica* L. is the only self-fertile, allotetraploid species (2n = 4x = 44) while all other species including *C. canephora* are diploids (2n = 22) and are generally self-incompatible (Charrier and Berthaud, 1985). Based on molecular–cytogenetic analysis, Lashermes *et al.* (1999). established that *C. eugenioides* and *C. canephora* or ecotypes related to these diploid species were the likely progenitors of *Coffea arabica*.

2.3. Cytogenetics and Genome Structure

Like several other members the family Rubiaceae, the basic chromosome number of the genus *Coffea* is 11 (Krug, 1934). Nevertheless, all known species of the genus *Coffea* are diploids (2n = 2x = 22) and generally self-incompatible, except *C. arabica*, which is the only tetraploid species (2n = 4x = 44) and self-compatible in nature. Several studies have established the allotetraploid origin of *C. arabica*, which has evolved from a cross between two ancestral ecotypes of the diploid species *C. eugenioides* and *C. canephora*. Further, the molecular as well as cytogenetic studies provided complementary evidence on the existence of two less differentiated subgenomes (designated as Ea and Ca) which was a consequence of a recent speciation event (Lashermes *et al.*, 1999; Herrera *et al.*, 2007; Tesfaye *et al.*, 2007; Clarindo and Carvalho, 2008). Molecular studies have also confirmed that *C. arabica* exhibit diploid-like meiotic behaviour, which allows preferential pairing between homologous chromosomes (Lashermes *et al.*, 2000).

Interestingly, the genome structure among various *Coffea* species is very similar. Studies conducted earlier on several species revealed that somatic chromosomes are rather small (1–3 μm) and morphologically similar (Krug, 1937; Bouharmont, 1963). Thus, the similarities in chromosome size and shape have been reported to be the most important limitation to identify and distinguish individual species based on chromosome length or arm ratios (Herrera *et al.*, 2011). Experiments using several chromosome banding methods have also established that coffee chromosomes exhibit uniform pericentromeric patterns of heterochromatin, suggesting that most of the coffee species retained not only the same number but also the same (or a very similar) pattern of repetitive sequences during their evolution (Pierozzi *et al.*, 2001). However, recent advancements in cytological methods have enabled high-quality chromosome preparations in *C. canephora* and *C. arabica* and these studies have revealed clear morphological differences between the chromosomes of these two species and enhanced the possibility to assemble karyograms in other coffee species (Clarindo and Carvalho, 2006, 2008). The flow cytometric analysis also revealed important differences in DNA content among different species of *Coffea* (Cros *et al.*, 1995). The 2C values of DNA content was reported to vary from 0.95 ± 0.13 pg to 1.78 ± 0.33 pg among diploid species and 2.6 ± 0.23 pg in *C. arabica*. Based on geographic distribution as well as phenology and excluding the differences due to ploidy level, the variation in DNA content among coffee species might be due to changes in the copy number of repeated DNA sequences along its evolutionary history (Herrera *et al.*, 2011). Cytomolecular techniques, such as fluorescent *in situ* hybridization (FISH), have been successfully employed for studying genome organization of *C. arabica* (Raina *et al.*, 1998, Lashermes *et al.*, 1999), chromosome differentiation between species (Pinto-Maglio *et al.*, 2001), detection of alien chromatin in interspecifc hybrids (Barre *et al.*, 1998, Herrera *et al.*, 2007) and more recently, the localization of introgressed fragments (Herrera *et al.*, 2007). Recent evidence of high similarity in genome as well as chromosome structures between coffee and other solanaceous species (Lin *et al.*, 2005, Mueller *et al.*, 2005) will provide additional genomic information useful for development of the future physical and cytogenetic maps for coffee (Herrera *et al.*, 2011).

2.4. Genetic Resources and Diversity

As per the current inventory, the genus *Coffea* comprises of over 100 species; the taxonomic classification has, however, become increasingly complex since new species are still being discovered from West Africa, Central Africa, Madagascar and East Africa (Charrier and Eskes, 2004; Stoffelen *et al.*, 2008, 2009). As majority of these species represent high similarity at genome level, cross breeding between the species has been reported to be possible to a large extent and wild coffee trees, therefore, constitute a valuable resource for breeding. Although, nearly 50 per cent of known species are conserved worldwide in field gene banks, their precise genetic evaluation has been weakly documented. On the other hand, the natural habitats of wild coffee are threatened by deforestation and other human activities and of the 103 *Coffea* species now identified, 73 (70.9 per cent) are considered critically endangered, 36 are endangered, and 23 species including *C. arabica* are vulnerable (Davis *et al.*, 2006).

In general, *C. arabica* is characterized by low genetic diversity, attributed to the allotetraploid origin, reproductive biology and evolution process of this species (Lashermes *et al.*, 1996b). Genetic diversity within *C. arabica* cultivars and wild collections was extensively studied using various DNA marker approaches *viz.* random amplified polymorphic DNA (RAPD) (Lashermes *et al.*, 1996b; Zezlina *et al.*, 1999; Anthony *et al.*, 2001; Aga *et al.*, 2003; Sera *et al.*, 2003; Cristancho *et al.*, 2004). The amplified fragment length polymorphism (AFLP) and simple sequence repeat (SSR) markers were also employed for the analysis (Anthony *et al.*, 2002a, 2000b, Steiger *et al.*, 2002, Prakash *et al.*, 2002, Aggarwal *et al.*, 2004) and all these analyses indicated low genetic diversity within arabica varieties.

With respect to cultivated species, the primary genepool comprising of wild and cultivated varieties of *C. arabica* and *C. canephora* forms the major genetic resources. The genetic resources of *C. arabica* comprise wild plants collected in the centre of diversity *i.e.*, Ethiopia, Kenya, and Sudan, landraces conserved on farm holdings in Ethiopia (Labouisse *et al.*, 2008). Moreover, plants grown in the primary centre of dispersion *i.e.*, Yemen (Eskes 1989), varieties and mutants selected worldwide in the populations of Typica and Bourbon that spread across the world in the eighteenth century (Krug *et al.*, 1939) and spontaneous hybrids resulted from co-cultivation of *C. arabica* and other diploid species also constitute genepool. Apart from the field gene bank with over 4000 collections of *C. arabica* established in Ethiopia the centre of diversity, several coffee growing countries participated in germplasm exchange programmes as well as expeditions/collecting missions that enabled the establishment of field gene banks in few countries like Brazil, Costa Rica, Côte d'Ivoire Kenya, Madagascar, Cameroon, Colombia and India.

Coffea canephora has one of the widest areas of natural distribution of the sub-genus *Coffea* as it extends West to East from Guinea to Sudan and north to south from Cameroon to Angola (Berthaud, 1986). In the global context, the robusta coffee genepool is conserved in *ex-situ* collection plots, in different countries *viz.*, Côte d'Ivoire, Cameroon, Uganda, India, Indonesia, Brazil *etc.* and the wealthiest of which is the Divo collection in Côte d'Ivoire. It is reported to contain more than 700 wild genotypes collected by ORSTOM (presently IRD – Institut de Recherche pour le Développement), France in collaboration with CIRAD (Centre de Cooperation International Enrecherche Agronomique pour le Développement), France; IBPGR (International Board for Plant Genetic Resources), Italy; FAO (Food and Agriculture Organization), Italy and MNHN (Muséum national d' histoire naturelle), France, between 1975 and 1987, in five African countries *i.e.*, Côte d'Ivoire and Guinea in

West Africa and Cameroon, Congo and Central African Republic in Central Africa. Besides the above, CIRAD made a collection of more than 600 accessions of diverse origin; local varieties and populations, forms from village plantations and selected materials developed at other research stations, and the same are also conserved in the Experimental Station of Divo (Dussert *et al.*, 1999).

The primary analysis of genetic diversity in robusta based on enzymatic polymorphism (Berthaud, 1986, Montagnon *et al.*, 1992) enabled the definition of the structure of the species. It has been reported that the collections of *C. canephora* could be distinguished into two groups, the "Guinean group", with wild populations from the Côte d'Ivoire, and the "Congolese group", with wild as well as cultivated types from the Central African Republic and Cameroon. Based on restriction fragment length polymorphism (RFLP) the wild and cultivated forms of robusta coffee were grouped into five diversity groups: A, B, C, D, and E (Dussert *et al.*, 1999). The genetic diversity of 40 robusta accessions from India was compared with 14 representative samples of core collections of *C. canephora* using AFLP and SSR markers (Prakash *et al.*, 2005). All the studies revealed high diversity in robusta collections compared to arabica.

The genetic resources of *C. liberica,* the third cultivated species of *Coffea,* comprise wild plants collected from the center of diversity *i.e.,* West and Central Africa. Besides, the collections from plantations covering the two bio-geographical groups *i.e.,* EA group (*C. liberica* var. liberica) and EB group (*C. liberica* var. dewevrei), reported to be distributed in Guinea and Central African Republic respectively, are also included (Stoffelen, 1998).

Apart from the cultivated species, efforts towards collection and conservation of non-cultivated and wild coffee species have been limited. Major explorations taken up in forests of Madagascar during 1960s, jointly by the French Museum of Natural History, CIRAD and IRD (Charrier, 1978), have resulted in establishment of collections representing about 50 species in a field gene bank at Kianjavato, Madagascar. Subsequently, survey missions were organized in African countries such as Guinea (Le Pierrès *et al.*, 1989), Côte d'Ivoire (Berthaud 1986), Cameroon (Anthony *et al.*, 1985), Central African Republic (Berthaud and Guillaumet 1978), Republic of Congo (de Namur *et al.*, 1987), and Tanzania and Kenya (Anthony *et al.*, 1987). Thus, collections were introduced and maintained in field gene banks in Côte d'Ivoire and few other countries.

The molecular phylogeny and phylogenetic relationships of *Coffea* species were inferred using the internal transcribed spacer (ITS) 2 region of the nuclear ribosomal DNA by Lashermes *et al.* (1996a) as well as the chloroplast DNA variation by Cros *et al.* (1998). The results suggested a radial mode of speciation and a recent origin in Africa for the genus *Coffea* (Etienne *et al.*, 2002). Several major clades were identified with strong geographical correspondence (*i.e.*, West Africa, Central Africa, East Africa, and Madagascar).

In addition to the accessions belonging to cultivated and wild species, the collections grouped under the subgenera *Psilanthus* provides additional genepool. The sub genus *Psilanthus* includes approximately 20 species with wide distribution, covering the tropical humid regions of Africa, India, South East Asia, and the Pacific. In India, several indigenous coffee species, earlier grouped under the genus *Coffea* (currently regrouped under the sub genus *Psilanthus*), are reported to be distributed (Narasimhaswamy and Vishweswara, 1963). The *Psilanthus* species belong to two distinct phytogeographical groups, with the first group comprising of *P. bababudanii*, *P. bengalensis* (Figures 17.3a and b), *P. khasiana*, and *P. fragrans*, all distributed mostly in the eastern Himalayas. The second group includes *P. malabaricus*, *P. travancorensis*, and *P. wightianus*, which are all confined to Western and Eastern Ghats in southern India. Some of these species, *P. bengalensis*, *P. travancorensis*, *P. wightianus*, and *P. khasiana* have been collected and established in coffee gene banks in India.

Figure 17.3: Flowering (Left) and Bearing Pattern (Right) in Wild Species *Psilanthus bengalensis* Endemic to India. The fruits are less in number at each node, mostly single seeded, and beans are known to contain low caffeine or no caffeine.

Thus, considerable variability has been reported among diploid species and some of the diploid species form valuable gene reservoir for different breeding purposes (Berthaud and Charrier, 1988). Nevertheless, some diploid species of *Coffea* are cross compatible and produce relatively fertile progeny (Anthony, 1992, Louarn, 1993).

3. Genetic Improvement by Conventional Breeding

In natural habitat, arabica grows into a small tree but looks like a bushy shrub on plant training. On the other hand, robusta coffee trees as the name denotes are big and robust in stature with broader leaves than arabica trees. Arabica is adaptable to high altitudes while robusta is a low land coffee. Additionally, the arabica and robusta coffee trees differ in plant morphology, vigour, ploidy status, breeding behaviour, genome diversity, yield potential and also in resistance genes. The bean as well as liquor quality of arabica is superior to that of robusta but arabica plants are more susceptible to all major pests and diseases of coffee such as coffee leaf rust (*Hemileia vastatrix Berk and Br*), coffee berry disease (*Colletotrichum kahawae* Waller et Bridge), white stem borer (*Xylotrechus quadripes* Chevrolat) and nematodes (*Meloidogyne* sp. and *Pratylenchus* sp.). Robusta is more tolerant to these diseases and pests with high yield potential. Thus, the main focus of genetic improvement has been on high production coupled with a broad spectrum of resistance in arabica and high production with improved bean and liquor quality in robusta. Further, based on the objective and breeding behaviour of the two commercially important species of coffee, appropriate strategies have been used for breeding new coffee varieties. A comprehensive account of coffee breeding has been published by Herrera *et al.* (2012), while Montagnon *et al.* (2012) have reviewed the status of coffee breeding for quality and recently. Recently, Vander vossen *et al.* (2015) reviewed the prospects of breeding and disseminating next generation (hybrid) cultivars of arabica coffee for sustainable coffee production under changing conditions of diseases, pests and climate while Prakash *et al.* (2016) have summarized a brief overview of coffee genetic improvement in India

3.1. Arabica Coffee Improvement

Though the spread of coffee cultivation across the continents started in 16th century, organized efforts towards development of new arabica coffee varieties were initiated only in early 19th century. Worldwide, the first phase of arabica breeding was undertaken during the period from 1920 to 1950s, with the main focus on yield, quality improvement and adaptability to local conditions. The main strategies adopted were selection and crossing within the base populations. However, in countries like India, Indonesia and Sri Lanka, the primary focus of arabica coffee breeding was on leaf rust resistance because of the onslaught of this disease in Asian sub-continent during late 18th century. The early breeding efforts for improvement of arabica resulted in development of vigorous and productive cultivars like 'Mundo Novo', 'Caturra' and 'Catuai' from Brazil, 'Kents', S.288 and S.795 from India, 'Blue Mountain' from Jamaica and several others. Some of these varieties are still under commercial cultivation not only in countries of their origin but also in other countries.

Subsequently, because of the gradual spread of leaf rust disease to various other coffee growing countries across Africa, Central and South America, and also the spread of coffee berry disease (CBD) in Africa, the focus of arabica coffee breeding was shifted towards disease resistance especially coffee leaf rust (CLR) and coffee berry disease (CBD). This second phase of arabica breeding undertaken during the period from 1960s to 1990s was very successful and resulted in several high yielding varieties with broad spectrum of host resistance in a relatively short period. The success was mainly because of the coordination between the coffee breeding groups across the continents that were involved in in exchange of coffee genetic resources (Meyer *et al.*, 1968) and exploitation of new genetic diversity by application of advanced selection and breeding methods (Van der Vossen, 1985). In addition, a wealth of basic information on coffee genetics generated in Brazil (Sybenga, 1960; Carvalho *et al.*, 1969) and also the establishment of Coffee Rust Research Centre (CIFC) in Oeiras, Portugal to work exclusively on various aspects of coffee leaf rust pathogen (*Hemileia vastatrix*), have contributed significantly in taking forward the rust resistance breeding programmes in many countries. Concurrently, identification of several dwarf mutants of arabica *viz.*, 'Caturra', 'San Ramon' and 'Villasarchi' with compact bush stature and high yielding potential and spontaneous hybrids of robusta and Arabica such as Hibrido de Timor (HDT) and Devamachy with high vigour, coupled with disease resistance, facilitated the development of several high yielding and disease tolerant cultivars with compact bush stature. The varieties like

Catimor, Ruiru 11, Colombia, Sarchimor *etc.* belongs to this group and have been extensively cultivated in different countries.

3.1.1 Arabica Coffee Improvement in India

According to legend, a holy saint Baba Budan went on a pilgrimage to Mecca and is reported to have brought seven seeds from Yemen, and planted on his hermitage located in Chandragiri hills now named after the saint as Baba Budan Giri in Chikmagalur District of Karnataka state. The arabica coffee plants remained as back yard plants for long time and it was in late 1820s, British entrepreneurs ventured upon coffee cultivation by exploring the ideal locations in South India. Coffee cultivation progressed rapidly during the next 40 years and coffee trade flourished. In later part of 18th century, coffee leaf rust disease caused by an obligate parasitic fungus, *Hemileia vastatrix*, started devastating arabica coffee plantations in India and other South-East Asian countries. In India, efforts were made by some enterprising planters to select the disease tolerant plants from existing populations. Thus the cultivars 'Coorgs', 'Chicks' and 'Kents', named after the place of selection or the person responsible, became popular and were largely used for cultivation in 1920s. In a short time, these varieties also succumbed to rust disease and it was during this time that rust resistant diploid *Coffea* species like *C. liberica* and *C. canephora* were introduced. Some efforts were also made to develop hybrids between arabica and diploid species but with little success due to variation in ploidy level (tetraploid vs diploid). Nevertheless, hybrids like 'Hamiltons', 'Jacksons', 'Netrakonda' and 'Chandrapore' were developed and used for cultivation. In spite of the best efforts, the leaf rust disease could not be effectively managed and has remained the major concern for arabica coffee cultivation. Hence, systematic research on coffee was the felt need and the Mysore Coffee Experiment Station was established at Balehonnur in Karnataka state, during 1925 with the primary mandate of developing rust resistant varieties and to address the problems in coffee cultivation (Anonymous, 2014). The Coffee Board of India, took over the reigns of the 'Mysore Coffee Experiment Station' in the year 1946 and rechristened it as Central Coffee Research Institute (CCRI).

Towards accomplishing its mandate, the research group primarily focused on collection of the existing variability from among the arabica plantations and established an indigenous gene bank in the research station during 1925-1940. In early breeding programmes, these indigenous collections were exploited and used for the development of rust tolerant varieties like S.288 and S.795. Subsequently, the exotic germplasm that was introduced during 1954 to 1965 from different countries including Ethiopia, the centre of diversity for arabica coffee, as well as few spontaneous tetraploid interspecific hybrids, were used for breeding programmes most importantly as sources of resistance. Though arabica is known for its narrow genetic base, CCRI successfully utilized the available variability in line with the scope and objectives of breeding and developed 13 improved arabica genotypes for commercial cultivation by employing proven breeding methods (Anonymous, 2014; Srinivasan and Narasimhaswamy, 1975). The improved arabica genotypes developed at CCRI, were designated as Sln.1 to Sln.13 and the breeding strategies commonly used for self pollinating species such as inter-varietal hybridization followed by pedigree selection (Sln. 3), pure line selection (Sln.1, Sln. 4, Sln. 8), interspecific hybridization followed by back-cross breeding (Sln. 6), double crosses (Sln. 10), multiple crosses (Sln. 7.3) and introgressive breeding (Sln. 5A, Sln.5B, Sln. 9, Sln. 12 and Sln.13) using spontaneous tetraploid interspecific hybrids were successfully exploited for arabica coffee improvement. Depending on the adaptability to the different agro-climatic conditions, the station bred selections are recommended for cultivation across the coffee tracts. At present, the improved selections are planted in over 90 per cent of the area of which Sln.3 (S.795), Sln.5A, Sln.5B, Sln.6, Sln.9 and Sln.13 (Chandragiri) (Figure 17.4) are very popular among the coffee growers.

Figure 17.4: Sln.13 (Chandragiri), an Improved Variety of Arabica Coffee.

3.2. Robusta Coffee Improvement

As in case of arabica, robusta coffee improvement was also pursued in early 19th century, with main objectives of yield and bean quality improvement. The pioneering work on coffee biology and selection, carried out in East Java in the early years of 19th century [reviewed by Cramer (1957)] formed the basis for robusta coffee improvement programmes taken up subsequently in India and in Africa (Van der Vossen, 1985). It was reported that the nucleus robusta stock introduced into Java in 1901 came from trees already under cultivation in Zaire in 1895, originating in Lomani river region. The material selected in Java was reintroduced in the Belgian Congo around 1916 at INEAC (Institut National pour l'EtudeAgronomique du Congo Belge), which has become the major selection centre of *C. canephora* from 1930 to 1960 (Montagnon *et al.*, 1998). It is interesting to note that the improved seed from Java was mainly used to establish robusta plantations in India, Uganda, Côte d'Ivoire and also in Zaire, from where robusta coffee originated (Van der Vossen, 1985). Thus, the cultivated robusta across the world have been reported to belong to diversity group 'E' (Dussert *et al.*, 1999).

Robusta coffee being allogamous in nature, breeding methods commonly employed for cross-pollinated crops, such as mass selection and intra as well as interspecific hybridization, have been used for its improvement. In mass selection strategy, elite plants with respect to plant vigor, yield potential and bean quality characters are selected in base populations and advanced through open pollinated seed. Some of the well known varieties like Apoata of Brazil, S.274 of India, and Nemaya of Central America were derived using this strategy. In another strategy, single plant progenies were subjected to selection and plants that yielded higher than the family mean yields were selected and used for bi- or polyclonal gardens. Either seed mixture or mixture of clones were released for commercial cultivation as clonal varieties in different countries like Balehonnur robusta clones (BR series) of India, SA and BP selections of Indonesia and IF clones of Côte d'Ivoire. The intra-specific hybridization strategy was primarily based on exploiting the available genetic diversity within the species. In robusta, two major diversity groups—the Guinean group (from West Africa) and the Congolese group (from central African countries) have been identified. The Congolese coffee types are known for better agronomic value than Guinean types and majority of the cultivated *C. canephora* populations in the world belong to Congolese group. The Guinean types are limited to Côte d'Ivoire and Guinea in either wild or cultivated forms. Dussert *et al.* (1999). conducted extensive studies on the robusta populations in Côte d'Ivoire based on morphological characters and molecular markers and reported two genetic groups (SG 1 and SG 2) and four subgroups within the Congolese types. Berthaud (1986) also emphasized that Guinean genotypes are of great value for robusta coffee breeding and in fact it was reported that the most productive clones selected in Côte d'Ivoire during 1960s were the hybrids between Congolese and Guinean types. Leroy *et al.* (1993) reported that high amount of heterosis for plant vigour and yield was achieved in inter-group hybrids compared to intra-group hybrids.

Inter-specific hybridization strategy was also exploited for robusta improvement. At the beginning of 20th century, a spontaneous diploid interspecific hybrid between *C. canephora* var. *ugandae* and *C. congensis*, called Congusta or Conuga was identified which subsequently proved to be of considerable commercial value (Cramer, 1957). This strategy was successfully exploited for development of a fertile hybrid variety C x R in India. In Brazil and Côte d'Ivoire, the tetraploid breeding strategy was adopted to develop tetraploid interspecific hybrids between arabica and tetraploid robusta coffee (colchicines-induced autotetraploids) with the twin objectives of improving quality of robusta and transferring the vigour and disease resistance into arabica (Capot, 1972; Monaco *et al.*, 1974). Popularly called "arabustas," these hybrids exhibited superior performance with respect to vigour, adaptation to tropical lowlands, and cup quality. A selection program undertaken in Madagascar also resulted in clonal hybrid varieties HA, HB, H865, *etc.*

3.2.1. Robusta Coffee Improvement in India

The epidemics of *Hemileia vastatrix* (orange rust) in 1860s threatened the arabica coffee cultivation in India that paved way for other tolerant species like *Coffee canephora* (Robusta) and *Coffea liberica*. These diploid species especially *C. canephora* adapted well to the low altitude regions. It is believed that, ancient introductions into India might be from the initial nucleus material of Java and the primary genepool comprised of the descendants of these introductions under cultivation in farmers fields. This was confirmed by the genetic similarity exhibited by the Indian genepool with the collections of diversity group 'E' (Prakash *et al.*, 2005). Ceylon was also reported to be the other source of robusta, introduced into the country during early 19th century, which is popularly known as Peradeniya Robusta. At present, both these robusta types still occupy larger area under robusta and is broadly known as Old robusta. The bushes of old Robusta grow vigorously into moderately large trees. Fruits are red, round to oblong with pronounced navel, small to medium in size, 30-40 per node and borne in tight clusters. Beans are smaller in size, comprises of less than 50 per cent AB grade and liquor quality rated as Fair Average Quality (FAQ) to Good. Hence, robusta improvement programmes in India were mainly directed towards bean quality improvement and development of compact bush types with early bearing habit. Breeding methods common to cross pollinated crops, such as mass selection and intra as well as inter-specific hybridization, have been followed and three superior robusta selections have been developed by CCRI for commercial cultivation in India. The first selection popularly known as S.274 was developed by following mass selection strategy. The plants outstanding in terms of vigour, yield and bean quality characters were selected in base populations and progenies were advanced through open pollinated seed. In case of second selection, known as BR series, the superior plants, which yielded higher than the family mean yields in single plant progenies, were selected and used for establishing biclonal or polyclonal gardens. Seed mixture or mixture of clones of these bi or polyclonal gardens was given for commercial cultivation. The third selection popularly called as C x R was developed by inter-specific hybridization between *C. canephora* and *C. congensis*. The F_1 hybrid between of *C.congensis* and robusta was back crossed to robusta and BC progeny was subjected to mass selection followed by sib mating, that resulted in a highly fertile and compact C x R hybrid variety. The C x R variety (Figure 17.5) exhibits good vegetative vigour and compact bush stature and is, therefore, suitable for planting at closer spacing than other robusta varieties. Further, the bean size is bold in C x R coupled with superior liquor characteristics than other robusta varieties, that contributes for value addition.

Figure 17.5: Sln.3R (CxR), an Improved Variety of Robusta Coffee.

The current thrust of coffee improvement in India is on maximizing productivity through development of high yielding hybrids coupled with durable host resistance in arabica and evolving drought tolerant robusta genotypes to cope with the changing climate, more efficiently.To accomplish these objectives, systematic breeding programmes are being pursued with emphasis on development of heterotic F_1 hybrids in both arabica and robusta, using genetically distant genotypes and male sterile lines identified in wild genepool/land races from Ethiopia. Towards breeding for durable rust resistance, pyramiding of resistance genes in selected arabica cultivars has been pursued using marker assisted selection. Search for new sources of tolerance/resistance to white stem borer (*Xylotrechus quadripes*) in coffee gene pool for exploitation in breeding and improving the efficiency of conventional breeding through integration of genomic information are some of the priorities of coffee genetic improvement in India.

3.3. Coffee Genetic Improvement: Scope and Limitations

Inspite of the limited genetic diversity available in arabica and several other inherent constraints, remarkable successes have been achieved with respect to the development of new varieties that contributed significantly for sustainable growth of the world coffee industry. Nevertheless, the changing climatic conditions such as rise in temperatures and erratic rainfall patterns, disease and pest outbreaks due changes in climate are posing new challenges for coffee cultivation. Hence, there is a constant need for genetically improved coffee varieties to meet the ever-changing demands of the growers, markets and environment. On the other hand, the available opportunities are limited for conventional genetic improvement of coffee. The problem is more pronounced in arabica coffee because of its tetraploid status and very narrow genetic base, making genetic improvement more difficult to realize. Moreover, the long generation cycles, difficulties in realizing homozygous lines, variation in ploidy level, incompatibility barriers, reproductive barriers for interspecific crossing, low frequency of inter-genome crossing over are some of the well known constraints for conventional breeding. The situation called for newer, easy and efficient alternatives and the recent advances in DNA markers/technologies and coffee genomics provide new opportunities to overcome some of limitations of conventional coffee improvement with requisite speed and efficiency. Recently, Van der vossen *et al.* (2015) reviewed the prospects of breeding and disseminating next generation (hybrid) cultivars of arabica coffee for sustainable coffee production under changing conditions of diseases, pests and climate. It was suggested that international networking on coffee breeding facilitate sharing of financial and genetic resources as well as scientific information, application of genomics-assisted selection technologies and pre-breeding for specific characters. Thus, there exists a great promise for integration of conventional breeding approaches with advanced genotyping methods based on DNA markers and genomics, to hasten the genetic improvement of coffee.

4. Molecular Markers in Coffee

The economic significance of coffee crop has attracted considerable interest towards application of a wide range of DNA marker approaches for polymorphism studies and other molecular genetic analysis. Various DNA markers such as restriction fragment length polymorphism (RFLP), random amplified polymorphic DNA (RAPD), amplified fragment length polymorphism (AFLP), cleaved amplified polymorphism (CAP), inverse sequence-tagged repeat (ISTR), simple sequence repeat (SSR) or microsatellites and SNPs have been successfully used in coffee. Initial studies in this regard were mainly focused on genetic diversity analysis of coffee germplasm, origin of tetraploid *C. arabica* species and analysis of alien genome introgression. Subsequently, efforts were also made towards development of coffee specific genomic SSR markers and identification of markers linked to the genes of interest. Further, during the last decade there has been a considerable increase in number of codominant markers available from SSR mining in coffee expressed sequence tag (EST) databases that provided new opportunities for application of these markers for molecular-genetic analysis of coffee.

Initial efforts with biochemical markers such as isozymes in coffee revealed that these markers failed to distinguish between different genotypes within the species *C. arabica* (Berthou and Trouslot, 1977; Berthaud and Charrier, 1988; Steiger *et al.*, 2002). The advent of DNA marker technologies soon proved to be more efficient and facilitated overcoming the limitations of the conventional markers. To begin with, use of RAPD markers in different coffee accessions representing major *Coffea* species was successful in detecting inter-specific variation between *C. canephora* and *C. liberica* (Lashermes *et al.*, 1993). Subsequently, RFLPs (Paillard *et al.*, 1996, Lashermes *et al.*, 1999, 2000; Hererra *et al.*, 2002a; Dussert *et al.*, 2003) and multilocus RAPDs (Orozco-Castillo *et al.*, 1994; Lashermes *et al.*, 1996b; Agwanda *et al.*, 1997; Anthony *et al.*, 2001; Aga *et al.*, 2003; Sera *et al.*, 2003;Chaparro *et al.*, 2004) were extensively used in coffee, for diversity analysis and introgressive breeding. Later, the AFLP marker approach (amplified fragment length polymorphism) proved to be more efficient for diversity studies and introgression analysis (Prakash *et al.*,

2002, 2005; Coulibaly *et al.*, 2003a; Maluf *et al.*, 2005) and also for development of trait linked marker for rust resistance (Prakash *et al.*, 2004; Mahe *et al.*, 2008). AFLP marker system was also found to be the most-efficient method to estimate genetic diversity in comparative studies using RFLP, RAPD, AFLP and microsatellites (Powell *et al.*, 1996). Microsatellites or SSR (Simple Sequence Repeat) markers became a marker of choice (Powell *et al.*, 1996) due to its reproducibility, multi-allelic nature, co-dominant inheritance, relative abundance and wide genome coverage. The advantage of SSR markers was effectively demonstrated in coffee for genetic mapping, diversity assessment, population genetics, and marker assisted selection (de Vienne, 1998; Mettulio *et al.*, 1999; Gupta and Varshney, 2000; Combes *et al.*, 2000; Anthony *et al.*, 2002; Baruah *et al.*, 2003; Moncada and McCouch, 2004; Poncet *et al.*, 2004). Further, more recent marker approach, the Single Nucleotide Polymorphisms (SNPs) was also applied successfully for analysis of gene expression in *Coffea* (Vidal *et al.*, 2010).

The development and use of molecular markers in coffee were comprehensively reviewed by Hendre and Aggarwal (2007) and Herrera *et al.* (2011). The updated information on application of DNA markers in coffee is furnished in Table 17.1. Further, in the following sections of this chapter, the efforts towards development of DNA markers as well as their application for coffee genetic analysis with special reference to origin of *Coffea arabica*, analysis of genetic diversity, analysis of alien genome introgression, development of linkage maps, QTL analysis and marker assisted selection are briefly highlighted.

4.1. Development of DNA Markers

Sequencing projects for gene discovery in different plant species during last decade has resulted in establishment of huge sources of EST databases as well as DNA sequence information generated and deposited in online databases (Rudd, 2003; Varshney *et al.*, 2005b). *Coffea* genomes, being relatively large, partial sequencing of anonymous cDNA clones (Expressed sequence tags - ESTs) is most common method followed for generating data on the coding regions of genomes. Thus, EST databases have become the fastest growing segment of the online DNA databases in coffee (Wolfsberg and Landsman, 1997; Fernandez *et al.*, 2004; Lin *et al.*, 2005; Vieira *et al.*, 2006; Poncet *et al.*, 2006; Barbosa *et al.*, 2008).

EST databases were initially developed by suppression subtractive hybridization (SSH) and Fernandez *et al.* (2004) created a library of 527 non-redundant (NR) EST sequences associated with reactions to the rust fungus. Lin *et al.* (2005) generated an EST database based on sequences of cDNA clones derived from different stages/tissues of five *C. canephora* varieties. About 33,000 different unigenes (http://www.lge.ibi.unicamp.br/cafe; Vieira *et al.*, 2006) were discovered under Brazilian Coffee Genome Project from clones from 37 cDNA libraries of *C. arabica*, *C. canephora* and *C. racemosa*, representing specific stages of cells and plant development. Barbosa *et al.* (2008) developed a web interface to the EST sequence database on Coffee at https://alanine.cenargen.embrapa.br/CoffEST, originally developed for the Brazilian Coffee Genome EST project and later incorporated *Coffea canephora* EST data of Cornell University (59,718 raw EST sequences, Lin *et al.*, 2005) and Institut

Table 17.1: Update on Application of Various Types of DNA Marker Approaches in Coffee

Marker Technique	*Scope/Application*	*Reference*
RFLP	Diversity Studies, development of Linkage maps	Lashermes *et al.* (1996a, 2001), Paillard *et al.* (1996)., Herrera *et al.* (2002b), Crouzillat *et al.* (2010).
RAPD	Inter-specific variation, genetic diversity analysis and gene introgression, genetic characterization, development of linkage maps	Orozco-Castillo *et al.* (1994).), Paillard *et al.* (1996)., Berthaud and Charrier (1988)., Lashermes *et al.* (1993, 1996b; 2001), Anthony *et al.* (2001)., Masumbuko *et al.* (2003)., Silveira *et al.* (2003)., Sera *et al.* (2003)., Aggarwal (2005)., Hendre and Aggarwal (2007)., Sakiyama *et al.* (2008)., Setotaw *et al.* (2010)., Kathurima *et al.* (2012)., Bigirimana *et al.* (2013)., Lashermes *et al.* (2001).
AFLP	Genetic diversity studies and introgression analysis	Powell *et al.* (1996)., Lashermes *et al.* (2000)., Prakash *et al.* (2002, 2004, 2005), Coulibaly *et al.* (2003a)., Maluf *et al.* (2005)., Aggarwal (2005)., Setotaw *et al.* (2010)., Herrera *et al.* (2002a)
	Development of Linkage maps and identification of QTLs	Ky *et al.* (2000)., Coulibaly *et al.* (2003a)., Lashermes *et al.* (2001)., Pearl *et al.* (2004)., Hendre and Aggarwal (2007)., Priolli *et al.* (2008).
	Development of trait linked marker for rust resistance	Prakash *et al.* (2004)., Mahe *et al.* (2008)., De Brito *et al.* (2010).
SSR	Genetic mapping, diversity assessment, population genetics, and marker assisted selection, Development of Linkage maps and identification of QTLs	de Vienne (1998), Gupta and Varshney (2000)., Anthony *et al.* (2002b), Herrera *et al.* (2002a, 2002b), Coulibaly *et al.* (2003b)., Poncet *et al.* (2004, 2006), Aggarwal(2005)., Tornincasa *et al.* (2006)., Musoli *et al.* (2006)., Hendre *et al.* (2008)., Cubry *et al.* (2008)., Sakiyama *et al.* (2008)., Cubry *et al.* (2006 a, b), Hendre and Aggarwal (2007, 2014), Priolli *et al.* (2008)., Setotaw *et al.* (2010)., Priyono *et al.* (2010)., Bigirimana *et al.* (2013)., Al-Murish *et al.* (2013)., Devasia (2011), Devasia *et al.* (2014)., Leroy *et al.* (2014)., Lashermes *et al.* (2001)., Crouzillat *et al.* (2010)., Priyono and Sumirat (2012)., Missio *et al.* (2009).
ISSR	Evolutionary studies, genetic diversity	Tesfaye *et al.* (2006)., Aga *et al.* (2005).
ISTR	Genetic diversity	Aga *et al.* (2006).
SNPs	Analysis of gene expression in *Coffea;* diversity studies	Vidal *et al.* (2010).; Priyono *et al.* (2010).; Crouzillat *et al.* (2010).; Priyono and Sumirat (2012).
SRAP, TRAP	Diversity Studies	Al-Murish *et al.* (2013).

de Recherche pour le Développement – IRD (8,782 raw EST sequences, Poncet *et al.*, 2006). Plechakova *et al.* (2009) developed a comprehensive web data source on annotated or mapped microsatellite markers in Rubiaeceae published by MoccaDB in an online database.

Genic and Genomic Microsatellite markers specific to *Coffea canephora* have been developed using data available on these web sites (EST –SSRs) and by following pre-cloning enrichment stratergy (Poncet *et al.*, 2007; Hendre *et al.*, 2008; Missio *et al.*, 2009; Devasia *et al.*, 2013; Hendre and Agarwal, 2014). The high cross-species transferability of the newly developed markers revealed the potential of these markers for use in evolutionary and diversity studies (Poncet *et al.*, 2004, 2007; Devasia *et al.*, 2014; Hendre and Agarwal, 2014; Cubry *et al.*, 2008).

4.2. Use of DNA Markers

Identification of molecular markers linked to traits of interest is an important pre requisite for efficient Marker Assisted Selection (MAS) to complement the conventional breeding programmes. The availability of highly saturated linkage maps paved way for identification of QTLs and trait linked markers in various crops. DNA markers provide powerful and reliable tools for evaluating genetic variation both within and between populations and genetic analysis of crop plants (Qamaruz *et al.*, 1998; Powell *et al.*, 1996; Moncada and McCouch, 2004; Varshney, 2005a). In coffee, substantial progress has been achieved with respect to use of DNA markers for diverse applications and some of the salient leads are summarized as follows.

4.2.1. Origin of *Coffea arabica*

Coffea arabica, the most economically important species of the genus *Coffea*, is an allotetraploid species while all other species are diploids. Hence, the origin of arabica coffee has always been an intriguing subject and the classical cytogenetic studies suggested that *C. arabica* might have originated from the union of unreduced gametes from a cross involving *C. eugenioides* and any one of the species of the genus such as *C. canephora*, *C. congensis* and *C. liberica* (Cramer, 1957; Carvalho, 1952; Narasimhaswamy, 1962), followed by genetic regulation of synapsis leading to the progressive diploidisation and evolution of present day amphidiploid *C. arabica* (Charrier and Berthaud, 1985). Subsequently, DNA marker based analysis using conserved regions from nuclear and/or chloroplast genome have not only substantiated some of the above inferences but also provided better understanding of the evolution and origin of arabica coffee. Based on DNA marker analysis, it was reported that *C. eugenioides* and *C. canephora* or ecotypes related to these diploid species (Lashermes *et al.*, 1993, 1996a, 1996d, 1999) as the likely progenitors of *C. arabica*. Chloroplast (cp) genome analysis and RFLP allelic diversity across different genera suggested a relatively recent speciation of *C. arabica* over 1 MYA (Lashermes *et al.*, 1999). The ISSR (Inter Simple Sequence Repeat) fingerprint data generated in the forest populations of *Coffea arabica* in Ethiopia provided strong evidence on evolution of *C. arabica* through a single allopolyploidization event (Tesfaye *et al.*, 2006). Phylogenetic analyses of chloroplast sequence data (also depict *C. eugenioides* as sister to *C. arabica*) suggested that an ancestor of *C. eugenioides* as the maternal parent of *C. arabica*,

4.2.2. Genetic Diversity Analysis

The coffee genetic resources includes primary genepool comprising the wild and cultivated varieties of *C. arabica* and *C. canephora* and secondary genepool comprising of over 100 related diploid species of *Coffea* and *Psilanthus*. Initialy, the genetic diversity among coffee species and cultivars was determined with morphological, biochemical or isozyme markers. The analysis of six isozyme patterns in different *C. arabica* accessions with high level of morphological variation revealed absence of polymorphism in the isozyme pattern suggesting that isozymes were not appropriate for genetic diversity studies and for identification of *C.arabica* accessions (Berthou and Trouslot, 1977). The genetic diversity of *C. canephora* Pierre ex A. Froehner was also assessed by enzymatic polymorphism (Berthaud, 1986; Montagnon *et al.*, 1992). However, these markers were often not very efficient to measure genetic variation. Isoenzyme evidence for inter specific origin of Piata coffee was reported by Medina *et al.* (1995).

Subsequently, with the advent of DNA marker technologies, various marker approaches were used in germplasm characterization and management in coffee. Genetic diversity within *C. arabica* cultivars and wild collections has been extensively analysed using various DNA marker approaches such as RAPD (Lashermes *et al.*, 1996b; Zezlina *et al.*, 1999; Anthony *et al.*, 2001; Aga *et al.*, 2003; Sera *et al.*, 2003; Cristancho *et al.*, 2004), AFLP (Anthony *et al.*, 2002a; Steiger *et al.*, 2002; Prakash *et al.*, 2002; Aggarwal *et al.*, 2004) and SSR markers (Anthony *et al.*, 2002a, 2002b; Aggarwal *et al.*, 2004). The polymorphism in *C. arabica* was first studied using RFLP (Lashermes *et al.*, 1996a) and RAPD (Orozco-Castillo *et al.*, 1994; Lashermes *et al.*, 1996b). Bekele (2005) analysed the genetic diversity of *C. arabica* genotypes collected from the northwestern and southwestern parts of Ethiopia using morphological, biochemical and molecular (AFLP and SSR) markers. More recently, the genetic variability in cultivated coffee was studied using dominant and co dominant marker systems of RAPDs, ISSRs, SSRs and AFLPs (Tran, 2005; Hendre *et al.*, 2008). Genetic diversity studies of cultivated varieties and wild collections by various authors have revealed higher genetic diversity of wild germplasm. Based on RAPD analysis, Lashermes *et al.* (1996d) confirmed the narrow genetic base of commercial cultivars and a relatively higher genetic diversity within wild collected germplasm collections of *C. arabica*. The study also demonstrated an East–West differentiation in Ethiopia. Similar findings were reported by Aga *et al.* (2003) and Montagnon *et al.* (1996) reported that accessions from the western part of the Great Rift Valley showed higher genetic variability within a population while accessions from the eastern part were genetically closer to coffee cultivars from other parts of the world. The use of molecular markers in assessing population genetic diversity can serve as a tool in identifying and selecting priority sites for *in situ* conservation as well as developing conservation management strategies (Krishnan, 2013). In general, SSR and AFLP data revealed higher degree of genetic variability among individuals within and between coffee varieties compared to the other marker systems. Nevertheless, irrespective of the marker approach, majority of the genetic variability studies indicated low genetic diversity within *C. arabica*, which has been attributed to the allotetraploid origin, reproductive biology and evolution process of this species. Maluf *et al.* (2005) suggested a combination of botanical, agronomic and molecular descriptors complemented with gene polymorphism for the purpose of cultivar identification. There were several reports on diversity analysis of field gene banks established in different coffee growing countries using DNA markers.

4.2.3. Molecular Characterization of Representative Coffee Germplasm Available in India

The genetic diversity of representative coffee germplasm collections available in India was analysed using four marker approaches *viz.*, RAPD, ISSR, f-AFLP and SSR markers (Aggarwal, 2005). The collections included a set of 25 accessions representing different provinces of Ethiopia, 33 rust differential clones, 16 superior selections (14 arabica and 2 robusta) developed by the Central Coffee Research Institute (CCRI) for commercial cultivation and 16 different species of *Coffea*.

Diversity among Ethiopian collections: A set of 25 accessions representing the collections from different provinces of Ethiopia *viz.*, Shoa, Illubabor, Gojam, Kaffa, Erytrea, Sidamo and Harar provinces was analysed using 25 RAPD and 15 ISSR markers. The data generated using both the multilocus markers, RAPD and ISSR suggested a very narrow genetic base of the collection. Further, the analysis revealed no distinct generic affinities/clusters between Ethiopian arabicas and provincial distinction of genotypes thus geographical isolation does not translate into genetic variation among the collections. In RAPD analysis, combinations of two random decamer primers were also employed, but it did not increase polymorphic bands. Interestingly, the grouping pattern of Ethiopian arabicas was found similar with both the marker approaches.

Characterization of selections released by Central Coffee Research Institute (CCRI): Coffee breeding programmes undertaken at CCRI since 1925 resulted in the development of 13 improved Arabica selections and three superior robusta selections. These improved selections have been cultivated across the Indian coffee tracts depending on their agro-climatic suitability. All the 16 selections were fingerprinted using high-resolution DNA marker techniques involving nuclear genomic markers such as RAPD, f-AFLP, SSR and IRAP (Aggarwal, 2005). All the marker systems generated distinct profiles for tetraploid arabicas and diploid robustas, with more polymorphism among robusta samples. The Arabica selections indicated limited variability and average percent polymorphic markers using the different marker approaches ranged from 51 to 62 per cent for diploid selections and only 18-23 per cent for tetraploid arabicas. All the selections grouped in two distinct clusters, representing arabica genotypes and robusta as per their genetic origin. Most of the semi-dwarf types grouped together and the low variability among the arabica-based selections confirmed the narrow genetic base of arabica genome (~20 per cent polymorphism). In spite of the low variability, the cultivated arabica selections could be discriminated from each other but by using a large number of DNA markers. Based on fingerprint data generated by use of the various marker systems, a reference DNA polymorphism panel was prepared for distinguishing the CCRI Selections (Aggarwal, 2005).

Molecular characterization of coffee rust differential clones: A set of 33 rust differential clones, comprises of 29 different clones of *C. arabica* and four clones of diploid species, one each of *C. racemosa, C. excelsa, C. canephora* and *C. congensis* maintained in the coffee germplasm bank at CCRI were subjected to genetic analysis using 35 RAPD primers, 7 AFLP primer combinations and over 150 in-house developed SSR primers (Aggarwal, 2005). The analysis indicated presence of only subtle variation among the arabica-based rust differentials while it was substantial between arabica and diploid rust differentials. Among the rust differentials, all the arabica genotypes formed one cluster and the four diploid rust differentials clustered out as per their genetic origin. The grouping of the arabica-based rust differentials was however not in accordance with rust resistance factors. Eighteen clones of differential coffee tree hosts for *Hemileia vastatrix* Berk. et Br. were characterized with 12 RAPD markers (Teixeira *et al.*, 2004) and the genetic distances was obtained by cluster analysis based on the UPGMA method which clearly defined the clones into three groups. Group A, consisted of one clone (Kawisari Hybrid CIFC 644/18), the most divergent clone; group B, five clones of Híbrido de Timor including Clone CIFC 4106 the plant selected from Timor Island, CIFC 1343/269 and CIFC 832/1 seed introductions from Híbrido de Timor. Group C consisted of nine materials of arabica and three *C. liberica* introgressed arabicas.

Molecular characterization of Coffea species: The taxonomy of coffee species was first described mainly based on morphological characters of the specimens preserved in different herbaria. Subsequently, in order to establish molecular phylogeny and phylogenetic relationships among *Coffea* species, DNA sequence data of the internal transcribed spacer ITS (2) region of the nuclear ribosomal DNA (Lashermes *et al.*, 1997), as well as the chloroplast DNA variation (Lashermes *et al.*, 1996a, Cros *et al.*, 1998) were successfully used. Based on the results, a radial mode of speciation and a recent origin in Africa for the genus *Coffea* was suggested (Etienne *et al.*, 2002). In addition, several major clads were also identified, which reveal a strong geographical correspondence *i.e.*, West Africa, Central Africa, East Africa and Madagascar. In the Indian context, 14 different species of *Coffea* and four species belongs to sub genus *Psilanthus* present in gene bank were analyzed using mobility-based DNA markers such as RAPD, ISSR and SSR (Aggarwal, 2005). Further, direct sequencing of three phylogenetically informative domains of nuclear and organelle genomes *viz.*, internal transcribed spacer regions ITS1-5SrDNA-ITS2 of the nuclear ribosomal DNA, 16S rDNA domain of mitochondrial genome and intergenic 'trnL' region of chloroplast genome was also undertaken. Individual plant sample as well as pooled samples were used for analysis to define the sampling strategy for DNA marker based analysis as the plant material is inherently heterozygous. The intra-species variation was also analyzed using four of the coffee species *viz.*, *C. euginioides, C. stenophylla, C. dewevrei* and *C. salvatrix* using RAPDs/ISSRs and ITS1-5SrDNA-ITS2 sequencing. The analysis revealed the presence of high variability between the species and out of 625 reproducible amplicons generated using 20 RAPD primers, 96 per cent were polymorphic. Similarly, all the 162 reproducible amplicons generated using nine selected ISSR primers were polymorphic and informative for the purpose of species relationships. Comparative analysis of pooled or individual samples using different DNA marker systems established that multiple individual samples are not necessary to draw valid inferences on sample size/type for inferring genetic affinities between coffee species. Sequencing of around 30 kb of the three genomic domains of the 19 representative genotypes covering all the coffee species and related genera revealed significant variation in the form of both Indels as well as base substitutions, across species. The average sequence sizes of the ITS 1 and ITS 2 of the ITS1-5SrDNA-ITS2 nuclear ribosomal domain was found to be 247 and 231 bp, respectively. Similarly, average sequence sizes of the partial Mt16S rDNA conserved domain and the intergenic 'trnL' region of chloroplast genome were 756 bp and 532 bp for all the species. Among the diploid species, *C. canephora* was found to be phylogenetically most close to *C. arabica* followed by the cluster comprising of *C. congensis, C. liberica* and *C. dewevrei*. The ITS sequence based analysis defined the utility of nuclear ITS domain in sequence based studies to derive realistic affinities between different *Coffea* species. It was also established that analysis of whole domain is necessary than confining to one of the variable segment *i.e.*, ITS 1 or ITS 2 for reliable inference in phylogenetic reconstruction (Hendre and Aggarwal, 2007). Furthermore, the sequence based genomic analysis of the three organellar compartments *viz.*, nucleus, chloroplast, and mitochondrion suggests that organelle DNA may not be the ideal candidate for phylogenetic analysis of coffee species (Aggarwal, 2005). The molecular data of the indigenous species, *Psilanthus bengalensis, P. travencorensis, P. khasiana* and *P. wightiana* validated the placement of these species under the related Paracoffea genus *Psilanthus.* Undertaking genetic diversity analysis of two wild populations of a coffee species endemic to the littoral forests of southeastern Madagascar, *Coffea commersoniana* (Baill.) A. Chev., using SSR markers, Krishnan *et al.* (2013). found high genetic partitioning among the two *in situ* populations, necessitating the need to keep two populations separately for restoration purposes.

Molecular characterization of cultivated varieties and wild collections of arabica: Genetic diversity analysis of commercially cultivated *Coffea arabica* varieties from America, India and Africa and native Ethiopian populations with microsatellites markers indicated wide genetic variability in native Ethiopian samples compared to the other commercial varieties (Tornincasa *et al.*, 2006). The Ethiopian Arabicas revealed complex geographical patterns of genetic diversity, with most regions possessing their own genotypes (Tesfaye *et al.*, 2006). Genetic diversity among 115 coffee accessions from the *Coffea* germplasm collection of IAC, Brazil was assessed using SSR markers (Silvestrini *et al.*, 2007). The germplasm represented 73 accessions of *Coffea arabica* derived from spontaneous and subspontaneous plants in Ethiopia and Eritrea, 13 commercial cultivars of *C. arabica* developed by the Breeding Program of IAC, one accession of *C. arabica* cv. 'Geisha', 13 accessions of *C. arabica* from Yemen, five accessions of *C. eugenioides*, four accessions of *C. racemosa* and six accessions of *C. canephora*. The study revealed differentiation between the cultivated plants of *C. arabica* and accessions derived from spontaneous and subspontaneous plants from Ethiopia. The results agreed with previously reported narrow genetic basis of cultivated plants of *C. arabica* and supported the hypotheses about domestication of the species. The study also showed significant genetic diversity among accessions from Ethiopia and Eritrea present in the germplasm collection of IAC, Brazil. The genetic variability and population structure of 68 accessions of *C. arabica* (wild and cultivated) and of three diploid species available in Colombia were evaluated using 47 SSR markers (López-Gartner *et al.*, 2009). The structure analysis inferred nine subpopulations (*k*=9), for which the greatest values of probability were obtained. Three of the groups corresponded to the three diploid species as expected. There were six groups identified within *C. arabica*. The genetic subdivisions within *C. arabica* were based on geographical origin, degree of domestication and dispersal history of coffee. The analysis provided a strong evidence of population structure in *C. arabica*. Genetic diversity studies on Rwandan Germplasm using RAPD and SSR markers (Bigirimana *et al.*, 2013) showed distict genetic differences between local cultivars such as BM 139, BM 71 and Mibilizi when compared to CBD and CLR resistant varieties like Rume Sudan, Sln. 6 and HDT. The *C. arabica* genotypes collected between and within different valleys of Yafea City, Yemen also exhibited diversity as analysed by using SRAP, TRAP and SSR primers (Al-Murish *et al.*, 2013). The genetic diversity studies carried out in Híbrido de Timor germplasm lines from the germplasm bank of UFV (Universidade Federal de Viçosa)/EPAMIG (Empresa de Pesquisa Agropecuária de Minas Gerais) using AFLP, RAPD and SSR molecular markers indicated considerable genetic diversity in these lines (Setotaw *et al.*, 2010).

More recently, Lin Zhou *et al.* (2016) developed 7538 single nucleotide polymorphism (SNP) markers using expressed sequence tags (EST) of *Coffea arabica*, *C. canephora* and *C. racemosa* from public databases. Among these, 180 SNPs were selected for validation using 25 *C. arabica* and *C. canephora* accessions from Puerto Rico. Based on the validation data, a panel of 55 SNP markers was found polymorphic across the two species. This panel enabled the differentiation of all tested accessions of *C. canephora*, which accounted for 79.2 per cent of the total polymorphism in the samples. Only 21.8 per cent of the polymorphic SNPs were detected in the 12 *C. arabica* cultivars. It was reported that this coffee SNP panel provided robust and universally comparable DNA fingerprints and can be used as a genotyping tool to assist coffee germplasm management and coffee cultivar authentication.

In case of the second commercial species *C. canephora* (Robusta), the primary analysis of genetic diversity was studied by enzymatic polymorphism (Berthaud, 1986, Montagnon *et al.*, 1992). These studies, although limited to few samples, provided basic information for defining the structure of the species, with two distinct groups. The first one was the 'Guinean Group' consisting of wild populations of Côte d'Ivoire and the the second one was 'Congolese group' comprising of wild material of Central African Republic and Cameroon and also the cultivated materials. Subsequent studies using RFLP polymorphism (Dussert *et al.*, 1999) grouped the wild and cultivated forms of robusta coffee into five diversity groups (A, B, C, D, E). All the studies established high genetic diversity in robusta compared to arabica.

Subsequently, more efficient marker approaches *viz.*, AFLP, SSRs and SNPs were used to assess the genetic diversity among *Coffea canephora* accessions including wild and germplasm collections (Prakash *et al.*,2005; Musoli *et al.*, 2006; Leroy *et al.*, 2006; Cubry *et al.*, 2006a, 2006b; Ferrão *et al.*, 2013; Ogutu *et al.*, 2016). All these studies revealed high variability among the accessions which was attributed to allogamous nature of the species. Prakash *et al.* (2005) assessed genetic diversity among *C. canephora* genotypes available in germplasm collection of India in comparison with 14 representative samples of core collections of *C. canephora* and three accessions of *C. congensis* using AFLP and SSR marker approaches. The accessions of Indian genepool grouped together with the robusta types identified as diversity group 'E'. The study clearly revealed high amount of diversity in the representative accessions of a core collection from Africa, the centre of genetic diversity of robusta coffee, which was not present in the cultivated Robusta genotypes. A study on the genetic differentiation of wild populations of *C. canephora* from Uganda revealed high diversity and differentiation (Musoli *et al.*, 2009). The study also revealed that the Ugandan populations of wild origin are different from other known genetic diversity groups of western and central Africa. Considering the high level of differentiation of wild populations of *C. canephora* from different centres of origin, there is an imminent need to develop a comprehensive conservation strategy to protect the valuable genetic resources from rapid extinction of primary forests in Africa due to climate change effects. Leroy *et al.* (2014) characterised *C. canephora* germplasm collections with SSRs and proposed the construction of genetic core collections for improvement of coffee germplasm management.

More recently, Oguttu *et al.* (2016) analyzed the genome-wide distribution of microsatellites in the *Coffea canephora* genome and a set of 100 SSRs were selected to characterize 96 coffee accessions, including 10 wild accessions collected from Mt. Marsabit (Kenya). Wild coffee species from Mt. Marsabit showed a close genetic similarity with cultivated accessions in Kenya, suggesting that the wild species in Mt. Marsabit played an important role in the domestication of cultivated coffee in Kenya. Significantly, low pairwise genetic divergence was observed between cultivated and wild accessions in Kenya, suggesting a relatively narrow level of genetic basis among coffee germplasm in Kenya. In addition, cultivated and wild coffee accessions in Kenya showed a great divergence from those in other countries.

Thus, in contrast to the tetraploid *C. arabica*, the diploid species of the genus *Coffea*, were found to exhibit considerable variability and some of the diploid species form valuable gene reservoir, for various breeding programmes (Berthaud and Charrier, 1988). The diploid *Coffea* varieties were reported to interbreed freely with each other and produce relatively fertile progeny (Anthony, 1992, Louarn, 1993).

4.2.4. Analysis of Alien Genome Introgression

The efficiency of DNA molecular marker technologies in detecting the alien genome introgression in coffee has been successfully demonstrated through many studies (Lashermes *et al.*, 2000, Prakash *et al.*, 2002, 2004, Herrera *et al.*, 2002a, 2002b, 2004). These studies were complemented by use of fluorescence *in situ* hybridization (FISH) towards establishing the genome introgression in coffee (Herrera *et al.*, 2007). Lashermes *et al.* (2000) analysed 19 arabica coffee introgression lines (BC_1F_4) and two accessions derived from a spontaneous interspecific cross (Hibrido de Timor) between tetraploid *Coffea arabica* (2n=4x=44) and diploid *C. canephora* (2n=2x=22) for the introgression of *C. canephora* genetic material, using AFLP marker approach.

The study established that AFLP technique was extremely efficient for DNA marker generation in coffee as well as for the detection of introgression in *C. arabica*. The genetic diversity observed in the Timor hybrid-derived genotypes appeared to be approximately double than that in *C. arabica*. Although representing only a small proportion of the genetic diversity available in *C. canephora,* the Timor Hybrid obviously constitutes a considerable source of genetic diversity for arabica breeding. Analysis of genetic relationships among the Timor Hybrid-derived genotypes suggested that introgression was not restricted to chromosome substitution but also involved chromosome recombinations. The Timor Hybrid-derived genotypes varied considerably in the number of AFLP markers attributable to introgression and were estimated to represent from 9 per cent to 29 per cent of the *C. canephora* genome.

AFLP analysis of early Indian arabica varieties derived from a putative spontaneous hybrid between *C. arabica* and *C. liberica* (Prakash *et al.*, 2002) provided the molecular evidence of natural hybridization between *C. arabica* and *C. liberica* and extent of *C. liberica* genome introgression in different *C. arabica* populations. *C. liberica* accessions of EA group (*C. liberica* var liberica of Guinean origin) seemed to be the likely progenitor in the origin of natural hybrid. Analysis of genetic relationships in the introgressed lines suggested that introgression was limited to few fragments. In *C. arabica* accessions, only 35 polymorphic bands were seen confirming the low genetic diversity. On the contrary, although representing a small amount of alien genome introgression, the Liberica-introgressed genotypes provided notable genetic diversity. Herrera *et al.* (2002a) investigated the behaviour of the *C. canephora* genome and its interaction with the *C. arabica* genome in tetraploid interspecific hybrids (*C. arabica* × *C. canephora 4x*) resulting from a cross between an accession of *C. arabica* and a tetraploid plant of *C. canephora* obtained following colchicine treatment. Segregation and co-segregation of restriction fragment length polymorphism (RFLP) and microsatellite loci-markers were studied in two BC_1 populations. At almost all loci analysed, the segregation of *C. canephora* alleles, transmitted by the (*C. arabica* × *C. canephora* 4*x*) hybrids, conformed to the expected ratio assuming random chromosome segregation and the absence of selection. The recombination frequencies estimated in both plant materials were rather similar, suggesting that recombination in the (*C. arabica* × *C. canephora* 4*x*) hybrid is not significantly restricted by the genetic differentiation between chromosomes belonging to the different genomes. The hybrid (*C. arabica* × *C. canephora* 4*x*) therefore appeared particularly favourable to intergenomic recombination events and gene introgressions. Recombination fractions of *C. canephora* chromosome segments using RFLP and microsatellites indicated that the recombination was not restricted by the genetic differentiation between chromosomes belonging to the different genomes (Herrera *et al.*, 2002b). The hybrid (*C. arabica x C. canephora*) showed inter genomic recombination events and gene introgressions of *C. canephora* alleles indicating a severe counter selection against the introgression of genetic material from *C. canephora* into *C. arabica* in form of triploid hybrids. Anthony *et al.* (2002b) characterized genetic groups within coffee species and identified genes introgressed from diploid species into interspecific hybrid, for the coffee leaf rust (*Hemileia vastatrix*), coffee berry disease

(*Colletotrichum kahawae*) and root-knot nematodes using SSR markers. Herrera *et al.* (2004) analysed factors controlling gene introgression into cultivated arabica coffee (*Coffea arabica* L.) by species-specific microsatellite markers. Back cross progenies of the interspecific triploid hybrid plants between the tetraploid species *C. arabica* (2n = 44) and a diploid species (2n = 22), either *Coffea canephora* or *Coffea eugenioides*, were studied for the extent of introgression. The study established that although the frequency of introgressed markers seemed as expected, assuming random chromosome segregation and diploid gamete formation, in the BC1 derived from triploid hybrids involving *C. canephora*, this frequency appeared significantly lower in the BC1 derived from triploid hybrids involving *C. eugenioides*. The comparison of reciprocal progenies between *C. arabica* and triploid interspecific hybrids (*C. arabica* x *C. canephora*) used as male or female parent revealed a very strong effect of the backcross direction. Sakiyama *et al.* (2008) reported the genome of Hı́brido de Timor plants to be mainly composed of *C. arabica* and a smaller portion of *C. canephora* genome. Based on SSR and AFLP analysis of an F_2 population derived from a cross between T5296 and Et6, Lashermes *et al.* (2011) investigated the number and mode of inheritance of *canephora*-introgressed segments as well as their sub-genome localisation and rate of recombination. The results suggested that the transfer of desirable genes into *C. arabica* from *C. canephora* is not limited by the ploidy level differences or the suppression of recombination between the different genomes. Genetic diversity analysis of 24 genotypes by using Random Amplified Polymorphic DNA (RAPD) primers and microsatellites (Kathurima *et al.*, 2012), grouped the genotypes into three clusters with *C. eugenioides* in the first cluster, un-introgressed arabica genotypes in the second and the third cluster with *Coffea canephora* (Robusta) and robusta introgressed genotypes, Ruiru 11, Hibrido de Timor and Catimor. Thus, the prospects of molecular markers in analyzing genome introgressions and evolution of species in coffee have been successfully demonstrated.

4.2.5. Development of Linkage Maps

Genetic maps have been proved as the valuable resources for applying genetic technologies towards crop improvement such as marker-assisted selection and gene pyramiding (Brar and Dhaliwal, 1997). The available linkage maps for *Coffea* species are based on random amplified polymorphic DNA (RAPD), RFLP, AFLP and microsatellite markers (Coulibaly *et al.*, 2003a; Hendre and Aggarwal, 2007; Ky *et al.*, 2000; Lashermes *et al.*, 2001; Paillard *et al.*, 1996) Interestingly, most of the maps developed for coffee were mainly on *Coffea canephora* (Hendre and Aggarwal, 2007, Paillard *et al.*, 1996, Lashermes *et al.*, 2001) and interspecific crosses (Ky *et al.*, 2000; Coulibaly *et al.*, 2003a) and only a few for *Coffea arabica* (Pearl *et al.*, 2004; Moncada 2016), mainly due to the low polymorphism of *Coffea arabica* (Herrera *et al.*, 2011). The number of mapped loci ranged from 147 to 464.

Linkage maps developed for coffee (*C. canephora*) was initially based on doubled haploids (DH) populations using RFLP and RAPD markers (Paillard *et al.*, 1996). The study indicated a low polymorphism rate and mapped a total of 1402 cM length and consisted of 15 linkage groups. Lashermes *et al.* (2001). constructed the linkage map of *C. canephora* consisting of 162 DNA markers (AFLP, RAPD, SSR and RFLP) spanning 1041 cM of the genome, and the number of linkage groups equivalent to the haploid number of chromosomes in *C. canephora*. The study established the cross compatibility behavior of doubled haploid and hybrid genotypes and the self-incompatibility system in *C. canephora* was confirmed to be gametophytic and controlled by a single locus (s-locus). Ky *et al.* (2000) published the first *Coffea* interspecific linkage map based on a backcross progeny of two diploid parental lines (*C. pseudozanguebariae* x *C. liberica* var. Dewevrei) x *C. liberica* var. Dewevrei) using AFLP and RFLP markers. The map comprised 181 loci distributed over 14 linkage groups covering 1144 cM. Coulibaly *et al.* (2003b). developed linkage map in a *Coffea* interspecific backcross progeny (*Coffea heterocalyx* x *C. canephora*) x *C. canephora*) using AFLP and SSR markers and mapped 190 loci in 15 linkage groups covering a total map length was 1,360 cM. Hendre and Aggarwal (2007) developed a relatively well populated framework linkage map for cultivated diploid robusta, using a pesudo-testcross population. This first generation map developed using a trait-specific (drought tolerance) mapping population has a total of 374 mapped markers (comprising 185 RAPDs, 118 AFLPs, 71 SSRs, and a moderate marker density of 3.3 cM centimorgans) spread over 11 major and 5 minor linkage groups.

The linkage maps developed for Arabica are very few due to the requirement of large number of molecular markers to detect polymorphism in Arabica. Pearl *et al.* (2004) constructed a genetic linkage map on a pseudo-F_2 population of arabica coffee derived from a cross between the cultivars Mokka hybrid and Catimor using AFLP markers with a total map length of 1,802.8 cM and average distance of 10.2 cM between adjacent markers. Nagai *et al.* (2006) constructed a prilimary linkage map in a segregating mapping population of *C. arabica* from a cross between Tall Mokka and Catimor, distinctively different in cupping quality as well as leaf, bean characteristics using 699 AFLP markers. The map proved useful in mapping QTL controlling source and sink traits. AFLP and SSR markers were used to build a genetic map of an interspecific F_2 population between *C. arabica* and *C. canephora* (Priolli *et al.,* 2008). Marker trait associations for sugar content, caffeine, CGA and total production per plant was reported. The study was successful in QTL detection for coffee quality and productivity. Lashermes *et al.* (2010) described the construction of a physical map in *C. arabica* spanning the resistance locus of S_H3, a major dominant gene that has been introgressed from a wild coffee species *Coffea liberica* (genome L) into the allotetraploid cultivated species, *Coffea arabica* (genome C^aE^a) by using a bacterial artificial chromosome (BAC) library. Genetic analysis was performed using a single nucleotide polymorphism detection assay based on Sanger sequencing of amplicons. The *C. liberica*-derived chromosome segment that carries the S_H3 resistance gene appeared to be introgressed on the sub-genome C^a. The position of the S_H3 locus was delimited within an interval of 550 kb on the physical map. Recently, Moncada *et al.* (2016) used 338 SSRs on a F_2 mapping population consisting of 278 individuals developed from a cross between Caturrax CCC1046 to construct a framework linkage map. Subsequently, SNP markers were added to construct a more robust genetic map. The integrated linkage map consists of 22 linkage groups populated by 848 SSR and SNP markers, with a total map length of 3800 cM.

4.2.6. QTL Analysis in Coffee

It is well established that, most of the agronomically important traits in plants are quantitative in nature, controlled by multiple genes. Molecular markers are powerful means for studying genetic basis of traits which are polygenic in nature (Tanksley, 1993) and molecular markers have been successfully used to identify quantitative trait loci (QTL) for complex traits like rooting ability. These could be selected more easily in a breeding programme than the traits themselves (Wang and Paterson, 1994). QTL mapping facilitates the screening for specific traits that are difficult to quantify and the traits influenced by environment (Hanson *et al.*, 1990). The genetic map provides means for studying number of genes controlling the trait, the location of the genes on chromosomes, the effect of variation in gene expression on the trait and the evaluation of genes regulating phenotype in different populations. In case of coffee, the first report was on linkage of RFLP marker to self-incompatibility on robusta doubled haploid map (Lashermes *et al.*, 1996c). Later, identification of few RAPD markers linked with resistance to coffee berry disease (CBD) caused by *Colletotrichum kahawae* were reported (Agwanda *et al.*, 1997). The study revealed 24 candidate markers linked to targeted trait, of which only three showed strong association with putative CBD resistance *T* gene (from Catimor). Subsequently, Coulibaly *et al.* (2002) validated the RFLP marker linked to self-incompatibility (Lashermes *et al.*, 1996c) on the intraspecific AFLP linkage map. Further, Coulibaly *et al.* (2003b). identified three fertility restorer QTLs using the partial AFLP based interspecific map. Similarly, QTL responsible for trigonelline content was identified on another interspecific linkage map developed by Ky *et al.* (2000). Further, putative QTLs were suggested for caffeine, chlorogenic acid and sucrose content in diploid coffee (Ky *et al.*, 2000). In Arabica, Pearl *et al.* (2004) identified QTLs with respect to source-sink traits. In addition to these earlier reports, some of the successful efforts on QTL mapping reported in recent years are highlighted in this chapter.

4.2.6.1. QTL for Yield and Quality Traits

Development of reference genetic map in coffee is an important step towards the management of key quantitative trait loci (QTLs). Crouzillat *et al.* (2010) used RFLPs, SSRs and SNP markers for development of Robusta reference genetic map. The mapping population consisted of *Coffea canephora* (2n = 2x = 22), from two locations in Indonesia, consisting of two Indonesian elite clones BP409 and Q121. Numerous QTLs for the agronomic and biochemical traits of interest were identified by transposition of the reference genetic map to other Robusta progenies. The genetic map of Arabica was also developed using F_2 mapping population and 277 SSRs were successfully mapped covering 1860 cM. Results suggested higher recombination rate in arabica than robusta genome. A QTL comparison study was performed based on the arabica draft map in conjunction with the different robusta maps on agronomic, technological and biochemical quantitative traits. Initial results indicated major differences in the genetic control of key quality characters for both arabica and robusta coffee species.

QTLs for yield and quality-related traits were identified in a genetic map of *Coffea canephora* constructed using 236 markers (Leroy *et al.*, 2011). QTLs were identified for yield, bean size, chlorogenic acid content, sucrose and trigonelline and acidity and bitterness of coffee beverages and regions of the *C. canephora* genome influencing beverage quality were identified. Five QTL zones were co-localized with candidate genes related to the biosynthesis of biochemicals: two genes coding for caffeine biosynthesis, one gene implicated in the biosynthesis of chlorogenic acids, and two genes implicated in sugar metabolism. QTLs on cherry and green bean traits were identified in the Robusta Coffee (*Coffea canephora* Pierre) using single nucleotide polymorphisms (SNPs) and simple sequence repeats (SSRs) markers (Priyono and Ucu Sumirat, 2012). Two QTLs for caffeine content, was identified by Ky *et al.* (2013) using an interspecific cross between two highly differentiated species of *Coffea liberica dewevrei* and *Coffea pseudozanguebariae.* Very recently, Moncada *et al.* (2016) identified QTLs associated with yield, plant height, and bean size. F_2 mapping population consisting of 278 individuals was developed from a cross between Caturra×CCC1046 and subjected for analysis with 338 SSRs and the data was used to construct a framework linkage map. Subsequently, SNP markers were utilized to construct a more robust genetic map. F_3 progenies of the mapping population were planted in five locations and evaluated for yield, plant height, and bean size. A major QTL for yield was significant at two locations and a second was significant at one location. Two QTLs governing plant height and bean size were detected.

4.2.6.2. QTLs for other Agronomical Traits

The QTLs identified for other agronomically important traits in coffee include QTLs for fertility restorer, disease resistance and QTLs for biotic and abiotic resistance traits. (Coulibaly *et al.*,2003b; Agwanda *et al.*, 1997; Herrera *et al.*, 2009; Devasia, 2011; Romero *et al.*, 2014). Three fertility restorer QTLs were identified using partial AFLP based interspecific map (Coulibaly *et al.*, 2003b). QTL mapping in coffee has been successful in identifying few RAPD markers linked to Coffee Berry Disease (CBD) resistance (Agwanda *et al.*, 1997). The study was carried out using the indirect approach of diverse genotypes (five susceptible and eight resistant Arabica cultivars/selections) and unmapped RAPD markers. Genetic analysis of partial resistance to coffee leaf rust (*Hemileia vastatrix* Berk and Br.) introgressed into the cultivated *Coffea arabica* L. from the diploid *C. canephora* species was reported by Herrera *et al.* (2009).

QTLs for drought adaptive traits in coffee was identified in the trait specific mapping population of *C. canephora* (L1 Valley x S.3334) developed in India- QTLs were identified for eight traits spread over on two parent specific linkage maps. Four major QTLs were located for transpiration rate *qR-Trr1*, Instantaneous WUE (*qR-iWUE2*), root: shoot ratio (*qR-RSR4*) and root length (*qR-Rl1*) (Devasia J, 2011). One major QTL for adult plant resistance to coffee leaf rust (*Hemileia vastatrix*), spanning a region of 2.5 cM designated Q_{CLR_4} was identified on Chromosome 4 in a study in the natural Timor hybrid (*Coffea arabica* x *C. canephora*) (Romero *et al.*, 2014).

5. Markers Linked to the Agronomic Traits and Marker Assisted Selection in Coffee

Development of improved varieties of coffee by conventional breeding is a rather a slow process and requires approximately 25 years due to the long generation time (5–6 years) and it takes at least five generations of selection to obtain superior plants. Molecular markers serve as useful tools in mapping traits of agronomic importance and holds great promise for hastening the development of improved varieties, through marker assisted selection (MAS). It also helps in improving our understanding of physiological and molecular aspects behind biological phenomenon (Wang and Paterson, 1994). Therefore, for the last couple of decades, conscious efforts have been made globally to integrate molecular markers based technologies in order to provide impetus and to improve the efficiency of genetic improvement of coffee. Some of the leads obtained with special reference to identification of markers linked to the agronomically useful traits and their exploitation for marker assisted selction in coffee are summarized in the following sections.

5.1. Disease Resistance

5.1.1. Coffee Leaf Rust (CLR)

Coffee leaf rust, caused by the fungus *Hemileia vastatrix*, is the most devastating disease of arabica coffee (*Coffea arabica*). Therefore, development of arabica coffee varieties resistant to leaf rust pathogen has been the main focus of breeding in India and many other countries. Resistance to coffee leaf rust is reported to be determined by atleast nine resistance genes, S_H1 through S_H9, either singly or in combination. Of these resistance genes, S_H1, S_H2, S_H4 and S_H5 are identified in the tetraploid species *C. arabica* where as S_H6, S_H7, S_H8 and S_H9 are present in the diploid species *C. canephora*, while S_H3 is present in another diploid species *C. liberica* (Bettencourt and Wagner, 1971; Bettencourt and Rodrigues, 1988; Prakash *et al.*, 2004). The resistance genes present in *C. liberica* and *C. canephora* have been successfully introgressed into *C. arabica* cultivars, after several decades of breeding and selection efforts and coffee varieties that manifest a broad spectrum of resistance were developed for commercial cultivation. Nevertheless, the major challenge being faced by the coffee breeders is achieving durability of resistance in commercial cultivars because of the adaptive capacity of the rust pathogen to mutate into new virulent races with ability to overcome the resistance in the improved varieties. Pyramiding of resistance genes from diploid coffee species in a selected arabica genotype is one of the promising approaches to achieve durable rust resistance.

In early breeding programmes of India, S.26, a spontaneous hybrid between arabica and a diploid species was used as a donor for rust resistance and popular arabica cultivars like S.288 and S.795 were developed. On detailed investigation of

S.26 derived lines using AFLP markers, Prakash *et al.* (2002) established the *C. liberica* introgression in S.26 and its derivatives. Further, Prakash *et al.* (2004) nvestigated the mechanism of introgression of a leaf rust resistance gene (S_H3) into *C. arabica* from *C. liberica*. An F_2 progeny of 101 individuals derived from a cross between Matari, an arabica accession and liberica-introgressed line S.288, was evaluated for resistance against three different races of *H. vastatrix*. The progeny segregated for S_H3 gene in a 3:1 ratio, as expected for a single dominant gene. AFLP analysis of a population with 15 primer combinations that generated candidate marker bands associated with the S_H3 resistance gene, resulted in identification of 21 markers strongly linked to S_H3 gene. All the 21 markers were grouped together in a single linkage group of 6.3 cM. Consequently, Mahé *et al.* (2008) developed ten sequence-characterized genetic markers closely associated with the S_H3 leaf rust resistance gene which included sequence characterized amplified regions (SCAR) markers resulting from the conversion of (AFLP) markers previously identified (Prakash *et al.*, 2004), simple sequence repeats (SSR) markers and SCAR markers derived from end-sequences of bacterial artificial chromosome (BAC) clones.

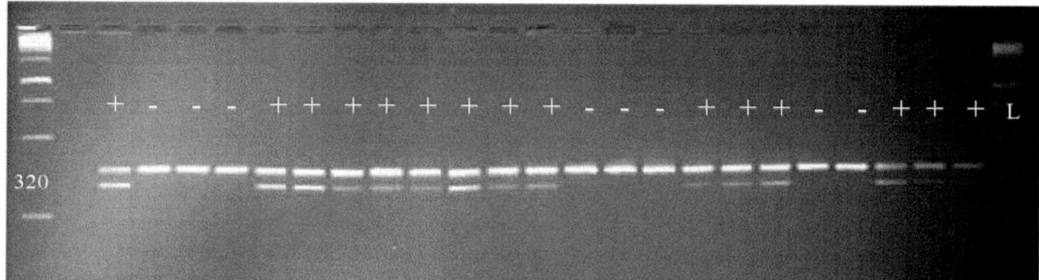

Figure 17.6: Amplification Profile of F_1 Hybrid Progeny with SCAR - BA 124 12K Showing Presence (+) and Absence (-) of S_H3.

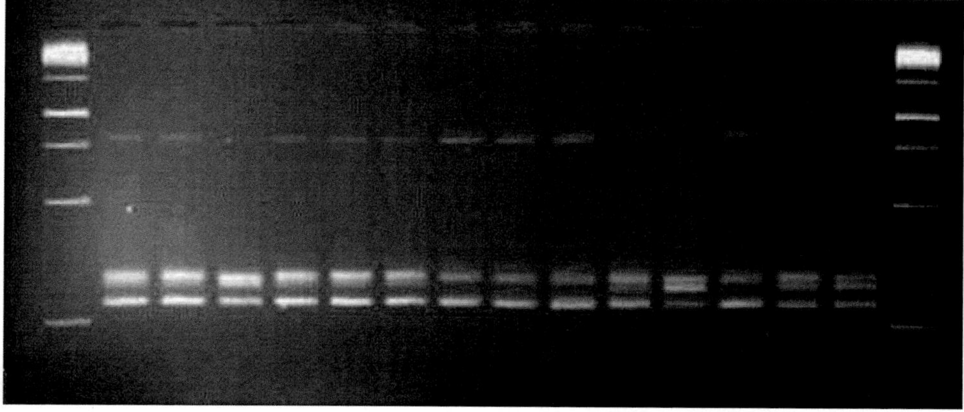

Figure 17.7: Amplification Profile of F1 Hybrid Progeny with SCAR Sat.244 Showing Homozygous and Heterozygous Status of S_H3.

Availability of SCAR markers for S_H3 gene (Figures 17.6 and 17.7) has facilitated the marker assisted selection (MAS) in breeding populations and for pyramiding of S_H3 gene with robusta genes (S_H6, S_H7, S_H8, S_H9) in the hybrid genotypes (Prakash *et al.*, 2012) (Figure 17.8). This strategy is being used in arabica breeding in several countries like India, Colombia and Brazil.

Figure 17.8: A New Breeding Line of Arabica (S.4817) with Integration of SH3 Gene by MAS.

Herrera *et al.* (2009) analysed a segregating F_2 population derived from a cross between the susceptible *Coffea arabica* cv. Caturra and a *C. canephora* introgressed arabica line exhibiting high partial resistance using AFLP and SSR markers. Molecular analysis enabled identification of seven polymorphic markers (5 AFLP and 2 SSR) exhibiting significant association with partial resistance. Coexistence of resistance homozygous alleles (RR) at codominant SSR loci was correlated with high resistance. This study was the first attempt to develop PCR-based sequence specific markers linked to partial rust resistance in coffee.

De Brito *et al.* (2010) reported three AFLP markers linked to one of the resistance genes of *C. canephora* origin (S_H7, S_H8, S_H9). Studies on Timor hydrid derievatives, F_2 population from artificial cross between Hý´brido de Timor UFV 427-15, the resistant parent and the susceptible genotype Catuaý´ Amarelo UFV 2143-236 (IAC 30) using AFLP primers resulted in identification of three AFLP markers linked to the resistant gene (De Brito *et al.*, 2010). The inheritance study of the Hý´brido de Timor UFV- 427-15 to *H. vastatrix* race II, conducted with three coffee populations (F_2, BCr and BCs) also confirmed that the resistance was monogenic and dominant. Diola *et al.* (2011). studied the same population derived from the Híbrido de Timor UFV 42715 and identified 25 closely related markers, and obtained the first saturated and high-density genetic map obtained from the region containing the resistance gene derived from the Híbrido de Timor.

Alvarenga *et al.* (2011) identified DNA sequences in arabica coffee potentially involved in rust resistance by data mining and *in silico* analysis, from data obtained in the Brazilian coffee genome project. Testing of 59 primer pairs in resistance and susceptible plants resulted in amplification of the region of DNA that corresponded to a partial open reading frame encoding a disease resistance protein. Talhinhas *et al.* (2014) carried out a 454-pyrosequencing transcriptome analysis of *H. vastatrix* germinating urediniospores (gU) and appressoria (Ap) and compared to previously published *in planta* haustoria-rich (H) data. The study contributed to characterization of molecular processes that lead to appressoria-mediated infection by rust fungi. The results from the study also pointed towards the identication of novel candidate virulence factors.

Genetic diversity and population structure of *Hemileia vastatrix*, with respect to the host and geographical origin, was studied using AFLP markers in 112 monopustule isolates of the pathogen from ve major coffee producing states in Brazil (Cabral *et al.*, 2016). The isolates collected on coffee genotypes of *C. arabica, C. canephora*, derivatives of Hýbrido de Timor and Icatu indicated a low level of genotypic diversity and low level of population differentiation. The genetic diversity did not represent any apparent geographic pattern.

5.1.2. Coffee Berry Disease

Coffee berry disease (CBD) caused by *Colletotrichum kahawae* is another serious disease of concern and a major constraint to Arabica coffee (*Coffea arabica*) production in Africa. It was reported that the Hibrido de Timor, a spontaneous interspecific hybrid of *C. arabica* and *C. canephora* and its derivatives possess resistance to CBD. Agwanda *et al.* (1997) reported three RAPD markers to be closely associated with T gene of the three genes (T, R and k) found in Hibrido de Timor (T gene), Catimor (T gene), Rume Sudan (R and k genes) and K7 (k gene), known to show resistance to CBD disease. Gichuru *et al.* (2008) screened two F_2 populations derived from crosses of cv. Catimor (resistant) and cv. SL28 (susceptible) for resistance to CBD and subjected them to marker analysis with 57 microsatellites and 31 AFLP markers. The analysis resulted in identification of eight AFLP and two microsatellites markers linked tightly to the resistant phenotype and mapped to one unique chromosomal fragment introgressed from *C. canephora*. The gene conferring the resistance was localized within an 11 cM segment and it was reported that the locus carried a major resistance gene designated *Ck*-1, which is likely to be synonymous to the *T* gene described in previous studies by Agwanda *et al.* (1997). Recently, Kiguongo *et al.* (2014). identified microsatellite markers that co-segregated with resistance to CBD in a F_2 population. The study also confirmed correlation between phenotypic data and molecular data with regard to resistance to *Colletotrichum kahawae* infection. Two alleles/loci of SSR markers, M24 (~210 bp) and Sat 227 (~200 bp) was reported to be putatively linked to resistance to *C. kahawae* which are useful as diagnostic markers in breeding programs to develop CBD resistant lines.

5.2. Nematode Resistance

Root characteristics and molecular polymorphisms associated with resistance to *Pratylenchus coffea* in robusta was reported by Touran *et al.* (1995). The genetic variation among the group was evaluated with SDS-PAGE analysis of the protein and using 17 RAPD primers on susceptible and resistance clones. Differences between the groups were found in root anatomy, total polyphenol content, protein and DNA patterns. A specific protein marker of molecular weight 29 kDa was found in the resistant group, indicating that the resistant clones had specific enzyme as product of DNA associated with resistance.

Meloidogyne exigua, the root-knot nematode, is a major agronomic constraint in all major coffee growing (*Coffea arabica*) areas especially in Latin America. Noir *et al.* (2003) studied the mode of inheritance of the *M. exigua* resistance transferred into *C. arabica* from a related species, *Coffea canephora* and identified 14 AFLP markers, associated with the resistance to *M. exigua*. A localized genetic map of the chromosome segment carrying *Mex*-1 was constructed. The association of the identified AFLP markers with *Mex*-1 was confirmed by analysis of a set of genotypes involving 28 arabica introgression lines either resistant or susceptible to *M. exigua* in field conditions. Identification of markers was considered as an important starting point to perform an early selection of resistant seedlings.

Recently, Pereira *et al.* (2016) identified microsatellite markers associated with resistance to *Meloidogyne exigua* in coffee. F_5 progenies derived from a cross between Híbrido de Timor 440-10 and Catuaí Amarelo IAC 86 were subjected to marker assays with 44 SSR markers. Of the evaluated markers, 11 showed a polymorphic pattern with a mean number of 4.5 alleles per marker. SSRCafé 13 allele 2, SSRCafé 19 allele 3, SSRCafé 40 allele 2, SSRCafé 15 allele 3, and SSRCafé 20 allele 3 were correlated with the root gall index of *M. exigua*. The SSR markers validated in this population provides scope to select progenies resistant to nematodes in coffee-breeding programmes.

5.3. Markers Associated with Root Characteristics and Physiological Traits for Drought Tolerance

Between the two commercial species of coffee, *Coffea canephora* is characterized by shallow root system and is, therefore, highly sensitive to drought stress that affects growth and production. Achar *et al.* (2015) identified markers linked to high root types in *Coffea canephora* using F_1 population of 134 individuals developed from two contrasting parents, L1 valley with low root and S.3334 with high root type at Central Coffee Research Institute, India. All the 134 F_1 individuals were phenotyped for root and associated physiological traits (29 traits) and genotyped with 41 of the 320 RAPD and nine of the 55 SSR polymorphic primers. The study resulted in identification of 13 putative RAPD markers associated with root traits such as root length, secondary roots, root dry weight, and root to shoot ratio, in which root length associated marker $OPS1_{850}$ showed high phenotypic variance of 6.86 per cent. Two microsatellite markers linked to root length ($CPCM13_{400}$) and root to shoot ratio ($CM211_{300}$). Besides, 25 markers were associated with more than one trait and few of the markers were associated with positively related physiological traits with potential scope for marker assisted trait selection.

5.4. Scope and Limitations for Marker Assisted Selection in Coffee

Although started with a modest beginning, the information generated on DNA markers in coffee has helped in generation of very useful information. Further, the potential of DNA markers has been successfully demonstrated for a wide range of applications in coffee, from diversity analysis to QTL mapping with ease and accuracy. Nevertheless, information on markers linked to various agronomic traits and their application for MAS are very limited. To our knowledge, only the markers linked to specific genes that impart tolerance/resistance for coffee leaf rust (CLR); coffee berry disease and root knot nematode have been successfully used for MAS in countries like India, Colombia, Brazil and Kenya. Hence, there is a need for development of coffee specific PCR based efficient DNA markers for rational application in genetic-linkage analysis. Development of dense and well covered reference molecular linkage maps would facilitate MAS based molecular breeding that greatly helps in complementing the conventional genetic improvement. The success in development of these tools would strengthen the efforts to identify QTLs linked to gene(s) of agronomic importance that provide much needed impetus to the coffee breeding programmes.

Efforts to generate ideal mapping population in Arabica and construction of linkage map of Arabica has limited success, even by using large number of molecular markers due to the narrow genetic base of Arabica. The reports on linkage maps for tetraploid Arabica were based on introgressed populations and so the marker coverage is likely to be more in introgressed regions. In *C. canephora*, substantial progress has been achieved where linkage maps have been successfully evolved for diploid *C. canephora* and other diploid interspecific hybrids. With the advent of affordable sequencing technologies, there has been a shift in focus towards whole genome sequencing and huge repository of data has been generated under global initiatives, in recent past. Significant leads have been made under the Brazilian Coffee Genome Project initiated during 2002 (Vieira *et al.*, 2006) to generate sequence information on *Coffea arabica*, *Coffea canephora* and *Coffea racemosa* (Brazilian Coffee Genome Project). Similarly, the genome sequence of *Coffea canephora* has been published recently (Denoeud *et al.*, 2014) provided new insights into the convergent evolution of caffeine biosynthesis (Denoeud *et al.*, 2014) Of late, the impact of climate change effects is strikingly visible especially in flareup of diseases and pests as well as drought in several coffee growing countries posing new challenges to the coffee breeders. Hence, there exists a great scope and promise for exploitation of the new developments in molecular tools for marker-assisted breeding aimed at focused genetic improvement of coffee.

Acknowledgements

The authors wish to acknowledge each and every scientist of coffee improvement group, since the establishment of erstwhile Mysore Coffee Experimental Station to the present Central Coffee Research Institute. The administrative support from authorities of Centre for Cellular and Molecular Biology (CCMB), Hyberabad and the financial support from Department of Biotechnology, New Delhi, Govt. of India, for the collaborative projects on 'Development and Application of DNA markers in coffee, is gratefully acknowledged.

References

Achar, D., Awati, M.G., Udayakumar, M. and Prasad, T.G. (2015). Identification of putative molecular markers associated with root traits in *Coffea canephora* Pierre ex Froehner. *Molecular Biology International*, **2015**: 11.

Aga, E. and Bryngelsson, T. (2006). Inverse sequence-tagged repeat (ISTR) analysis of genetic variability in forest coffee (*Coffea arabica* L.) from Ethiopia. *Genetic Resources and Crop Evolution*, **53**: 721–728.

Aga, E., Bekele, E. and Bryngelsson, T. (2005). Inter-simple sequence repeat (ISSR) variation in forest coffee trees (*Coffea arabica* L.) *populations from Ethiopia. Genetica*, **124**: 213-221.

Aga, E., Bryngelsson, T., Bekele, E. and Salomon, B. (2003). Genetic diversity of forest arabica coffee (*Coffea arabica* L.) in Ethiopia revealed by random amplified polymorphic DNA (RAPD) analysis. *Hereditas*, **138**: 36–46.

Aggarwal, R.K. (2005). DNA fingerprinting of coffee germplasm for conservation and better management and development of molecular map of coffee:Molecular characterization and linkage analysis. Completion report submitted to *Department of Biotechnology*, New Delhi, India, Project No. BT/PR1302/AGR/08/095/98. pp. 48.

Aggarwal, R.K., Rajkumar, R., Rajendrakumar, P., Hendre, P.S., Baruah, A., Phanindranath, R., Annapurna, V., Prakash, N.S., Santaram, A., Srinivasan, C.S. and Singh, L. (2004). Fingerprinting of Indian coffee selections and development of reference DNA polymorphism panels for creating molecular IDs for variety identification. In: *Proceedings of 20th International Scientific Colloquium on Coffee Paris*, France, ASIC. pp. 751–755.

Agwanda, C., Lashermes, P., Trouslot, P., Combes, M.C. and Charrier, A. (1997). Identification of RAPD markers for resistance to Coffee Berry Disease, *Colletotrichum kahawae*, in *Arabica coffee. Euphytica*, **97**: 241–248.

Al-Murish, T.M., Elshafei, A.A., Al-Doss, A.A. and Barakat, M.N. (2013). Genetic diversity of coffee (*Coffea arabica* L) in Yemen via SRAP, TRAP and SSR markers. *Journal of Food Agriculture and Environment*, **11(2)**: 411-416.

Alvarenga, S.M., Caixeta, E.T., Hufnagel, B., Thiebaut, F., Maciel-Zambolim, E., Zambolim, L. and Sakiyama, N.S. (2011). Molecular markers from coffee genome expressed sequences potentially involved in resistance to rust. *Pesquisa Agropecuária Brasileira*, **46:** 890–898.

Anonymous (2014). Coffee Guide, *Coffee Board of India*, Bangalore.

Anthony, F. (1992). Les ressources génétiques des caféiers: Collecte, gestion d'un conservatoire et évaluation de la diversité génétique. *z*, ORSTOM, Paris.

Anthony, F., Berthaud, J., Guillaumet, J.L. and Lourd, M. (1987). Collecting wild *Coffea* species in Kenya and Tanzania. *Plant Genetic Resources Newsletters*, **69**: 23–29.

Anthony, F., Bertrand, B., Quiros, O., Wilches, A., Lashermes, P., Berthaud, J. and Charrier, A. (2001). Genetic diversity of wild coffee (*Coffea arabica* L.) using *molecular markers. Euphytica*, **118**: 53–65.

Anthony, F., Couturon, E. and de-Namur, C. (1985). Les caféiers sauvages du Cameroun. Résultats d'une mission de prospection effectuée par l'ORSTOM en 1983. In *Proceedings of XI Colloque Scientifique International sur le Café*, ASIC, Lausane. pp. 495–505

Anthony, F., Combes, M.C., Astorga, C., Bertrand, B., Graziosi, G. and Lashermes, P. (2002a). The origin of cultivated *Coffea arabica* L. varieties revealed by AFLP and SSR markers. *Theoretical and Applied Genetics*, **104**: 894–900.

Anthony, F., Quiros, O., Topart, P., Bertrand, B and Lasherms, P. (2002b). Detection by simple sequence repeat markers of introgression from *Coffea canephora* in *Coffea arabica* cultivars. *Plant Breeding*, **121**: 542-544.

Barbosa, A.V., Sales, R.M.O.B. and da Silva, F.R. (2008). CoffEST is the complete resource for *Coffea spp*. EST analysis. Paper presented at 22nd International Conference on Coffee Science, Campinas, Brazil. Programme abstracts. Montpellier, France: *Association for Science and Information on Coffee*, PB635.

Barre, P., Laissac, M., D'Hont, A., Louarn J., Charrier A., Hamon S. and Noirot M. (1998). Relationship between parental chromosomic contribution and nuclear DNA content in the *Coffea* interspecific hybrid: *C. pseudozanguebariae* × *C. liberica* var. *dewevrei. Theoretical and Applied Genetics*, **96**: 301–305.

Baruah, A., Naik, V., Hendre, P.S., Rajkumar, R., Rajendrakumar, P. and Aggarwal, K. (2003). Isolation and characterization of nine microsatellite markers from *Coffea arabica* L., showing wide cross-species amplification. *Molecular Ecology Notes*, **3**: 647-650.

Bekele, Y.D. (2005). Assessment of cup quality, morphological, biochemical and molecular diversity of *Coffea arabica* L. ***genotypes of Ethiopia,*** University of Free State, Ethiopia, Doctoral Thesis.

Berthaud, J. (1986). Les resources g_en_etiques pour l' am_elioration des caf_eiers africains diploides. ORSTOM, collection Travaux et documents, Montpellier, France. pp. 379.

Berthaud, J. and Charrier, A. (1988). Genetic Resource of *Coffea*. In: Clarke, R.J. and Macrae, R. (Eds.), *Coffea- Agronomy* (Vol. 4). *Elsivier Applied Science Publishers Ltd.*, Amsterdam. pp. 1-42.

Berthaud, J. and Guillaumet, J.L. (1978). Les caféiers sauvages en Centrafrique : Résultats d'une mission de prospection (Janvier–Février 1975). *Café Cacao Thé*, **3**: 171–186.

Berthou, F. and Trouslot, P. (1977). Analysis of enzymatic polymorphism in the genus *Coffea*: adaptation of an electrophoresis method in series: first results. In: *VIIIth Colloque Scientifique International sur le Cafe*. Abidjan. pp. 373–383

Bettencourt, A.J. and Rodriguez Jr. C.J. (1988). Principles and practice of coffee breeding for resistance to rust and other diseases. In: Clarck, R.J. and Macrae, R. (Eds.), *Coffee* (Vol. 4). *Agronomy, Elseviers Applied Science*. pp. 199-235.

Bettencourt, A.J. and Noronha-Wagner, M. (1971). Genetic factors conditioning resistance of *Coffea arabica* L. to *Hemileia vastatrix* Berk and Br. *Agronomia Lusitana*, **31**: 285–292.

Bigirimana, J., Njoroge, K., Muthomi, J.W., Gahakwa1, D., Phiri, N.A., Gichuru, E.K. and Walyaro. D.J. (2013). Genetic diversity among disease resistant coffee varieties and cultivars in Rwanda based on RAPD and SSR marker. *Journal of Renewable Agriculture*, **1(6)**: 106-112.

Bouharmont, J. (1963). Somatic chromosomes of *Coffea* species. *Euphytica*, **12**: 254–257.

Brar, D.S. and Dhaliwal, H.S. (1997). Molecular markers and their application in crop improvement. *Third Agricultural Science Congress. Punjab Agricultural University*, Ludhiana, March 12-15, 1997.

Cabral, P.G.C., Maciel-Zambolim, E.M., Oliveira, S.A.S., Caixeta, E.T. and Zambolim, L. (2016). Genetic diversity and structure of *Hemileia vastatrix* populations on *Coffea* spp. *Plant Pathology*, **65(2)**: 196–204.

Capot, J. (1972). L'amélioration du caféier en Cote d'Ivoire. Les hybrides Arabusta. *Café Cacao Thé*, **16**: 3–18.

Carvalho, A. (1952). Taxonomia de *Coffea arabic*a L.: *Caracteres morfologicas dos haploids*. *Bragantia*, **12**: 201–212.

Carvalho, A., Ferwerda, F.P., Frahm-Leviveld, J.A., Medina, P.M., Mendes, A.J.T. and Monaco, L.C. (1969). Coffee. In: Ferwerda, F. P. and Wit, F. (Ed.), *Outlines of Perennial Crop Breeding in the Tropics*. Veenman and Zonen NV, Wageningen. pp.189- 241

Chaparro, A.P., Cristancho, M.A., Cortina. H.A. and Gaitán, A.L. (2004). Genetic variability of *Coffea arabica* L. accessions from Ethiopia evaluated with RAPDs. *Genetic Resources and Crop Evolution*, **51**: 291–297.

Charrier, A. (1978). La structure génétique des caféiers spontanés de la région Malgache (Mascarocoffea). Leurs relations avec les caféiers d'origine Africaine (Eucoffea). Mémoires ORSTOM 87. Paris.

Charrier, A. and Berthaud, J. (1985). Botanical classification of coffee. In: Cliffort, M. N. and Wilson, K.C. (Ed.), *Coffee: Botany, Biochemistry and Production of Beans and Beverage.* Croom Helm, London. pp. 13–47.

Charrier, A. and Eskes, A.B. (2004). Botany and Genetic of Coffee. In: Wintgens, J. N. (Ed.), *Coffee: Growing, Processing, Sustainable Production, a Guidebook for Growers, Processors, Traders, and Researchers*. Wiley-VCH, Weinhelm, Wiley. pp. 25-56.

Clarindo, W.R. and Carvalho, C.R. (2006). A high quality chromosome preparation from cell suspension aggregates culture of *Coffea canephora*. *Cytologia*, **71**: 243–249.

Clarindo, W.R. and Carvalho, C.R. (2008). First *Coffea arabica* karyogram showing that this species is a true allotetraploid. *Plant Systematics and Evolution*, **274**: 237–341.

Combes, M.C., Andrzejewski, S., Anthony, F., Bertrand, B., Rovelli, P., Graziosi, G. and Lashermes, P. (2000). Characterization of microsatellite loci in *Coffea arabica* and related coffee species. *Molecular Ecology*, **9**: 1178–1180.

Coulibaly, I., Louarn, J., Lorieux, M., Charrier, A., Hamon, S. and Noirot, M. (2003b). Pollen viability restoration in a *Coffea canephora* P and *C. heterocalyx* Stoffelen backcross. QTL identification for marker-assisted selection. *Theoretical and Applied Genetics*, **106**: 311–316.

Coulibaly, I., Noirot, M., Lorieux, M., Charrier, A., Hamon, S. and Louarn, J. (2002). Introgression of selfcompatibility from *Coffea heterocalyx* to the cultivated species *Coffea canephora*. *Theoretical and Applied Genetics*, **105**: 994–999.

Coulibaly, I., Revol, B., Noirot, M., Poncet, V., Lorieux, M., Carasco-Lacombe, C., Minier, J., Dufour, M. and Hamon, P. (2003a). AFLP and SSR polymorphism in a *Coffea* interspecific backcross progeny, (*C. heterocalyx* x *C. canephora*) x *C. canephora*. *Theoretical and Applied Genetics*, **107**: 1148–1155.

Cramer, P. J. S. (1957) Review of literature of coffee research in Indonesia. In: Wellman, F. L. (Ed.), Turrialba, Costa Rica. *SIC International American Institute of Agricultural Sciences*. pp. 262

Cristancho, M.A., Chaparro, A.P., Cortina, H.A. and Gaitan, A.L. (2004). Genetic variability of *C. arabica* accessions from Ethiopia evaluated with RAPDs. In: *Proceedings of 20th International Scientific Colloquium on Coffee*, Bangalore. Paris, France: ASIC. pp. 672–675.

Cros, J., Combes, M.C., Chabrillange, N., Duperray, C., Monnot des Angles, A. and Hamon, S. (1995). Nuclear DNA content in the subgenus Coffea (Rubiaceae): Inter- and intraspecific variation in African species. *Canadian Journal of Botany*, **73**: 14–20.

Cros. J., Combes, M.C, Trouslot, P., Anthony, F., Hamon, S., Charrier, A. and Lashermes, P. (1998). Phylogenetic analysis of chloroplast DNA variation in *Coffea* L. *Molecular Phylogenetics and Evolution*, **9**: 109–117.

Crouzillat, D., Rigoreau, M., Lefebvre-Pautigny, F., Priyono, P., Broun, P. and Lambot, C. (2010). A coffee high density genetic map for quantitative trait loci analysis on agronomical, technological and biochemical characteristics in Robusta and Arabica. In: *Proceedings of 23rd ASIC International Conference on Coffee Science*, Bali, Indonesia.

Cubry, P., Musoli, P., Legnaté, H., Aluka, P., Dufour, M., Pot, D., De Bellis, F. and Leroy, T. (2006a). Genetic diversity within *Coffea canephora* germplasm maintained in RCI using SSR markers, results and future prospects. In: *Proceedings of the 21st International Conference on Coffee Science. Paris, France: ASIC. Colloque Scientifique International sur le Café, Montpellier, France.*

Cubry, P., Musoli, P., Legnate, H., Pot, D., de Bellis, F., Poncet, V., Anthony, F., Dufour, M. and Leroy, T. (2008). Diversity in coffee assessed with SSR markers: structure of the genus Coffea and perspectives for breeding. *Genome*, **51**: 50- 63.

Cubry, P., Pot, D., de Bellis, F., Musoli, P., Aluka, P., Legnate, H., Dufour, M. and Leroy, T. (2006b). Genetic diversity in coffea genus using microsatellites loci. *Proceedings of the 21st International Conference on Coffee Science*. Paris, France: ASIC. Colloque Scientifique International sur le Café, Montpellier, France.

Davis, A.P., Govaerts, R., Bridson, D.M. and Stoffelen, P. (2006). An annotated taxonomic conspectus of the genus *Coffea* (Rubiaceae). *Botanical Journal of the Linnean Society*, **152**: 465–512.

De Brito, G.G., Caixeta, E.T., Gallina, A.P., Zambolim, E.M., Zambolim, L., Diola, V. and Loureiro, M.E. (2010). Inheritance of coffee leaf rust resistance and identification of AFLP markers linked to the resistance gene, *Euphytica*, **173**: 255–264.

de Namur, C., Couturon, E., Sita, P. and Anthony, F. (1987). Résultats d'une mission de prospection des caféiers sauvages du Congo. In : *Proceedings of XII Colloque Scientifique International sur le Café,* ASIC, Lausane. pp. 397–404.

de Vienne, D. (1998). In : de Vienne, D. (Ed.), *Les marqueurs moléculaires en génétique et biotechnologies végétales*. INRA, Paris, France.

Denoeud, F., Carretero-Paulet, L., Dereeper, A., Droc, G., Guyot, R., Pietrella, M., Zheng, C., Alberti, A., Anthony, F., Aprea, G. and Aury, J.M. (2014). *The coffee genome provides insight into the convergent evolution of caffeine biosynthesis. Science*, **345**: 1181-1184.

Devasia, J. (2011). Development of genomic resources, linkage map and identification of QTLs in a mapping population (L1 Valley x S.3334) of Robusta Coffee (*Coffea canephora* Pierre ex A. Froehner). *Unpublished doctoral dissertation*, University of Agricultural Sciences, GKVK, Bangalore, Karnataka, India.

Devasia, J., Madhura, J.N., Prakash, N.S., Divya, K.D., Udayakumar, M. and Raghuramulu, Y. (2014). Cross species transferability of novel SSR markers and genetic relationships of selected *Coffea* species. *Paper presented in PLACROSYM XXI, Calicut*, Kerala, 10- 12 December 2014.

Devasia, J., Madhura, J.N., Raj Kumar, R., Sheshasayee, M.S., Kulkarni, R.S., Udayakumar, M. and Jayarama. (2013). Development of SSR markers for Robusta coffee (*Coffea canephora*). *Journal of Plantation Crops*, **41(3)**: 277-286.

Diola, V., de Brito, G.G., Caixeta, E.T., Zambolim, E.M., Sakiyama, N.S. and Loureiro, M.E. (2011). High-density genetic mapping for coffee leaf rust resitance. *Tree Genetics and Genomes*, **7**: 1199–1208.

Dussert, S., Lashermes, P., Anthony, F., Montagnon, C., Trouslot, P., Combes, M. C. *et al.* (1999). Le caféier, *Coffea canephora*. In: Hamon, P., Seguin, M., Perrier, X. and Glaszmann, C. (Eds.), *Diversité Génétique des Plantes Cultivées*. Repères, CIRAD, France. pp. 175–194.

Dussert, S., Lashermes, P., Anthony, F., Montagnon, C., Trouslot, P., Combes, M. C. *et al.* (2003). Coffee (*Coffea canephora*). In: Hamon, P., Seguin, M., Perrier, X. and Glaszmann, J. C. (Eds.), *Genetic Diversity of Cultivated Tropical Crops*. Collection Repères. CIRAD, Montpellier, France. pp. 239–258

Eskes, A.B. (1989). Identification, d*escription and collection of coffee types in P.D.R.* Yemen. Biodiversity International, Rome.

Etienne. H., Anthony, F., Dussert, S., Fernandez, D., Lashermes, P. and Bertrand, B. (2002). Biotechnological applications for the improvement of coffee (*Coffea arabica* L) In *Vitro Cellular and Developmental Biology–Plant*, **38**: 129-138.

Fernandez, D., Santos, P., Agostini, C., Bon, M.C., Petitot, A.S., Silva, M.C., Guerra-Guimarães, L.E., Ribeiro, A.N., Argout, X. and Nicole, M. (2004). Coffee (*Coffea arabica* L.) genes early expressed during infection by the rust fungus (*Hemileia vastatrix*). *Molecular Plant Pathology*, **5**: 527–536.

Ferrão, L.F.V., Caixeta, E.T., Souza, F.D., Zambolim, E.M., Cruz, C.D., Zambolim, L. and Sakiyama, N.S. (2013). Comparative study of different molecular markers for classifying and establishing genetic relationships in *Coffea canephora*. *Plant Systematics and Evolution*, **299(1)**: 225-238.

Gichuru, E.K., Agwanda, C.O., Combes M.C, Mutitu, E.W., Ngugi, E.C.K., Bertrand Benoît *et al.* (2008). Characterisation and genetic mapping of a gene conferring resistance to coffee berry disease (*Colletotrichum kahawae*) in Arabica coffee (*Coffea arabica* L.). In: *21st International Conference on Coffee Science*, Montpellier (France), 11th - 15th September 2006.

Gupta, P.K. and Varshney, R.K. (2000). The development and use of microsatellite markers for genetic analysis and plant breeding with emphasis on bread wheat. *Euphytica*, **113 (3):** 163-185.

Hanson, A.D., Peacock, W.J., Evans, L.T., Arntzen, C.J. and Khush, G.S. (1990). Drought resistance in rice. *Nature*, **234**: 2.

Hendre, P.S. and Aggarwal, R.K. (2007). DNA markers: development and application for genetic improvement of coffee. In: Varshney, R. K and Tuberosa, R. (Eds.), *Genomic Assisted Crop Improvement: Genomics Applications in Crops* (Vol. 2). Springer-Verlag, Germany. pp. 399-434.

Hendre, P.S. and Aggarwal, R.K. (2014). Development of genic and genomic SSR markers of Robusta Coffee (*Coffea canephora* Pierre Ex *A. Froehner*). *PLoS ONE*, **9(12):** e113661.

Hendre, P.S., Phanindranath, R., Annapurna, V., Lalremruata, A. and Aggarwal, R.K. (2008). Development of new genomic microsatellite markers from robusta coffee (*Coffea canephora* Pierre ex A. Froehner) showing broad cross-species transferability and utility in genetic studies. *BMC Plant Biology*, **8**: 51.

Herrera, J.C., Cortina, H.A., Anthony, F., Prakash, N.S., Lashermes, P., Gaitan, A.L., *et al.* (2011). COFFEE (*Coffea spp.*). In: Singh Ram, J. (Ed.), *Genetic Resources, Chromosome Engineering and Crop Improvement - Medicinal Crops* (vol. 6). CRC press. pp. 589–640.

Herrera, J.C., Alvarado, A.V., Cortina, G.H.A., Combes, M.C., Romero, G.G. and Lashermes, P. (2009). Genetic analysis of partial resistance to coffee leaf rust (*Hemileia vastatrix* Berk and Br.) introgressed into the cultivated *Coffea arabica* L. from the diploid *C. canephora* species. *Euphytica*, **167 (1)**: 57-67.

Herrera, J.C., Cortina, H.A., Anthony, F., Prakash, N.S., Lashermes, P., Gaitan, A.L., *et al.* (2012). COFFEE'. In: Genetic Resources, Chromosome Engineering and Crop improvement - *Medicinal Crops (Vol. 6). CRS Press, Boca Raton*, Florida, USA. pp. 589-640.

Herrera, J.C., D'Hont, A. and Lashermes, P. (2007). Use of fluorescence *in situ* hybridization as a tool for introgression analysis and chromosome identification in coffee (*Coffea arabica* L.). *Genome,* **50**: 619–626.

Herrera, J.C., Combes, M.C., Anthony, F., Charrier, A. and Lashermes, P. (2002b) Introgression into the allotetraploid coffee (*Coffea arabica* L.): Segregation and recombination of the *C. canephora* genome in the tetraploid interspecific hybrid (*C. arabica* × *C. canephora*). *Theoretical and Applied Genetics*, **104**: 661–668.

Herrera, J.C., Combes, M.C., Cortina, H. and Lashermes, P. (2004). Factors influencing gene introgression into the allotetraploid *Coffea arabica* L. from its diploid relatives. *Genome*, **47(6)**: 1053-1060.

Herrera, J.C., Combes, M.C., Cortina, H., Alvarado, G. and Lashermes, P. (2002a) Gene introgression into *Coffea arabica* by way of triploid hybrids (*C. arabica* × *C. canephora*). *Heredity*, **89**: 488–494.

Kathurima, C.W., Kenji, G.M., Muhoho, S.M., Boulanger, R., Gichimu, B.M. and Gichuru, E.K. (2012). Genetic diversity among commercial coffee varieties, advanced selections and museum collections in Kenya using molecular markers. *International Journal of Biodiversity and Conservation*, **4**: 39-46.

Kiguongo, A.P.K., Omondi, C.O., Gichuru, E.K. and Kasili, R.W. (2014). Analysis of simple sequence repeat markers linked to coffee berry disease resistance genes in a segregating population of arabica coffee (*Coffea arabica L.*). *International Journal of Biotechnology and Food Science*, **2(8)**: 156-166.

Krishnan, S. (2013). Current status of coffee genetic resources and implications for conservation. *CAB Reviews*, **8(016)**: 1-9.

Krishnan, S., Ranker, T., Davis, A.P., Rakotomalala, J.J. (2013). The study of genetic diversity patterns of *Coffea commersoniana*, an endangered coffee species from Madagascar: a model for conservation of other littoral forest species. *Tree Genetics and Genomes*, **9**: 179–87.

Krug, C.A. (1934) Contribuçao para o estudo da citologia do gênero Coffea. *Campinas, Instituto Agronomico, Boletim Técnico*, **11**: 1–10.

Krug, C.A. (1937) Observaçoes citologicas em Coffea. III. Campinas, Instituto Agronomico, Boletim, **37**: 1–19.

Krug, C.A., Mendes, J.E.T. and Carvalho, A. (1939) *Taxonomia de Coffea arabica L. Bolétim Técnico* no. 62. *Instituto Agronômico do Estado, Campinas.*

Ky, C.L., Barre, P. and Noirot, M. (2013). Genetic investigations on the caffeine and chlorogenic acid relationship in an interspecific cross between *Coffea liberica dewevrei* and *C.* Pseudozanguebariae. *Tree Genetics and Genomes,* **9**: 1043–1049.

Ky, C.L., Barre, P., Lorieux, M., Trouslot, P., Akaffou, S., Louarn, J., Charrier, A., Hamon, S. and Noirot, M. (2000). Interspecific genetic linkage map, segregation distortion and genetic conversion in coffee (*Coffea sp.*). *Theoretical and Applied Genetics*, **101**: 669–676.

Labouisse, J.P., Bellachew, B., Kotecha, S. and Bertrand, B. (2008). Current status of coffee (*Coffea arabica* L.) genetic resources in Ethiopia: Implications for conservation. *Genetic Resources and Crop Evolution*, **55**: 1079–109.

Lashermes, P., Andrzejewski, S., Bertrand, B., Combes, M.C., Dussert, S., Graziosi, G., Trouslot, P. and Anthony, F. (2000). Molecular analysis of introgressive breeding in coffee (*Coffea arabica* L.). *Theoretical and Applied Genetics*, **100**: 139-146.

Lashermes, P., Combes, M.C., Robert, J., Trouslot, P., D'Hont, A, Anthony, F. and Charrier, A. (1999). Molecular characterization and origin of the *Coffea arabica* L. genome. *Molecular Genomics and Genetics*, **261**: 259–266.

Lashermes, P., Combes, M.C., Ansaldi, C., Gichuru, E. and Noir, S. (2011). Analysis of alien introgression in coffee tree (*Coffea arabica* L.). *Molecular Breeding*, **27**: 223-232.

Lashermes, P., Combes, M.C., Prakash, N.S., Trouslot, P., Lorieux, M. and Charrier, A. (2001). Genetic linkage map of *Coffea canephora*: effect of segregation and analysis of recombination rate in male and female meioses. *Genome*, **44**: 589-596.

Lashermes, P., Combes, M.C., Ribas, A., Cenci, A., Mahé, L. and Etienne, H. (2010). Genetic and physical mapping of the S_H3 region that confers resistance to leaf rust in coffee tree (*Coffea arabica* L.). *Tree Genetics and Genomes*, **6**: 973–980.

Lashermes, P., Combes, M.C., Trouslot, P. and Charrier, A. (1997). Phylogenetic relationships of coffee tree species (*Coffea* L.) As inferred from ITS sequence of nuclear ribosomal DNA. *Theoretical and Applied Genetics*, **94**: 947-955.

Lashermes, P., Cros, J., Combes, M.C., Trouslot, P., Anthony, F., Hamon, S. and Charrier, A. (1996a). Inheritance and restriction fragment length polymorphism of chloroplast DNA in the genus *Coffea* L. *Theoretical and Applied Genetics*, **93**: 626–632.

Lashermes, P., Coutoron, E. and Charrier, A. (1996c) Inheritance and genetic mapping of self incompatibility in *Coffea canephora* Pierre. *Theoretical and Applied Genetics*, **93**: 458–462.

Lashermes, P., Trouslot, P., Anthony, F., Combes, M.C. and Charrier A. (1996b) Genetic diversity for RAPD markers between cultivated and wild accessions of *Coffea arabica*. *Euphytica*, **87**: 59-64.

Lashermes, P., Cros, J., Combes, M.C., Trouslot, P., Anthony, F. and Charrier, A. (1996d) Molecular analysis of the origin and genetic diversity of Coffea arabica L. In: Implications for Coffee Improvement. *Proceedings of EUCARPIA Meeting on Tropical Plants*. Montpellier. pp. 23–29.

Lashermes, P., Cros, J., Marmey, P. and Charrier, A. (1993). Use of random amplified DNA markers to analyse genetic variability and relationships of *Coffea* species. *Genetic Resources and Crop Evolution*, **40**: 91–99.

Lashermes, P., Paczek, V., Trouslot, P., Combes, M.C., Couturon, E. and Charrier, A. (2000). Single-locus inheritance in the allotetraploid *Coffea arabica* L. and interspecific hybrid *C. arabica* × *C. canephora*. *Journal of Heredity*, **91**: 81–85.

Le Pierrès, D., Charmetant, P., Yapo, A., Leroy, T., Couturon, E., Bontemps, S. and Tehe, H. (1989). Les caféiers sauvages de Côte d'Ivoire et de Guinée: Bilan des missions de prospection effectuées de 1984 à 1987. In : *Proceedings of XIII Colloque Scientifique International sur le Café, ASIC, Paipa, Colombia*. pp. 420–428.

Leroy, T., Cubry, P., Durand, N., Dufour, M., De Bellis, F., Jourdan, I., *et al.* (2006). *Coffea spp.* and Coffea canephora diversity evaluated with microsatellites and SNPs. Lessons from comparative analysis. In: Proceedings of the 21st International Conference on Coffee Science. Paris, France: ASIC. *Colloque Scientifique International sur le Café,* Montpellier, France.

Leroy, T., De Bellis, F., Legnate, Y., Kananura, E., Gonzales, G., Pereira, L., Andrade, A.C., Charmetant, P., Montagnon, C., Cubry, P. and Marraccini, P. (2011). Improving the quality of African robustas: QTLs for yield and quality-related traits in *Coffea canephora*. *Tree Genetics and Genomes*, **7**: 781–798.

Leroy, T., Montagnon, C., Eskes, E.B. and Charrier, A. (1993). Reciprocal recurrent selection applied to *Coffea canephora* Pierre. Characterisation and evaluation of breeding populations and value of intergroup hybrids. *Euphytica*, **67**: 113-125.

Leroy, T., De Bellis, F., Legnate, H., Musoli, P., Kalonji, A., Loor Solórzano, and Cubry, P. (2014). Developing core collections to optimize the management and the exploitation of diversity of the coffee *Coffea canephora*. *Genetica*, **142**: 185-199.

Lin Zhou, Vega, F.E., Tan, H., Ramirez Lluch, A.E., Meinhardt, L.W., Fang, W., Mischke, S., Irish, B. and Zhang, D. (2016). Developing single nucleotide polymorphism (SNP) markers for the identification of coffee germplasm. *Tropical Plant Biology*, **9**: 82-95.

Lin, C., Mueller, L.A., McCarthy, J., Crouzillat, D., Pétiard, V. and Tanksley, S.D. (2005). Coffee and tomato share common gene repertoires as revealed by deep sequencing of seed and cherry transcripts. *Theoretical and Applied Genetics*, **112**: 114–130.

López-Gartner, G., Cortina, H., McCouch, S.R. and Moncada, M.D.P. (2009). Analysis of genetic structure in a sample of coffee (*Coffea arabica* L.) using fluorescent SSR markers. *Tree Genetics and Genomes*, **5(3)**: 435-446.

Louarn, J. (1993). La fertilitédes hybrides intrspécifique setles relations génomiques entreles caféiers diplod' origines africaines (Genre *Coffea* L. sous-genre *Coffea*). Thèse del. Université Paris XI, Orsay, France.

Mahe, L., Combes, M.C., Várzea, V.M.P., Guilhaumon, C. and Lashermes, P. (2008). Development of sequence characterized DNA markers linked to leaf rust (*Hemileia vastatrix*) resistance in coffee (*Coffea arabica* L.). *Molecular Breeding*, **21**: 105-113.

Maluf, M.P., Silvestrini, M., Ruggiero, L.M. de. C., Guerreiro Filho, O. and Colombo, C.A. (2005). Genetic diversity of cultivated *Coffea arabica* inbred lines assessed by RAPD, AFLP and SSR marker systems. *Scientia Agricola*, **62(4)**: 366-373.

Masumbuko, L.I., Bryngelsson, T., Mneney, E.E. and Salomon, B. (2003). Genetic diversity in Tanzanian Arabica coffee using random amplified polymorphic DNA (RAPD) markers. *Hereditas*, **139**: 56 – 63.

Medina, H.P., Carvalho, A., Ballve, R.M.L., Bordignon, R., Silvarolla, M.B. and Lima, M.M.A. (1995). Isozyme evidence for the interspecific origin of Piata coffee. *Bragantia*, **54**: 263-273.

Mettulio, R., Rovelli, P., Anthony, F., Anzueto, F., Lashermes, P. and Graziosi, G. (1999). Polymorphic microsatellites in *Coffea arabica*. In: *Proceedings of XVIII International Conference on Coffee Science. ASIC*, Lausane. pp. 344–347.

Meyer, F.G., Fernie, L.M., Narasimhaswamy, R.L., Monaco, L.C. and Greathead, D.J. (1968). FAO Coffee Mission to Ethiopia, *Food and Agricultural organisation of the United Nations*, Rome.

Missio, R.F., Caixeta, E.T., Zambolim, E.M., Pena, G.F., Ribeiro, A.P., Zambolim, L., Pereira, A.A. and Sakiyama, N.S. (2009). Assessment of EST-SSR Markers for genetic analysis on Coffee. *Bragantia*, **68(3)**: 573-581.

Monaco, L.C., Carvalho, A. and Fazuoli, L.C. (1974). Melhoramento do cafeeiro. Germoplasma do café Icatu e seu potencial no melhoramento. In: *Resumos II Cogresso Brasileiro de pesquisas cafeeiras. IBC/MIC*, Rio de Janeiro, Brazil.

Moncada, M., Tovar, E., Montoya, J.C., González, A., Spindel, J. and McCouch, S. (2016). A genetic linkage map of coffee (*Coffea arabica* L.) and QTL for yield, plant height and bean size. *Tree Genetics and Genomes*, **12**: 5.

Moncada, P. and McCouch, S. (2004). Simple sequence repeat diversity in diploid and tetraploid Coffea species. *Genome*, **47**: 501-509.

Montagnon, C. and Bouharmont, P. (1996). Multivariate analysis of phenotype diversity of *Coffea arabica*. *Genetic Resources and Crop Evolution*, **43(3)**: 221–227.

Montagnon, C., Leroy, T. and Yapo, A. (1992). Etude compl_ementaire de la diversit_e genotypique et phenotypique des cafeiers de l' espece *C. canephora* en collection en Cote d Ivorie. In: *XIVe colleque scientifique international sur le cafe, Paris, France*, ASIC. pp. 444–450.

Montagnon, C., Leroy, T. and Eskes, A.B. (1998). Amélioration variétale de Coffea canephora. II. Les programmes de sélection et leurs résultats. *Plantations, Recherche, Développement*, **5**: 89–98.

Montagnon, C, Marraccini, P. and Bertrand, B. (2012). Breeding for coffee quality. In: Oberthür Thomas, Läderach Peter, and Cock James H. (Ed.), *Specialty coffee: Managing Quality*. Penang, IPNI [Southeast Asia]. pp. 89-117.

Mueller, L.A., Solow, T.H., Taylor, N., Skwarecki, B., Buels, R., Binns, J., Lin, C., Wright, M.H., Ahrens, R., Wang, Y. and Herbst, E.V. (2005). The SOL genomics network. A comparative resource for Solanaceae biology and beyond. *Plant Physiology*, **138**: 1310–1317.

Musoli, P., Aluka, P., Cubry, P., Dufour, M., de Bellis, F., Nakendo, S., Nabaggala, A., Ogwang, J., Kyetere, D., Leroy, T. and Bieysse, D. (2006). Fighting against Coffee Wilt Disease: Uganda wild *C. canephora* Genetic Diversity and its Usefulness In: *Proceedings of the 21st International Conference on Coffee Science. Paris, France: ASIC. Colloque Scientifique International sur le Café.* 21, 2006/09/11-15, Montpellier, France.

Musoli, P., Cubry, P., Aluka, P., Billot, C., Dufour, M. and De Bellis, T. (2009). Genetic differentiation of wild and cultivated populations: diversity of *Coffea canephora* Pierre in Uganda. *Genome*, **52**: 634–646.

Nagai, C., Jones, M.R., Byers, A.E., Adamski, D.J. and Ming. R. (2006). Development and Characterization of a True F_2 population for Genetic and QTL Mapping in Arabica. In: *Proceedings of 21st ASIC International Conference on Coffee Science,* 11-15 September, Montpellier, France.

Narasimhaswamy, R.L. (1962). Some thoughts on the origin of *C. arabica* L. Coffee. *Turrialba*, **4**: 1–5.

Narasimhaswamy, R.L. and Vishweswara S. (1963). A note on the occurrence of three species of coffee indigenous to India, *Coffee*, **5(19)**: 65-71.

Noir, S., Anthony, F., Bertrand, B., Combes, M.C. and Lashermes, P. (2003). Identification of a major gene (*Mex*-1) from *Coffea canephora* conferring resistance to *Meloidogyne exigua* in *Coffea arabica*. *Plant Pathology*, **52(1)**: 97–103.

Oguttu, C., Fang, T., Yan, L., Wang, L., Huang, L., Wang, X., Ma, B., Deng, X., Owiti, A., Nyende, A. and Han, Y. (2016). Characterization and utilization of microsatellites in the *Coffea canephora* genome to assess genetic association between wild species in Kenya and cultivated coffee. *Tree Genetics and Genomes*, **12(3)**: 54.

Orozco-Castillo, C., Chalmers, K.J., Waugh, R. and Powell, W. (1994).) Detecion of genetic diversity and selective gene introgression in coffee using RAPD markers. *Theoretical and Applied Genetics*, **87**: 934-940.

Paillard, M., Lashermes, P. and Pétiard, V. (1996). Construction of a molecular linkage map in coffee. *Theoretical and Applied Genetics*, **93(1-2)**: 41-47.

Pearl, H.M., Nagai, C., Moore, P.H., Steiger, D.L., Osgood, R.V. and Ming, R. (2004). Construction of a genetic map for arabica coffee. *Theoretical and Applied Genetics*, **108**: 829–835.

Pereira, T.B., Setotaw, T.A., Santos, D.N., Mendes, A.N., Salgado, S.M, Carvalho, G.R. and Rezende, R.M. (2016). Identification of microsatellite markers in coffee associated with resistance to *Meloidogyne exigua*. *Genetics and Molecular Research*, **29**: 15(3).

Pierozzi, N.I., Pinto-Maglio, C.A.F., Silvarola, B. and Barbosa, R.L. (2001). Karyotype characterization in some diploid coffee species by acetic orcein, C-band and AgNO3 techniques. *Chromosome Research*, **9**: 99.

Pinto-Maglio, C.A.F., Barbosa, R.L., Cuéllar, T., Maluf, M.P., Pierozzi, N.I., Silvarolla, M.B. and Orsi, C.H. (2001). Chromosome characterization in *Coffea arabica* L. using cytomolecular techniques. In: 14th International Chromosome Conference, Würzburg, Alemanha, *Chromosome Research*. pp.100

Plechakova, O., Tranchant-Dubreuil, C., Benedet, F., Couderc, M., Tinaut, A., Viader, V., De Block, P., Hamon, P., Campa, C., de Kochko, A. and Hamon, S. (2009). MoccaDB-an integrative database for functional, comparative and diversity studies in the Rubiaceae family. *BMC Plant Biology*, **9(1)**: 123.

Poncet, V., Hamon, P., Minier, J., Carasco, C., Hamon, S. and Noirot, M. (2004). SSR cross-amplification and variation within coffee trees (*Coffea* spp.). *Genome*, **47**: 1071-1081.

Poncet, V., Rondeau, M., Tranchant, C., Cayrel, A., Hamon, S., de Kochko, A. and Hamon, P. (2006). SSR mining in coffee tree EST databases: potential use of EST-SSRs as markers for the Coffea genus. *Molecular Genetics and Genomics*, **276(5)**: 436-449.

Poncet, V., Dufour, M., Hamon, P., Hamon, S., Alexandre de Kochko, and Leroy, T. (2007). Development of genomic microsatellite markers in *Coffea canephora* and their transferability to other coffee species. *Genome*, **50(12)**: 1156–1161.

Powell, W., Morgante, M., Andre, C., Hanafey, M., Vogel, J., Tingey, S. and Rafalsky, A. (1996). The comparison of RFLP, RAPD, AFLP, and SSR (microsatellite) markers for germplasm analysis. *Molecular Breeding*, **2**: 225–238.

Prakash, N.S., Combes, M.C., Dussert, S., Naveen, S. and Lashermes, P. (2005). Analysis of genetic diversity in Indian robusta coffee genepool (*Coffea canephora*) in comparison with a representative core collection using SSRs and AFLPs. *Genetic Resources and Crop Evolution*, **52**: 333–343.

Prakash, N.S., Combes, M.C., Somanna, N. and Lashermes, P. (2002). AFLP analysis of introgression in coffee cultivars (*Coffea arabica* L.). *Euphytica*, **124**: 265–27.

Prakash, N.S., Yashoda, M.H., Suresh, N., Mishra, M.K., Hanumantha, B.T., Devasia, J. *et al*. (2012). Pyramiding of Rust Resistance Genes as an Approach for Durable Resistance in Arabica – Indian experience, In: *Proceedings of the 24th International Conference on Coffee Science (ASIC)*, 11-17 November.

Prakash, N.S., Devasia, J., Raghuramulu, Y. and Aggarwal, R.K (2016). Coffee Improvement in India. In: Rajpal, V.R.R., Rama S., Raina, S.N. (Eds.), *Molecular Breeding for Sustainable Crop Improvement*. Springer International Publishing Switzerland. pp. 231-268.

Prakash, N.S., Marques, D.V., Varzea, V.M.P., Silva, M.C., Combes, M.C. and Lashermes, P. (2004). Introgression molecular analysis of a leaf rust resistance gene from *Coffea liberica* into *C. arabica* L. *Theoretical and Applied Genetics*, **109(6)**: 1311-1317.

Priolli, R.H.G., Ramos, L.C.S., Pot, D., Moller, M., Gallo, P.B., Pastina, M.M., Garcia, A.A., Yamamoto, P.Y., Lannes, S.D., Ferreira, L.P. and Scholz, M.B. (2008). Construction of a genetic map based on an interspecific F_2 population between

Coffea arabica and *Coffea canephora* and its usefulness for quality related traits. In: *22nd International Conference on Coffee Science, ASIC 2008, Campinas, SP, Brazil, Association Scientifique Internationale du Café (ASIC)*. pp. 882-890.

Priyono, M. Rigoreau. and Crouzillat, D. (2010). Evaluation of transferability of simple sequence repeats (SSR) and single nucleotide polymorphisms (SNP) marker for genetic map development in Robusta coffee (*Coffea canephora* Pierre). *Pelita Perkebunan*, **26**: 126–141.

Priyono and Ucu Sumirat (2012). Mapping of quantitative trait Loci (QTLs) controlling cherry and green bean characters in the Robusta Coffee (*Coffea canephora* Pierre). *Journal of Agricultural Science and Technology*, **A2**: 1029-1039.

Qamaruz-Zaman, F., Fay, M.F., Parker, J.S. and Chase, M.W. (1998). Molecular techniques employed in the assessment of genetic diversity: a review focusing on orchid conservation. *Lindleyana*, **13**: 259–283.

Raina, S.N., Mukai, Y. and Yamamoto, M. (1998). *In situ* hybridization identifies the diploid progenitor species of *Coffea arabica* (*Rubiaceae*). *Theoretical and Applied Genetics*, **97**: 1204–1209.

Romero, G., Vásquez, L.M., Lashermes, P. and Herrera, J.C. (2014). Identification of a major QTL for adult plant resistance to coffee leaf rust (*Hemileia vastatrix*) in the natural Timor hybrid (*Coffea arabica* x *C. canephora*). *Plant Breeding*, **133(1)**: 121-129.

Rudd, S. (2003). Expressed sequence tags: alternative or complement to whole genome sequences?. *Trends in Plant Sciences*, **8**: 321–329.

Sakiyama, N.S., Caixeta, E.T., Setotaw, T.A., Zambolim, E.M., Zambolim, L. and Pereira, A.A. (2008). Híbrido de Timor: a valuable source of genetic variability for arabica coffee. In: *22nd International Conference on Coffee Science, ASIC 2008, Campinas, SP, Brazil. Association Scientifique Internationale du Café (ASIC)*. pp. 930-932.

Sera, T., Ruas, P. M., Ruas, C. F., Diniz, L. E. C., Carvalho, V. P., Rampim, L., Ruas, E. A. and Silveira, S.R. (2003). Genetic polymorphism among 14 elite *Coffea arabica* L. cultivars using RAPD markers associated with restriction digestion. *Genetics and Molecular Biology*, **26**: 59–64.

Setotaw, T.A., Caixeta, E.T., Pena, G.F., Maciel-Zambolim, E., Pereira, A.A. and Sakiyama, N.S. (2010). Breeding potential and genetic diversity of "Híbrido de Timor" coffee evaluated by molecular markers. *Crop Breeding and Applied Biotechnology*, **10**: 289–304.

Silveira, S.R., Ruas, P.M., Ruas, C.F., Sera, T., Carvalho, V.P. and Coelho, A.S.G. (2003). Assessment of genetic variability within and among coffee progenies and cultivars using RAPD markers. *Genetics and Molecular Biology*, **26(3)**: 329-336.

Silvestrini, M., Junqueira, M.G., Favarin, A.C., Guerreiro-Filho, O., Maluf, M.P., Silvarolla, M.B. and Colombo, C.A. (2007). Genetic diversity and structure of Ethiopian, Yemen and Brazilian *Coffea arabica* L. accessions using microsatellites markers. *Genetic Resources and Crop Evolution*, **54(6)**: 1367-1379.

Srinivasan, K.H. and Narasimhaswamy, R.L. (1975). A review of coffee breeding work done at the Government Coffee Experiments Station, Balehonnur. *Indian Coffee*, **34**: 311–321.

Steiger, D.L., Nagai, C., Moore, P.H., Morden, C.W., Osgood, R.V. and Ming, R. (2002). AFLP analysis of genetic diversity within and among Coffea arabica cultivars. *Theoretical and Applied Genetics*, **105**: 209–215.

Stoffelen, P. (1998). Coffea and Psilanthus in tropical Africa: A systematic and palynological study, including a revision of the west and central African species. PhD Thesis, Katholieke Universiteit, Leuven. pp. 187–209.

Stoffelen, P., Noirot, M., Couturon, E. and Anthony, F. (2008). A new caffeine-free coffee species from Cameroon. *Botanical Journal of the Linnean Society*, **158**: 67–72.

Stoffelen, P., Noirot, M., Couturon, E., Bontems, S., De Block, P. and Anthony, F. (2009). *Coffea anthonyi* Stoff. and F. Anthony, a new self-compatible Central African coffee species, closely related to an ancestor of *C. arabica* L. *Taxon*, **58**: 133–140.

Sybenga, J. (1960) Genetics and cytology of coffee: A literature Review. *Bibliography Genetics*, **19**: 217-316.

Talhinhas, P., Azinheira, H.G., Vieira, B., Loureiro, A., Tavares, S., Batista, D. Morin, E., Petitot, A.S., Paulo, O.S., Poulain, J. and Da Silva, C. (2014). Overview of the functional virulent genome of the coffee leaf rust pathogen *Hemileia vastatrix* with an emphasis on early stages of infection. *Frontiers in Plant Science*, **5**: 88.

Tanksley, S.D. (1993). Mapping polygenes. *Annual Review of Genetics*, **27**: 205-233.

Teixeira-Cabral, T.A., Sakiyama, N.S., Zambolim, L., Pereira, A.A., Barros, E.G., and Silva, D.G. (2004). Characterization of differential coffee tree hosts for Hemileia vastatrix Berk. et Br. with RAPD markers. *Crop Breeding and Applied Biotechnology*, **4:** 68-73.

Tesfaye, G.K., Govers, K., Oljira, T., Bekele, E. and Borsch, T. (2006). Genetic diversity of wild Coffea arabica in Ethiopia: analyses based on plastid, ISSR and microsatellite markers. In: 21st International Conference on Coffee Science, Montpellier, France. *Association Scientifique Internationale du Café (ASIC)*. pp. 802-810.

Tesfaye, K., Borsch, T., Govers, K. and Bekele, E. (2007). Characterization of *Coffea* chloroplast microsatellites and evidence for the recent divergence of *C. arabica* and *C. eugenioides* chloroplast genomes. *Genome*, **50**: 1112–1129.

Tran, T.M.H. (2005). Genetic variation in cultivated coffee (*Coffea arabica L.*) accessions in Northern New South wales, Australia, Southern Cross University, Masters Thesis, 2005.

Tornincasa, P., Dreos, R., De Nardi, B., Asquini, E., Devasia, J., Mishra, M.K., Terra, L.D., Crisafulli, P., Pallavivini, A. and Graziosi, G. (2006). Genetic diversity of commercial coffee (*C. arabica L.*) from America, India and Africa assessed by simple sequence repeats (SSR), In: *Proceedings of 21th ASIC International Conference on Coffee Science*, Montpellier, France.

Touran, M.N., Pancor Ludlow, O.A., Sudamadji, D., Mawardi, S., Udayakumar, M. and Prasad, T.G. (1995). $\Delta^{13}C$ isotope discrimination in plants- A potential technique to determine WUE. In: Rao, R. C. N. and Wright, G. C. (Eds.), *Selection for WUE in Grain Legumes:* A report of workshop held at ICRISAT Center. Andhra Pradesh, India. pp. 42-45.

Van der Vossen, H.A.M. (1985). Coffee selection and breeding. In: *Coffee: Botany, biochemistry and production of beans and beverage*, Croom Helm, London. pp. 48–96

Van der Vossen, H., Bertrand, B. and Charrier, A. (2015). Next generation variety development for sustainable production of arabica coffee (*Coffea arabica* L.): A review. *Euphytica*, **204**: 243-256.

Varshney, R.K., Graner, A. and Sorrells, M.E. (2005a). Genic microsatellite markers in plants: features and applications. *Trends in Biotechnology*, **23**: 48-55.

Varshney, R.K., Balyan, H.S. and Langridge, P. (2005b). Wheat. In: Kole, C. (Ed.), *The Genome: Cereals and Millets*. Science Publishers, Inc., Enfield (NH), USA. pp. 121–219.

Vidal, R.O., Mondego, J.M.C., Pot, D., Ambrósio, A.B., Andrade, A.C., Pereira, L.F.P., Colombo, C.A., Vieira, L.G., Carazzolle, M.F. and Pereira, G.A. (2010). A high-throughput data mining of single nucleotide polymorphisms in *Coffea* species expressed sequence tags suggests differential homeologous gene expression in the allotetraploid *Coffea arabica*. *Plant Physiology*, **154(3)**: 1053–1066.

Vieira, L.G.E., Andrade, A.C., Colombo, C.A., Moraes, A.H., Metha, Â., Oliveira, A.C., Labate, C.A., Marino, C.L., Monteiro-Vitorello, C.D., Monte, D.D. and Giglioti, É. (2006). Brazilian coffee genome project: an EST-based genomic resource. *Brazilian Journal of Plant Physiology*, **18(1)**: 95-108.

Wang, G.G. and Paterson, A.H. (1994).) Assessment of DNA pooling strategies for mapping QTLs. *Theoretical and Applied Genetics*, **88**: 355-361.

Wolfsberg, T.G. and Landsman, D. (1997). A comparison of expressed sequence tags (ESTs) to human genomic sequences. *Nucleic Acids Research*, **25**: 1626-1632.

Zezlina, S., Soranzio, M., Rovelli, P., Krieger, M.A., Sodhal, M.R. and Graziosi, G. (1999). Molecular characterization of the cultivar Bourbon LC. In: *Proceedings of 18th International Scientific Colloquium on Coffee*. Paris, France: ASIC; 1999. pp. 314–329.

Chapter 18

Tea

☆ *Sangeeta Borchetia, Sagarika Das, Afruza Zaman, Madhurjya Gogoi, Pritom Choudhury, Hemanta Saikia and Tanoy Bandyopadhyay*

1. Introduction

Tea is one of the most popular non-alcoholic beverages across the world. Cultivation of tea worldwide comprises three main natural hybrids (Banerjee, 1992a) *viz., C. sinensis* (L.) O. Kuntze or China type, *C. assamica* (Masters) or Assam type and *C. assamica* sub spp. *lasiocalyx* (Planchon ex Watt.) or Cambod or Southern type. The genus *Camellia* presently comprises more than 300 species (Mondal *et al.*, 2004). The two main types of tea, China and Assam, differ from one another in their growth habits and biochemical attributes. Since members of the genus *Camellia* are self incompatible, highly heterozygous and out crossing, clonal introgressants from these two extremes are frequently observed and still referred to as China or Assam depending on their morphological proximity with the main taxa (Wight, 1962; Banerjee, 1992b). Tea diversity has suffered severe erosion over the years due to intensive selection and breeding for desirable traits (yield and quality) and absence of proper conservation programs in the producing countries. Knowledge of genetic variability and relationships between genotypes is essential in breeding programmes (Bhagawati *et al.*, 2015). In tea, various techniques have been applied from time to time to investigate the genetic relationships; these include the use of morphological, physiological, biochemical and molecular marker techniques. Molecular information is an absolute measure to understand the genetic differences in a detailed manner without the involvement of environmental interferences. Molecular markers are used to study both genetic diversity and genetic fingerprinting (Abdel-Mawgood, 2012). With the advent of Polymerase Chain Reaction (PCR), DNA marker technology has gained a new dimension facilitating the development of marker-based gene tags, map-based cloning of important genes, variability and phylogenetic studies. The modern biological techniques based on genomic sequencing, DNA molecular markers and genetic maps play a pivotal role to understand genetic structures of agronomic traits for rapid and precise crop improvement. In plants like rice, tomato, soybean and *Medicago,* availability of functional markers is comparatively easier as information on genome sequence is available. The basic genomic information regarding tea is at developing stage as tea has a large diploid genome (~4 Gb, 2n=30), long juvenile period and it takes 22–25 years to breed a new tea cultivar through traditional methods (Chen *et al.*, 2007) which are hurdles for marker development. Tan *et al.* (2013). reported availability of 49760 expressed sequence tag (ESTs) and ~200 single sequence repeat (SSR) or microsatellite markers with respect to tea. More effective markers are required for genetic mapping and molecular breeding programs for *C. sinensis.* Genetic maps are essential tools for implementing quantitative trait loci (QTL) analysis and marker assisted selection (MAS) breeding in tea.

2. Tea Breeding

Wide genetic variability present in early tea population is due to open pollination among the cultivated and non-cultivated teas. Seeds produced from Assam tea seed baris are called as *jats,* which refers to its pedigree. The contribution of *jats* towards improvement of yield and quality has also been quite significant (Wight and Gilchrist, 1961). Selection of promising genetic materials from these heterogeneous populations, having natural variability, was utilized in the tea improvement programmes (Wight, 1956; Barua, 1963; Bezbaruah, 1969; Satyanarayana and Sharma, 1986). Breeding objectives, with time, have transformed from high yield to high quality and then to high quality, high yield and tolerance to various types of stress (Liang *et al.,* 2007). To obtain uniform quality and yield, vegetative propagation technique was initiated which brought the concept of cloning the selected genetic materials. Clonal propagation of tea replaced propagation through seeds in the 1960s. Today, vegetatively propagated clones and F_1 progenies of clonal crosses are gradually replacing the old *jats* and in the due course of time, *jats* are likely to be completely eliminated. In true sense, a *jat bari* consists of a group of different genotypes having a lesser degree of variability among themselves. *Jat* populations are used as foundation stocks for clonal selection and other tea breeding works. The parents of most of the biclonal seed stocks are either clones or seedling progenies selected and maintained through vegetative propagation (Figure 18.1) (Konwar, 1999).

The genetic diversity of crops tends to shrink with time and eventually the propagated elite clones end up representing only a fraction of the genetic diversity in the entire gene pool. However, dependence on limited number of clones for productivity may lead to decrease in genetic diversity and, in turn, may also increase their susceptibility to infectious pests and diseases (Bandyopadhyay, 2011).

There are limitations to the success in developing hybrid progenies of desired characteristics through conventional breeding. The characteristics of the plant species are governed by the genes of the chromosomes. In gamete formation through

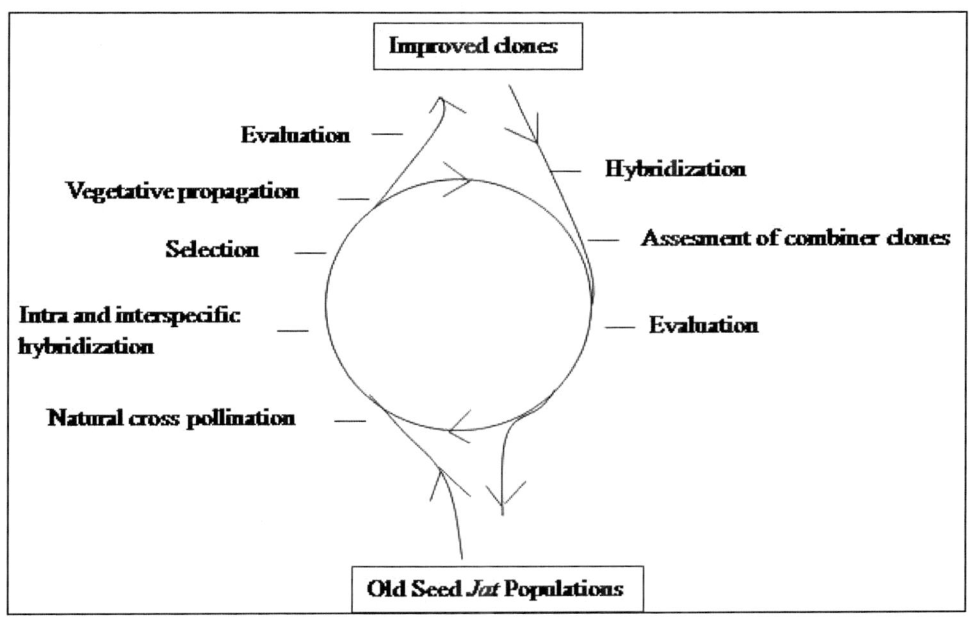

Figure 18.1: Plant Breeding Programmes for Improvement in *Jat* Tea. (Courtsey: Konwar, 1999).

meiosis during the sexual reproduction phase, some genes remain linked which never separate and cannot be brought together in the offspring through crossing over. Therefore, it is not possible to eliminate some undesired characteristics or combine some desired characteristics in the progeny (Das *et al.*, 2012). The new technologies developed through molecular marker assisted breeding are likely to help in overcoming these limitations and provide the means for developing plants of desired characteristics. Expression of morphological characters is influenced by environmental factors which misleads in selection of particular trait in tea plant. The molecular characters remain unchanged with changes in the environment and use of trait related DNA markers can provide a better and reliable means for selecting germplasm with desired trait. The theoretical advantages and potential applications of genetic markers in plant breeding were reported (Crouch and Ortiz, 2004). The feasibility and practical utility of genetic markers gained importance with the advent of DNA marker technology in the 1980s and DNA based molecular markers were used since then in various fields of taxonomy, plant breeding, genetic engineering (Jonah *et al.*, 2011). It is important to understand the genome of a plant to exploit the benefits of modern technologies. Tea in Assam acquires special characteristics due to genetic and environmental interactions. To develop superior planting materials, priority is given to selection of plants having potential for high yield and quality of tea with very high rooting ability and tolerance to stress. Tocklai Tea Research Institute, Jorhat, India has developed 33 TV clones, 153 TRA/Garden series clones and 15 biclonal seed stocks which now occupy more than 48 per cent of the total tea acreage of North Eastern India. Few of them are listed below (Tables 18.1 and 18.2)

Table 18.1: List of some Tea Cultivars (TV clones) and their Traits

Tea Cultivars	*Traits*	*Tea Cultivars*	*Traits*
TV 1	Drought tolerant	TV 23	Very high yield, drought tolerant, vigorous growth, termite resistant, good for interplanting
TV 6	High flavour Assam variety, resistance to red rust	TV 24	Cold resistant, high leaf proline
TV 9	Tolerant to water logging, Early flusher, drought tolerant, good for interplanting	TV 25	High yield, drought tolerant, high leaf area index, high water use efficiency, resistance to cockchafer grub, good for interplanting
TV 10	Assam flavour and bright liquor, resistance to Pink mite infestation, high leaf chlorophyll	TV 26	High yield, drought tolerant, low transpiration, resistance to root knot nematode, good for interplanting
TV 11	High flavour Assam variety and bright liquor, resistance to throny stem blight infection	TV 27	High yield, high shoot density, resistance to black rot
TV13	Large leaf Assam flavour variety, soft shoot and less fibre	TV 28	High yield, drought tolerant, resistance to blister blight
TV 18	Minty flavour, high yield, good for interplanting	TV 29	High yield, triploid (2n = 45), resistance to root diseases
TV 19	High yield	TV 30	High yield, resistance to blister blight and tea mites
TV 20	Good for interplanting	TV 31	Resistance to tea mosquito bug, Coppery yellow shoot
TV 21	Quality clone with unique Assam flavour, light leaf Assam variety	TTRI 1	Quality clone
TV 22	Very high yield, drought tolerant, high photosynthesis, resistance to Eelworm infestation, good for interplanting		

Table 18.2: List of some TRA/Garden Series Clones

Tea Cultivars	*Traits*	*Tea Cultivars*	*Traits*
Phoobsering 1258	Blister blight resistant, thickest cuticle (4.5 μm)	Sikim 1	Resistant to pink and scarlet mite, high flavour
Phoobsering 1404	Thickest epidermal cell layer (21 μm), yellowish green leaf	Tukdah 383	Highest Darjeeling flavour clone (Flavour index 147:100), high pubescence in bud, flowers profusely
Tukdah 135	Early flusher, high rooting ability in nursery	Phoobsering 312	High Darjeeling flavour (Flavour index 141:100), drought tolerant, resistant to blister blight, suitable for high elevation
Bannockburn 777	Resistant to red rust disease, best clone for southern hill slope of Darjeeling	Tukdah 246	Resistant to red spider mite, unique Darjeeling flavour (Flavour index 138:100)
Sundaram (B/5/63)	Triploid (2n = 45), resistant to red spider mite, more leaf hair per unit area (6.26 mm^{-2}), Yield clone	Ambari Vallai 2	Hifh flavour index (Flavour index 134:100), very high nursery rooting, spreading and dense frame
Balasam 7/1A/76	Resistant to Jassids and to green fly	Bannockburn 157	High Darjeeling flavour (Flavour index 131:100), resistant to purple mite
Badamtam 15/263	Cold resistant, resistant to mites	Kopati 1/1	High Darjeeling flavour (Flavour index 129:100), early flusher, Assam hybrid, good for mid elevation
Rungli Rungliot 4/5	High yield, Highest number of stomata per unit area (271 mm^{-2}), good nursery rooter	Bannockburn 688	High yield, Good flavour Darjeeling clone (Flavour index 128:100), high Geraniol (20.74) and Linalool (9.23) content, deep leaf bullation
Rungli Rungliot 17/144	Early flusher	Tukdah 78	High Darjeeling flavour clone (Flavour index 121:100), resistant to pink and scarlet mites, suitable for all elevation of Darjeeling, dark green leaf, compact frame
Thurbo 9	Resistant to blister blight, Thick leaf blade, unserrated leaf margin	Teesta Valley 1	Good Darjeeling flavour (Flavour index 117:100), drought tolerant China hybrid clone, suitable for mid and low elevation of Darjeeling
Happy Valley 39	Rose like sweetie aroma producing cultivar, synthesised highest Hexanol and t-2 Hexanol, Yield clone	Tukdah 145	Darjeeling flavour above average (Flavour index 110:100), Assam hybrid, high Geraniol content (16.87)

Contd...

Table 18.2–*Contd...*

Tea Cultivars	*Traits*	*Tea Cultivars*	*Traits*
Charaidew Parbat 1	Rose like sweetie aroma, Assam hybrid for Darjeeling plantation, Yield clone	Lengri 51, Lengri 56, Bormajan 19	Drought tolerant
Koomsong 23, Koomsong 29, Baghmari 10, Gohpur 33, Huplongcherra 18	Quality clone	Nagrijuli 6/24, Seajuli 16, Huldibari 19, Leesh River 9/34, Mohargung and Gulma 25, Sanyasithan 10, Sukna 7	Yield clone

with their traits which can be used in further breeding program to develop elite planting materials. DNA fingerprinting as well as trait specific genetic markers are required for meaningful selection of tea plants with desired characteristics. Success in these aspects may help in developing the modalities of marker-assisted selection (MAS) for quick identification of desired plants.

3. Applications of Molecular Marker Assisted Studies in Tea

Molecular markers have a wide range of applications in crop breeding and have several advantages over morphological and biochemical markers. Molecular markers are preferred than other descriptors due to their unlimited number and as they remain unaffected by environmental factors or developmental stages of the plant (Winter and Kahl, 1995). DNA-based molecular markers have proved to be quite efficient in analyzing genetic diversity and variation, molecular identification, molecular phylogenetics of tea germplasm as well as construction of the genetic linkage map of tea (Ni *et al.*, 2008). The common molecular marker technologies applied in tea breeding and diversity studies include: Restriction Fragment Length Polymorphism (RFLP) (Botstein *et al.*, 1980), Simple Sequence Repeats(SSR) (Tautz, 1989), Randomly Amplified Polymorphic DNA (RAPD) (Williams *et al.*, 1990), Cleaved Amplified Polymorphic Sequence (CAPS) (Konieczny and Ausubel, 1993), Inter Simple Sequence Repeats (ISSR) (Zietkiewicz *et al.*, 1994), Amplified Fragment Length Polymorphism (AFLP) (Vos *et al.*, 1995) and Single Nucleotide Polymorphism (SNP) (Chen and Sullivan, 2003).

3.1. Genetic Variability Studies

Genetic diversity of species is important for its long-term viability and adaptability to the changing environment and hence for its fitness. To preserve, evaluate and utilize germplasm efficiently, knowledge on the extent of genetic diversity of the germplasm is essential (Smith and Smith, 1989; Gepts, 1993; Zahid *et al.*, 2009). Genetic diversity (variation) studies are highly significant in order to conserve the old germplasm. World tea production has increased by an average of 25 per cent in the last two decades as a result of widespread replacement of the varied local seedling populations with superior and identical clonally propagated tea cultivars in many countries (Yao *et al.*, 2008). Various morphological, physiological and biochemical methods were used to assess genetic diversity in tea but plasticity of these descriptors has given preference to molecular markers in recent approaches.

Genetic diversity in Assam, China and Cambod varieties of tea was first characterized by RAPD markers by Wachira *et al.* (1995). Later, several related studies were carried out to establish the genetic diversity among different tea cultivars of different regions (Wachira *et al.*, 1997; Kaundun *et al.*, 2000; Chen and Yamaguchi, 2002, 2005; Young-Goo *et al.*, 2002; Hui *et al.*, 2004, Mewan *et al.*, 2005; Chen *et al.*, 2005; Afridi *et al.*, 2011; Vijayan *et al.*, 2013).

Genetic diversity of different tea cultivars and wild relatives was also assessed through AFLP analysis (Wachira *et al.*, 2001; Lai *et al.*, 2001; Kaundun and Matsumoto, 2002; Balasaravanan *et al.*, 2003; Mishra and Sen-Mandi, 2001; Mishra *et al.*, 2009, Kafkas *et al.*, 2009).

Differentiation of Japanese green tea cultivars and 463 local tea plants was revealed by RFLP analysis of phenylalanine ammonia lyase DNA (Matsumoto *et al.*, 2002). The results obtained from RFLP analysis indicated a process of differentiation from the ancestral material probably introduced from China to the local tea plants and, finally, cultivars which were produced by selecting from local tea plants and crossing. RFLP was also utilized in delineating genetic relationship of Korean tea germplasm with the Japanese tea germplasm (Matsumoto *et al.*, 2004). Genetic diversity studies were also performed for *Camellia sinensis* and *C. assamica* varieties based on PCR- RFLP analysis of *PAL, CHS2* and *DFR* genes (Kaundun and Matsumoto, 2003a). PCR-RFLP analysis of chloroplast DNA was performed using cpDNA universal primers to study the genetic diversity among 30 different cultivars of *Camellia* in China (Chen *et al.*, 2012). Genetic relationship between tea plants rich in catechins was established using morphological characters and RAPD primers establishing that DNA based markers and ultimately genetic characters could be correlated to secondary metabolites of tea (Kim *et al.*, 2011).

Genomic microsatellite markers were identified and characterized for both wild and cultivated tea using different approaches. RAPD-PCR based isolation of microsatellites (Ueno *et al.*, 1999; Hung *et al.*, 2008), the fast isolation by AFLP of sequences containing repeats (FIASCO) method for wild tea (Yang *et al.*, 2009) were employed to detect the heterozygosity and genetic relationships among tea cultivars for population studies and DNA fingerprinting aimed at the genetic improvement of tea (Freeman *et al.*, 2004; Ohsako *et al.*, 2008; Bali *et al.*, 2013; Bhardwaj *et al.*, 2013; Mewan *et al.*, 2012, Bhardwaj *et al.*, 2014; Tan *et al.*, 2015). EST-SSR markers prove to be advantageous over the genomic markers in trait associated studies and construction of linkage maps. Such markers have been characterized in the tea plant as reported from a number of works (Sharma *et al.*, 2011; Ma *et al.*, 2012; Tan *et al.*, 2013).

Molecular cloning of phenylalanine ammonia-lyase cDNA and classification of varieties and cultivars of tea plants (*Camellia sinensis*) was done by Matsumoto *et al.* (1994).). Cultivar identification of tea and genetic diversity analysis of tea germplasm using CAPS marker developed from EST sequences was performed in Japan and Taiwan (Ujihara *et al.*, 2011; Hu *et al.*, 2014). In tea, CAPS markers analysis was mostly concentrated on three crucial genes of the phenylpropanoid pathway, *i.e.* phenylalanine ammonia lyase (*PAL*), chalcone synthase (*CHS*) and dihydroflavanol 4-reductase (*DFR*). DNA polymorphism between var. *sinesis* and var. *assamica* was studied using PCR-RFLP analysis (Kaundun and Matsumoto, 2003a).

ISSR markers were used for genetic diversity and fingerprinting study of tea clones (Lai *et al.*, 2001; Mondal, 2002; Roy and Chakraborty, 2009; Ji *et al.*, 2011; Liu *et al.*, 2012a, b). Genetic diversity in tea has apparently been found to be quite high when evaluated using ISSR technique. This was evident in the estimation of genetic diversity in ancient tea populations of *Camellia assamica* in Yunnan (Lai *et al.*, 2001, Devarumath *et al.*, 2002, Mondal, 2002). ISSR markers were also used to establish genetic diversity between six Chinese tea cultivars (Yao *et al.*, 2005) and differentiate tea germplasm at the inter-specific level (Liu *et al.*, 2012b). It was also observed that 35-40 per cent diversity was present among *C. assamica* germplasm in China with the help of ISSR markers (Ji *et al.*, 2011).

SNP markers were developed from EST sequences derived from *Camellia sinensis* which effectively established DNA fingerprints for most of the varieties (Fang *et al.*, 2014a; Zhang *et al.*, 2014). SNPs along with SSRs derived from the EST sequences of tea were mined and polymorphism was validated through sequencing (Yuan, 2011).

3.2.Identification of Elite Clones

Traditionally, selection of tea plants with desirable traits was carried out from variants produced by natural or artificial crossing (Barua, 1963) or by somaclonal variation (Arulpragasam, 1963). Later, vegetative propagation of the selected plants was done which ultimately gave required number of desirable plants. However, this is a prolonged and labour intensive process as tea is a perennial slow-growing crop. Alternatively, molecular markers are used these days for the selection of elite or superior clones from a population. Estimation of the genetic relatedness and/ or variability among genotypes in tea using different molecular markers can be beneficial in selecting desirable parents for breeding purposes.

Fifteen Chinese elite tea genetic resources were identified (Chen *et al.*, 2005) by the presence of 20 unique RAPD markers and absence of 11 unique markers. EST-SSR markers were also developed for the identification of elite clones for various tea breeding programmes (Zhao *et al.*, 2008; Ma *et al.*, 2010). A number of microsatellite markers were developed and used for tea varietal identification (Kaundun and Matsumoto, 2004; Sharma and Kumar, 2005;Chen *et al.*, 2007; Ujihara *et al.*, 2009; Yang *et al.*, 2009; Taniguchi *et al.*, 2012b; Ujihara *et al.*, 2012). Tea varietal identification was attempted by using ISSR markers (Thomas *et al.*, 2006) in somaclonal variants of South Indian tea germplasm. STS and CAPS markers were also used for the identification of tea varieties (Ujihara *et al.*, 2011; Hu *et al.*, 2014) because of their advantages over other molecular markers including co-dominance, locus specificity and higher reproducibility.

3.3. Varietal Protection

Tea is produced in more than 44 countries of the world including China, India, Kenya, Sri Lanka, Korea, Thailand and Myanmar. India is the second largest producer of tea with a total contribution of 22 per cent of the world's production after China, which contributes around 31 per cent. There are four main types of commercial tea: black tea, green tea, oolong tea and white tea. India is the largest producer of black tea, whereas, China produces maximum of the world's green tea in addition to oolong tea and white tea. As tea is the second most widely consumed beverage in the world after water, there is a huge demand for good quality tea world over.

Among the various methods of tea varietal authentication, near-infrared spectroscopy developed in 1988 has been the most widely studied technique. Being a rapid and non-invasive method, near-infrared spectroscopy has been utilized in a number of authentication studies (Luypaert *et al.*, 2003, Zhang *et al.*, 2004, Chen *et al.*, 2008, He *et al.*, 2007, Tan *et al.*, 2012). However, this technique has some limitations when large numbers of genotypes need to be tested. Molecular markers such as microsatellites (Kaundun and Mastsumoto, 2004; Ujihara *et al.*, 2009; Ma *et al.*, 2010; Ujihara *et al.*, 2011), STS and CAPS (Ujihara *et al.*, 2011; Hu *et al.*, 2014) were also used for varietal identification in tea. Fingerprinting of improved cultivars using cultivar-specific molecular marker can be helpful to resolve problems related to clonal cultivar certifications and protecting intellectual property rights for breeders.

Molecular markers can also be used for the detection of adulteration in commercial plant products. The first successful study on the detection of cashew (*Anacardium occidentale*) husk in tea was reported by Dhiman and Singh (2003) where PCR primers developed from intergenic spacer regions of 5S ribosomal RNA genes were applied. STS-RFLP markers were also developed for varietal authentication of green tea made in Japan (Kaundun and Matsumoto, 2003b).

With technological advancement, SNP markers have become the molecular marker of choice for varietal authentication as these are abundant in the genome, biallelic, codominant and does not require electrophoretic separation on the basis of size. Fang *et al.* (2014b) generated SNP fingerprints of cacao beans which were later used to unambiguously distinguish adulterants from authentic beans. Based on this study, an attempt of EST-derived SNP validation in tea was performed (Fang *et al.*, 2014a), where DNA from a variety of samples (fresh tea leaves and processed commercial loose-leaf tea) was genotyped. These SNP markers proved to be successful in tea variety authentication as well as quality control and can be used in future for germplasm management and tea breeding programs.

3.4. Genetic Fidelity of Micropropagated Plants

The occurrence of cryptic genetic instability due to somaclonal variation in tissue cultured plants can limit the broader utility of the micropropagation system (Salvi *et al.*, 2001). It is important to establish genetic uniformity of micropropagated plants to confirm the quality of the plantlets for its commercial utility. Different types of marker systems such as RAPD (Mondal and Chand, 2002), ISSR (Devarumath *et al.*, 2002) and SSR (Borchetia *et al.*, 2009) have been used to test the genetic fidelity of micro-propagated tea plants and are found suitable for the assessment of the genetic fidelity of *in vitro*-raised clones.

3.5. Functional Genomics to Develop Molecular Markers in Tea

Functional genomics focuses on the function of DNA by characterizing genes, RNA transcripts, and their protein products. It aims to understand the relationship between an organism's genome and its phenotype. Thus, functional genomics involves cloning of genes and studying differential expression of genes associated with a particular trait (Mondal, 2014). Tea functional genomics was initiated with the isolation of the chalcone synthase gene from the Japanese cultivar 'Yabukita' (Takeuchi *et al.*, 1994). Differential gene expression is one such technique whereby the differences in transcript levels of a genes can be used for screening molecular markers associated with the trait. Molecular markers associated with QTL/major genes for desired traits are being developed and routinely used in many plants (Ruan, 2010). cDNA-AFLP method has been widely used to study the differential gene expression in ESTs obtained from an SSH (suppression subtractive hybridization) library in tea (Mondal and Sutoh, 2013; Das *et al.*, 2013). Gupta *et al.* (2013). identified 108 transcript fragments obtained from an SSH library using cDNA-AFLP approach and found to be overexpressed in drought tolerant tea cultivars. cDNA-AFLP approach was also followed by Yang *et al.* (2012) to identify 68 differentially expressed genes related to the morphological variation of two contrasting callus lines. SNPs were identified from the EST database of tea from which 253 putative SNPs were verified in 17 tea varieties through sequencing. Further, 34 sequences containing SNP with functional roles have been identified (Zhang *et al.*, 2014). Heat shock proteins (HSP70) and superoxide dismutase (SOD) were reported to be highly expressed in drought tolerant cultivars of tea, which may be potent loci for developing molecular markers through the application of functional genomics (Maritim *et al.*, 2015). Molecular markers derived from functionally characterized sequences provide advantage over random DNA markers due to their linkage with specific trait loci which may be associated with major agronomic and economic traits of the tea plant.

3.6. Taxonomic and Interspecies Relationships

The selection and use of proper genetic markers for taxonomic classification is a major bottleneck in molecular phylogenetics. Chloroplast DNA (cpDNA) and mitochondrial DNA (mtDNA) sequences had been widely used to investigate interspecific relationships (Waugh *et al.*, 1990). Non-coding regions are targets for phylogenetic studies as they show higher rates of evolution than coding regions. Internal Transcribed Spacer (ITS) region of the nuclear rDNA cistron was used as a suitable marker for taxonomic classification and phylogenetic analysis of the genus *Camellia* (Vijayan *et al.*, 2009). ITS2 primary sequence and its secondary structure in phylogenetic analysis has enabled deciphering the phylogenetic grouping and resolving the taxonomic classification of species within the *Camellia* genus. Combination of ITS2 rRNA gene marker and a multiple global alignment based on sequence and secondary structures put necessitated revision of the existing classification (Gogoi *et al.*, 2015).

4. Marker Assisted Selection in Tea Breeding

Any DNA marker, which is tightly associated with an agronomically important trait (with a distance <10cM), may be used as a molecular tool for marker assisted selection (Ribaut and Hoisington, 1998; Staub and Serquen, 1996). Tea breeding depends mainly on the conventional methods like individual selection and hybridization. Most of the released tea cultivars were developed through individual selection from natural population or progeny of uncontrolled pollination (Yao *et al.*, 2008). As a result, the pedigree data of most of the tea cultivars, which is a prerequisite for genetic improvement programs, is non-existent.

4.1. Characterization of Breeding Population

Tea cultivars from different regions of the world were compared for their utility in tea breeding (Yao *et al.*, 2008). Studies involving germplasm of China (*Camellia sinensis*), Assam (*C. assamica*) and Cambod (*C. assamica* ssp. *lasiocalyx*) varieties have been carried out using molecular markers (Mondal, 2002, Balasaravana *et al.*, 2003,

Paul *et al.*, 1997). These markers were able to group the germplasm according to their geographical area and provide details of the extent of genetic diversity present within these germplasm. The China variety was found to exhibit highest genetic diversity, the Cambod variety exhibited the least genetic variation while the Assam variety showed intermediate genetic diversity (Roy and Chakraborty, 2009). Yao *et al.* (2008) studied the genetic diversity of 48 tea cultivars from Japan, Kenya and different regions of China using ISSR markers which distinguished the cultivars into distinct groups based on their area of cultivation. Population structure and diversity was also studied using 449 EST-SSR markers on tea accessions collected from different provinces of China (Yao *et al.*, 2012) which grouped the accessions based on their eco-geographic regions. The variability in the repeats of 5S rDNA with specific restriction endonucleases (*Bam*H I, *Sau*3AI and *Apo*I) in 28 tea cultivars collected from different regions in India were evaluated. The results unequivocally demonstrated the applicability of 5S rDNA gene as a molecular marker for the differentiation of Chinary tea (*Camellia sinensis*) from Assam (*C. assamica*) and Cambod (*C. assamica* ssp. *lasiocalyx*) types (Singh andAhuja, 2006). Recently, a RAPD marker was also developed which could specifically distinguish Assam type tea from others. This RAPD amplicon was eventually used to develop a SCAR marker for identification of 'Assam' type tea (Kalita *et al.*, 2014).

4.2. Breeding for Abiotic Stress

Tea plants, like other plantation crops, experience abiotic stresses (drought, salinity, water logging, extreme temperature changes, heavy metals, *etc.*) which negatively influence the growth and productivity of tea. Global warming and climate change are major threats to crop production due to water and temperature stresses. Drought and winter dormancy have been observed to affect tea plants to a great extent in terms of productivity and survival. Crop loss due to drought is estimated to be up to 50 per cent (Bray *et al.*, 2000). Drought also leads to 6-9 per cent plant deaths depending upon the cultivar and severity of conditions (Burgess and Carr, 1993). In north-east India, which is the main tea producing region of the country, 30-40 per cent of tea yield is affected by drought (Singh and Handique, 1993; Barua, 1989; Jain, 1999). Tea productivity is likely to be affected in the coming years which may lead to greater economical and social problems. To overcome this situation, screening of inherent tolerant cultivars is essential. These cultivars can be used in conventional breeding practices to develop germplasm with desired yield and quality. Research on other abiotic stresses like soil salinity, heavy metal toxicity, water logging, heat stress, *etc.* which affects tea yield are yet to be initiated.

Molecular marker studies related to drought stress in tea are less compared to other crops. An important study involving molecular markers was performed by Mishra and Sen-Mandi (2004 a, b). They carried out genetic evaluation of 10 Darjeeling tea clones known for drought tolerant/susceptibility, based on field trials, using RAPD markers, which were co-related with drought-specific superoxide dismutase (SOD) and ascorbate peroxidase (APX). Drought-specific RAPD markers capable of screening drought-tolerant cultivars were identified in this experiment. Similar experiments were carried out independently by Wium (2009) and Malebe (2011) using cultivars from Malawi. Mphangwe *et al.* (2013) also carried out RAPD analysis on 32 germplasm from Malawi for the identification of trait-related markers and identified two drought related markers (bands) which could be used to screen drought-tolerant cultivars from among germplasm. Molecular markers associated with drought tolerance in tea were identified from the tea transcriptome using the cDNA–AFLP approach (Gupta *et al.*, 2010). Sharma *et al.* (2010) showed that substantial genetic diversity exists among tea germplasm associated with various trait attributes, including drought tolerance, resistance to frost, tolerance to water logging and tolerance to wind, using AFLP markers. These inferences can further be validated using previously reported 324 microsatellite (SSR) markers in tea. With the advent of next-generation sequencing, sequencing of core collections would generate high-density genome-wide SNP data enabling genome-wide association studies in tea for traits, including drought tolerance and other abiotic stresses.

Winter dormancy is another abiotic stress that affects tea yield. During winter time, tea plants (or any other crop) are unable to initiate growth from meristematic tissues or any other cells or organs that otherwise have the ability to grow under favourable conditions (Rohde and Bhalerao, 2007). Winter dormancy is characterized by reduced growth of the apical bud and the associated two leaves used for commercial tea production. Plant's growth rate may be reduced upto 20 per cent as compared to that during active growing season of the year (Nandi *et al.*, 1995).

Mphangwe *et al.* (2013) in their study using RAPD markers with respect to drought also identified two specific bands associated with extreme temperature tolerance in tea (one specific for low temperature tolerance and the other for heat tolerance). Extensive work related to tolerance to winter dormancy is lacking which needs to be taken up as a priority research.

4.3. Breeding for Pest and Pathogen Resistance

Biotic stress in tea includes diseases and pests which affects tea productivity. Diseases may either be bacterial (crown gall, bacterial shoot blight, *etc.*) or fungal (blister blight, black rot, red rust, *etc.*). Among these, blister blight has been one of the most devastating diseases in southern and eastern Asia, leading to an annual loss of up to 43 per cent of tea crop (Ordish, 1952). It is caused by the fungus *Exobasidium vexans* and affects the young leaves and shoots, which is used for commercial tea production. Other diseases which infect stems and roots also affect the productivity. Various insects and pests also affect tea yield by hampering the growth of the plant. Tea mosquito bug (*Helopeltis theivora*) is the most significant pest which causes around 7-15 per cent loss of crop annually (Gurusubramanian and Bora, 2008). Other insects such as aphids, red spider mites, loopers, *etc.* also cause a considerable amount of crop loss per year.

Chemical pesticides, insecticides and biological agents are being used to control the menace of insects, pests and various diseases on tea crop. Tea breeding efforts and development of tolerant cultivars can partly help to curb their infestation. Screening for tolerant tea cultivars is needed using molecular markers. EST-based markers associated with resistances of blister blight (Agarwal and Das, 2009) and insect (*Helopeltis theivora*) (Ahmed and Das, 2009) were developed to aid the marker assisted selection process in tea breeding. Mphangwe *et al.* (2013) also developed

RAPD marker associated with tolerance to stem canker (*Phomopsis theae*). However, this marker was negatively associated to tolerance, and was present in the susceptible cultivars and absent from the tolerant ones. Recently, Suganthi *et al.* (2014) used 138 RAPD markers to screen genomic DNA from eight tea cultivars of which four were relatively tolerant to *Helopeltis theivora* attack and four were susceptible to it. Out of these, three RAPD markers were linked to tea mosquito resistance. These three RAPD markers were then converted into SCAR markers which can be used as an important tool to screen tea cultivars for *Helopeltis theivora* resistance.

Despite the efforts for development of trait-associated molecular markers for cultivar screening, the numbers of molecular markers developed are very few. The tea genome is about 4 Gb (Tanaka and Taniguchi, 2006) and the unsaturated genetic maps limit the practical exploitation of molecular markers in tea breeding and selection and, thus, demands additional research in this area.

4.4. QTL Mapping and Genetic Maps in Tea

Genetic maps are constructed for the effective analysis of Quantitative trait loci (QTL) and application of marker assisted selection (MAS) in plant breeding, particularly for perennial crops. They are vital tools for producing improved cultivars through breeding of plants with desired traits. Identification of agronomically important QTLs and their utilization in crop improvement requires the mapping of these QTLs using molecular markers. The first genetic map of tea was constructed using RAPD markers related to theanine content, date of bud sprouting, resistance to anthracnose and tolerance to cold (Tanaka, 1996). Genetic maps of *Camellia sinensis* has been created mainly based on dominant markers such as RAPD, AFLP and ISSR markers (Hackett *et al.*, 2000; Huang *et al.*, 2005; Huang *et al.*, 2004). QTLs were identified in tea for yield and drought tolerance using RAPD, AFLP and SSR markers which showed segregation of the progeny in 1:1 ratio after backcrossing were used for construction of linkage map of tea (Kamunya *et al.*, 2009; Kamunya *et al.*, 2010). Construction of a reference linkage map of tea was also done using the co-dominant SSR and CAPS markers as landmark markers developed from both genomic DNA and EST sequences (Taniguchi *et al.*, 2012a). Similarly a number of polymorphic molecular markers such as SSR, CAPS, STS, AFLP, ISSR and RAPD were utilized for construction of an integrated genetic map of tea based on their maternal and paternal lineages (Hu *et al.*, 2013). SSR-based linkage map was constructed for the floral transcriptome of tea that consisted of 15 linkage groups (Tan *et al.*, 2013). *Camellia sinensis* varieties with different catechin content were interbred to characterize the QTLs for catechin content in the tender shoots of the progeny population using SSR markers from which a moderately saturated genetic map was constructed (Ma *et al.*, 2014).

4.5. Limitation of Marker Assisted Selection in Breeding Programs

Improvement of polygenic traits through MAS is a complex endeavour as most of the agronomically important traits are complex and regulated by several genes. The difficulty in manipulating quantitative traits is related to their genetic complexity, mainly the number of genes involved in their expression and interactions among genes (epistasis). Several genes are involved in expression of a quantitative trait but have smaller individual effects on the phenotype and the individual genes effects cannot be identified. This warrants repetitions of field tests to characterize accurately the effects of QTLs and evaluate their stability across environments. The evaluation of QTL by environment interaction (Q × E) continues to be a major constraint on the efficiency of MAS.

Marker Assisted Selection (MAS) in tea has huge potential but it has not been applied practically in tea crop improvement programmes, the main reasons being the high exorbitant cost of MAS as a large amount of marker loci are needed to be scored in breeding material. With the rapid development of sequencing technology and molecular marker technology undoubtedly the cost is decreasing with time but cost of some stages involved in MAS requires significant financial investment. It requires huge funding source and considerable time to complete all the stages of MAS to achieve fruitful results. Understanding of majority of argonomically important genes with respect to phenotypic trait largely remain unknown (Collard and Mackill, 2008). The other factors which hinder the integration of modern and conventional breeding techniques are:

a. The necessity to re-examine the marker–trait associations on multiple mapping populations in the absence of close gene–trait linkage and considering the accuracy of QTLs/mapping genes.
b. The markers need to be validated to ascertain its consistency in the identification of QTLs/genes in wide range of samples.
c. Limited or insufficient number of co-dominant and polymorphic markers for testing of planting material for breeding which is very important for crosses between closely related species and for interspecific hybridization
d. The problem with the introgression of "minor" QTLs for traits with multigene control, where environment and epistatic interactions have a direct or indirect effect.
e. The requirement of trained manpower or trained human factor: highly skilled persons are required in projects concerned with MAS.

Thorough knowledge of molecular methods and analysis of plant genomes along with understanding of statistical programs for mapping genes/QTLs is a must. Additionally along with molecular biology knowledge, the familiarity with methodological basis of phenotypic selection is important which is seen to be often lacking.

5. Advances in Molecular Markers for Gene Assisted Plant Breeding Practices

Next generation sequencing (NGS) assisted marker selection has the potential to generate large data set in shorter duration to select them as effective markers. The main limitations of molecular markers like RAPDs are their repeatability across plants which do not permit easy comparisons of maps at interspecific or intergeneric levels. NGS can produce millions of sequence reads at a relatively low cost and facilitate the functional characterization of genes/genomes to provide a more comprehensive view of diversity and gene function in plants responsible for many traits (Brady and Provart, 2007). Recently, various research groups have resorted to NGS to generate *C. sinensis* transcriptome data, which offered a substantial increase in DNA sequence availability for this species (Shi *et al.*, 2011; Wu *et al.*, 2013). Transcriptome generates enormous data to identify a substantial number of putative SSRs sequences to develop novel SSR markers and construct a moderately saturated SSR-based genetic linkage map for QTL mapping and MAS breeding in *C. sinensis* (Tan *et al.*, 2013). Compared to leaf and root research of *C. sinensis* plant, reproductive organs research is underrepresented in public databases. It is important to elucidate the gene expression profiling of these organs to expand species genetic resources and will enable the investigation of molecular mechanisms of flowering and pollination, such as self-incompatibility and various fruit-bearing traits in different cross-parent combinations (Takeda, 1990). Floral transcriptome analysis generated 431 novel polymorphic SSR markers and a moderately saturated linkage map for *C. sinensis* was constructed (Tan *et al.*, 2013). SSR markers are robust tools for genetic mapping and molecular breeding in crops. SSR marker development feasibility for non-model organisms has been enhanced with NGS technology further, it has also led to development and extensive use of single nucleotide polymorphism (SNP) markers (Weising *et al.*, 2005). Technology development in the field of chip based assay platforms enable SNPs to be determined more accurately and precisely. SNPs are biallelic and codominant markers and the error rate in allele calling is much lower than the SSRs. SNPs based selection are quick, low-cost and multiplex genotyping techniques can be employed. These advantages have made SNPs, the markers of choice, for accurate genotype identification in crop improvement. Recently, SNP fingerprint patterns for single cacao beans where DNA extracted from the seed coat were analysed using a nanofluidic system. Based on the SNP profiles, an assumed adulterant variety was unambiguously distinguished from the authentic beans by multilocus matching (Fang *et al.*, 2014a). Developments of SNP markers through the data mining of EST databases of tea plants and assessment of their potential applications for tea varietal identification were done. Mining of putative SNPs were done from the EST database at the NCBI GenBank (http://www.ncbi.nlm.nih.gov/) and validated for genotyping in nanofluidic 96.96 Dynamic Array TM IFC system (Integrated Fluidic Circuit; Fluidigm Corp.). This technology enabled development of SNPs markers for both fresh and processed tea products, including white, green, oolong and black tea. The SNP resources reported by Fang *et al.* (2014b) represent the first study of EST-derived SNP validation in tea and demonstrate the utility of EST databases as an alternative approach for *de novo* SNP identification in species whose genome sequences are not yet available.

These technological advances have facilitated the functional characterization of genes and provided a more comprehensive view of diversity and gene function in plants. Plant breeding requires genetic variability to be selected in order to increase the frequencies of favorable alleles and genetic combinations. Evaluation of germplasm resources is required for the continuous improvement of crop plants, including the analysis of variation within and among germplasm (Hodgkin *et al.*, 2001). In this direction, Targeting Induced Local Lesions in Genomes (TILLING) and a similar procedure called ecotype TILLING (EcoTILLING) was developed to identify all allelic variants of a DNA region present in an artificial mutant collection which can be induced by transferred DNA tagged lines and transposon tagged lines (Perez-de-Castro *et al.*, 2012). The availability of sequences coming from NGS and the information provided by gene expression studies is significantly increasing the number and quality of candidates for TILLING and EcoTILLING studies. These procedures were successfully used in many crops (Colbert *et al.*, 2001; Perry *et al.*, 2003; Caldwell *et al.*, 2004; Weil *et al.*, 2007; Triques *et al.*, 2007; Dahmani-Mardas *et al.*, 2010). Rice was the first crop for which EcoTILLING was applied (Kadaru *et al.*, 2006). Subsequently, EcoTILLING has been used in other crops and wild relatives, like barley (Mejlhede *et al.*, 2006), wheat (Wang *et al.*, 2008), or the wild peanut *Arachis duranensis* (Ramos *et al.*, 2009).

Novel methodologies *viz.*, cytochrome P450 based analogues (PBA) (Yamanaka *et al.*, 2003), tubulin based polymorphism (TBP) (Bardini *et al.*, 2004), intron-targeting polymorphism (ITP) (Liu *et al.*, 2015), transposable element based markers (TEMs) (Yadav *et al.*, 2015), resistance-gene based markers (RGMs) (Poczai *et al.*, 2013) and targeted fingerprinting markers (TFMs) have been devised and applied in major plant groups. These are mainly PCR based multilocus techniques that either incorporate modifications to existing methods or rely on new principles, and belong to the class of gene-targeted and/or functional markers. However, these novel methods have large potential application in the molecular breeding programmes of tea.

Although these techniques have potential to provide new sources of marker system but are still under-utilized and not popular like RAPD or AFLP. Efforts are being made to develop new and more efficient markers for agricultural and economically important plants but are hindered due to the lack of genome information in tea plants. Development of high throughput technique like NGS can be expected to provide more structured datasets which could be used alone or in combination with sequence level characters in developing markers for varied traits in tea plant.

6. Conclusion

Genetic marker development is an important area of research for molecular breeding in tea. The ideal genetic marker to be used in highly heterozygous tea is yet to be developed. Highly informative markers, like AFLPs, offer potential to assay large numbers of markers very quickly. Co-dominance markers (*e.g.* SSR and SNP) can allow effective selection of recessive alleles of desired traits in the heterozygous status. These markers are, however, difficult and expensive to develop. MAS is used more widely for simply inherited traits than for polygenic traits. The development and access to reliable PCR-based markers such as SSRs and SNPs will significantly increase efficiency of genotyping large tea populations. The utilization of MAS for improving polygenic traits in a quick time-frame and cost-effective manner is still elusive and at an initial stage. The extensive use of molecular markers for various applications, *e.g.* germplasm evaluation, genetic mapping, gene discovery, characterization of traits and tea improvement has been initiated to facilitate molecular breeding in tea. High costs and technical demands of molecular breeding will continue to be a major obstacle for its large-scale use. Other challenges includes identification of markers that can be applicable across populations and having reliable marker-trait association, requirement of multiple mapping populations and validation of QTL positions and effects in the germplasm of interest. High throughput transcriptome sequencing and mapping will lead to identification of large number of structural gene loci from where markers can be generated. The large number of markers will complement and develop the power of quantitative genetics to define and locate functional genes for complex or polygenic traits. Fine structure mapping of complex traits will ultimately lead to identification of genes determining phenotypes in tea. Functional genomics has marked a significant paradigm shift in the field of molecular breeding and in future, quantitative genetics will look towards genomics for information involving complex traits. Bioinformatics will also play an important role to integrate with genomics and develop comprehensive research strategies towards more efficient molecular breeding for future tea improvement.

References

Abdel-Mawgood, A.L. (2012). DNA based techniques for studying genetic diversity. In: Caliskan, M. (Ed.), *Genetic Diversity in Microorganisms.* InTech Publishers. pp. 95-122.

Afridi, S.G., Ahmad, H., Alam, M., Khan, I.A. and Hassan, M. (2011). DNA landmarks for genetic diversity assessment in tea genotypes using RAPD markers. *African Journal of Biotechnology*, **10**: 15477-15482.

Agarwal, N. and Das, S. (2009). Dissecting molecular mechanisms related to blister blight disease in tea; An EST approach. *Proceedings of the Indo-US workshop on Epigenetic Regulation and Genome Control*, Hyderabad, India. pp. 99.

Ahmed, P. and Das, S. (2009). A functional genomic approach to analyze herbivory related transcripts in tea. *Proceedings of the Indo-US workshop on Epigenetic Regulation and Genome Control*, Hyderabad, India. pp.100.

Arulpragasam, P.V. (1963) Assam Review and Tea News, **80**: 13–20.

Balasaravanan, T., Pius, P.K., Raj Kumar, R., Muraleedharan, N. and Shasany, A.K. (2003). Genetic diversity among south Indian tea germplasm (*Camellia sinensis, C. assamica* and *C. assamica* spp. *lasiocalyx*) using AFLP markers. *Plant Science*, **165**: 365-372.

Bali, S., Raina, S.N., Bhat, V., Aggarwal, R.K. and Goel, S. (2013). Development of a set of genomic microsatellite markers in tea (*Camellia* L.) (Camelliaceae). *Molecular Breeding*, **32**: 735-741.

Bandyopadhyay, T. (2011). Molecular marker technology in genetic improvement of tea. *International Journal of Plant Breeding and Genetics*, **5**: 23-33.

Banerjee, B. (1992a). Botanical classification of tea. In: Wilson, K.C. and Clifford, M.N. (Eds.), *Tea Cultivation to Consumption*. Chapman and Hall, London. pp. 25–51.

Banerjee, B. (1992b) Selection and breeding of tea. In: Willson, K.C. and Clifford, M.N. (Eds.), *Tea Cultivation to Consumption.* Chapman and Hall, London. pp. 53–86.

Bardini, M., Lee, D., Donini, P. and Mariani, A. (2004). Tubulin-based polymorphism (TBP): a new tool, based on functionally relevant sequences, to *assess genetic diversity in plant species. Genome*, **291**: 281–291.

Barua, D.N. (1989). Propagation technique. In: Barua, D.N. (Ed.), *Science and Practice in Tea Culture. Tea Research Association*, Calcutta. pp. 223–264.

Barua, D.N. (1963) Characteristics of the Tocklai released clones. *Two and A Bud*, **10(1)**: 26-28.

Barua, P.K. (1963) Tea breeding at Tocklai. *Two and a Bud*, **10**: 7–11.

Bezbaruah, H.P. (1969) Economics of a selection scheme. In: *Proceedings of 24th Tocklai Conference.*

Bhagawati, P., Singh, H.R., Borchetia, S., Chowdhury, P., Saikia, H. and Bandyopadhyay, T. (2015). *Comparative study on morphological and molecular genetic diversity analysis*. In: Das, K. (Ed.), *Biotik*. Maharathi Prakashan, India. pp. 66-79.

Bhardwaj, P., Kumar, R., Sharma, H., Tewari, R., Ahuja, P.S. and Sharma, R.K. (2013). Development and utilization of genomic and genic microsatellite markers in Assam tea (*Camellia asamica* ssp. *assamica*) and related *Camellia* species. *Plant Breeding*, **132**: 748-763.

Bhardwaj, P., Sharma, R. K., Kumar, R., Sharma, H. and Ahuja, P.S. (2014). SSR marker based DNA fingerprinting and diversity assessment in superior tea germplasm cultivated in Western Himalayas. *Proceedings of the Indian National Science Academy*, **80**: 157-162.

Borchetia, S., Das, S.C., Handique, P.J. and Das, S. (2009). High multiplication frequency and genetic stability for commercialization of the three varieties of micropropagated tea (*Camellia* spp.). *Scientia Horticulturae*, **120**: 544–550.

Botstein, D., White, R.L., Skolnick, M. and Davis, R.W. (1980). Construction of a genetic linkage map in man using restriction fragment length polymorphism. *American Journal of Human Genetics*, **32**: 314-331.

Brady, S.M. and Provart, N.J. (2007). Extreme breeding: leveraging genomics for crop improvement. *Journal of the Science of Food and Agriculture*, **87**: 925–929.

Bray, E.A., Bailey-Serres, J. and Weretilnyk, E (2000). Responses to abiotic stresses. In: Gruissem, W., Buchannan, B. and Jones, R. (Eds.), *Biochemistry and Molecular Biology of Plants. American Society of Plant Physiologists*, Rockville, MD. pp. 1158-1249.

Burgess, P.J. and Carr, M.K.V. (1993). Response of tea (*Camellia sinensis*) clones to drought. I. Yield, dry matter production and partitioning. *Aspects of Applied Biology*, **34**: 249–258.

Caldwell, D.G., McCallum, N., Shaw, P., Muehlbauer, G.J., Marshall, D.F. and Waugh, R.A. (2004). Structured mutant population for forward and reverse genetics in barley (*Hordeum vulgare* L.). *Plant Journal*, **40**: 143-150.

Chen, L. and Yamaguchi, S. (2002). Genetic diversity and phylogeny of tea plant (*Camellia sinensis*) and its related species and varieties in the section *Thea* genus *Camellia* determined by randomly amplified polymorphic DNA analysis. *Journal of Horticultural Science and Biotechnology*, **77**: 729-732.

Chen, L. and Yamaguchi, S. (2005). RAPD markers for discriminating tea germplasm on the inter-specific level in China. *Plant Breeding*, **124(4)**: 404–409.

Chen, L., Gao, Q., Chen, D. and Xu, C. (2005). The use of RAPD markers for detecting genetic diversity, relationship and molecular identification of Chinese elite tea genetic resources preserved in tea germplasm repository. *Biodiversity and Conservation*, **14**: 1433-1444.

Chen, L., Zhou, Z.X. and Yang, Y.J. (2007). Genetic improvement and breeding of tea plant (*Camellia sinensis*) in *China: From individual selection to hybridization and molecular breeding*. Euphytica, **154**: 239-248.

Chen, Q., Zhao, J., Liu, M., Cai, J. and Liu, J. (2008). Determination of total polyphenols content in green tea using FT-NIR spectroscopy and different PLS algorithms. *Journal of Pharmaceutical and Biomedical Analysis*, **46**: 568–573.

Chen, S., Qi, G., Li, H., Zou, Y. and Shan, H. (2012). PCR-RFLP analysis of cpDNA in tea cultivars (*Camellia sinensis* L.) in Sichuan of China. *Journal of Agricultural Science*, **4(5)**: 25-30.

Chen, X. and Sullivan, P.F. (2003). Single nucleotide polymorphism genotyping: Biochemistry, protocol, cost and throughput. *Pharmacogenomics Journal*, **3**: 77-96.

Colbert, T., Till, B.J., Tompa, R., Reynolds, S., Steine, M.N., Yeung, A.T., McCallum, C.M., Comai, L. and Henikoff, S. (2001). High throughput screening for induced point mutations. *Plant Physiology*, **126**: 480-484.

Collard, B.C. and Mackill, D.J. (2008). Marker-assisted selection: an approach for precision plant breeding in the twenty-first century. Philosophical Transactions of the Royal Society B: *Biological Sciences*, **363**: 557-572.

Crouch, J.H. and Ortiz, R. (2004). Applied genomics in the improvement of crops grown in Africa. *African Journal of Biotechnology*, **3**: 489-496.

Dahmani-Mardas, F., Troadec, C., Boualem, A., Lévêque, S., Alsadon, A.A., Aldoss, A.A., Dogimont, C. and Bendahmane, A. (2010). *Engineering melon plants with improved fruit shelf life using the TILLING approach. PLoS ONE*, **5**: e15776.

Das, A., Saha, D. and Mondal, T.K. (2013). An optimized method for extraction of RNA from tea roots for functional genomics analysis. *Indian Journal of Biotechnology*, **12**: 129-132.

Das, S.C, Das, S. and Hazarika, M. (2012). Breeding of the Tea Plant (*Camellia sinensis*) in India. In: *Global Tea Breeding: Achievements, Challenges and Perspectives*. Zhejiang University Press, Springer Berlin Heidelberg. pp. 69-124.

Devarumath, R.M., Nandy, S., Rani, V., Marimuthu, S., Muraleedharan, N. and Raina, S.N. (2002). ISSR and RFLP fingerprints as useful markers to evaluate genetic integrity of micropropagated plants of three diploid and triploid elite tea clones representing *Camellia sinensis* (China type) and *C. assamica* ssp. *assamica* (Assam-India type). *Plant Cell Reports*, **21**: 166-173.

Dhiman, B. and Singh, M. (2003). Molecular detection of cashew husk (*Anacardium occidentale*) adulteration in market samples of dry tea (*Camellia sinensis*). *Planta Medica*, **69**: 882-884.

Fang, W.P., Meinhardt, L.W., Tan, H.W., Zhou, L., Mischke, S. and Zhang, D. (2014a) Varietal identification of tea (*Camellia sinensis*) using nanofluidic array of single nucleotide polymorphism (SNP) markers. *Horticulture Research*, **1**: 14035.

Fang, W.P., Meinhardt, L.W., Mischke, S., Bellato, C.M., Motilal, L. and Zhang, D. (2014b) Accurate determination of genetic identity for a single cacao bean, using molecular markers with a nanofluidic system, ensures cocoa authentication. *Journal of Agriculture Food Chemistry*, **62**: 481–487.

Freeman, S., West, J., James, C., Lea, V. and Mayess, S. (2004). Isolation and characterization of highly polymorphic microsatellites in tea (*Camellia sinensis*). *Molecular Ecology Notes*, **4**: 324-326.

Gepts, P. (1993). The use of molecular and biochemical markers in crop evolution studies. In: Hecht, M.K., Maclntyre, R.J. and Clegg, M.T. (Eds.), *Evolutionary Biology*. Plenum Press, New York. pp. 51-94.

Gogoi, M., Zaman, A., Borchetia, S. and Bandyopadhyay, T. (2015). Phylogenetic relationship of *Camellia* genus based on ITS2 primary sequence and secondary structure. In: Das, K. (Ed.), *Biotik*. Maharathi Prakashan, India. pp. 36-50.

Gupta, S., Bharalee, R., Bhorali, P., Das, S.K., Bhagawati, P., Bandyopadhyay, T., Gohain, B., Agarwal, N., Ahmed, P., Borchetia, S., Kalita, M.C., Handique, A.K. and Das, S. (2013). Molecular analysis of drought tolerance in tea by cDNA-AFLP based transcript profiling. *Molecular Biotechnology*, **53**: 237-248.

Gupta, S., Bhorali, P., Bharalee, R., Bandyopadhyay, T., Gohain, B., Das, S., Agarwal, N., Singh, H.R., Ahmed, P., Bhagawati, P., Bhattacharyya, N., Borchetia, S. and Das, S. (2010). Transcript profiling and identification of molecular markers associated with drought tolerance in tea. Paper presented at the Plant Genomics in China XI, Changchun, China.

Gurusubramanian, G. and Bora, S. (2008). Insecticidal resistance to tea mosquito bug *Helopeltis theivora* waterhouse (Miridae: Heterptera) in Northeast India. *Journal of Environmental Research and Development*, **2**: 560-567.

Hackett, C.A., Wachira, F.N., Paul, S., Powell, W. and Waugh, R. (2000). Construction of a genetic linkage map for *Camellia sinensis* (tea). *Heredity*, **85**: 346-355.

He, Y., Li, X.L. and Deng, X.F. (2007). Discrimination of varieties of tea using near infrared spectroscopy by principal component analysis and BP model. *Journal of Food Engineering*, **79**: 1238–1242.

Hodgkin, T., Roviglioni, R., De Vicente, M.C. and Dudnik, N. (2001). Molecular methods in the conservation and use of plant genetic resources. *Acta Horticulture*, **546**: 107–118.

Hu, C.Y., Lee, T.C., Tsai, H.T., Tsai, Y.Z. and Lin, S.F. (2013). Construction of an integrated genetic map based on maternal and paternal lineages of tea (*Camellia sinensis*). Euphytica, **191**: 141-152.

Hu, C.Y., Tsai, Y.Z. and Lin, S.F. (2014). Development of STS and CAPS markers for variety identification and genetic diversity analysis of tea germplasm in Taiwan. *Botanical Studies*, **55**: 12.

Huang, F.P., Liang, Y.R., Lu, J.L., Chen, R.B. and Mamati, G. (2004). Evaluation of genetic diversity in Oolong tea germplasm by AFLP fingerprinting. *Journal of Tea Science*, **24**: 183–189.

Huang, J.A., Li, J.X., Huang, Y.H., Luo, J.W., Gong, Z.H. and Liu Z.H. (2005). Construction of AFLP molecular markers linkage map in tea plant. *Journal of Tea Science*, **25**: 7-15.

Hui, L.X., Lin, L.C., Peng, S.Z., Wu, L.J., Wen, S.C., Hua, G.Z., Xuan, C., Li, X.H., Liu, C.L., Shi, Z.P., Luo, J.W., Shen, C.W., Gong, Z.H. and Chen, X. (2004). Analysis of genetic relationships among "Rucheng Baimao Cha" plants with RAPD method. *Journal of Tea Science*, **24**: 33–36.

Hung, C.Y., Wang, K.H., Huang, C.C., Gong, X., Ge, X.J. and Chiang, T.Y. (2008). Isolation and characterization of 11 microsatellite loci from *Camellia sinensis* in Taiwan using PCR-based isolation of microsatellite arrays (PIMA). *Conservation Genetics*, **9**: 779-781.

Jain, N.K. (1999). *Global advances in tea science*. Aravali Books International, New Delhi. pp. 882.

Ji, P.Z., Li, H., Gao, L.Z., Zhang, J., Cheng, Z.Q. and Huang, X.Q. (2011). ISSR diversity and genetic differentiation of ancient tea (*Camellia sinensis* var. *assamica*) plantations from China: implications for precious tea germplasm conservation. *Pakistan Journal of Botany*, **43(1)**: 281-291.

Jonah, P.M., Bello, L.L., Lucky, O., Midau, A. and Moruppa, S.M. (2011). The importance of molecular markers in plant breeding programmes. *Global Journal of Science Frontier Research*, **11(5)**: 5-12.

Kadaru, S.B., Yadav, A.S., Fjellstrom, R.G. and Oard, J.H. (2006). Alternative EcoTILLING protocol for rapid, cost-effective single-nucleotide polymorphism discovery and genotyping in rice (*Oryza sativa* L.). *Plant Molecular Biology Reporter*, **24**: 3-22.

Kafkas,S., Ercisli, S., Dogan, Y., Erturk, Y., Haznedar, A. and Sekban, R. (2009). Polymorphism and genetic relationships among tea genotypes from Turkey revealed by Amplified Fragment Length Polymorphism markers. *Journal of the American Society for Horticultural Science*, **134**: 428-434.

Kalita, R., Barooah, M., Modi, M.K. and Sen, P. (2014). Development of 'Assam' type tea specific SCAR marker from RAPD products. *Indian Journal of Biotechnology*, **13**: 376-380.

Kamunya, S.M., Wachira, F.N., Pathak, R.S., Korir, R., Sharma, V., Kumar, R., Bhardwaj, P., Chalo, R., Ahuja, P.S. and Sharma, R.K. (2010). Genomic mapping and testing for quantitative trait loci in tea (*Camellia sinensis* (L.) O. Kuntze). *Tree Genetics and Genomes*, **6**: 915-929.

Kamunya, S.M., Wachira, F.N., Pathak, R.S., Sharma, R K., Korir, R., Kinyangi, T., Kiplang'at, J., Sharma, V., Chalo, R. and Ahuja, P.S. (2009). Genetic mapping and identification of quantitative trait loci for yield and drought tolerance in tea (*Camellia sinensis* (L.) O. Kuntze). *Tea*, **30**: 19-41.

Kaundun, S.S. and Matsumoto, S. (2002). Heterologous nuclear and chloroplast microsatellite amplification and variation in tea, *Camellia sinensis*. *Genome*, **45**: 1041–1048.

Kaundun, S.S. and Matsumoto, S. (2003a).. Development of CAPS markers based on three key genes of the phenylpropanoid pathway in tea, *Camellia sinensis* (L.) O. Kuntze and differentiation between *assamica* and *sinensis* varieties. *Theoretical and Applied Genetics*, **106**: 375–38.

Kaundun, S.S. and Matsumoto, S. (2003b).. Identification of processed Japanese green tea based on polymorphisms generated by STS-RFLP analysis. *Journal of Agriculture and Food Chemistry*, **51**: 1765–1770.

Kaundun, S.S. and Matsumoto, S. (2004). PCR-based amplicon length polymorphisms (ALPs) at microsatellite loci and indels from non-coding DNA regions of cloned genes as a means of authenticating commercial Japanese green teas. *Journal of the Science of Food and Agriculture*, **84**: 895–902.

Kaundun, S.S., Zhyvoloup, A. and Park, Y. (2000). Evaluation of the genetic diversity among elite tea accessions using RAPD markers. *Euphytica*, **115**: 7–16.

Kim, Y.D., Jeong, M.J., Song, H.J., Kim, J.C. and Choi, M.S. (2011). Morphological characters and genetic relationship between catechins-rich and –poor tea tree (*Camellia sinensis* L.) lines. *Forest Science and Technology*, **8**: 28-33.

Konieczny, A. and Ausubel, F.M. (1993). A procedure for mapping *Arabidopsis* mutations using co- dominant ecotype-specific PCR-based markers. *The Plant Journal*, **4**: 403-410.

Konwar, B.K. (1999). *Biodiversity of tea in North East India and their conservation at Tocklai. Two and a Bud*, **46(2)**: 7-12.

Lai, J.A., Yang, W.C. and Hsiao, J.Y. (2001). An assessment of genetic relationships in cultivated tea clones and native wild tea in Taiwan using RAPD and ISSR markers. B*otanical Bulletin of Academia Sinica*, **42**: 93-100.

Liang, C., Zhi-Xiu, Z. and Ya-Jun, Y. (2007). Genetic improvement and breeding of tea plant (*Camellia sinensis*) in China: from individual selection to *hybridization and molecular breeding*. *Euphytica*, **154**: 239-248.

Liu, B., Sun, X., Wang, Y., Li, Y., Cheng, H., Xiong, C. and Wang, P. (2012a). Genetic diversity and molecular discrimination of wild tea plants from Yunnan Province based on Inter simple sequence repeats (ISSR) markers. *African Journal of Biotechnology*, **11**: 11566-11574.

Liu, B.Y., Cheng, H., Li, Y.Y., Wang, L.Y., He, W. and Wang, P.S. (2012b). Fingerprinting for discriminating tea germplasm using Inter simple sequence repeat (ISSR) markers. *Pakistan Journal of Botany*, **44**: 1247-1260.

Liu, J., Li, J., Qu, J. and Yan, S. (2015). Development of genome-wide insertion and deletion polymorphism markers from next-generation sequencing data in rice. *Rice*, **8**: 27.

Luypaert, J., Zhang, M.H. and Massart, D.L. (2003). Feasibility study for the use of near infrared spectroscopy in the qualitative and quantitative analysis of green tea, *Camellia sinensis* (L.). *Analytica Chimica Acta*, **478**: 303–312.

Ma, J.Q., Ma, C.L., Yao, M.Z., Jin, J.Q., Wang, Z.L., Wang, X.C. and Chen, L. (2012). Microsatellite markers from tea plant expressed sequence tags (ESTs) and their applicability for cross-species/genera amplification and genetic mapping. *Scientia Horticulturae*, **134**: 167-175.

Ma, J.Q., Yao, M.Z., Ma, C.L., Wang, X.C., Jin, J.Q., Wang, X.M. and Chen, L. (2014). Construction of a SSR-based genetic map and identification of QTLs for *catechins content in tea plant* (*Camellia sinensis*). *PLoS ONE*, **9**: e93131.

Ma, J.Q., Zhou, Y.H., Ma, C.L., Yao, M.Z., Jin, J.Q, Wang, X.C. and Chen, L. (2010). Identification and characterization of 74 polymorphic EST-SSR markers in the tea plant, *Camellia sinensis* (*Theaceae*). *American Journal of Botany*, **97**: 153-156.

Malebe, P. (2011). Development of molecular marker assisted selection for drought tolerance in tea *Camellia sinensis* (L.) O. Kuntze. Unpublished M.Sc. dissertation, University of Pretoria, Pretoria, South Africa.

Maritim, T.K., Wachira, F.M., Kamunya, S.M., Mireji, P., Mwendia, C., Muoki, R.C., Wamalwa, M., Stomeo, F. and Martina, K.P. (2015). Transcriptome based identification of drought responsive genes in the tea plant. *Proceedings of the First International Conference on Tea Science and Development. Advances in Tea Research*, Karatina, Kenya. pp. 82-93.

Matsumoto, S., Kiriiwa, Y. and Takeda, Y. (2002). Differentiation of Japanese green tea cultivars as revealed by RFLP analysis of phenylalanine ammonia-lyase DNA. *Theoretical and Applied Genetics*, **104**: 998-1002.

Matsumoto, S., Kiriiwa, Y. and Yamaguchi, S. (2004). The Korean tea plant (*Camellia sinensis*): RFLP analysis of genetic diversity and relationship to Japanese tea. *Breeding Science*, **54**: 231-237.

Matsumoto, S., Takeuchi, A. and Hayatsu, M. (1994).) Molecular cloning of phenylalanine ammonia-lyase cDNA and classification of varieties and cultivars of tea plants (*Camellia sinensis*) using the tea PAL cDNA probe. *Theoretical and Applied Genetics*, **89**: 671–675.

Mejlhede, N., Kyjovska, Z., Backes, G., Burhenne, K., Rasmussen, S.K. and Jahoor, A. (2006). EcoTILLING for the identification of allelic variation in the powdery mildew resistance genes *mlo* and *Mla* of barley. *Plant Breeding*, **125**: 461-467.

Mewan, K.M., Abeysinghe, I.S.B., Saha, M.C., Pang, Y., Karunanayake, E.H., Tirimanne, T.L.S., Everard, J.M.D.T. and Dixon, R.A. (2012). Development of genomic and EST microsatellite markers and their usefulness in genetic conservation and improvement of tea (*Camellia sinensis* L.). *Sri Lanka Journal of Tea Science*, **77**: 22-44.

Mewan, K.M., Liyanage, A.C., Everard, J.M., Gunasekare, M.T.K. and Karunanayaka, E. (2005). Studying genetic relationship among tea accessions in Sri Lanka using RAPD. *Sri Lanka Journal of Tea Science*, **70**: 42-53.

Mishra, R.K. and Sen-Mandi, S. (2004a) Genetic diversity estimates for Darjeeling tea clones based on AFLP markers. *Journal of Tea Science*, **24**: 86-92.

Mishra, R.K. and Sen-Mandi, S. (2004b) Molecular profiling and development of DNA marker associated with drought tolerance in tea clones growing in Darjeeling. *Current Science*, **87**: 60-66.

Mishra, R.K., Chaudhury, S., Ahmad, A., Pradhan, M. and Siddiqi, T.O. (2009). Molecular analysis of tea clones (*Camellia sinensis*) using AFLP markers. *International Journal of Integrative Biology*, **5(2)**: 130-136.

Mishra, R.K. and Sen-Mandi, S. (2001). DNA fingerprinting and genetic relationship study of tea plants using Amplified Fragment Length Polymorphism (AFLP) technique. *Indian Journal of Plant Genetic Resources*, **14(2)**: 148-149.

Mondal, T.K. (2002). *Assessment of genetic diversity of tea* (*Camellia sinensis* (L.) O. Kuntze) by inter-simple sequence repeat polymerase chain reaction. *Euphytica*, **128**: 307-315.

Mondal, T.K. (2014). Functional genomics. In: *Breeding and Biotechnology of Tea and its Wild Species*. Springer, India. pp. 149-167.

Mondal, T.K. and Sutoh, K. (2013). Applications of Next- Generation Sequencing for abiotic stress tolerance. In: Barh, D., Zambare, V. and Azevedo, V. (Eds.), *The World of Omics and Applications in Plant and Agricultural Sciences*. CRC Press, USA. pp. 383-402.

Mondal, T.K. and Chand, P.K. (2002). Detection of genetic instability among the micropropagated tea (*Camellia sinensis*) plants. In *Vitro Cellular and Developmental Biology-Plant*, **37**: 1–5.

Mondal,T.K., Bhattacharya, A., Laxmikumaran, M. and Ahuja, P.S. (2004). Recent advance in tea biotechnology. *Plant Cell Tissue and Organ Culture*, **75**: 795–856.

Mphangwe, N.I., Vorster, J., Steyn, J.M., Nyirenda, H.E., Taylor, N.J. and Apostolides, Z. (2013). Screening of tea (*Camellia sinensis*) for trait-associated molecular markers. *Applied Biochemistry and Biotechnology*, **171**: 437-449.

Nandi, S.K., Palni, L.M.S. and Rashmi (1995). Chemical manipulation of dormancy in tea shoots and associated biochemical changes. *Journal of Plantation Crops*, **23**: 52–58.

Ni, S., Yao, M., Chen, L, Zhao, L. and Wang, X. (2008). Germplasm and breeding research of tea plant based on DNA marker approaches. *Frontiers of Agriculture in China*, **2**: 200-207.

Ohsako, T., Ohgushi, T., Motosugi, H. and Oka, K. (2008). Microsatellite variability within and among local landrace populations of tea, *Camellia sinensis* (L.) O. Kuntze, in Kyoto, Japan. *Genetic Resources and Crop Evolution*, **55**: 1047-1053.

Ordish, G. (1952). Untaken harvest. Man's loss of crops from pest, weed and disease. *Untaken harvest. Man's loss of crops from pest, weed and disease.* Constable and Company Ltd., London. pp. 171

Paul, S., Wachira, F.N., Powell, W. and Waugh, R. (1997). Diversity and genetic differentiation among population of Indian and Kenyan tea (*Camellia sinensis* (L.) O. Kuntze) revealed by AFLP markers. *Theoretical and Applied Genetics*, **94**: 255-263.

Perez-de-Castro, A.M., Vilanova, S., Canizares, J., Pascual, L., Blanca, J.M., Diez, M.J., Prohens, J. and Pico, B. (2012). Application of genomic tools in plant breeding. *Current Genomics,* **13**: 179-195.

Perry, J.A., Wang, T.L., Welham, T.J., Gardner, S., Pike, J.M., Yoshida, S. and Parniske, M.A. (2003). TILLING reverse genetics tool and a web-accessible collection of mutants of the legume *Lotus japonicus*. *Plant Physiology*, **131**: 866-871.

Poczai, P., Varga, I., Laos, M., Cseh, A., Bell, N., Valkonen, J.P.T. and Hyvonen, J. (2013). Advances in plant gene-targeted and functional markers: A review. *Plant Methods*, **9**: 6.

Ramos, M.L., Huntley, J.J., Maleki, S.J. and Ozias-Akins, P. (2009). Identification and characterization of a hypoallergenic ortholog of Ara h 2.01. *Plant Molecular Biology*, **69**: 325-335.

Ribaut, J.M. and Hoisington, D. (1998). Marker assisted selection: New tools and strategies. *Trends in Plant Science*, **3**: 236-239.

Rohde, A. and Bhalerao, R.P. (2007). Plant dormancy in the perennial context. *Trends in Plant Science,* **12**: 217–223.

Roy, S.C. and Chakraborty, B.N. (2009). Genetic diversity and relationships among tea (*Camellia sinensis*) cultivars as revealed by RAPD and ISSR based fingerprinting. *Indian Journal of Biotechnology*, **8**: 370-376.

Ruan, C. (2010). Germplasm-regression-combined marker-trait association identification in plants. *African Journal of Biotechnology*, **9**: 573-580.

Salvi, N.D., George, L. and Eapen, S. (2001). Plant regeneration from leaf base callus of turmeric and random amplified polymorphic DNA analysis of regenerated plants. *Plant Cell, Tissue and Organ Culture*, **66**: 113–119.

Satyanarayana, N. and Sharma, V.S. (1986). Tea (*Camellia* L. spp.) germplasm in South India. In: Srivastava, H.C. (Ed.), *Plantation Crops* (Vol. II.). Oxford and IBH Publishing Co. Pvt. Ltd., New Delhi. pp. 173-179.

Sharma, H., Kumar, R., Sharma, V., Kumar, V., Bhardwaj, P., Ahuja, P.S. and Sharma, R.K. (2011). Identification and cross-species transferability of 112 novel unigene-derived microsatellite markers in tea (*Camellia sinensis*). *American Journal of Botany*, **98**: e133-e138.

Sharma, P. and Kumar, S. (2005). Differential display-mediated identification of three drought responsive expressed sequence tags in tea [*Camellia sinensis*(L.) O. Kuntze]. *Journal of Biosciences,* **30**: 231–235.

Sharma, R.K., Negi, M.S., Sharma, S., Bhardwaj, P., Kumar, R., Bhattachrya, E., Tripathi, S.B., Vijayan, D., Baruah, A.R., Das, S.C., Bera, B., Rajkumar, R., Thomas, J., Sud, R.K., Muraleedharan, N., Hazarika, M., Lakshmikumaran, M., Raina, S.N. and Ahuja, P.S. (2010). AFLP-based genetic diversity assessment of commercially important tea germplasm in India. *Biochemical Genetics,* **48**: 549–564.

Shi, C.Y., Yang, H., Wei, C.L., Yu, O., Zhang, Z.Z., Sun, J., Li, Y.Y., Chen, Q., Xia, T. and Wan, X.C. (2011). Deep sequencing of the *Camellia sinensis* transcriptome revealed candidate genes for major metabolic pathways of tea-specific compounds. *BMC Genomics,* **12**: 131.

Singh, D. and Ahuja, P.S. (2006). 5S rDNA gene diversity in tea (*Camellia sinensis* (L.) *O. Kuntze*) and its use for variety identification. *Genome,* **49(1)**: 91-96.

Singh, I.D. and Handique, A.C. (1993). Breeding for resistance to water stress in tea (*Camellia sinensis* L.). *Two and a Bud,* **40**: 41-49.

Smith, J.S.C. and Smith, O.S. (1989). The description and assessment of distances between inbred lines of maize: The utility of morphological, biochemical and genetic descriptors and a scheme for the testing of distinctiveness between inbred lines. *Maydica,* **34**: 151-161.

Staub, J.E. and Serquen, F.X. (1996). Genetic markers, map construction and their application in plant breeding. *HortScience,* **31**: 729-740.

Suganthi, M., Senthilkumar, P., Arvinth, S. and Rajkumar, R. (2014). RAPD and SCAR markers linked to tea-mosquito resistance in tea. *Journal of Crop Improvement,* **28**: 795-803.

Takeda, Y. (1990). Cross compatibility of tea (*Camellia sinensis*) and its allied species in the Genus *Camellia*. Japan International Research Centre for Agricultural Sciences, **24**: 111-116.

Takeuchi, A., Matsumoto, S. and Hayatsu, M. (1994).) Chalcone synthase from *Camellia sinensis* isolation of the cDNAs and the organ specific and sugar-responsive expression of the genes. *Plant and Cell Physiology,* **35**: 1011-1018.

Tan, L.Q., Wang, L.Y., Wei, K., Zhang, C.C., Wu, L.Y., Qi, G.N., Cheng, H., Zhang, Q., Cui, Q.M. and Liang, J.B. (2013). Floral transcriptome sequencing for SSR marker development and linkage map construction in the tea plant (*Camellia sinensis*). *PLoS ONE,* **8:** e81611.

Tan, L.Q., Zhang, C.C., Qi, G.N., Wang, L.Y., Wei, K., Chen, S.X., Zou, Y., Wu, L.Y. and Cheng, H. (2015). Heterozygosities and genetic relationship of tea cultivars revealed by simple sequence repeat markers and implications for breeding and genetic mapping populations. *Genetics and Molecular Research,* **14**: 1557-1565.

Tan, S.M., Luo, R.M., Zhou, Y.P., Xu, H., Song, D.D., Tan, Z., Yang, T.M. and Nie, Y. (2012). Boosting partial least-squares discriminant analysis with application to near infrared spectroscopic tea variety discrimination. *Journal of Chemometrics*, **26**: 34–39.

Tanaka, J. (1996). RAPD linkage map of tea plant and the possibility of application in tea genetics and breeding. *Tea Research Journal*, **84**: 44-45.

Tanaka, J. and Taniguchi, F. (2006). Estimation of the genome size of tea (*Camellia sinensis*), *Camellia* (*Camellia japonica*), and their interspecific hybrids by flow cytometry. *Tea Research Journal*, **101**: 1-7.

Taniguchi, F., Fukuoka, H. and Tanaka, J. (2012b). Expressed sequence tags from organ-specific cDNA libraries of tea (*Camellia sinensis*) and polymorphisms and transferability of EST-SSRs across *Camellia* species. *Breeding Science*, **62**: 186–195.

Taniguchi, F., Furukuwa, K., Ota-Metoku, S., Yamaguchi, N., Kono, I., Fukuoka, H. and Tanaka, J. (2012a). Construction of a high-density reference linkage map of tea (*Camellia sinensis*). *Breeding Science*, **62**: 263-273.

Tautz, D. (1989). Hypervariability of simple sequences as a general source for polymorphic DNA markers. *Nucleic Acids Research*, **17**: 6463-6471.

Thomas, J., Vijayan, D., Joshi, S.D., Lopez, J. and Kumar, R.R. (2006). Genetic integrity of somaclonal variants in tea (*Camellia sinensis* (L.) O Kuntze) as revealed by inter simple sequence repeats. *Journal of Biotechnology*, **123(2)**: 149-154.

Triques, K., Sturbois, B., Gallais, S., Dalmais, M., Chauvin, S., Clepet, C., Aubourg, S., Rameau, C., Caboche, M. and Bendahmane, A. (2007). Characterization of *Arabidopsis thaliana* mismatch specific endonucleases: Application to mutation discovery by TILLING in pea. *Plant Journal*, **51**: 1116-1125.

Ueno, S., Yoshimaru, H., Tomaru, N. and Yamamoto, S. (1999). Development and characterization of microsatellite markers in *Camellia japonica* L. *Molecular Ecology Resources*, **8**: 335–336

Ujihara, T., Hayashi, N. and Tanaka, J. (2012). Identification of material cultivar of green tea infusions by simple sequence repeat markers. *Food Science and Technology Research*, **18**: 209–217.

Ujihara, T., Ohta, R., Hayashi, N., Kohata, K. and Tanaka, J. (2009). Identification of Japanese and Chinese green tea cultivars by using simple sequence repeat markers to encourage proper labeling. *Bioscience Biotechnology and Biochemistry*, **73**: 15–20.

Ujihara, T., Taniguchi, F., Tanaka, J.I. and Hayashi, N. (2011). Development of expressed sequence tag (EST)-based cleaved amplified polymorphic sequence (CAPS) markers of tea plant and their application to cultivar identification. *Journal of Agricultural and Food Chemistry*, **59**: 1557–1564.

Vijayan, D., Thomas, J., John, K.M.M., Pius, P.K. and Muraleedharan, N. (2013). Characterization of selected South Indian tea (*Camellia* ssp.) germplasm using morphological traits and RAPD markers. *Kennean Journal of Science*, **2**: 45-52.

Vijayan, K., Zhang, W.J. and Tsou, C.H. (2009). Molecular taxonomy of *Camellia* (*Theaceae*) inferred from NRITS sequences. *American Journal of Botany*, **96**: 1348–1360.

Vos, P., Hogers, R., Bleeker, M., Reijans, M., Van der, Lee, Hornes, M., Fritjers, A., Pot, J., Peleman, J., Kuiper, M. and Zabeau, M. (1995). AFLP: a new technique for DNA fingerprinting. *Nucleic Acids Research*, **23**: 4407-4414.

Wachira, F.N., Powell, W. and Waugh, R. (1997). An assessment of genetic diversity among *Camellia sinensis* L. (cultivated tea) and its wild relatives based on randomly amplified polymorphic DNA and organelle-specific STS. *Heredity*, **78**: 603-611.

Wachira, F.N., Tanaka, J. and Takeda, Y. (2001). Genetic variation and differentiation of tea (*Camellia sinensis*) germplasm revealed by RAPD and AFLP variation. *Journal of Horticultural Science and Biotechnology*, **76**: 557-563.

Wachira, F.N., Waugh, R., Hackett, C.A. and Powell, W. (1995). Detection of genetic diversity of tea (*Camellia sinensis* L.) using RAPD markers. *Genome*, **38**: 201–210.

Wang, J., Sun, J.Z., Liu, D.C., Yang, W.L., Wang, D.W., Tong, Y.P. and Zhang, A.M. (2008). Analysis of *Pina*and *Pinb*alleles in the microcore collections of Chinese wheat germplasm by EcoTILLING and identification of a novel *Pinb*allele. *Journal of Cereal Science*, **48**: 836- 842.

Waugh, R., Vande Ven, W.T.G., Phillips, M.S. and Powell, W. (1990). Chloroplasts DNA diversity in the genus *Rubus* (*Rosaceae*) revealed by southern hybridization. *Plant Systematics and Evolution*, **172**: 65–75.

Weil, C.F. and Monde, R.A. (2007). Getting to the point - mutations in maize. *Crop Sciences*, **47**: S60-S67.

Weising, K., Nybom, H., Pfenninger, M., Wolff, K. and Kahl, G. (2005). DNA Fingerprinting in Plants: *Principles, Methods, and Applications*. CRC Press, Boca Raton, Florida.

Wight, W. (1956). Commercial selection and breeding of tea in India. *World Crop*, **8**: 263-268.

Wight, W. (1962). Tea classification revised. *Current Science*, **31**: 298-299.

Wight. W. and Gilchrist, R.C.J.H. (1961) The concept of kind of tea. *Nature*, **161**: 14-16.

Williams, J.G.K., Kubelik, A.R., Livak, K.J., Rafalski, J.A. and Tingey, S.V. (1990). DNA polymorphisms amplified by arbitrary primers are useful as genetic markers. *Nucleic Acids Research*, **18**: 6231-6235.

Winter, P. and Kahl, G. (1995). Molecular marker technologies for plant improvement. *World Journal of Microbiology and Biotechnology*, **11**: 438-448.

Wium, M, (2009). Characterization of genetic diversity in selected cultivars and identification of a possible molecular marker for drought tolerance in tea *Camellia sinensis* (L.) O. Kuntze. Unpublished M.Sc. dissertation, University of Pretoria, Pretoria, South Africa.

Wu, H., Chen, D., Li, J., Yu, B., Qiao, X., Huang, H. and He, Y. (2013). *De novo* characterization of leaf transcriptome using 454 sequencing and development of EST-SSR markers in tea (*Camellia sinensis*). *Plant Molecular Biology Report*, **31**: 524-538.

Yadav, C.B., Bonthala, V.S, Muthamilarasan, M., Pandey, G., Khan, Y. and Prasad, M. (2015). Genome-wide development of transposable elements-based markers in foxtail millet and construction of an integrated database. *DNA Research,* **22:** 79-90.

Yamanaka, S., Suzuki, E., Tanaka, M., Takeda, Y., Watanabe, J.A. and Watanabe, K.N. (2003). Assessment of cytochrome P450 sequences offers a useful tool for determining genetic diversity in higher plant species. *Theoretical and Applied Genetics,* **108**: 1–9.

Yang, D., Liu, Y., Suna, M., Zhao, L., Wang, Y., Chen, X., Wei, C., Gao, L. and Xia, T. (2012). Differential gene expression in tea (*Camellia sinensis* L.) calli with different morphologies and catechin contents. *Journal of Plant Physiology,* **169**: 163-175.

Yang, J.B., Yang, J., Li, H.T., Zhao, Y. and Yang, S.X. (2009). Isolation and characterization of 15 microsatellite markers from wild tea plant (*Camellia taliensis*) using FIASCO method. *Conservation Genetics,* **10**: 1621–1623.

Yao, M.Z., Chen, L. and Liang, Y.R. (2008). Genetic diversity among tea cultivars from China, Japan and Kenya revealed by ISSR markers and its implication for parental selection in tea breeding programmes. *Plant Breeding,* **127**: 166-172.

Yao, M., Huang, H., Yu, J. and Chen, L. (2005). Analysis on applicability of ISSR in molecular identification and relationship investigation of tea cultivars. *Journal of Tea Science,* **25**: 153-157.

Yao, M.Z., Ma, C.L., Qiao, T.T., Jin, J.Q. and Chen, L. (2012). Diversity distribution and population structure of tea germplasm in China revealed EST-SSR markers. *Tree Genetics and Genomes,* **8**: 205-220.

Young-Goo, P., Kaundun, S.S. and Zhyvoloup, A. (2002). Use of bulked genomic DNA-based RAPD methodology to assess the genetic diversity among abandoned Korean tea plantations. *Genetic Resources and Crop Evolution,* **49**: 1–7.

Yuan, W.L. (2011). Mining and application of molecular markers from EST database and transcriptome sequencing in tea plant (*Camellia sinensis*). PhD thesis, *Chinese Academy of Agricultural Sciences,* China.

Zahid, N.Y., Abbasi, N.A., Hafiz, I.A. and Ahmad, Z. (2009). Genetic diversity of indigenous Fennel (*Foeniculcum vulgare* Mill) germplasm Pakistan assessed by RAPD markers. *Pakistan Journal of Botany,* **41(4)**: 1759-1767.

Zhang, C.C., Wang, L.Y., Wei, K. and Cheng, H. (2014). Development and characterization of single nucleotide polymorphism markers in *Camellia sinensis* (*Theaceae*). *Genetics and Molecular Research,* **13**: 5822-5831.

Zhang, M.H., Luypaert, J., Ferna´ndez Pierna, J.A., Xu, Q.S. and Massart, D.L. (2004). Determination of total antioxidant capacity in green tea by near-infrared spectroscopy and multivariate calibration. *Talanta,* **62**: 25–35.

Zhao, L.P., Liu, Z., Chen, L., Yao, M.Z. and Wang, X.C. (2008). Generation and characterization of 24 novel EST derived microsatellites from tea plant (*Camellia sinensis*) and cross- species amplification in its closely related species and varieties. *Conservation Genetics*, **9**: 1327-1331.

Zietkiewicz, E., Rafalski, A. and Labuda, D. (1994).) Genomic fingerprinting by simple sequence repeat (SSR) – anchored polymerase chain reaction amplification. *Genomics*, **20(2)**: 176-183.

In vitro Conservation

Chapter 19

Cryopreservation Techniques

☆ *Haeng-Hoon Kim, Elena Popova and Praveen K. Saxena*

1. Introduction

Secure long-term storage collections of crop germplasm is one of basic requirements for national food security (Wang *et al.*, 2014). Accessibility of plant genetic resources, including both popular commercial and heirloom cultivars, is of key importance for successful breeding programs (Panis and Lambardi, 2006; Benelli *et al.*, 2013). *In vitro* storage and cryopreservation are considered excellent options for long-term storage of clonal plant genetic resource collections (Engelmann, 1997; Pence, 2011). Cryogenic banks, where multiple cultivars are stored in liquid nitrogen (LN) or its vapor phase (-150 to -196°C) have been established for apple, banana, mulberry, potato, mint and garlic among others (Panis *et al.*, 2005; Fukui *et al.*, 2011; Kim *et al.*, 2012a; Wang *et al.*, 2014; Panta *et al.*, 2015; Volk *et al.*, 2016). However, there are some major challenges specific to cryogenic collections that remain unresolved despite the attention they have received in the scientific literature on cryopreservation. Apart from the obvious logistical challenges in handling a large numbers of samples, there are methodological problems that need to be overcome such as high sensitivity of some types of plant materials to dehydration and/or toxic effects of cryoprotectant solutions (Kim *et al.*, 2009a). Cryopreservation protocols have been reported for over 200 species (Sakai and Engelmann, 2007; Gonsalez-Arnao *et al.*, 2014), yet cryopreservation of a new species may require resource-expensive optimization procedures. Moreover, when a cryopreservation protocol which has been tested with a few genotypes is applied to the whole collection, it may result in variable regrowth (Kim *et al.*, 2012a; Panta *et al.*, 2015). In this case, heirloom domestic cultivars, that have limited stress tolerance and/or unusual physiology or storage behavior, will be at highest risk. In this review, we briefly discuss the existing methods for crop cryopreservation and propose a workable model that has the potential to be tested in cryopreserving different plant materials based on their tolerance to dehydration and toxicity of cryoprotectant solutions.

2. A Practical Classification of Plant Cryopreservation Methods

2.1. Slow-cooling (Two-step/programmed/conventional cooling)

During pre-freezing to about -40°C with a slow cooling rate (-0.1-1°C/min), cells and tissues endure freeze-induced inter-cellular crystallization, while intra-cellular crystallization can be limited. Developed in 1970-1980s, this method was the most efficient for cryopreserving undifferentiated cell cultures, dormant buds and *in vitro* apices of cold hardy plants. Most of the materials required pre-freezing equilibration with low concentration (<20 per cent) cryoprotectant solution, *e.g.* glycerol, dimethylsulfoxide (DMSO), sucrose or proline applied alone or in combination. Some examples of method application for cryopreservation of cell cultures include: wheat (Chen *et al.*, 1985), ginseng (Butenko *et al.*, 1984), soybean (Engelmann, 1992), rice (Anthony *et al.*, 1997), *Hevea* (Engelmann *et al.*, 1997), cassava (Escobar *et al.*, 1997), citrus (Pérez *et al.*, 1997) and alfalfa (Volkova *et al.*, 2015) (see Nosov *et al.*, 2014 for a review). The method was also effective for cryopreservation of organized tissues such as dormant/winter buds of pear (Oka *et al.*, 1991) and plum (Brison *et al.*, 1995) as well as *in vitro* grown apical meristems of strawberry, blackberry, raspberry, apple and grape (Chang and Reed, 1999; Reed, 2008).

2.2. Rapid-Cooling

The methodology of rapid-cooling is based upon the premise of inter- and intra-cellular solidification of cells devoid of crystallization (refers to 'vitrification') during cooling and warming. This method has been used to successfully cryopreserve diverse plant materials ranging in their sensitivity to dehydration, from sensitive (shoot tips, somatic embryos) to relatively tolerant (seeds, zygotic embryos, pollen) materials in both the presence or absence of highly concentrated cryoprotectant solutions, depending on modification (see below).

2.2.1. Desiccation

This method is suitable for dehydration of tolerant materials such as orthodox seeds and zygotic embryos, pollen, spores *etc.* (Pritchard, 2007). In the 'preculture-desiccation' method, the materials are incubated in presence of high concentrations (0.3 to 1.2 M) of osmotically active chemicals such as sugars or sugar alcohols to increase their desiccation tolerance before drying and cryopreservation. The preculture-desiccation approach was efficient for cryopreserving recalcitrant seeds and their excised embryos as well as *in vitro* cultured somatic embryos (Lambardi *et al.*, 2008; Gonzalez-Arnao *et al.*, 2008). An alternative method, encapsulation-desiccation (or encapsulation-dehydration) is based on the technology of synthetic seeds and operates with plant materials encapsulated in calcium-alginate gel (Fabre and Dereuddre, 1990). This method can be applied to vegetative organs which possess inherent or inducible dehydration tolerance. Highly efficient post-cryopreservation regrowth using this method has been demonstrated for hop

(Martínez *et al.*, 1999), grape (Wang *et al.*, 2000), strawberry (Medina *et al.*, 2007), potato (Kaczmarczyk *et al.*, 2011), apple (Li *et al.*, 2015) and a wide range of *Ribes* and *Rubus* germplasm (see Reed, 2008 for a review) and several other crops (Engelmann *et al.*, 2008).

2.2.2. Vitrification Method

The solution-based vitrification method incorporates cryoprotection of samples, using highly concentrated vitrification solutions (VSs), before cooling. It is suitable for small and desiccation sensitive vegetative organs, such as in vitro shoot-tips, bulbs, *etc.* in association with or without encapsulation. Diverse terminology has been used to describe the many options for containers/vehicles/devices in which to perform cooling and warming: vials and straws (vitrification), aluminum foil strips (droplet-vitrification), and cryo-plates (V-plate) (Sakai and Engelmann, 2007; Kaczmarczyk *et al.*, 2011; Yamamoto *et al.*, 2015; Vujoviæ *et al.*, 2015). For a facilitated cooling and warming, other options for containers can be tested, including open-pulled straw, cryoloops, cryohooks, grids that are used for cryopreservation of animal cells and tissues (Kuwayama, 2005; Lee *et al.*, 2006; Nakashima *et al.*, 2010; Wang *et al.*, 2008).

In the vitrification procedure, the material is subjected to a series of predetermined treatments such as preculture, osmoprotection (loading) and cryoprotection (dehydration) in concentrated VSs followed by quick immersion into liquid nitrogen (Figure 19.1).

After rapid rewarming, samples are rehydrated in so called unloading solutions while removing the cryoprotectants then placed on the regrowth medium. This method, originally developed by Sakai *et al.* (1990) has been successfully applied to cryopreserve many species of both tropical and temperate origin, using different types of plant materials (Sakai and Engelmann, 2007). The most prominent examples of the use of vitrification-based methods for large-scale cryopreservation of crops include potato, banana, and garlic collections (Panis *et al.*, 2005; Kim *et al.*, 2012a; Panta *et al.*, 2015).

3. Cryoinjury

Plant cell injury during the cryopreservation procedure has been studied and reviewed extensively (Mazur, 2004; Benson, 2008, Table 1). In the two-factor hypothesis, Mazur *et al.* (1972) suggested that cells can be damaged by the formation of intracellular ice. Even in the absence of intracellular ice crystallization, cells suffer from dramatic dehydration, resulting in concentration of potentially toxic solutes and increasing cytoplasm viscosity (Lovelock, 1953). Both extensive dehydration and low temperatures may cause phase transition in membrane lipids, thus leading to the loss of membrane integrity and semi-permeable properties, and, consequently, cell death (Steponkus, 1984; Uemura and Steponkus, 1999). Reduction of plasma membrane ATPase activity and perturbation of membrane polypeptide profiles, as well as depolymerization of microtubules, have been recorded in freeze-thawed cells (Arora and Palta, 1991; Bartolo and Carter, 1991). Cytotoxicity of concentrated cryoprotectant solutions is the most common injury incurred in solution-based

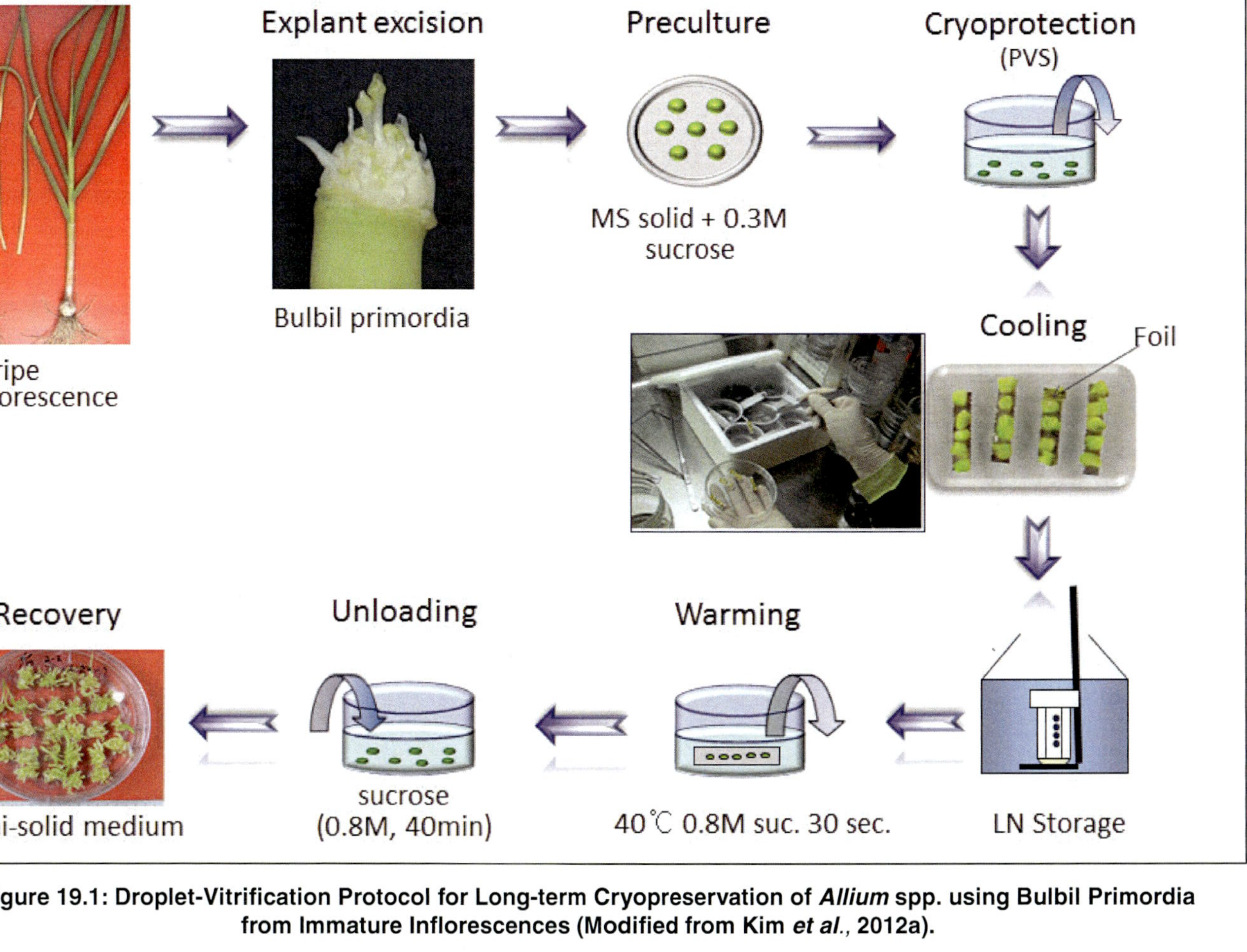

Figure 19.1: Droplet-Vitrification Protocol for Long-term Cryopreservation of *Allium* spp. using Bulbil Primordia from Immature Inflorescences (Modified from Kim *et al.*, 2012a).

vitrification procedures (Fahy *et al.*, 1990, 2004). Recently, the generation of reactive oxygen species in plant cells in response to multiple stresses associated with cryopreservation has been recognized and considered as one of the most hazardous factors (Benson and Bremner, 2004). Some of the most common strategies to overcome the injuries caused by cryopreservation are summarized in Table 19.1.

Table 19.1: Cryoinjuries and Strategies to Overcome them

Classes	*Injuries*	*Mechanism/Procedure*	*Strategies to Overcome*
Two factor hypothesis (Mazur *et al.*, 1972)	Intracellular ice crystallization/ recrystallization	Mechanical cell damage during cooling/warming	VS with high ice blocking efficiency, sample dehydration
	Extensive cellular dehydration	Increasing cytoplasm viscosity, increasing concentration of potentially toxic solutes,conformation changes in macromolecules, phase transition in membranes, and many others	Induction of dehydration tolerance, Determination of safe range of water content for plant materials before cryopreservation
Cytotoxicity (Fahy *et al.*,1990)	Osmotic stress and chemical toxicity	Toxicity of cryoprotectant solution during cryoprotection	Induction of tolerance during preculture; Balancing of cryoprotection and cytotoxicity
Oxidative stress (Thomson, 2009)	Oxidative stress	Reactive oxygen species (ROS) during excision, cryoprotection, unloading, *etc.*	Antioxidants (vitamins, polyphenols, indole amines); Optimization of regrowth conditions
Apoptosis (Xu, 2010)	Programmed cell death	Physiological mechanisms	Optimization of regrowth conditions (antioxidants, subculture, *etc.*)

4. Development of the Cryopreservation Protocol

4.1. Selection and Optimization of an Appropriate Method

Suitable methods can be chosen based on the characteristics of the material, *i.e.* its desiccation tolerance and size/permeability. However, simple comparison of cryopreservation techniques using the same experimental conditions, *e.g.* same preculture, osmoprotection treatment, cryoprotection duration, *etc.*, may result in difficulties in data interpretation and rationalization of the protocols. For example, over 14 reports are available on cryopreservation of isolated root cultures of commercially important crops and medical plants (reviewed by Popova *et al.*, 2015); however, the physiological differences in the material, different composition of the regrowth medium and the diversity of procedures used, do not allow any generalization on the most suitable protocol for root cultures. Each cryopreservation method has advantages and disadvantages that should be considered, and preparation of plant material for each method may require different approaches.

Existing literature suggests that the preculture-desiccation method or slow-cooling method is appropriate for desiccation of tolerant or frost hardy materials, respectively. For fragile and desiccation sensitive materials, such as cultures of undifferentiated plant cells, both slow-cooling method and vitrification methods can be applied. The vitrification approach is also suitable for tiny and desiccation sensitive shoot tips. For larger and desiccation sensitive shoot tips, encapsulation-desiccation is a preferable option. In practice, however, there are many factors favoring the application of one or another technique in different laboratories, including availability of cooling and warming containers and equipment, training and personal preference, cost of chemicals, and many others. In the laboratories at the Sunchon National University (Republic of Korea) and Gosling Research Institute for Plant Preservation (GRIPP, Canada), the authors of this review have adopted a series of preliminary tests that can be applied to plant materials in order to evaluate their sensitivity to dehydration and toxic effects of cryoprotectant solutions before cryopreservation. Based on the droplet-vitrification technique, this approach involves a series of preculture treatments followed by exposure of specimens to modified cryoprotectant solutions. This strategy and was effectively applied to develop cryopreservation protocols for meristematic tissues of chrysanthemum, lily, garlic and potato (Lee *et al.*, 2011; Kim *et al.*, 2012a; Popova *et al.*, 2015), somatic embryos of forest trees (Shin *et al.*, 2012) and hairy root cultures of various medicinal plants (Kim *et al.*, 2012b).

4.2. Development of 'Standard' Vitrification Procedure for the Species

4.2.1. Pre-conditioning (Preculture, cold acclimation)

Pre-conditioning is critically important for the acquisition of dehydration tolerance. For some plants such as chrysanthemum, potato or date palm, cold acclimation can be replaced by preculture in the presence of elevated concentrations of sucrose (Lee *et al.*, 2011; Fki *et al.*, 2013; Wang *et al.*, 2014; Folgado *et al.*, 2015). Preculture using sucrose-enriched medium has proven effective in enhancing post-cryopreservation regrowth of different plant materials, including shoot tips, somatic embryos, isolated root and cell cultures (Sakai and Engelmann, 2007; Popova *et al.*, 2011; Nosov *et al.*, 2014). Concentration of sucrose in the preculture medium can be selected based on the tolerance of plant material to osmotic stress. It is also important to consider compatibility with the selected cryoprotectant solution. For example, preculture with 0.7-1.0 M sucrose is preferable before applying highly concentrated plant vitrification solution 3 (PVS3) composed of, w/v, 50 per cent glycerol and 50 per cent sucrose (Nishizawa *et al.*, 1993), which is used to dehydrate large, compact and dehydration tolerant samples. By contrast, lower concentrations of sucrose (0.3-0.5 M) may be used to preculture small and dehydration-sensitive samples before applying PVS2 (30 per cent glycerol + 15 per cent dimethyl sulfoxide (DMSO) + 15 per cent ethylene glycol (EG) + 13.7 per cent sucrose, w/v) as it induces lower level of osmotic stress compared to PVS3 (Kim *et al.*, 2009a).

In most cases, dehydration tolerance is dependent on the inherent characteristics of the selected plant material. Hence, at the initial stage of protocol development, characterization of the material according to its dehydration tolerance is important. Based on our experience, the material can be considered tolerant if it tolerates preculture with 25 per cent sucrose for 17-24 hours. By contrast, very sensitive materials show significant decrease in regrowth after being exposed to 17.5 per cent sucrose solution for 17-24 hours. Materials with medium tolerance withstand 17-24 hours preculture with 17.5 per cent sucrose but show noticeable decrease in regrowth after treatment with 25 per cent sucrose for 17-24 hours. The preculture treatments that are used in the laboratories of the authors of this review for different materials depending on their dehydration tolerance are the following:

a. No-preculture
b. S-10 per cent for 30h, transfer to S-25 per cent O/N - for tolerant materials
c. S-10 per cent for 30h, transfer to S-17.5 per cent O/N, (optionally transfer to S-25 per cent 5h) - for moderate tolerant materials
d. S-10 per cent for 1-5 days, (optionally transfer to S-17.5 per cent for 5h) - for sensitive materials

Where S-10 per cent; S-17.5 per cent and S-25 per cent are preculture solutions composed of basal MS medium salts and vitamins (Murashige and Skoog, 1962) supplemented with 10, 17.5 and 25 per cent (w/v) sucrose, respectively (pH = 5.7). O/N – overnight (approx. 17 h).

4.2.2. Osmoprotection (loading)

The main purpose of the osmoprotection step is to assist in osmotic adaptation of the material before applying highly concentrated VSs. This step is of crucial importance for cryopreservation of osmotically sensitive materials such as shoot tips. In most reports on plant cryopreservation, the osmoprotection solution is composed of 2.0 M glycerol and 0.4 M sucrose (Nishizawa *et al.*, 1993). Alternatively, a set of osmoprotection treatments using binary mixtures of glycerol and sucrose has been adapted in our laboratories (Kim *et al.*, 2009b).

a. No osmoprotection
b. C4-35 per cent (G 17.5 per cent + S 17.5 per cent)/C6-40 per cent for 30~50 min
c. C7-32.1 per cent (G 2M + 0.4M S)for 30~50 min

Where C4-35 per cent is composed of, w/v, 17.5 per cent glycerol and 17.5 per cent sucrose; C6-40 per cent is composed of, w/v, 20.0 per cent glycerol and 20.0 per cent sucrose and C7-32.1 per cent is composed of, w/v, 18.4 per cent glycerol and 13.7 per cent sucrose. All solutions are prepared on basal MS medium (pH = 5.7).

A two-step osmoprotection solution composed of C4-35 per cent and C10-50 per cent (25 per cent glycerol + 25 per cent sucrose) can be tested when the material is extremely sensitive to osmotic stress. In this case the duration of cryoprotection step (see below) should be shortened since the samples are sensitive to cytotoxicity of VSs.

4.2.3. Cryoprotection (Dehydration)

Cryoprotection is the most critical step in the majority of vitrification-based cryopreservation protocols. At this stage, plant material is dehydrated by highly concentrated cryoprotectant solutions (also called vitrification solutions, VSs) before being exposed to LN. Sufficient cryoprotection is required to avoid ice crystallization in the living cells upon rapid immersion into LN. Other beneficial properties of VSs are based on their ability to inhibit ice nucleation in the tissues, maintain the stability of membranes and macromolecules during dehydration and reduce freezing temperature of the intercellular solutions (Samygin, 1994; Volk and Walters, 2006). Some cryoprotectants exhibit anti-oxidative properties (Benson and Bremner, 2004).

Both composition of the VS and duration of treatment should be optimized to ensure that samples are sufficiently dehydrated and cryoprotected, while minimizing the toxic effects of the solutions. Two VSs traditionally used in plant cryopreservation are PVS3 and PVS2 (see above). These solutions are different in their action. Application of PVS3 results in severe osmotic stress in plant tissues, while the use of PVS2 is restricted by high chemical toxicity of its constituents: permeating cryoprotectants ethylene glycol (EG) and DMSO. As a result, PVS3 is very effective for cryoprotection of large and osmotically tolerant materials such as micro-bulbs (*e.g.* lily), large shoot tips and rhizomes, while PVS2 is primarily used for cryopreservation of tiny and dehydration-sensitive explants (callus, small meristems) that are tolerant to chemical toxicity (*e.g.* some potato cultivars). In the search for balance between protective and toxic effects of VSs and to broaden their application to samples with moderate and high sensitivity, we developed a series of VSs based on glycerol and sucrose (B-type solutions) or glycerol and sucrose combined with EG and DMSO (A-type solutions). The information presented in Table 19.2 may be helpful while selecting the VS for a specific explant.

Table 19.2: Alternative Plant Vitrification Solutions (PVS) based on the Cytotoxicity of Materials to VSs and Size/Permeability of the Materials (Modified from Popova *et al.*, 2015)

Cytotoxicity		*Size and Permeability of Materials*			
Osmotic	*Chemical*	*Tiny (callus)*	*Small (meristem, 1mm)*	*Medium (shoot tips, ~2mm)*	*Large/semi-permeable (Bulb, rhizome)*
T*	T	B5-80%, A3-70~80% (RT)#	B3-90%, A3-70~80% (RT)	B1-100%, A3-90% (RT)	B1-100%
T	S	B5-80%	B3-90%	B1-100%	B1-100%
S	T	A3-70~80% (RT)	A3-80~90% (RT)	A3-90% (RT)	A3-90% (ice)
S	S	A3-70~80% (ice), B5-80%, B3-90%	A3-80~90% (ice), B3-90%	A3-90% (ice), B1-100%	??

* T: tolerant, S: sensitive, # RT: at room temperature, ice: on ice bath. BI-100 per cent (PVS3): 50 per cent glycerol + 50 per cent sucrose, B5-80 per cent : 40 per cent glycerol + 40 per cent sucrose, A3-90 per cent : 37.5 per cent glycerol + 15 per cent DMSO + 15 per cent EG + 22.5 per cent sucrose, A3-70 per cent : 29.2 per cent glycerol + 11.7 per cent DMSO + 11.7 per cent EG + 17.4 per cent sucrose.

Similar to their prototypes PVS3 and PVS2, solutions of A- and B-type induce different types of stress in plant tissues. Cryoprotection of samples with A3-90 per cent induced mainly chemical toxicity with additional osmotic stress due to the presence of DMSO and EG, while solutions of B-type primarily induce osmotic stress (Kim *et al.*, 2009; Popova *et al.*, 2015). Hence, a comparison of these two groups of

VSs will help reveal the characteristics of the materials in terms of their sensitivity to toxicity of the VSs. For cryopreserving sensitive materials, VSs need to be diluted. For example, for hairy roots, total concentration of cryoprotectants in the solution should not exceed 70-80 per cent (Kim *et al.*, 2012b).

When the samples are tolerant to osmotic stress, B1-100 per cent and respective dilutions are recommended, regardless of size. When samples are tolerant to chemical toxicity with small size, A3-90 per cent and respective dilutions are preferable. However, A3-90 per cent may not work well with medium and large samples, such as shoot tips, bulbs, and rhizomes. If the samples are sensitive to both osmotic stress and chemical toxicity, cryoprotection with A3-90 per cent and respective dilutions appliedin an ice bath is preferable. The same applies for tiny and fragile explants. However, both types of the VSs may be ineffective if the samples are medium or large in size, and are sensitive to both osmotic and chemical toxicity. In this case, alternative techniques, such as encapsulation-desiccation can be applied.

4.2.4. Cooling and Warming

In the vitrification method, fast cooling and rewarming is important to decrease the chance of ice crystallization and growth between 0 and -130°C (Samygin, 1994). The cooling and rewarming rate during cryopreservation is highly dependent on the container, the volume and composition of cryoprotectant solution (if used), size of the explants and many other factors. For example, cooling and rewarming rates has been determined to be, respectively, 3.2 and 1.8 °C/s for cryovials containing 0.5 ml PVS2 and 106 and 49 °C/s for aluminum foil strips with 5 of 2 ml drops of PVS2 (Teixeira *et al.*, 2014). With most species and material types, cryopreservation using aluminum foils strips resulted in higher regrowth than cryopreservation in cryovials (*e.g.* Kim *et al.*, 2012 a,b). A recently developed protocol using aluminum cryo-plates also ensured high rates of cooling and rewarming, thus resulting in high post-cryopreservation regrowth of chrysanthemum, potato and cherry (Yamamoto *et al.*, 2011, 2015; Vujovic *et al.*, 2015). Interestingly, for materials that cannot be sufficiently dehydrated before cryopreservation, due to their extreme sensitivity, high speed rewarming may be more important than rapid cooling (Kim *et al.*, 2009a).

4.2.5. Unloading

Unloading serves to remove toxic cryoprotectants and allows gradual rehydration of the rewarmed samples before placing them on regrowth medium. Traditionally, 0.8 to 1.2 M sucrose solutions are used (Sakai and Engelmann, 2007). In our laboratories, we employed unloading with 35 per cent sucrose solution (S-35 per cent) for 30-50 min or with 41 per cent sucrose solution (S-41 per cent) for shorter periods. The solution can be refreshed once or twice during unloading process.

4.2.6. Regrowth

During cryopreservation plant tissues experience tremendous stress caused by dehydration, mechanical damage, toxic effect of chemicals, *etc.* As a result, plant materials can be severely damaged even when the optimized cryopreservation protocol is used. To help the recovery systems overcome the challenge, after cryopreservation, the explants should be treated similar to premature babies and provided with the most possible comfort and supportive conditions, essential nutrients and chemicals to help mitigate the consequences of the stress. In general, dark condition for a minimum of three days after cryopreservation followed by 2-5 days under dim light is recommended to avoid accelerating of oxidative stress processes (Sakai and Engelmann, 2007). Recent studies have also suggested that the use of antioxidants, such as reduced glutathione, glycine betain, ascorbic acid, indole amines, *etc.* may help plant tissues to alleviate stress and increase regrowth (Uchendu *et al.*, 2010; Ren *et al.*, 2014). Manipulation of ammonium nitrate concentration in the regrowth medium may improve post-cryopreservation survival in some species (Decruse *et al.*, 2004). Plant growth regulators are also important, for example, supplementing GA_3 alone or in combination with cytokinins (Lee *et al.*, 2011; Wang *et al.*, 2014) was found to be beneficial to successful regrowth of cryopreserved chrysanthemum shoot tips, while increasing BA concentration promoted regrowth of orchid materials (Popova *et al.*, 2010).

4.2.7. Biochemical and Thermal Analyses

Development and optimization of cryopreservation protocols can be facilitated by using a range of biochemical and thermal analyses in parallel with regrowth assessment. The composition of the VS and duration of cryoprotectant treatment can be optimized based on the information of in-flow and out-flow of VS constitutes in samples during the osmoprotection and cryoprotection, and unloading steps respectively. For example, concentration of glycerol in garlic shoot tips reached its maximum of ca. 180 mg g-1 FW after 150 min of treatment with PVS3 (Kim *et al.*, 2004). By contrast, DMSO content was maximized after 40 min of PVS2 treatment (Kim, 2004). During unloading, concentration of DMSO in shoot tips decreased two-fold within 5 min while concentration of glycerol decreased gradually over 30 min (Kim *et al.*, 2004; Kim, 2004). Thermal analysis can be used to determine cooling and rewarming rates in the developed protocol and can evaluate the effectiveness of the cryoprotectant solution to decrease the content of freezable water and reduce crystallization temperature in plant tissues (Volk and Walters, 2006; Bruòáková *et al.*, 2011; Teixeira *et al.*, 2014).

4.3. The Proposed Steps in a Droplet-Vitrification Protocol using Modified VSs

The following procedures can be followed, with further optimization, for cryopreserving plant materials depending on their sensitivity to osmotic stress and chemical toxicity caused by cryoprotectant solutions.

4.3.1. Plant Materials with High Tolerance to Osmotic and Chemical Stress (*e.g.* garlic shoot tips)

Cold acclimation: for temperate crops, duration and temperature depending on species

Preculture: 0.3-0.5 M sucrose for 1 day

Osmoprotection: no

Cryoprotection: PVS3 at room temperature for 60-180 min

Cooling - Rewarming*

Unloading: 0.8-1.2 M sucrose, 50 min

* Cooling - Rewarming: After treatment with VS, explants are placed in 5μM drops of the same VS on aluminum foil strips (2 cm × 0.8 cm) and plunged directly to LN ('cooling'). Foils can be rewarmed in unloading solution preheated to 40°C for 20 s, followed by unloading at room temperature ('rewarming').

4.3.2. Plant Materials with Moderate Tolerance to Osmotic and Chemical Stress (*e.g.* embryogenic callus cultures, somatic embryos)

Preculture: 0.3 M sucrose for 1 day or no preculture

Cryoprotection: C4-35 per cent or classic loading solution (2.0 M glycerol and 0.4 M sucrose) for 20 min at room temperature

Cryoprotection: PVS3 at room temperature for ca. 40 min, or A3-80 per cent at 0°C for 40 min

Cooling - Rewarming*

Unloading: 0.8 M sucrose, 30 min

4.3.3. Plant Materials with Moderate Tolerance to Osmotic Stress and High Sensitivity to Chemical Stress (*e.g.* chrysanthemum shoot tips)

Preculture: 0.3 M sucrose for 31 h, then 0.5 M sucrose for 17 h, then 0.7 M sucrose for 5 h

Cryoprotection: C4-35 per cent for 40 min at room temperature

Cryoprotection: PVS3 at room temperature for 40-60 min, or A3 at 0°C for 50-60 min

Cooling - Rewarming*

Unloading: 0.8 M sucrose, 40 min

4.3.4. Plant Materials with High Sensitivity to both Osmotic and Chemical Stress (*e.g.* hairy root cultures)

Preculture: step-wisely 0.3 M sucrose for 24-48 h, then 0.5 M sucrose for 5 h

Cryoprotection: C4-35 per cent for 30 min at room temperature

Cryoprotection: B5-80 per cent at room temperature for 10-20 min, or A3-70 per cent at 0°C for 20 min

Cooling - Rewarming*

Unloading: 0.8 M sucrose, 30 min

5. Conclusion

Successful cryopreservation has been reported for over 200 plant species including many economically important crops and fruit trees. However, the researchers and curators of the collections may face multiple problems when initiating cryogenic storage *de novo*. One of such problems is selection of the proper cryopreservation protocol. The choice of the most effective protocol to ensure high regrowth rate of cryopreserved materials is dependent on many factors. The most critical limitations for successful cryopreservation using vitrification and droplet-vitrification methods are the dehydration sensitivity of selected plant material and toxic effects of traditionally used cryoprotectant solutions PVS2 and PVS3. These limitations may be overcome through optimization of the combination of various steps in the cryopreservation process. At every step, treatments can be selected based on the initial characteristics of plant material such as size, permeability to cryoprotectants, and tolerance/sensitivity to osmotic and chemical stress. By applying specific preculture and osmoprotection treatments, and a range of modified VSs, the time required for protocol development may be reduced, and the existing protocols can be easily adapted to new cultivars with enhanced regrowth rate after cryopreservation.

References

Anthony, P., Lowe, K.C., Power, J.B. and Davey, M.R. (1997). Synergistic enhancement of the post-thaw growth of cryopreserved rice cells by oxygenated perfluorocarbon and pluronic F-68. *Cryobiology*, **35**: 201-208.

Arora, R. and Palta, J.P. (1991). A loss in the plasma membrane ATPase activity and its recovery coincides with incipient freeze-thaw injury and post-thaw recovery in onion bulb scale tissue. *Plant Physiology*, **95**: 846-852.

Bartolo, M.E. and Carter, J.V. (1991). Effect of microtubule stabilization on the freezing tolerance of mesophyll cells of spinach. *Plant Physiology*, **97**: 182-187.

Benelli, C., de Carlo, A. and Engelmann, F. (2013). Recent advances in the cryopreservation of shoot-derived germplasm of economically important fruit trees of *Actinidia, Diospyros, Malus, Olea, Prunus, Pyrus* and *Vitis*. *Biotechnology Advances*; **31**: 175-185.

Benson, E.E. and Bremner, D. (2004). Oxidative stress in the frozen plant: a free radical point of view. In: Fuller, B.J., Lane, N. and Benson, E.E. (Eds.), *Life in the Frozen State*. CRC Press, USA. pp. 205-241

Benson, E.E. (2008). Cryopreservation theory. In: Reed, B.M. (Ed.), *Plant Cryopreservation: A Practical Guide*. Springer, Berlin, Germany. pp. 15-32

Brison, M., de Boucaud, M.T. and Dosba, F. (1995). Cryopreservation of *in vitro* grown shoot tips of two interspecific *Prunus* rootstocks. *Plant Science*, **105**: 235-242.

Bruòáková, K., Zámeèník, J., Urbanová, M. and Èellárová, E. (2011). Dehydration status of ABA-treated and cold-acclimated *Hypericum perforatum* L. shoot tips subjected to cryopreservation. *Thermochimica Acta*, **525**: 62–70.

Butenko, R.G., Popov, A.S., Volkova, L.A., Chernyak, N.D. and Nosov, A.M. (1984). Recovery of cell cultures and their biosynthetic capacity after storage of *Dioscorea deltoidea* and *Panax ginseng* cells in liquid nitrogen. *Plant Science Letters*, **33**: 285-292.

Chang, Y. and Reed, B.M. (1999). Extended cold acclimation and recovery medium alteration improve regrowth of *Rubus* shoot tips following cryopreservation. *CryoLetters*, **20**: 371-376.

Chen, T.H.H., Kartha, K.K. and Gusta, L.V. (1985). Cryopreservation of wheat suspension culture and regenerable callus. *Plant Cell, Tissue and Organ Culture*, **4**: 101-109.

Decruse, S.W., Seeni, S. and Nair, G.M. (2004). Preparative procedures and culture medium affect the success of cryostorage of *Holostemma annulare* shoot tips. *Plant Cell, Tissue and Organ Culture*, **76**: 179-182.

Engelmann, F. (1992). Effects of freezing in liquid nitrogen on the properties of a soybean (*Glycine max* L. var. acme) callus strain used as a bioassay for cytokinin activity. *CryoLetters*, **13**: 331-336.

Engelmann, F. (1997). In vitro conservation methods. In: Ford-Lloyd, B.V., Newburry, H. J. and Callow, J.A. (Eds.), *Biotechnology and Plant Genetic Resources Conservation and Use*. CABI, Wallingford, UK. pp. 119-162

Engelmann, F., Gonzalez-Arnao, M.T., Wu, Y. and Escobar, R. (2008). Development of encapsulation dehydration. In: Reed, B.M. (Ed.), *Plant Cryopreservation: A Practical Guide*. Springer, Berlin, Germany. pp. 59-75

Engelmann, F., Lartaud, M., Chabrillange, N., Carron, M.P. and Etienne, H. (1997). Cryopreservation of embryogenic calluses of two commercial clones of *Hevea brasiliensis*. *CryoLetters*, **18**: 107-116.

Escobar, R.H., Mafla, G. and Roca, W.M. (1997). A methodology for recovering cassava plants from shoot tips maintained in liquid nitrogen. *Plant Cell Reports*, **16**: 474-478.

Fabre, J. and Dereuddre, J. (1990). Encapsulation-dehydration: A new approach to cryopreservation of *Solanum* shoot tips. *CryoLetters*, **11**: 413-426.

Fahy, G.M., Lilley, T.H., Linsdell, H., Douglas, M.S. and Meryman, H.T. (1990). Cryoprotectant toxicity and cryoprotectant toxicity reduction: In search of molecular mechanisms. *Cryobiology*, **27**: 247-268.

Fahy, G.M., Wowk, B., Wu, J. and Paynter, S. (2004). Improved vitrification solutions based on the predictability of vitrification solution toxicity. *Cryobiology*, **48**: 22-35.

Fki, L., Bouaziz, N., Chkir, O., Benjemaa-Masmoudi, R., Rival, A., Swennen, R. (2013). Cold hardening and sucrose treatment improve cryopreservation of date palm meristems. *Biologia Plantarum*, **57**: 375-379.

Folgado, R., Panis, B., Sergeant, K., Renaut, J., Swennen, R. and Hausman, J.F. (2015). Unravelling the effect of sucrose and cold pretreatment on cryopreservation of potato through sugar analysis and proteomics. *Cryobiology*, **71**(3):432-41.

Fukui, K., Shirata, K., Niino, T. and Kashif, I.M. (2011). Cryopreservation of mulberry winter buds in Japan. *Acta Horticulturae*, **908**: 483-488.

Gonzalez-Arnao, M.T., Martinez-Montero, M.E., Cruz-Cruz, C.A. and Engelmann, F. (2014). Advances in Cryogenic Techniques for the Long-Term Preservation of Plant Biodiversity. In: Ahuja, M.R. and Ramawat, K.G. (Eds.), *Biotechnology and Biodiversity* (*Sustainable Development and Biodiversity, Vol. 4*). Springer, Netherlands. pp. 129-170

Gonzalez-Arnao, M.T., Panta, A., Roca, W.M., Escobar, R.H. and Engelmann, F. (2008). Development and large scale application of cryopreservation techniques for shoot and somatic embryo cultures of tropical crops. *Plant Cell, Tissue and Organ Culture*, **92**: 1-13.

Kaczmarczyk, A., Rokka, V.M. and Keller, E.R.J. (2011). Potato shoot tip cryopreservation. A review. *Potato Research*, **54**: 45–79.

Kim, H.H. (2004). Analysis of factors affecting the cryopreservation of garlic shoot tips. Unpublished doctoral dissertation, Seoul National University, Seoul, Republic of Korea.

Kim, H.H., Lee, Y.G., Shin, D.J., Ko, H.C., Gwag, J.G., Cho, E.G. and Engelmann, F. (2009a). Development of alternative plant vitrification solutions in droplet-vitrification procedures. *CryoLetters*, **30**: 320-334.

Kim, H.H., Lee, Y.G., Park, S.U., Lee, S.C., Baek, H.J., Cho, E.G. and Engelmann, E. (2009b). Development of alternative loading solutions in droplet-vitrification procedures. *CryoLetters*, **30**: 291-299.

Kim, H.H., Popova, E., Shin, D.J., Yi, J.Y., Kim, C.H. and Lee, J.S. (2012a). Cryobanking of Korean *Allium* germplasm collections: results from a 10 year experience. *CryoLetters*, **33**: 45-57.

Kim, H.H., Popova, E.V., Shin, D.J., Bae, C.H., Baek, H.J. and Park S.U. (2012b). Development of a droplet-vitrification protocol for cryopreservation of *Rubia akane* (Nakai) hairy roots using a systematic approach. *CryoLetters*, **33**: 506-517.

Kim, J.B., Kim, H.H., Cho, E.G. Kim, H.I., Kim, Y.H. and Engelmann, F. (2004). Evolution of sucrose and glycerol concentration in garlic shoot tips during a vitrification procedure. *CryoLetters*, **25**: 91-100.

Kuwayama, M. (2005). Highly effective vitrification method for cryoptreservation of human oocytes. RBM Online **11**: 300-308.

Lambardi, M., Ozudogru, E.A. and Benelli C. (2008). Cryopreservation of embryogenic cultures. In: Reed, B.M. (Ed.), *Plant Cryopreservation: A Practical Guide*. Springer, Berlin, Germany. pp. 177-210

Lee, J.H., Kim, G.J.,Kim, S., Lee, W.W., Lee, H.J., Lee, D.R. and Chung, H.M. (2006). Comparison of different vehicles on human embryonic stem cells using vitrification. *Reproductive Developmental Biology* **30**: 278-285.

Lee, Y.G., Popova, E., Cui, H.Y., Kim, H.H., Park, S.U., Bae, C.H. (2011). Improved cryopreservation of chrysanthemum (*Chrysanthemum morifolium*) using droplet-vitrification. *CryoLetters*, **32**: 487-497.

Li, B.Q., Feng, C.H., Wang, M.R., Hu, L.Y., Volk, G. and Wang Q.C. (2015). Recovery patterns, histological observations and genetic integrity in *Malus* shoot tips cryopreserved using droplet-vitrification and encapsulation-dehydration procedures. *Journal of Biotechnology*, **214**: 182-191.

Lovelock, J.E. (1953). The mechanism of the cryoprotective effect of glycerol against haemolysis by freezing and thawing. *Biochimica et Biophysica Acta*, **11**: 28-36.

Martínez, D., Tamés, R.S. and Revilla, M.A. (1999). Cryopreservation of in vitro-grown shoot-tips of hop (*Humulus lupulus* L.) using encapsulation/dehydration. *Plant Cell Reports*, **19**: 59-63.

Mazur, P. (2004). Principles of cryobiology. In: Fuller, B.J., Lane, N. and Benson, E.E. (Eds.), *Life in the Frozen State*. CRC Press, USA. pp. 3-65

Mazur, P., Leibo, S.P. and Chu, E.H.Y. (1972). A two-factor hypothesis of freezing injury. Evidence from Chinese hamster tissue culture cells. *Experimental Cell Research*, **71**: 345-355.

Medina, J.J., Clavero-Ramírez, I., González-Benito, M.E., Gálvez-Farfán, J., López-Aranda, J.M. and Soria, C. (2007). Field performance characterization of strawberry (*Fragaria* × *ananassa* Duch.) plants derived from cryopreserved apices. *Scientia Horticulturae*, **113**: 28-32.

Murashige, T. and Skoog, F. (1962). A revised medium for rapid growth and bio assays with tobacco tissue cultures. *Physiologia Plantarum*, **15**: 473-497.

Nakashima,A., Ino, N., Kusumi, M., Ohgi, S., Ito, M., Horikawa, T., Nakagawa, K., Saito, T., Kamura, T. and Saito, H. (2010). Optimization of a novel nylon mesh container for human embryo ultrarapid vitrification. *Fertility and Sterility*, **93**: 2405-2410.

Nishizawa, S., Sakai, A., Amano, A.Y. and Matsuzawa, T. (1993). Cryopreservation of asparagus (*Asparagus officinalis* L.) embryogenic suspension cells and subsequent plant regeneration by vitrification. *Plant Science*, **91**: 67-73.

Nosov, A.M., Popova, E.V. and Kochkin, D.V. (2014). Isoprenoid production *via* plant cell cultures: biosynthesis, accumulation and scaling-up to bioreactors. In: Paek, K.Y., Murthy, H.N. and Zhjong, J.J. (Eds.), *Production of Biomass And Bioactive Compounds Using Bioreactor Technology*. Springer, Dordrecht, Netherlands. pp. 563-623

Oka, S., Yakuwa, H., Sate, K. and Niino, T. (1991). Survival and shoot formation *in vitro* of pear winter buds cryopreserved in liquid nitrogen. *Horticultural Science*, **26**: 65-66.

Panis, B. and Lambardi, M. (2006). Status of cryopreservation technologies in plants (crops and forest trees).In: Ruane, J. and Sonnino, A. (Eds.), *The Role of Biotechnology in Exploring And Protecting Agricultural Genetic Resources. United Nations Food and Agriculture Organization (FAO)*, Rome, Italy. pp. 61-78

Panis, B., Piette, B. and Swennen, R. (2005). Droplet vitrification of apical meristems: A cryopreservation protocol applicable to all *Musaceae*. Plant Science, **168**: 45–55.

Panta, A., Panis, B., Ynouye, C., Swennen, R., Roca, W. and Tay, D. (2015). Improved cryopreservation method for the long-term conservation of the world potato germplasm collection. *Plant Cell, Tissue and Organ Culture*, **120**: 117–125.

Pence, V.C. (2011). Evaluating costs for *in vitro* propagation and preservation of endangered plants. *In Vitro Cellular and Developmental Biology - Plant*, **47**: 176–187.

Pérez, R.M., Navarro, L. and Duran-Vila, N. (1997). Cryopreservation and storage of embryogenic callus cultures of several *Citrus* species and cultivars. *Plant Cell Reports*, **17**: 44-49.

Popova, E., Bukhov, N., Popov, A. and Kim, H.H. (2010). Cryopreservation of protocorm-like bodies of the hybrid orchid *Bratonia* (*Miltonia flavescens* × *Brassia longissima*). *CryoLetters*, **31**: 426-37.

Popova, E., Shukla, M., Kim, H.H. and Saxena, P.K. (2015). Plant cryopreservation for biotechnology and breeding. In: Al-Khayri, J.M. (Ed.), *Advances in Plant Breeding Strategies: Breeding, Biotechnology and Molecular Tools*. Springer, Switzerland. pp. 63-93.

Popova, E.V., Paek, K.Y. and Kim, H.H. (2011). Cryopreservation of medicinal plants: the case of *in vitro* cultures. In: Kumar, A. and Roy, S. (Eds.), *Plant Tissue Culture and Applied Plant Biotechnology*. Pointer publishers, Jaipur, India. pp. 153-196.

Pritchard, H.W. (2007). Cryopreservation of desiccation-tolerant seeds. In: Day, J.G. and Stacey, G.N. (Eds.), *Methods in Molecular Biology, V. 368* (*Cryopreservation and Freeze-Drying Protocols. 2nd Ed.*). Humana Press Inc., Totowa, New Jersey. pp. 185-201.

Reed, B.M. (2008). Cryopreservation of temperate berry crops. In: Reed, B.M. (Ed.), *Plant Cryopreservation: A Practical Guide*. Springer, Berlin, Germany. pp. 333-364.

Ren, L., Zhang, D., Shen, X.H. and Reed, B.M. (2014). Antioxidants and anti-stress compounds improve the survival of cryopreserved *Arabidopsis* seedlings. *Acta Horticulturae*, **1039**: 57-60.

Sakai, A. and Engelmann, F. (2007). Vitrification, encapsulation-vitrification and droplet-vitrification: A review. *CryoLetters*, **28**: 151-172.

Sakai, A., Kobayashi, S. and Oiyama, I. (1990). Cryopreservation of nucellar cells of navel orange (*Citrus sinensis* Osb. var. *brasiliensis* Tanaka) by vitrification. *Plant Cell Reports*, **9**: 30-3.

Samygin, G.A. (1994).) The process responsible for plant cell damage by extracellular ice. *Russian Journal of Plant Physiology*, **41**: 539-549.

Shin, D.J., Kong, H.J., Popova, E.V., Moon, H.K., Park, S.Y. and Park, S.U. (2012). Cryopreservation of *Kalopanax septemlobus* embryogenic callus using vitrification and droplet-vitrification. *CryoLetters*, **33**: 402-410.

Steponkus, P.L. (1984). Role of the plasma membrane in freezing injury and cold acclimation. *Annual Review of Plant Physiology*, **35**: 543-584.

Teixeira, A.S., Gonzalez-Benito, M.E. and Molina-Garcia, A.D. (2014). Measurement of cooling and warming rates in vitrification-based plant cryopreservation protocols. *Biotechnology Progress*, **30**: 1177-1184.

Thomson, L.K., Fleming, S.D., Aitken, R.J., De Iuliis, G.N., Zieschang, J-A and Clark, A.M. (2009). Cryopreservation-induced human sperm DNA damage is predominantly mediated by oxidative stress rather than apoptosis. *Human reproduction*, **24**: 2016-2070.

Uchendu, E.E., Muminova, M., Gupta, S. and Reed, B.M. (2010). Antioxidant and anti-stress compounds improve regrowth of cryopreserved *Rubus* shoot tips. *In Vitro Cellular and Developmental Biology- Plant*, **46**: 386-393.

Uemura, M. and Steponkus, P.L. (1999). Cold acclimation in plants: relationship between the lipid composition and the cryostability of the plasma membrane. *Journal of Plant Research*, **112**: 245-254.

Volk, G.M. and Walters, C. (2006). Plant vitrification solution 2 lowers water content and alters freezing behavior in shoot tips during cryoprotection. *Cryobiology*, **52**: 48-61.

Volk, G.M., Jenderek, M. and Chao, C.T. (2016). Prioritization of *Malus* accessions for collection cryopreservation at the USDA-ARS National Center for Genetic Resources Preservation. *Acta Horticulturae*, In press.

Volkova, L.A., Urmantseva, V.V., Popova, E.V. and Nosov, A.M. (2015). Physiological, cytological and biochemical stability of *Medicago sativa* L. cell culture after 27 years of cryogenic storage. *CryoLetters*, **36**: 252-263.

Vujovic, T., Chatelet, P., Ružiæ, Đ. and Engelmann, F. (2015). Cryopreservation of *Prunus spp.* using aluminium cryo-plates. *Scientia Horticulturae*, **195**: 173-182.

Wang, B., Wang, R.R., Cui, Z.H., Bi, W.L., Li, J.W., Li, B.Q. (2014). Potential applications of cryogenic technologies to plant genetic improvement and pathogen eradication. *Biotechnology Advances*, **32**: 583-595.

Wang, Q., Tanne, E., Arav, A. and Gafny, R. (2000). Cryopreservation of *in vitro*-grown shoot tips of grapevine by encapsulation-dehydration. *Plant Cell Tissue and Organ Culture*, **63**: 41-46.

Wang, R.R., Gao, X.X., Chen, L., Huo, L.Q., Li, M.F. and Wang, Q.C. (2014). Shoot recovery and genetic integrity of *Chrysanthemum morifolium* shoot tips following cryopreservation by droplet-vitrification. *Scientia Horticulturae*, **176**: 330-339.

Wang, Y., Xiao, Z., Li, L., Fan, W. and Li, S.W. (2008). Novel needle immersed vitrification: Apractical and convenient method with potential advantages in mouse and human ovarian tissue cryopreservation. *Human Reproduction*, **23**: 2256-2265.

Xu, F., Moon, S., Zhang, X., Shao, L., Song, Y.S. and Demirci, U. (2010). Multi-scale heat and mass transfer modelling of cell and tissue cryopreservation. *Mathematical Physical and Engineering Sciences*, **368**: 561-583.

Yamamoto, S., Rafique, T., Priyantha, W.S., Fukui, K., Matsumoto, T. and Niino, T. (2011). Development of a cryopreservation procedure using aluminium cryo-plates. *CryoLetters*, **32**: 256-265.

Yamamoto, S., Wunna, T., Rafique, T., Arizaga, M.V., Fukui, K., Gutierrez, E.J.C. (2015). The aluminum cryo-plate increases efficiency of cryopreservation protocols for potato shoot tips. *American Journal of Potato Research*, **92**: 250-257.

Chapter 20

Coconut

☆ *Anitha Karun, K.K. Sajini, V. Aparna and M.K. Rajesh*

1. Introduction

The coconut palm, *Cocos nucifera* L. (2n=32), is a woody perennial monocotyledonous tree that belongs to the family Arecaceae that includes 27 genera and 600 species. Gunn *et al.* (2011) proposed the South East Asian Island and Southern margins of Indian subcontinent as the ancestral homes of coconut. Coconut is distributed mainly in coastal regions between 20° N and 20° S from sea level to 1000 m above sea level in more than 86 countries which can be grouped into eight distinct coastal/oceanic regions on four continents (Harris, 2001). The ideal growing temperature for coconut ranges between 24 and 30°C (Woodroof, 1979). It is grown on more than 12.19 million ha, 94 per cent of which are in Asia and the South Pacific (APCC, 2014).

The genus *Cocos* is monotypic and has no known truly wild forms. The varieties in coconut fall into two main groups: the tall *typica*, which is commonly cultivated for commercial production and the dwarf *nana*, grown for ornamental and breeding purpose (Narayana and John, 1949). The palm grows from a single terminal vegetative bud with a juvenile period of 3-4 years for dwarfs and 8-10 years for talls. The palm is monoecious with large number of male flowers and few female flowers. The seed is a large one seeded drupe. The economically productive period for talls is 80-90 years and that of dwarfs are 30-40 years (Menon and Pandalai, 1958).

While considering the conservation objectives for coconut genetic diversity, *ex situ* conservation in field gene bank becomes the choice for conserving and utilizing maximum genetic diversity in coconut. The field gene banks provide easy and ready access to conserved material for research as well as for use. But its major drawbacks are threat from pests, diseases, natural calamities, availability of land resources, urbanization that results in loss of valuable collections and high maintenance costs (Withers and Engels, 1990). Moreover, diseases such as root wilt and incidence of insects and pest attacks necessitates the conservation of target genes present in the existing varieties that becomes inevitable for future use in breeding programmes. This envisages the development of a complementary conservation strategy integrating *in situ* and *ex situ* approaches (Maxted *et al.*, 1997) and also utilization of alternate techniques of conservation such as *in vitro* conservation.

2. Coconut Genetic Resources

Plant genetic resources are the most valuable and essential basic raw material of any crop improvement programme and its conservation is highly essential to safeguard against from genetic erosion, genetic vulnerability and genetic wipeout for a sustainable agricultural production. Being a monotypic genus, the genetic diversity in coconut is mainly found in different ecotypes/landraces. The genetic diversity in coconut is gradually getting reduced due to various reasons. The high production cost, market fluctuations in price, labour requirement, incidence of pests and diseases compel coconut farmers to shift the land use for more profitable crops. Consequently, some of the traditional, locally adapted populations are gradually getting diminished or lost. Environmental disasters as well as developmental activities also pose risk and cause the loss of diversity in coconut. Hence there is urgent need to collect and conserve coconut genetic resources for its efficient utilization as unexplored genes or alleles may have better use in future breeding programmes. The large stature and recalcitrant (Chin and Roberts, 1980) or homoiohydrous behaviour of coconut seeds makes it impossible for seed storage by conventional methods. At present, *ex situ* conservation of coconut genetic diversity is through field gene banks whereas *in situ* conservation is mainly on farm or in home gardens or on remote islands and atolls (Sthapit *et al.*, 2005). The ICAR-Central Plantation Crops Research Institute (CPCRI) located at Kasaragod in India maintains the world's largest collection of coconut germplasm constituting a total of 455 accessions (ICAR-CPCRI, 2015-16). The National and multi-site International Gene Banks for the conservation of coconut genetic diversity have been developed by the coconut growing countries world over under ADB-funded project of COGENT (Coconut Genetic Resources Network) of the Bioversity International (Batugal and Jayashree, 2005). The highest priority is to duplicate field collections *in vitro* as embryos and pollen (Ramanatha Rao and Batugal, 1998) and to facilitate international exchange of germplasm. In India the Central Plantation Crops Research Institute, Kasaragod hosts the International Coconut Genebank for South Asia (ICG-SA) that is located at Kidu, Karnataka. In ICG-SA India has conserved 91 accessions of indigenous and exotic coconut germplasm. The other multi-site gene banks are located in Indonesia (for Southeast and East Asia), Papua New Guinea (for the South Pacific), Cote d' Ivorie (for Africa and the Indian Ocean) and Brazil (for Latin America Caribbean).

3. Conservation Strategies

The concept of germplasm conservation demands that collection methods initially capture maximum variation and subsequently, conservation and regeneration techniques minimize losses through time (Astley, 1992). The two approaches for conservation of plant genetic resources are *in situ* and *ex situ*. In the case of *in situ* conservation, the genetic resources are maintained in the natural habitat in which the diversity has evolved *e.g.* natural reserves and on-farm conservation. The advantage of in *situ* conservation is that it facilitates the ongoing processes of natural evolution (Krogstrup *et al.*, 1992). It is probably the best way to preserve biodiversity. Nevertheless, it is expensive in terms of the area required and in the cost of establishment and maintenance. Conserving plant genetic resources outside their original habitat is termed as *ex situ* conservation *e.g.* botanic gardens, seed banks, field gene banks, *in vitro* collections, pollen and DNA storage. It is generally used to protect populations in danger of destruction, replacement or deterioration. For any given gene pool, a complementary conservation strategy involving *in situ* and *ex situ* approaches and methods are necessary for efficient and cost effective conservation (Maxted *et al.*, 1997).

4. Conservation of Coconut Genetic Resources

Coconut is a monotypic genus without any known wild forms. Consequently, the conservation of coconut diversity is within a single species that includes many ecotypes. As seeds are large in size and highly recalcitrant (Chin and Roberts, 1980), *ex situ* conservation of seeds under conventional dry and low temperature conditions is not feasible in coconut. In coconut, the *in situ* conservation method presently being utilized is on farm conservation in the farmer's fields as components of the traditional agricultural systems and in remote islands and atolls. *Ex situ* conservation is mainly in field gene banks. Field gene banks provide easy access to conserved material for use. But they confer the risk of destruction by natural calamities, pests, diseases and are costly for maintenance. Hence there is urgent need to develop safety duplicates of the living collections using alternate strategies of conservation such as *in vitro* conservation.

In vitro culture techniques provide some important tools for improved conservation and management of plant genetic resources (Ramanatha and Riley, 1994). In view of high maintenance cost of field gene banks and loss of important genetic material due to diseases and pest attacks, conservation of coconut genetic diversity can be complemented utilizing other systems like *in vitro* conservation. The advantages of *in vitro* conservation are maintenance of material in pathogen-free condition that facilitates safer distribution, conservation of vegetatively propagated plants, somaclonal and gametoclonal variants, rare germplasm arising through somatic hybridization and genetic engineering, storage of pollen enhancing longevity, meristem culture, conservation of plants from endangered species, storage of cell cultures for industrial applications, and those that produces recalcitrant seeds. Further, the cultures are not subjected to environmental disturbances (Withers and Engelmann, 1997). *In vitro* options provide short to medium (slow growth) and long-term (cryopreservation) conservation of coconut genetic resources.

For short and medium term storage, the objective is to increase the intervals between subcultures by reducing growth. This is achieved by modifying the environmental conditions and/or the culture medium. The advantages of this method are saving space and other resources, cost effective exchange and distribution of material at any times across the world and rescue of germplasm in the event of outbreak of diseases. In coconut, it is especially beneficial for germplasm collection, exchange and surveillance of diseases before the introduction of collected material (embryo or pollen) for further evaluation (Assy-Bah and Engelmann, 1993). The various methods for limiting *in vitro* growth are addition of osmotic agent like sucrose (Sarkar and Naik, 1998), use of growth retardants (Tahtamouni *et al.*, 2001), storage at low temperature (Bertrand-Desbrunais *et al.*, 1992), and encapsulation in alginate beads (Karun *et al.*, 1996). Short to medium term *in vitro* conservation of coconut has been mainly utilized for collection and international exchange of germplasm. Karun and Sajini (1994).) reported short-term storage of coconut zygotic embryos in sterile water without losing its viability for two months. This technique has been utilized in one indigenous and five international expeditions conducted by ICAR-CPCRI for the collection of coconut genetic diversity (Karun *et al.*, 2002). Mature zygotic embryos encapsulated in alginate beads and stored in sterile water at room temperature for two months showed normal germination upon transfer to the retrieval medium. But none germinated after storage in a refrigerator at 10° C for the same period (Karun *et al.*, 1996). Assy-Bah and Engelmann (1993) reported medium term storage (6 to 12 months) of coconut zygotic embryos. They identified the balance of sucrose and activated charcoal for the slow growth action.

Cryopreservation is the *ex situ* storage of biological materials at ultra low temperature (-196 °C) using liquid nitrogen (Withers and Engelmann, 1997) at which all metabolic process within them are completely arrested allowing conservation for a theoretically unlimited period of time (Engelmann, 2004). Even though the initial inputs for cryogene banking are high, its maintenance requires minimal resources. Therefore, it is cost effective (Reed *et al.*, 2004) and requires limited space (De-Zu and Prichard, 2009). Additional advantages are maintenance of cultures in disease-free state without genetic alteration or modification.

It has emerged as an important biotechnological tool for storing wide range of plant germplasm especially those which are difficult to store due to large size, high moisture content, high desiccation/freezing sensitivity (recalcitrant) and those propagated vegetatively and endangered plant species (Nikishina *et al.*, 2007). This technique ensures safe and efficient long term conservation of different types of seeds (Chaudhury and Chandel, 1994), zygotic embryos (Engelmann, 2000), somatic embryos (Gonzalez-Arnao *et al.*, 2003), embryonic axes (Laura *et al.*, 1996), shoots tips (Escobar and Roca, 1997) and pollen (Zhang *et al.*, 2009). Cryopreservation is also employed in *in vitro* propagation systems based on somatic embryogenesis, to avoid somaclonal variation and loss of embryogenic potential that can occur during long-term maintenance of actively growing embryogenic cultures (Silvia *et al.*, 2004). Cryopreservation is also be utilized for virus eradication (cryotherapy) (Brison *et al.*, 1997; Helliot *et al.*, 2002).

The standard techniques of cryopreservation are the classical controlled rate cooling and new vitrification based ones. Classical and new cryopreservation approaches are different based on the techniques employed and the physical mechanisms upon which they are attained (Withers and Engelman, 1997). The classical cryopreservation techniques involve slow cooling in a programmable freezer to a defined pre-freezing temperature (- 40° C) followed by plunging in liquid nitrogen. The material, after treatment with cryoprotectant mixture to a certain period of time, was slowly cooled (from 0.1 to 0.5 °C/min) so that ice is formed in the extracellular spaces. Depending upon the rate of cooling and pre-freezing temperature, most or all intracellular water is removed that will avoid the detrimental intracellular ice formation upon subsequent immersion of specimen in liquid nitrogen. As freezable water is reduced the cytoplasm becomes concentrated and vitrifies upon exposure to liquid nitrogen. In order to avoid recrystallization in which ice melts and reforms larger and more damaging crystal size, rapid rewarming was performed (Mazur, 1984). The classical approach to cryopreservation was most successful in preserving protoplast culture, cell suspensions and callus cultures. It was not much valuable in shoot tips and mature zygotic and somatic embryo cultures that comprise mixture of cell sizes and types (Withers and Engelmann, 1997; Fukai, 1995).

The new cryopreservation techniques are based on vitrification. In vitrification based procedure, cell dehydration is performed prior to freezing by exposure of samples to concentrated cryoprotective media and/or air desiccation followed by rapid cooling. Vitrification is defined as the transition of the liquid phase to an amorphous glassy solid at the glass transition (T_g) temperature (Fahy *et al.*, 1984). The biological material stored in this stable condition may be maintained for a long time without alteration or modification (Burke, 1986). Vitrification-based procedures offer practical advantages in comparison with classical freezing techniques (Steponkus *et al.*, 1992). This procedure is simple than classical ones, and requires minor modifications for different cell types. It is more suitable for complex organs like shoot tips and embryos. Here the critical step to achieve survival is the dehydration step and not the freezing step as in classical protocols. Therefore if samples to be frozen are amenable to desiccation to critical water content (which vary depending on the procedure employed and type and characteristics of the propagule to be frozen) then further decline in survival was not observed after cryopreservation (Engelmann, 1997).

The development of an effective cryopreservation protocol requires the optimization of numerous variables such as type of explants, size, water content, and concentration of cryoprotectant (if utilized) and freezing behavior of the species. Before cryopreservation, the cells have to be dehydrated artificially to protect them from damage caused by the crystallization of intracellular water into ice (Mazur, 1984). As membranes are the primary site of injury (Fujikawa *et al.*, 1999), tolerance mechanisms for desiccation and freezing mainly depends on membrane stabilization through changes in lipid composition, accumulation of sugars and production of membrane protecting polypeptides, induction of anti-oxidative mechanisms, and the transcription of genes coding molecular chaperones (Thomashow, 1999). Engelmann (2000) described seven vitrification-based procedures for cryopreservation those were air desiccation, pre-growth, pre-growth desiccation, encapsulation dehydration, encapsulation vitrification, vitrification and droplet freezing.

5. Steps Involved in Cryopreservation

The successive steps involved in cryopreservation are selection of material, pretreatment, freezing, storage, thawing, and post retrieval treatments.

5.1. Selection of Material

It is preferable to select young and meristematic tissues for cryopreservation studies. The cells of these tissues are small and are characterized by low water contents, few vacuoles, dense cytoplasm, high nucleo-cytoplasmic ratio that enables them to withstand freezing (Engelmann, 1991). The physiological stage of the material is also vital for the successful recovery after cryopreservation (Berjak *et al.*, 1993).

5.2. Pre-treatment

The material is to be pretreated for a specific period for preparing it for the freezing process. The samples are cultured in a medium enriched with various cryoprotective substances like monosaccharides, oligosaccharides, polysaccharides, mannitol, sorbitol, DMSO etc (Benson and Withers, 1988) for different durations. Cryoprotective substances like sucrose act as an osmotic agent in dehydrating the samples and also protect the membrane (Fujikawa *et al.*, 1999). The nature of cryoprotectants, their concentration and duration will have to be determined on each species basis.

5.3. Freezing

The two types of freezing processes followed are slow and ultra rapid freezing. For slow freezing, a programmable freezer is required to obtain precise and reproducible freezing condition. For rapid freezing, the specimens have been enclosed in cryovials and rapidly immersed in liquid nitrogen as done in case of coconut (Assy-Bah and Engelmann, 1992b) or oil palm (Engelmann *et al.*, 1995).

5.4. Storage

The samples have to be kept at the temperature of liquid nitrogen for storage for theoretically unlimited period of time (Engelmann, 2004).

5.5. Thawing

Immersing the cryotubes containing the samples in a water bath set at 37-40°C carries out rapid thawing. It reduces the chance of ice micro crystals formed during freezing to larger crystals of a size, which would cause damage to cellular integrity (Engelmann, 1991). Fast rewarming of coconut zygotic embryos at 40±1°C for 3 min produced regrowth after cryopreservation (Sisunandar *et al.*, 2010).

5.6. Viability Assessment

Viability tests are conducted for quick and accurate prognosis of the efficiency of cryoprotection and to forecast the chances for recovery of explants after cryopreservation. Viability tests are usually correlated with the survival and regrowth of the plant material subsequent to cryopreservation (Verleysen *et al.*, 2004). Different techniques for viability testing include staining and meristem electrolyte leakage (MEL) tests. The commonly used staining tests are with FDA (flurescein diacetate) and TTC (triphenyl tetrazolium chloride). FDA is absorbed by the living cells and transformed into flurescein, and measured in UV (Windholm 1972). In TTC test, the tetrazolium salt solution is reduced to red colored formazan by the activity of dehydrogenase ions released by respiration of the viable plant cells in the mitochondria (Stenponkus and Lanpher, 1967). In non-optimized cryopreservation protocols, the primary cause of cell death is due to damaged cell membranes (Senaratna and McKersie, 1986). This led to cell lysis and leakage of electrolytes that can be quantified in terms of cell critical to survival and also injured cells (Ketchie *et al.*, 1972; Becwar *et al.*, 1982; Kioko *et al.*, 2006; Pammenter. *et al.*, 1991; Malik and Chaudhury, 2006; Savage, 1992). But the actual assessement of viability is regrowth upon transfer into the medium after cryostorage. Gomes-Copeland *et al.* (2012) used electrolytic conductivity and potassium leaching tests to assess the viability of cryopreserved embryos of coconut cultivar 'Anão Verde do Brasil de Jiqui' (AVeJBr).

5.7. Post-retrieval Treatment

The samples after cryopreservation are to be set for recovery in best conditions (Benson and Withers, 1988). To eliminate the toxic effect of cryoprotective substances, the samples are rinsed and diluted in the culture medium. By successive transfer to progressively less concentrated media avoided osmotic shock for the samples (Benson and Withers, 1988). For better regrowth the nature of the medium (solid/ liquid) can also be changed. To avoid photo oxidation, the samples are placed initially in dark condition for better regrowth. Growth hormones may also be supplemented to the medium for the stimulation of growth (Sajini, 2011). Addition of non vitamin antioxidant and anti-stress compounds (LA: lipoic acid, GSH: glutathione, GB: glycine betaine) during the PVS 2 vitrification protocol was found to be effective for the improvement of regrowth (Esther *et al.*, 2010).

6. Cryopreservation of Coconut Zygotic Embryos

Bajaj (1984) suggested the possibility of long term conservation of coconut zygotic embryos when the embryos resumed growth after freezing at -196°C. Here, the immature embryos (1-1.5 cms) of West Coast Tall cultivar of coconut were partially dehydrated and cut into transverse halves. They were treated with a cryoprotectant solution [7 per cent dimethyl sulfoxide (DMSO) and 7 per cent sucrose in MS liquid medium] blotted dry and wrapped in a single layer of sterile aluminum foil. It was then frozen by gradually lowering into liquid nitrogen and kept for five minutes. The frozen samples were thawed in warm water (35 to 40° C), washed and cultured on MS medium containing 2, 4-D (0.2 mg L^{-1}) NAA (0.5mg L^{-1}) and kinetin (0.1 mg L^{-1}). The retrieved embryos and their segments in cultures showed a lag period of up to 4 months without any sign of growth. In some cultures, the embryo subsequently showed an overall swelling and elongation. Chin *et al.* (1989). reported the survival of one single coconut embryo, 15 months after freezing using a classical protocol (cryoprotection with DMSO and slow freezing).

Using immature embryos of coconut (7-8 months after pollination), Assy-Bah and Engelmann (1992a) could successfully establish rooted plantlets from embryos after freezing in liquid nitrogen. The immature embryos from the coconut hybrid PB 121 were placed for 4 hours in Petri dishes on standard medium containing 600 g L^{-1} glucose. Pregrowth on this medium was compared with pregrowth on medium supplemented with the cryoprotectants glycerol, sorbitol or polythyleneglycol

(PEG) 6000 at 5, 10 or 15 per cent. Thereafter, the embryos were immersed rapidly in liquid nitrogen. Thawing was carried out by immersion of the cryotubes for 30 sec in a water bath at 40 °C. After freezing in liquid nitrogen, survival was obtained in three conditions only: pregrowth with 10 and 15 per cent glycerol (25 and 10 per cent survival respectively) and 10 per cent sorbitol (43 per cent survival). PEG showed no cryoprotective effect at the concentrations tried. But only one rooted plantlet could be obtained from embryos pre-treated with 15 per cent glycerol after 2.5 months (Assy-Bah and Engelmann, 1992a). Embryo pre-treatment with cryoprotectant solution containing, 1.75 mol L^{-1} sucrose + 15 per cent glycerol for 12 and 16 hours presented lower moisture content and larger viability by tetrazolium test in coconut cultivar Brazil Green Dwarf (BGD) (Gomes-Copeland *et al.*, 2015).

Assy-Bah and Engelmann (1992b) reported cryopreservation of mature embryos of four varieties of coconut (hybrid PB 121, Cameroon Red Dwarf, Indian Tall, Renell tall). The embryos were pretreated in the laminar air current for 4 hours and subsequently incubated in medium containing 600 g L^{-1} glucose and 15 per cent glycerol for 11-20 hours. After rapid freezing and thawing a recovery rate of 33 and 93 per cent was observed depending on the variety. Karun *et al.* (2005) reported cryopreservation of mature embryos of West Coast Tall variety of coconut after desiccation pretreatments. Maximum retrieval of healthy plantlets was obtained from the embryos subjected to 18 h silica gel or 24 h laminar air flow desiccation treatment. Irreversible damage of shoot meristem was observed when the moisture content of the embryo was reduced below 20 per cent.

Sisunandar *et al.* (2010a) reported an improved cryopreservation protocol for a wide range of coconut cultivars from Indonesia. The method included four optimized steps *viz*: rapid dehydration, rapid cooling, rapid warming and recovery *in vitro* and acclimatization and soil supported growth. For rapid dehydration, the embryos were placed in a glass jar equipped with a stainless steel mesh (1-2 mm) platform and an autoclavable fan placed below the platform and housed in a polycarbonate tube. Activated silica gel (680 g) was placed into the lower portion of the glass jar in two separate zones. Following this protocol, 20 per cent (when cryopreserved 12 days after harvesting) and 40 per cent (when cryopreserved at the time of harvest) of all Malayan Yellow Dwarf embryos cryopreserved could be retrieved to normal seedlings. Differential scanning calorimetric studies showed that this protocol induced a drop in embryo fresh weight to 19 per cent and significantly reduced the amount of water remaining that could produce ice crystals (0.1 per cent). Of the 20 cultivars tested, 16 were found to produce between 10 per cent and 40 per cent normal seedlings, while four cultivars generated between 0 per cent and 10 per cent normal seedlings after cryopreservation.

Sisunandar *et al.* (2010b) conducted morphological, cytological or molecular studies in coconut plantlets recovered after cryopreservation. The embryos from four different cultivars were subjected to rapid dehydration in a drying chamber containing activated silica gel for 8 h to decrease the moisture content from 78-80 per cent to 19-20 per cent. The plants recovered from cryopreservation showed no morphological variation through measurement of shoot elongation rates, production of opened leaves and number and total length of primary roots. There was no variation in chromosome number (2n=32), type of chromosomes, the length of the long and short arms, the arm length ratio, and centromeric index in all studied cultivars independently of cryopreservation as revealed by karyotype analysis. Genetic and epigenetic fidelity of coconut plants recovered from cryopreservation was assessed through microsatellite (SSR) analysis and global DNA methylation rates. There were no significant genetic or epigenetic differences between the seedlings originating from cryopreserved embryos and respective control.

Sajini *et al.* (2011) studied the effect of preculture conditions, vitrification and unloading solutions on survival and regeneration of coconut zygotic embryos after cryopreservation. Among the seven plant vitrification solutions tested, PVS3 was found to be the most effective for regeneration of cryopreserved embryos (Figure 20.1). The optimal protocol involved preculture of embryos for 3 days on medium with 0.6 M sucrose, PVS3 treatment for 16 h, rapid cooling and rewarming and unloading in 1.2 M sucrose liquid medium for 1.5 h. Under these conditions, 70-80 per cent survival (corresponding to size enlargement and weight gain) was observed with cryopreserved embryos and 20-25 per cent of the plants regenerated (showing normal shoot and root growth) from cryopreserved embryos were successfully established in pots.

Figure 20.1: Plantlet Formation from Coconut Zygotic Embryo following Cryopreservation in PVS3.

7. Cryopreservation of Coconut Plumular Tissues

Hornung *et al.* (2001) followed encapsulation dehydration protocol for the cryopreservation of plumular tissues of coconut. The encapsulated plumules were precultured for 72-96 h in medium with 0.75 M sucrose and desiccated with silica gel to around 30 per cent moisture content. Callus growth was observed from the plumule after freezing in liquid nitrogen. Bandupriya *et al.* (2007) studied the effect of abscisic acid in the encapsulation-dehydration method for cryopreservation of coconut plumules. The survival and recovery rate of frozen plumules was significantly increased after the addition of ABA (40 μM) to the sucrose pretreatment medium. Post cryopreservation, 84 per cent of plumules showed survival 39 per cent of which had been recovered.

Coconut plumule (apical dome with 3-4 leaf primordial) extracted from mature embryos (11-12 months after pollination) was also utilized for cryopreservation studies by Nan *et al.* (2008). Plumules excised from Malayan Yellow Dwarf embryos were first precultured on standard medium with 0.12 M sucrose for three days. Thereafter, for encapsulation, the plumular tissues were suspended in standard medium containing 3 per cent (v/v) sodium alginate and 0.15 M sucrose. After making beads in 0.1 M $CaCl_2$ containing 0.15 M sucrose, it was pretreated for 2-3 days sequentially in standard medium containing various sucrose concentrations (0.5 M, 0.75 M and 1 M). Thereafter, the beads were dried for 6 to 24 h on sterile filter paper over 40 g silica gel in 125 ml air tight boxes. After freezing, regrowth of plumules was obtained for plumules dehydrated for 14 h (21 per cent) and 16 h (20 per cent). Pregrowth of encapsulated plumule beads in 1 M sucrose for 16 hours resulted in 20 per cent leafy shoot production from the cryopreserved samples. The histological studies in recovered plumules after cryopreservation showed structural similarity with control cells.

Bandupriya *et al.* (2010) investigated the most suitable method to transport/store mature zygotic embryos of coconut (for excision of plumules) for cryopreservation work. Three different conditions tested were transportation as solid endosperm cores containing embryos (refrigerated for 10 days), embryos in solidified agar or KCl solution (stored at 27°C in dark for 10days). Following encapsulation-dehydration technique, plumular tissues, excised from embryos stored in KCl and solidified agar showed significantly higher rate of recovery than embryos in albumen cores in unfrozen samples. In frozen plumules, there was no significant difference in recovery under three conditions tested. In frozen plumules pretreated with 1.0 M sucrose, the rate of recovery (40 per cent) was significantly higher in ones excised from embryos stored in solidified agar when compared to other two conditions.

8. Conservation of Coconut Pollen

Pollen represents the haploid phase in the life cycle of plants. At maturity, the pollen grains are dispersed either in two celled (vegetative and generative cell) or three celled stage (vegetative and two sperm cells). Under natural conditions the two-celled pollen has much longer life span because of their protective structure, low water content, and reduced metabolic activity. The trinucleate pollen is short lived due to its wall characteristics, high moisture content, and high rate of metabolism. The life span of pollen is primarily determined by the plant genome and also influenced by environmental conditions (Kozaki, 1975).

Based on pollen longevity, the plant taxa are classified into three main groups (Harrington, 1970) such as, long-lived pollen (6 months to 1 year), pollen with a medium life span (approximately 1-3 months) and short-lived pollen (few minutes to few days). Depending on water content, pollen having more than 30 per cent water is classified as partially hydrated like recalcitrant seeds and if it is less than 30 per cent as partially dehydrated like orthodox seeds (Kermode and Finch-Savage, 2002). Jain and Shivanna (1989) reported a positive correlation between the loss of viability and a reduction in the amount of membrane phospholipids irrespective of pollen storage conditions. The desiccated pollen is to be rehydrated in humid air for restoration of membrane integrity and viability of pollen (Shivanna and Heslop-Harrison, 1981; Bernard, 1973; Karipidis *et al.*, 2007; Hoekstra, 1984; Hoekstra and Van der Wal, 1988; Crowe *et al.*, 1989 a, b).

Maintenance of pollen viability for breeding and research purposes necessitate its storage (Nath and Anderson, 1975). Apart from the use of stored pollen in breeding programmes like storage of multiple genotypes in small space, its transportation, use of individual genotype over a long period of time, pollination in asynchronously flowering plants, it also offers successful exchange and conservation of plant genetic resources (Ganeshan *et al.*, 2008; Panella *et al.*, 2009; Tandon *et al.*, 2007). Pollen conservation provides a viable backup method for conserving genes (Towil and Walters, 2000).

The viability of pollen is to be ensured for effective use in hybridization as well as conservation studies. The pollen viability can be assessed by utilizing staining methods (Panella *et al.*, 2009) or by germination on artificial media (Verdeil and Pannetier 1990; Muccifora *et al.*, 2003; Nandakumar *et al.*, 1988) and was found to correlate with its fertilizing ability (Visser, 1955; Aloni *et al.*, 2001). Temperature is one of the most important environmental factors affecting pollen germination, pollen tube growth and fruit set (Kakani *et al.*, 2002; Burke *et al.*, 2004; Lee *et al.*, 1985). Karim *et al.* (2011) reported variation for cardinal temperatures (T_{min}, T_{opt} and T_{max}) of pollen germination percentage and pollen tube growth. The principal component analysis (PCA) identified maximum percentage pollen germination and pollen tube length of the species, and T_{max} for the two processes as the most important pollen parameters in describing a species tolerance to high temperature.

The percentage of *in vitro* germination of stored pollen has been correlated with its fertilizing ability (Visser, 1955). But the ultimate test for viability is the ability for fruit and seed set after pollination. There are reports that storage and other stresses affect vigour (vigorous pollen tubes) before affecting viability of pollen (Shivanna *et al.*, 1991). So in addition to pollen viability, the study of pollen vigour (pollen tube growth) is also of much importance for pollen storage studies (Page *et al.*, 2006; Harding and Tucker, 1969; David, 1971, 1974; Ottaviano *et al.*, 1980, 1982,).

In the case of coconut, the microspores inside microsporangia undergo nuclear division to give rise to a lenticular shaped generative cell and large tube nucleus (Menon and Pandalai, 1958). When the anthers become fully mature, the pollen sacs burst and shed their pollen before the opening of the male flower. The mature coconut pollen grain contains three nuclei (Patel, 1938). The male flowers on the upper spikes of the inflorescence and those situated on the sides of the female flower open first and commence opening from the apex of the spike and extends downwards. Even though the male flowers open throughout the day, most of them open from 8 A.M. to 10 A.M. Each male flower carries about 272 million pollen grains (Aldaba, 1921). Fresh pollen grains are spherical and smooth while within the anther, but after few seconds of exposure to dryness, they turn ellipsoidal with a longitudinal groove in the middle and measure about 0.063 mm in length and 0.020 mm in breadth (Menon and Pandalai, 1958). Pollen production in individual male flowers in coconut was significantly higher in summer and cold seasons than in the rainy season (Gangolly *et al.*, 1961). Under natural condition, the life span of fresh coconut pollen is only few days (Patel, 1938). The pollen must be conditioned to prolong and maintain high viability. Coconut pollen is desiccation tolerant and can be dried to 5 per cent moisture content without any lozss in viability (Whitehead, 1963). Franchi *et al.* (2011) classified coconut seed and pollen as recalcitrant seed and orthodox pollen (RSOP). Under Indian conditions, Patel (1938) observed 25 per cent of infertile pollen grains in coconut. Aldaba (1921) observed 3 to 33 per cent of infertile pollen grains in Philippines. Nair and Sharma (1963) reported the occurrence of coconut pollen variations comprising the trichotomocolpate, porate, operculate and I-furrowed forms in coconut. Several workers have reported *in vitro* pollen germination in coconut (Aldaba, 1921; Marechal, 1928; Patel, 1938; Nambiar, 1960; Nampoothiri, 1970). Several workers have noticed the importance of boron for germination of pollen in the stigmatic secretions (Gaugh and Ouggar, 1953) or in the germination medium (Johri and Vasil, 1961; Richards, 1986; Shivanna and Johri, 1985; Asif *et al.*, 1983; Leduc *et al.*, 1990; Muccifora *et al.*, 2003; Mortazavi *et al.*, 2010).

8.1. Short Term Storage of Coconut Pollen

The pollen after dehiscence from the anther can be stored through the manipulation of moisture content, storage temperature and humidity. Patel (1938) observed that fresh coconut pollen wrapped in bloting paper become nonviable after ninth day at room temperature. The longevity of pollen was prolonged for 16 days by storage over 35 per cent sulphuric acid (Marechal, 1928) wherein the viability estimates was 35-60 per cent. Desiccation to 5 per cent moisture content by oven drying at 40°C and storage over damp calcium chloride (RH 40 per cent at 5°C) resulted in considerable reduction of viability during the first month of storage, but there was very little further reduction after seven months after collection (Whitehead, 1963). By freeze-drying and storage at room temperature, coconut pollen could be conserved for short periods (Whitehead, 1965).

8.2. Coconut Pollen Cryopreservation

Even though partial dehydration can prolong coconut pollen viability for short periods, cryopreservation is the only available option for its long term storage (Engelmann, 1997; Towill and Walters, 2000). Coconut pollen cryopreservation could be utilized for long-term storage of pollen from palms of distinguishable characteristics that can be utilized for future breeding programmes, facilitating crosses for the production of hybrids and also for the conservation of diverse genotypes. It can also be advantageous in distributing and exchanging germplasm and also for the continuous availability for pollination over extended periods of time. The viability of pollen was studied by *in vitro* germination on media supplemented with sucrose (Kakani *et al.*, 2005).

Karun *et al.* (2006) reported pollen cryopreservation in West Coast Tall (WCT) accession of coconut. The authors accounted medium comprising 8 per cent sucrose, 1 per cent gelatin, 1 per cent agar and 0.01 per cent boric acid for best pollen germination. The variation observed for pollen germination among WCT genotypes was reduced after cryopreservation. The overall pollen tube growth was significantly more when pollen was incubation dried or cryopreserved. Karun *et al.* (2014). studied the effect of cryopreservation and cryostorage on pollen germination and fertility of coconut cultivars, WCT and Chowghat Orange Dwarf (COD). Germination and vigour of cryopreserved pollen were generally higher compared to that of pollen dried in oven and non-cryopreserved. After 4 years and 6 years of cryopreservation, 29 per cent and 32 per cent germination was reported in COD and WCT cultivars (Figure 20.2). Cryopreserved pollen of five Tall and five Dwarf accessions also displayed 24-31 per cent and 25-49 per cent germination, respectively (Karun *et al.*, 2014). Normal nut set was also observed when WCT pollen, cryostored for six years, was used for hybridization in COD palms (Figure 20.3).

9. Light and Ultrastructural Studies Conducted in Cryopreserved Materials

In coconut, Nan *et al.* (2008) reported cellular changes with respect to dehydration duration and pregrowth treatment during cryopreservation of plumular tissues of coconut. For shorter dehydration of plumules (6 hours), the cells in the first level of meristematic zone recovered from its shorted size whereas underlying layers showed high plasmolysis with significant periplasmic space. For intermediate dehydration, apart from this, the occurrences of starch grains were also

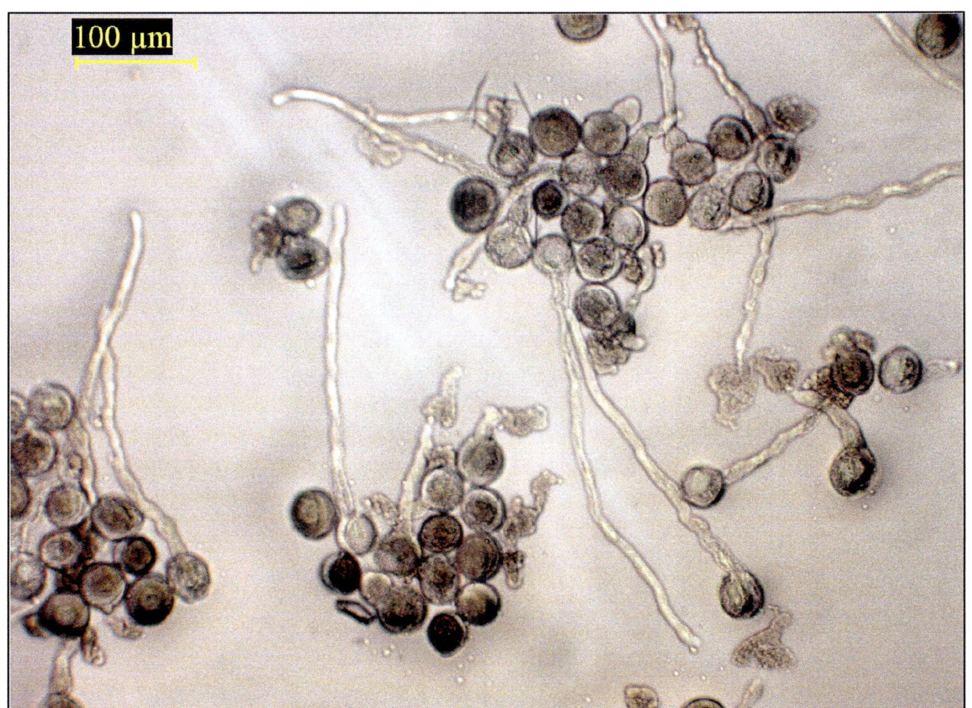

Figure 20.2: Germination in WCT Pollen Cryostored for Four Years.

Figure 20.3: Nut Set in COD Palms Hybridized with WCT Pollen Cryostored for Six Years.

observed. When long dehydration durations (16 hours) were used, the cells appeared of normal isodiametric shape, spherical nuclei with high nucleo-cytoplasmic ratio. There was increase in starch grains from 0.5 M to 1 M sucrose in pregrowth medium. The occurrence of active dividing cells in the plumular tissues at the periphery of the meristem was in accordance with the actual regrowth of plumular explants of coconut after cryopreservation. After studying different dehydration durations of cryopreserved coconut embryos and plumules, Nan *et al.* (2014) reported three categories of cell ultra structural damages during cryopreservation. The first stage of changes concerned the plasmolysis of cells with small vacuoles, condensation of chromatin, changing in the conformation of the DNA and the nucleus and arrest of mitosis. These types of changes are described in general in the context of a desiccation tolerance. The second degree of the changes was the retraction of the cytoplasm inside the cell, the increase in the periplasmic volume. The third degree of modification concerned the deformation of the walls, the invagination or the lysis of the plasma membrane resulting in the observation of distorted cells and the bursting of the nucleus. These two types of modifications are irreversible and correspond to an absence of regrowth of the samples.

10. Conclusion

It has been observed that survival and germination after cryopreservation of coconut zygotic embryos is mainly dependent on genotype as well as cryopreservation methods. Hence it is necessary to develop an efficient cryopreservation protocol independent of genotypic effect so that maximum diversity can be conserved with ease. The studies presented in this chapter elucidate the effectiveness of various cryopreservation methods in coconut zygotic embryo cryopreservation and efficacy of pollen cryostorage on viability and fertility of stored pollen. Though cryopreservation of coconut zygotic embryos, plumule and pollen is feasible, post-retrieval percentage cryopreserved material is still reported as low. Perfect protocols for different cryopreservation methods need to be standardized for different explant material including embryogenic calli, somatic embryos in addition to embryos, plumule and pollen for long term maintenance coconut germplasm.

References

Aldaba, V.C. (1921). The pollination of coconut. *The Philippine Agricultural Scientist*, **10**: 195-210.

Aloni, B., Mary, P., Mason, P. and Leah, K. (2001). The effect of high temperature and high atmospheric CO_2 on carbohydrate changes in bell pepper (*Capsicum annum* L.) pollen in relation to its germination. *Physiologia Plantarum*, 112: 505-512.

Asif, M.I., Al-Tahir, O.A. and Farah, A.F. (1983). The effects of some chemicals and growth substances on pollen germination and tube growth of date palm. *Horticulture Science*, **18**: 479-480.

Assy-Bah, B. and Engelmann, F. (1993). Medium-term conservation of mature embryos of coconut. *Plant Cell, Tissue and Organ Culture*, **33**: 19-24.

Assy-Bah, B. and Engelmann, F. (1992a). Cryopreservation of immature embryos of coconut (*Cocos nucifera* L.). *Cryoletters*, **13**: 67-74.

Assy-Bah, B. and Engelmann, F. (1992b) Cryopreservation of mature embryos of coconut (*Cocos nucifera* L.) and subsequent regeneration of plantlets. *Cryoletters*, **13**: 117-126.

Astley, D. (1992). Preservation of genetic diversity and accession integrity. *Field Crops Research*, **29**: 205-224.

Bajaj, Y.P.S. (1984). Induction of growth in frozen embryos of coconut and ovules of citrus. *Current Science*, **53**: 1215-1216.

Bandupriya, H.D.D., Fernando,.S.C., Verdil J.L. and Malaurie, B. (2010). Cryopreservation of encapsulated plumules of coconut: Effect of transport/ store conditions. *Asia-Pacific Journal of Molecular Biology and Biotechnology*, **18**: 135-137.

Bandupriya, H.D.D., Fernando, S.C., Verdeil, J.L. and Malaurie, B. (2007). Effect of abscisic acid on survival and recovery of cryopreserved plumule explants of coconut (*Cocos nucifera* L.). *Cocos*, **18**: 58-68.

Batugal, P., and Jayashree, K. (2005). COGENT's Multi-site International Coconut Genebank. In: Batugal, P., Ramanatha Rao, V. and Oliver, J. (Eds.), *Coconut Genetic Resources. International Plant Genetic Resources Institute*, Regional Office for Asia, the Pacific and Oceania (IPGRI-APO), Serdang, Selangor DE, Malaysia. pp. 92-106.

Becwar, M.R., Stanwood, P.C. and Roos, E.E. (1982). Dehydration effects on imbibitional leakage from desiccation sensitive seeds. *Plant Physiology*, **69**: 1132-1135.

Benson, E.E and Withers, L.A (1988). The applications of germplasm storage in biotechnology. In: Pais, M.S.S., Mavituna, F. and Novais, J.M. (Eds.), *Plant Cell Biotechnology*. Springer-Verlag, Berlin, Heidelberg. pp. 426-431.

Berjak, P., Vertucci, C.W. and Pammenter, N.W. (1993). Effects of developmental status and dehydration rate on characteristics of water and desiccation-sensitivity in recalcitrant seeds of *Camellia sinensis*. *Seed Science Research*, **3**: 155-166.

Bernard, G. (1973). Quelques aspects the la lypholysation du pollen de cocotier. *Oleagineux*, **28**: 447-551.

Bertrand-Desbrunais, B., Noirot. and Charrier, A. (1992). Slow growth of *in vitro* conservation of coffee (*Coffea* spp.). *Plant Cell, Tissue and Organ Culture*, **31**: 105-110.

Brison, M., de Boucaud, M.T., Pierronnet, A. and Dosba, F. (1997). Effect of cryopreservation on the salinity state of a cv. of *Prunus* rootstock experimentally infected with Plum Pox Potyvirus. *Plant Science*, **123**: 189-196.

Burke, J.J., Velten, J. and Oliver, M.J. (2004). *In vitro* analysis of cotton pollen germination. *Agronomy Journal*, **96**: 359-368.

Burke, M. J. (1986). The glassy state and survival of anhydrous biological systems. In: Leopold, A C. (Ed.), *Membranes, Metabolism and Dry Organisms*. Cornell University Press, Ithaca. pp. 358-363.

Chaudhury, R and Chandel, K.P.S. (1994).) Germination studies and cryopreservation of seeds of black pepper (*Piper nigrum* L.) -A recalcitrant species. *Cryoletters*, **15**: 145-150.

Chin, H.F. and Roberts, E.H. (1980). *Recalcitrant Crop Seeds*. Tropical Press. SDN, BDH, Kuala Lumpur, Malaysia.

Chin, H.F., Krishnapillay, B. and Hor, Y.I. (1989). A note on the cryopreservation of embryos from young coconut (*Cocos nucifera* var MAWA). *Pertanika Journal*, **12**: 183-186.

Coconut Statistical Year Book (2014). *Alit Pirmansah, Asian and Pacific Coconut Community (APCC)*, Jakarta, Indonesia. pp. 6-7.

Croew, J.H., Hoskstra, F.A. and Crowe, L.M. (1989a). Membrane phase transitions are responsible for imbibitional damage in dry pollen. *Proceedings of the National Academy of Sciences*, **86**: 520-525.

Croew, J.H., Hoskstra, F.A., Crowe, L.M., Anchordoguy, T.J. and Drobnis, E. (1989b). Lipid phase transitions measured in intact cells with Fourier Transform infrared spectroscopy. *Cryobiology*, **26**: 76-83.

David, L.M. (1971). A correlation between gametophytic and sporophytic characteristics in *Zea mays* L. *Science*, **171**: 1155-1156.

David, L.M. (1974). Correlation between speed of pollen tube growth and seedling height in *Zea mays* L. *Nature*, **249**: 491-492.

De-Zhu, L. and Pritchard, H.W. (2009). The science and economics of *ex situ* plant conservation. *Trends in Plant Science*, **14**: 614-621.

Engelmann, F. (1991). *In vitro* conservation of tropical plant germplasm- a review. *Euphytica*, **57**: 227-243.

Engelmann, F. (1997). *In vitro* conservation methods. In: Ford-Lloyd, B.V., Newbury, H.J. and Callow, J.A. (Eds.), *Biotechnology and Plant Genetic Resources: Conservation and Use*. CABI, UK. pp. 119-162.

Engelmann, F. (2000). Importance of cryopreservation for the conservation of plant genetic resources. In: Engelmann, F and Takagi, H. (Eds.), *Cryopreservation of Tropical Plant Germplasm: Current Research Progress and Applications*. JIRCAS, Tsukuba and IPGRI, Rome, Italy. pp. 8-20.

Engelmann, F. (2004). Plant cryopreservation: Progress and prospects. In *Vitro Cellular and Developmental Biology- Plant*, **40**: 427-433.

Engelmann, F., Chabrillange, N. and Duval, Y. (1995). Cryopreservation of zygotic embryos and kernels of oil palm (*Elaeis guineensis* Jacq.). *Seed Science Research*, **5**: 81-86.

Escobar, R.H. and Roca, W.M. (1997). Cryopreservation of cassava shoot tips through rapid freezing. *African Journal Root and Tuber Crops*, **2**: 214-215.

Esther, E.U., Magfrat, M., Sandhya, G. and Reed, B.M. (2010). Antioxidant and anti-stress compounds improve regrowth of cryopreserved *Rubus* shoot tips. *In Vitro Cellular and Developmental Biology- Plant*, **46**: 386-393.

Fahy, G.M., MacFarlane, D.R., Angell, C.A. and Meryman, H.T. (1984). Vitrification as an approach to cryopreservation. *Cryobiology*, **21**: 407-426.

Franchi, G.G., Piotto, B., Nepi, M., Baskin, C.C., Baskin, J.M and Pacini, E. (2011). Pollen and seed desiccation tolerance in relation to degree of developmental arrest, dispersal and survival. *Journal of Experimental Botany*, **62**: 5267-81.

Fujikawa, S., Jitsuyama, Y. and Kuroda, K. (1999). Determination of the role of cold acclimation induced diverse changes in plant cells from the viewpoint of avoidance of freezing injury. *Journal of Plant Research*, **112**: 237-244.

Fukai, S. (1995). Cryopreservation of germplasm of *Chrysanthemums*. In: Bajaj, Y.P.S. (Ed.), *Cryopreservation of Plant Germplasm. Biotechnology in Agriculture and Technology* Vol. 32. Springer- Verlag, Berlin. pp. 447-457.

Ganeshan, S., Rajasekharan, P.E., Shashikumar, S. and Decruze, W. (2008). Cryopreservation of pollen. In: Reed, B.M (Ed.), *Plant Cryopreservation: A Practical Guide*, Springer Science Business Media, LLC, New York. pp. 443-464.

Gangolly, S.R., Kamalakaran, A.K., Balakrishnan T. K. and Pandalai, K.M. (1961) Studies on pollen in the coconut (*Cocos nucifera* L.). 1. Its importance, output in different varieties and composition in the still air. *Indian Coconut Journal*, **14**: 49-66.

Gaugh, H.G. and Ouggar, W.M. (1953). The role of boron in the translation of sucrose. *Plant Physiology*, **28**: 457-466.

Gomes-Copeland, K.K.P., Silva Lédo, A.D., Almeida, F.T.C.D., Miranda, R.P. and Santos, I.R.I. (2012). Assessing the viability of cryopreserved coconut zygotic embryos by electrolytic conductivity and potassium leaching. Pesquisa Agropecuária Brasileira, **47**: 8-13.

Gomes-Copeland, K.K.P., Santos, I.R.I., Almeida, F.T.C. and Lédo, A.S. (2015).. Performance of cryoprotectants, dehydration methods and tetrazolium test on the zygotic embryos of BGD coconut. *Scientia Plena*, **11**: 2-7.

Gonzalez-Arnao, Juárez, M.T.J., Ortega, C., Navarro, L. and Duran-Vila, N. (2003). Cryopreservation of ovules and somatic embryos of citrus using the encapsulation-dehydrtion technique. *Cryoletters*, **24**: 85-94.

Gunn, B.F., Baudouin, L. and Olsen, K.M (2011). Independent origins of cultivated coconut (*Cocos nucifera* L.) in the old world tropics. *PLoSONE*, **6**: e21143.

Harding, J. and Tucker, C.I. (1969) Quantitative studies on mating systems. III. Methods for the estimation of male gametophytic selective values and differential out crossing rates. *Evolution*, **23**: 85-95.

Harries, H.C. (1990). Malesian origin for a domestic *Cocos nucifera*. In: Baas, P., Kalkman, K. and Geesink, R. (Ed.), *The Plant Diversity of Malesia*. Kluwer Academic Publishers, Dordrecht. pp. 351- 357.

Harries, H.C. (2001). The coconut palm (*Cocos nucifera* L.). In: Last, F.T. (Eds.), *Tree crop Ecosystems*. Elsevier, New York. pp. 321-338.

Harrington, J.F. (1970). Seed and pollen storage for conservation of plant gene resources. In: Frankel, O.H. and Bennet, E. (Eds.), *Genetic Resources in Plants: Their Exploration and Conservation*. Oxford and Edinburgh, Blackwell. pp. 501-521.

Helliot, B., Panis, B., Poumay, Y., Swennen, R., Lepoivre, P. and Frison, E. (2002). Cryopreservation for the elimination of cucumber mosaic and banana streak viruses from banana (*Musa* spp). *Plant Cell Reports*, **20**: 1117-1122.

Hoekstra, F.A. (1984). Imbibitonal chilling injury in pollen. *Plant Physiology*, **74**: 815-821.

Hoekstra, F.A. and Van der Wal, E.W. (1988). Initial moisture content and temperature of imbibition determine extent of imbibitional injury in pollen. *Plant Physiology*, **133**: 257-262.

Hornung, R., Domas, R. and Lych, P.T. (2001). Cryopreservation of plumular explants of coconut (*Cocos nucifera* L.) to support programmes for mass clonal propagation through somatic embryogenesis. *Cryoletters*, **22** : 211-220.

Jain, A. and Shivanna, K.R. (1989). Loss of viability during storage is associated with changes in membrane phospholipids. *Phytochemistry*, **28**: 999-1002.

Johri, B.M. and Vasil, I.K. (1961) Physiology of pollen. *The Botanical Review*, **27**: 325-381.

Kakani, V.G., Reddy, K.R., Koti, S., Wallace, T.P., Prasad, P.V.V., Reddy, V.R. and Zhao, D. (2005). Differences in *in vitro* pollen germination and pollen tube growth of cotton cultivars in response to high temperature. *Annals of Botany*, **96**: 59-67.

Kakani, V.G., Prasad, P.V.V., Craufurd, P.Q. and Wheeler, T.R. (2002). Response of *in vitro* pollen germination and pollen tube growth of groundnut (*Arachis hypogaea* L.) genotypes to temperature. *Plant, Cell and Environment*, **25**: 1651-1661.

Karim, S., Behrouz, S., Vahid, R and Mahmood, K. (2011). Influence of temperature on the *in vitro* pollen germination and pollen tube growth of various native Iranian almonds (*Prunus* L. spp.) species. *Trees*, **25**: 809-822.

Karipidis, C., Olympios, C., Passam, H.C. and Savvas, D. (2007). Effect of moisture content of tomato pollen stored cryogenically on *in vitro* germination, fecundity and respiration during pollen tube growth. *The Journal of Horticultural Science and Biotechnology*, **82**: 29-34.

Karun, A. and Sajini, K.K. (1994).) Short-term storage of coconut zygotic embryos. *Current Science*, **67**: 118-120.

Karun, A., Parthasarathy, V.A., Kumaran, P.M., Iyer, R.D. and Sajini, K.K. (2002). *Coconut Embryo Culture Protocol for Germplasm Collection*, Technical Bulletin No. 45, CPCRI, Kasaragod, India.

Karun, A., Sajini, K.K. and Iyer, R.D. (1996). *In vitro* active conservation of coconut zygotic embryos. *Journal of Plantation Crops*, **24**: 586-593.

Karun, A., Sajini, K.K. and Parthasarathy, V.A. (2005). *Cryopreservation of mature coconut embryos by desiccation method*. CORD, **21**: 13-19.

Karun, A., Sajini, K.K., Meera, N., Kumaran, P.M. and Samsudheen. (2006). Cryopreservation of coconut (*Cocos nucifera* L.) pollen. *Journal of Plantation Crops*, **34**: 568-571.

Karun, A., Sajini, K.K., Niral, V., Amarnath, C.H., Remya, P., Rajesh, M.K., Samsudeen, K. and Jerard,B.A. (2014). Coconut (*Cocos nucifera* L.) pollen cryopreservation. *Cryoletters*, **35**: 407-417.

Kermode, A.R. and Finch-Savage, B.E. (2002). Desiccation sensitivity in orthodox and recalcitrant seeds in relation to development. In: Black, M. and Pritchard, H.W. (Eds.), *Desiccation and Survival of Plants: Drying Without Dying*. CABI Publishing, Reading. pp. 149-184.

Ketchie, D.O., Beeman, C.H. and Ballard, A.L. (1972). Relationship of electrolytic conductance to cold injury and acclimation in fruit trees. *Journal of the American Society of Horticultural Science*, **97**: 403-406.

Kioko, J.I., Berjak, P. and Pammenter, N.W. (2006). Viability and ultra structural responses of seeds and embryonic axes of *Trichilia emetica* to different dehydration and storage conditions. *South African Journal of Botany*, **72**: 167-176.

Kozaki, I. (1975). Storage method of pollen. In: Matsuo, T. (Ed.). JIBP Synthesis V, Gene conservation, exploration, collection, preservation, and utilization. University of Tokyo Press, Tokyo.

Krogstrup, P., Baldursson, S. and Norgaavd, J.V. (1992). *Ex situ* genetic conservation by use of tissue culture. *Opera Botanica*, **113**: 49-53.

Laura, V.J., Ana, M.V. and Antonio, B. (1996). Cryopreservation of somatic embryos and embryonic axis of *Camellia japonica* L. *Plant Cell Reports*, **15**: 699-703.

Leduc, N., Monnier, M. and Douglas, G.C. (1990). Germination of trinucleate pollen formulation of a new medium for *Capsella bursapastoris*. *Sexual Plant Reproduction*, **3**: 228-235.

Lee, C.W., Thomas, J.C. and Buchmann, S.L. (1985). Factors affecting *in vitro* germination and storage. of jojoba pollen. *Journal of American Society for Horticultural Science*, **110**: 671-676.

Malik, S.K. and Chaudhury, R. (2006). The cryopreservation of embryonic axes of two wild and endangered *Citrus* species. *Plant Genetic Recourses: Characterization and Utilization*, **4**: 204-209.

Marechal, H. (1928) Observations and preliminary experiments on the coconut palm with a view to developing improved seed nuts for Fiji. *Fiji Agriculture Journal*, **1**: 16-45.

Maxted, N., Ford-Loyd, B.V. and Hawkes, J.G. (1997). Complementary conservation strategies. In: Maxted, N., Ford-Lloyd, B.V. and Hawkes, J.G. (Eds.), *Plant Genetic Conservation: The In Situ Approach*. Chapman and Hal, London. pp. 15-39.

Mazur, P. (1984). Freezing of living cells: mechanisms and applications. *American Journal of Physiology*, **247**: 125-142.

Menon, K.P.V. and Pandalai, K.M. (1958). Floral Biology. In: The coconut palm: A monograph. *Indian Central Coconut Committee*, Ernakulam, Kerala, India. pp. 39-85.

Mortazavi, S.M.H., Arzani, K. and Moieni. (2010). Optimizing storage and *in vitro* germination of date palm (*Phoenix dactylifera*) pollen. *Journal of Agricultural Science and Technology*, **12**: 181-189.

Mucciflora, S., Bellani, L.M. and Gori, P. (2003). Ultrastructure, viability, and *in vitro* germination of the tricellular *Sambucus nigra* L. pollen. *International Journal of Plant Science*, **164**: 855-860.

Nair, P.K.K. and Sharma, M. (1963). Pollen grains of *Cocos nucifera* L. *Grana Palynologica*, **4**: 373-379

Nambiar, M.C. (1960). Pollen irradiation and culture of pollen tubes of coconut. *Current Science*, **29**: 317-318.

Nampoothiri, K.U.K. (1970). Pollen studies in coconut (*Cocos nucifera* L.) with special reference to a sampling procedure. *Indian Journal of Agricultural Sciences*, **40**: 457-460.

Nan, O.N., Hocher, V., Verdeil, J.L., Konan, J.L., Ballo, K., Mondeil, F. and Malaurie, B. (2008). Cryopreservation by encapsulation dehydration of plumules of coconut. *Cryoletters*, **29**: 339-350.

Nan. O.A., Tiécoura, K., Bi, S.G., Verdeil, J.L. and Malaurie, B. (2014). Ultrastructural changes during cryopreservation of plumules and embryos of coconut (*Cocos nucifera* L.). I, **5**: 103-115.

Nandakumar, P.B.A., Chaudhury, R. and Shivanna, K.R. (1988). Effect of storage on pollen germination and pollen tube growth. *Current Science*, **57**: 557-559.

Nath, J. and Anderson, J.O. (1975). Effect of freezing and freeze drying on the viability and storage of *Lilium longiflorum* L. and *Zea mays* L. pollen. *Cryobiology*, **12**: 81-88.

Narayana, G.V. and John, C.M. (1949). Varieties and forms of the coconut. *Madras Agriculture*, **36**: 349-366.

Nikishina, T.V., Popova, E.V., Vakhrameeva, M.G., Varlygina, T.I., Kolomeitseva, G.l., Burov, A.V., Popovich, E.A., Shirokov, A.I., Shumilov,V-YU. and Popov, A.S. (2007). Cryopreservation of seeds and protocorms of rare temperate orchids. *Russian Journal of Plant Physiology*, **54**: 121-127.

Ottaviano, E., Sari-Gorla, M. and Mulcahy, D.L. (1980). Pollen tube growth rate in *Zea mays*: Implications for genetic improvement of crops. *Science*, **210**: 437-438.

Ottaviano, E., Sari-Gorla, M. and Pe', M.E. (1982). Male gametophytic selection in maize. *Theory of Applied Genetics*, **63**: 249-254.

Ottaviano, E., Sari-Gorla, M. and Villa, M. (1983). Pollen competitive ability in maize: Within population variability and response to selection. *Theoretical and Applied Genetics*, **76**: 601- 608.

Page, T., Moore, G.M., Will. J. and Halloran, G.M. (2006). Pollen viability in *Kunzea pomifera* (Myrtaceae) as influenced by sucrose concentration and storage. *Australian Journal of Botany*, **54**: 553-558.

Pammenter, N.W., Vertucci, W.C.W. and Berjak, P. (1991). Homeohydrous (Recalcitrant) seeds: Dehydration, the state of water and viability characteristics in *Landolphia kirkii*. *Plant Physiology*, **96**: 1093-1098.

Panella, L., Wheeler, L. and McClintock, M. E. (2009). Long term survival of cryopreserved sugar beet pollen. *Journal of Sugar Beet Research*, **46**: 1-9.

Patel, J.S. (1938). The coconut: A monograph. Govt. Press. Madras. pp. 90-133.

Ramanatha Rao, V. and Riley, K.W. (1994). The use of biotechnology for conservation and utilization of plant genetic resources. *Plant Genetic Recourses Newsletter*, **97**: 3-20.

Ramanatha Rao, V and Batugal, P. (1998). *Proceedings of the COGENT Regional Coconut Genebank Planning Workshop*. IPGRI-APO, Serdang.

Reed, B.M., Engelmann, F., Dulloo, M.E. and Engles, J.M.M. (2004). Technical guidelines for the Management of field and *in vitro Germplasm collection*. IPGRI handbook for Gene banks. No.7. IPGRI, Rome, Italy. pp.106.

Richards, A.J. (1986). *Plant Breeding Systems.* George Allen and Unwin, London. pp. 529.

Sajini, K.K., Karun, A., Amarnath, C.H. and Engelmann, F. (2011). Cryopreservation of coconut (*Cocos nucifera* L.) zygotic embryos by vitrification. *Cryoletters*, **32**: 317-328.

Sarkar, D. and Naik, P.S. (1998). Effect of inorganic nitrogen nutrition on cytokinin-induced potato micro tuber production *in vitro*. *Potato Research*, **41**: 211-217.

Savage, W.E. (1992). Embryo water status and survival in the recalcitrant species *Quercus robur* L. Evidence for critical moisture content. *Journal of Experimental Botany*, **43**: 663- 669.

Senaratna, T. and McKersie. (1986). Loss of desiccation tolerance during germination. In: Leopold, A.C. (Ed.), *Membranes, Metabolism and Dry Organisms*. Cornell University Press, Ithaca. pp. 85-101.

Shivanna, K.R. and Heslop-Harrison, J. (1981). Membrane state and pollen viability. Annals of Botany, **47**: 759-770.

Shivanna, K.R. and Johri, B.M. (1985). *The Angiosperm Pollen: Structure and Function.* Wiley Eastern Publishing, New Delhi. Wiley Eastern. 374p.

Shivanna, K.R., Linskens, H.F. and Cresti, M. (1991). Pollen viability and pollen vigor. *Theoretical and Applied Genetics*, **81**: 38-42.

Silvia, V., Mariano, T., Cristina, C. and Ana, M.V. (2004). Cryopreservation of embryogenic cultures from mature *Quercus suber* trees using vitrification. *Cryoletters*, **25**: 177-186.

Sisunandar, Peter A.S., Yohannes M.S., Samosir, Rival, A. and Adkins, S.W. (2010). Dehydration improves cryopreservation of coconut (*Cocos nucifera* L.). *s*, **61**: 289-296.

Sisunandar, Rival, A., Turquay, P., Samosir, Y. and Adkins, S.W. (2010b). Cryopreservation of coconut (*Cocos nucifera* L.) zygotic embryos does not induce morphological, cytological or molecular changes in recovered seedlings. *Planta*, **232**: 435-447.

Steponkus, P.L. and Lamphear, F.O. (1967) Refinement of the triphenyl tetrazolium chloride method of determining cold injury. *Plant Physiology*, **42**: 1423-1426.

Steponkus, P.L., Langis, R. and Fujikawa, S. (1992). Cryopreservation of plant tissues by vitrification. In: Steponkus. P.L. (Ed.), *Advances in Low Temperature Biology. Vol.1*. JAI Press Ltd. London. pp.1-61.

Sthapit, B., Ramanatha Rao, V. and Jarvis, D. (2005). *In situ* conservation of coconut diversity. In: Batugal, P., Ramanatha Rao, V. and Oliver, J. (Eds.), *Coconut*

Genetic Resources. International Plant Genetic Resources Institute, Regional Office for Asia, the Pacific and Oceania (IPGRI-APO), Serdang, Selangor DE, Malaysia. pp. 143-149.

Tahtamouni, R.W., Shibli, R.A. and Ajlouni, M.M. (2001). Growth responses and physiological disorders in wild pear (*Pyrus syriaca* Boiss) during slow growth *in vitro* preservation on osmostressing media. *Plant Tissue Culture,* **11**: 15-23.

Tandon, R., Chaudhury, R. and Shivanna, K.R. (2007). Cryopreservation of oil palm pollen. *Current Science,* **92**: 182-183.

Thomashow, M.F. (1999). Plant cold acclimation: freezing tolerance genes and regulatory mechanisms. *Annual Review of Plant Physiology and Plant Molecular Biology,* **50**: 571-599.

Towill, L.E. and Walters, C. (2000). Cryopreservation of pollen. In: Engelmann, F. and Takagi, H. (Eds.), *Cryopreservation of Tropical Plant Germplasm-Current Research Progress and Applications. Japan International Research Center for Agricultural Sciences,* Tsukuba, Japan/*International Plant Genetic Resources Institute,* Rome, Italy. pp. 115-129.

Verdeil, J.L. and Pannetier, C. (1990). Optimisation of *in vitro* coconut (*Cocos nucifera* L.) pollen germination conditions in order to develop a viability test. *Oleagineux,* **45**: 179-181.

Verleysen, H., Samyn, G., Van Bockstaele, E. and Debergh, P. (2004). Evaluation of analytical techniques to predict viability after cryopreservation. *Plant Cell Tissue and Organ Culture,* **77**: 11-21.

Visser, T. (1955). Germination and storage of pollen. Mededelingen van de Landbouwhogesschool, Wageningen, The Netherland, **55**: 1-68.

Whitehead, R.A. (1965). Freeze-drying and room temperature storage of coconut pollen. *Economic Botany,* **19**: 267-275.

Whitehead, R.A. (1963). The processing of coconut pollen. Euphytica, **12**: 167-177.

Widholm, J.M. (1972). The use of uorescein diacetate and phenosafranine for determining viability of cultured plant cells. *Stain Technology,* **47**: 189-194.

Withers, L.A. and Engelmann, F. (1997). *In vitro* conservation of plant genetic resources. In: Altman, A. (Ed.), *Biotechnology in Agriculture.* Marcel Dekker, Inc. New York. pp. 57-88.

Withers, L.A. and Engels, J.M.M. (1990). The test tube gene bank: A safe alternative to field conservation. *IBPGR Newsletter from Asia and the Pacific,* **3**: 1-2.

Woodroof, J.G. (1978).*Coconuts: Production, Processing, Products.* AVI Publishing, Westport, Connecticut. 307p.

Zhang, Y.L., Chen, R.D., Huang, C.J. and Liu, Y. (2009). Cryo-banking of *Prunus mume* pollen and its application in cross-breeding. *Cryoletters,* **30**: 165-170.

Chapter 21

Cocoa

✰ *Anitha Karun, V. Aparna, M.K. Rajesh and P. Chowdappa*

1. Introduction

Cocoa (*Theobroma cocoa* L.) is an important tropical plantation crop worldwide. Cocoa seed is important raw material for the production of cocoa powder and butter; these are important ingredients for preparation of chocolate as well as various confectionery products (Li *et al.*, 1998). Cocoa is a native of Tropical America (Cheesman, 1944); later cultivation of this tropical crop spread to the countries in Asia and Africa (Zhang and Motilal, 2016). Cocoa requires a temperature of 21-32°C, well distributed rainfall of 100-250 cm for its optimal growth. It grows only below 1000 m of elevation, ideally below 300 m from the mean sea level. The main growing areas of the crop are situated approximately within 20° North and South latitude of the equator (http://www.cacaoweb.net/cacao-tree.html).

2. Cocoa Genetic Diversity

Genetic resources are a crucial element for the development of new and improved cultivars to achieve a more sustainable and cost-effective means of cocoa production. Cocoa genetic resources comprise the range of genetic variability that provides the raw material for breeding new and improved cultivars. The primary gene pool of cocoa is situated in the Amazon basin, ranging from French Guiana to Bolivia, where a large number of wild populations still exist. The centre of diversity of cocoa is in the upper Amazonian rainforest. Recent molecular studies suggest that the diversity of natural cocoa populations might be stratified by the major river basins (Thomas *et al.*, 2012). Within each river basin, wild cocoa is grouped in patches and separated by large spatial distances between patches. Gene flow is limited and mating is likely confined within patches due to short-distance seed

Table 21.1: Various Sections of *Theobroma* Species According to the Classification Proposed by Cuatrecasas (1964)

Sections and Theobroma Species	*Common Name*
Section *Andropetalum*	
T. mammosum Cuatr. and León	
Section *Glossopetalum*	
T. angustifolium Moçiño and Sessé	'cacao de mico'
T. canumanense Pires et Fróes	
T. chocoense Cuatr.	
T. cirmolinae Cuatr.	
T. grandiflorum (Willd. ex Spreng.) Schum.	'cupuassu'/'cupuaçu'/´Copoasu´/´Cupuasu´
T. hylaeum Cuatr.	
T. nemorale Cuatr.	
T. obovatum Klotzsch ex Bernoulli	' cabeça de urubu'/´Cacahuillo´/´Ushpa cacao´
T. simiarum Donn. Smith.	
T. sinuosum Pavón ex Hubber	
T. stipulatum Cuatr.	
T. subincanum Mart.	'cupuí'/´Macambillo´/´Macambo Sacha
Section *Oreanthes*	
T. bernouillii Pittier	
T. glaucum Karst.	
T. speciosum Willd.	'cacaui'
T. sylvestre Mart.	'cacau azul'
T. velutinum Benoist	
Section Rhytidocarpus	
Theobroma bicolor Humb. and Bonpl.	'mocambo'/'patashte'/´macambo´
Section *Telmatocarpus*	
T. gileri Cuatr.	
T. microcarpum Mart	'cacaurana'
Section *Theobroma*	
Theobroma cacao	'cacao'

and pollen dispersal. Only a very small fraction of the diversity was dispersed from the Amazon to Mesoamerica and thus the ancient cultivated materials have a narrow genetic background (CacaoNet, 2012). This genus *'Theobroma'* contains 22 species; out of which 15 species are edible (CacaoNet, 2012). All these related species possess potential commercial value, mainly because of the sweet seed-surrounding pulp of their fruits in addition to the beans and other crop related characteristics. *T. grandiflorum* (cupuassu), *T. bicolor* and *T. angustifolium* are generally cultivated in native areas of this crop. *T. grandiflorum* is considered an important fruit crop in various Amazonian countries and its cultivation has been increasing, especially in Brazil (CacaoNet, 2012). Considerable work has been done on intergeneric and interspecific crosses involving these species and cocoa with variable levels of success. The 'tertiary genepool' germplasm consists mainly of various species of the genus *Herrania*.

3. Germplasm Conservation Strategies

For many crop plants, germplasm can be stored in the form of dried seeds at low temperature (*i.e.* so-called 'orthodox' seed storage) but this is not possible with cocoa because of its recalcitrant nature. The hygroscopic nature of the seeds further worsens possibility of seed storage in cocoa (Chandel *et al.*, 1995). Cocoa seeds are not even suitable for cold storage conditions; with increasing maturity of the seeds, freezing sensitivity also increases. Fully matured seeds survived desiccation up to 35 per cent of moisture level. But embryo axis was fully sensitive to freezing conditions at all physiological maturity stages (Chandel *et al.*, 1995). Cocoa germplasm can be conserved in two ways:

1. ***In situ:*** The genetic resources are maintained in the natural habitat in which the diversity has evolved *viz.*, natural reserves and on-farm conservation. The advantage of in *situ* conservation is that it facilitates the ongoing processes of natural evolution.
2. ***Ex situ:*** Comprising all cocoa germplasm currently maintained in field genebanks as living trees and/or in *in vitro* collections as tissues, embryos pollen and DNA *etc.*

Ex situ collections play a crucial role in the conservation of many varieties, particularly those that have already disappeared from farmers fields. New germplasm can be introduced into a field genebank as seedling trees. *Ex situ* field genebank collections have the advantages that once the trees are established they can remain in the ground for many decades and can readily provide the bud wood, seed or pollen needed for evaluation and incorporation into breeding programmes. But they are under the risk of destruction due to natural calamities, pests, diseases and are costly for maintenance. Hence there is urgent need to develop safety duplicates of the living collections using alternate strategies of conservation such as *in vitro* conservation. There are 40 genebanks for cocoa around the world. Most of them are supported by public-private funding. There are currently only two international collections managed by the Cocoa Research Unit of the University of the West Indies (CRU/UWI), Trinidad and the Centro Agro-nómico Tropical de Investigacióny Enseñanza (CATIE), Costa Rica. These collections are called as the International Cocoa Genebank, Trinidad (ICG). These institutes concluded agreements with the International Treaty for Plant Genetic Resources for Food and Agriculture (ITPGRFA) to maintain their respective collections as global collections of cocoa genetic resources for the long term and to make the germplasm freely available to any professionally qualified institution or individual. The safe movement of germplasm at the global level, including virus indexing, is achieved through the International Cocoa Quarantine Centre (ICQC), an intermediate quarantine facility supported by the chocolate industry and USDA, at the University of Reading, UK. The USDA/ARS facility in Miami, USA, offers quarantine facilities for regional transfers. The benefits of conserving and utilizing the cocoa genetic diversity will be realized if this diversity is of interest and is made available to researchers engaged in breeding programmes.

Despite the existence of over 24,370 cocoa germplasm accessions in *ex situ* collections worldwide, including 3500 accessions that are held in the two international collections, much of this germplasm is under-utilized or at risk, due to the lack of adequate long-term funding to conserve or utilize the existing germplasm effectively. Furthermore, genetic studies suggest that the material held in *ex situ* genebanks, particularly the international genebanks, does not fully represent the known range of diversity and still there are potential genetic variations yet be to be discovered in the rainforests and farmers fields of the Amazonian region. It has been estimated that even in Brazilian Amazon, where the greatest collecting activity has taken place, only some 20 per cent of the potential diversity has been explored (Bartley, 2005). Central American countries, especially Bolivia, Colombia, Ecuador, Peru and Venezuela, remain largely unexplored for cocoa diversity. With the rapid deforestation in this region, drastic changes in land use and replacement of traditional cocoa varieties with modern ones, both in the Amazon region and in other regions where cocoa is grown, there is the likelihood of irreversible genetic erosion unless further steps are taken to conserve materials *in situ*, or to collect and conserve them *ex situ*.

4. *In vitro* Conservation

Development of a cryopreservation technique for long term storage of cocoa germplasm is very important for preserving biological diversity and genetic fidelity. Cryopreservation is the preservation of viable cells, tissues and organ in liquid nitrogen (LN) at -196° C (Engelmann, 1991; Benson, 1999) and can be stored for indefinite periods without genetic erosion (Golmirzaie *et al.*, 1999). Cryopreservation method involves a sequence of treatments including encapsulation-dehydration, sucrose preculture, silica gel desiccation and liquid nitrogen storage *etc* (Fang *et al.*, 2004).

Cryopreservation of cocoa using zygotic embryos, embryonic axes, callus and somatic embryos as explants has been successfully demonstrated. Grout *et al.* (1983). tried desiccation storage for embryo axis and seeds in liquid nitrogen. Embryos are processed and frozen after aseptically removed from surface sterilized immature fruits. Extracted embryos are placed into a basal media containing cryoprotectant. Treated embryos are transferred to cryovials and then stored in liquid nitrogen container. Cryopreserved embryos were tested after certain storage by transferring the thawed embryos into a germination media (Duhem *et al.*, 1988). During desiccation and freezing, embryo axis was found to be severely damaged and fails to germinate on tissue culture medium. The immature zygotic embryos retains the ability to produce callus and undergo somatic embryogenesis after slow hydrated freezing and desiccated fast freezing in liquid nitrogen were reported by Pence (1991). Immature embryos of cocoa could be able to survive exposure to liquid nitrogen either in hydrated or desiccated state (Pence, 1991). Florin *et al.* (2000) standardized a cryopreservation technique for embryogenic callus of cocoa.

Cryopreservation using encapsulation-dehydration technique for somatic embryos was developed by Fang *et al.* (2004). for long-term conservation of cocoa (*Theobroma cacao* L.) germplasm. Survival of individually encapsulated somatic embryos after desiccation and cryopreservation was achieved through optimization of concentration of cryoprotectant (abscisic acid and sugar), duration of osmotic and evaporative dehydration and development stage of the embryo. Up to 63 per cent somatic embryos of the genotype SPA4, in early-cotyledonary stage, survived cryopreservation following seven days preculture with 1 M sucrose and silica exposure for 4 hours (16 per cent moisture content in bead). Recovered SPA4 somatic embryos were converted to plantlets at a rate of 33 per cent and the regenerated plants were phenotypically comparable to non-cryopreserved somatic embryo-derived plants. This optimized protocol was successfully applied to three other genotypes *e.g.* EET272, IMC14 and AMAZ12 with recovery frequencies of 25, 40 and 72 per cent, respectively (but the latter two genotypes using 0.75M sucrose).

Fang *et al.* (2008) also evaluated the role of ethylene and oxidative stress for the recovery of cryopreserved embryos. Dimethyl sulfoxide (DMSO), a free radical scavenger, and an anti-oxidant compound, quercetin, were supplemented to media for enhancing the recovery of cryopreserved embryos. Wetten *et al.* (2008) tested somaclonal variations in cryopreserved somatic embryos with the help of microsatellite markers and none of the embryos exhibited the aberrations in their DNA profiles. Electron microscopic studies revealed that, primary somatic embryos arise from intermediary callus unlike secondary somatic embryos which were originated from the cells of epidermal calli (Wetten *et al.,* 2008). Secondary somatic embryos, thus pose low risk of undergoing mutations than primary somatic embryos, since they are derived from epidermal cell rather than callus cells. No polymorphism was observed in cryopreserved secondary somatic embryos when they are tested for genetic fidelity with microsatellite markers (Fang *et al.,* 2009).

Quainoo (2009) investigated the effect of liquid nitrogen storage time on survival and regeneration of somatic embryos of cocoa (*Theobroma cacao* L.). Somatic embryos from different cocoa genotypes (AMAZ 3-2, AMAZ 10-1, AMAZ 12, SIAL 93, and IMC 14), at 15.45 per cent moisture content, were cryopreserved in LN for one hour, four and eight weeks. Somatic embryos of the genotypes emerged from the alginate beads at different periods 4 to 12 week's post-cryopreservation. Individual genotypes subjected to low temperature storage time did not show significant differences in post-cryopreservation survival, although different genotypes responded differently with AMAZ 12 and IMC 14 recorded the highest and lowest mean survival rates of 58 per cent and 35 per cent, respectively. Plantlets originating from five genotypes were weaned and these plantlets developed normally and were comparable to non-cryopreserved somatic embryo-derived plantlets in the glasshouse.

Fang and Wetten (2011) studied the structural integrity of cryopreserved somatic embryos of cocoa following encapsulation dehydration method. Results showed that the parenchyma cells of the hypocotyls and radicle were the major sites of injury possibly due to their large size and non-cytoplasmic nature whereas the shoot meristem and provascular strand were well preserved throughout the treatments. The extent of damage increased with each step of the encapsulation dehydration procedure. Even though post thaw regrowth of injured embryos was possible, it often failed to convert into plantlets. The authors suggested the maintenance of structural integrity of the somatic embryos at each treatment step for the successful cryopreservation of cocoa germplasm.

Adu-Gyamfi and Wetten (2012) used vitrification technique for cryopreservation of secondary somatic embryos derived from floral organs. About 74.5 per cent post-cryostorage survival rate of secondary somatic embryos (SSE) could be obtained by pre-culturing SSE on 0.5 M sucrose medium followed by 60 minutes dehydration in cold PVS2. Cation sources were removed from the embryo development solution or the recovery medium in order to reduce the free radical related cryo-injury to the material. This treatment gave a significant benefit during recovery stage. After optimization of this protocol with cocoa genotype AMAZ 15, the same protocol was tested in five other additional lines and successful results were obtained with this vitrification procedure

Adu-Gyamfi *et al.* (2016) studied the epigenetic variations among *in vitro* multiplied somatic embryos, cryopreserved and post- cryopreservation generated somatic embryos along with the ortet trees, using methylation sensitive amplified polymorphism technique. He observed higher level of epigenetic changes in post-cryopreservation generated somatic embryos compared to *in vitro* multiplied and cryopreserved somatic embryos. Furthermore, the passage of cryopreserved embryos through another embryogenic stage led to further increase in variations. Interestingly, these epigenetic variations were reversible to a certain extent.

5. Conclusion

Early attempts for *in vitro* conservation of cocoa started during early 1980s and most of the cryopreservation work was carried out using somatic embryos as well as immature embryos. Though cryopreservation of cocoa tissues appears feasible, a protocol with good recovery of plantlets, along with lowers levels of phenotypic, genetic and epigenetic variation, needs to be standardized for proper maintenance as well as sustainable use of cocoa germplasm. Development of new protocols is also required to conserve different explant material like shoot meristem, embryos, somatic embryos, pollen and DNA samples of wild as well as cultivated species of cocoa.

References

Adu-Gyamfi, R. and Wetten, A. (2012). Cryopreservation of cocoa (*Theobroma cacao* L.) somatic embryo vitrification. *Cryoletters*, **33**: 494-505.

Adu-Gyamfi, R., Wetten, A. and Lopez, C.M.R. (2016). Effect of cryopreservation and post cryopreservation on somatic embryogenesis and epigenetic fidelity of cocoa. *PLoS ONE*, **11**: 0158857.

Bartley, B.G.D. (2005). The foundations of diversity. In: Bartley, B.G.D. (Ed.), *The Genetic Diversity of Cacao and its Utilization*. CABI Publishing, Wallingford, U.K. pp. 68-196.

Benson, E.E. (1999). Cryopreservation. In: Benson, E.E. (Ed.), *Plant Conservation Biotechnology*. T.J. International Ltd, Padstow, U.K. pp. 83-95.

CacaoNet (2012). A Global Strategy for the Conservation and Use of Cacao Genetic Resources, as the Foundation for a Sustainable Cocoa Economy (B. Laliberté, compiler). *Bioversity International*, Montpellier, France. 174p.

Chandel, K.P.L., Chaudhury, R., Radhmani, J. and Malik, S.K. (1995). Desiccation and freezing sensitivity in recalcitrant seeds of tea, cocoa and jackfruit. *Annals of Botany*, **76**: 443-450.

Cheesman, E. (1944). Notes on the nomenclature, classication and possible relationships of cocoa populations. *Tropical Agriculture*, **21**: 145-146.

Cuatrecasas, J. (1964). Cacao and its allies: A taxonomic revision of the genus *Theobroma*. *Contributions from the United States National Herbarium*, **35**: 379-614.

Duhem, K., Le Mercier, N. and Boxus, P. (1988). Difficulties in the establishment of axenic *in vitro* cultures of field collected coffee and cacao germplasm. *Acta Horticulture*, **225**: 67-76.

Engelmann, F. (1991). *In vitro* conservation of *tropical plant germplasm*-a review. *Euphytica*, **57**: 227-243.

Fang, J.Y. and Wetten, A. (2011). Importance of structural integrity of somatic embryos for long-term cryopreservation of cocoa (*Theobroma cacao* L.) germ plasm. *African Journal of Agricultural Research*, **6**: 395-396.

Fang, J.Y., Wetten, A. and Hadley, P. (2004). Cryopreservation of cocoa (*Theobroma cacao* L.) somatic embryos for long-term germplasm storage. *Plant Science*, **166**: 669-675.

Fang, J.Y., Wetten, A. and Johnston, J. (2008). Headspace volatile markers for sensitivity of cocoa (*Theobroma cacao* L.) somatic embryos to cryopreservation. *Plant Cell Reports*, **27**: 453-461.

Fang, J.Y., Wetten, A. and Hadley P. (2004). Cryopreservation of cocoa (*Theobroma cacao* L.) somatic embryos for long-term germplasm storage. *Plant Science*, **166**: 669-675.

Fang, J.Y., Wetten, A., Raphael, A.G., Wilkinson, M. and Lopez, R.C. (2009). Use of secondary embryos promotes genetic fidelity in cocoa (*Theobroma cacao* L). *Agricultural and Food Science*, **18**: 152-159.

Florin, B., Brulard, E. and Pétiard, V. (2000). *In vitro* cryopreservation of cacao genetic resources. In: Engelman, F. and Tagaki, H. (Eds.), *Cryopreservation of Tropical Plant Germplasm*. *Japanese International Research Centre for Agricultural Sciences and IPGRI*, Rome, Italy. pp. 344-347.

Golmirzaie, A.M., Panta, A. and Toledo, J. (1999). Biotechnological advances in the conservation of root and tuber crops. In: Benson, E.E. (Ed.), *Plant Conservation Biotechnology*. Taylor and Frances, London, UK. pp. 165-178.

Grout, B.P.W., Shelton, K. and Pritchard, W. (1983). Orthodox behaviour of plant seed and cryopreservation of excised embryos for genetic conservation. *Annals of Botany*, **52**: 381-384.

Li, Z., Traore, A., Maximova, S. and Guiltinan, M. (1998). Somatic embryogenesis and plant regeneration from floral explants of cacao (*Theobroma cacao* L.) using thidiazuron. In *Vitro Cellular and Developmental Biology-Plant*, **34**: 293-299.

Pence, V.C. (1991). Cryopreservation of immature embryos of *Theobroma cacao* L. *Plant Cell Reports*, **10**: 144-147.

Quainoo, A.K. (2009). Effect of liquid nitrogen storage time on the survival and regeneration of somatic embryos of cocoa. *African Crop Science Journal*, **17**: 17-24.

Thomas, E., Zonneveld, V.M., Loo, J., Hodgkin, T., Galluzi, G., and Etten V.J. (2012). Present special diversity pattern of *Theobroma cacao* L. In the neotropics reflect genetic differentiation in Pleistocene refugia followed by human influenced dispersal. *PLoS ONE*, **7**: e47676.

Wetten, A., Fang, J.Y. and Lopez, C.R. (2008). Cryopreservation of cocoa (*Theobroma cacao* L.) somatic embryos- is somaclonal variation an issue? Society for Low *Temperature Biology Meetings*, University of Copenhagen, Denmark. pp. 152-159.

Zhang, D. and Motilal, L. (2016). Origin, dispersal and current global distribution of cacao genetic diversity In: Baily, L.W. and Meinhardt, L.W. (Eds.), *Cocoa Diseases*. Springer International Publishing, Switzerland. pp. 1-31.

Chapter 22

Spices

☆ *K. Nirmal Babu, Minoo Divakaran, S.P. Geetha, G. Yamuna, K.V. Saji, P.N. Ravindran and K.V. Peter*

1. Introduction

Plant genetic resources, a subset of biodiversity, contain the genetic material which holds key to food security and sustainable agricultural development. Direct threats to plant survival are a combination of habitat loss, aggressive alien species, over-exploitation and climate change. Due to the destruction of their natural habitat, spread of high yielding varieties and selection pressure, genetic variability of crops is gradually eroded leading to large-scale depletion of variability. Loss of plants leads to worsening conditions such as food insecurity, increasing vulnerability to disease, lower material wealth, deteriorating social relations and restricted freedom of choice and action. This demands urgent action to conserve germplasm be it at species, genepool or ecosystem level, for posterity (Frankel, 1975).

Conservation of species, particularly the agriculturally important ones which have an impact on human well being, has now attained paramount importance, in our efforts to achieve sustainable utilization of biological resources, by preventing further loss. Plant genetic resources, constituting genotypes or populations of cultivars (landraces, advance/improved cultivars), genetic stocks, wild and weedy species, which are maintained in the form of plants, seeds, tissues, *etc.*, hold key to food security and sustainable agricultural development (Iwananga, 1994). The opportunities offered by conservation biotechnology should not be neglected or restricted by lack of interconnectivity between traditional and contemporary conservation practitioners.

Spices and herbs are aromatic plants–fresh or dried plant parts like foliage, young shoots, roots, bark, buds, seeds, berries and other fruits of which are mainly used to flavour our culinary preparations, confectionary. They are also major ingredients in indigenous medicine and perfumery. Spices and herbs are grown throughout the world–different plant species in different regions. Peninsular India is a rich repository of spices and over 100 species of spices and herbs are grown. Black pepper, cardamom, ginger, turmeric, vanilla, capsicum, cinnamon, clove, nutmeg, tamarind, coriander, cumin, fennel, fenugreek, dill, caraway, anise and herbs like saffron, lavender, thyme, oregano, celery, anise, sage and basil are important as spices. India being the native home of many spices, their conservation and characterization are one of the priority programmes.

2. Conservation Strategies

Conventionally, *in situ* conservation allows evolution to continue within the area of natural occurrence, and *ex situ* conservation provides a better degree of protection to germplasm compared to *in situ* conservation. *Ex situ* conservation includes germplasm banks, common garden archives, seed banks, DNA banks and techniques involving tissue culture, cryopreservation; incorporation of disease, pest and stress tolerance traits through genetic transformation and ecological restoration of rare plant species and their populations. In many crop spices conventional seed storage can satisfy most of the conservation requirements.

But crops with recalcitrant seeds, and those whose conservation needs cannot be satisfied by seed storage, have to be stored *in vitro.* Most field gene banks are prone to high labour cost, vulnerability to hazards like natural disasters, pests and pathogens attack (especially viruses and systemic pathogens), to which they are continuously exposed and required large areas of space (Chandel and Pandey, 1991). This supports *in vitro* and cryo conservation. In addition, other resources like continuous supply of standard stock cultures for experiments to examine physiological and biochemical processes, cell and callus lines developed for *in vitro* synthesis of valuable secondary products, flavours and other important compounds will benefit strongly from *in vitro* cultures. Thus *in vitro* storage system becomes an important starategy of conserving genepool. Cryopreservation is, so far, the only viable procedure for long-term germplasm conservation of many species and, therefore, can be used for base collections. Each technology should be chosen on the basis of utility, security and complementarity to the other components of strategy.

However, a balance needs to be struck between seed, field genebank, *in vitro* conservation of propagules, tissues, pollen, cell lines and DNA storage for overall objective of conserving gene pool (Figure 22.1, Withers, 1991). The different systems of *ex-situ* conservation of biodiversity, according to targeted germplasm and period of conservation are listed in Table 22.1. Conservation of the germplasm in *in vitro* gene bank and cryo bank is a viable method and a safe alternative to augment the conventional conservation strategies.

3. Technologies for *In vitro* and Cryo Conservation

3.1. Micropropagation

Plant regeneration and successful cloning of genetically stable plantlets in tissue culture is an important pre-requisite in any *in vitro* conservation effort. These

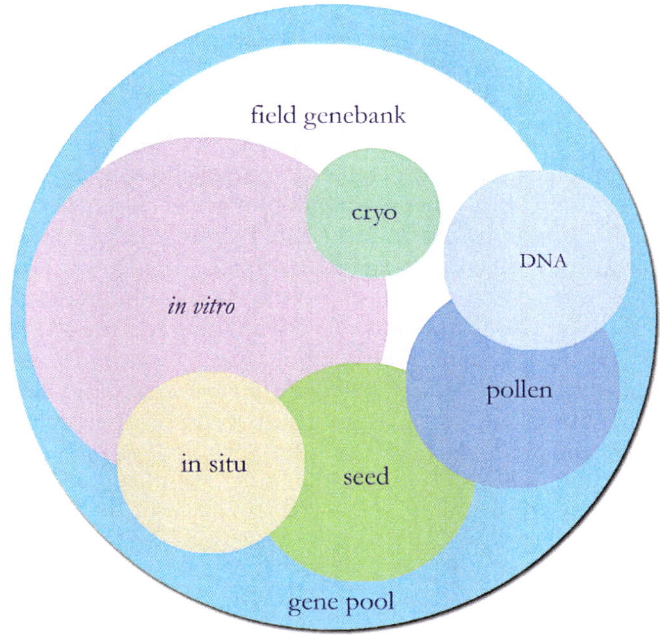

Figure 22.1: A Balanced Strategy for Gene Pool Conservation (Withers, 1991).

Table 22.1: Systems of *ex-situ* Conservation of Biodiversity, According to Targeted Germplasm and Period of Conservation

Conservation Systems		*Targeted Germplasm*	*Term of Conservation*
Seed Type	*System*		
Orthodox	Seed bank	Cultivated species Gene pools	Medium or long term
	Botanic garden	Species for classification Flora	Depends on life cycle and no. of sub-culture cycle
	Cryopreservation bank	Cultivated species	Long term
	DNA bank	Special sequences	
Recalcitrant	*In vitro* bank	Cultivated species Wild species	Short term
	In vivo bank	Cultivated species	Varies with life cycle and no. of subculture
	Botanic garden	ClassificationFlora	Varies with life cycle and no. of subculture
	Pollen bank	Cultivated plants (male)	Long term
	Cryopreservation bank	Cultivated species	Long term
	DNA bank	Special sequences	Medium to long term

Adopted from Benjamin P, Daniel D and Mariano M; MiDL course on *ex-situ* conservation of plant genetic resources.

techniques form the base for establishing tissue cultures and developing *in vitro* technology for conservation. Simultaneously these tissue-cultured plants should be evaluated for their morphological and genetic stability in culture. The *in vitro* storage experiments, as much as possible, should use using growth regulators free media to reduce the rate of multiplication which in turn will reduce the extent of variation.

Micropropagation protocols for stable cloning of elite genotypes of spice crops have been standardised. Protocols are available for black pepper and its related species cardamom, ginger, turmeric and related genera, large cardamom, kasturi turmeric, mango ginger, *Kaempferia galanga, K. rotunda, Alpinia* spp, large cardamom, vanilla and related species, cinnamon, camphor, cassia seed and herbal spices like lavender, celery, thyme, mint, anise, savory, spearmint and oregano (Nirmal Babu *et al.*, 1999, 2005; Peter *et al.*, 2002; Minoo, 2002). These techniques formed the basis for establishing tissue cultures and developing *in vitro* technology forconservation. Protocols are available for micropropagation and multiplication of many endangerd species like *Piper hapnium, P. silent vallyensis, P. schmidtii, P. wightii, P. barberi, Vanilla aphylla, V. pilifera, V. walkyrie, V. wightiana, K. rotunda* and *Alpinia galanga* are available (Nirmal Babu *et al.*, 2012).

3.2. Synthetic Seeds

Synthetic seeds include encapsulated buds, bulbs or any form of meristem that can develop into a plant. Encapsulation is technique used in the production of 'synthetic seeds' by coating somatic embryos/shoot tips in alginate beads. Several gels like agar, alginate, polyco 33 (Bordon Co), Carboxy methylcellulose, Carrageenan, gelrite (Kelco Co), Guargum, sodium pectate, tragacanth gum *etc.* have been tested for synthetic seed production. Alginate hydrogel is frequently selected as a matrix for synthetic seed because of its moderate viscosity and low spinnability of solution, low toxicity for somatic embryos and quick gellation, low cost and biocompatibility characteristics. Encapsulation protects both embryos and micro-plantlets. It provides favorable conditions for handling, storage and mechanical seeding. Artificial or 'synthetic seeds' could serve as an ideal system for low cost plant movement, propagation, conservation and exchange of germplasm.

Reports on synthetic seeds are available in ginger, cardamom and piper by encapsulating the somatic embryos and shoot buds in calcium or sodium alginate (Nirmal Babu *et al.*, 2012) (Figure 22.2).

3.3. *In vitro* Conservation

In vitro conservation involves maintenance of explants in a sterile, pathogen-free environment and is widely used for the conservation of species that produce recalcitrant seeds, or do not produce seeds (Engelmann, 1997). The principle of slow growth storage is that the safety of *in vitro* culture be ensured without disadvantages of frequent subculturing. Thus the risk of contamination at each transfer interval, inputs in terms of labour and consumables are reduced.

Various *in vitro* conservation methods can be used. For short-and medium-term storage, the aim is to increase the intervals between subcultures by reducing growth.

Figure 22.2: Shoot Buds of Ginger Encapsulated in Na-alginate (*Source*: Nirmal Babu *et al*., 2012).

This is achieved by modifying the environmental conditions, including the culture medium to realize slow-growth conservation. The most widely applied technique is temperature reduction (varying from 0-5°C for cold tolerant species to 9-18°C for tropical species) that can be combined with a decrease in light intensity or storage in the dark (Engelmann, 1997) and adjustment of the growth medium. Though growth reduction can generally be achieved by lowering the culture temperature, but the scope for temperature reduction depends upon the species to be conserved. Several methods, such as temperature reduction, medium modification, use of osmoticums, *etc.*, have been found to reduce the rate of growth of tissue cultures, so that it can be kept unattended for moderate length of time (Withers, 1980, 1987, 1991; Withers and Williams, 1986; Ashmore 1997). Alternatives to standard slow-growth conservation include modification of the gaseous environment of cultures, desiccation and encapsulation of explants.

A schematic diagram of *in vitro* conservation starategy for conserving spices germplasm is given in Figure 22.3.

Conservation of pepper, cardamom, herbal spices, vanilla and ginger germplasm in *in vitro* gene banks by slow growth has been reported by Nirmal Babu *et al.* (1999, 2007), Peter *et al.* (2002), Geetha (2002) and Ravindran *et al.* (2004) Protocols for *in vitro* conservation by slow growth of black pepper and its related species *viz., P. barberi, P. colubrinum, P. betle* and *P. longum* were standardized by maintaining cultures at low temperatures, in the presence of osmotic inhibitors, at reduced nutrient levels, or minimizing evaporation loss by using closed containers (Nirmal Babu *et al.*, 2012). Black pepper cultures could be maintained in half strength WPM supplemented with 15 gl^{-1} each of sucrose and mannitol for one year with 85 per cent survival. In *P. barberi,* full strength WPM with 25 gl^{-1} sucrose and 5 gl^{-1} mannitol was suitable for storage of cultures up to one year with 80 per cent survival (Nirmal Babu *et al.*, 2012). Shoot tips of *P. longum* and *P. colubrinum* could be stored up to one year in

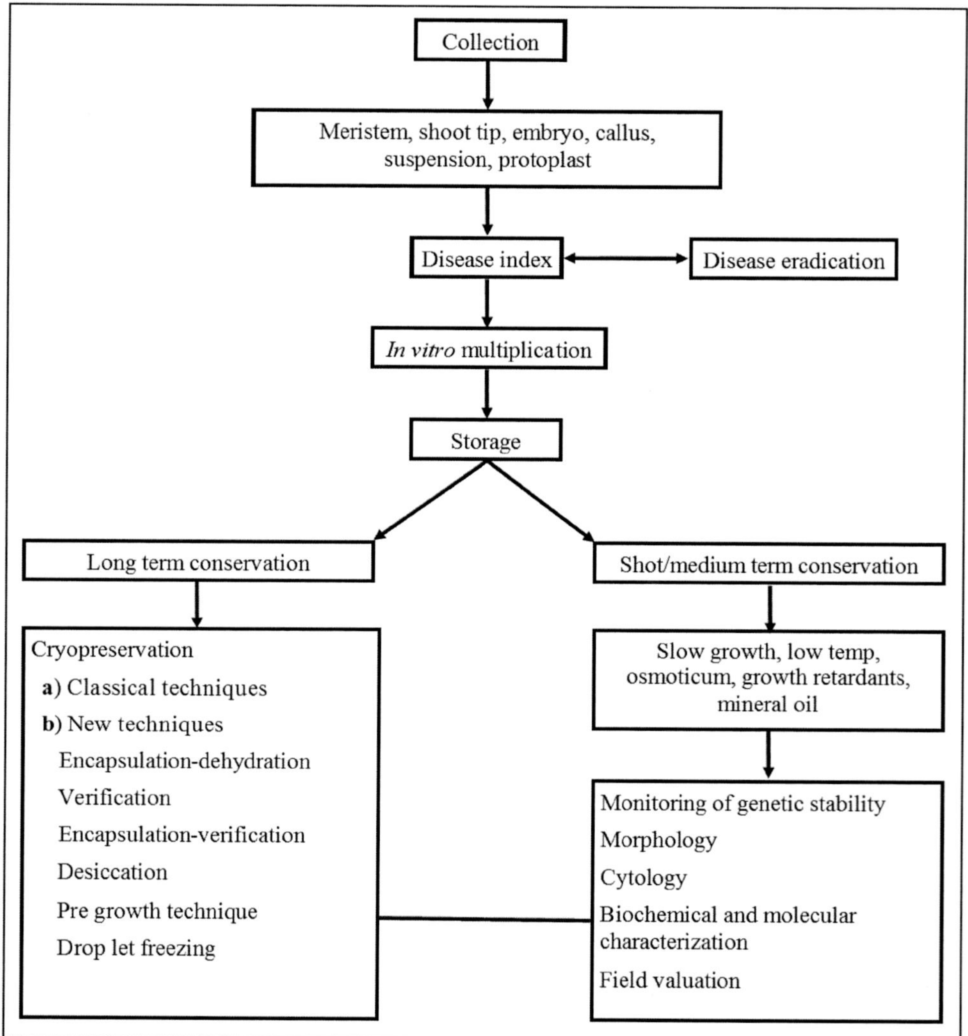

Figure 22.3: Schematic Representation of Important Steps for *In vitro*/Cryo conservation of Spice Germplasm.

full strength WPM with 20 gl^{-1} sucrose and 10 gl^{-1} mannitol with 75 per cent and 70 per cent survival respectively (Nirmal Babu *et al.,* 2012). *P. betle* cultures could be stored in half strength WPM supplemented with 20 gl^{-1} sucrose for one year (Nirmal Babu *et al.,* 2012). Technology for *in vitro* conservation of Zingeberaceous crops like ginger, turmeric, Kaempheria, cardamom and their related species was standardized by Geetha (2002) and vanilla by Minoo (2002). The other reports on *in vitro* conservation were made by Peter *et al.* (2002) and Ravindran *et al.* (2004). The conserved materials of all the species showed normal rate of multiplication when transferred to multiplication medium after storage. The normal sized plantlets, when transferred to soil, established with over 80 per cent success. They developed into normal plants without any deformities and were morphologically similar to mother plants. RAPD profiling of these conserved plants also showed their genetic integrity (Nirmal Babu *et al.*, 2012).

Conservation of genetic resources in *in vitro* gene banks is now an established convention and two gene banks for conservation of spices germplasm functions at ICAR-Indian Institute of Spices Research,Kozhikode and ICAR-National Bureau of Plant Genetic Resources, New Delhi.

4. Cryopreservation

For long-term conservation of the problem species, cryopreservation is the only effective method currently available. Dramatic progress has been made in recent years in the development of new cryopreservation techniques and cryopreservation protocols have been established for over 100 different plant species.

Cryopreservation is an attractive option for long-term storage. Liquid nitrogen (–196°C) is routinely used for cryogenic storage, since it is relatively cheap and safe, requires little maintenance and is widely available. Below –120°C the rate of chemical or biophysical reactions is too slow to cause biological deterioration (Kartha, 1985). Only in the long term, there might be a little risk of ionising radiation causing genetic changes in materials stored at cryogenic temperatures (Grout, 1995). There are two main types of new cryopreservation techniques, although combinations of them have also been used. The first one is actually termed vitrification (*sensu stricto*) and the second one, encapsulation-dehydration.

Plant germplasm stored in liquid nitrogen (-196°C) does not undergo cellular divisions. In addition, metabolic and most physical processes are stopped at this temperature and plant material can be indefinitely stored without alteration or modification. As such, plants can be stored for very long time periods and both the problem of genetic instability and the risk of loosing accessions due to contamination or human error during subculturing are overcome. Most cryopreservation endeavours deal with recalcitrant seeds, *in vitro* tissues from vegetatively propagated crops, species with a particular gene combination (elite genotypes) and dedifferentiated plant cell cultures. Care must be taken to avoid ice crystallization during the freezing process, which otherwise would cause physical damage to the tissues. The existing cryogenic strategies rely on air-drying, freeze dehydration, osmotic dehydration, addition of penetrating cryoprotective substances and adaptive metabolism (hardening), encapsulation, vitification or combinations of these processes. An array of plant material could be considered for cryopreservation that include meristems, cell, callus and protoplast cultures, somatic and zygotic embryos, anthers, pollen or microspores and whole seeds (Withers, 1985; Kartha, 1985). At the Federal Centre for Breeding Research on Cultivated Plants (BAZ, Germany), the viability of potato apices after long-term storage in liquid nitrogen has been studied. The cryopreservation technique employed was the droplet method and storage was carried out in liquid nitrogen. Plant regeneration from long-term (3 to 8 years) *versus* short-term stored apices was studied in 51 cultivars. A significant decrease in recovery of the long-term stored apices was observed only in three of 51 cultivars. Cryopreservation methods have been developed for more than 80 different plant species in various forms like cell suspensions, calluses, apices, somatic and zygotic embryos (Kartha and Engelmann, 1994; Engelmann, 1997, 2000, Engelmann *et al.*, 1994; Ashmore 1997). However, routine utilization of cryopreservation technique is still restricted almost exclusively to the conservation of cell lines in research laboratories.

5. Cryo-conservation in Spices

Reports on cryopreservation of spices genetic are of recent origin. In black pepper, cryopreservation of seeds in liquid nitrogen (LN) was reported by Choudhary and Chandel, (1994), and Choudhury and Malik (2004) by desiccating the seeds to 12 per cent moisture content with a survival rate of 45 per cent. Yamuna (2007) reported a vitrification procedure in which the somatic embryos were precultured for three days on SH basal medium containing 0.3 M sucrose and subjected to vitrification treatment for 60 minutes at 25°C resulting in 71 per cent survival after cryopreservation. Cryopreservation of encapsulated shoot buds of endangered *Piper barberi* using vitrification technique was reported by Peter *et al.* (2002) and Ravindran *et al.* (2004) with 60 per cent success (Figure 22.4). Studies on cryopreservation of critically endangered *P.barberi* revealed that, the encapsulation-vitrification procedure produced 70 per cent survival of cryopreserved shoot tips (Yamuna, 2007). Genetic fidelity studies showed that the regenerated plants were similar to the controls.

Figure 22.4. Shoot Tips of *Piper barberi* Encapsulated in Na-alginate, arrow Indicates Shoot Tip Used as Explants (Source: Nirmal Babu *et al.*, 2012).

In cardamom, Choudhary and Chandel (1995) reported that seeds possessing 7.7-14.3 per cent moisture content could be successfully cryo-preserved with 80 per cent germination when tested after one-year storage in vapor phase of liquid nitrogen (at - 150°C). Ravindran *et al.* (2004) reported that 70 per cent success was observed when *in vitro* grown cardamom shoot tips (1.0-2.0 mm) when subjected to progressive increase of sucrose concentrations (0.1, 0.3, 0.5, 0.7, 0.9, and 1.0) for two days and transferred to 1.8 ml cryotube containing ice cold PVS2 solution [30 per cent (v/v) glycerol + 15 per cent (v/v) ethylene glycol + 15 per cent (v/v) DMSO in culture medium with 0.4 M sucrose, pH (5.8)] at 0°C for three hours and were directly immersed into liquid nitrogen for one hour. The vials were thawed in 40°C water for one minute, cryoprotectant removed and the shoot tips were washed 2-3 times in 1.2 M sucrose solution and cultured on MS medium supplemented with BAP and NAA. Yamuna (2007) tested the effect of encapsulation-dehydration, encapsulation-vitrification and vitrification methods on cryopreservation of cardamom (Figure 22.5). In the vitrification treatment, to enhance tolerance to vitrification solution (PVS2), a two-step sucrose preculture with 0.3 M and 0.75 M sucrose for one day each and an osmo-protection step with a loading solution (LS) of 2 M glycerol and 0.4 M sucrose were performed prior to PVS2 treatment. The shoots that were dehydrated with PVS2 for 60 minutes retained a high level of shoot formation (70 per cent). The vitrification procedure resulted in higher regrowth (70 per cent) when compared to encapsulation vitrification (62 per cent) and encapsulation dehydration (60 per cent). In all the three cryopreservation procedures tested, shoots grew after cryopreservation without intermediary callus formation. The genetic stability of cryopreserved cardamom shoots were confirmed using ISSR and RAPD based molecular markers.

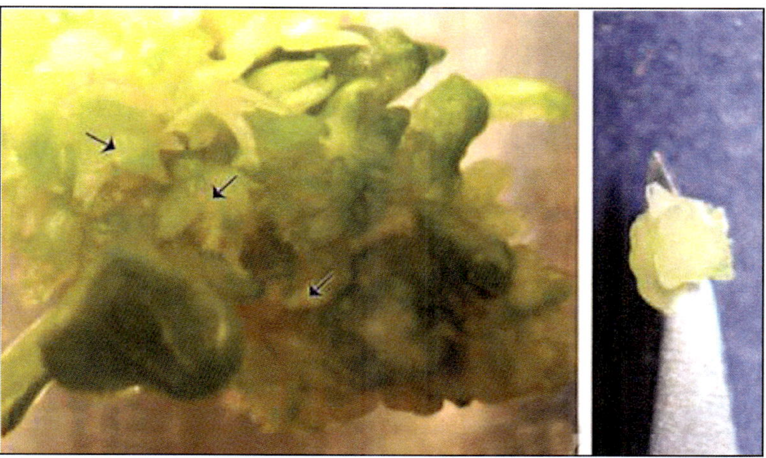

Figure 22.5: Plant Regeneration from Cryopreserved Miniature Shoots of Cardamom by Vitrification. Left: cardamom culture with miniature shoots; Right: excised meristematic clumps used for cryopreservation (Source: Nirmal Babu *et al.*, 2012).

Cryopreservation of ginger and turmeric shoot tips was successfully done with 80 per cent of recovery using vitrification method (Figure 22.6). But the rate of recovery was only 40 per cent when encapsulated shoot tips were dehydrated in progressive increase of sucrose concentration together with 4- 8 hours of desiccation (Peter *et al.*, 2002; Ravindran *et al.*, 2004). Efficient cryopreservation techniques were developed for *in vitro* grown shoots of ginger based on encapsulation-dehydration, encapsulation-vitrification and vitrification procedures (Yamuna *et al.*, 2007; Yamuna, 2007; Nirmal Babu *et al.*, 2012). The vitrification procedure resulted in higher regrowth (80 per cent) when compared to encapsulation-vitrification (66 per cent) and encapsulation-dehydration (41 per cent). The genetical stability of shoot apices was confirmed by RAPD and ISSR assays based molecular profiling and it suggested that no genetic aberrations originated in ginger plants during culture and cryopreservation.

Figure 22.6: Plantlets Regenerating from Cryopreserved Shoot Bud of Ginger (Source: Nirmal Babu *et al.*, 2012).

Technology for cryopreservation of vanilla germplasm, using encapsulation and vitrification methods, are available. Encapsulated *in vitro* grown shoot tips of vanilla could be cryopreserved with 70 per cent success when pretreated with progressive increase of sucrose concentration (0.1 M-1.0 M) for one day each and dehydrated for eight hours (Peter *et al.*, 2002; Minoo, 2002; Ravindran *et al.*, 2004; Minoo and Babu, 2009 a; Minoo *et al.*, 2010). Attempts have also been made to cryopreserve *V. planifolia* Andr. using *in vitro* fragmented explants (IFEs) and the apices derived from them. Cryopreservation of apices from *in vitro* grown plants was achieved using the droplet vitrification protocol. Maximum survival (30 per cent) and further regeneration (10 per cent) of new shoots were obtained for apices derived from clusters of *in vitro* plantlets produced from microcuttings through a three-step droplet vitrification protocol. Plants could be successfully regenerated from cryopreserved seeds of capsicum (Peter *et al.*, 2002; Ravindran *et al.*, 2004). The regrowth after cryopreservation and average number of new embryos developed from cryopreserved ECs were comparable to that of untreated control (98 per cent and 13 embryos per clump, respectively). Both normal and abnormal plants were produced from control and cryopreserved cultures, indicating that appearance of abnormalities was not related to cryopreservation. The regenerants with normal phenotype showed the same peaks of relative DNA content regardless of cryopreservation. The results suggest that simple desiccation method is effective for cryopreservation of coriander somatic embryos with subsequent regeneration. Plants could be regenerated from cryopreserved seeds of anise (Peter *et al.*, 2002). Successful cryopreservation of seeds, meristems, somatic or zygotic embryos were reported in *Allium* spp. (Niwata, 1995). Preliminary success was reported in cryopreservation of mint (Leigh and Remi, 2003). Most of the reports are confined to a few genotypes and hence the techniques standardized needs to be extended to more genotypes before adopting them for routine conservation. Reports of cryoconservation of spices like *Ocimum, Lavendula, Salvia* are available from ICAR-National Bureau of Plant Genetic Resources (ICAR-NBPGR), New Delhi. Mandal *et al.* (2000) reported propagation and conservation of four pharmaceutically important herbs hoary basil, sweet basil, shrubby basi; and sacred basil using encapsulating axillary vegetative buds harvested from garden-grown plants of these four *Ocimum* species in calcium alginate gel. The gel contained Murashige and Skoog (MS) nutrients and 1.1-4.4 μM benzyladenine (BA). The encapsulated buds could be stored for 60 days at 4°C. Plants retrieved from the encapsulated buds could be hardened-off and established in soil.

Shoot tips of *in vitro*-grown plants *Syzygium francissi* were successfully cryostoraged at –196°C by the encapsulation-dehydration method. A preculture of formed beads on MS medium containing 0.75 *M* sucrose for 1 day, followed by 6 hours dehydration (20 per cent moisture content) led to the highest survival rate after cryostorage for 1 hour. This method is a promising technique for *in vitro* propagation and cryopreservation of shoot tips from *in vitro*-grown plantlets of *S. francissi* germplasm (Shatnawi *et al.*, 2004).

Efficient vitrification, encapsulation and dehydration of calli was reported in two species of *Crocus* (Chand *et al.*, 2000). After cryopreservation, the highest survival (55.6 - 61.1 per cent) and regrowth (16.7 - 27.8 per cent) rates respectively were achieved when calluses of *C. hyemalis* and *C.moabiticus* were pretreated with 0.5 M sucrose for two days after two hours of dehydration.

6. Pollen Storage

Pollen storage can be of considerable value in supplementing the germplasm conservation strategies by facilitating hybridization between plants with different periods of flowering and to transport pollen across the globe for various crop improvement programmes in addition to developing haploid or homozygous lines. The technique of pollen storage is simple since pollen can be dried (less than 5 per cent moisture content on a dry weight basis) and stored below 0°C. Pollen storage might represent an interesting alternative for the long-term conservation of problematic species However, pollen has a relatively short life compared with seeds. An advantage is that pests and diseases are rarely transferred by pollen (excepting some virus diseases). This allows safe movement and exchange of germplasm as pollen. (Hoekstra,1995, IPGRI, 1996). Alexander *et al.* (1991) and Rajasekharan and Ganeshan (2003) reported freeze preservation of capsicum pollen in liquid nitrogen (–1960C) for 42 months. Pollen from two asynchronously flowering species of *Vanilla viz.*, cultivated *V. planifolia* and its wild relative *V. aphylla,* were cryopreserved after desiccation, pretreatment with cryoprotectant dimethyl sulphoxide (5 per cent) and cryopreserved -196°C in liquid nitrogen (Figure 22.7). The cryopreserved pollen was latter thawed and tested for their viability both *in vitro* and *in vivo.* A germination percentage of 82.1 per cent and 75.4 per cent in *V. planifolia* and *V. aphylla* pollen respectively were observed indicating their viability. The cryopreserved pollen of *V. planifolia* was used successfully to pollinate *V. aphylla* flowers resulting in fruit set the seeds thus obtained were successfully cultured to develop hybrid plantlets (Minoo, 2002). This system is of great importance and can be used for conserving the haploid gene pool of *Vanilla* in cryo- banks and their subsequent utility in crop improvement.

Figure 22.7: Germination of Cryopreserved Vanilla Pollen (Source: Nirmal Babu *et al.*, 2012).

7. DNA Storage

Concurrent with the advancements in gene cloning and transfer has been the development of technology for the removal and analysis of DNA. DNAs from the nucleus, mitochondria and chloroplast are now routinely extracted and immobilized onto nitrocellulose sheets where the DNA can be probed with numerous cloned genes. In addition, the rapid development of polymerase chain reaction (PCR) now means that one can routinely amplify specific oligonucleotides or genes from the entire mixture of genomic DNA.These advances, coupled with the prospect of the loss of significant plant genetic resources throughout the world, have led to the establishment of DNA bank for the storage of genomic DNA.The conserved DNA will have numerous uses *viz.*, molecular phylogenetics and systematics of extant and extinct taxa, production of previously characterized secondary compounds in transgenic cell cultures, production of transgenic plants using genes from gene families, *in vitro* expression and study of enzyme structure and function and genomic probes for research laboratories.

The vast resources of dried specimens in the world's herbaria may hold considerable DNA that would be suitable for PCR. It is highly likely that the integrity of DNA would decrease with the age of specimens. Because there are many types of herbarium storage environments, preservation and collections, there is a need for systematic investigations of the effect of modes of preparation, collection and storage on the integrity of DNA in the world's major holdings.

The advantage of storing DNA is that it is efficient and simple and overcomes many physical limitations and constraints that characterize other forms of storage (Adams, 1988, 1990, 1997; Adams and Adams, 1991; Adams *et al.*, 1994). The disadvantage lies in problems with subsequent gene isolation, cloning and transfer but, most importantly, it does not allow the regeneration of live organisms (Maxted *et al.*, 1997). DNA banking is yet to catch up in spices. DNA samples of over 600 genotypes of spices is stored in the DNA bank of ICAR-IISR, Kozhikode, Kerala, India.

8. Callus and Cell Culture Systems

Quatrano (1968) and Nag and Street (1975) reported the first successful experiments on cryopreservation of plant cells. Since then, a large number of cell suspension and calli cultures have been successfully cryopreserved (Engelmann *et al.*, 1994). In general, callus cultures are more difficult to cryopreserve than cell suspensions, because of the relative volume of the callus, its slow growth rate and cellular heterogeneity (Withers, 1987). Plant cells cultured *in vitro* produce wide range of primary and secondary metabolites of economic value. Production of phytochemicals from plant cell cultures has been presently used for pharmaceutical products. Production of flavour components and secondary metabolites *in vitro* using immobilized cells is an ideal system for aromatic and medicinally important crops.

9. Genetic Stability of Conserved Materials

An important prerequisite for any conservation technique is that the regenerants produced from the conserved material should be true-to-type. There are ample evidences to indicate that under certain culture conditions the materials undergo genetic changes (somaclonal variations) and as a consequence, lose their integrity and uniformity. This would be highly undesirable in spices varieties where the purpose is not only to conserve a genotype but also retain its specific quality traits. Thus testing for the genetic stability of *in vitro* conserved materials is of utmost importance. Besides morphology, cytology and isozyme profiling, sophisticated biochemical and DNA-based techniques have enabled more critical analysis of the genetic stability of *in vitro* materials. RAPD, ISSR and SSR analyses can be done to evaluate genetic fidelity of the cryopreserved lines of spices. RAPD and ISSR, SSR profiles have been developed in spices by Nirmal Babu *et al.* (2003, 2007) and Ravindran *et al.* (2004). Morphological characters, coupled with RAPD profiles using 24 RAPD primers, have indicated genetic fidelity among randomly selected micropropagated plants of Subhakara and Aimpiriyan, indicating that micropropagation protocol can be used for commercial cloning of black pepper (Nirmal Babu *et al.*, 2003). Genetic uniformity of micropropagated *Piper longum* using RAPD profiling was reported by Ajith (1997) and Parani *et al.* (1997) for conservation. Peter *et al.* (2002) and Ravindran *et al.* (2004) reported that the conserved materials of species showed normal rate of multiplication when transferred to multiplication medium after storage. The normal sized plantlets when transferred to soil were found to be established with over 80 per cent success rate. Thus conserved materials developed into normal plants without any deformities and were morphologically similar to mother plants. RAPD profiling of these conserved plants also showed their genetic uniformity.

10. Germplasm Exchange

Germplasm exchange using *in vitro* culture is to certain extent practiced in potato, other tuber crops, orchids, anthuriums *etc.* In view of the necessity of reducing the possibility of introduction of new pathogens and pests along new plant introductions, it is imperative that we use *in vitro* technology for plant introduction wherever possible, especially in spices. Utilization of microrhizomes in ginger and turmeric can be a positive development in this regard.

11. Conclusion

In vitro and cryo-conservation can form an ideal supplement for field gene banks in conserving the active germplasm in vegetatively propagated crop species and form a source for multiplication and exchange of disease free material of various genotypes as and when required. Tissue culture techniques, together with cryopreservation, are of great interest for the medium and long-term conservation of plant germplasm, particularly that of tropical species. However, before routinely using *in vitro* techniques for germplasm preservation, various problems have to be encountered and suitable amendments in the existing protocol requires to be worked out. The germplasm has to be evaluated in order to store a representative sample of the variability of the species, a minimal knowledge of the biology and physiology of the species is required. Trials must be carried out in order to determine precisely the conditions for slow growth storage as well as for cryopreservation. Species specific *in vitro* culture conditions have to be developed for effective and efficient conservation.

References

Adams, R.P. and Adams, J.E. (1991). *Conservation of Plant Genes: DNA Banking and In Vitro Biotechnology.* Academic Press, New York. 345p.

Adams, R.P. (1988). The preservation of genomic DNA: DNA Bank Net. American *Journal of Botany*,75: 156 (*Abst. Suppl.*).

Adams, R.P. (1990). The preservation of Chihuahuan plant genomes through *in vitro* biotechnology: DNA Bank–Net, a genetic insurance policy. In: Powell, A.M., Hollander, R.R., Barlow, J.C., McGillivray, W.B. and Schmidly, D.J. (Eds.), *Third Symposium on Resources of the Chihuahuan Desert Region*. Printech Press, Lubbock, TX. pp. 1-9.

Adams, R.P. (1997). Conservation of DNA: DNA Banking. In: Callow, J.A., Ford–Loyd, B.V. and Newbury, H.J. (Eds.), *Biotechnology and Plant Genetic Resources: Conservation and Use. Biotechnology in Agriculture Series*, No. 19: CAB International. pp. 163-174.

Adams, R.P., Miller, J.S., Golenberg, E.M. and Adams, J.E. (1994).) *Conservation of Plant Genes 11: Utilization of Ancient and Modern DNA*: Missouri Botanical Garden Press. St. Louis, MI. 276p.

Ajith, A. (1997). Micropropagation and genetic fidelity studies in *Piper longum* L. In: Edison, S., Ramana, K.V., Sasikumar, B., Nirmal Babu, K. and Santhosh, J.

E. (Eds.), *Biotechnology of Spices, Medicinal and Aromatic Plants. Indian Society for Spices*, Calicut, India. pp. 94–97.

Alexander, M.P., Ganeshan, S. and Rajasekharan, P.E. (1991). Freeze preservation of capsicum pollen (*Capsicum annuum*) in liquid nitrogen (–196°C) for 42 months: Effect on viability and fertility. *Plant Cell Incompatibility Newsletter*, **23**: 1–4.

Ashmore, S.E. (1997). Status report on the development and application of *in vitro* techniques for the conservation and use of plant genetic resources. *International Plant Genetic Resources Institute*. Rome, Italy. 67p.

Chand, P.K., Mandal, J. and Pattnaik, S. (2000). Alginate encapsulation of axillary buds of 24 *Ocimum americanum* L. (hoary basil), *O. basilicum* L. (sweet basil), *O. gratissimum* L. 25 (shrubby basil), and *O. sanctum* L. (sacred basil). In *Vitro Cellular and Developmental Biology (Plant)*, **36(4)**: 287–292.

Chandel, K.P.S. and Pandey, R. (1991). Plant Genetic Resources Conservation – Recent Approaches. In: Paroda, R.S. and Arora, A.K. (Eds.), *Plant Genetic Resources Conservation and Management – Concepts and Approaches*. New Delhi, IBPGR, Regonal Office for South and Southeast Asia (ROSSEA). pp. 248 – 272.

Choudhury, R. and Chandel, K.P.S. (1994).) Germination studies and cryopreservation of seeds of black pepper (*Piper nigrum* L.), *a recalcitrant species. CryoLetters*, **15**:145-150.

Choudhary, R. and Chandel, K.P.S. (1995). Studies on germination and cryopreservation of cardamom (Elletaria cardamomum Maton.) seeds. *Seed Science and Biotechnology*, **23(1)**: 235-240.

Choudhury, R. and Malik, S.K. (2004). Genetic conservation of plantation crops and spices using cryopreservation. *Indian Journal of Biotechnology*, **3**: 348–358.

Engelmann, F. (1997). *In vitro* conservation methods. In: Callow, J.A., Ford–Loyd, B.V. and Newbury, H.J. (Eds.), *Biotechnology and Plant Genetic Resources: Conservation and Use. Biotechnology in Agriculture Series*, No. 19: CAB International. UK. pp. 119–161.

Engelmann, F. (2000). Importance of cryopreservation for the conservation of plant genetic resources. In: Engelmann, F. and Takagi, H. (Eds.), *Cryopreservation of Tropical Plant Germplasm: Current Research Progress and Application. Japan International Research Centre for agricultural Sciences*, Japan/International Plant genetic Resources Institute: Rome, Italy. pp. 8-20.

Engelmann, F., Benson, E.E., Chabrillange, N., Gonzalez-Arnao, M.T., Mari, S., Michaux-Ferriere, N., PauletGlazmann, J.C. and Charrier, A. (1994).) Cryopreservation of several tropical plant species using encapsulation/ dehydration of apices. In: *Proceedings of VIII APTC Meeting, Firenze*, Italy.

Frankel, O.H. (1975). Genetic resources survey as a basis for exploration. In: Frankel, O.H. and Hawkes, J. G. (Eds.), *Crop Genetic Resources for Today and Tomorrow*. Cambridge University Press, Cambridge. pp. 99.

Geetha, S.P. (2002). *In vitro* technology for genetic conservation of some genera of Zingiberaceae. Ph.D Thesis. Calicut University

Grout, B.W.W. (1995). Introduction to the *in vitro* preservation of plant cells, tissues and organs. In: Grout, B.W.W. (Ed.), *Genetic Preservation of Plant Cells In Vitro.* Springer–Verlag, Berlin. pp.1–20.

Hoekstra, F.A. (1995). Collecting pollen for genetic resources conservation. In: Guarino, L., Rao, V.R. and Reid, R. (Eds.), *Collecting Plant Genetic Diversity: Technical guidelines*.CAB International, Wallingford, UK. pp.527-550.

IPGRI (1996). Programme activities, germplsm maintenance and usein Annual Report, IPGRI: Rome, pp.56-65.

Iwananga, M. (1994).) Role of International organisations in global genetic resource management. In: Plant Genetic Resource Management in the Tropics. JIRCAS International Symposium Series. No. 2. P. 1–6. Proc. 27th International Symposium on Tropical Agriculture Research, *Japan International Research Centre for Tropical Agricultural Sciences,* Ministry of Agriculture, Forestry and Fisheries, Tsukuba, Japan, August, 25 – 26, 1993.

Kartha, K. K. and Engelmann, F. (1994).) Cryopreservation and germplasm storage. In: Vasil, I. K. and Thorpe, T. A. (Eds.), *Plant Cell and Tissue Culture*. Kluwer Academic Publishers, Dordrecht/Boston/London. pp. 195-230.

Kartha, K.K. (1985). *Cryopreservation of Plant Cells and Organs*. CRC Press, Boca Raton, Florida. 288p.

Leigh, E.T. and Remi, B. (2003). Cracking in a vitrification solution during cooling or warming does not effect growth of cryopreserved mint shoot tips. *CryoLetters*, **24**: 341–346.

Mandal, J., Pattnaik, S. and Chand, P.K. (2000). Alginate encapsulation of axillary buds of *Ocimum americanum* L. (hoary basil), *O. Basilicum* L. (sweet basil), *O. Gratissimum* L. (shrubby basil), and *O. sanctum* L. (sacred basil). In *Vitro Cellular and Developmental Biology – Plant*, **36(4)**: 287-292.

Maxted, N., Brian, F. and Hawkes, J.G. (1997). *Plant Genetic Conservation : the In Situ Approach.* Springer. 446p.

Minoo, D. (2002). Seedling and somaclonal variation and their characterization in Vanilla. Ph.D Thesis. Calicut University, Kerala, India.

Minoo, D. and Babu, K.N. (2009). Micropropagation and *in vitro* conservation of vanilla (*Vanilla planifolia* Andrews) In: Jain, S.M. and Saxena, P.K. (Eds.), *Springer Protocols, Methods in Molecular Biology, Protocols for In Vitro Cultures and Secondary Metabolite Analysis of Aromatic and Medicinal Plants*: The Humana Press, Springer, USA. pp. 129-138.

Minoo, D., Nirmal Babu, K. and Michel, G. (2010). Biotechnological applications in Vanilla, In: Odoux E. and Michel G. (Eds.) Vanilla, CRC Press, Boca Raton, USA.pp. 51-73.

Nag, K.K. and Street, H.E. (1975). Freeze preservation of cultured plant cells, 11. The freezing and thawing phases. *Physiologia Plantarum,* **340**: 254-260.

Nirmal Babu, K., Geetha, S.P., Minoo, D., Ravindran, P.N. and Peter, K.V. (1999). *In vitro* conservation of germplasm. In: Ghosh, S.P. (Ed.), *Biotechnology and its Application in Horticulture.* Narosa Publishing House, New Delhi. pp. 106–129.

Nirmal Babu, K., Geetha, S.P., Minoo, D., Yamuna, G., Praveen, K., Ravindran, P.N. and Peter, K.V. (2007). Conservation of spices genetic resources through *in vitro* conservation and cryopreservation. In: Peter, K.V. and Abraham, Z. (Eds.), *Biodiversity in Horticultural Crops- Vol. 1*, Daya Publishing house, New Delhi. pp. 210-233.

Nirmal Babu, K., Minoo, D., Geetha, S.P., Ravindran, P.N. and Peter, K.V. (2005). Advances in biotechnology of spices and herbs. *Indian Journal of Experimental Biology*, **1(2)**: 155–214.

Nirmal Babu, K., Ravindran, P.N. and Sasikumar, B. (2003). Field evaluation of tissue cultured plants of spices and assessment of their genetic stability using molecular markers. *Final Report submitted to Department of Biotechnology*, Government of India. pp. 94.

Nirmal Babu, K., Yamuna, G., Praveen, K., Minoo, D., Ravindran, P.N. and Peter, K.V. (2012). Cryopreservation of Spices Genetic Resources, In: Igor I. Katkov (Ed). *Current Frontiers in Cryobiology*, ISBN- 978-953-51-0191-8, InTech-Open Access Publisher. (Croatia) pp. 457- 484.

Nirmal Babu, K., Yamuna, G., Praveen, K., Minoo, D., Ravindran, P.N. and Peter, K.V. (2012). Cryopreservation of spices genetic resources. In: Katkov, I. (Ed.), *Current Frontiers in Cryobiology*, InTech, doi: 10.5772/35401.

Niwata, E. (1995). Cryopreservation of apical meristems of garlic (*Allium sativum* L.) and high *subsequent plant regeneration. CryoLetters*, **16**: 102–107.

Parani, M., Anand, A. and Parida, A. (1997). Application of RAPD finger printing in selection of micropropagated plants of *Piper longum* for conservation. *Current Science*, **73(1)**: 81–83.

Peter, K.V., Ravindran, P.N., Nirmal Babu, K., Sasikumar, B., Minoo, D, Geetha, S.P. and Rajalakshmi, K. (2002). Establishing *in vitro* conservatory of spices germplasm. ICAR project report. *Indian Institute of Spices Research*, Calicut, Kerala, India. pp. 131.

Quatrano, R.S. (1968) Freeze preservation of cultured flax cells using DMSO. *Plant Physiology*, **43**: 2057– 2061.

Rajasekharan, P.E. and Ganeshan, S. (2003). Pollen cryopreservation in capsicum species: A feasibility study. Capsicum and Eggplant Newsletter, **22**: 87–90.

Ravindran, P.N., Nirmal Babu, K., Saji, K.V., Geetha, S.P., Praveen, K. and Yamuna, G. (2004). Conservation of spices genetic resources in *in vitro* gene banks. ICAR project report. *Indian Institute of Spices Research*, Calicut, Kerala, India. pp. 81.

Shatnawi, M.A., Johnson, K.A. and Torpy, F.R. (2004). *In vitro* propagation and cryostorage of *Syzygium francissi* (*Myrtaceae*) by the encapsulation-dehydration method. *In Vitro Cellular and Developmental Biology-Plant*,**40(4)**: 403-407.

Withers, L.A. and Williams, J.T. (1986). *In vitro* conservation; *International Board of Plant Genetic Resources – Research Highlights* 1984-85. IBPGR, Rome. pp. 21.

Withers, L.A. (1980). *Tissue Culture Storage for Genetic Conservation*. IBPGR Technical Report. *International Board for Plant Genetic Resources, Rome*. pp. 91.

Withers, L.A. (1985). Cryopreservation of cultured cells and meristems. In: Vasil, I.K. (Ed.), *Cell Culture and Somatic Cell Genetics of Plants* (*Vol.2*)*: Cell Growth, Nutrition, Cytodifferentiation and Cryopreservation.* Academic Press, Orlando, Florida. pp. 253-316.

Withers, L.A. (1987). Long–term preservation of plant cells, tissues and organs. *Oxford Surveys of Plant Molecular and Cell Biology*, **4**: 221–272.

Withers, L.A. (1991). Biotechnology and plant genetic resources conservation. In: Paroda, R.S. and Arora, R.K. (Eds.), *Plant Genetic Resources Conservation and Management - Concepts and Approaches*. IBPGR, New Delhi. pp. 273-297.

Yamuna, G. (2007). Studies on cryopreservation of spices genetic resources. Ph. D Thesis, University of Calicut, Kerala, India.

Yamuna, G., Sumathi V., Geetha, S.P., Praveen, K., Swapna, N. and Nirmal Babu, K. (2007). Cryopreservation of *in vitro* grown shoot of Ginger (*Zingiber officinale* Rosc). *CryoLetters,* **28(4)**: 241-252.

Chapter 23

Rubber, Tea, Coffee, Oil Palm and Arecanut

☆ *Rekha Chaudhury and S.K. Malik*

1. Introduction

The conservation of plant genetic resources (PGR) is essentially to protect and safeguard invaluable genetic resources that are fast depleting due to various biotic and abiotic stresses and have large potential future use. Utilization of these PGRs is linked to food security and agrobiodiversity. Conservation strategies are dependent on plant species and explants to be conserved, availability of best techniques and resources and feasibility. *In situ* and *ex situ* conservation methods that can complement one another based on specific needs are essential for holistic efficient and cost-effective conservation.

Over the past six decades, biotechnological research has focused on developing and improving protocols for (i) embryo culture; (ii) clonal propagation *via* somatic embryogenesis; (iii) homozygote production *via* anther culture; (iv) germplasm conservation *via* cryopreservation; and (v) genetic transformation. Cryopreservation, in particular, has shown enormous potentiality in effecting safe long-term conservation of PGRs over the past 40 years in view of simplicity and the applicability to a wide range of genotypes and explants like seeds, embryos, embryonic axes, meristems, pollen, dormant buds and genomic resources (Chaudhury *et al.*, 2015).

Plantation crops need to be propagated and maintained clonally in field, however with several limitations. Field maintenance proves most expensive due to (1) high labour costs (2) vulnerability to environmental hazards and pest and pathogens and (3) large space requirement. This can lead to sudden loss of valuable germplasm or accumulation of systemic pathogens, especially viruses. It can be overcome using *in vitro* repositories or seed genebanks or cryobanks for short-, medium- and long-term conservation. *In vitro* conservation using mainly normal and slow growth conditions, induction of storage organs *etc.* have been reported for several plantation crops (Mandal, 1999; Mandal *et al.*, 2000; Manivel, 2000; Sreenath, 2000; Thulaseedharan *et al.*, 2000). Arecanut is the least worked out species in this regard.

Cryopreservation technology, the *ex situ* storage of biological materials at ultra-low temperatures, usually between -150 to -196°C using liquid nitrogen (LN) as a safe and efficient long-term technology, has remained an important biotechnological tool for storing wide range of plant germplasm (Chaudhury *et al.*, 2015) and for pathogen-free status of germplasm (Wang *et al.*, 2014). Cryopreservation is the only technique available presently, for long-term storage of vegetatively propagated species (in the form of shoot tips, meristems, somatic embryos and cells), difficult-to-store non-orthodox seed species (comprising intermediate and recalcitrant seeds), to which several plantation crops belong, for pollen to meet breeders need. For plantation crops, cryopreservation of vegetative tissues such as shoot apices and somatic embryos and zygotic tissues like seeds, embryos and embryonic axes and pollen have been reported from several laboratories.

Seeds are the most preferred explants for storage due to ease of handling. However, many plantation crops, which are indigenous to the tropics belong mainly to the category of recalcitrant seeds (Chaudhury and Malik, 2004). These cannot be desiccated to low moisture content without substantial loss of viability. In addition, seeds with intermediate seed storage behavior can withstand desiccation up to 10-14 per cent, however with viability decline. Careful controlled processing of such seeds result in extension of their storage life at low and ultra-low temperatures.

Storage of pollen is another option which is comparable to seed storage. Pollen of several species is desiccation-tolerant (*i.e.* can be dried to less than 5 per cent moisture content on a dry weight basis) when shed at binucleate state and can be stored below 0°C in desiccated state. In contrast, trinucleate pollen produced from species like arecanut and oil palm, have high moisture content at anthesis, survive only limited desiccation and have short life span ranging from a few hours to days (Barnabas and Kovacs, 1997). Pollen storability trends are not correlated with seed storability trends and hence pollen storage of recalcitrant seed species can be undertaken to complement their conservation efforts.

The chapter deals with current status of use of various cryotechniques to achieve *ex situ* conservation of some important plantation crops, *Camellia sinensis* L. (O.) Kuntze (tea), *Elaeis guineensis* Jacq. (oil palm), *Areca catechu* L. (arecanut), *Hevea brasiliensis* Muell.-Arg. (rubber) and several spp. of *Coffea* (coffee) using explants like seeds, zygotic embryos, somatic embryos, embryonic axes, shoot apices and pollen.

2. Cryopreservation Methods

Cryopreservation techniques commonly used for various species are as follows:

2.1. Air Desiccation-Freezing

In this technique, the seeds are desiccated to different target moisture contents by maintaining over charged silica gel in a desiccator for 4-48 hrs before freezing in LN. Aseptically excised embryos and embryonic axes are desiccated in sterile air under laminar airflow cabinet up to 5 hours, to around 11-16 per cent moisture level depending on critical moisture content of each, and then frozen by rapid plunging in liquid nitrogen.

2.2. Pregrowth-Desiccation

Explants like zygotic embryos, somatic embryos, shoot apices and embryonic axes are pregrown on media containing different cryoprotectants in order to impart greater desiccation tolerance and homogeneity in the behaviour. The explants are then dehydrated under the laminar airflow cabinet or over silica gel or processed by encapsulation-dehydration or vitrification.

2.3. Encapsulation-Dehydration

Explants are suspended in 3-5 per cent sodium alginate solution and picked up to dispense individually into 100 mM $CaCl_2$ followed by shaking to obtain beads. The beads are then dehydrated in sucrose-enriched media for a minimum period of 17 hours followed by desiccation and freezing.

2.4. Vitrification

It involves the treatment of tissues with cryoprotectants in vitrification solutions followed by fast freezing. Most commonly used is Plant Vitrification Solution 2 (PVS2). Recovery growth of explants is tested after thawing and removal of cryoprotectants. Initially developed for shoot apices, cell suspensions and somatic embryos, it has been lately applied to zygotic embryos and embryonic axes.

2.5. Droplet Vitrification

Developed by Panis *et al.* (2005), this technique follows a normal vitrification protocol except during immersion in LN, droplets of vitrification solution enclosing the explants are placed on sterile aluminum strips. A very high cooling and rewarming rates are ensured as explants are in direct contact with LN and with unloading solution.

2.6. Pollen Storage

Pollen samples are desiccated using various desiccants for few hours before sealing in cryovials and storing at temperatures at or below -150°C.

2.7. Thawing

For thawing of frozen samples, polypropylene cryovials containing the explants are dipped into a water bath (37-40°C) to warm the specimen rapidly.

3. Success using Cryotechniques in different Genera of Plantation Crops

3.1. Rubber (*Hevea brasiliensis* L.)

3.1.1. Cryostorage of Zygotic Embryos and Embryonic Axes

Cryopreservation of rubber was reported to be successful by Normah *et al.* (1986) using air desiccation-freezing method for embryonic axes with 50 per cent survival after cryoexposure. In later studies (Normah, 1987), higher survival rates (69-71 per cent) were achieved when axes were desiccated for 2 and 3 hours, cooled by stepwise method or by direct immersion and thawed rapidly. A detailed review on storage attempts on rubber seeds and embryonic axes was provided by Normah and Chin (1995). It was concluded that best cryoprotocol was desiccation of rubber axes to 19 per cent followed by stepwise cooling and rapid thawing.

In later studies by Sam and Hor (1999), zygotic embryos were first precultured for 16 hours on 0.3 M sucrose medium before subjecting to desiccation by PVS2 treatment for 70 to 120 min which led to 57 per cent survival after rapid freezing and thawing. In another study, using encapsulation-dehydration technique, rubber embryos were encapsulated in beads and then precultured on medium enriched with sucrose (0.3, 0.5, 0.7, 0.9 M) for 24 hours followed by desiccation in laminar flow to obtain 15 per cent moisture in the explant (Yap *et al.*, 1999). Preculture in 0.3 M sucrose, after LN exposure, led to 45-70 per cent viability, and 30-60 per cent survival.

3.1.2. Cryostorage of Pollen

Pollen of rubber has been reported to be successfully cryostored for 3-7 weeks as viability was confirmed by seed set resulting from field pollination (Hamzah *et al.*, 1999). It demonstrated the full functionality of cryostored pollen when used for hand-pollination.

3.2. Tea (*Camellia sinensis* L. (O.) Kuntze)

3.2.1. Cryostorage of Seeds and Embryonic Axes

Fully mature tea seeds were reported to be highly desiccation sensitive (Kim *et al.*, 2005) and could not survive LN exposure (Chaudhury *et al.*, 1990) with success later achieved with embryonic axes excised from seeds (Chaudhury *et al.*, 1991). In later studies, seeds were partially desiccated up to 14 per cent (Hu *et al.*, 1994) and 8-10 per cent (Wang *et al.*, 1999) and could be successfully cryostored using simple protocol. No loss of germinability of tea seeds, cryostored for two months, was recorded (Wang *et al.*, 1999). Simultaneously, tea axes could be successfully cryostored with high survival values of up to 95 per cent (Chaudhury *et al.*, 1991) following desiccation-fast freezing. Kim *et al.* (2002) used cotyledonary embryonic axes at moisture contents between 21.5 and 15.0 per cent for cryopreservation and highest percentages of plantlet production from cryopreserved explants ranged between 75.1 and 80.4 per cent. In studies by Kaviani (2010), success in cryopreservation was achieved with alginate encapsulated embryonic axes of *C. sinensis* which were kept in MS liquid medium with 0.75 M sucrose for 2 h followed by air dehydration with moisture contents of about 15-20 per cent.

3.2.2. Cryostorage of Shoot Tips

Cryostorage of shoot tips have been successful in tea. In *Camellia sinensis* cv. Yabukita, two month old *in vitro* grown plantlets were cold-hardened for five weeks. Vitrification of shoot tips, precultured for two days on sucrose-enriched media, was then carried out. The normal shoot formation rate of cryopreserved apices was 60 per cent. Using the encapsulation technique, 40 per cent shoot formation was reported (Kuranuki and Sakai, 1995). In both cases, apices developed shoots directly within 2 or 3 weeks without intermediary callus formation producing true-to-type plants (Kuranuki and Sakai, 1995).

3.3. Coffee (*Coffea* spp.)

Different *Coffea* species exhibit non-orthodox (intermediate or recalcitrant) seed storage behaviour (Dussert *et al.*, 2002; 2007). Extensive cryopreservation studies have been carried out on different *Coffea* species with diverse explants (Dussert *et al.*, 2002).

3.3.1. Cryostorage of Seeds and Zygotic Embryos

In *Coffea arabica* seeds, optimal water content for cryopreservation was found to be 0.2 g H_2O g^{-1}dry weight. Seedlings after cryopreservation could be successfully recovered if seeds were rehydrated above water immediately after warming (Dussert *et al.*, 1997). Extraction of zygotic embryos from whole seeds after cryopreservation led to higher (80-90 per cent) survival values (Dussert *et al.*, 1998). In subsequent studies, four coffee species, desiccated to different moisture contents followed by fast and slow freezing, showed variable success rates (Dussert *et al.*, 1998). Using air desiccation–freezing method, seeds of *C. liberica* showed about 53 per cent survival after cryopreservation (Normah and Vengadasalam, 1992). Zygotic embryos of *C. canephora, C. arabica,* hybrid *arabusta* (*C. arabica x C. canephora*) and *C. liberica* have been able to withstand freezing after partial desiccation (survival, 41-95 per cent) (Abdelnour-Esquivel *et al.*, 1992). At IRD-Montpellier (France) since 1997, cryopreservation efforts have been made to investigate basic principles in desiccation and freezing sensitivity in seeds of different *Coffea* spp. to enable cryobanking (Dussert et al., 2001). Continued efforts led to establishment of world's first coffee cryobank (Dussert *et al.*, 2007). Optimization of rewarming and rehydration protocols led to 100 per cent survival of seeds when recovered after cryostorage.

3.3.2. Cryostorage of Somatic Embryos

Somatic embryos have also been attempted for cryostorage. In coffee (*C. arabica* and *C. canephora*), using conventional methods, slow freezing globular embryos were cultivated on a medium enriched with sucrose, pretreated later with sucrose and DMSO before slow freezing (Bertrand-Desbrunais *et al.*, 1988). A 50 per cent recovery of cryopreserved samples was achieved through secondary embryogenesis. Higher recovery (100 per cent through adventitious embryos) was achieved using simplified freezing of heart shaped somatic embryos at –20°C (Tessereau, 1993). Heart and torpedo shaped embryos, processed through encapsulation-dehydration technique, were later recovered directly (Hatanaka *et al.*, 1994). After cryopreservation, 63 per cent of embryos were viable and half of them developed into whole plantlets. Using desiccation method, *C. canephora* somatic embryos were subjected to 12-week freeze-hardening on high sucrose and abscisic acid (ABA) before seven day desiccation and fast freezing (Tessereau *et al.*, 1994). Approximately 64 per cent cryopreserved embryos developed directly into plantlets. Using a similar methodology, 70 per cent recovery of plantlets was reported in *C. arabica* (Mycock *et al.*, 1995).

Abdelnour-Esquivel (2000) used method of pregrowth-desiccation followed by cryoprotectant treatment and slow freezing to somatic embryos. Embryos were pretreated with increasing concentrations of sucrose up to 0.75 M followed by incubation with 5 per cent DMSO and slow freezing at 0.5°C min^{-1} to -40°C before plunging in to LN. Variable success (9-61 per cent) was observed in *C. arabica* and *C. canephora*.

3.3.3 Cryostorage of Shoot Tips

In *Coffea sessiliflora* and *C. racemosa*, shoot apices excised from three week old nodal cuttings were cultured overnight on standard medium and then processed through the standard encapsulation technique. In the former species, a 3-10 day treatment of beads in 0.75 M sucrose before cryopreservation led to 38 per cent survival and in the later, treatment of beads in increasing sucrose concentration from 0.5 to 1 M led to 27 per cent survival (Mari *et al.*, 1995).

3.3.4. Cryostorage of Pollen

No efforts have been reported for long term coffee pollen cryostorage however, 1-2 months longevity by storing at low humidity and temperature (Carvalho and Monaco, 1969; Ferwerda, 1969) and by more than two years by storing under vacuum at -18°C (Walyaro and van der Vossen, 1977) have been reported.

3.4. Oil Palm (*Elaeis guineensis* Jacq.)

3.4.1. Cryostorage of Zygotic Embryos

Oil palm seeds, earlier classified as orthodox, were later defined as intermediate in seed storage behaviour. In studies by Grout *et al.* (1983), oil palm embryos desiccated to 10.4 per cent moisture content were frozen in LN with recovery rate of 75 per cent and there was no loss in viability even after 8 months storage. Before cryopreservation, partial rehydration of oil palm embryos was reported useful (Engelmann *et al.*, 1995b). Zygotic embryos were extracted from dehydrated or rehydrated seeds and desiccated to 0.12-0.3 g H_2O g^{-1} dry weight before freezing in LN rapidly. When extracted from rehydrated kernels, 65 per cent of the embryos desiccated to around 0.3 g H_2O g^{-1} dry weight developed into plantlets after cryopreservation. In contrast, only 25 per cent of embryos at 0.12 g H_2O g^{-1} dry weight extracted from cryopreserved dry kernels developed into plantlets indicating a role of rehydration of kernels before embryo extraction. Rajanaidu and Ainul (2013) reported cryobanking of 33,250 diverse accessions of oil palm as zygotic embryos by the Malaysia Palm Oil Board, Malaysia.

3.4.2. Cryostorage of Somatic Embryos

In oil palm, finger-shaped somatic embryos, pregrown with 0.75 M sucrose followed by rapid freezing and rapid thawing (Engelmann and Dereuddre, 1988), led to 20 per cent success. Later, pregrown somatic embryos subjected to 16 hours desiccation in silica gel before fast freezing led to 80-90 per cent survival (Dumet *et al.*, 1993a, b). Inclusion of the desiccation step allowed the use of standard somatic embryos. Later Dumet *et al.* (1993c) applied this procedure to diverse clones (39) of somatic embryos resulting in highest survival of 31 per cent for clones in optimal physiological state and lowest survival of 12 per cent for those in a poor state.

For the first time, droplet vitrification method was applied to polyembryoids of oil palm which led to 68 per cent survival after cryopreservation (Gantait *et al.*, 2015). Polyembryoids (3-5 mm diameter) were pre-cultured before processing using PVS2 and were frozen on aluminium strips. It proved to be a better method in terms of higher regeneration in oil palm in comparison to the earlier reported conventional vitrification method. Polyembryoids with haustoria bearing a number of embryoids were the best stage of explants for cryopreservation as judged by morphological examination. These explants exhibited maximum morphogenetic competency in terms of developing plantlets after treatments.

3.4.3. Cryostorage of Meristem

There is one report on success in cryostorage of apical meristems of oil palm (Ainul *et al.*, 2009) resulting in 45 per cent recovery growth when tested after 24 hours.

3.4.4. Cryostorage of Pollen

Oil palm pollen could be cryogenically stored for periods beyond eight years without any significant loss in their viability and germinability and hence can be used effectively for pollinations (Tandon *et al.*, 2007). Oil palm pollen cryostored for up to eight years retained as high as 54 per cent viability as judged by fluorescein diacetate reaction (FCR) test and 49 per cent by *in vitro* germinability. This was comparable to 52 per cent germination before storage (Tandon *et al.*, 2007).

3.5. Arecanut

3.5.1. Cryostorage of Zygotic Embryos

Arecanut is propagated through seeds which are recalcitrant (Raja *et al.*, 2002), and desiccation sensitive, with critical moisture content of 32.8 per cent and hence, Raja *et al.* (2014) attempted to cryopreserve zygotic embryos. The fully developed zygotic embryos (4.0-4.5 mm L X 3.0-3.5 dia) were desiccated to different moisture levels for 1-5 hours under laminar air flow. Desiccated embryos were subsequently fast frozen to -196°C. Recovery, after an eight hour cryoexposure, was 85 per cent under *in vitro* when moisture content of the axes was 21.8 per cent on four hours of drying. After cryostorage at this moisture, maximum survival (70 per cent) was observed after 30 days of culture. Recently pollen collection and *in vitro* germination method has been standardized in arecanut (Anonymous, 2014, www. cpcri.gov.in).

4. Factors Determining Successful Cryopreservation

There are several pre- and post cryoexposure factors determining the success and recovery of plantlets which fall under following themes:

4.1. Collection and Handling of Fruits and Seeds

Seeds of non-orthodox species possess high moisture contents and are short-lived necessitating collection, handling and processing for storage readily. Fresh harvests are required to be transported in shortest possible time through courier. For retaining healthy viable condition for subsequent cryopreservation experiments, freshly harvested fruits, seeds and vegetative materials are to be processed for storage within a few days or weeks of harvest.

4.2. Developmental Stage

The maturity level of explants is of immense importance during cryopreservation. Recalcitrant embryos and embryonic axes acquire some degree of desiccation tolerance at a particular physiological maturity stage, which is variable for each species. As the seeds develop towards maturity, a gradual decrease in desiccation tolerance has been observed in non-orthodox species which necessitates selection of the right maturity stage of explant for cryopreservation.

4.3. Desiccation and Freezing Rates

For most of the species, fast drying (3-5 hours in laminar air flow) followed by fast freezing has shown to cause minimum damage to cells and hence is widely adopted. Flash drying, using forced drying rapidly in a stream of dry air flowing at rates of approximately 10 $lmin^{-1}$, have proved advantageous in retention of viability (Wesley-Smith *et al.*, 1992).

The direct plunging of small explants into LN, causing freezing rates of 200°C min^{-1}, has been the best. Keeping the explant size and cryo-containers to a minimum and sub-cooling of LN under vacuum to cause quenching increases the rate of freezing.

4.4. Recovery Growth

The suitability of a protocol for cryopreservation can be assessed only in terms of good recovery growth of cryopreserved explants. Whole seeds are germinated in Petri plates in controlled lab conditions. In case of embryos, embryonic axes, somatic embryos and shoot apices recovery growth is to be assessed under *in vitro* conditions. Optimised culture media and culture conditions are essential for achieving normal recovery growth from cryopreserved tissues, especially during the first weeks of culture. In coffee, coconut and oil palm, the use of the optimal recovery medium enhances not only the survival percentage of embryos but increases normal growth without callusing after cryopreservation (Engelmann *et al.*, 1995a). Addition of 2,4-D (0.2 mgl^{-1}) for short period increased the recovery growth after cryopreservation in oil palm (Engelmann *et al.*, 1985). IBA (0.1 mgl^{-1}) and BA (0.5 mgl^{-1}) improved the growth of cryopreserved embryonic axes of *Coffea liberica* (Normah and Vengadasalam, 1992). Improved growth of immature embryonic axes of *C. arabica* was observed after culturing them on a recovery media supplemented with GA_3 (Abdelnour Esquivel *et al.*, 1992). Cryopreservation in some cases affected normal growth of embryonic axes as in *Theobroma cacao* where only regenerative callus could be obtained after cryopreservation (Pence, 1991).

5. Cryobanking

Large-scale testing experiments for practical and feasible establishment of cryobanking has been successful for major important crops and are being continued

(N'Nan *et al.*, 2012; Benelli *et al.*, 2013). Establishment of cryobanks ensure availability of diverse germplasm, thus consistently providing valuable genes for breeding more productive elite cultivars by traditional breeding and genetic transformation. Reviews enlisting operational risk and safety issues in cryobanking have been published (Benson, 2008; Keller *et al.*, 2005; Benelli *et al.*, 2013). Cryo-banking cannot replace conventional conservation methods, but does serve as a complementary conservation strategy to field gene banking and *in vitro* conservation, *i.e.*, a backup and safety storage technique (Benson, 2008).

6. Prospects and Future Strategies

Variable success in plantation crops, after cryoexposure of explants using different cryotechniques, has been reported. Limitations such as low survival percentage and extent and mode of regrowth in few of the species have led to non-uniform applicability of these to diverse genotypes. In addition, insufficient research efforts for understanding of mechanisms underlying the desiccation and freezing sensitivity have been undertaken.

In vitro culture methods especially using zygotic embryos has been standardized for several plantation crops. This aids in conservation due to ease in handling seeds and embryos for various manipulations. The varying results obtained after seed/embryo/plumule cryopreservation can be attributed to heterogeneity of the explants, availability of limited quantity of explants resulting in less replication, non-optimization of recovery conditions, suitable *in vitro* recovery media, strong link between amenability of a genera/accession to in vitro conditions and their recovery post-cryoexposure. However, there is still scope for various technical improvements in the current cryopreservation protocols as percentage survival in several cases is less than 50 per cent.

Overall cryopreservation, where biological materials are conserved for long-term when stored below the glass transition temperatures, has proven to be the exclusive cost-effective method for non-orthodox species to which most plantation crops belong. In fact cryobanking has proven its worth for germplasm belonging to plants, animals, fishes and microbes (Chaudhury *et al.*, 2015). In several cryolabs around the world, cryopreservation costs have been worked out to be much cheaper than that of *in vitro* slow growth and field genebanks. Cost of coffee field collection was compared with coffee cryo-selection at CATIE in Costa Rica (Dulloo *et al.*, 2009) and found it to be the cheapest method.

Except studies on oil palm and coffee, where large number of diverse germplasm have been cryobanked, there is no other large-scale experimentation on range of genotypes of other species of plantation crops.

At present, pollen storage is not often used in genebanks and the technique needs to be thoroughly assessed based on needs for a species. It is essential to undertake this methodology to supplement the seed or clonal genebanks. Since the technique is simple, effective and relatively inexpensive, it should be widely adopted for plantation crop species. This would have selective advantage in these species since several of these are not amenable to *in vitro* conditions and also produce difficult-to-store intermediate and recalcitrant seeds. During storage the viability and fertilizing ability of pollen should be maintained. Pollen used for germplasm conservation should remain viable for many years and this is achievable only through cryogenic storage. Herrera *et al.* (2002) has reported protocol for the induction of androgenensis and plant regeneration in *Coffea arabica* from isolated microspores *in vitro* using colchicine pretreatment. With the possibility of haploid and diplodized plant regeneration directly from pollen assuring pure gametophytic origin their storage assumes a new significance. Hopefully, in near future, it would be feasible to raise whole plants from cryostored pollen.

References

Abdelnour-Esquivel, A., Villalobos, V. and Engelmann, F. (1992). Cryopreservation of zygotic embryos of *Coffea* spp. *CryoLetters*, **13:** 297-302.

Ainul, M.M., Tarmizi, A.H. and Kushairi, A. (2009). Isolation and characterization of genes involved in fatty acid biosynthesis from cryopreservation of oil palm (*Elaeis guineensis*) apical meristem on tissue culture clonal materials. In: *Proceedings of the 8th Malaysia Congress on Genetics, 4-6 August 2009*, Genting Highlands, Malyasia. pp. 478.

Barnabas, B. and Kovacs, G. (1997). Storage of pollen. In: Shivanna, K. R. and Sawhney V.K. (Eds.), *Pollen Biotechnology for Crop Production and Improvement*. Cambridge University Press, UK. pp. 293-314.

Benelli, C., de Carlo, A. and Engelmann, F. (2013). Recent advances in the cryopreservation of shoot-derived germplasm of economically important fruit trees of *Actinidia, Diospyros, Malus, Olea, Prunus, Pyrus* and *Vitis*. *Biotechnology Advances*, **31:** 175–185.

Benson, E.E. (2008). Cryopreservation of phytodiversity: A critical appraisal of theory and practice. *Critical Reviews in Plant Sciences*, **27**: 141-219.

Bertrand-Desbrunais, A., Fabre, J., Engelmann, F., Dereuddre, J. and ET Charrier, A. (1988). Reprise de l'emryogenese adventive a partir d'embryons somatiques de cafeier (*Coffea arabica* L.) après leur congelation dans l'azote liquide. *Comptes Rendus de l'Academie des Science*, **307:** 795-801.

Carvalho, A. and Monaco, L. C. (1969) The breeding of arabica coffee. In: Ferwerda, F. P. and Wit, F. (Eds.), *Outlines of Perennial Crop Breeding in the Tropics*. Misc. *Paper Agricultural University*, Wageningen. pp. 198-216.

Chaudhury, R, Malik, S.K. and Bansal, K.C. (2015). Cryobanking of bioresources –Cross linking plant, animal, fish, microbe and medical sciences In : *Book of Abstracts of International Conference on 'Low Temperature Science and Biotechnological Advances', 27-30th April, 2015*, New Delhi, India. pp. 20-23.

Chaudhury, R. and Malik, S.K. (2004). Genetic conservation of plantation crops and spices using biotechnological approaches. *Indian Journal of Biotechnology*, **3**: 348-358.

Chaudhury, R., Lakhanpaul, S. and Chandel, K.P.S. (1990). Germination and desiccation tolerance of tea [*Camellia sinensis* (L.) O. Kuntze] seeds and feasibility of cryopreservation. *Sri Lanka Journal of Tea Science,* **59**: 89-94.

Chaudhury, R., Radhamani, J. and Chandel, K.P.S. (1991). Preliminary observations on the cryopreservation of desiccated embryonic axes of tea [*Camellia sinensis* (L.) O. Kuntze] seeds for genetic conservation. *CryoLetters,* **12:** 31-36.

Dulloo, M. E., Ebert, A. W., Dussert, S., Gotor, E., Astorga, C., Vasquez, N., Rakotomalala J. J., Rabemiafara, Eira, A. M., Bellachew, B., Omondi, C., Engelmann, F., F. Anthony, Watts, J., Qamar Z. and Snook L. (2009). Cost efficiency of cryopreservation as a long-term conservation method for coffee genetic resources. *Crop Science* **49:** 2123-2138.

Dumet, D., Engelmann, F., Chabrillange, N. and Duval, Y. (1993b). Cryopreservation of oil palm (*Elaeis guineensis* Jacq.) somatic embryos involv-ing a desiccation step. *Plant Cell Report,* 12: 352-355.

Dumet, D., Engelmann, F., Chabrillange, N. and Duval, Y. (1993c). Development of cryopreservation for oil palm somatic embryos using an improved process. *Oleagineux* 48: 273-278.

Dumet, D., Engelmann, F., Chabrillange, N., Duval, Y. and Dereuddre, J., (1993a). Importance of sucrose for the acquisition of tolerance to desiccation and cryopreservation of oil palm somatic embryos. *CryoLetters,* **14:** 243-250.

Dussert S, Vasquez, N., Salazar, K., Anthony, F. and Engelmann F. (2007).. Cryopreservation of coffee genetic resources. In: Engelmann, F., Dulloo, M.E., Astorga, Dussert, S. and Anthony, F. (Eds.), *Conserving Coffee Genetic Resources, Complementary Strategies for Ex Situ Conservation of Coffee* (*Coffea arabica L.*) *Genetic Resources- A case study in CATIE, Costa Rica.* pp. 49-58.

Dussert, S., Chabrillange, N., Engelmann, F., Anthony, F., Hamon, S. (1997). Cryopreservation of coffee (*Coffea arabica* L.) seeds: Importance of the precooling temperature. *CryoLetters* 18: 269-276.

Dussert, S., Chabrillange, N., Engelmann, F., Anthony, F., Louarn, J., Hamon, S. (1998). Cryopreservation of seeds of four coffee species (*Coffea arabica, C. costatifructa, C. racemosa* and *C. sessiliflora*): Importance of seed water content and cooling rate. *Seed Science Research,* **8**: 9-15.

Dussert, S., Chabrillange, N., Engelmann, F., Anthony, F., Vasquez, N. and Hamon, S. (2002). Cryopreservation of *Coffea* (Coffee). In: Towill, L.E. and Bajaj, Y.P.S. (Eds.), *Biotechnology in Agriculture and Forestry, Vol 50. Cryopreservation of Plant Germplasm II.* Springer-Verlag Berlin Heidelberg. pp. 220-233.

Dussert, S., Chabrillange, N., Rocquelin, G., Engelmann, F., Lopez, M., Hamon, S. (2001). Tolerance of coffee (*Coffea* spp) seeds to ultra-low temperature exposure in relation to calorimetric properties of tissue water, lipid composition, and cooling rate, *Plant Physiology,* **122**: 495-504.

Engelmann, F. and Dereuddre, J. (1988). Effect du milieu de culture sur la production d'embryoides destines a la cryoconservation chez le palmier a huile (*Elaeis guineensis* Jacq.). Comptes Rendus de I'Academie des Science 306 ser III : 515-520.

Engelmann, F., Dumet, D., Chabrillange, N., Abdelnour-Esquivel, A., Assy-Bah, B., Dereuddre, J. and Duval, Y. (1995a). Factors affecting the cryopreservation of coffee, coconut and oil palm embryos. *IPGRI/FAO Plant Genetic Resource Newsletter*, **103:** 27-31.

Engelmann, F., Chabrillange, N., Dussert, S. and Duval, Y. (1995b). Cryopreservation of zygotic embryos and kernels of oilpalm (*Elaeis guineensis* Jacq.). *Seed Science Reserach*, **5:** 81-86.

Engelmann, F., Duval, Y. and Dereuddre, J. (1985). Survie et proliferation d'embryons somatiques de palmier a' huile (*Elaeis guineensis* Jacq.) après congelation dans l'azote liquide. *Comptes Rendus de I'Academie des Science III*, **301:** 111-116.

Ferwerda, F.P. (1969). The breeding of canephora coffee. In: Ferwerda, F.P. and Wit, F. (Eds.), *Outlines of Perennial Crop Breeding in the Tropics.* Misc. *Paper, Agricultural University*, Wageningen. pp. 216-235.

Gantait, S., Sinniah, U.R., Suranthran, P., Palanyandy, S.R. and Subramaniam, S. (2015). Improved cryopreservation of oil palm (*Elaeis guineensis* Jacq.) polyemryoids using vitrification approach and assessment of genetic fidelity. *Protoplasma*, **252**: 89-101

Grout, B.W.W., Shelton, K. and Pritchard, H.W. (1983).. Orthodox behaviour of oil palm seed and cryopreservation of the excised embryo for genetic conservation. *Annals of Botany*, **52:** 381-384.

Hamzah, S., Chan, J.L. and Yeang, H.Y. (1999). *Hevea brasiliensis* seedlings from hand pollination using cryostored pollen validated by isozymes. *Journal of Rubber Research*, **2**: 62-68.

Hatanaka, T., Yasuda, T., Yamaguchi, T. and Sakai, A. (1994).) Direct regrowth of encapsulated somatic embryos of coffee (*Coffea canephora*) after cooling in liquid nitrogen. *CryoLetters*, **15:** 47-52.

Herrera, J.C., Moreno, L.G., Acuna, J.R., De Pena, M. and Osorio, D. (2002). Colchicine-induced microspore embryogenesis in coffee. *Plant Cell Tissue and Organ Culture*, **71**: 89-92.

Hu, J., Guo, and Shi, S.X. (1994).) Partial drying and post-thaw conditioning improve the survival and germination of cryopreserved seeds of tea (*Camellia sinensis*). IPGRI/FAO *Plant Genetic Resources Newsletter*, **98:** 25-28.

Kaviani, B. (2010). Cryopreservation by ED for long-term storage of some important germplasm: Seed of lily (*Lilium ledebourii* (Baker) Bioss.), embryonic axes of Persian lilac (*Melia azedarach* L.) and tea (*Camellia sinensis* L.) *Plant Omics Journal*, **3**: 177-182.

Keller, E.R., Grübe, M., Senula, A. (2005). Cryopreservation in the Gatersleben genebank - state of the art in potato, garlic and mint. Memorias Congreso Intercional Biotecnología y Agricultura (Bioveg 2005) *Centre de Bioplantas, Ciego de Avila*, CU. pp. 97-105.

Kim, H.H., Cha, Y.S., Baek, H.J., Cho, E.G., Chae, Y.A. and Engelmann, F. (2002). Cryopreservation of tea (*Camellia sinensis* L.) seeds and embryonic axes. *CryoLetters*, **23(4)**: 209–216

Kim, H.H., Yoon, J.W., Park, S.U., Kim, J.H., Cho, E.G. and Engelmann, F. (2005). Assessment of desiccation sensitivity of tea embryos for cryopreservation. *CryoLetters*, **26(4)**: 269–276.

Kuranuki, Y. and Sakai, A. (1995). Cryopreservation of *in vitro*-grown shoot tips of tea (*Camellia sinensis*) by vitrification. *CryoLetters*, **16**: 345-352.

Mandal, B.B. (1999). Conservation biotechnology of endemic and other economically important plant species of India. In: Benson, E.E. (Ed.), *Plant Conservation Biotechnology*. T J International Ltd, Padstow, UK. pp. 211-223.

Mandal, B.B., Tyagi, R.K., Pandey, R., Sharma, N. and Agrawal, A. (2000). *In vitro* conservation of germplasm of agri-horticultural crops at NBPGR: an overview. In: Razdan, M.K. and Cocking, C. (Eds.), *Conservation of Plant Genetic Resources In Vitro, Vol 2: Applications and Limitations.* Science Publishers Inc., Oxford and IBH Publishing Co. Pvt. Ltd., New Delhi. pp. 279-308.

Manivel, L., (2000). Biotechnology in the improvement of tea. In: Chadha, K.L. (Ed.), *Biotechnology in Horticultural and Plantation Crops.* Malhotra Publishing House, New Delhi. pp. 303-309.

Mari, S., Engelmann, F., Chabrillange, N., Huet, C. and Michaux-Ferriere, N. (1995). Histo-cytological study of apices of coffee (*Coffea racemosa* and *C. sessiliflora*) *in vitro* plantlets during their cryopreservation using the encapsulation-dehydration technique. *CryoLetters*, **16**: 289-298.

Mycock, D.J., Wesley-Smith, J. and Berjak, P. (1995). Cryopreservation of somatic embryos of four species with and without cryoprotectant pre-treatment. Annals of Botany, **75:** 331-336.

N'Nan, O., Borges, M., Konan, KJ-L., Hocher, V., Verdeil, J-L., Tregear, J., N'guetta, A.S.P. and Malaurie, B. (2012). Simple cryopreservation process of zygotic embryos of ten accessions of coconut (*Cocos nucifera* L.). *In Vitro Cellular and Developmental Biology - Plant.* **48:** 160-166.

Normah, M. N. (1987). Effects of temperature on rubber (*Hevea brasiliensis* Muell-Arg.) seed storage. Ph.D. thesis, Universiti Pertanian Malaysia, Malaysia.

Normah, M.N. and Chin, H.F. (1995). Cryopreservation of germplasm of rubber (*Hevea brasiliensis* Muell.-Arg.). In: Bajaj Y.P.S. (Ed.), *Biotechnology in Agriculture and Forestry, Vol 32- Cryopreservation of Plant Germplasm I.* Springer-Verlag, Berlin, Heidelberg. pp. 180-189.

Normah, M.N. and Vengadasalam, M. (1992). Effect of moisture contents on cryopreservation of *Coffea* and *Vigna* seeds and embryos. *CryoLetters*, **13**: 199-208.

Normah, M.N., Chin, H. F. and Hor, Y. L. (1986). Desiccation and cryopreservation of embryonic axes of *Hevea brasiliensis* Muell. -Arg. Pertanika, **9**: 299-303.

Panis, B, Piette, B. and Swennen, R. (2005). Droplet vitrification of apical meristems: A cryopreservation protocol applicable to all Musaceae. *Plant Science*, **168**: 45-55.

Pence, V.C. (1991). Cryopreservation of immature embryos of Theobroma cacao. *Plant Cell Reports*, **10:** 144-147.

Raja, K., Palanisamy, V. and Selvaraju, P. (2014). Desiccation and cryopreservation of recalcitrant arecanut (*Areca catechu* L.) embryos. *Plant Genetic Resources Newsletter*, **133**: 16-18.

Raja, K., Palaniswami, V. and Selvaraja, P. (2002). Studies on the seed viability influenced by the storage temperature and moisture in the recalcitrant arecanut, *Areca catechu* L. In: *Abstracts of PLACROSYM- XV*, 10-13th Dec, 2002, Mysore. pp. 35

Rajanaidu, N. and Ainul, M.M. (2013). Conservation of oil palm and coconut genetic resources. In: Normah, M. N., Chin, H. F. and Reed B.M. (Eds.), *Conservation of Tropical Plant Species.* Springer Science + Business Media New York. pp. 189-212.

Sam, Y.Y. and Hor, Y.L. (1999). Effect of vitrification solution on survival of zygotic embryos of Rubber (*Hevea brasiliensis*) in liquid nitrogen. In: Marzalina, M. (Ed.), *Recalcitrant Seeds.* FRIM, Malaysia. pp. 146-152.

Sreenath, H.L. (2000). Biotechnology in the improvement of coffee. In: Chadha K.L. (Ed.), *Biotechnology in Horticultural and Plantation Crops*. Malhotra Publishing House, New Delhi. pp. 310-322.

Tandon, R., Chaudhury, R and Shivanna, K.R. (2007). Cryopreservation of oil palm pollen. *Current Science* **92**: 182-183.

Tessereau, H. (1993). Development of a simplified method for the cryopreservation of plant tissues and somatic embryos and study of the acquisition of freezing tolerance. Ph D Thesis. University of Paris VI, France.

Tessereau, H., Florin, B., Meschine, C., Thierry, C. and Petiard, V. (1994).) Cryopreservation of somatic embryos a tool for germplasm storage and commercial delivery of selected plants. *Annals of Botany*, **74**: 547-555.

Thulaseedharan, A., Jayasree, P.K. and Venkatachalam, P. (2000). Biotechnological approaches for crop improvement in rubber. In: Chadha, K.L. (Eds.), *Biotechnology in Horticultural and Plantation Crops*. K. L. Malhotra Publishing House, New Delhi. pp. 323-351.

Walyaro, D.J. and van der Vossen, H.A.M. (1997). Pollen longevity and artificial cross-pollination in Coffea arabica L. Euphytica, 26: 225-231.

Wang, L., Ming, De, Z. and Quin, Z. (1999). Study on storage of tea seed. *Journal of Tea Science*, 19: 25-28.

Wang, B., Wang, R., Cui, Z., Bi, W., Li, J., Li, B., Ozudogru, E.A., Volk, G.M. and Wang, Q. (2014). Potential applications of cryogenic technologies to plant genetic improvement and pathogen eradication. *Biotechnology Advances*, **32:** 583-595.

Wesley-Smith, J., Vertucci, C.W., Berjak, P., Pammenter, N.W. and Crane, J. (1992). Cryopreservation of desiccation-sensitive axes of Camellia sinensis in relation to dehydration, freezing rate and the thermal properties of tissue water. *Journal of Plant Physiology*, 140: 596-604.

Yap, L.V., Hor, Y.L. and Normah, M.N. (1999). Effects of sucrose preculture and subsequent desiccation on cryopreservation of alginate-encapsulated Hevea brasiliensis embryo. In: Marzalina, M. (Eds.), *Recalcitrant Seeds*. FRIM, Malaysia. pp. 140-145.

Transgenics

Chapter 24

Rubber

☆ *A. Thulaseedharan, R.G. Kala, R. Jayashree, K. Rekha and S. Sobha*

1. Introduction

Hevea brasiliensis (Wild.ex A. Juss.) Mull. Arg. (rubber tree), belonging to the family Euphorbiaceae, is the major commercial source of natural rubber. Natural rubber produced in the milky cytoplasm (latex) of specialized cells called latex vessels is one of the most important biological macromolecules used for the manufacture of more than 35,000 variety products. Latex vessels originate from the cambial cells and they are articulated and anastomosing. Latex vessels are present in all parts of the plant except the heartwood. Latex is collected by the controlled wounding of the bark, called tapping. Rubber is separated from the latex upon coagulation and further processing. Although, natural rubber has been found in the latex of over 2000 plant species belonging to 311 genera of 79 families, *H. brasiliensis* remains as the only cultivated species as a source of commercial natural rubber because of its abundance in the latex, better quality, convenience of harvesting and perennial nature. *H. brasiliensis* is predominantly cultivated in the tropics where an equatorial monsoon climate prevails.

1.1. Geographic Origin and Domestication of Rubber

The geographical origin of *H. brasiliensis* is the Amazon River basin of South America (Schultes, 1977). Rubber is one of the recently domesticated crops in the world. Before the commercial cultivation of *H. brasiliensis*, the major sources of natural rubber were *Ficus elastica* and *Castilla elastica,* which grow wild in the forests of Central and South America, India, Africa, Madagascar *etc.* The flourishing rubber industry in various parts of the world found it difficult to sustain itself with the limited supply of wild rubber from South America. The growing demand for this industrial raw material prompted successful transfer of *H. brasiliensis* to Asia and the subsequent establishment of commercial rubber plantations. *H. brasiliensis* was introduced to Tropical Asia in 1876 through Kew Gardens from the seeds collected from the RioTapajos region of the upper Amazon region of Brazil by Sir Henry Wickam (Dijkman, 1951). The planting materials were brought to Kew Gardens from the center of origin, propagated, and further distributed to other botanical gardens around the world (Baulkwill, 1989). *H. brasiliensis* is now commercially cultivated in the tropical regions of Asia, Africa, and South America, in countries like Indonesia, Thailand, Malaysia, India, China, Sri Lanka, the Philippines, Vietnam, Nigeria, Cameroon, Ivory Coast, Liberia, Brazil, Mexico, *etc.* However, the major share of natural rubber is from Tropical Asia.

1.2. Botany

H. brasiliensis is a quick-growing, erect tree with a straight trunk and the bark is usually gray and fairly smooth. In the wild, the trees may grow to over 40 m with a life span of more than 100 years (Figure 24.1). However, the cultivated plants rarely grow beyond 25 m height because of the growth reduction due to harvesting of latex by tapping. Moreover, the trees will be replanted after about 30 years, when yield fall to an uneconomic level (Webster and Paardekooper, 1989). The young plants show characteristic growth pattern of alternating periods of rapid elongation and consolidation. The tree is deciduous with annual leaf fall followed by refoliation and flowering. The leaves are arranged in groups or storey's. The petioles are long, usually about 15 cm, with extra floral nectaries present in the region of insertion of the leaflets (Premakumari and Saraswathyamma, 2000).

H. brasiliensis is monoecious with diclinous flowers arranged in a pyramid-shaped panicle. The flowers are short-stalked and fragrant. Male flowers are smaller in size but more in number than the female flowers. In the male flower, there are 10 sessile anthers arranged on a slender staminal column in two whorls of five each. Each anther contains two pollen sacs that split longitudinally on dehiscence. Female flowers are present at the tip of the panicle and its branchlets. The mature female flowers are recognized by their relatively bigger size and the green torus basal disc. The gynoecium is tricarpellary and syncarpous with an ovule in each locule. Pollination is mediated by insects. After fertilization the ovary will develop into a three-lobed dehiscent capsule, regma, with three large mottled seeds. Fruits ripen 5-6 months after fertilization (Mydin and Saraswathyamma, 2005). The chromosome number of *H. brasiliensis* is $2n=2x=36$ (Saraswathyamma *et al.*, 1984). Although, *Hevea* behaves as a diploid, some investigators consider it as an amphidiploids, with $2n=36$ and $x=9$, that stabilized during the course of evolution (Raemer, 1935; Ong, 1976; Whycherly, 1976).

1.3. Economic Importance

Natural rubber is an important biological macromolecule used as an industrial raw material for the manufacture of more than 35,000 rubber based products. Chemically natural rubber is *cis*-1,4-polyisoprene, having molecular weight of 2,00,000 to 80,00,000 and with good viscoelastic properties. The flexibility of natural

Figure 24.1: A Mature Rubber Plantation Showing an Enlarged View of the Tapping Area at the Inset.

rubber to undergo vulcanization with sulphur under high temperature is an important attribute giving better physical properties for the finished products. The higher strength, low heat buildup and better resistance to wear and flex cracking made natural rubber a suitable raw material for the manufacture of automobile tires including aircrafts. A major share of natural rubber produced is consumed in the automobile tire industry. Natural rubber is water resistant and a good insulator also. The global area under rubber cultivation is about 11.2 million hectares (Natural Rubber Trends and Statistics, 2016) with an annual production of 12.3 million tones (Rubber Statistical Bulletin, 2015) with a value of about US $ 16 billion for the raw material alone. The major rubber producing countries are Thailand, Indonesia, Malaysia, India, China, Vietnam *etc.*

2. Genetic Transformation in *Hevea*

2.1. Rationale for Transgenic Development

In order to meet the ever increasing demand for natural rubber, the production also has to be increased. The prime objective of rubber tree breeding is the yield improvement combined with other characters like abiotic stress tolerance, high initial vigour in growth, good latex vessel system, high growth rate after initiation of latex harvest, tolerance to major diseases and wind damage, development of latex-timber clones *etc.* (Varghese, 1992; Thulaseedharan *et al.*, 2000). Tremendous increase in yield has been achieved over the last century through conventional breeding. Conventional methods of genetic improvement involve introduction, selection and hybridization. Introduction of planting materials enriches the genetic diversity of the species and accelerates the process of genetic improvement. Ortet selection, mother tree selection or plus tree selection, is the oldest selection method adopted in rubber trees. Perfection of bud grafting as a propagation method facilitated the fixation of desired characters and the development of early primary clones through ortet selection. Progress made in the ortet selection between 1919 and 1926 in Indonesia and Malaysia resulted in the development of many classical primary clones commercially very promising (Marattukalam *et al.*, 1980). Traditional breeding, such as hand-pollination between selected parent clones, evaluation of F1 hybrids, selection of promising recombinants from the progeny, and multiplication by bud grafting are still the most important methods of conventional breeding (Varghese and Mydin, 2000). Progress in yield improvement in rubber tree resulted in a gradual increment from 650 kg/ha in unselected seedlings during 1920's to 1600 kg/ha in the best clones during the 1950's. The yielding potential was further enhanced to 2500 kg/ha with many of the promising clones developed as a result of rigorous breeding and selection (Clement-Demangne *et al.*, 2000; Priyadarshan, 2003).

As rubber is a perennial tree with a long life span, the conventional breeding strategies for crop improvement are long drawn process. The major constraints in conventional breeding includes highly heterozygous nature of the crop, long gestation period (6 - 7 years), seasonal flowering, low fruit set, lack of early selection parameters, very long period of field experiments for estimating traits such as latex yield, susceptibility to TPD and wind damage, pronounced interaction of genotype X environment *etc.* The major advantages are monoecious nature of the tree, which makes hybridization easy and the amenability to vegetative propagation (Saraswathyamma, 2002). The steadily increasing demand for natural rubber has compelled researchers to explore the possibility of recombinant DNA and *in vitro* techniques for rubber tree genetic improvement. The transfer of selected genes in a single generation by genetic transformation and further vegetative multiplication by bud grafting is especially interesting for this species, since its improvement is limited by long breeding cycles and high levels of heterozygosity.

2.2. Genetic Transformation Methods in *Hevea*

The recent developments in recombinant DNA and *in vitro* plant regeneration techniques have opened new avenues for the direct introduction of specific genes controlling important agronomic traits into crop plants. Different techniques for gene transfer into plant systems have been developed, which includes *Agrobacterium tumefaciens*-mediated transformation, transformation through micro projectile bombardment, electroporation, polyethylene glycol (PEG)-mediated direct gene transfer to protoplasts, microinjection and fibre mediated transformation. Among these, genetic transformation in *Hevea* has been brought about by employing *A. tumefaciens* and microprojectile bombardment. The possibility of genetic transformation in rubber tree was first explored in 1991 by *A. tumefaciens*-mediated transformation of callus derived from *in vitro* and *in vivo* seedling cultures (Arokiaraj and Rahaman, 1991).

2.3. *Agrobacterium tumefaciens*-Mediated Genetic Transformation in *Hevea*: Protocol

In rubber, *Agrobacterium*- mediated genetic transformation is the widely used and effective method of gene transformation (Arokiaraj *et al.*, 1996, 1998; Jayashree *et al.*, 2003; Sobha *et al.*, 2003a; Kala *et al.*, 2003; Montoro *et al.*, 2000). The natural capacity of the gram negative soil bacterium, *Agrobacterium tumefaciens* to introduce a segment of DNA present in the Ti plasmid makes it an efficient vector system in genetic transformation. During infection by *A. tumefaciens*, a piece of DNA is transferred from the bacterium to the plant cell. This piece of DNA is a copy of a segment called the T-DNA (transfer DNA) present in a specific plasmid, the Ti-plasmid which is found in *A. tumefaciens*. The efficiency of T-DNA transfer varies with plant species, cultivars and target tissues. Several factors such as difference in tissue culture regime, physiological state of the tissue and *Agrobacterium* strains used are influencing the efficiency of transformation and regeneration.

Different laboratories have developed protocol for *Agrobacterium*-mediated genetic transformation in *Hevea*. Essentially the protocols are similar with slight modifications. An efficient *Agrobacterium* protocol for genetic transformation in *Hevea* developed at Rubber Research Institute of India (RRII) with *Hb*.MnSOD gene is described below and summarized in Figure 24.2. *Agrobacterium* strain EHA 101 was transformed with the binary vector pDU 96.2144, harboring the 702-nucleotide Mn.superoxide dismutase cDNA from *H. brasiliensis* (*Hb.MnSOD*). Apart from *MnSOD* gene, the binary vector also contained *uid*A as the reporter gene and *npt*II as the selectable marker gene. Two *Hb.MnSOD* gene constructs under the control of CaMV 35S promoter (Jayasree *et al.*, 2003) and under the control of FMV 34S promoter (Sobha *et al.*, 2003a) were tried separately. *Agrobacterium* harboring the binary vector was grown initially in AELB agar medium. Single colonies from these were transferred to liquid AELB medium in the presence of 50 mg/l kanamycin and 20 mg/l gentamycin. The cultures were grown overnight at 28°C in a shaker with 200 rpm for 24 hours to obtain an optical density of 0.5 OD at 420 nm. The bacteria were pelleted by centrifugation at 5000 rpm for 10 minutes and re-suspended in the induction medium so as to get a bacterial density of 5×10^8 cells/ml. MS medium containing 100 µm acetosyringone and 1.0 mM each of proline and betaine hydrochloride was used as induction medium. The pH of the medium was adjusted to 5.2 with 1N KOH, filter sterilized and used for tissue infection. *Hevea* anther calli precultured on callus proliferation medium was used as target

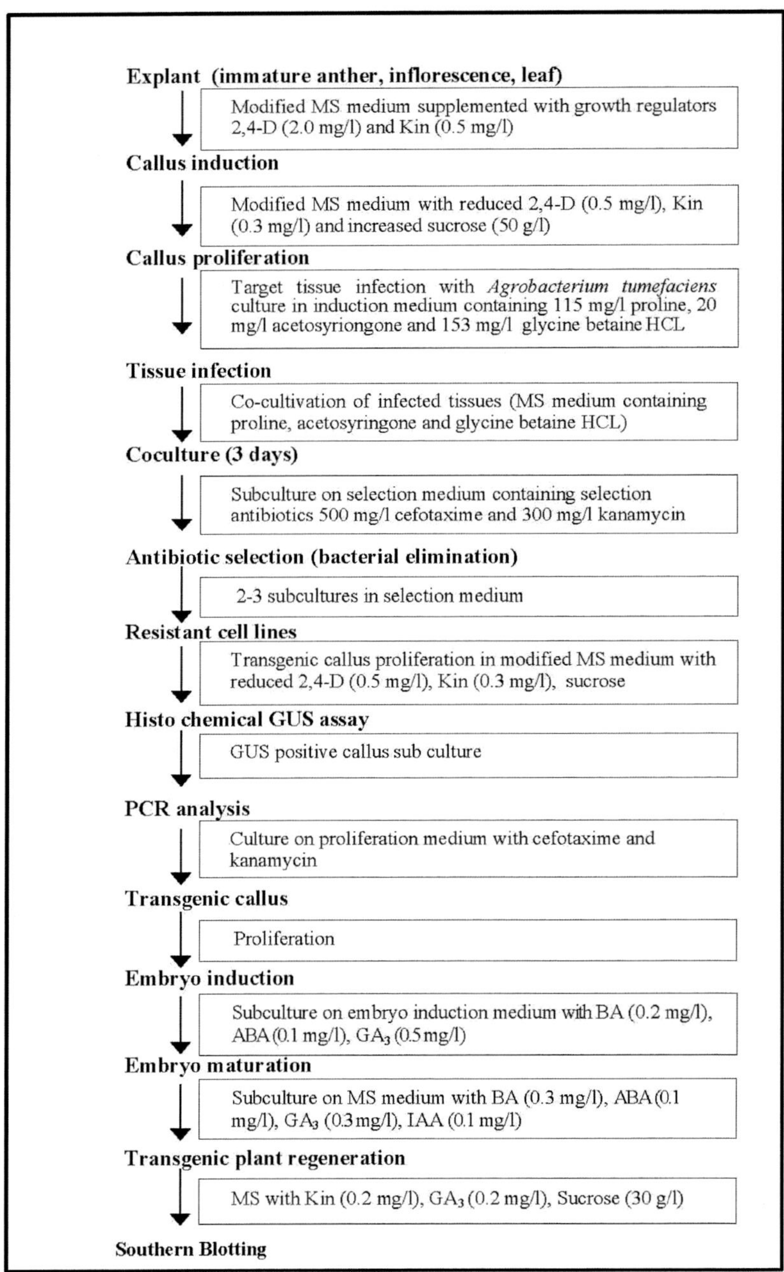

Figure 24.2: Flow Chart of *Agrobacterium* Mediated Genetic Transformation and Plant Regeneration in *H. brasiliensis*.

tissues for *Agrobacterium* infection. Approximately 2.0 g calli were immersed in 5 ml of *Agrobacterium* culture and the callus was wounded with a scalpel blade. *Agrobacterium* infection was given for 10 minutes and the infected calli were blotted dry and transferred to filter paper placed above the co-cultivation medium. After 3 days of co-culture, the explants were subcultured on a selection medium containing 500 mg/l cefotaxime and 300 mg/1 kanamycin and maintained at 25 ± 2°C in the dark. Subculture to fresh selection medium was carried out every 3 weeks. After 8 weeks of culture, kanamycin-resistant callus lines obtained were subjected to histochemical GUS assay according to Jefferson (1987). GUS positive lines obtained were proliferated and embryo induction tried. After two months of culture, the embryogenic calli obtained were subcultured over embryo-induction medium. Transgenic plant regeneration was also obtained from the somatic embryos. The gene integration in the transgenic plants was confirmed by PCR analysis and Southern hybridization. The different hybridization patterns obtained for the transgenic plants indicate random integration and multiple insertions of the T-DNA in the genome of the plants.

A protocol optimized by Rekha (2013) also resulted in high transformation frequency. In this protocol, *Agrobacterium* glycerol stock 50 µl was added to 5 ml MGL medium with pH 7.0 (Jones *et al.*, 2005) containing 50 mg/l kanamycin, 75 mg/l rifampicin and 100 mg/l carbenicillin. The culture was grown at 23°C at 250 rpm. After 24 hours of growth, the bacterial culture was subjected to a pH shock by diluting 2.5 ml of the culture with 7.5 ml of TY medium (pH 5.5) containing the respective antibiotics and 200 µM acetosyringone. The cultures were incubated overnight at 23°C at 250 rpm. Next day, 1.5 ml of the bacterial culture was diluted to 20 ml with TY medium (pH 5.5), containing 200 µM acetosyringone. The OD was adjusted to the optimal level (0.1-0.2 at 600 nm) and used for *Agrobacterium* infection (Rekha *et al.*, 2013; Rekha *et al.*, 2014b).

2.4. Factors Influencing *Heavea agrobacterium* Mediated Transformation

Several factors were found to influence effective *Agrobacterium* infection, transgene integration and further plant regeneration. *Agrobacterium* strain, target tissue for *Agrobacterium* infection and the media components used are the major factors. Therefore, efforts were made by different investigators to optimize these parameters and the technique was improved considerably.

2.4.1. *Agrobacterium* Strain

Agrobacterium strain is an important factor determining the efficiency and frequency of genetic transformation. Species specificity is also observed for many of the *Agrobacterium* strains. *Agrobacterium* strains so far used for *Hevea* transformation include EHA101, EHA 105, LBA 4404, pGV2260 and pGV3850. These strains have been reported to show high efficiency genetic transformation in *Hevea* (Arokiaraj *et al.*, 1996, 1998; Jayashree *et al.*, 2003; Sobha *et al.*, 2003a; Kala *et al.*, 2014; Rekha, 2013; Rekha *et al.*, 2014b; Jayasree *et al.*, 2013).

Figure 24.3: Development of *Hevea* Transgenic Plants Integrated with *Hb. Mn SOD* Gene using Immature Anther-Derived Callus as initial Explants.

(A) emergence of transgenic callus lines on selection medium containing kanamycin (300 mg/l) and cefotaxime (500 mg/l) after *Agrobacterium* infection; (B) globular embryos from transgenic callus; (C) transformed embryos showing histochemical GUS expression; (D) regenerated transgenic plantlets; (E) PCR confirmation of transgenic plants using SOD gene specific primers: the upper bands indicates amplification of native SOD and lower bands indicates transgene (cDNA); (F) Southern blot analysis with *nptII* specific probe and (G) hardened transgenic plants growing in containment facility.

2.4.2. Antibiotic Selection Markers

In most of the reports on *Hevea* genetic transformation using *Agrobacterium*, *neomycin phosphoransferase* II gene for tolerance to the antibiotic, kanamycin has been used as the selection marker (Arokiaraj *et al.*, 1996; Jayashree *et al.*, 2003; Rekha *et al.*,

2014b). Blanc *et al.* (2006) successfully used paromomycin to select transformants. There are recent reports on the use of *hpt* as the selection gene for the antibiotic hygromycin (Jayashree *et al.*, 2014). Qualitative marker commonly used in *Hevea* is the GUS reporter (Jayashree *et al.*, 2003; Sobha *et al.*, 2003a). Leclercq *et al.* (2010) has used green fluorescent protein as an efficient selection marker for *Agrobacterium* mediated transformation in *Hevea*. In recent reports on transformation, most binary vectors were found to avoid the qualitative markers since these were found to hamper plant regeneration (Rekha *et al.*, 2014b; Jayashree *et al.*, 2014). Different investigators used different antibiotics at variable concentration based on the target tissues used. With anther derived callus kanamycin concentration ranging from 100- 350 mg/l has been used by different workers (Jayashree *et al.*, 2003; Rekha *et al.*, 2014b; Jayashree *et al.*, 2015). When intact tissues such as leaf and root sections and somatic embryos were used for *Agrobacterium* infection, kanamycin concentration below 100 mg/l was found suitable for transgenic tissue regeneration (Kala *et al.*, 2014; Huang *et al.*, 2015). With hygromycin, of 40 mg/l was optimal for selection of *Hevea* transformants (Jayashree *et al.*, 2014).

2.4.3. Vacuum Infiltration

Vacuum infiltration was also attempted for enhancing the transformation frequency. The conditions for vacuum infiltration *viz.*, vacuum pressure and period of infiltration were standardized. Vacuum infiltration at 30 psi pressure for 10 minutes was found to be ideal for obtaining high frequency transformation in *Hevea brasiliensis*. Irrespective of the gene constructs experimented, transformation frequency was significantly improved by adopting vacuum infiltration. Employing this technique, the transformation efficiency of *MnSOD* gene construct with FMV34S promoter could be enhanced from 14 to 50 percent (Sobha *et al.*, 2013).

2.4.4. Target Tissue

Target tissue for *Agrobacterium* infection is an important factor determining the transformation frequency. Arokiaraj *et al.* (1996) has successfully used anther callus for *Agrobacterium* mediated transformation of *Hevea.* Explants have a significant role in transformation efficiency as different explants behaved differently. Among the different types of explants used for transformation, highest transformation frequency (76.4 per cent) was obtained for the embryogenic calli derived from immature zygotic embryo (Rekha *et al.*, 2014b), 63 per cent for embryogenic calli derived from anther (Rekha *et al.*, 2006) followed by 60 per cent for embryogenic calli derived from leaf (Kala *et al.*, 2006). Transformation frequency was found to be low with two-month-old primary calli from anther (Jayashree *et al.*, 2003).The feasibility of using intact explants directly as target tissue for *Agrobacterium* infection was also explored. Three target tissues, *viz.*, leaf explants from glass house and pre-cultured in modified MS medium for one week, leaf and root explants from *in vitro* developed somatic plants were used in the study. Different explant pre-treatments such as air drying in laminar air flow hood, soaking in sterile water, sterile water containing acetosyringone (40 mg/l) sterile water containing acetosyringone (40 mg/l) and picloram (2.0 mg/l) for 20 minutes were given prior to infection with *Agrobacterium*. Both pre-cultured leaf and *in vitro* root explants soaked in sterile water containing acetosyringone (40 mg/l) and picloram (2.0 mg/l) responded well to bacterial infection with *in vitro* root explants giving maximum transformation efficiency (67 per cent) (Kala *et al.*, 2012). Huang *et al.* (2015) used somatic embryos as target tissues for *Agrobacterium* mediated transformation and a transformation frequency of 4 per cent was obtained.

2.4.5. Medium Components

The influence of $CaCl_2$ on *Agrobacterium tumefaciens*-mediated gene transfer in *H. brasiliensis* friable calli was investigated by Montoro *et al.* (2000). In this experiment, the calli was normally proliferated in maintenance medium (MM) containing 9mM $CaCl_2$. For *Agrobacterium* infection, the calli were precultured on MM containing a range of $CaCl_2$ concentrations and inoculated with *Agrobacterium* suspension. Transfer of friable calli from MM containing 9 mM $CaCl_2$ to calcium-free medium significantly enhanced the transient β-glucuronidase activity. Interestingly, the use of calcium-free *Agrobacterium* re-suspension medium to inoculate friable calli again dramatically increased the transformation efficiency (Montoro *et al.*, 2000). The infection, co-culture, selection medium components and *Agrobacterium* strains significantly influenced T-DNA delivery, integration and stable transformation. Kala *et al.* (2012) observed that addition of silver nitrate in the infection (10.0 mg/l), co-cultivation (10.0 mg/l) and selection (20.0 mg/l) medium significantly suppressed bacterial overgrowth and improved the texture of callus in newly emerged lines. Improvement in the transformation frequency by 7 per cent was obtained by the inclusion of either of the thiol compounds lipoic acid (50 mg/l) or L–cysteine (100 mg/l) in the infection, co-cultivation and selection medium and addition of the surfactant pluronic F68 (300 mg/l) in the infection medium.

2.5. Biolistic Transformation System

Biolistic technique utilizes high velocity micro projectiles to deliver DNA directly in to cells and tissues. Gene delivery into intact tissues by DNA coated micro projectiles allows genetic transformation of several recalcitrant species. The major advantage of this system is that a wide variety of explants can be used for bombardment. There is no need of specialized vectors and this is the only reliable method of chloroplast transformation. Usually this technique results in transient gene expression and chimera formation. Effective gene transfer using this system are found to be influenced by several factors including nature and size of particles, target tissues, growth stage, and media supplements that help cell survival.

In *Hevea*, Arokiaraj *et al.* (1994) used biolistic technique for transforming anther derived callus with vectors harboring the *gus* gene, *npt* II and *cat* gene. Plasmid DNA was precipitated on to tungsten particles loaded on to a micro projectile and accelerated towards the target placed 5 cm below the stopping plate with a biolistic particle gun. The calli were then dark incubated for 24 hours after which they were transferred to incubation medium containing antibiotics from which kanamycin resistant transformants were obtained. Optimization of parameters which would influence DNA delivery such as micro projectile velocity, coating mixture and particle dispersal were carried out which proved that micro projectiles can deliver DNA into *Hevea* cells and helped recovery of kanamycin resistant transformants.

Wang *et al.* (2013) has also successfully used this technique for transforming *Hevea* calli, for introduction of an *Arabidopsis* gibberellic acid insensitive (GAI) gene to induce the dwarfing phenotype.

2.6. Transgenic Plant Regeneration

Efficient plant regeneration protocols through somatic embryogenesis are an essential prerequisite for crop improvement through transgenic approaches (Thulaseedharan *et al.,* 2000). The main challenge with genetic transformation of tree species is achievement of high transformation efficiency for desired cultivars combined with efficient plant regeneration (Arokiaraj *et al.,* 2009). Wang *et al.* (1980) and Wan *et al.* (1981) successfully regenerated rubber plants through somatic embryogenesis from anther walls. Carron (1981) used inner integument tissue of seeds for somatic embryogenesis and was successful in plantlet development. Extensive experiments were carried out by several investigators to enhance the frequency of somatic embryo induction and plant regeneration. Studies were also conducted to optimize cultural conditions, nutritional and hormonal requirements during somatic embryogenesis. In order to identify the suitable explant source, a variety of explants such as leaf, tender shoots, integumental tissues of immature fruit, immature anther, immature inorescence *etc.* were tried. Extensive optimization experiments were carried out to improve the plant regeneration efficiency through somatic embryogenesis for the Indian clones of rubber. Immature anther (Kumari Jayasree *et al.,* 1999), immature inorescence (Sushamakumari *et al.,* 2000) and leaf (Kala *et al.,* 2005, 2006) were identied as suitable explant sources and protocols were developed for high-frequency somatic embryo induction and plant regeneration. Kumari Jayasree *et al.* (1999) reported a standardized protocol for the induction of friable embryogenic callus, somatic embryogenesis, and further plant regeneration from the immature anthers. Culture conditions and other nutritional requirements for improving the efficiency of somatic embryo induction and germination were also investigated (Kumari Jayasree *et al.,* 2001). Dark incubation favoured callus induction and proliferation as well as embryo induction, whereas plantlet regeneration was found to be light dependent. Embryo germination percentage was significantly enhanced by incorporation of higher concentrations of GA_3 up to 2.0 mg/l; however, further plant development was affected by increasing GA_3 levels (Kumari Jayasree and Thulaseedharan, 2004). Among the various cytokinins such as BA, zeatin (ZEA), kinetin and thidiazuron (TDZ) tried on germination of somatic embryos derived from immature anther explants, TDZ was found to be superior to BA and ZEA while kinetin showed the least response. Maximum embryo germination and plantlet regeneration was 80 and 82 per cent respectively, when the medium was supplemented with TDZ (Kumari Jayasree and Thulaseedharan, 2005). Sushamakumari *et al.* (2000) studied the role of sucrose and abscisic acid (ABA) on somatic embryogenesis from immature inflorescence explants. A higher sucrose level was found to be essential for effective embryo induction as well as maturation. Callus induction could be obtained from leaf explants in MS medium with enhanced calcium nitrate along with organic supplements and phytohormones BA (1.0 mg/l), 2,4-D (1.5 mg/l) and NAA (0.2 mg/l). Embryo induction was obtained in modified MS medium by the addition of amino acids, organic supplements like casein hydrolysate, coconut water and phytohormones, 2.0 mg/l BA, 1.0 mg/l GA_3, 0.2 mg/l NAA and 0.1 mg/l 2.4-D. Maturation and apex induction of embryos could be obtained in woody plant medium. Plant regeneration was obtained in MS medium in the presence of activated charcoal (Kala *et al.*, 2005, 2006).

Most of the investigators regenerated transgenic rubber plants integrated with different genes following minor modifications in the procedure reported for non-transgenic callus. At RRII, regeneration of transgenic plants integrated with the *Hb.MnSOD* gene under the control of different promoters were developed (Jayashree *et al.*, 2003; Sobha *et al.*, 2003a) following the method reported by Kumari Jayasree *et al.* (1999) with modifications. A combination of ABA (0.1mg/l) and phytagel (0.4 per cent) promoted the frequency of embryo induction. A high sucrose level was beneficial for both embryo induction as well as maturation in rubber. Addition of organic supplements and polyamines played a significant role in the induction and maturation of the embryos. Inclusion of spermine (2.0 mg/l) in the embryo-induction medium had a positive effect on embryogenesis. Although, casein hydrolysate (200 mg/l) was good for the embryo induction, maturation was favoured by the addition of 150 mg/l banana powder. Addition of amino acids like glutamine and proline influenced the maturation frequency dramatically. Plant regeneration was promoted in the medium with reduced levels of sucrose (20 g/l) and phytagel (0.2 per cent).

3. Progress of *Hevea* Transgenic Breeding

The first transgenic rubber plant was developed in 1994 incorporated with β-glucuronidase (*GUS*) gene (Arokiaraj *et al.*, 1994). A critical step in *Agrobacterium*-mediated transformation and transgenic tissue regeneration in *H. brasiliensis* is the establishment of optimal conditions for T-DNA delivery into infected tissue and tissue recovery without bacterial overgrowth. In order to optimize the *Agrobacterium* and biolistic-mediated genetic transformation systems in rubber trees, different investigators employed *GUS* as the marker gene under the control of Cauliower Mosaic Virus (CaMV) 35S promoter (Arokiaraj *et al.*, 1994, 1996, 1998; Montoro *et al.*, 2000, 2003; Blanc *et al.*, 2006). The selectable marker gene used was neomycin phosphotransferase II (*npt*II) for resistance to the antibiotic kanamycin. Different laboratories were successful in developing transgenic plants for increased tolerance to abiotic stresses and TPD, enhanced rubber yield, recombinant protein production and altered phenotype (Venkatachalam *et al.*, 2006, 2007; Thulaseedharan *et al.*, 2008).

3.1. Transgenics for Abiotic Stress Tolerance

Climate change and global warming are generating rapid changes in the atmosphere that are unprecedented. Worldwide, it has been estimated that approximately 70 per cent of yield reduction is the direct result of abiotic stresses. The adverse environmental conditions such as drought, temperature extremes, high solar radiation, low atmospheric humidity, poor soils, *etc.*, limit the expansion of rubber cultivation to marginal areas in several rubber producing countries. After developing efficient plant regeneration protocols through somatic embryogenesis, transgenic breeding was attempted at RRII for introgression of genes for desirable agronomic traits. Initial focus was to develop transgenic plants tolerant to abiotic stresses such as drought, temperature extremes (high and low) and TPD. Plants exposed to environmental stress generate excess reactive oxygen species (ROS) such as superoxide (O_2^-), hydrogen peroxide (H_2O_2) and hydroxyl radicals (OH^-) (Asada and Takahashi, 1987). Superoxide dismutase (SOD) is the first enzyme involved in the detoxifying process of reactive oxygen species (Fridovich, 1986). Significant yield loss occurs in plantations across the rubber-growing countries due to TPD, a physiological disorder, characterized by the browning of bark followed by the cessation of latex flow. Increased free radical accumulation and a reduced level of SOD enzyme and cytokinins were reported in the TPD-affected trees (Das *et al.*, 1998). Therefore, the gene encoding *MnSOD* was identified to generate transgenic rubber plants for tolerance to abiotic stress and tapping panel dryness. Transgenic rubber plants were developed using the gene coding for superoxide dismutase (Jayashree *et al.*, 2003). Immature anther-derived calli were transformed through *Agrobacterium* mediated genetic transformation. The binary plasmid vector contained *MnSOD* cDNA isolated from *H. brasiliensis* under the control of the CaMV 35S promoter. The binary vector also contained GUS gene (*uid*A) for screening and the neomycin phosphotransferase gene (*npt*II) for selection of the transformed cell lines. The morphology of the transgenic plants was similar to that of untransformed plants. Histochemical GUS assay revealed the expression of the *uid*A gene in transgenic calli at different developmental stages as well as in the hardened plants. The presence of the *uid*A, *npt*II and *Hb.MnSOD* genes in the hardened transgenic plants were confirmed by polymerase chain reaction and genomic Southern blot hybridization analysis (Jayashree *et al.*, 2003). Transgenic plants integrated with *MnSOD* gene were also developed with a different gene construct with figwort mosaic virus 34S (FMV 34S) promoter for the expression of *MnSOD* along with the selectable marker gene *npt*II and the reporter gene GUS (Sobha *et al.*, 2001, 2003b).

Over-expression of SOD, peroxidase and catalase enzymes in response to abiotic stresses like water stress, osmotic stress and under different light regimes were studied in the *MnSOD* transgenic cultures under *in vitro* conditions. Water stress was induced by the addition of different concentrations (0.2 - 1.0 per cent) of phytagel and osmotic stress using polyethylene glycol (PEG), mannitol and sorbitol (2-10 per cent) in the culture medium. More than 50 per cent over-expression of SOD was observed when 0.4 per cent phytagel was added to the medium and 40 per cent over-expression when the culture medium was supplemented with 4.0 per cent PEG, with a corresponding over-expression of catalase and peroxidase (Sobha *et al.*, 2001). Jayashree *et al.* (2011) studied the transgene expression and other drought tolerant traits in *MnSOD* transgenic plants. Six month old bud grafted plants grown in polybags were subjected to water stress and the drought tolerant traits were evaluated through molecular, physiological and biochemical tools. A higher SOD transcript level was observed in the transgenic plants. The SOD enzyme activity showed 35 and 31 per cent higher under normal and drought conditions respectively compared with their controls. The drought mediated reduction in photosystem II and mid-day leaf water potential and photosynthetic oxygen evolution were also found to be lower in the transgenic plants compared with their control. The physiological performance of one year old *MnSOD* transgenic bud grafted plants was evaluated in a dry sub humid environment by withholding irrigation to assess the recovery after rewatering. The parameters for drought tolerance studied *viz.*

the pre-dawn leaf water potential, relative water content, net photosynthesis rate, stomatal conductance, antioxidant enzyme levels *etc.* were found to better in the transgenic plants compared with their controls (Sumesh *et al.*, 2014).

Leclercq *et al.* (2012) regenerated transgenic *Hevea* plants over-expressing a *Hevea brasiliensis* cytosolic *CuZnSOD* gene. They studied the physiological parameters related to drought tolerant traits after subjecting to water deficit treatments. Lower stomatal conductance and proline content are observed in the transgenic lines. Over-expression of the *HbCuZnSOD* gene and activation of all ROS-scavenging enzymes also suggested more efficient protection against ROS in the transgenic line.

In order to achieve tolerance to a variety of stresses Rekha *et al.* (2013b; 2014b) attempted to transfer the gene encoding tobacco osmotin (Tb osm) to *Hevea*. Osmotin is a stress responsive multifunctional protein belonging to PR-5 protein family, providing osmotolerance to plants (Amjad and Malik, 2008). It is reported that its transcription can be activated by several factors like sodium chloride, desiccation, ethylene, wounding, abscisic acid, tobacco mosaic virus, fungi and UV light (Liu *et al.*, 1994). Transgenic plants were developed through *Agrobacterium* mediated genetic transformation. Plants were germinated, hardened and established in polybags. The molecular confirmation of gene integration was done by PCR, Southern blot and RT-PCR (Rekha *et al.*, 2013b, 2014a). The stress tolerance studies conducted using transgenic calli integrated with osmotin gene, showed better expression for stress tolerance in the transgenic calli as indicated by the higher accumulation of proline under stress in the transgenic calli (Rekha *et al.*, 2013b). Similarly, the transgenic calli could survive and proliferate in culture media containing 150 mM NaCl, whereas the non-transgenic calli showed retarded growth even in the presence of 50 mM NaCl (Rekha *et al.*, 2013a, 2016).

Recently, efforts have been made for the introduction of sorbitol-6-phosphate dehydrogenase gene, isolated from apple, into the *Hevea* genome for imparting drought tolerance. Sorbitol serves as a compatible solute, under conditions of decreasing water, anti-freezing agent for chilling or as a scavenger of free radicals. Embryogenic callus from immature anther was used as the target tissue for *Agrobacterium* mediated transformation and plantlets were regenerated (Kumari Jayasree *et al.*, 2015).

In certain rubber growing areas, wind damage seriously affects rubber cultivation. Pruning is carried out in such areas to escape from wind damage. In order to develop dwarf phenotypes for imparting tolerance to wind damage, Wang *et al.* (2013) attempted introduction of an *Arabidopsis* gibberellic acid insensitive (GAI) gene into *Hevea* to induce dwarfing phenotype. The genetic transformation was carried out through microparticle bombardment using 40 day old calli and plantlets were regenerated.

Since tapping panel dryness (TPD) is a major problem especially in high yielding clones and reduced levels of cytokinins were reported in the TPD-affected trees (Das *et al.*, 1998), an attempt was made to overproduce cytokinin in rubber tissues by incorporating the gene coding for isopentenyl transferase (*ipt*), to combat TPD. The putative transformed calli were able to grow without external supply of hormones and showed increased cytokinin levels compared to the controls. However, the embryos showed developmental abnormalities and most of the transformants were severely deformed (Kala *et al.*, 2003). Isopentenyl transferase gene has also been used in selection of marker-free transgenic plants (Ebinuma *et al.*, 1997). Alternatively, moderate ipt expression using an alternative approach by fusing ipt at a distance from the constitutive promoter showed only a 2–3 fold increase of cytokinin levels. This resulted in improved stress tolerance in transgenic plants supporting the notion that moderate increases in cytokinin can be an effective strategy for improving stress tolerance (Peleg and Blumwald, 2011).

3.2. Transgenic Rubber Plants for Enhanced Rubber Yield

As rubber is a perennial tree with a long life span, conventional methods take a long time to release a clone with desirable characters. The efforts made by conventional breeding has improved rubber productivity considerably (Mydin and Gireesh, 2016), but a stage will be reached when the rubber biosynthetic rate of the tree itself become a limiting factor. At this juncture, yield enhancement can be made only by manipulating the factors influencing the rate of rubber biosynthesis. Yield improvement through transgenic attempts, the transfer of key regulatory genes associated with rubber biosynthesis. Another important factor influencing latex yield is the number of laticifers (Gomez, 1982). Jasmonic acid is the signalling molecule involved in the laticifer differentiation (Hao and Wu, 2000). Introduction of genes encoding enzymes involved in the jasmonic acid biosynthesis can bring about latex vessel differentiation in *Hevea*. A key enzyme in this pathway, allene oxide synthase (AOS) has been cloned and work has been initiated in this direction (Arokiaraj *et al.*, 2002).

Rubber biosynthesis occurs in the latex, mainly by the mevalonate pathway (MVA), where acetyl Co A is converted to poly isoprene. One of the key regulatory points in the isoprenoid pathway is the irreversible conversion of HMG-CoA to mevalonate catalyzed by 3-Hydroxy-3-Methylglutaryl Coenzyme A Reductase (HMGR). This enzyme activity was reported to be low (0.078 nmol MVA/ml of latex) compared to that of other enzymes up to IPP (isopentenyl pyrophosphate), suggesting its rate limiting role in the pathway (Lynen, 1969). The association between *hmgr* enzyme activity and rubber biosynthesis in *Hevea* is well established (Nair and Kurup, 1990). HMGR is encoded by a group of three genes namely *hmgr*1, *hmgr*2 and *hmgr*3 which form a small gene family and *hmgr*1 is involved in rubber biosynthesis (Chye *et al.*, 1992). The regulatory role of *hmgr*1 gene in rubber biosynthesis was proved in the Malaysian clone RRIM 600 and PB 235 where a positive correlation was observed between the dry rubber content and *hmgr*1 mRNA (Nuntanuwat, 2006; Suwanmanee *et al.*, 2007).

The initial attempts to transform *Hevea* with *hmgr1* gene was made by Arokiaraj *et al.* (1995) using *Hevea* anther callus by particle bombardment where the activity in the transformed callus ranged from 70-410 per cent and the enzyme level in the somatic embryos were 250-300 per cent compared with the wild type control (Arokiaraj *et al.*, 1995). However, the transgenic embryos failed to germinate. Transgenic *Hevea* plants integrated with laticifer specific *hmgr*1 gene were developed from the embryogenic callus of zygotic origin and the plantlets were acclimatized and maintained in the containment facility (Jayashree *et al.*, 2010). The *Agrobacterium* strain employed in the study was EHA 105 which produced a good transformation frequency with all type of target tissues used (Jayashree *et al.*, 2012, 2013). The acclimatized plants were PCR screened for the presence of the transgene and the integration and expression of the transgene was confirmed by Southern and northern hybridization respectively. HMGR activity was semi quantified in selected plants using ELISA and the activity was higher in the transgenic plants compared to control wild type plant (Jayashree *et al.*, 2014).

3.3. Stacking Genes for Multiple Traits

Gene stacking is a viable approach for introducing multiple genes and it involves the integration of two or more genes simultaneously or sequentially into the target tissue by standard delivery systems such as *Agrobacterium* mediated transformation. In *Hevea brasiliensis*, genes coding for different agronomic traits *viz.* manganese superoxide dismutase (MnSOD) for enhanced environmental stress tolerance and 3-hydroxy-3-methyl-glutaryl-CoA reductase (*hmgr*1) for improved latex yield were integrated in *Hevea* callus by *Agrobacterium* mediated repeated transformation. Initially *Agrobacterium*-mediated gene integration was carried out with the binary vector harboring MnSOD gene and *npt*II as the selectable marker gene using embryogenic callus derived from immature zygotic embryo as the target tissue. Kanamycin resistant callus lines were selected and MnSOD gene integration was detected by GUS histo-chemical staining and PCR using *MnSOD* gene specific primer. This MnSOD transgenic callus was used as the target tissue for the integration of *hmgr*I gene containing *hpt* as the selectable marker gene. Multiple gene integration was ascertained in the hygromycin resistant callus lines by performing PCR using *MnSOD* and *hpt* gene specific primer pairs. Somatic embryo induction and maturation was reported from the multiple gene integrated transgenic callus lines (Sobha *et al.*, 2014).

3.4. Opportunities of *Hevea brasiliensis* for Biopharming

Using biotechnological tools, transgenic plants have been used as living factories for the production of a variety of recombinant proteins including edible vaccines, therapeutic proteins, and antibodies for immunotherapy. For commercial production of diagnostic and therapeutic products, the pharmaceutical industry mainly depends upon microorganisms involving sophisticated bioreactors. Plants for production of recombinant proteins are cost effective and can be easily scaled up for large volume production. The rubber tree has many unique advantages over other plants for biopharming. The bark of the rubber tree contains a complex network of articulated latex vessels, notably in the soft bark of the trunk from which rubber is collected. The latex is a cytoplasm that contains rubber particles, micro vacuoles known as lutoids, and double-membrane organelles rich in carotenoids assimilated to plastids, the Frey-Wysling particles (Pardekooper, 1989). It means latex contains all the machinery for the protein synthesis, if the desired gene is inserted into rubber plants. In rubber tree, latex harvesting is by non-destructive method and a continuous process throughout the year for a period of over 25 year and the latex replenishment after each tapping is rapid. Moreover, once the tree is genetically transformed, the trait could be xed in the T1 generation itself with large-scale clonal propagation. Therefore, rubber tree is the most suitable candidate for biopharming (Yeang *et al.*, 1998). Laticiferous specific promoters would be more useful for incorporating genes for biopharming rather than constitutive promoters. Montorro *et al.* (2009) have studied the expression of a laticiferous promoter of hevein gene *(Hev.2.1)* in transgenic *Hevea*.

3.4.1. Transgenic Rubber Plants for Recombinant Protein

The heterologous expression of proteins in the latex vessels was first explored by Arokiaraj (2000) and Arokiaraj *et al.* (2002). They developed transgenic rubber plants that secrete human serum albumin (HSA) in the serum fraction, of rubber latex, through *Agrobacterium*-mediated genetic transformation using a binary vector pLGMR HSA, which contained a 1.8-kb HSA.cDNA fused with the CaMV 35S promoter and polyA tail to control the expression. HSA in the leaf and latex extracts was detected using an antibody coupled to a protein chip array. In the latex an expression level of upto 24 µg of HSA per milliliter of latex extract was observed. Further, Yeang *et al.* (2002) reported the expression of a functional recombinant single-chain variable fragment (ScFv) antibody in the latex of transgenic rubber. The binary vector used for *Agrobacterium* mediated genetic transformation contained the gene coding for a mouse immunoglobulin ScFv with specificity for the dental bacterium *Streptococcus gordonii*. Gene expression was controlled by the 35S CaMV promoter and nos terminator regulatory sequences, together with the tobacco pathogenesis-related protein *prla* signal sequence. The ScFv antibody concentration detected in the latex was upto 3 µg ml/llatex serum for the most productive plants. The expression of the protein varied between individual plants in quantity and functionality. The concentration of the protein increased as the transgenic plants aged. Kala *et al.* (2006) used a cDNA sequence coding for a 10.8 kDa TB antigen protein isolated from *Mycobacterium tuberculosis* for *Hevea* genetic transformation.

Transgenic *H. brasiliensis* was also developed for the expression of a gene encoding human atrial natriuretic factor (HANF), a peptide hormone that is involved in regulating cardiac blood pressure (Sunderasan *et al.*, 2012). *Agrobacterium*-mediated genetic transformation was performed on anther callus with a constructs bearing HANF gene in the pGPTV-Kan vector. The *HANF* gene was fused with hevein promoter for the over-expression of the interested gene in the latex vessels (Arokiaraj and Shuib, 2011). The presence of the *HANF* transcript in leaf samples was detected by RT-PCR in a number of original transformants and their vegetative generations. Western-immunoblot using polyclonal anti-atrial natriuretic peptide also detected a band in the low molecular weight region in the leaf protein of an original transformed plant, confirming the production of functional HANF protein.

4. Antibiotic Marker-free Transgenic Plants

In genetic transformation experiments, selectable marker genes are widely used for the selection of transformed cell lines, since they confer resistance to an antibiotic or herbicide that kills non-transformed cells, but they serve no purpose once transgenic plants have been regenerated. The presence of these marker genes, especially antibiotic marker genes in the genetically modified (GM) plants has raised considerable public concern regarding their potential transfer to other organisms and their safety (Thomson *et al.*, 2001), the environmental implications of growing GM crops (Ho, 2001) and the health aspects of consuming GM foods. Further herbicide resistance genes might be transferred by out crossing into weeds (Dale *et al.*, 2002). Gene stacking by using different selectable marker genes will lead to duplication of promoters and polyA signals that may cause an unanticipated gene silencing (Hohn *et al.*, 2001, Ebinuma and Komamine, 2001). Therefore, researchers working in the area of genetic engineering tried to develop technologies for the excision of the antibiotic marker genes, conferring resistance to clinically relevant antibiotics along with its promoter from the transgenic cells/plants.

Several strategies were employed for excising selectable marker genes from transgenic plants. In the co-transformation method, two independent T-DNAs, one with selection marker and the other with the gene of interest, offers a simple approach for marker gene elimination in annual/cereal crops (Depicker *et al.*, 1985; McKnight *et al.*, 1987; De Block and Debrouwer, 1991). If the two DNAs integrate at different loci in the T_0 plant, they will get segregated in the T_1 progeny leading to the generation of marker-free transgenic plants (Veluthambi *et al.*, 2003; Daley *et al.*, 1998; Komari *et al.*, 1996). In the marker elimination by site-specific recombination system, excision of the selectable marker gene is done by using site specific recombination system (Dale and Ow, 1990; Qin *et al.*, 1994; Srivastava *et al.*, 1999; Srivastava and Ow, 2002), The Cre-loxP site-specific recombination system is based on the ability of microbial site- specific recombinase (*cre*) gene to cleave DNA at specific sites and ligate it to the cleaved DNA at a second target sequence (loxP site). In this system the Cre (cause recombination) is the recombinase enzyme and loxP (locus of x- over P) is the respective recombination site. The *cre* gene expression can be controlled by specific gene promoters, such as inducible gene promoters, including chemical inducible promoters and heat shock protein promoter (Dale and Ow, 1991; Bryant and Leather, 1992; Rao *et al.*, 2011; Nandy and Srivastava, 2011, 2012).

Hevea brasiliensis, being a perennial tree crop with a long breeding cycle, it will take several years to produce marker-free transgenic plants by co-transformation. Further, in positive selection using *pmi* gene, preliminary studies revealed that *Hevea* normal callus can grow in medium containing high concentration of mannose indicating that it is insensitive to mannose. Therefore it is presumed that the positive selection (*pmi*) method is not a viable approach in *Hevea* genetic transformation for producing marker-free transgenic plants. For the removal of the antibiotic marker gene from transgenic *Hevea* plants, a binary vector with heat inducible Cre-loxP system (HSP:Cre:loxP) was developed by Sobha *et al.* (2015b). The vector contained transcription units *npt11* under CaMV35S promoter and *cre* gene with inducible heat shock promoter (HSP). Both units were flanked by loxP sites to allow excision of *npt*II gene by Cre-loxP recombination (Sobha *et al.*, 2015a). Functional validation of the binary vector in tobacco plants was carried out by *Agrobacterium* mediated transformation with the binary vector using tobacco leaf disc and callus and the transgene integration was ascertained by performing PCR using *npt*II and *cre* gene specific primers. Successful excision of the marker genes was achieved by imparting heat shock treatment at 42°C for 3 hrs. (Sobha *et al.*, 2015a). Functional validation of the construct (pNS14) in tobacco plants ascertained that the same construct could be employed for producing marker-free transgenic *Hevea brasiliensis* plants. Work has been initiated in this direction for developing antibiotic marker-free transgenic rubber plants. This clean vector technology mediated by the Cre/loxP system is a novel approach for producing marker-free transgenic *Hevea* plants. An added advantage of this technique is that the removal of antibiotic marker gene after each transformation step enables the use of the same antibiotic for gene stacking without crossing which is highly desirable in a tree crop like *Hevea*, which is propagated mainly through bud grafting.

5. Biosafety Issues

Genetic engineering (GE) and genetically modified organisms (GMOs) provide powerful tools for sustainable development in agriculture, healthcare and many other industries. Use of GMOs is common now in healthcare industry and agriculture, often for the production of better quality products. While healthcare industry is highly regulated and the products are generally life-saving drugs, in agriculture it is more open and deals with crops, their protection from insect pests and diseases, improving their taste, quality and acceptability to the consumers. Since GM crops and animals are grown under open environment and they can interact with other organisms in the surrounding environment, GMOs in agriculture have become a more sensitive issue than they are in the healthcare industry. With the adoption of GM crops in many developing and developed countries, the global area of biotech crops has increased to 179.7 million hectares during 2015 (James, 2015). The introduction of the transgenic crop to the natural system has generated serious public concerns particularly about damage to human health and the natural environment. Risks to human health are mainly related to toxicity, allergenicity and resistance to antibiotics used for the development of the GM crop. The toxicity depends upon the specific product produced in the GM crop as a result of transgene expression. The introduction of specific proteins into the crop sometimes produces allergic response even if it is not a food crop. However, there is no evidence proving transgenic crops triggering more risks than conventional products in triggering allergies. The use of antibiotic resistance genes as selectable markers have also raised concerns about the transfer of such genes to microorganisms and developing antibiotic resistance in the disease causing organisms, although, the probability of such transfer is extremely rare. The concerns about environmental risks due to the introduction of transgenic crops include impact of the transgenic trait on other related species, the potential build-up of resistance in specific insect populations affecting the biodiversity and unintended effects on non-targeted organisms. Accidental cross-breeding between transgenic crops and traditional varieties through pollen transfer can contaminate the traditional local varieties. Since, rubber is not a food crop there is not such food safety issue with transgenic rubber plant. Gene flow from transgenic crops to others requires, the presence of sexually compatible wild relatives close to the crop, an overlap of flowering time between the transgenic crop and the wild relatives and the presence of pollinating agents. Since, *Hevea brasiliensis* is a native of South

America and rubber cultivation is predominantly in the South Asian countries where the presence of sexually compatible wild relatives are rare, the risk of gene flow is also very remote.

6. Conclusion and Future Prospects

Conventional agriculture relying on selection and traditional breeding programmes is a slow process for deploying new plant varieties with improved or modified traits quickly in a tree species like *H. brasiliensis*. New systems are essential for delivery of next generation crop plants with novel traits to exploit the potential of crop based and bio products industry to meet social and environmental needs. Optimization of the protocol by identifying the ideal explant source and conditions would speed up genetic modification in *Hevea*. At present there are several constraints in transformation such as low frequency plant regeneration and hardening. Use of constitutive promoters, selection antibiotics *etc.* poses hindrance to field planting of the developed transgenics. Engineering the chloroplast genome is emerging as a successful approach towards this end. Manipulation of the chloroplast genome has become routine in model system tobacco and has been extended to other solanaceous crops. Expression of the transgene in this organelle offers unique advantages that render this technology safe and acceptable to the public. Gene containment is the most notable advantage that this technology offers, because chloroplast genomes are maternally inherited in most plant species. In addition, chloroplast engineering overcomes the challenges of low level expression, gene silencing, positional effect and multistep engineering of multiple genes, which are current limitations of nuclear transformations. In rubber, experiments are underway to screen a number of chosen endogenous *Hevea* chloroplast promoters to drive the expression of the reporter gene, *uidA* for *Hevea* specific chloroplast transformation vector. Available literature strongly supports the possibility of the transgenic technology to produce new varieties of *Hevea* with improved abiotic and biotic stress tolerance, increased latex and timber yield, altered phenotypes for protection from wind damage and production of recombinant proteins. Most of the biosafety issues can be overcome by the use of marker-free and chloroplast transformation technology.

References

Amjad, M.H. and Malik, Z.A. (2008). Over expression of tobacco osmotin gene leads to salt stress tolerance in strawberry (*Fragaria* x *ananassa* Duch.) plants. *Indian Journal of Biotechnology*, **7**: 465-471.

Arokiaraj, P. and Shuib, S.S.S. (2011). Functional analysis of latex-specific promoters (Hevein).*Paper presented in the IRRDB Biotechnology Workshop*, Kuala Lumpur, Malaysia.

Arokiaraj, P. and Wan Abdul Rahaman, W.Y. (1991). *Agrobacterium*-mediated transformation of *Hevea* cells derived from *in vitro* and *in vivo* seedling cultures. *Journal of Natural Rubber Research*, **6**: 55 - 61.

Arokiaraj, P. (2000). Genetic Transformation of *Hevea Brasiliensis* (Rubber Tree) and its applications towards crop improvement and production of recombinant proteins of commercial value. In: Mohan Jain, S. and Subhash C. Minocha. (Eds.),

Molecular Biology of Woody Plants (vol 66: 305). *Forestry Sciences*. Springer, Netherlands. pp. 305-325.

Arokiaraj, P., Jaafar, H., Hamzah, S., Yeang, H.Y. and Wan, A.R. (1995). Enhancement of *Hevea* crop potential by genetic transformation: HMGR activity in transformed tissue. Aspects of the Breeding of *Hevea brasiliensis*.In: *Proceedings of IRRDB Symposium on Physiology Molecular Biology*. Penang, Malaysia. pp. 74-82.

Arokiaraj, P., Jones, H., Cheong, K.F., Coomber, S. and Charlwood, B.V. (1994).) Gene insertion into *Hevea brasiliensis. Plant Cell Reports*, **13**: 425- 431.

Arokiaraj, P., Jones, H., Jaafar, H., Coomber, S. and Charlwood, B.V. (1996). *Agrobacterium*- mediated transformation of *Hevea* anther calli and their regeneration into plantlets. *Journal of Natural Rubber Research*, **11**: 77-87.

Arokiaraj, P., Leelawathy, R. and Yeang, H.Y. (2009). The Super virulence Plasmid pToK47 from *Agrobacterium tumefaciens* A281 improves transformation efficiency of *Hevea brasiliensis*. *American Journal of Biochemistry and Biotechnology*, **5(3)**: 137-141.

Arokiaraj, P., Ruker, F., Obermayer, E. and Yeang, H.Y. (2002). Expression of human serum albumin in transgenic *Hevea brasiliensis. Journal of Rubber Research*, **5(3)**: 157-166.

Arokiaraj, P., Yeang, H.Y., Cheong, K.F., Hamzah, S., Jones, H., Coomber, S. and Charlwood, B.V. (1998). CaMV 35S promoter directs β-glucuronidase expression in the laticiferous system of transgenic *Hevea brasiliensis* (rubber tree). *Plant Cell Reports*, **17(8)**: 621-625.

Asada, K. and Takahashi, M. (1987). Production and scavenging of active oxygen in photosynthesis. In: Kyle, D.J., Osmond, C.B. and Arntzen, C.J. (Eds.), *Photoinhibition Topics in Photosynthesis*. pp. 227-287.

Baulkwill, W.J. (1989). The history of natural rubber production. In: Webster, C.C. and Baulkwill, W.J. (Eds.), *Rubber*. Longmann Scientific and Technical, Essex, UK. pp. 1–56.

Blanc, G., Baptiste, C., Oliver, G., Martin, F. and Montoro, P. (2006). Efficient *Agrobacterium tumefaciens* mediated transformation of embryogeneic calli and regeneration of *Hevea brasiliensis* Mull. Arg. *Plant Cell Reports*, **24(12)**: 724-733.

Bryant, J. and Leather, S. (1992). Removal of selectable marker genes from transgenic plants: Needless sophistication or social necessity?.*Trends in Biotechnology*, **10**: 274-275.

Carron, M.P. (1981). Germination *in vitro* d'embryons immatures d'*Hevea. Caoutch Plast*, **612**: 93.

Chye, M.L., Tan, C.T. and Chua, M.N. (1992). Three genes encode 3-hydroxy-3-methylglutaryl Coenzyme A reductase in *Hevea brasiliensis*: hmg1 and hmg3 are differentially expressed. *Plant Molecular Biology*, **19**: 473-484.

Clement Demange, A., Legnate, H., Seguin, M., Carron, M.P., Guen, V., Le. and Chapuset T. (2000). Rubber Tree. In: Charrier, A., Jacquot, M., Hamon, S. and

Nicolas, D. (Eds.), *Tropical Plant Breeding*. Collection Reperes, CIRADORSTOM, Montpellier, France. pp. 455–480.

Dale, E.C. and Ow, D.W. (1990). Intra and intermolecular site-specific recombination in plant cells mediated by bacteriophage P1 recombinase. *Gene*, **91(1)**: 79-85.

Dale, E.C. and Ow, D.W. (1991). Gene transfer with subsequent removal of the selection gene from the host genome. *Proceedings of the National Academy of Sciences*, **88**: 10558-10562.

Dale, P.J., Clarke, B. and Fontes, E.M.G. (2002). Potential for the environmental impact of transgenic crops. *Nature Biotechnology*, **20**: 567-74.

Daley, M., Knauf, V.C., Summerfelt., K.R. and Turner, J.C. (1998). Co-transformation with one *Agrobacterium tumefaciens* strain containing two binary plasmids as a method for producing marker- free transgenic plants. *Plant Cell Reports*, **17**: 489-96.

Das, G., Raj, S., Pothen, J., Sethuraj, M.R., Sinha, T.P. and Sen-Mandi, S. (1998). Status of free radical and its scavenging system with stimulation in *Hevea brasiliensis. Plant Physiology and Biochemistry*, **25**: 47-50.

DeBlock, M. and Debrouwer, D. (1991). Two T-DNA's co-transformed into *Brassica napus* by a double *Agrobacterium* infection are mainly integrated at the same locus. *Theoretical and Applied Genetics*, **82**: 257-63.

Depicker, A., Herman, L., Jacobs, A., Schell, J. and Van Montague, M. (1985). Frequencies of simultaneous transformation with different TDNAs and their relevance to the *Agrobacterium* plant cell interaction. *Molecular Gene Genetics*, **201**: 477-84.

Dijkman, M.J. (1951). *Hevea*: Thirty years of research in the Far East. University of Miami Press, Florida. pp. 5–7.

Ebinuma, H. and Komamine, A. (2001). MAT (Multi-Auto-Transformation) vector system. The oncogenes of *Agrobacterium* as positive markers for regeneration and selection of marker-free transgenic plants. In *Vitro Cell Developmental Biology-Plant*, **37**: 103–113.

Ebinuma, H., Sugitha, K., Matsunaga, E. and Yamakado, M. (1997). Selection of marker-free transgenic plants using the isopentenyl transferase gene. *Proceedings of the National Academy of Science*, **94**: 2117-2121.

Fridovich, I. (1986). Biological effects of the superoxide radical. *Biochimica et BiophysicaActa*, **247**: 1–11.

Gomez, J.B. (1982). Anatomy of *Hevea* and its influence on latex production. MRRDB Monograph No 7, *Rubber Research Institute*, Kuala lumpur.

Hao, B.Z. and Wu, J.L. (2000). Laticifer differentiation in *Hevea brasiliensis*: Induction by exogenous jasmonic acid and linolenic acid. *Annals of Botany*, **85**: 37–43.

Ho, M.W. (2001). Horizontal gene transfer-the hidden hazards of genetic engineering. In: *Biotechnology and Biosafety Series- IV*. Third World Network, Penang, Malaysia. pp. 1-30.

Hohn, B., Levy, A. and Puchta, H. (2001). Elimination of selection markers from transgenic plants.*Current Opinion in Biotechnology*, **12(2)**: 139-143.

Huang, T.D., Li, J., Li, Y.T., Huang, H.S. and Hua, Y.W. (2015). Somatic embryo, an alternative target tissue for *Agrobacterium*-mediated transformation in *Hevea brasiliensis*. *Journal of Rubber Research*, **18(3)**: 171–188.

James, C. (2015). Global status of commercialized biotech/GM Crops. Brief 51: ISAAA: Ithaca, New York.

Jayashree, R., Nazeem, P.A., Venkatachalam, P., Rekha, K., Ambily, P.K., Kala, R.G., Suni, A.M., Krishnakumar, R., Sobha, S., Leda, P. and Thulaseedharan, A. (2014). Integration and expression of *hmgr*1 gene in the transgenic plants of *Hevea brasiliensis* (clone RRII 105). *International symposium on Plantation Crops*, Kozhikode.

Jayashree, R., Rekha, K., Sushamakumari, S., Sobha, S., Jayasree, P.K., Kala, R.G., Jiji, M. and Thulaseedharan, A. (2012). Effect of salinity stress and osmtica on embryogenesis and plant regeneration from MnSOD transgenic cell lines in *Hevea brasiliensis*. *Journal of Plantation Crops*, **40(3)**: 203-207.

Jayashree, R., Rekha, K., Venkatachalam, P., Sushamakumari, S., Sobha, S.,Vineetha, M., Suni, A.M., Leda, P., Thulaseedharan, A. and Nazeem, P.A. (2013). Role of bacterial strain in determining the efficiency of genetic transformation in *Hevea brasiliensis*. *Rubber Science*, **26(2)**: 217-227.

Jayashree, R., Rekha, K., Venkatachalam, P., Uratsu, S.L., Dandekar, A.M., Kumari Jayasree, P., Kala, R.G., Priya, P., Sushamakumari, S., Sobha, S., Asokan, M.P., Sethuraj, M.R. and Thulaseedharan, A. (2003). Genetic transformation and regeneration of rubber tree (*Hevea brasiliensis* Muell. Arg) transgenic plants with a constitutive version of an anti-oxidative stress superoxide dismutase gene. *Plant Cell Reports*, **22**: 201-209.

Jayashree, R., Sobha, S., Rekha, K., Supriya, R., Vineetha, M., Ancy, S.I., Sushamakumari, S., Kala, R.G., Jayasree, P.K., Thulaseedharan, A., Annamalainathan, K., Nair, D.B., Sreelatha, S., Krishnakumar, R. and Jacob, J. (2010). Development of stress tolerance in *Hevea brasiliensis* through transgenic approaches. In: *International workshop on "Climate Change and Rubber Cultivation: R and D Priotities"*. *Rubber Research Institute of India*, Kottayam, India. pp. 46-47.

Jayashree, R., Sobha,S., Rekha, K., Supriya, R., Vineetha, M., Sushamakumari, S., Kala, R.G., Jayasree, P.K. and Thulaseedharan, A. (2011). Over expression of MnSOD and drought related traits in MnSOD transgenic *Hevea brasiliensis*. *Natural Rubber Research*, **24(1)**: 18-27.

Jefferson, R.A. (1987). Assaying Chimeric Genes in Plants, the GUS Gene Fusion System. *Plant Molecular Biology*, **5(4)**: 387– 405.

Jones, H.D., Doherty, A. and Wu, H. (2005). Review of methodologies and a protocol for the *Agrobacterium*-mediated transformation of wheat. *Plant Methods*, **1**: 5.

Kala, R.G., Abraham, V., Sobha, S., Jayasree, P.K., Suni, A.M. and Thulaseedharan, A. (2012). *Agrobacterium* mediated genetic transformation and somatic embryogenesis from leaf callus of *Hevea brasiliensis*: effect of silver nitrate. In: *Prospects of Bioscience: Addressing the* Issues. Springer, India. pp. 303.

Kala, R.G., Jayasree, P.K., Sobha, S., Rajani Joseph. and Thulaseedharan, A. (2006). Plant regeneration from leaf derived embryos of *Hevea brasiliensis*. International Symposium on Frontiers in Genetics and Biotechnology - Retrospect and Prospect. Osmania University, Hyderabad.

Kala, R.G., Jayasree, P.K., Sobha, S., Sushamakumari, S., Jayasree, R., Rekha, K., Venkatachalam, P. and Thulaseedharan, A. (2003). Introduction of the gene coding for isopentenyl transferase into *Hevea brasiliensis*: Effect on plant regeneration. 10th Congress of FAOBMB, Bangalore.

Kala, R.G., Jayasree, P.K., Sushamakumari, S., Jayashree, R., Rekha, K., Sobha, S. and Thulaseedharan, A. (2005). *In vitro* regeneration of *Hevea brasiliensis* from leaf explants. In: *Proceedings of the ICAR National Symposium on Biotechnological Interventions for Improvement of Horticultural Crops: Issues and Strategies*. Trichur, India. pp. 105-106.

Kala, R.G., Reshmi, J., Sobha, S., Jayashree, R. and Thulaseedharan, A. (2014). Genetic transformation of *Hevea brasiliensis* Muell.Arg. using intact explants as target tissues for *Agrobacterium* infection. *Journal of Tropical Agriculture*, **52(1)**: 21-30.

Komari, T., Hiei, Y., Saito, N. and Kumashino, T. (1996). Vectors carrying two separate T-DNAs for co-transformation of higher plants mediated by *Agrobacterium tumefaciens* and segregation of transformants free from selection markers. *Planta Journal*, **10(1)**: 165-174.

Kumari Jayasree, P. and Thulaseedharan, A. (2004). Initiation and maintenance of long term somatic embryogenesis in *Hevea brasiliensis*. In: *IRRDB Biotechnology Workshop*. Kuala Lumpur, Malaysia.

Kumari Jayasree, P. and Thulaseedharan, A. (2005). *In vitro* germination of *Hevea* somatic embryos: Effect of cytokinins. *Plant Cell Biotechnology and Molecular Biology*, **6**: 61-64.

Kumari Jayasree, P., Asokan, M.P., Sobha, S., Sankariammal, L., Rekha, K., Kala, R.G., Jayashree, R. and Thulaseedharan, A. (1999). Somatic embryogenesis and plant regeneration from immature anthers of *Hevea brasiliensis* (Muell.Arg.). *Current Science*, **76**: 1242-1245.

Kumari Jayasree, P., Divya, S., Supriya, R. and Thulaseedharan, A. (2015). *Agrobacterium* mediated transformation of *Hevea brasiliensis* with apple cDNA encoding sorbitol-6-hosphate dehydrogenase. *Rubber Science*, **28(1)**: 31-39.

Kumari Jayasree, P., Sunitha, S.S. and Thulaseedharan, A. (2001). Effect of cytokinins on *in vitro* germination of *Hevea* somatic embryos. In: *Proceedings of IRRDB Symposium on Biotechnology and Rubber Tree. Cirad-Montpellier*, France.

Leclercq, J., Lardet, L., Martin, F., Chapuset, T., Oliver, G. and Montoro, P. (2010). The green fluorescent protein as an efficient selection marker for *Agrobacterium tumefaciens*-mediated transformation in *Hevea brasiliensis* (Müll. Arg). *Plant Cell Reports*, **29**: 513-22.

Leclercq, J., Martin F.S., Sanier, C., Clement -Vidal, A., Fabre, D., Oliver, G., Lardet, L., Ayar, A., Peyramard, M. and Montoro, P. (2012). Overexpression of a cytosolic isoform of the *Hb* CuZnSOD gene in *Hevea brasiliensis* changes its response to a water deficit. *Plant Molecular Biology*, **80(3)**: 255-72.

Liu, D., Raghothama, K.G., Hasegawa, P.M. and Bressan, R.A. (1994).) Overexpression of osmotin in potato delays development of disease symptoms. *Proceedings of National Academy of Sciences*, **91**: 1888–1892.

Lynen, F. (1969) Biochemical problems of rubber synthesis. *Journal of Rubber Research Institute of Malaysia*, **21**: 389-406.

Marattukalam, J.G., Saraswathyama, C.K. and George, P.J. (1980). Crop improvement through ortet selection in India. In:*International Rubber Conference*, Kottayam, India.

McKnight, T.D., Lillis, M.T. and Simpson, R.B. (1987). Segregation of genes transferred to one plant cell from two separate *Agrobacterium* strains. *Plant Molecular Biology*, **8**: 439-45.

Montoro, P., Lagier, S., Baptiste, C., Marteaux, B., Pujade-renaud, V., Leclercq, J. and Alemanno, L. (2009). Expression of the HEV2.1 Gene Promoter in Transgenic *Hevea brasiliensis*. *Plant Cell Tissue Organ Culture*, **94**: 55–63.

Montoro, P., Rattana, W., Pujade-Renaud, V., Michaux-Ferriere, N., Monkolsook, Y., Kanthapura, R. and Adunsadthapong, S. (2003). Production of *Hevea brasiliensis* transgenic embryogenic callus lines by *Agrobacterium tumefaciens*: Roles of calcium. *Plant Cell Reports*, **21**: 1095–1102.

Montoro, P., Teinseree, N., Rattana, W., Kongsawadworakul, P. and Michaux Ferriere, N. (2000). Effect of exogenous calcium on *Agrobacterium tumefaciens*-mediated gene transfer in *Hevea brasiliensis* (rubber tree) friable calli. *Plant Cell Reports*, **19**: 851–855.

Mydin, K.K. and Gireesh,T. (2016). Diversity and heterosis by recombination breeding of *Hevea brasiliensis* in India. *Rubber Science*, **29(1)**: 20-35.

Mydin, K.K. and Saraswathyamma, C.K. (2005). *A Manual of Breeding of Hevea brasiliensis*. *Rubber Research Institute of India*, Kottayam. pp. 30–31.

Nair, N.U. and Kurup, P.A. (1990). Possible application of the ratio of HMG CoA to mevalonate in the bark of *Hevea brasiliensis* as an indicator of rubber yield. Physiology and Exploitation of *Hevea brasiliensis*. In: *Proceedings of IRRDB Symposium*. Kunming, China.

Nandy, S. and Srivastava, V. (2011). Site-specific gene integration in rice genome mediated by the FLP-FRT recombination system. *Journal of Plant Biotechnology*, **9(6)**: 713-721.

Nandy, S. and Srivastava, V. (2012). Marker-free site specific gene integration in rice based on the use of two recombination systems. *Journal of Plant Biotechnology*, **10(8)**: 904-912.

Natural Rubber Trends and Statistics (2016). ANRPC; http://anrpc.org/html/member_nrts.aspx. Accessed in September 2016,

Nuntanuwat, W. (2006). Relationship between the expression of 3-hydroxy-3-methylglutaryl coenzyme A reductase (HMGR-1) and rubber content in *Hevea brasiliensis* (Willd. ex A. Juss.) Muell.Arg. cultivars RRIM 600, BPM24 and PB235. *An Abstract presented to the Thaksin University.*

Ong, S.H. (1976). Chromosome morphology at the pachytene stage in *Hevea brasiliensis*: A preliminary report. In: *Proceedings of International Rubber conference. Rubber Research Institute of Malaysia*, Kuala Lumpur. pp. 3-12.

Pardekooper, E.C. (1989). Exploitation of the rubber tree. In: Webster, C.C. and Baulkwill, W.L. (Eds.), *Rubber*. Longman Scientific and Technical, Singapore. pp. 349–414.

Peleg, Z. and Blumwald, E. (2011). Hormone balance and abiotic stress tolerance in crop plants. *Current Opinion in Plant Biology*, **14**: 290–295.

Premakumari, D. and Saraswathyamma, C.K. (2000). The para rubber tree. In: George, P.J. and Jacob, C.K. (Eds.), *Natural Rubber: Agromanagement and Crop Processing. Rubber Research Institute of India*, Kottayam. pp. 29–35.

Priyadarshan, P.M. (2003). Breeding *Hevea brasiliensis* for environmental constraints. *Advances in Agronomy*, **79**: 351-400.

Qin, M., Bayley, C., Stockton, T. and Ow, D. W. (1994).) Cre recombinase-mediated site-specific recombination between plant chromosomes. *Proceedings of the National Academy of Sciences*, **91**: 1706-1710.

Raemer, H. (1935). Cytology of *Hevea. Genetics*, **17**: 193-194.

Rao, M.V., Parameswari, C., Sripriya, R. and Veluthambi, K. (2011). Transgene stacking and marker elimination in transgenic rice by sequential *Agrobacterium* mediated co-transformation with the same selectable marker gene. *Plant Cell Reports*, **30**: 1241-1252.

Rekha, K. (2013). *Agrobacterium* mediated molecular breeding in *Hevea brasiliensis for crop improvement*. Ph D Thesis. M. G. University, Kerala.

Rekha, K., Jayashree, R., Sushamakumari, S., Sobha, S., Jomini, T., Supriya, R. and Nazeem, P.A. (2013b). Integration and expression of osmotin gene in *Hevea brasiliensis via. Agrobacterium* mediated transformation. *Journal of Plantation Crops*, **40(1)**: 80-85.

Rekha, K., Jayasree, R., Kumari Jayasree, P., Venkatachalam, P., Jinu, C. and Thulaseedharan, A. (2006). High frequency *Agrobacterium* mediated genetic transformation in *Hevea brasiliensis*. In: *International symposium on "frontiers in genetics and biotechnology- retrospect and prospect"*, Hyderabad, India.

Rekha, K., Nazeem, P.A., Venkatachalam, P., Jayashree, R., Akshara, R., Sobha, S. and Sushamakumari, S. (2016). Expression of stress tolerance in transgenic cell lines. *Rubber Science*, **29(2)**: 40-55.

Rekha, K., Nazeem, P.A., Venkatachalam, P., Jayashree, R., Sobha, S. and Sushamakumari, S. (2014a). Development of transgenic plants of *Hevea brasiliensis* integrated with osmotin gene. In: *National seminar on GM crops- issues and prospects*. KAU, Vellanikkara.

Rekha, K., Nazeem, P.A., Venkatachalam, P., Jayasree, R., Sobha, S. and Sushama kumari, S. (2014b). Development of osmotin transgenics in *Hevea brasiliensis* Muell.Arg. using explants of zygotic origin. *Journal of Tropical Agriculture*, **52(1)**: 7-20.

Rekha,K., Jayashree, R., Sushamakumari, S., Sobha, S., Jomini, T., Supriya, R. and Nazeem, P.A. (2013a). Expression of stress tolerance in transgenic cell lines in *Hevea brasiliensis*.In: *National Seminar on Tree Biotechnology: Emerging opportunities in forestry and tree science*, At IFGTB, Coimbatore, Tamil Nadu, India.

Rubber Statistical Bulletin 2015, IRSG; http://www.rubberstudy. com/documents/WebSiteData_Aug2016.pdf)

Saraswathyamma, C.K. (2002). Advances in crop improvement in *Hevea* in the traditional rubber growing tract of India. In: *Proceedings of the Rubber Planters' Conference*. Kottayam, India. pp. 101–116.

Saraswathyamma, C.K., Markose, V.C., Licy, J., Annamma, Y. and Panikkar, A.O.N. (1984). Cytomorphological studies in an induced polyploidy of *Hevea brasiliensis* Muell. *Arg. Cytologia*, **49**: 725-729.

Schultes, R.E. (1977). Wild *Hevea:* An untapped source of germplasm. *Journal of the Rubber Research Institute of Sri Lanka*, **54**: 227-257.

Sobha, S., Rekha, K., Sushamakumari, S., Jayashree, R., Kala, R.G., Jayasree, P., Deepa, K. and Thulaseedharan, A. (2013). High frequency *Agrobacterium* mediated genetic transformation in rubber tree *via*. Vacuum infiltration. *Journal of Plantation Crops*, **41(3)**: 300-306.

Sobha, S., Rekha, K., Sushamakumari, S., Jayasree, R., Kala, R.G., Deepa, K. and Thulaseedharan, A. (2014). *Agrobacterium* mediated multiple gene integration in *Hevea brasiliensis. Journal of Tropical Agriculture*, **52(1)**: 31-38.

Sobha, S., Sushamakumari, S., Rekha, K., Jayashree, R., Kala, R. G., Jayasree P. K., Asokan, M.P., Sethuraj, M.R. and Thulaseedharan, A. (2001). Abiotic stress induced over-expression of superoxide dismutase enzyme in transgenic *Hevea brasiliensis. Symposium on "Natural Rubber Research"*, CIRAD, France.

Sobha, S., Sushamakumari, S., Rekha, K., Jayashree, R., Kala, R.G., Supriya, R. and Thulaseedharan, A. (2012). Factors promoting *Agrobacterium* mediated genetic transformation efficiency in *Hevea brasiliensis. Rubber Science*, **25(2)**: 173-182.

Sobha, S., Sushamakumari, S., Thanseem, I., Jayasree, P.K., Rekha, K., Jayashree, R., Kala, R.G., Asokan, M.P., Sethuraj, M.R., Dandekar, A.M. and Thulaseedharan, A. (2003a).. Genetic transformation of *Hevea brasiliensis* with the gene coding for superoxide dismutase with FMV 34S promoter. *Current Science*, **85**: 1767-1773.

Sobha, S., Sushamakumari, S., Thanseem, I., Rekha, K., Jayashree, R., Kala, R.G., Kumari Jayasree, P., Asokan, M.P., Sethuraj M.R., Dandekar, A.M.

and Thulaseedharan, A. (2003b).. Abiotic stress induced over-expression of superoxide dismutase enzyme in transgenic *Hevea brasiliensis. Indian Journal of Natural Rubber Research*, **16**: 45-52.

Sobha, S., Thulaseedharan, A., Nandy, S. and Srivastava, V. (2015a). Synthesis of a Clean vector based on heat-inducible Cre-loxP system for production of marker-free transgenic rubber tree. *Proceedings of the National Conference on "New Frontiers in Plant Sciences and Biotechnology"*, Goa University.

Sobha, S., Thulaseedharan, A., Nandy, S. and Srivastava, V. (2015b). Towards the development of marker-free transgenic rubber tree by synthesizing a "Clean Vector" utilizing heat-inducible Cre-loxP system In: Krishnan, S. and Rodrigues, B.F. (Eds.), *Advances in Plant Sciences and Biotechnology*. pp. 198-210.

Srivastava, V. and Ow, D. W. (2002). Biolistic mediated site-specific gene integration in rice. *Molecular Breeding*, **8**: 345-350.

Srivastava, V., Anderson, O.D. and Ow, D.W. (1999). Single copy transgenic wheat generated through the resolution of complex integration patterns. *Proceedings of National Academy of Science*, (**96**): 11117-11121.

Sumesh, K.V., Satheesh, P.R., Sreelatha, S., Ravichandran, S., Thulaseedharan, A., Jayashree, R., Krishnakumar, R., Annamalainathan, K., Singh, M. and Jacob, J. (2014). Drought tolerance in MnSOD transgenic *Hevea brasiliensis* in a dry sub-humid environment. *Journal of Plantation Crops*, **42(1)**: 70-77.

Sunderasan, E., Badaruddin, B.E., Azharuddin, A. and Arokiaraj, P. (2012). Genetic Transformation of *Hevea brasiliensis* with Human Atrial Natriuretic Factor. *Journal of Rubber Research*, **15** (**4**): 255–264.

Sushamakumari, S., Sobha, S., Rekha, K., Jayasree, R. and Asokan, M.P. (2000). Influence of growth regulators and sucrose on somatic embryogenesis and plant regeneration from immature inflorescence of *Hevea brasiliensis. Indian Journal of Natural Rubber Research*, **13**: 19-29.

Suwanmanee, P., Sirinupong, N., Nunthanuwat, W., Chatree, T., Chocknukul, S., Seedungnu, N.Y. and Suwachitanont, W. (2007). Expression of HMG-CoA synthase (*hmgs*) and HMG-CoA reductase-1 (*hmgr-1*) reveal coordinated regulation of rubber biosynthesis in *Hevea brasiliensis* (B.H.K.) Mull. *Arg. TERPNET 30, Strasbourg*, France.

Thomson, L.C., Calendar, R. and Ow, D.W. (2001). Gene insertion and replacement in *Sachizo-saccharomyces pombe* mediated by the *Streptomyces* bacteriophage *phi*C31 site-specific recombination system. *Molecular Genetics Genomics*, **265**: 1031-1038.

Thulaseedharan, A., Kumari Jayasree, P. and Venkatachalam, P (2000). Biotechnological approaches for crop improvement in rubber. In: Chadha, K.L., Ravindran, P.N. and Sahijram,L. (Eds.), *Biotechnology in Horticultural and Plantation Crops*. Malhotra Publishing, Calcutta. pp. 323–351.

Thulaseedharan, A., Venkatachalam, P., Jayashree, R., Kala, R.G., Rekha, K., Sobha, S., Jayasree, P.K., Sushamakumari, S. and Saleena, A. (2008). Rubber Tree. In: Kole, C. and Hall, T.C. (Eds.), *Compendium of Transgenic Crop Plants: Transgenic*

Plantation Crops, Ornamentals and Turf Grasses, Blackwell Publishing, Oxford, UK.

Varghese, Y.A. and Mydin, K.K. (2000). Genetic improvement. In: George, P.J. and Jacob, C.K. (Eds.), *Natural Rubber. Agromanagement and Crop Processing. Rubber Research Institute of India*, Kottayam. pp. 36–46.

Varghese, Y.A. (1992). Germplasm resources and genetic improvement. In: Sethuraj, M. and Mathew, N.M. (Eds.), *Natural Rubber Biology, Cultivation and Technology*. Elsevier, Amsterdam. pp. 88-115.

Veluthambi, K., Gupta, A.K. and Sharma, A. (2003). The current status of plant transformation technologies. *Current Science*, **84(3)**: 368-380.

Venkatachalam, P., Jayasrhee, R., Rekha, K., Sushamakumari, S., Sobha, S., Kumari Jayasree, P., Kala, R.G. and Thulaseedharan, A. (2006). Rubber Tree (*Hevea brasiliensis* Muell. Arg). In: Wang, K.W. (Ed.), *Methods in Molecular Biology – Protocols*. Humana Publishing House, USA. pp. 143-64.

Venkatachalam, P., Kumari Jayasree, P., Sushamakumari, S., Jayashree, R., Rekha, K., Sobha, S., Priya, P., Kala, R.G. and Thulaseedharan, A. (2007). Current Perspectives on application of Biotechnologyto Assist the Genetic Improvement of Rubber Tree(*Hevea brasiliensis* Muell. Arg.): An Overview. *Functional Plant Science and Biotechnology*, **1(1)**: 1-17.

Wan Abdul Rahaman., Ghandimathi, W.Y., Rohani, H., Paranjothy, K. (1981). Recent developments in tissue culture of *Hevea*. In: Rao, A.N. (Ed.), *Tissue Culture of Economically Important Plants*. Costed, Singapore. pp. 152-158.

Wang, Y., Chen, X., Peng, S., Wu, K. and Hong, L. (2013). Genetic transformation and regeneration of *Hevea brasiliensis* transgenic plant with GAI gene by microparticle bombardment. *Romanian Biotechnological Letters*, **18**: 7910-7919.

Wang, Z., Zeng, X., Chen, C., Wu, H., Li, Q., Fan, G. and Lu, W. (1980). Induction of rubber plantlets from anther of *Hevea brasiliensis* Muell. Arg. *In vitro Chinese Journal of Tropical Crops*, **1**: 25-26.

Webster, C.C. and Paaradekooper, E.C. (1989). The Botany of the Rubber Tree. In: Webster,C.C. and Baulkwill,W.J. (Eds.), *Rubber*. *Longman Scientific and Technical*, Essex. pp. 572-84.

Wycherley, P.R. (1976). Rubber. In: Simmonds, N.W. (Ed.), *Evolution of Crop Plants*. Longman, London. pp. 77–80.

Yeang, H.Y., Arokiaraj, P., Jaafar, H., Arif, S.A.M., Rajamanikam, S., Chan, J.L., Sharib, J., Leelavathi, R., Samsidar, H. and Van der logt, C.P.E. (2002). Expression of a functional recombinant antibody fragment in the latex of transgenic *Hevea brasiliensis*. *Journal of Rubber Research*, **5(4)**: 215–225.

Yeang, H.Y., Arokiaraj, P., Jaafar, H., Hamzah, S., Arija, M.A.S. and Jones, H. (1998). Rubber Latex as an Expression System for High-value Proteins. In: Shewry, P.R., Napier, J.A. and Davis, P.J. (Eds.), *Engineering Crop Plants for Industrial End Uses*. London, Portland Press. pp. 55–64.

Chapter 25

Spices

☆ *A.I. Bhat*

1. Introduction

Conventional plant breeding has contributed significantly to crop improvement over past fifty years. However, there is intense pressure to produce further improvements in crop quality and quantity as a result of population growth, social demands, health requirements, environmental stress and ecological considerations. Traditional plant breeding has not able to withstand this increasing demand due to the limited gene pool and restricted range of organism between which genes can be transferred due to the species barriers. Genetic transformation holds great promise for overcoming these major constraints to crop productivity as it involves recombination of an efficient cell or tissue culture regeneration system with recombinant DNA technology, which would transfer specific genes from other taxa, unrelated plants, microbes, and animals. Because of these possibilities, it is now feasible to introduce genes that have previously been inaccessible to the conventional plant breeder or which did not exist in the crop of interest to impart desired traits, such as high yield, insect resistance, disease resistance and herbicide resistance, without disrupting their fundamental genetic background within a short period of time. The first stably transformed plants were reported in early 1980s (Herrera-Estrella *et al.*, 1983). The production of novel chimeric genes in 1980s resulted in the expansion of plant transgenic technologies, transformation vectors, DNA delivery systems, combined with plant regeneration systems.

The major components for the development of transgenic plants are: (1) the development of reliable tissue culture regeneration systems; (2) efficient transformation techniques for the introduction of genes into the crop plants; (3) preparation of gene constructs in suitable vectors; (4) selection and multiplication of transgenic plants; (5) molecular and genetic characterization of transgenic plants for stable and efficient gene expression; and (6) evaluation of transgenic plants. Genetic transformation methods in plants are mainly classified into direct and indirect transformation systems. The majority of gene transfer experiments have focused on maximizing the efficiencies for the recovery of stably transformed plants, and also extending the range of genotypes that could be engineered using a specific procedure. Plant transformation is performed using a wide range of tools such as *A. tumefaciens* Ti plasmid vectors, biolistic bombardment, micro injection, chemical (PEG) treatment of protoplasts and electroporation of protoplasts. Though all methods have advantages that are unique to each of them, transformation using *Agrobacterium* and biolistic bombardment are the most extensively used methods. In general, *Agrobacterium* has been used as the vector for genetic transformation of diverse dicotyledonous species, but biolistic bombardment has been a very useful technique to introduce foreign DNA into plant cells of monocotyledons and dicotyledonous plants.

India is known as "The Home of Spices". Spices have been considered important in the culinary art from time immemorial. They are used for flavouring, seasoning and imparting aroma in foods. Besides, some of them are known to be fungistatic, antimicrobial or antibiotic. Their antioxidative activity helps to preserve foods from oxidative deterioration, increasing their shelf life. They are also processed into numerous 'value added' attractive spice products of importance such as spice oleoresins, essential oils, curry powder *etc.* Spices may comprise different plant components or parts such as floral parts or fruits or berries or seeds or rhizomes or roots or leaves or kernel or aril or bark or bulbs *etc.* (Pruthi, 1992). There is a constant mention of the important role spices have played in the life of the ancient people and also in changing the course of the world history. It was the lure of the exotic spices that inspired the European navigators and explorers that brought them to the shores of India and South-East Asia. Even today, spices are of considerable economic importance for all spice producing, exporting and importing countries of the world. The global spice trade is expected to increase with the growing consumer demand in importing countries for more exotic, ethnic tastes in food (Peter, 2001).

International Organization for Standardization (ISO) lists 109 herb and spice plant species useful as ingredients in food. India is known as one of the largest producers, consumers and exporters of spices and spice products. India grows over 50 spices in different parts. However, the country has one of the lowest productivity in many of the spice crops. Abiotic and biotic factors are among the important factors that contribute for the lower productivity of spices in India. Conventional breeding programmes are cumbersome and time consuming. Besides, in many spice crops source of resistance to biotic and abiotic stresses are not available or not characterized in the available germplasm collections. Hence transgenic approaches are one of the alternatives to impart abiotic and biotic stresses and to enhance quality and yield of spice crop. In this chapter, developments in the transgenics of different spice crops are discussed.

2. Black Pepper (*Piper nigrum*) (Family: Piperaceae)

Black pepper (*Piper nigrum*), known as the 'King of Spices', originated in the tropical evergreen forests of Western Ghats of India (Ravindran, 2000). It is one of the most ancient crops cultivated in India. It is used for a variety of purposes including in medicine. It constitutes an important component of culinary seasoning and an essential ingredient of numerous commercial foodstuffs. India has the largest area and production of black pepper in the world. India is a leading exporter of black pepper that accounts for about 40 per cent of export earnings through spices (Ravindran, 2000).

2.1. Development of Transformation System

There are two preliminary reports on *Agrobacterium* mediated transformation of black pepper. Sasikumar and Veluthambi (1996) reported black pepper transformation using cotyledon and primary leaves as explants and *Agrobacterium* strain LBA4404 harbouring binary vector PGA 472 with β-glucuronidase (*GUS*) and neomycin phosphotransferase (*npt*II) as the selectable markers. They could get callusing (20 per cent) in infected cotyledon explants where as there was no appreciable callusing in infected primary leaves under kanamycin selection. Transformed calli remained fresh and proliferated in medium containing up to 150 μg/ml of kanamycin but in the case of control calli, no callus proliferation was observed above 50 μg/ml of kanamycin. Sim *et al.* (1998). cultured leaf, petiole and stem explants from axenic seedlings of black pepper and inoculated with *Agrobacterium* strain LBA4404 containing plasmids pMOG23 and pTOK47 containing *GUS* and *nptI*I genes in callus inducing medium in the dark at 28° C. After co cultivation for two days, the explants were transferred to Murashige and Skoog (MS) based medium supplemented with carbenicillin and cefotaxime to kill *Agrobacterium*, and kanamycin at 75 mg/l for selection of transformed tissue. After 10 days, the leaf explants tested showed expression *GUS* gene but PCR analysis using *npt*II specific primers failed to detect the presence of transgene. Both these reports on black pepper transformation did not report regeneration of transformed tissue into plantlets.

Nair and Gupta (2006) reported a very efficient micropropagation strategy through cyclic secondary somatic embryos obtained from secondary embryos that were borne from the root pole region of primary somatic embryos derived from micropylar region of germinating mature seeds of black pepper. Using these somatic embryos as explants, Jiby and Bhat (2011) developed an efficient *Agrobacterium* mediated transformation protocol for black pepper. Cyclic secondary somatic embryos (embryogenic mass), co-cultivated with *Agrobacterium* carrying the *GUS* reporter gene, were cultured on plant growth regulator free Schenk and Hildebrandt (SH) medium and transformants were selected in the medium containing cefotaxime and step wise increase in kanamycin concentration from 25 to 100 μg/ml (Fig. 25.1). The transient GUS gene expression was determined histochemically. Transformants that survived in the selection medium were hardened in the green house. An average of nine hardened putative plantlets was obtained per gram of embryogenic mass. Presence of transgene in these plantlets was assayed by PCR, dot blot and Southern blot hybridization. Maju and Sonia (2012) reported direct regeneration of shoots from the bulged portion of shoot tip and nodal explants cultured on SH and MS medium with various combinations of cytokinins and auxins. They also reported genetic transformation of Panniyur 1 variety of black pepper by infecting seedling

Figure 25.1: Different Stages of Transgenesis in Black Pepper.

(a) Co-cultivated embryogenic mass, (b) Growing points under kanamycin selection, (c) Embryo clusters formed under kanamycin selection, (d) Embryogenic mass in liquid SH medium, (e) Fully developed plantlets in liquid SH medium, (f) Hardened plants maintained in green house (Source: Jiby and Bhat, 2011).

derived explants with *Agrobacterium* carrying pCAMBIA 1301 vector with GUS and HPT markers. Shoot regenerated from explants in the presence of hygromycin were analysed by PCR and GUS histochemical assay for confirmation of transgenic nature of plants.

Sasi *et al.* (2015) reported a loop-mediated isothermal amplification (LAMP) and real-time LAMP based assays as an alternative to PCR for quick and sensitive detection of transgenic black pepper plants. Primers (six each) were designed based on the nucleotide sequence of two target regions [kanamycin and *Cauliflower mosaic virus* (CaMV) 35 S promoter] integrated into the genome of transgenic black pepper. The assay successfully detected the transgenic plants whereas no cross-reaction was recorded with non-transgenic plants. The detection limit for LAMP was up to 10000 times that for conventional PCR and 1/1000 times that for real-time LAMP. The assays were validated by testing putative transformants of black pepper.

2.2. Development of Transgenic Black Pepper for Fungal and Viral Resistance

Babu *et al.* (2013) reported delayed response to infection by the fungus, *Phytophthora capsici* causing foot rot disease in black pepper plants transformed with *osmotin* gene. Bhat *et al.* (2014) attempted pathogen derived resistance approach to get virus resistant plants by transforming black pepper with sequences from *Cucumber mosaic virus* (CMV) and *Piper yellow mottle virus* (PYMoV) sequences. Three constructs (CMV coat protein gene in sense orientation, reverse transcriptase region of PYMoV in sense and antisense orientations) were prepared in the binary vector, pBI121, mobilized into *A. tumefaciens* and were used for transforming embryogenic mass of four varieties of black pepper. After co-cultivation, explants were selected in a step wise increase in the concentration of kanamycin from 25 to 100 μg/ml in the SH medium. Of the 23 transgenic lines of PYMoV in the sense orientation and 40 transgenic lines in antisense orientation, 15 and 21 transgenic lines respectively showed presence of transgene when tested through PCR. The integration of the transgene was ascertained through Southern hybridization using transgene specific probe while the transcript production was confirmed though RT-PCR and northern hybridization. Out of 67 transgenic plants carrying the CMV coat protein gene when subjected to PCR, 12 plants showed the presence of transgene. Southern analysis of these plants using transgene specific probe confirmed the presence of transgene in nine plants with three insertions in all plants. All nine plants showed transcript and protein production when subjected to northern and western blotting. Short listed transgenic lines of PYMoV and CMV when subjected to challenge inoculation with respective viruses, four plants with PYMoV sequences showed symptom remission and two plants showed resistance to CMV.

Piper colubrinum, an exotic species distantly related to black pepper, shows high degree of resistance to the oomycete pathogen *Phytophthora capsici*, which causes the devastating 'foot rot' disease in black pepper. *In planta* transformation of black pepper *via* pollen tube pathway using total exogenous DNA of *P. colubrinum* was reported (Asha and Rajendran, 2010). When resultant seeds were germinated and tested *in vitro* and *ex vitro* against *P. capsici*, 39 per cent of plants did not take up the infection. Enhanced expression of serine/threonine protein kinase (*PcSTPK*) gene in *P. colubrinum* was noticed upon infection by *P. capsici* (Krishnan *et al.*, 2015). A *Tobacco Rattle Virus* (TRV)-based virus-induced gene silencing (VIGS) construct was established for functional validation of PcSTPK in *Piper colubrinum* by Krishnan *et al.* (2015). The construct TRV:PcSTPK VIGS vector was inltrated into young leaves of *Piper colubrinum* and the time course study revealed that STPK transcript levels was signicantly down regulated. Knock-down of *PcSTPK* by VIGS increased the susceptibility to *P. capsici* infection, as evidenced by the appearance of foliar necrotic lesions and increased proliferation and sporulation of *P. capsici* on the leaf surface indicating the possible role of PcSTPK in modulating antifungal defense response in the plant. The osmotin *PR 5* gene homologue from *Piper colubrinum* showed significant over expression in response to *P. capsici* infection. The functional validation of the same was reported by Anu *et al.* (2015). using a TRV based VIGS in *Piper colubrinum*. The TRV construct carrying *Piper colubrinum* osmotin (PCOSM) was infiltrated into *Piper colubrinum* plants. Three weeks post infiltration, significant down regulation of PCOSM was observed. The silenced plants when challenged with *P. capsici*, showed increased *P. capsici* growth accompanied by decreased accumulation of H_2O_2 indicating that osmotin gene is required for resisting *Piper colubrinium* infection and has possible role in hypersensitive cell death response and oxidative burst signaling during infection.

3. Caraway (*Carum carvi*) (Family: Umbelliferae)

Caraway or 'caraway seed' of commerce is the fruit of a biennial herb known botanically as *Carum carvi*. It is native to North and Central Europe, and is extensively cultivated in Holland, Russia, Poland, Bulgaria, Rumania, Syria and Morocco. The seeds, on steam distillation, yield an aromatic essential oil which finds greater use in medicine (Pruthi, 1992).

Krens *et al.* (1997) reported a high-frequency direct regeneration when cotyledonary node explants were used. Transient expression of *GUS* was obtained when cotyledonary node explants were used. This explant type proved to be the best for stable transformation resulting in transgenic plants. Parameters determining regeneration and transformation efficiency were optimized. The percentage of explants giving transgenic plants was as high as 13 per cent. This system for the rapid production of transgenic caraway plants opens up possibilities for studying metabolic engineering with this crop.

4. Cardamom (*Elettaria cardamomum*) (Family: Zingiberaceae)

Cardamom is the second most important national spice of India and hence known as "Queen of Spices". Cardamoms of commerce are the dried capsules of these plants, which contain seeds possessing a pleasant characteristic aroma. The plant is indigenous to India and Sri Lanka and more than 70 per cent of the world production is from India. It is cultivated in evergreen rain forests at 760-1500 m above sea level (Pruthi, 1992). Low production of cardamom is mainly due to the onslaught of pests and diseases caused by viruses and fungi.

In order to produce transgenic cardamom resistant to *Cardamom Mosaic Virus* (CdMV), Backiyarani *et al.* (2005) cloned coat protein gene of CdMV in plant expression vector, pAHC 17 under the ubiquitin promoter. Based on the *in vitro* studies, Josephrajkumar *et al.* (2006) suggested production of transgenic cardamom expressing protease inhibitor, such as aprotinin, for the management of cardamom shoot and capsule borer, *Conogethes punctiferalis*. A preliminary standardization of biolistic based transformation of cardamom embryogenic callus using GUS reporter gene was reported (Babu *et al.*, 2013).

5. Celery (*Apium graveolens*) (Family: Umbelliferae)

Celery seed is the dried ripe fruit of the umbelliferous herb. The native habitat of celery extends from Sweden to Egypt, Algeria and Ethiopia, and in Asia. It is one of the important minor spices of India. The dried ripe fruits are used as spice. Leaves and stalks are used as salads in soups. It is rich in beta- carotene, folic acid, vitamin C, calcium, magnesium, potassium, and fiber. The crop is susceptible to abiotic and biotic stresses. Thus, genetic engineering strategies could be a powerful means for introducing agriculturally valuable traits such as herbicide, disease or pest resistance.

Only a few reports to date are available concerned transformation efforts of celery. Catlin *et al.* (1988) obtained a total of 20 kanamycin resistant plants using *A. tumefaciens* mediated transformation of petiole explants of cv. PI257228. Celery callus and seedlings were also used as models for transient transformation by Liu *et al.* (1992). Glufosinate herbicide resistant celery plants were obtained from cvs. XP85 and XP166 using *A. tumefaciens* with the bar transgene (Loskutov *et al.*, 2008). Morphologically typical and atypical transgenic plants were regenerated from both leaf and seedling explants using a selection system of 0.25 mg of ammonium glufosinate.

Song *et al.* (2007) reported a protocol for rapid and efficient production of transgenic celery plants *via* somatic embryo regeneration from *A. tumefaciens* inoculated leaf sections, cotyledons and hypocotyls. Co-cultivation was carried out for four days in the dark on callus induction medium supplemented with acetosyringone followed by selection in the medium containing kanamycin and timentin. Explants that survived for 12 weeks in the selection were regenerated *via* somatic embryogenesis on Gamborg B5 + 4.92 µM 6 (γ-γ-dimethyl allyl amino) purine (2iP) + 1.93 µM α-naphthaleneacetic acid (NAA) + 25 mg/l kanamycin + 250 mg/l timentin after 8 weeks. Using this protocol, the transformation frequency of 5 per cent for leaf sections, 18 per cent for cotyledons, and 16 per cent for hypocotyl explants were obtained. Stable integration of the model transgenes with 1–3 copy numbers was confirmed in selected transgenic events by Southern blot analysis of *gusA*. Progeny analysis by histochemical GUS assay showed stable Mendelian inheritance of the transgenes.

Loskutov *et al.* (2008 evaluated methods for celery transformation using *A. tumefaciens* and the *bar* gene as selectable marker. Callus selection (CS) and the flamingo-bill explant (FB) methods were evaluated for efficacy in transformation. *A. tumefaciens* strains EHA105 and GV3101, each with the *bar* gene under the promoters NOS (pGPTV-BAR) or 35S (pDHB321.1), were used. Leaf explants were inoculated and co-cultivated for two days in the dark. Calluses emerged on the explants after 4-6 weeks were selected for glufosinate (GS) resistance by a two step method with increasing concentration of glufosinate. The explants that survived the selection were allowed to shoot on Gamborg B5 medium + 2iP (4.9 µM) + NAA (1.6 µM) and rooted on MS in 5–6 months time. Conversely, using FB with inoculation by GV3101/pDHB321.1, putative transgenic celery plants were obtained in just six weeks. Southern blot analyses indicated 1–2 copies in CS lines and 1 copy in FB lines. Herbicide assays on whole plants with 100 and 300 mg/l glufosinate indicated a range of low to high tolerance for lines derived by both methods. The *bar* gene was found to be Mendelian inherited in one self-fertile CS derived line.

6. Coriander (*Coriandrum sativum*) (Family: Umbellifereae)

The green leaves and seeds of coriander are used as spice. Mexico and USA are the major growers of coriander. It is a native of Mediterranean region and India is one of the major producers of coriander in the world. Stem, leaves and fruits of coriander have a pleasant aromatic odour. The entire plant is used in preparing chutneys and sauces, and leaves are used for flavouring curries and soups. The fruits are used as condiment in the preparation of several items (Pruthi, 1992).

The phytohormone ethylene is involved in many developmental processes, including leaf and flower senescence. Ethylene is perceived by plants through receptors that trigger the downstream signal transduction pathway. The mutated ethylene receptor *ERS1* (ethylene response sensor) from *Arabidopsis* is of a dominant negative nature and confers ethylene insensitivity in *Arabidopsis.* To investigate if the altered *ERS1* gene can affect the tissue senescence in heterologous plants, *ERS1*was introduced into coriander by *Agrobacterium*-mediated transformation (Wang and Kumar, 2004). Transgenic plants were regenerated by co-cultivating hypocotyl segments with *A. tumefaciens* harboring binary vector pCGN1547 that carried the *ERS1* gene. The presence and expression of the transgene was confirmed by genomic Southern blot and reverse transcriptase-PCR analyses. Leaf and flower senescence were delayed significantly in the transgenic plants. The ability of the mutated ERS1 gene to confer the ethylene- insensitive phenotype can be exploited for extending the shelf-life of leafy vegetables.

7. Cumin (*Cuminum cyminum*) (Fam: Apiaceae)

Cumin is an annual herbaceous plant cultivated in tropical regions of the world including Egypt, India, China, Pakistan, Spain and the Eastern Mediterranean. India is the largest producer of cumin, contributing about 70 per cent of the total world production. Its root, stem, leaf and flower are a rich source of essential oils and polyphenols with antioxidant properties. However, cumin production has declined in recent years and the area of cumin cultivation is becoming depleted due to various factors that include abiotic stress such as salinity. The development of an abiotic stress-tolerant plant that can cope with adverse conditions might be an alternative so that unsuitable land is used for sustainable development. Conventional breeding methods have provided limited scopes to improve traits in cumin due to its low-genetic diversity and also due to inefficient and time-consuming approaches. Recent advances in gene manipulation, DNA technology, and genetic transformation provide a potential approach for the development of transgenic cumin.

A simple and efficient method was developed for multiple shoot regeneration of cumin from imbibed embryo cultures (Ebrahimie *et al.*, 2003). The method yielded a large number of shoots within short period of time (30–50 days) without any subculturing. The effects of different media, different embryo explants and various combinations of plant growth regulators on callus formation and shoot regeneration were also studied. Simultaneous callus formation and shoot regeneration was obtained. The best response for multiple shooter generation was observed on B5 medium containing 1.0 mg/l BAP, 0.2 mg/l NAA and 0.4 mg/l IAA, with an average of 140 shoots per explants. A microprojectile bombardment-mediated genetic transformation of embryo axes and plant regeneration in cumin was reported by Singh *et al.* (2010). Pre-cultured cumin embryos were bombarded under 27 inches Hg vacuum, 25 mm distance from rupture disc to macrocarrier, 10 mm macrocarrier flight distance using 1100 psi rupture disc and 9 cm microprojectile travel distance. About 91 per cent embryos showed transient GUS expression after 24 hours. Shoot tips and roots of T0 plantlets exhibited GUS expression done after three months of bombardment. Transformation was conrmed by performing PCR detection of *nptII* and *GUS* genes respectively from T0 transgenics and Southern blot analysis using PCR amplied DIG labeled *nptII* gene as probe.

Pandey *et al.* (2013) reported an efficient and reproducible method of *Agrobacterium*-mediated genetic transformation for this crop. A direct regeneration method without callus induction was optimized using embryos as explants material in Gamborg's B5 medium supplemented with 0.5 μM 6 BA and 2.0 μM NAA. Pre-cultured elongated embryos wounded with razor blade were co-cultivated with *Agrobacterium* carrying the binary construct for 72 hours in the medium containing 300 μM acetosyringone. About 95 per cent embryos showed transient *GUS* expression after co-cultivation. Putative transformed embryos were cultured on B5 medium for shoot proliferation and regenerated plants were allowed to root. T0 plantlets showed *GUS* expression and gene integration was confirmed *via* PCR amplification. A transformation efficiency of 1.5 per cent was obtained and gene copy number analysed by Southern blot analysis indicated single-copy gene integration.

8. Fenugreek (*Trigonella foenum-graecum*) (Family: Papilionaceae)

Fenugreek is an annual plant extensively cultivated in India and northern Africa, the seeds and leaves of which are used as food and also as traditional medicines. Two biologically active alkaloid metabolites, galactomannan and diosgenin, are found in the seed of fenugreek plants. Diosgenin is found in roots, stems, and young leaves and used for the synthesis of oral contraceptives, hormones, and other steroids. Most of the studies concerning the production of diosgenin have focused on root and also hairy root cultures. Genetically transformed hairy roots are highly differentiated and cause stable and extensive production of secondary metabolites.

A. rhizogenes induced hairy roots is ideal for production of secondary metabolites such as diosgenin in fenugreek. In order to evaluate the transformation frequency and the efficiency of transgenic hairy root induction, leaf and stem explants from two fenugreek ecotypes, Karaj and Bushehr, were infected with *A. rhizogenes* strain K599 harboring a *GFP* (green fluorescent protein) gene (Shahabzadeh *et al.*, 2013). Regardless of ecotype, the ability of stem explants for the induction of hairy roots (8.09) and the transformation frequency (81.3 per cent) was higher compared with leaf explants with the values of 5.97 and 71.88 per cent, respectively. The number of transgenic GFP-positive hairy roots ranged from 4.2 to 13.5 in the Karaj ecotype and 3.8 to 9.9 in Bushehr. The effect of genotype, type of explants and bacterial concentration on the hairy root production, transformation frequency and rate of growth of transgenic roots were also studied (Shahabzadeh *et al.*, 2013).

9. Garlic (*Allium sativum*) (Family: Lilliaceae)

Garlic is an important minor spice or condiment known all over the world as a valuable condiment for foods, and a popular remedy or medicine for various ailments and physiological disorders. It is a hardy bulbous perennial with narrow flat leaves and bears small white flowers and bulbils. Garlic does not readily produce seeds and is propagated vegetatively thus preventing improvement of garlic by conventional breeding methods.

9.1. Development of Transformation System

The temperature and number of days of co-cultivation with *Agrobacterium* was shown to be an important factor in transient expression of the *uidA* in garlic (Kondo *et al.*, 2000). After a culture period of five months in selection medium containing hygromycin, 20 shoots were induced from ca. 1000 calluses among which 15 plants expressed *GUS* gene and shoots developed into transgenic garlic plants. Integration of the *uidA* was confirmed by Southern blot analyses. Robledo-Paz *et al.* (2004) used microprojectile bombardment to introduce DNA into embryogenic callus of garlic and produce stably transformed garlic plants. Embryogenic calluses were bombarded with plasmid DNA containing genes coding for *hpt* and *gus*. Putative transformed calluses were indentured in the bombarded tissue after four months of selection on 20 mg/l hygromycin B. The transgenic nature of the selected material was demonstrated by *GUS* histochemical assay and Southern blot hybridization analysis.

Eady *et al.* (2005) reported *A. tumefaciens*-mediated transformation of leek (*Allium porrum*) and garlic using immature leek and garlic embryos as explants. The method involved the use of a binary vector containing the m-GFP- ER reporter gene and *npt* II selectable marker. The presence of transgenes in the genome of the plants was confirmed using TAIL-PCR and Southern analysis. Kenel *et al.* (2010) reported efficient *A. tumefaciens*-mediated transformation and regeneration of garlic immature leaf tissue. The method involved the use of immature embryos and a binary vector containing the m-GFP-ER reporter gene and *hpt* selectable marker. The presence of transgenes in the genome of the plants was confirmed using Southern analysis.

9.2. Development of Transgenic Garlic Resistant to Herbicide and Insects

Park *et al.* (2002) reported generation of chlorsulfuron (a sulfonylurea herbicide) resistant transgenic garlic plants by particle bombardment. The callus tissue from the apical meristem of garlic cloves were subcultured and repeatedly selected calli with brittle, non-mucilaginous surfaces were selected for over six months was used as explants for transformation. Recombinant DNA that contained the acetolactate synthase (*ALS*) gene from a chlorsulfuron-resistant *Arabidopsis* mutant, the Cauliflower Mosaic Virus 35S promoter, the *GUS* reporter gene, and the hygromycin phosphotransferase (*hpt*) selectable marker gene was used for transformation. The explants were bombarded twice with tungsten particles coated with the DNA constructs. Transformed calli were selected in hygromycin B and regenerated into plants were confirmed using PCR, Southern and northern blot analyses. The regenerated plants survived in the presence of 3 mg/l chlorsulfuron, demonstrating that their *ALS* was insensitive to the herbicide.

Zheng *et al.* (2004) reported the development of a reliable *Agrobacterium* mediated transformation system for garlic and its application in producing insect resistant transgenic garlic lines. Callus induced from apical and non-apical root segments of *in vitro* plantlets, true garlic seeds and bulbils were used as explants, *gusA* and the *GFP* gene coding for green fluorescent protein were used as reporter genes. The protocol required short time period of about six months. The highest transformation frequency was 1.47 per cent in the cv. 'Printanor'. PCR and Southern hybridization showed that the reporter gene *gusA* and the selectable marker gene *hpt* were stably integrated into the garlic genome. Using this protocol transgenic garlic resistant to beet armyworm using the cry1Ca and H04 resistance genes from *Bacillus thuringiensis* was developed. After transfer of the transgenic *in vitro* garlic plants to the green house, the cry1Ca plants developed normally and grew well to maturity with normal bulbs. However, all transgenic *in vitro* H04 garlic plants did not survive after transfer to the greenhouse. Transgenic cry1Ca garlic plants proved completely resistant to beet armyworm in a number of *in vitro* bio-assays.

10. Ginger (*Zingiber officinale*) (Family: Zingiberaceae)

Ginger is an herbaceous perennial, the rhizome of which is used as spice. India is a leading producer of ginger in the world. The aroma of ginger is pleasant and spicy, and flavours penetrating. Hence it is used in the manufacture of a number of food products and in medicine. Ginger oil is a food flavorant in soft drinks. The absence of seed set in ginger makes conventional breeding methods inapplicable warranting genetic modification through biotechnological means. Suma *et al.* (2008). reported development of *A. tumefaciens* based genetic transformation and regeneration of ginger using *GUS* reporter gene. Optimum concentration of bacteria, co-cultivation period, concentration of acetosyringone and kanamaycin required were standardized. Transformants were recovered on selection media containing 100 mg/l kanamycin and a combination of 2,4-D 1.0 mg/l and BA 0.5 mg/1, and regenerated in half strength MS media of BA 3.0 mg/1 and 2,4-D 0.5 mg/l. Successful transformation was confirmed by histochemical GUS assay and PCR analysis.

11. Onion (*Allium cepa*) and Shallot (*Allium cepa* var. *aggregatum*) (Family: Lilliaceae)

Onion and shallot are used both for cooking and as a condiment for flavouring or for pickling. The flavour of shallot is somewhat milder than that of onions and is used for flavouring curries. Mild onions are used for cooking or as salad. Pungent varieties are used as condiment for flavouring a number of foods. Dehydrated onions, onion flakes, kibbled onions and onion powder are used for flavouring ketchups, sauces *etc.* There is a need to improve the production and nutritive value of these economically important crops and genetic engineering techniques can be exploited as an additional method for introduction of useful traits into established cultivars (Pruthi, 1992).

11.1. Development of Transformation System

Zheng *et al.* (2001) described a reliable transformation protocol which could be used year-round for onion and shallot. It was based on *Agrobacterium tumefaciens* as a vector, with three-week old callus, induced from mature zygotic embryos, as target tissue. Subspecies (onion and shallot) and cultivar were important factors for successful transformation as shallot was responsive to transformation than onion and shallot cv. Kuning gave the best results. Also, it was found that constant reduction of the size of the calli during subculturing and selection by chopping, enhances exposure to the selective agent hygromycin, thereby improved the selection efficiency significantly. Callus induction medium and co-cultivation period showed a significant effect on successful stable transformation. The usage of different *Agrobacterium* strains, callus ages, callus sources and osmotic treatments during co-cultivation did not influence transformation efficiency. The highest transformation frequency (1.95 per cent) was obtained with shallot cv. Kuning. PCR and Southern hybridization were used to confirm transgene integration and its copy number. FISH performed on 12 plants from two different lines representing two integration events showed that original T-DNA integration had taken place on the distal end of chromosomes 1 or 5. A total of 83 transgenic plants were transferred to the greenhouse and these plants appeared to be diploid and normal in morphology.

A new selection system for onion transformation by *Agrobacterium* and biolistic, that does not require the use of antibiotics or herbicides, was developed by Aswath *et al.* (2006). The selection system used the *Escherichia coli* gene that encodes phosphomannose isomerase (*pmi*). Transgenic plants carrying the *manA* gene that codes for *pmi* could detoxify mannose-6-phosphate by conversion to fructose-6-phosphate, an intermediate of glycolysis. Six week old embryogenic callus initiated from seedling radicle was used for transformation. Transgenic plants were produced effciently with transformation rates of 27 and 23 per cent using *Agrobacterium* and biolistic system, respectively. Untransformed shoots were eliminated by a stepwise increase from 10 g/l sucrose with 10 g/l mannose in the first selection to only 10 g/l mannose in the second selection. Integrative transformation was confirmed by PCR, RT-PCR and Southern hybridization. Cheng *et al.* (2009) reported transient expression of linear gene cassettes containing a *GUS* reporter gene in onion epidermal cells *via* direct transformation. The basic transformation solution used was MS liquid medium. Hypertonic pretreatment of explants and transformation cofactors, including Ca2+, surfactant assistants, *Agrobacterium* LBA4404 cell culture on transformation efficiency were evaluated. Prior to the incubation of the explants and target linear cassette in transformation solution for 3 hours, the onion lower epidermal explants were pre-cultured in darkness for 48 hours and then transferred to MS solid media for 72 h. FITC-labeled linear DNA was used to trace the delivery of DNA entry into the cell and the nuclei. By GUS staining and flow-cytometry- mediated fluorescent detection, a significant increase of the ratios of fluorescent nuclei as well as expression of the GUS reporter gene was observed by each designed transformation solution.

Sandhu and Gosal (2009) reported transient GUS expression in onion epidermal layer cells through particle bombardment. The onion epidermal layer(s) were bombarded with tungsten coated plasmid pWRG2426 containing *GUS* gene under the control of CaMV 35S promoter. The *GUS* gene expression was assessed within 24 hours following particle bombardment by counting the number of islands of cells showing indigo colouration and the measure of intensity was based on relative indigo colour development among the bombarded cells. Mythili *et al.* (2012) studied factors influencing *in vitro* regeneration such as the age of the embryogenic callus, the salt content in the basal medium (MS or BDS), gelling agent strength, plant growth regulators (picloram, Kinetin, 2,4-D) in the regeneration medium and aeration of the culture vessels. They also studied factors such as method of inoculation, influence of acetosyringone; and selection agents (kanamycin or geneticin) influencing *Agrobacterium m*ediated transformation of onion var. 'Arka Niketan'. Inoculation of the explants with *Agrobacterium* under vacuum filtration was suitable only for immature embryos while inoculating with bacterial suspension was suitable for the embryogenic callus. The effect of acetosyringone was influenced by basal medium and plant growth regulators. Geneticin was found to be a better selectable agent.

Callus obtained from basal meristem plate and twin scale leaves of two cultivars (Bellary and CO3) were used by Malla *et al.* (2015) for *Agrobacterium* mediated transformation using binary vector pCAMBIA 1301 containing the *GUS* gene with different concentrations of acetosyringone. The study revealed that the frequency of callus induction was the maximum in MS fortified with B5 vitamins supplemented with 0.5 and 1 mg/l picloram. Regeneration of plantlets from the callus was observed on MS supplemented with 0.5 mg/1 each of BAP and KIN and 0.1 mg/1 NAA. *In vitro* bulb response was observed on MS with B5 vitamins supplemented with 2.0 mg/l BAP. *GUS* gene integration in the transgenic plants was confirmed by PCR. The maximum *GUS* gene expression was observed at a concentration of 150 µM acetosyringone in the transformed plants by histochemical assay.

11.2. Development of Transgenic Onion and Shallot Resistant to Herbicide and Insects

Eady *et al.* (2003a). produced transgenic onion plants tolerant to herbicides containing active ingredients *viz.*, glyphosate and phosphinothricin from immature embryos of open pollinated and hybrid parent onion lines were used as explants and a maximum transformation frequency of 0.9 per cent was obtained. Transformants of different onion cultivars, grown on different selective agents and confirmed by Southern analysis, thrived with no apparent ill effects when sprayed with the respective herbicides at double the recommended field dosage for weed eradication. Eady *et al.* (2003b). studied inheritance and expression of introduced DNA in transgenic onion plants. Transgenic onion plants containing the *Cauliflower Mosaic Virus* 35S promoter (CaMV35S) and *GFP* gene construct encoding the visual green fluorescent reporter protein from pBin m GFP ER and the CaMV35S-bar gene construct encoding resistance to the herbicide phosphinothricin from pCAMBIA3301 were produced by *Agrobacterium*-mediated transformation were used in the study. These plants were grown to maturity and selfed in order to determine the expression and inheritance of the transgenes. Both the expression of GFP and tolerance to phosphinothricin inherited in a Mendelian fashion. Levels of expression in F1 offspring varied due to environmental and genetic factors and copy number did strongly influence GFP protein production and expression. In the majority of plants there were no obvious detrimental phenotypic effects caused by the transgene, the integration event, or somaclonal variation.

Agrobacterium-mediated genetic transformation was applied to produce beet armyworm (*Spodoptera exigua*) resistant tropical shallots. A cry1Ca or a H04 hybrid gene from *Bacillus thuringiensis,* driven by the chrysanthemum ribulose-1,5-bisphosphate carboxylase/oxygenase small subunit (Rubisco SSU) promoter, along with the HPT driven by the CaMV 35S promoter, was employed for genetic transformation. An average transformation frequency of 3.68 per cent was obtained from two shallot cultivars, Tropix and Kuning. After transfer of the *in vitro* plants to the greenhouse 69 per cent of the cry1Ca and 39 per cent of the H04 transgenic shallots survived the first half year. After one year of cultivation in the greenhouse, the remaining cry1Ca and H04 transgenic plants grew vigorously and had a normal bulb formation, although the cry1Ca transgenic plants (and controls) had darker green leaves compared to their H04 counterparts. PCR and Southern analysis conrmed integration of T-DNA into the shallot genome. Northern blot and ELISA analyses revealed expression of the *cry1Ca* or *H04* gene in the transgenic plants. The amount of Cry1Ca expressed in transgenic plants was higher than the expression levels of H04 (0.39 vs. 0.16 per cent of the total soluble leaf proteins, respectively). There was a good correlation between protein expression and beet armyworm resistance. Cry1Ca or H04 gene expression of at least 0.22 or 0.08 per cent of the total soluble protein in shallot leaves was sufficient to confer complete resistance against beet armyworm. The study also confirmed earlier findings that H04 toxin was more effective in controlling *S. exigua* than Cry1Ca toxin.

12. Paprika/Chilli (*Capsicum annuum*) (Family: Solanaceae)

Chilli or Paprika or Hungarian paprika, also called sweet pepper or Spanish pimento, is the mild or non-pungent variety of chilli capsicum. The dried ripe red paprikas are valued chiefly for their brilliant red colour and mild flavour. The crop is susceptible for attack by several fungi and viruses. The most transformation studies in chilli refer to the use of marker (*npt*II) or reporter gene (*GUS*) in order to establish adequate protocols; however some genes have also been utilized to generate transgenic plants with tolerance to viruses and fungi.

12.1. Development of Transformation System

The first report of *Agrobacterium* mediated genetic transformation of chilli using explants such as hypocotyls, cotyledons and leaves was reported by Liu *et al.* (1990). Although kanamycin resistant shoot buds were obtained, no further elongation and plant formation occurred. Manoharan *et al.* (1998) reported transformation using *A. tumefaciens* carrying a binary vector plasmid pBI 121and cotyledonary explants. Shoot buds produced were elongated and rooted in the presence of kanamycin (25 mg/l). The transgenic nature of regenerated plants was confirmed by histochemical staining of *GUS*, PCR and Southern hybridization analyses of *NPT*II gene.

Kim *et al.* (2002) developed a transformation system for paprika based on mannose as selection agent. A dosage curve for optimizing the selection conditions was established by mixing mannose and sucrose. They found that mannose selection has an increased transformation frequency compared to kanamycin selection. Nianiou *et al.* (2002) established a regeneration and transformation protocol for the sweet red pepper type 'Florinis' and for two pepper hybrids PO1 and C using hypocotyl explants. The rate of plant regeneration was found to depend on the types of explants cultured and the media used. Shoot bud initiation is more effective on MS media supplemented with IAA and BAP and shoot bud development with addition of GA3. Rooted shoots are successfully established in soil. *Agrobacterium* and the particle gun were used for transformation. *A. tumefaciens* strain LBA4404 harboring a plasmid containing GUS reporter gene and the NPT II selection gene or a plasmid with chloroplastic Cu/Zn SOD gene of tomato. Pepper hypocotyls were bombarded with plasmid that contained the GUS reporter gene driven by the

CaMV-35S promoter employing particle gun mediated transformation. Of the two methods, more transgenic plants were obtained in the case of particle gun.

A highly efficient transformation system using cotyledons as explants was reported by Li *et al.* (2003). Cotyledon explants were preconditioned for transformation for two days in a medium supplemented with sucrose. After two days of co-cultivation, explants were selected on 500 mg/l carbenicillin for two days. Explants were then placed on medium containing $AgNO_3$, kanamycin sulfate and carbenicillin. After 4-5 weeks, the explants with buds were transferred to medium supplemented with sucrose, $AgNO_3$ and hormones such as IAA, BA, gibberellic acid along with kanamycin and carbenicillin for elongation of buds. After 1-6 weeks, 1-2 cm long elongated shoots were excised and placed in MS medium containing NAA, IAA, kanamycin and carbenicillin for rooting. All four tested genotypes showed a high differentiation efficiency (81 per cent), elongation rate (61 per cent) and rooting efficiency (90 per cent). PCR results showed that 41 per cent of the plants were transgenic. High frequency shoot regeneration and *Agrobacterium* mediated transformation using shoot tips, cotyledons and hypocotyls explants was reported by Sobhakumari and Lalithakumari (2005).

A tissue culture independent *Agrobacterium* mediated *in planta* transformation was attempted in two varieties of chilli (Kumar *et al.*, 2009). In order to establish a reliable and highly efficient method for genetic transformation of pepper, a monitoring system featuring *GFP* as a marker was applied to *Agrobacterium*-mediated transformation by Jung *et al.* (2011). A callus-induced transformation (CIT) system was used to express *GFP* gene. Expression of GFP was observed in all tissues of T0, T 1 and T2 peppers in which the whole pepper plant exhibited GFP fluorescence. The transformation rate ranged from 0.47 to 1.83 per cent depending on the genotype. This technique could enhance selection power by monitoring GFP expression at the early stage of callus *in vitro*. The detection of GFP expression in the callus led to successful identification of shoots that contained the transgene.

Kumar *et al.* (2012) reported an effective and reproducible auxin free regeneration method for six different red pepper cultivars (ACA-10, Kashi Anmol, LCA-235, PBC-535, Pusa Jwala and Supper) using hypocotyl explants and an efficient *Agrobacterium*-mediated transformation protocol. The explants (hypocotyls, cotyledonary leaves and leaf discs) collected from axenic seedlings were cultured on either hormone free MS medium or MS medium supplemented with BAP alone or in combination with IAA. Inclusion of IAA in the regeneration medium resulted in callus formation at the cut ends of explants, formation of rosette leaves and ill defined shoot buds. Regeneration of shoot buds was achieved from hypocotyls grown in MS medium supplemented with different concentrations of BAP unlike other explants which failed to respond. Incorporation of GA3 in shoot elongation medium at 0.5 mg/l concentration enhanced the elongation in two cultivars, LCA-235 and Supper, while other cultivars showed no significant response. Chilli cultivar, Pusa Jwala was transformed with βC1 ORF of satellite DNA β molecule associated with Chilli leaf curl Joydebpur virus through *A. tumefaciens*. Transgene integration in putative transformants was confirmed by PCR and Southern hybridization analysis.

12.2. Development of Transgenic Paprika Resistant to Herbicide, Fungi and Viruses

Yamakawa *et al.* (1998) produced transgenic chilli expressing phenylalanine ammonia lyase (*PAL*) gene from parsley using *A. rhizogenes* strain A13 harbouring recombinant binary vector pBI121 and hypocotyls as explants. After four weeks, the hairy roots produced were transferred onto medium supplemented with cefotaxime and subsequently selected in kanamycin. The integration of the transgene was confirmed through Southern hybridization. Hairy roots containing *PAL* showed different *PAL* activity, slow growth and altered morphology.

In another study, *Cucumber mosaic virus* (CMV) and *Tobacco mosaic virus* (TMV) coat protein (*CP*) genes have been transferred to chili pepper cultivar 8212 by a modified procedure of *A. tumefaciens*-mediated transformation using hypocotyl as the explants (Cai *et al.*, 2003). PCR analysis revealed the presence of both CMV and TMV CP genes in at least 11 primary transformants out of 49 kanamycin-resistant chili pepper plants. Ten T1 lines, from five independent transformation events, were identified as putative homozygous transgenic lines based on the rooting assay of their T2 seedlings on the kanamycin-containing media. Integration and expression of CMV CP and TMV CP transgenes in one of the homozygous line, 16-13, were confirmed by Southern blot, RT-PCR and western blot analyses. Line 16-13 was highly resistant to infection of homologous CMV and TMV strains in greenhouse conditions when successively challenged with CMV and TMV or challenged with TMV alone. Furthermore, field trials on T2, T3 and T4 progenies of Line 16-13 were performed on scales of 123, 300 and 10,000 plants, respectively, in consecutive years 1996, 1997 and 1998 with the permission of the Chinese government authority. The transgenic plants displayed delayed symptom development and signicantly milder disease severity in field conditions when compared to untransformed chili pepper plants, resulting in 47 and 110 per cent increase in pepper fruit yield in surveys conducted in 1997 and 1998 trials, respectively. Finally, quality analysis and biosafety assessment were performed on transgenic chili pepper fruit concurrently with the control fruit, and demonstrated that the transgenic chili pepper fruit is substantially equivalent to the non-transgenic pepper in terms of the quality and biosafety when consumed as a food additive.

Lee *et al.* (2004) used two genes, *TMV-CP* and *PPI1* (pepper-PMMV interaction 1 transcription factor), to transform commercially important chili pepper inbred lines (P915, P409) by means of *Agrobacterium* co-culture and obtained eighteen independently transformed T0 plants. They also reported that use of correct type of callus and selection of callus-mediated shoot formation are important in the transformation of chilli. Lee *et al.* (2009) produced transgenic peppers resistant to a new CMV pathotype, CMVP1 using coat protein gene from CMV pathotype, CMV P0 as transgene using *Agrobacterium* mediated transformation. Transgenic peppers tolerant to CMVP1 were selected in a plastic house as well as in the field. Three independent T3 pepper lines highly tolerant to the CMVP1 pathogen were found to also be tolerant to the CMVP0 pathogen. These selected T3 pepper lines were phenotypically identical or close to the non-transformed lines. However, after CMVP1 infection, the height and fruit size of the non-transformed lines became shorter and smaller, respectively, while the T3 pepper lines maintained a normal phenotype

Juan-Xu *et al.* (2009) reported an *Agrobacterium*-mediated transformation system with *Cre* and *Barnase* genes designed to control plant fertility by cell lethal gene Barnase expressing at specific developmental stage and in specific tissue of male organ under the control of Cre/lox system. PCR and Southern blotting analysis of kanamycin plantlet indicated that the foreign genes had been integrated into the genome of pepper. The transgenic plants with *Cre* gene developed well, blossomed out, and set fruit normally. The transgenic plants with *Barnase* gene grew well with normal appearance of flower, but they showed different fertility from complete sterility, partial sterility to complete fertility.

Aguilar-Barragán and Ochoa-Alejo (2014) used *Tobacco rattle virus* (TRV) based virus induced gene silencing system (VIGS) system to determine the role of MYB, MYC, and WD40 transcription factors (TFs) that regulate the expression of structural biosynthetic genes at different steps on chilli. The accumulation of anthocyanins in chilli pepper fruits of plants transformed with TRV2-MYB and TRV2-WD40 constructs was significantly reduced compared to the control or empty TRV2-transformed plants. A significant reduction in gene expression of both TFs was also detected. The expressions of the chalcone synthase (*CHS*), chalcone isomerase (*CHI*), flavonoid 3′,5′-hydroxylase (*F3´5´H*), dihydroflavonol 4-reductase (*DFR*), and UDP-glucose:flavonoid 3-O-glucosyltransferase (*3GT*) genes were decreased in the plants transformed with the TRV2-MYB construct but not the transcription of flavanone 3-hydroxylase (*F3H*). When chilli pepper plants were infected with the TRV2-WD40 construct, a significant reduction in *CHS, F3H, F3´5´H, DFR* and *3GT* expression, but not in *CHI* in the fruits was observed. Mythili *et al.* (2015). produced transgenic chili expressing baculovirus chitinase gene which showed *in vitro* inhibition against fungal pathogens such as *Colletotrichum capsici* and *Alternaria alternata* causing anthracnose disease.

13. Turmeric (*Curcuma longa*) (Family: Zingiberaceae)

Turmeric is a rhizomatous species known both for its culinary and medicinal uses. It is mainly cultivated in India, Pakistan, Sri Lanka, Bangladesh, and China. India is the largest producer and a major exporter of this spice. Turmeric powder is obtained from its boiled, dried and polished underground rhizomes. Cultivated turmeric is a sterile polyploid species that is propagated only clonally *via* rhizomes. Due to the vegetative propagation, the rhizome is susceptible to accumulation and transmittance of pathogens and soil-borne diseases. It is essential to develop new strategies that combine tissue culture and genetic engineering techniques to complement breeding programs, and effective transformation approaches for identication of gene function and improvement of physiological traits for this species. Preliminary studies in this direction included *in vitro* micropropagation and plant regeneration from callus culture. An efficient method for stable transformation for turmeric was developed by Shirgurkar *et al.* (2006) using particle bombardment. Callus cultures initiated from shoots were bombarded with gold particles coated with plasmid pAHC25 containing the *bar* and *gusA* genes each driven by the maize ubiquitin promoter. Transformants were selected on medium containing glufosinate. Transgenic shoots regenerated from these were multiplied and stably transformed plantlets were identified by PCR and histochemical *GUS* assay. Transformed plantlets were resistant to glufosinate. An efficient somatic embryogenesis system and a reliable *Agrobacterium*-mediated transformation protocol were developed for turmeric (He and Gang, 2014). Calli derived from turmeric inflorescences were used as source tissues for transformation. Factors affecting transformation and regeneration efficiency were evaluated, including callus induction and culture conditions, *Agrobacterium* strains, co-cultivation conditions, selection agent sensitivity and bacterial elimination, and transformant selection. Optimized transformation conditions were identied, including use of *Agrobacterium* strain EHA105 with plasmid pBISN1 for infection; a modied B5 medium system for callus induction, subculture, co-culture and selection; and MS media for transformant regeneration. Transgenic plants and their vegetative (clonal) progeny stably expressed the transgene as indicated by *GUS* assay, PCR and Southern blot analysis. In addition, a transient gene expression system was also developed that involved *Agrobacterium* infiltration of young turmeric leaves followed by *in vitro* regeneration of plantlets.

14. Vanilla (*Vanilla planifolia*) (Family: Orchidaceae)

Vanilla is the second most expensive spice on the world market. Natural vanillin obtained from the cured pods (fruits) of this plant is used for spicing a variety of food and confectionaries. There are three important cultivated species namely *V. planifolia* (Mexican vanilla), *V. pompona* (West Indian vanilla), and *V. tahitensis* (Tahitian vanilla). The important vanilla growing countries are Madagascar, Indonesia, Mexico, Comoro and Reunion. Vanilla is a native of Mexico and was introduced to India as early as 1835. Like other orchids, vanilla is also affected by fungal and viral diseases.

Wang *et al.* (1997) produced transgenic *Nicotiana benthamiana* plants using the coat protein gene of Vanilla necrosis potyvirus (VNV) (a virus known to infect vanilla) *via Agrobacterium tumefaciens*-mediated transformation. Four constructs contained either: sense (+) CP sequence, antisense (-) CP sequence, sense CP sequence with a Kozak's consensus ATG resulting in a change in the first amino acid, or antisense CP sequence with the Kozak's modification. When transgenic *N. benthamiana* plants were mechanically inoculated with a high concentration of VNV, one of the plant lines containing the full-length sense CP gene was highly resistant to virus infection. Plants from the resistant lines expressed the CP at a relatively low level compared to susceptible lines containing the same construct. Plants containing the other three constructs were either susceptible or showed delayed symptom expression in tobacco. However transgenic vanilla harbouring this sequence could not be developed.

Malabadi and Nataraja (2007) reported production of protocorm-like bodies (PLBs) in presence of putrescine from thin section culture of shoot tips of vanilla. Using PLBs as explants they established an *Agrobacterium* mediated genetic engineering of vanilla using *nptII* and *GUS* genes. The presence of transgene was confirmed by PCR followed by Southern and northern hybridization of PCR

Figure 25.2: ***Agrobacterium*** **Mediated Genetic Transformation in Vanilla.**

(a) Protocorm like bodies (PLBs) in selection medium after 30 days of co-cultivation (b, c, d, e and f) after 40, 50, 60, 90 and 120 days of culture in regeneration medium (g) hardened transgenic plants in green house.

products. Retheesh and Bhat (2011) established an efficient transformation protocol for vanilla using PLBs derived from shoot tips as explants (Figure 25.2). Of the ten media tested, MS medium containing 0.45 μM thidiazuron (TDZ) produced maximum PLBs per shoot tip. PLB's were co-cultured with *A. tumefaciens* strain

EHA105 harbouring the binary vector pBI121 containing the *GUS* and *NPTII* genes for three days in MS medium supplemented with acetosyringone and transferred to selective regeneration medium containing 4.43 µM BA, 2.68 µM NAA supplemented with 50 mg/l kanamycin and 250 mg/l cefotaxime. After 15 days of culture, the surviving explants were transferred to the same regeneration medium but with a higher concentration of kanamycin (75 mg/l). Finally, explants that survived after 30 days were subjected to more stringent selection in the regeneration medium supplemented with 100 mg/l of kanamycin (Fig. 25.2). Integration of T-DNA into nuclear genome of transgenic plant was confirmed by PCR and Southern hybridization while expression of transgene was confirmed by Northern hybridization.

References

Aguilar-Barragán, A. and Ochoa-Alejo, N. (2014). Virus-induced silencing of MYB and WD40 transcription factor genes affects the accumulation of anthocyanins in chilli pepper fruit. *Biologia Plantarum*, **58**: 567-574.

Anu, K., Jessymol, K.K., Chidambareswaren, M., Gayathri, G.S. and Manjula, S. (2015). Down regulation of *osmotin* (PRs) gene by virus-induced gene silencing (VIGS) leads to susceptibility of resistant *Piper colubrinum* Link. to the oomycete pathogen, *Phytophthora capsici* Leonian. *Indian Journal of Experimental Biology*, **53**: 329-334.

Asha, S. and Rajendran, P.C. (2010). Putative transgenic plants through *in planta* transformation against *Phytophthora* foot rot in black pepper (*Piper nigrum* L.). *Asian Journal of Biosciences*, **4**: 135-141.

Aswath, C.R., Mo, S.Y., Kim, D.H. and Park, S.W. (2006). *Agrobacterium* and biolistic transformation of onion using non-antibiotic selection marker phosphomannose isomerase. *Plant Cell Reports*, **25**: 92–99.

Babu, K.N., Suraby, E.J., Cissin, J., Minoo, D., Pradeepkumar, T., Parthasarathy, V.A. and Peter, K.V. (2013). Status of transgenics in Indian spices. *Journal of Tropical Agriculture*, **51**: 1-14.

Backiyarani, S., Manohari, C., Jebsingh, T., Jacob, T. and Usha, R. (2005). Cloning of coat protein gene of Kursuppara isolate of *Cardamom mosaic virus* for developing transgenic virus-resistant cardamom. In: *Recent trends in Horticultural Biotechnology* (Vol. II). New India Publishing Agency, New Delhi. pp. 693-698.

Bhat, A.I., Revathy, K.A., Sasi, S. and Jiby, M.V. (2014). Production and evaluation of transgenic black pepper expressing *Cucumber mosaic virus* and *Piper yellow mottle virus* sequences. Paper presented at the VIROCON 2014, *Tamil Nadu Agricultural University*, Coimbatore.

Cai, W.Q., Fang, R.X., Shang, H.S., Wang, X., Zhang, F.L., Li, Y.R., Zhang, J.C., Cheng, X.Y., Wang, G.L. and Mang, K.Q. (2003). Development of CMV- and TMV-resistant transgenic chili pepper: eld performance and biosafety assessment. *Molecular Breeding*, **11**: 25–35.

Catlin, D., Ochoa, O., McCormick, S.A. and Quiros, C.F. (1988). Celery transformation by *Agrobacterium tumefaciens*: cytological and genetic analysis of transgenic plants. *Plant Cell Reports*, **7**: 100-103.

Cheng, Y.Q., Yang, J., Xu, F.P., An, L.J., Liu, J.F. and Chen, Z.W. (2009). Transient expression of minimum linear gene cassettes in onion epidermal cells *via* direct transformation. *Applied Biochemistry and Biotechnology*, **159**: 739–749.

Eady, C., Davis, S., Catanach, A., Kenel, F. and Hunger, S. (2005). *Agrobacterium tumefaciens*-mediated transformation of leek (*Allium porrum* and garlic (*Allium sativum*). *Plant Cell Reports*, **24**: 209–215.

Eady, C., Davis, S., Farrant, F., Reader, J. and Kenel, F. (2003a). *Agrobacterium tumefaciens*-mediated transformation and regeneration of herbicide resistant onion (*Allium cepa*) plants. *Annals of Applied Biology*, **142**: 213-217.

Eady, C.C., Reader, J., Davis, S. and Dale, T. (2003b). Inheritance and expression of introduced DNA in transgenic onion plants (*Allium cepa*). *Annals of Applied Biology*, **142**: 219-224.

Ebrahimie, E., Habashi, A.A., Ghareyazie, B., Ghannadha, M. and Mohammadie, M. (2003). A rapid and efficient method for regeneration of plantlets from embryo explants of cumin (*Cuminum cyminum*). *Plant Cell Tissue and Organ Culture*, **75**: 19–25.

He, R. and Gang, D.R. (2014). Somatic embryogenesis and *Agrobacterium*-mediated transformation of turmeric (*Curcuma longa*). *Plant Cell Tissue and Organ Culture*, **116**: 333–342.

Herrera-Estrella, L., DeBlock, M., Messens, E., Hernalsteens, J.P., VanMontagu, M. and Schell, J. (1983). Chimeric genes as dominant selectable markers in plant cells. *European Molecular Biology Organization-EMBO Journal*, **2**: 987–996.

Jiby, M.V. and Bhat, A.I. (2011). An efficient *Agrobacterium*-mediated transformation protocol for black pepper (*Piper nigrum* L.) using embryogenic mass as explant. *Journal of Crop Science and Biotechnology*, **14**: 247-254.

Josephrajkumar, A., Chakrabarty, R. and Thomas, G. (2006). Midgut proteases of the cardamom shoot and capsule borer *Conogethes punctiferalis* (Lepidoptera: Pyralidae) and their interaction with aprotinin. *Bulletin of Entomological Research*, **96**: 91-98.

Juan-Xu, L., Yi-Xun, Y., Jian-Jun, L., Guo-Ju, C., Bi-Hao, C. and China, P.R. (2009). Study on *Agrobacterium*-mediated transformation of pepper with *Barnase* and *Cre* gene. *Agricultural Sciences in China*, **8**: 947-955.

Jung, M., Shin, S.H., Park, J.M., Lee, S.N., Lee, M.Y., Ryu, K.H., Paek, K.Y. and Harn, C.H. (2011). Detection of transgene in early developmental stage by GFP monitoring enhances the efciency of genetic transformation of pepper. *Plant Biotechnology Reporter*, **5**: 157–167.

Kenel, F., Eady, C. and Brinch, S. (2010). Efficient *Agrobacterium tumefaciens*-mediated transformation and regeneration of garlic (*Allium sativum*) immature leaf tissue. *Plant Cell Reports*, **29**: 223–230.

Kim, J.Y., Jung, M., Kim, H.S., Lee, Y.H., Choi, S.H., Lim, Y.P., Min, B.W., Yang, S.G. and Harn, C.H. (2002). A new selection system for pepper regeneration by mannose. *Journal of Plant Biotechnology*, **4**: 129-134.

Kondo, T., Hasegawa, H. and Suzuki, M. (2000). Transformation and regeneration and of garlic (*Allium sativum* L.) by *Agrobacterium* mediated gene transfer. *Plant Cell Reports*, **19**: 989-993.

Krens, F.A., Keizer, L.C.P. and Capel, I.E.M. (1997). Transgenic caraway, *Carum carvi* L.: a model species for metabolic engineering. *Plant Cell Reports*, **17**: 39-43.

Krishnan, A., Mahadevan, C., Mani, T. and Sakuntala, M. (2015). Virus-induced gene silencing (VIGS) for elucidation of pathogen defense role of serine/threonine protein kinase in the non-model plant *Piper colubrinum* Link. *Plant Cell Tissue and Organ Culture*, **122**: 269–283.

Kumar, A.M., Reddy, K.N., Srrevathsa, R., Ganeshan, G. and Udayakumar, M. (2009). Towards crop improvement in bell pepper (*Capsicum annuum* L.) transgenics (uidA: npt II) by a tissue culture independent *Agrobactreium* mediated *in planta* approach. *Scientia Horticulturae*, **119**: 362-370.

Kumar, R.V., Sharma, V.K., Chattopadhyay, B. and Chakraborty, S. (2012). An improved plant regeneration and *Agrobacterium*-mediated transformation of red pepper (*Capsicum annuum* L.). *Physiology and Molecular Biology of Plants*, **18**: 357-364.

Lee, Y.H., Jung, M., Shin, S.H., Lee, J.H., Choi, S.H., Her, N.H., Lee, J.H. and Harn, C.H. (2009). Transgenic peppers that are highly tolerant to a new CMV pathotype. *Plant Cell Reports*, **28**: 223–232.

Lee, Y.H., Kim, H.S., Kim, J.Y., Jung, M., Park, Y.S., Lee, J.S., Choi, S.H., Her, N.H., Lee, J.H., Hyung, N.I., Lee, C.H., Yang, S.G. and Harn, C.H. (2004). A new selection method for pepper transformation: callus-mediated shoot formation. *Plant Cell Reports*, **23**: 50–58.

Li, D., Zhao, K., Xie, B., Zhang, B. and Luo, K. (2003). Establishment of highly efficient transformation system for pepper (*Capsicum annuum* L.) *Plant Cell Reports*, **21**: 785-785.

Liu, C.N., Li, X.Q. and Gelvin, S.B. (1992). Multiple copies of *virG* enhance the transient transformation of celery, carrot and rice tissue by *Agrobacterium tumefaciens*. *Plant Molecular Biology*, **20**: 1071–1087.

Liu, W., Parrot, W.A., Hilderbrand, D.P., Collins, G.B. and Williams, E.G. (1990). *Agrobacterium* induced gall formation in bell pepper (*Capsicum annuum* L.) and formation of shoot like structures expressing introduced genes. *Plant Cell Reports*, **9**: 360-364.

Loskutov, A.V., Song, G.Q. and Sink, K.C. (2008). Evaluation of methods for celery (*Apium graveolens* L.) transformation using *Agrobacterium tumefaciens* and the bar gene as selectable marker. In *Vitro Cell Development Biology-Plant*, **44**: 239-245.

Maju, T.T. and Sonia, E.V. (2012). *In vitro* regeneration system for multiplication and transformation in *Piper nigrum* L. *International Journal of Medicinal and Aromatic plants*, **2**: 178-184.

Malabadi, R.B. and Nataraja, K. (2007). Genetic transformation of *Vanilla planifolia* by *Agrobacterium tumefaciens* using shoot tip sections. *Research Journal of Botany*, **2**: 86–94.

Malla, A., Srinivasan, B., Shanmugaraj, B.M. and Ramalingam, S. (2015). Micropropagation and DNA delivery studies in onion cultivars of Bellary, CO3. *Journal of Crop Science and Biotechnology*, **18**: 37- 43.

Manoharan, M., Vidya, C.S.S. and Sita, G.L. (1998). *Agrobacterium*-mediated genetic transformation in hot chilli (*Capsicum annuum* L. var. Pusa jwala). *Plant Science*, **131**: 77–83.

Mythili, J.B., Rajeev, P.R. and Chethana, B.S. (2012). Factors influencing *in vitro* regeneration and *Agrobacterium* mediated transformation of onion 'Arka Niketan'. *Acta Horticulturae*, **929**: 417-422.

Mythili, J.B., Rashmi, H.J., Suneetha, C., Saiprasad, G.V.S., Rajeev, P.R., Naveena, C., Anand, L., Ganeshan, G. and Riaz. (2015). Transgenic chili possessing *Baculovirus* chitinase gene exhibits *in vitro* fungal inhibition. *Journal of Crop Improvement*, **29**: 159-187.

Nair, R.R. and Gupta, S.D. (2006). High frequency plant regeneration through cyclic secondary somatic embryogenesis in black pepper (*Piper nigrum* L.). *Plant Cell Reports*, **24**: 699-707.

Nianiou, M., Zambounis, K.A. and Tsaftaris, A. (2002). Development of pepper transgenic plants via *Agrobacterium* and biolistic transformation. *Acta Horticulturae*, **579**: 83-87.

Pandey, S., Mishra, A., Patel, M.K. and Jha, B. (2013). An efficient method for *Agrobacterium*-mediated genetic transformation and plant regeneration in cumin (*Cuminum cyminum* L.). *Applied Biochemistry and Biotechnology*, **71**: 1–9.

Park, M.Y., Yi, N.R., Lee, H.Y., Kim, S.T., Kim, M., Park, J.H., Kim, J.K., Lee, J.S., Cheong, J.J. and Choi, Y.D. (2002). Generation of chlorsulfuron-resistant transgenic garlic plants (*Allium sativum* L.) by particle bombardment. *Molecular Breeding*, **9**: 171–181.

Peter, K.V. (Ed.) (2001). *Handbook of herbs and spices*. Cambridge, England: CRC Press.

Pruthi, J.S. (1992). *Spices and Condiments*. National Book Trust, New Delhi, India.

Ravindran, P.N. (2000). Black pepper: *Piper nigrum*, Series: *Medicinal and Aromatic Plants*- Industrial Profiles. Amsterdam, Netherlands. Harwood Academic Publishers.

Retheesh, S.T. and Bhat, A.I. (2011). Genetic transformation and regeneration of transgenic plants from protocorm-like bodies of vanilla (*Vanilla planifolia* Andrews) using *Agrobacterium tumefaciens*. *Journal of Plant Biochemistry and Biotechnology*, **20**: 262–269.

Robledo-Paz, A., Cabrera-Ponu, J.L., Villalobos-Aranbula, V.M., Herrer-Estrella, L. and Jofre-Garfias, A.E. (2004). Genetic transformation of garlic (*Allium sativum* L) by particle bombardment. *Horticulture Science*, **39**: 1208-1211.

Sandhu, J.S. and Gosal, S.S. (2009). A simple and rapid method for optimizing particle bombardment parameters through transient expression of gus reporter gene in onion epidermal layer cells. *Crop Improvement*, **36**: 6-8.

Sasi, S., Revathy, K.A. and Bhat, A.I. (2015). Rapid identification of transgenic black pepper using loop-mediated isothermal amplification (LAMP) and real-time LAMP assays. *Journal of Plant Biochemistry and Biotechnology*, **24**: 466-469.

Sasikumar, B. and Veluthambi, K. (1996). Transformation of black pepper (*Piper nigrum* L.) using *Agrobacterium* mediated Ti plasmid based vectors. *Indian Perfumer*, **40**: 13-16.

Shahabzadeh, Z., Heidari, B. and Hafez, R.F. (2013). Induction of transgenic hairy roots in *Trigonella foenum-graceum* co-cultivated with *Agrobacterium rhizogenes* harboring a GFP Gene. *Journal of Crop Science and Biotechnolgy*, **16**: 263 -268.

Shirgurkar, M.V., Naik, V.B., Arnold, S.V., Nadgauda, R.S. and Clapham, D. (2006). An efficient protocol for genetic transformation and shoot regeneration of turmeric (*Curcuma longa* L.) via particle bombardment. *Plant Cell Reports*, **25**: 112–116.

Sim, S.L., Jafar, R., Power, J. and Davey, M.R. (1998). Development of an *Agrobacterium* mediated transformation system for black pepper (*Piper nigrum* L.). *Acta Horticulturae*, **461**: 349-353.

Singh, N., Mishra, A., Joshi, M. and Jha, B. (2010). Microprojectile bombardment mediated genetic transformation of embryo axes and plant regeneration in cumin (*Cuminum cyminum* L.). *Plant Cell Tissue and Organ Culture*, **103**: 1–6.

Sobhakumari, V.P. and Lalithakumari, D. (2005). High frequency shoot regeneration and *Agrobacterium* mediated DNA transfer in red chilli (*Capsicum annuum* L.). *Plant Cell Biotechnology and Molecular Biology*, **6**: 9-16.

Song, G.Q., Loskutov, A.V. and Sink, K.C. (2007). Highly efcient *Agrobacterium tumefaciens*-mediated transformation of celery (*Apium graveolens* L.) through somatic embryogenesis. *Plant Cell Tissue and Organ Culture*, **88**: 193-200.

Suma, B., Keshavachandran, R. and Nybe, E.V. (2008). *Agrobacterium tumefaciens* mediated transformation and regeneration of ginger (*Zingiber officinale* Rosc.). *Journal of Tropical Agriculture*, **46**: 38–44.

Wang, Y. and Kumar, P.P. (2004). Heterologous expression of Arabidopsis ERS1 causes delayed senescence in coriander. *Plant Cell Reports*, **22**: 678-683.

Wang, Y.Y., Gardner, R.C. and Pearson, M.N. (1997). Resistance to vanilla necrosis potyvirus in transgenic *Nicotiana benthamiana* plants. *Journal of Phytopathology*, **145**: 7-15.

Yamakawa, T., Sekiguchi, S., Kodama, T., Smith, S.M. and Yeoman, M.M. (1998). Transformation of chilli pepper (*Capsicum frutescens*) with a Phenylalanine Ammonia Lyase gene. *Plant Biotechnology*, **15**: 189-193.

Zheng, S.J., Henken, B., Ahn, Y.K., Krens, F.A. and Kik, C. (2004). The development of a reproducible *Agrobacterium tumefaciens* transformation system for garlic

(*Allium sativum* L.) and the production of transgenic garlic resistant to beet armyworm (*Spodoptera exigua* Hübner). *Molecular Breeding*, **14**: 293–307.

Zheng, S.J., Khrustaleva, L., Henken, B., Soari, E., Jacobsen, E., Kik, C. and Krens, F.A. (2001). *Agrobacterium tumefaciens*-mediated transformation of *Allium cepa* L.: the production of transgenic onions and shallots. *Molecular Breeding*, **7**: 101–115.

Chapter 26

Cocoa

☆ *S.V. Ramesh, Ginny Antony, Tony Grace and M.K. Rajesh*

1. Introduction

Theobroma cacao L. (cocoa), a member of the family Sterculiaceae, is a tropical sub-canopy tree which has its origins in the rain forests of the Amazon basin (Hurst *et al.*, 2002). Cocoa derives its economic importance due to fat-rich seeds that are sole source of cocoa powder and cocoa butter which are raw materials for chocolate, cosmetic and pharmaceutical industries. Livelihood of more than two million growers spread over 50 countries worldwide is dependent on cocoa cultivation (Motamayor *et al.*, 2008).

Incidence of a number of serious pests and pathogens cause severe damage to the cultivation and production of cocoa thereby potential yields of cocoa plantations are reduced to great extent (Fulton, 1989). Traditional approaches of plant protection measures are being utilized to manage pest and diseases incidence in cocoa; however, practical difficulty in applying plant protection chemicals to cocoa plantations, remoteness of plantings and expensive application methods limits their use. Genetic improvement of cocoa to obtain strong resistance against pests and diseases have not yielded any significant outcome because of narrow genetic base and extended life cycle. Moreover, cocoa germplasm pool has not been explored extensively for novel genetic sources to confer disease and pest resistance. Notwithstanding serious impediments, some sources of partial resistance or tolerance to pests and diseases have been identified and utilized in crop improvement programmes (Eskes and Lanaud, 1997). However, till date, cocoa breeding programmes, which suffer from a long life cycle, funding constraints, and other limitations, have made limited progress with respect to the incorporation of durable, horizontal resistance to cocoa pests and diseases into widely used cocoa germplasm. Long-term endeavors towards breeding for disease resistance in cocoa have generated only limited success (Lopes *et al.*, 2011). The long generation time of cocoa plants, coupled with its large size, makes conducting requisite long-term and multi-location field trials cumbersome (Kennedy *et al.*, 1987). Moreover, as many of the cocoa pathogens are opportunistic, sources of resistance have not co-evolved for some of the most devastating of the major cocoa diseases (Maximova *et al.*, 2006).

Modern biotechnology based approaches equips researchers with an array of tools to complement and hasten conventional plant breeding programmes. Techniques and protocols have been standardized in many crops for incorporating novel sources of resistance and value-added traits to develop invaluable genetic resources that help accelerate breeding efforts. Among various techniques, genetic transformation is a very powerful tool in the hands of biotechnologists to alter genetic make-up of an organism not only to develop improved crop phenotype but also to perform gene-structure-function analysis. In the recent years, researchers have begun to apply the tools of genetic modification and molecular genetics to facilitate speeding up of cocoa breeding programmes (Maximova *et al.*, 2009).

2. Genetic Transformation Studies in Cocoa

Genetic transformation is a potential tool for performing basic research on functional genomics, and also provides means to introduce novel traits into cacao genome that are otherwise difficult to achieve through conventional or molecular breeding approaches. In the field of cocoa genetic transformation, Purdy and Dickstein (1989) made the first successful attempt by demonstrating the susceptibility of cocoa cells to *Agrobacterium*.

A major break-through was made when successful transformation of cocoa callus cells was established by Sain *et al.* (1994). Leaf strips from two cocoa clones *viz.* ICS-16 and SIC-5 were co-cultivated using the supervirulent *Agrobacterium* strain A281-kan. Accordingly, transformed cells were selected on a callusing medium containing 100 μg ml $^{-1}$ kanamycin. In addition, callus cultures of both the clones growing on kanamycin containing medium were analyzed for marker *npt II* (neomycin phosphotransferase II) gene expression by performing *npt II* assays and stable single site integration of *npt II* gene was demonstrated through Southern blotting studies.

Alongside *Agrobacterium*-mediated transformation, particle bombardment technique was also employed for introduction of foreign genes into cocoa (Perry *et al.*, 2000; Santos *et al.*, 2002). In both instances, reporter genes could be introduced into cocoa cells and visualized. In order to improve the transformation frequencies, optimization of osmotic adjustment was provided as a pre-treatment. Despite the successful regeneration of cacao plants through somatic embryogenesis, and introduction of alien DNA in to cocoa genome through *Agrobacterium*-mediated transformation or particle bombardment methodology, the development of transgenic cacao remained intangible. A highly efficient cocoa regeneration protocol was developed by Maximova *et al.* (2002) that allowed regeneration of secondary somatic embryos. The primary somatic embryo explants amenable for regeneration into secondary somatic embryos were originally initiated from floral tissue explants. The system developed has many advantages as more embryos per explants were produced than what was obtained in primary somatic embryo system. Further, greater proportion of embryos displayed morphological conformity and a significant leap was attained in percentage of conversion to plantlets. A set of stable transgenic cacao plants characterized with proper DNA integration and expression of visible marker green fluorescent protein (GFP), were obtained by Maximova *et al.* (2003). Furthermore, cacao transgenics were grown to maturity and efficient transmission of transgene to next generation was also proved. Co-cultivation of cocoa tissue with *Agrobacterium* was an issue in earlier attempts because growth of *Agrobacterium* overpowers cocoa explants and frequently damages the plant tissue. Furthermore, the commonly used antibiotic cefotaxime in tissue culture media to remove *Agrobacterium* cells post infection caused a severe reduction in cocoa somatic embryo production by 86 per cent. This situation prompted de Mayolo *et al.* (2003) to study and define suitable antibiotic systems and concentrations for cocoa transformation so that sufficient suppression of *Agrobacterium* is achieved without interfering cocoa somatic embryogenesis. Among the four antibiotics evaluated, moxalactam was not only very effective for *A. tumefaciens* counter-selection but also increased the regeneration frequency of secondary embryos.

Ultimately transgenic cocoa plants have been developed by Maximova *et al.* (2003) by combining the secondary somatic embryogenesis system developed earlier (Maximova *et al.*, 2002) with *A. tumefaciens* infection and co-cultivation (de Mayolo *et al.*, 2003). Three different plant transformation compatible binary vectors were constructed on the Bin19 T-DNA vector backbone system, with each plasmid containing *nptII* and *GFP* genes under the transcriptional control of the CaMV 35S promoter derivative E12-Ω. Successful regeneration of transformed secondary embryos results in transgenic cocoa plants carrying the visible marker gene green fluorescent protein (*EGFP*), the selectable marker gene neomycin phosphotransferase II (*nptII*), the class I chitinase gene from cocoa (*Chi*), and tobacco nuclear matrix attachment regions (MARs) in different combinations. Transgene, per se, did not influence number of transgenic plants produced as number of transgenics developed with marker gene and other transgenes such as *Chi* gene or MARs were found to be same. Nevertheless, incorporation of MARs gene in vector construct enhanced mean GFP expression in the transgenics and reduced the incidence of transgene induced gene silencing phenomenon in transgenic lines multiplied though reiterative somatic embryogenesis. The study was a milestone in cocoa genetic transformation as it yielded 94 transgenic plants and their growth and development are comparable to that of untransformed, control plants. Genetic crosses and segregation analysis of transgenic EGFP expression showed a near-perfect 1:1 thus unequivocally proving that the transgenic lines resulted from the insertion of a single locus of T-DNA.

Cocoa transgenic plants over expressing a cocoa class I chitinase gene (*TcChi1*) were developed with a view to obtain resistance against fungal pathogen *Colletotrichum gloeosporioides* causing anthracnose employing *Agrobacterium*-mediated transformation of somatic embryo cotyledons (Maximova *et al.*, 2006). Genetic transformation was done in genotype PSU-Scavina 6. The binary vector pGAM00.0511 (Maximova *et al.*, 2003) was modified to harbor cocoa *TcChi1* chitinase gene, EGFP, and the neomycin phosphotransferase II (NPTII) marker genes, under the control of constitutive E12-Ω CaMV-35S promoter. Southern blot studies revealed stable integration of transgene in eight independent cocoa lines. Further enhanced levels of expression of *TcChi1* transgene expression in genetically modified lines were confirmed by Northern blot analysis. *In vitro* fluorometric and quantified chitinase activity assays indicated that the expression of transgene varied in different transgenic lines; upto six fold increase of endochitinase activity was documented. In order to prove the utility of cocoa transgenic lines, leaf disc bioassay was carried out by evaluating antifungal activity of the transgene against the foliar pathogen *Colletotrichum gloeosporioides*. Bioassay revealed that expression of *TcChi1* in transgenic cocoa leaves significantly impeded the growth of *Colletotrichum* fungus and the development of leaf necrosis was also found to be reduced. These results established for the first time the value of the cocoa transformation system as a tool for gene functional analysis and the potential utility of the cocoa chitinase gene as a means of increasing resistance against fungal pathogens in cocoa.

Despite the availability of technique to genetically transform cacao, poor regeneration and transformation efficiencies hampered the progress. In order to refine the transformation protocol proposed by Maximova *et al.* (2003), various factors affecting somatic mebryogenesis and transformation efficiency *viz.*, concentration of hygromycin, β-lactam antibiotics, polyamines and composition of co-cultivation medium were evaluated. Besides these factors, concentration of *Agrobacterium* and sonication-assisted *Agrobacterium*-mediated transformation (SAAT) were also investigated (Silva *et al.*, 2009). Among the polyamines, spermine at 1,000 μM was found to improve embryogenic callus and increase the number of embryos per embryogenic callus. The study suggested use of β-lactam antibiotics such as timentin and meropenem because of its neutral effect on secondary somatic embryogenesis whereas the commonly used cefotaxime irrespective of the concentration used inhibits somatic embryogenesis. Authors also suggested sonication of explants and explants co-cultivation on tobacco feeder layers (Silva *et al.*, 2009). Thus this study is a comprehensive evaluation of factors affecting successful cacao genetic transformation and regeneration.

Understanding the molecular mechanism underlying somatic embryogenesis in cacao would help devise a better tool to enhance its regeneration potential and find ways to overcome major impediments in cocoa regeneration. *Arabidopsis thaliana* leafy cotyledon gene (*AtLEC2*) has been shown to enhance somatic embryogenic competency of plant cells (Karami *et al.*, 2009) hence, its functional ortholog *Theobroma cacao* leafy cotyledon gene *TcLEC2* was studied. Investigations exploring molecular role of *TcLEC2* gene in somatic embryogenesis and seed fatty acid biosynthesis revealed that *TcLEC2* is consistently expressed in cacao endosperm and cotyledon. Ectopic transient expression of *TcLEC2* activates expression of seed specific TF genes such as *TcAGL15, TcABI3* and *TcLEC1* in cacao. Further, expression levels were high in embryogenic calli implying that *TcLEC2* greatly enhances embryogenic competency. Thus constitutive over-expression of *TcLEC2* dramatically enhanced development of somatic embryos in cacao (Zhang *et al.*, 2014). Hence, it was suggested that expression of *TcLEC2* could be used as biomarker associated with efficient somatic embryogenesis and would help screening cacao genotypes for high embryogenic capacity (Zhang *et al.*, 2014). Along the similar lines, cacao baby boom transcription factor (*TcBBM*) an *Arabidopsis* BBM ortholog, was investigated for its potential in enhancing somatic embryogenesis (Florez *et al.*, 2015). Transient and constitutive expression analysis of *TcBBM* greatly improved proliferation of somatic embryos independent of exogenous hormone application (Florez *et al.*, 2015). These studies underlined the potential of transcriptional factors (TFs) expression based somatic cell reprogramming required to enhance regeneration of an elite cultivar and development of transgenic cacao. Embryogenic potential of cacao cultivars are highly genotypic dependent, hence somatic embryogenesis has not been employed as a tool of mass multiplication of elite cultivars. Hence, *Arabidopsis* ortholog TFs identified and characterized in cacao genotypes would not only serve as biomarker for somatic embryogenesis, but also help enhance its genetic transformation. In the present context, the methodology available for cacao genetic transformation is reproducible yet is marred by low recovery of transgenic embryo (Maximova *et al.*, 2003). Hence, TF expression based induction of somatic embryogenesis coupled with genetic transformation would serve as an invaluable tool in developing transgenic cacao genotypes.

Of late, a transient assay system has been developed for functional genomics study in cocoa genotypes (Fister *et al.*, 2016). The study utilizes, vacuum infiltration of induced *Agrobacterium* cells harboring binary vector into C stage cocoa leaf sections to transiently express visual marker gene. In addition, utility of the system for studying transgene cacao chitinase overexpression was demonstrated by performing bioassays for *Phytophthora tropicalis*. Thus it is is rapid method for gene function analysis in cacao.

3. Conclusion

Successful development of genetically modified cacao genotypes and developments of somatic embryogenesis-based regeneration techniques offers plethora of possibilities to obtain improved crop phenotypes. Experiments have established that the importance of inclusion of matrix attachment regions (MAR) in transformation vectors for stabilizing transgene expression during somatic embryogenesis and transgenic line establishment in cocoa. Future research which incorporates genes of interest into vectors containing MAR would further establish the applicability of the system for all alien genes. Such a system could then be used as a basis for analyzing gene structure and function in cocoa and also as a model system for testing the effectiveness of transgenes for enhancement of desired traits such as disease resistance or quality improvement. Developments in the field of cacao molecular biology and identification of cacao transcription factors, involved in somatic embryogenesis, offers a much needed fillip to develop efficient genetic transformation system. Exploration of role of such TFs, and demonstration of their utility in obtaining improved SE and genetic transformation could help bridge major crop improvement objectives in cacao.

Among the various objectives of improved crop phenotypes, development of resistance to pests and diseases is foremost important. In this context, the feasibility of expressing *Bt* toxins to confer cacao pod borer resistance could be explored. Developments in the field of transcriptomics, with the aid of next generation sequencing technologies, would help in deciphering genes responsible for fatty acid composition, cacao flavor that are potential target genes for genetic manipulation to develop specialty cacao products. The utility of genetic transformation system for crop improvement in cocoa, however, remains to be seen keeping in view the continued opposition by the public to crops which have been genetically modified. Transgenic cocoa material has not yet been released and its scope will depend upon economic, social, environmental, and political factors of the country in which transgenic cocoa is introduced. Nonetheless, the efficient tissue culture based somatic embryogenesis system on its own contributes to clonal propagation of elite cocoa germplasm. Genetic transformation protocol available also could contribute to functional genomics studies in cacao by identification of candidate resistance genes that in turn could accelerate molecular breeding of cacao. A recent potent addition to this arena of research is cacao transient assay system for functional genomics analysis (Fister *et al.*, 2016). It is anticipated that such a rapid system of gene-function analysis would be of immense help to accelerate functional genomics studies in cocoa as it precludes development of transgenic cocoa lines for analyzing gene functions.

References

de Mayolo, A.G., Maximova, S.N., Pishak, S. and Guiltinan, M.J. (2003). Moxalactam as a counter selection antibiotic for *Agrobacterium*-mediated transformation and its positive effects on *Theobroma cacao* somatic embryogenesis. *Plant Science*, **164**: 607–615.

Eskes, A. and Lanaud, C. (1997). Cocoa. *In*: Charrier, A., Jacquot, M., Hamon, S. and Nicolas, D. (Eds.), *L'Amelioration des Plantes Tropicales.* CIRAD/ORSTROM, Paris. pp. 623.

Fister, A.S., Shi, Z., Zhang, Y., Helliwell, E.E., Maximova, S.N. and Guiltinan, M.J. (2016). Protocol: transient expression system for functional genomics in the tropical tree *Theobroma cacao* L. *Plant Methods*, **12(1)**: 1.

Florez, S.L., Erwin, R.L., Maximova, S.N., Guiltinan, M.J. and Curtis, W.R. (2015). Enhanced somatic embryogenesis in *Theobroma cacao* using the homologous BABY BOOM transcription factor. *BMC Plant Biology*, **15(1)**: 1.

Fulton, R.H. (1989). The cacao disease trilogy: black pod, monilia pod rot, and witches' broom. *Plant Disease*, **73**: 601–603.

Hurst, W.J., Tarka, S.M., Powis, T.G., Valdez, F. and Hester, T.R. (2002). Cacao usage by the earliest Maya civilization. *Nature*, **418**: 289–290.

Karami, O., Aghavaisi, B. and Pour, A.M. (2009). Molecular aspects of somatic-to-embryogenic transition in plants. *Journal of Chemical Biology*, **2(4)**: 177-190.

Kennedy, A.J., Lockwood, G., Mossu, G., Simmonds, N.W. and Tan, G.Y. (1987). Cocoa breeding: past, present and future. *Cocoa Grower's Bulletin*, **38**: 5–22.

Lopes, U.V., Monteiro, W.R., Pires, J.L., Clement, D., Yamada, M.M. and Gramacho, K.P. (2011). Cacao breeding in Bahia, Brazil: strategies and results. *Crop Breeding and Applied Biotechnology*, **11**: 73-81.

Maximova, S.N., Alemanno, L., Young, A., Ferriere, N., Traore, A. and Guiltinan, M.J. (2002). Efficiency, genotypic variability, and cellular origin of primary and secondary somatic embryogenesis of *Theobroma cacao* L. *In Vitro Cellular and Developmental Biology-Plant*, **38**: 252–259.

Maximova, S.N., Guiltinan, M.J. and Lock, T.C. (2009). Cocoa. *Compendium of Transgenic Crop Plants*, **8(3)**: 85–98.

Maximova, S.N., Marelli, J.P., Young, A., Pishak, S., Verica, J.A. and Guiltinan, M.J. (2006). Over-expression of a cacao class I chitinase gene in *Theobroma cacao* L. enhances resistance against the pathogen, *Colletotrichum gloeosporioides*. *Planta*, **224**: 740–749.

Maximova, S.N., Miller, C., Antunez de Mayolo, G., Pishak, S., Young, A. and Guiltinan, M.J. (2003). Stable transformation of *Theobroma cacao* L. and influence of matrix attachment regions on GFP expression. *Plant Cell Reports*, **21**: 872–883.

Motamayor, J.C., Lachenaud, P., e Mota, J.W., Loor, R., Kuhn, D.N., Brown, J.S. and Schnell, R.J. (2008). Geographic and genetic population differentiation of the Amazonian chocolate tree (*Theobroma cacao* L.). *PLoS ONE*, **3(10)**: e3311.

Perry, M.D., Power, J.B., Lowe, K.C. and Davey, M.R. (2000). Biolistic (R) transformation of cacao [*Theobroma cacao* (L.)]. *Tropical Agriculture*, **77**: 64-66.

Purdy, L.H. and Dickstein, E.R. (1989). *Theobroma cacao*, a host for *Agrobacterium tumefaciens*. *Plant Disease*, **73**: 638–639.

Sain, S.L., Oduro, K.K. and Furtek, D.B. (1994). Genetic transformation of cocoa leaf cells using *Agrobacterium tumefaciens*. *Plant Cell Tissue and Organ Culture*, **37**: 342–351.

Santos, M., Albuquerque de Barros, E., Tinoco, M., Brasileiro, A. and Aragão, F. (2002). Repetitive somatic embryogenesis in cacao and optimisation of gene expression by particle bombardment. *Journal of Plant Biotechnology*, **4**: 71–76.

Silva, T.E., Cidade, L.C., Alvim, F.C., Cascardo, J.C. and Costa, M.G. (2009). Studies on genetic transformation of *Theobroma cacao* L.: evaluation of different polyamines and antibiotics on somatic embryogenesis and the efficiency of uidA gene transfer by Agrobacterium tumefaciens. *Plant Cell Tissue and Organ Culture*, **99(3)**: 287-298.

Zhang, Y., Clemens, A., Maximova, S.N. and Guiltinan, M.J. (2014). The *Theobroma cacao* B3 domain transcription factor *TcLEC2* plays a duel role in control of embryo development and maturation. *BMC Plant Biology*, **14**: 106.

Omics Applications

Chapter 27

Coffee

☆ *Alexandre de Kochko, Serge Hamon, Romain Guyot, Emmanuel Couturon, Valérie Poncet, Christine Dubreuil-Tranchant, Dominique Crouzillat and Michel Rigoreau and Perla Hamon*

1. Introduction

Despite the fact that coffee is the most traded crop and the second exported product by southern countries just after oil (Pendergrast, 2009), modern fields of research including genomics, transcriptomics and other "omics" were only recently initiated and are conducted by a relatively low number of laboratories and institutions worldwide. With the development of plant molecular biology, coffee genomics was mainly concentrated to genetic diversity studies (Anthony *et al.*, 2001; Steiger *et al.*, 2002; Poncet *et al.*, 2004; Tesfaye *et al.*, 2007; Cubry *et al.*, 2008) and construction of genetic maps (Paillard *et al.*, 1994; Paillard *et al.*, 1996; Dufour *et al.*, 2001; Pearl *et al.*, 2004). Later, construction of the first Bacterial Artificial Chromosome (BAC) libraries (Noir *et al.*, 2004; Leroy *et al.*, 2005; Cacao *et al.*, 2013) and establishment of first Expressed Sequence Tags (EST) libraries (Poncet *et al.*, 2006; Vieira *et al.*, 2006), in parallel, gained momentum. Advent of novel molecular markers such as Single Sequence Repeats (SSRs or microsatellites) and Single Nucleotide Polymorphism (SNPs) vastly improved genetic diversity analyses (Poncet *et al.*, 2006) and construction of saturated genetic maps. This was followed by gene expression studies utilizing newly developed expression profiling technologies. These studies either focus on global gene expression or on specific gene families or biosynthetic pathways involved in important traits for coffee cultivation and/or commercialization (Salmona *et al.*, 2008; Joët *et al.*, 2009; Privat *et al.*, 2011). These new tools, together with the sequencing of the first BACs, allowed the identification of genomic structural elements such as transposable elements (Lopes *et al.*, 2008; Hamon *et al.*, 2011; Dias *et al.*, 2015) and resulted in conduct of first synteny studies (Guyot *et al.*, 2009; Lefebvre-Pautigny *et al.*, 2010; Yu *et al.*, 2011; Guyot *et al.*, 2012). These developments were followed by whole genome sequencing efforts that led to deciphering of a complete genome of coffee. The first genome of a *Coffea* species sequenced was that of *C. canephora* (Denoeud *et al.*, 2014) as recommended by Alexandre de Kochko during the International Coffee Genomics Network (ICGN) inaugural meeting held in Paris in April 2007. This species was chosen for several reasons: it is cultivated, it is a diploid and a doubled haploid, produced by the IRD (Institut de Recherche pour le Développement, formerly ORSTOM) (Couturon, 1986) was available. *C. canephora* is also one of the parent of the other cultivated species *C. arabica* (Hamon *et al.*, 2009) (Figure 27.1). Unfortunately, the ICGN rapidly lost its credibility, but the first draft genome sequence of *Coffea* was established (Denoeud *et al.*, 2014) due to the joint efforts of two consortia, a French and an international one.The sequenced genotype was the doubled haploid kept in IRD green house in Montpellier (France). Recently, a new international consortium was established in order to achieve the sequencing of the *C. arabica* genome, the sole tetraploid species of the genus (2n=4x=44), as well as that of its diploid ancestral species *i.e.*

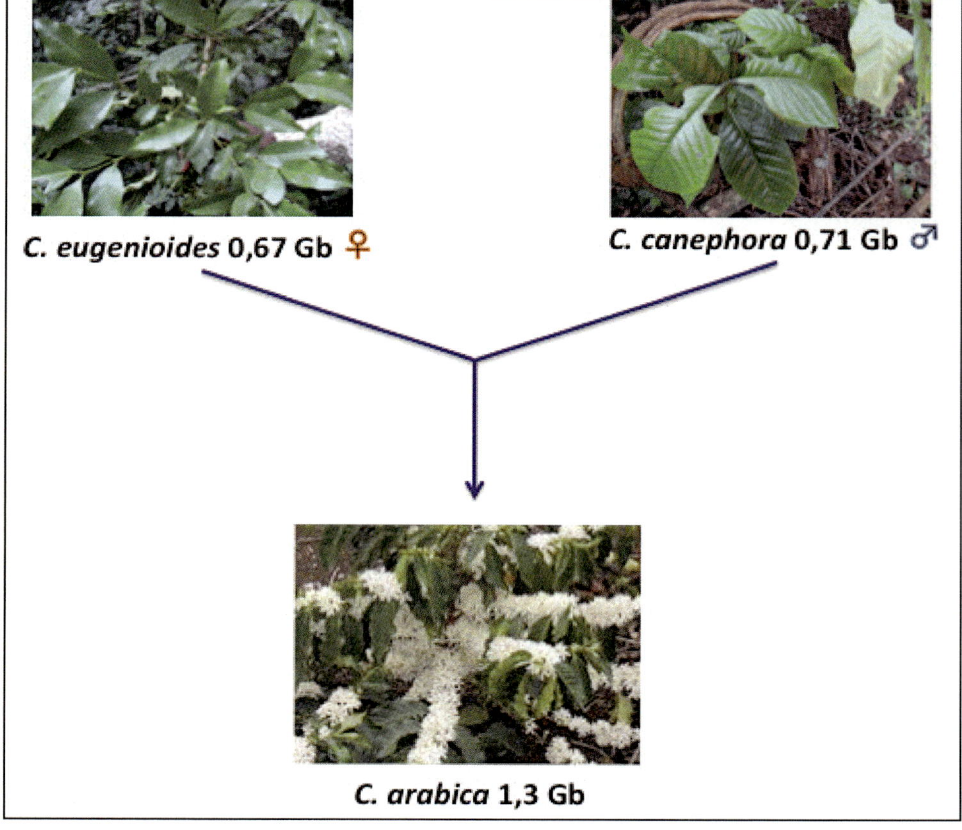

Figure 27.1: Origin of *Coffea arabica*.

C. canephora and *C. eugenioides*. Several initiatives, in different countries, also exist in order to sequence the genome of cultivated *C. arabica* varieties.

C. arabica originated from a spontaneous cross between *C. eugenioides* as the female parent and *C. canephora* as the male one. This cross had occurred 665,000 years ago maximum, probably much less (Yu *et al.*, 2011).

But coffee genomics is not limited to the two cultivated species. The genus *Coffea* extends all over sub-Saharan Africa, in some Indian Ocean Islands (Madagascar, Mauritius, Réunion and Comoros), and in Asia down to northern Australia. *Coffea* genus includes 125 described species including the species from a previously separated genus: *Psilanthus*, which was recently included in the genus *Coffea* (Davis *et al.*, 2011). Some additional taxons still remain to be described. Coffee genomics also encompasses genomics studies of some of the wild species as many questions regarding phylogenetic relationships among the genus and exact history of species diversification remain unanswered yet. In addtion, these species possess an extraordinary potential that could be extremely useful for the improvement of cultivated coffee trees especially when challenges such as climate change are a looming threat.

2. Early Steps in *Coffea* Genomics and Transcriptomics (before 2010)

If the definition of genomics is taken at its basic sense, *i.e.* the study of a genome (structurally and functionally) in its integrity instead of its components (mainly genes) independently, then one may consider that analyzing the genetic diversity with molecular tools and the establishment of genetic maps are a first genomic approach to study *Coffea* genomes.

Genetic diversity studies of *Coffea* using molecular tools started in the mid-1990s, immediately followed by the construction of the first genetic map of *C. canephora* (Paillard *et al.*, 1994; Paillard *et al.*, 1996). Molecular polymorphism and diversity studies were then carried out all over the world, both in coffee producing and consuming countries, using all the available tools, such as Restriction Fragments Length Polymorphism (RFLPs) through Random Amplified Polymorphic DNA (RAPDs), Inter Simple Sequence Repeat (ISSRs), Amplified Fragment length Polymorphism (AFLPs) and SSR among others (reviewed by de Kochko *et al.*, 2010) to SNPs. These studies not only concerned cultivated species, but wild coffee genotypes were also included in genetic mapping and diversity analyses (Ky *et al.*, 2000; Dufour *et al.*, 2001; Poncet *et al.*, 2004; Poncet *et al.*, 2007).

Genomic studies in their early stages invovled construction of BAC libraries. The first BAC libraries were constructed for the two cultivated species and strangely the more complicated one, *C. arabica*, preceded the simpler one, *C. canephora* (Noir *et al.*, 2004; Leroy *et al.*, 2005).

Global nuclear DNA content measurements and cytogenetic analyses were also performed during this period. It was shown that even if all the coffee species, but one (*C. arabica*), share the same chromosome number and all, but one (*C. arabica*), are diploid with 2n=2x=22, the DNA content was quite variable among species within the genus, both in Africa and in the Indian Ocean islands. As the ex-*Psilanthus* species were only recently added to the *Coffea* genus, little to no information is available for them. Furthermore a gradient of genome size was identified in Africa from East (small genomes) to West (larger genomes) and in Madagascar from North (small genomes) to South (larger genomes) (Cros *et al.*, 1995; Noirot *et al.*, 2003; Razafinarivo *et al.*, 2012). The chromosome structure was also analyzed and comparative studies were carried out (Pinto-Maglio and Da Cruz, 1987; Herrera *et al.*, 2007; Clarindo and Carvalho, 2008; 2009). In particular, a pioneering work involving 16 different species showed that species diversification among the genus was accompanied by the acquisition of a different chromosomal structure (Hamon *et al.*, 2009) in accordance with the botanical distinction between Eucoffea, Mozambicoffea and Mascarocoffea described by Chevalier as early as 1947 (Chevalier, 1947).

In parallel to these genetic/genomic approaches, coffee transcriptomics was also gaining momentum. The first Expression Sequence Tags (ESTs) were established in the early 2000 using the traditional Sanger sequencing protocol (Poncet *et al.*, 2006; Vieira *et al.*, 2006), but quite soon the Next Generation Sequencing (NGS) technologies took over; first using the Roche pyrosequencing approach (also known as 454) (Vieira *et al.*, 2006) and then the Illumina protocol (RNA-Seq) (Yuyama *et al.*, 2016). The first set of ESTs established by Nestlé R & D (France) and IRD (France) allowed the construction of the first RNA chip (Privat *et al.*, 2011).

Two main domains of coffee research have been investigated through transcriptomic studies: (i) the coffee cup quality: by analyzing the expression of genes involved in the biosynthetic pathways of compounds determining the organoleptic quality of coffee beverage such as sugars (Geromel *et al.*, 2006; Privat *et al.*, 2008; Joet *et al.*, 2014), phenolics (Joët *et al.*, 2010; Lepelley *et al.*, 2012) or caffeine (Dessalegn *et al.*, 2008; Barsalobres-Cavallari *et al.*, 2009; De Gaspari-Pezzopane *et al.*, 2012; Denoeud *et al.*, 2014) and (ii) the resistance to biotic and abiotic stresses, such as the resistance/tolerance to coffee rust; *Hemileia vastatrix* (Fernandez *et al.*, 2004) and climate change (Geromel *et al.*, 2008; Barsalobres-Cavallari *et al.*, 2009; Joët *et al.*, 2010; Marraccini *et al.*, 2011; Miniussi *et al.*, 2015). These previously published studies on gene expression only concerned a limited set of genes or a set of genes involved in different pathways (Joët *et al.*, 2009) or development stages (De Gaspari-Pezzopane *et al.*, 2012). One published study concerned the use of cDNA macro arrays, complemented by real time RT-PCRs, for the analysis of genes involved during the development of the coffee seed (Salmona *et al.*, 2008). This study highlighted the differential expression of genes according to the developmental stage, some of which are directly related to coffee quality.

3. More Recent Advances

The field of studies concerning the diversity, often conducted in parallel with evolutionary analyses, still continues nowadays using molecular markers, including novel ones such as transposable elements and SNPs. With the advent of improved technologies and the availability of computational analyzing tools, more genotypes are being analyzed through diversity studies. The *C. canephora* species is particularly interesting to study because it has the widest habitat range among all *Coffea* species.

It is found from South Guinea in North to Angola in South, and from Southern Guinea in the West to Uganda in the East. It is also found in lowland plains in Western Africa and up to the high plateaus in Uganda or Angola. Because of its widest range of natural habitat, *C. canephora* is expected to be genetically diverse. Indeed, utilizing a set of SSRs on a large sample of genotypes, the genetic structure of the species was established and seven diversity groups were identified (Gomez *et al.*, 2009). Wide cultivation and intervention through breeding programmes has introduced more dynamic but less structured diversification (Musoli *et al.*, 2009). *C. canephora* can be considered as being still undergoing diversification that may lead in a certain amount of time to the raise of new species. As a general rule, the border between species among the *Coffea* genus is very thin as almost all the species are inter-crossable (Louarn, 1992; Louarn, 1993).

Until very recently, very few studies have focused on Malagasy species despite the fact that Madagascar is a hotspot of *Coffea* diversification, with over 60 *Coffea* species are native to the Big Island. A recent study has finally overcome this lacuna by considering a large sample of African and Malagasy species in order to obtain a global view of the genus diversification and to gain understanding of origin and evolution of the genus. The study contradicted the widely held belief that Malagasy coffee trees have originated only due to migration from Eastern Africa (Razafinarivo *et al.*, 2013).

Transposable elements (TEs) are genetic mobile elements that constitute up to 80 per cent of plant genomes (Feschotte *et al.*, 2002) and they are considered the major factor governing genome size and genome structure evolution (Grover and Wendel, 2010). In *Coffea* genomes, preliminary observations explained the genome size differences among *Coffea* species based on the content of repetitive sequences including TEs (de Kochko *et al.*, 2004); however, no specific relations were found. This raises the question about the nature of events responsible for differences in genome size among the *Coffea* species.

Sequencing efforts of ESTs led to the development of a TEs repository in coffee genome. Studies of ESTs have further underlined that TEs could possibly play major roles in altering gene expression (Lopes *et al.*, 2008). BAC-based sequencing revealed more comprehensive directory of TEs from *C. canephora* genome (Dereeper *et al.*, 2013). Further, sequencing efforts have established that about 50 per cent of *C. canephora* genome is made of transposable elements of which 85 per cent are Long Terminal Repeat Retrotransposons (LTR-RTs). Among these, an outstanding conservation among plant genomes of several *Copia* groups was also noticed (Denoeud *et al.*, 2014). Active LTR-RTs, *i.e.* those undergoing transcription, might also be powerful markers to unravel the evolutionary history and genetic delimitation of closely related *Coffea* species. Investigations conducted on six species from the Millotii complex native to Madagascar suggested that the peak of transpositional activity of the *Gypsy* and *Copia* TEs occurred, respectively, before and after the speciation events, respectively (Roncal *et al.*, 2015).

Recently, a study on the composition of different *Coffea* species genomes was performed using partial sequencing representing as low as 3 per cent of genome coverage in length. Several interesting features were pointed out in this study: the TEs content of different species is variable according to their biogeographic location and genome size. Furthermore, a strong variation of LTR-RT was observed, suggesting differential dynamics of these elements in this group. Two LTR-RT lineages *viz.*, SIRE and Del, were clearly differentially accumulated between African and insular species, suggesting these lineages were associated to the genome divergence of *Coffea* species (Guyot *et al.*, 2016). This study is an important contribution to the understanding of the evolution and divergence of *Coffea* genomes.

Transcriptomics studies employing NGS technologies, mainly short read approaches like Illumina, have been conducted on few *Coffea* species. One of these studies, conducted on the *C. arabica* transcriptome, was in relation with an extremely important topic for coffee cultivation, *i.e.* global warming leading to water deficiency and drought situation in coffee growing regions (Mofatto *et al.*, 2016). Several differentially expressed genes under drought conditions were identified, some of these genes were known but some corresponded to uncharacterized genes. These findings signify the potency of the RNA-Seq approach in unraveling novel roles of genes that are previously unsuspected to act in biological processes. Other studies using RNA-Seq involved differential expression of genes in the polyploidy context of *C. arabica*. These studies have led to the identification of the effects of divergent trans-regulatory parental factors on the gene expression in polyploid hybrid and to the relative expression of homologues in different environmental conditions (Combes *et al.*, 2013; Combes *et al.*, 2015).

Another controversial study using RNA-Seq and whole genome sequencing tried to identify 'supposed' homeologous exchange events (HEE) between the two sub-genomes of *C. arabica* (Lashermes *et al.*, 2014). However, considering the high level of homologies between the two set of exonic sequences (>98 per cent, pers. com.) that are recovered in the RNA-Seq and short reads of RNA-Seq used in this study (72 nucleotides), the probability of misassembly or wrong mapping on to the reference genome, *C. canephora*, is very high. Also, the probability of mapping mismatches is very high leading to artificial recombination events. The fact remains that the possibility of recombination between chromosomes from different sub-genomes still seems likely.

The largest project dealing with *Coffea* genomics published to date, is that of the sequencing of the *C. canephora* genome (Denoeud *et al.*, 2014) conducted by a French group, headed by the Genoscope and IRD, along with an international consortium. The selected genotype was a doubled haploid from the E genetic diversity group of *C. canephora* (Central Africa) (Gomez *et al.*, 2009). This doubled haploid, developed by IRD, was obtained from an embryo that spontaneously appeared and then rescued before being allowed to grow before undergoing a chromosome doubling by the use of colchicine (Couturon, 1986). Sequencing, and overall assembling, of a doubled haploid is much easier than a highly polymorphic plant as *C. canephora* is allogamous.

Several important findings resulted from this work. It was shown that this genome has not undergone polyploidization events during its evolutionary history, outside the early hexaploidisation common to all angiosperms, making it similar to grape (Jaillon *et al.*, 2007). This also makes more comprehensible the results of synteny studies, which associated more closely coffee trees to grape that to Solanaceae (Guyot et al., 2012), the latter being phylogenetically closer to Rubiaceae (Bremer *et al.*, 2009). The *C. canephora* genome contains at least 25,574. This places C.

canephora among the plants with a relatively limited number of genes. But as only 80 per cent of the genome sequence has been assembled in scaffolds (*i.e.* with many gaps), it is highly likely that actual gene number could be higher. TEs, as already mentioned above, represent at least 50 per cent of the genome; among them, LTR-RTs are far the most represented with *Copia* families showing remarkably high homology with already described elements in sequenced plant genomes. The most noteworthy finding of this study was that the N-methyltransferase (NMT) genes, involved in the caffeine biosynthetic pathway, expanded through sequential tandem duplications, independently of genes from cacao and tea (other caffeine accumulating plants), suggesting that caffeine in eudicots is of polyphyletic origin.

4. Ongoing Projects

Several projects regarding both coffee transcriptomics and genomics are conducted in different laboratories worldwide. Some of the prominent studies are presented herewith.

4.1. Phylogeny of *Coffea* Genus

This project seeks to establish a definitive, complete and trustable phylogeny of the genus *Coffea*-using comparative genomics approaches. To date, phylogenetic relationships among the genus have been established only through the alignment of chloroplastic genes and/or a very limited number of nuclear genes. The resulting trees were badly resolved and could not provide any conclusive information on the origin of the genus and its path of diversification (Maurin *et al.*, 2007; Nowak *et al.*, 2012). The voluminous sequence data generated in this project is expected to provide crucial information.

4.2. Genome Comparison within the Genus *Coffea*

An ambitious project aims to compare genomes from several *Coffea* spp. following their sequencing and mapping on the reference *C. canephora* genome. The goal of this project is to establish the origin of genome size variations, the evolutionary dynamics of transposable elements and non-coding elements of *Coffea* genome. Besides this, other goals include studying overall allelic diversity and diversity of metabolic pathways that could be of interest for breeding targets especially to increase the sensory variability of the final product (Hamon *et al.*, 2015).

4.3. Diversity, Genetic Structure, Biogeography and Demographic History of *C. canephora*

C. canephora, as stated previously, is the most widely distributed species; its diverse range of habitat makes it quite interesting to analyze. Reconstructing its evolutionary history should give priceless clues regarding genome level mechanisms that allowed such a wider adaptation to different environments. Furthermore, it would also allow establishment of models that may mimic the impending genomic events that could possibly take place under the effect of climate change. The re-sequencing of several dozens of genotypes will give access to an almost unlimited number of SNPs not only for genomic studies, but also for molecular breeding purposes.

4.4. *C. arabica* Genome Sequencing

Arabica Coffee Genome Consortium
8 countries, 18 institutions, 55 researchers

An international consortium Arabica coffee genome consortium, which was established about three years ago, undertook the task of sequencing the tetraploid species *C. arabica*, the sole tetraploid of the genus and the most widely cultivated species worldwide. Among other goals of this consortium, resequencing of the *C. canephora* genome is also found, as the first version comprises many unfilled gaps and only about 30 per cent of the genome is correctly anchored (attribution of the scaffolds to pseudo molecules symbolizing the chromosomes). Furthermore, genome sequence of the other parental species, *C. eugenioides*, is also non-existent. In order to make a distinction between the two sub-genomes, sequencing of *C. eugenioides* genome is pertinent and it also constitutes one of the goals of the consortium. About 30 different *C. arabica* varieties (cultivated) and accessions (wild), including the lectotype described by Carl Linnaeus and conserved by the Linnean society at the Natural History Museum in London, are also being sequenced. The analysis of the genome sequence data should produce: (i) a high quality *C. arabica* genome, (ii) good quality *C. canephora* and *C. eugenioides* genomes, (iii) differentiate two parental sub-genomes, (iv) assess structural effects of polyploidization, (v) realize a fine annotation, (vi) catalog the genetic diversity in *C. arabica* and identify possible neo diversification induced by man, and (vii) make all these data easily accessible to researchers and breeders through web interfaces.

Sequencing a tetraploid genome is not an easy task, considering the additional diversity due to residual polymorphism (*C. arabica* cross-pollination varies from 7 to 13 per cent; Charrier *et al.*, 1978). The possible presence of paralogs and homeologs might also provide an additional source of variability which would make the assembly of the genome even more complicated. For this reason the chosen genotype is a dihaploid, *i.e.* it has only 22 chromosomes, a single set (11) from each parent (Figure 27.2).

The most advanced technologies are being utilized for performing this sequencing; among others, the use of the long reads technology provided by Pacific Biosciences (CA, USA) which delivers reads of several thousands of nucleotides and allows much more confident assemblies. As a result, only contigs are present in the final sequence assembly and this eliminates the gaps that are present when scaffolds are concerned. The Illumina RNA-Seq is an approach of choice to get a large amount of transcribed sequences representing a large amount of genes according to

Figure 27.2: Et39 the *C. arabica* Genotype Sequenced. A dihaploid has only one set of each parental chromosomes, the *C. arabica* dihaploid has 22 chromosomes instead of 44, 11 chromosomes from each parental species.

the numbers of cDNA libraries constructed from different tissues and organs. But the shortness of the reads may introduce some ambiguities overall in a tetraploid species and when the two parental genomes are highly similar as in *C. arabica*. In addition, it is now well established that a single gene does not produce only one transcript, as alternative splicing leads to production of several transcripts (Syed *et al.*, 2012). In *Arabidopsis*, about 60 per cent of the genes undergo alternative splicing (Marquez *et al.*, 2012), in order to better understand this issue, the sequencing of

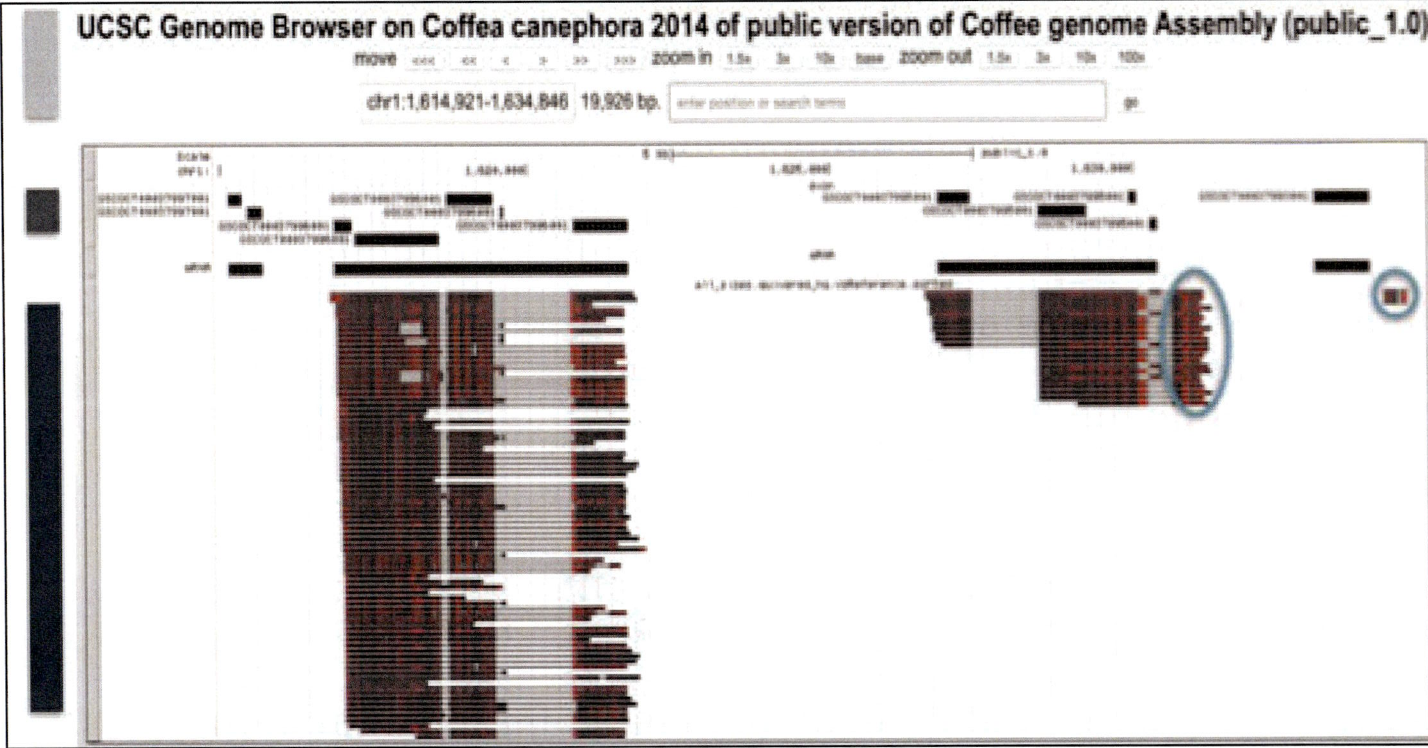

Figure 27.3: Example of Results Produced by the Iso-Seq Method. The different isoforms issued from a single gene are indicated by the black and grey lines. Blue circles indicate exons that were not identified in a previous annotation that used only RNA-Seq data.

full length transcripts, through the Iso-Seq method (Pacific Biosciences) has been performed. This method gives a variable number of isoforms, derived from the same gene sequence, revealing different types of alternative splicing as shown in Figure 27.3.

Finally, the Irys® system, furnished by Bionano-Genomics (CA, USA) for optical mapping, is also being used in this project to improve the final assembly which associates a structural (physical) mapping to the sequence.

5. Perspectives

Of late, coffee genomics, as well as that of many organisms, has undergone dramatic and rapid progress. The major difficulties encountered are capacities of analyzing voluminous data and obtaining useful information for usage in breeding programs. It is mandatory to associate the differences revealed in the DNA to their phenotypic expression and thus the big challenge to be addressed is to perform high quality phenotyping so as to associate numerous phenotypic traits at multi-location trials over a period of several years.

References

Anthony, F., Bertrand B., Quiros, O., Wilches, A., Lashermes, P., Berthaud, J. and Charrier, A. (2001). Genetic diversity of wild coffee (*Coffea arabica* L.) using molecular markers. *Euphytica*, **118(1)**: 53-65.

Barsalobres-Cavallari, C., Severino, F., Maluf, M. and Maia, I. (2009). Identification of suitable internal control genes for expression studies in *Coffea arabica* under different experimental conditions. *BMC Molecular Biology*, **10(1)**: 1.

Bremer, B., Bremer, K., Chase, M.W., Fay, M.F., Reveal, J.L., Soltis, D.E., Soltis, P.S., Stevens, P.F., Anderberg, A.A., Moore, M.J., Olmstead, R.G., Rudall, P.J., Sytsma, K.J., Tank, D.C., Wurdack, K., Xiang, J.Q.Y. and Zmarzty, S. (2009). An update of the Angiosperm Phylogeny Group classification for the orders and families of flowering plants: *APG III. Botanical Journal of the Linnean Society*, **161(2)**: 105-121.

Cacao, S.M.B., Silva, N.V., Domingues, D.S., Vieira, L.G.E., Diniz, L.E.C., Vinecky, F., Alves, G.S.C., Andrade, A.C., Carpentieri-Pipolo, V. and Pereira L.F.P. (2013). Construction and characterization of a BAC library from the *Coffea arabica* genotype Timor Hybrid CIFC 832/2. *Genetica*, **141(4-6)**: 217-226.

Charrier, A., Lanaud, C. and Guillet, J.M. (1978). Etude de la variabilité des descendances issues de pollinisations libres et contrôlées à Madagascar. *Bulletin – IFCC*, **14**: 89-98.

Chevalier, A. (1947). Recherche sur la classification et l'anatomie des Caféiers et de quelques Rubiacées. In: Lechevalier, P. (Ed.), *Les Caféiers Du Globe II Systématique Des Caféiers Et faux Caféiers*. Paris. pp. 356.

Clarindo, W.R. and Carvalho, C.R. (2008). First *Coffea arabica* karyogram showing that this species is a true allotetraploid. *Plant Systematics and Evolution*, **274(3-4)**: 237-241.

Clarindo, W.R. and Carvalho, C.R. (2009). Comparison of the *Coffea canephora* and *C. arabica* karyotype based on chromosomal DNA content. *Plant Cell Reports*, **28(1)**: 73-81.

Combes, M.C., Dereeper, A., Severac, D., Bertrand, B. and Lashermes, P. (2013). Contribution of subgenomes to the transcriptome and their intertwined regulation in the allopolyploid *Coffea arabica* grown at contrasted temperatures. *New Phytologist*, **200(1)**: 251-260.

Combes, M.C., Hueber, Y., Dereeper, A., Rialle, S., Herrera, J.C. and Lashermes, P. (2015). Regulatory divergence between parental alleles determines gene expression patterns in hybrids. *Genome Biology and Evolution*, **7(4)**: 1110-1121.

Couturon, E. (1986). The early sorting of spontaneous haploid plants of *Coffea canephora* pierre. *Cafe Cacao The*, **30(3)**: 171-176.

Cros, J., Combes, M.C., Chabrillange, N., Duperray, C., Monnot des Angles, A. and Hamon, S. (1995). Nuclear DNA content in the subgenus *Coffea* (*Rubiaceae*): Inter- and intra-specific variation in African species. *Canadian Journal of Botany*, **73**: 14-20.

Cubry, P., Musoli, P., Legnate, H., Pot, D., de Bellis, F., Poncet, V., Anthony, F., Dufour, M. and Leroy, T. (2008). Diversity in coffee assessed with SSR markers: Structure of the genus *Coffea* and perspectives for breeding. *Genome*, **51(1)**: 50-63.

Davis, A.P., Tosh, J., Ruch, N. and Fay, M.F. (2011). Growing coffee: *Psilanthus* (Rubiaceae) subsumed on the basis of molecular and morphological data; implications for the size, morphology, distribution and evolutionary history of *Coffea*. *Botanical Journal of the Linnean Society*, **167(4)**: 357-377.

de Gaspari-Pezzopane, C., Bonturi, N., Guerreiro Filho, O., Favarin, J.L. and Maluf, M.P. (2012). Gene expression profile during coffee fruit development and identification of candidate markers for phenological stages. *Pesquisa Agropecuaria Brasileira*, **47(7)**: 972-982.

de Kochko, A., Akaffou, S., Andrade, A.C., Campa, C., Crouzillat, D., Guyot, R., Hamon, P., Ming, R., Mueller, L.A., Poncet, V., Tranchant-Dubreuil, C., Hamon, S., Jean-Claude, K. and Michel, D. (2010). Advances in Coffea Genomics. In: Delseny,M. and Kader, J. (Eds.), *Advances in Botanical Research* (*Vol 53*). *Academic Press Ltd-Elsevier, Science Ltd*. pp. 23-63.

de Kochko, A., Boualem, A., Sarry, J., Hamon, P., Poncet, V. and Noirot, M. (2004). Coffee tree genome size variation: Its relation with retrotransposons copy numberASIC 2004 : *20th International Conference on Coffee Science ASIC*, Bangalore, India. pp. 696-698.

Denoeud, F., Carretero-Paulet, L., Dereeper, A., Droc, G., Guyot, R., Pietrella, M., Zheng, C., Alberti, A., Anthony, F., Aprea, G., Aury, J.M., Bento, P., Bernard, M., Bocs, S., Campa, C., Cenci, A., Combes, M.C., Crouzillat, D., Da Silva, C., Daddiego, L., De Bellis, F., Dussert, S., Garsmeur, O., Gayraud, T., Guignon, V., Jahn, K., Jamilloux, V., Joët, T., Labadie, K., Lan, T., Leclercq, J., Lepelley,

M., Leroy, T., Li, L.T., Librado, P., Lopez, L., Muñoz, A., Noel, B., Pallavicini, A., Perrotta, G., Poncet, V., Pot, D., Priyono, Rigoreau, M., Rouard, M., Rozas, J., Tranchant-Dubreuil, C., VanBuren, R., Zhang, Q., Andrade, A.C., Argout, X., Bertrand, B., de Kochko, A., Graziosi, G., Henry, R.J., Jayarama, Ming, R., Nagai, C., Rounsley, S., Sankoff, D., Giuliano, G., Albert, V.A., Wincker, P. and Lashermes, P. (2014). The coffee genome provides insight into the convergent evolution of caffeine biosynthesis. *Science*, **345**: 1181-1184.

Dereeper, A., Guyot, R., Tranchant-Dubreuil, C., Anthony, F., Argout, X., Bellis, F., Combes, M.C., Gavory, F., de Kochko, A., Kudrna, D., Leroy, T., Poulain, J., Rondeau, M., Song, X., Wing, R. and Lashermes, P. (2013). BAC-end sequences analysis provides first insights into coffee (*Coffea canephora* P.) genome composition and evolution. *Plant Molecular Biology*, **83(3)**: 177-189.

Dessalegn, Y., Labuschagne, M.T., Osthoff, G. and Herselman, L. (2008). Genetic diversity and correlation of bean caffeine content with cup quality and green bean physical characteristics in coffee (*Coffea arabica* L.). *Journal of the Science of Food and Agriculture*, **88(10)**: 1726-1730.

Dias, E.S., Hatt, C., Hamon, S., Hamon, P., Rigoreau, M., Crouzillat, D., Carareto, C.M.A., de Kochko, A. and Guyot, R. (2015). Large distribution and high sequence identity of a copia-type retrotransposon in angiosperm families. *Plant Molecular Biology*, **89(1-2)**: 83-97.

Dufour, M., Hamon, P., Noirot, M., Ristrerucci, A.M., Brottier, P., Vico, V. and Leroy, T. (2001). Potential use of SSR markers for *Coffea* spp. genetic mapping. In: ASIC, *19th Int Sci Colloq on Coffee. Trieste*, Italy.

Fernandez, D., Santos, P., Agostini, C., Bon, M.C., Petitot, A.S.C., Silva, M., Guerra-Guimaraes, L., Ribeiro, A.N.A., Argout, X. and Nicole, M. (2004). Coffee (*Coffea arabica* L.) genes early expressed during infection by the rust fungus (*Hemileia vastatrix*). *Molecular Plant Pathology*, **5(6)**: 527-536.

Feschotte, C., Jiang, N. and Wessler, S.R. (2002). Plant transposable elements: Where genetics meets genomics. *Nature Reviews Genetics*, **3(5)**: 329-341.

Geromel, C., Ferreira, L.P., Davrieux, F., Guyot, B., Ribeyre, F., Scholz, M.B.D., Luiz, F.P. P., Vaast, P., Pot, D., Leroy, T., Androcioli, A., Vieira, L.G.E., Mazzafera, P. and Marraccini, P. (2008). Effects of shade on the development and sugar metabolism of coffee (*Coffea arabica* L.) fruits. *Plant Physiology and Biochemistry*, **46(5-6)**: 569-579.

Geromel, C., Ferreira, L.P., Guerreiro, S.M.C., Cavalari, A.A., Pot, D., Pereira, L.F.P., Leroy, T., Vieira, L.G.E., Mazzafera, P. and Marraccini, P. (2006). Biochemical and genomic analysis of sucrose metabolism during coffee (*Coffea arabica*) fruit development. *Journal of Experimental Botany*, **57(12)**: 3243-3258.

Gomez, C., Dussert, S., Hamon, P., Hamon, S., de Kochko, A. and Poncet, V. (2009). Current genetic differentiation of *Coffea canephora* Pierre ex A. Froehn in the Guineo-Congolian African zone: cumulative impact of ancient climatic changes and recent human activities. *BMC Evolutionary Biology*, **9(1)**: 167.

Grover, C.E. and Wendel, J.F. (2010). Recent insights into mechanisms of genome size change in plants. *Journal of Botany*, 2010.

Guyot, R., Darré, T., Dupeyron, M., de Kochko, A., Hamon, S., Couturon, E., Crouzillat, D., Rigoreau, M., Rakotomalala, J.J., Raharimalala, N.E., Akaffou, S.D. and Hamon, P. (2016). Partial sequencing reveals the transposable element composition of *Coffea* genomes and provides evidence for distinct evolutionary stories. *Molecular Genetics and Genomics*, **291(5)**: 1979-1990.

Guyot, R., de la Mare, M., Viader, V., Hamon, P., Coriton, O., Bustamante-Porras, J., Poncet, V., Campa, C., Hamon, S. and de Kochko, A. (2009). Microcollinearity in an ethylene receptor coding gene region of the *Coffea canephora* genome is extensively conserved with *Vitis vinifera* and other distant dicotyledonous sequenced genomes. *BMC Plant Biology*, **9(1)**: 22.

Guyot, R., Lefebvre-Pautigny, F., Tranchant-Dubreuil, C., Rigoreau, M., Hamon, P., Leroy, T., Hamon, S., Poncet, V., Crouzillat, D. and de Kochko, A. (2012). Ancestral synteny shared between distantly-related plant species from the asterid (*Coffea canephora* and *Solanum* sp.) and rosid (*Vitis vinifera*) clades. *BMC Genomics*, **13**: 103.

Hamon, P., Duroy, P.O., Dubreuil-Tranchant, C., Mafra D'Almeida Costa, P., Duret, C., Razafinarivo, N., Couturon, E., Hamon, S., de Kochko, A., Poncet, V. and Guyot, R. (2011). Two novel Ty1-*copia* retrotransposons isolated from coffee trees can effectively reveal evolutionary relationships in the *Coffea* genus (Rubiaceae). *Molecular Genetics and Genomics*, **285(6)**: 447-460.

Hamon, P., Mueller, L., Strickler, S., Domingues, D., Pereira, L., Andrade, A., Ming, R., Wai, J., Albert, V., Giuliano, G., Pietrella, M., Aprea, G., Descombes, P., Moine, D., Guyot, R., Hamon, S., Couturon, E., de Kochko, A., Lepelley, M., Bellanger, L., Merot-L'Anthoene, V., Vandecasteele, C., Rigoreau, M., Crouzillat, D., Sankoff, D. and Zheng, C. (2015). The International Coffea Genome13 Project: A Way to Understand the Evolutionary History of Coffea Genomes and Unlock the Potential Use of Wild Species in Breeding? Plant and Animal Genome XXIII Conference. *Plant and Animal Genome*.

Hamon, P., Siljak-Yakovlev, S., Srisuwan, S., Robin, O., Poncet, V., Hamon, S. and de Kochko, A. (2009). Physical mapping of rDNA and heterochromatin in chromosomes of 16 *Coffea* species: A revised view of species differentiation. *Chromosome Research*, **17(3)**: 291-304.

Herrera, J.C., D'Hont, A. and Lashermes, P. (2007). Use of fluorescence *in situ* hybridization as a tool for introgression analysis and chromosome identification in coffee (*Coffea arabica* L.). *Genome*, **50(7)**: 619-626.

Jaillon, O., Aury, J.M., Noel, B., Policriti, A., Clepet, C., Casagrande, A., Choisne, N., Aubourg, S., Vitulo, N., Jubin, C., Vezzi, A., Legeai, F., Hugueney, P., Dasilva, C., Horner, D., Mica, E., Jublot, D., Poulain, J., Bruyere, C., Billault, A., Segurens, B., Gouyvenoux, M., Ugarte, E., Cattonaro, F., Anthouard, V., Vico, V., Del Fabbro, C., Alaux, M., Di, Gaspero, G., Dumas, V., Felice, N., Paillard, S., Juman, I., Moroldo, M., Scalabrin, S., Canaguier, A., Le Clainche, I., Malacrida,

G., Durand, E., Pesole, G., Laucou, V., Chatelet, P., Merdinoglu, D., Delledonne, M., Pezzotti, M., Lecharny, A., Scarpelli, C., Artiguenave, F., Pe, M.E., Valle, G., Morgante, M., Caboche, M., Adam-Blondon, A.F., Weissenbach, J., Quetier, F., Wincker, P. and French-Italian, P. (2007). The grapevine genome sequence suggests ancestral hexaploidization in major angiosperm phyla. *Nature*, **449**: 463-465.

Joët, T., Laffargue, A., Salmona, J., Doulbeau, S., Descroix, F., Bertrand, B., de Kochko, A. and Dussert, S. (2009). Metabolic pathways in tropical dicotyledonous albuminous seeds: *Coffea arabica* as a case study. *New Phytologist*, **182(1)**: 146-162.

Joet, T., Laffargue, A., Salmona, J., Doulbeau, S., Descroix, F., Bertrand, B., Lashermes, P. and Dussert, S. (2014). Regulation of galactomannan biosynthesis in coffee seeds. *Journal of Experimental Botany*, **65(1)**: 323-337.

Joët, T., Salmona, J., Laffargue, A., Descroix, F. and Dussert, S. (2010). Use of the growing environment as a source of variation to identify the quantitative trait transcripts and modules of co-expressed genes that determine chlorogenic acid accumulation. *Plant Cell and Environment*, **33(7)**: 1220-1233.

Ky, C.L., Barre, P., Lorieux, M., Trouslot, P., Akaffou, S., Louarn, J., Charrier, A., Hamon, S. and Noirot, M. (2000). Interspecific genetic linkage map, segregation distortion and genetic conversion in coffee (*Coffea* sp.). *Theoretical and Applied Genetics*, **101(4)**: 669-676.

Lashermes, P., Combes, M.C., Hueber, Y., Severac, D. and Dereeper, A. (2014). Genome rearrangements derived from homoeologous recombination following allopolyploidy speciation in coffee. *Plant Journal*, **78(4)**: 674-685.

Lefebvre-Pautigny, F., Wu, F., Philippot, M., Rigoreau, M., Priyono, Zouine, M., Frasse, P., Bouzayen, M., Broun, P., Pétiard, V., Tanksley, S. and Crouzillat, D. (2010). High resolution synteny maps allowing direct comparisons between the coffee and tomato genomes. *Tree Genetics and Genomes*, **6(4)**: 565-577.

Lepelley, M., Mahesh, V., McCarthy, J., Rigoreau, M., Crouzillat, D., Chabrillange, N., de Kochko, A. and Campa, C. (2012). Characterization, high-resolution mapping and differential expression of three homologous *PAL* genes in *Coffea canephora* Pierre (Rubiaceae). *Planta*, **236(1)**: 313-326.

Leroy, T., Marraccini, P., Dufour, M., Montagnon, C., Lashermes, P., Sabau, X., Ferreira, L. P., Jourdan, I., Pot, D., Andrade, A.C., Glaszmann, J.C., Vieira, L.G. and Piffanelli, P. (2005). Construction and characterization of a *Coffea canephora* BAC library to study the organization of sucrose biosynthesis genes. *Theoretical and Applied Genetics*, **111(6)**: 1032-1041.

Lopes, F.R., Carazzolle, M.F., Pereira, G.A.G., Colombo, C.A. and Carareto, C.M.A. (2008). Transposable elements in *Coffea* (Gentianales : Rubiacea) transcripts and their role in the origin of protein diversity in flowering plants. *Molecular Genetics and Genomics*, **279(4)**: 385-401.

Louarn, J. (1992). La fertilité des hybrides interspécifiques et les relations génomiques entre caféiers diploïdes d'origine africaine (Genre *Coffea* L., sous-genre *Coffea*). THE : Thèses, Université de Paris-Sud.

Louarn, J. (1993). Structure génétique des caféiers africains diploîdes basée sur la fertilité des hybrides interspécifiques. In: *ASIC 15 colloque. ASIC*. pp. 243-252.

Marquez, Y., Brown, J.W.S., Simpson, C., Barta, A. and Kalyna, M. (2012). Transcriptome survey reveals increased complexity of the alternative splicing landscape in *Arabidopsis*. *Genome Research*, **22(6)**: 1184-1195.

Marraccini, P., Freire, L., Alves, G., Vieira, N., Vinecky, F., Elbelt, S., Ramos, H., Montagnon, C., Vieira, L., Leroy, T., Pot, D., Silva, V., Rodrigues, G. and Andrade, A. (2011). RBCS1 expression in coffee: Coffea orthologs, *Coffea arabica* homeologs, and aexpression variability between genotypes and under drought stress. *BMC Plant Biology*, **11(1)**: 85.

Maurin, O., Davis, A.P., Chester, M., Mvungi, E.F., Jaufeerally-Fakim, Y. and Fay, M.F. (2007). Towards a phylogeny for *Coffea* (Rubiaceae): Identifying well-supported lineages based on nuclear and plastid dna sequences. *Annals of Botany*, **100(7)**: 1565-1583.

Miniussi, M., Del, Terra, L., Savi, T., Pallavicini, A. and Nardini, A. (2015). Aquaporins in *Coffea arabica* L.: Identification, expression and impacts on plant water relations and hydraulics. *Plant Physiology and Biochemistry*, **95**: 92-102.

Mofatto, L.S., Carneiro, F.D., Vieira, N.G., Duarte, K.E., Vidal, R.O., Alekcevetch, J.C., Cotta, M.G., Verdeil, J.L., Lapeyre-Montes, F., Lartaud, M., Leroy, T., De Bellis, F., Pot, D., Rodrigues, G.C., Carazzolle, M.F., Pereira, G.A.G., Andrade, A.C. and Marraccini, P. (2016). Identification of candidate genes for drought tolerance in coffee by high-throughput sequencing in the shoot apex of different *Coffea arabica* cultivars. *BMC Plant Biology*, **16**: 94. doi: 10.1186/s12870-016-0777-5

Musoli, P., Cubry, P., Aluka, P., Billot, C., Dufour, M., De Bellis, F., Pot, D., Bieysse, D., Charrier, A. and Leroy, T. (2009). Genetic differentiation of wild and cultivated populations: Diversity of *Coffea canephora Pierre in Uganda*. *Genome*, **52(7)**: 634-646.

Noir, S., Patheyron, S., Combes, M.C., Lashermes, P. and Chalhoub, B. (2004). Construction and characterisation of a BAC library for genome analysis of the allotetraploid coffee species (*Coffea arabica* L.). *Theoretical and Applied Genetics*, **109(1)**: 225-230.

Noirot, M., Poncet, V., Barre, P., Hamon, P., Hamon, S. and de Kochko, A. (2003). Genome size variations in diploid African *Coffea* Species. *Annals of Botany*, **92(5)**: 709-714.

Nowak, M.D., Davis, A.P. and Yoder, A.D. (2012). Sequence data from new plastid and nuclear *cosII* regions resolves early diverging lineages in *Coffea* (Rubiaceae). *Systematic Botany*, **37(4)**: 995-1005.

Paillard, M., Lashermes, P., Charrier, A. and Pétiard, V. (1994).) Study of polymorphism and construction of a molecular linkage map in coffee. In: *4th International Congress of Plant Molecular Biology*. Amsterdam, The Netherlands. pp. 39.

Paillard, M., Lashermes, P. and Pétiard, V. (1996). Construction of a molecular linkage map in coffee. *Theoretical and Applied Genetics*, **93(1)**: 41-47.

Pearl, H.M., Nagai, C., Moore, P.H., Steiger, D.L., Osgood, R.V. and Ming, R. (2004). Construction of a genetic map for arabica coffee. *Theoretical and Applied Genetics*, **108(5)**: 829-835.

Pendergrast, M. (2009). Coffee second only to oil? Is coffee really the second largest commodity? Mark Pendergrast investigates and finds some startling results. (Coffee Commodity) (Viewpoint essay). *Tea and Coffee Trade Journal*, pp. 38-41.

Pinto-Maglio, C.A.F. and Da Cruz, N.D. (1987). Pachytene chromosome morphology in *Coffea* L. I. *Nucleolar chromosomes. Caryologia*, **40**: 7-23.

Poncet, V., Dufour, M., Hamon, P., Hamon, S., de Kochk,o A. and Leroy, T. (2007). Development of genomic microsatellite markers in *Coffea canephora* and their transferability to other coffee species. *Genome*, **50(12)**: 1156-1161.

Poncet, V., Hamon, P., Minier, J., Carasco, C., Hamon, S. and Noirot, M. (2004). SSR cross-amplification and variation within coffee trees (*Coffea* spp.). *Genome*, **47(6)**: 1071-1081.

Poncet, V., Rondeau, M., Tranchant, C., Cayrel, A., Hamon, S., de Kochko, A. and Hamon, P. (2006). SSR mining in coffee tree EST databases: potential use of EST–SSRs as markers for the *Coffea* genus. *Molecular Genetics and Genomics*, **276(5)**: 436-449.

Privat, I., Bardil, A., Gomez, A., Severac, D., Dantec, C., Fuentes, I., Mueller, L., Joet, T., Pot, D., Foucrier, S., Dussert, S., Leroy, T., Journot, L., de Kochko, A., Campa, C., Combes, M.C., Lashermes, P. and Bertrand, B. (2011). The 'PUCE CAFE' Project: The first 15K coffee microarray, a new tool for discovering candidate genes correlated to agronomic and quality traits. *BMC Genomics*, **12(1)**: 5. doi: 10.1186/1471-2164-12-5.

Privat, I., Foucrier, S., Prins, A., Epalle, T., Eychenne, M., Kandalaft, L., Caillet, V., Lin, C.W., Tanksley, S., Foyer, C. and McCarthy, J. (2008). Differential regulation of grain sucrose accumulation and metabolism in *Coffea arabica* (Arabica) and *Coffea canephora* (Robusta) revealed through gene expression and enzyme activity analysis. *New Phytologist*, **178(4)**: 781-797.

Razafinarivo, N., Rakotomalala, J.J., Brown, S., Bourge, M., Hamon, S., de Kochko, A., Poncet, V., Dubreuil-Tranchant, C., Couturon, E., Guyot, R. and Hamon, P. (2012). Geographical gradients in the genome size variation of wild coffee trees (*Coffea*) native to Africa and Indian Ocean islands. *Tree Genetics and Genomes*, **8(6)**: 1345-1358.

Razafinarivo, N.J., Guyot, R., Davis, A.P., Couturon, E., Hamon, S., Crouzillat, D., Rigoreau, M., Dubreuil-Tranchant, C., Poncet, V., De Kochko, A., Rakotomalala, J.J. and Hamon, P. (2013). Genetic structure and diversity of coffee (*Coffea*) across Africa and the Indian Ocean islands revealed using microsatellites. *Annals of Botany*, **111(2)**: 229-248.

Roncal, J., Guyot, R., Hamon, P., Crouzillat, D., Rigoreau, M., Konan, O.N.G., Rakotomalala, J.J., Nowak, M.D., Davis, A.P. and de Kochko, A. (2015). Active transposable elements recover species boundaries and geographic structure in Madagascan coffee species. *Molecular Genetics and Genomics*, **291(1)**: 155-168.

Salmona, J., Dussert, S., Descroix, F., de Kochko, A., Bertrand, B. and Joët, T. (2008). Deciphering transcriptional networks that govern *Coffea arabica* seed development using combined cDNA array and real-time RT-PCR approaches. *Plant Molecular Biology*, **66(1)**: 105-124.

Steiger, D.L., Nagai, C., Moore, P.H., Morden, C.W., Osgood, R.V. and Ming, R. (2002). AFLP analysis of genetic diversity within and among *Coffea arabica* cultivars. *Theoretical and Applied Genetics*, **105**: 209-215.

Syed, N.H., Kalyna, M., Marquez, Y., Barta, A. and Brown, J.W.S. (2012). Alternative splicing in plants- coming of age. *Trends in Plant Science*, **17(10)**: 616-623.

Tesfaye, K., Borsch, T., Govers, K. and Bekele, E. (2007). Characterization of *Coffea* chloroplast microsatellites and evidence for the recent divergence of *C. arabica* and *C. eugenioides* chloroplast genomes. *Genome*, **50(12)**: 1112-1129.

Vieira, L.G.E., Andrade, A.C., Colombo, C.A., Araújo, A.H., Metha, A., Oliveira, A.C., Labate, C.A., Marino, C.L., Monteiro, C.B., Castro Monte, D., Giglioti, E., Kimura, E., Romano, E., Kuramae, E., Lemos, *E.G.*, Almeida, E.R., Jorge, E.C., Albuquerque, E.V.S., da Silva, F.R., Vinecky, F., Sawazaki, H.E., Dorry, H.F.A., Carrer, H., Abreu, I.N., Batista, J.A.N., Teixeira, J.B., Kitajima, J.P., Xavier, K., Lima, L., Camargo, L.E., Pereira, L.F., Coutinho, L., Lemos, M.V., Romano, M., Machado, M.A., Costa, M.M., Sá, M.F., Goldman, M.H., Ferro, M., Tinoco, M.L., Oliveira, M.B., Van Sluys, M.A., Shimizu, M., Maluf, M., Eira, M.T., Filho, O., Arruda, P., Mazzafera, P., Mariani, P., Oliveira, R., Harakava, R., Balbao, S., Tsai, S., di Mauro, S., Santos, S., Siqueira, W., Costa, G.G., Fernandes, E., Formighieri, E.F., Carazzolle, M.F. and Pereira, G.A. (2006). Brazilian coffee genome project: An EST-based genomic resource. *Brazilian Journal of Plant Physiology*, **18(1)**: 95-108.

Yu, Q., Guyot, R., de Kochko, A., Byers, A., Navajas-Pérez, R., Langston, B.J., Dubreuil-Tranchant, C., Paterson, A.H., Poncet, V., Nagai, C. and Ming, R. (2011). Micro-collinearity and genome evolution in the vicinity of an ethylene receptor gene of cultivated diploid and allotetraploid coffee species (*Coffea*). *The Plant Journal*, **67(2)**: 305-317.

Yuyama, P.M., Reis, O., Ivamoto, S.T., Domingues, D.S., Carazzolle, M.F., Pereira, G.A.G., Charmetant, P., Leroy, T. and Pereira, L.F.P. (2016). Transcriptome analysis in *Coffea eugenioides*, an Arabica coffee ancestor, reveals differentially expressed genes in leaves and fruits. *Molecular Genetics and Genomics*, **291(1)**: 323-336.

Chapter 28

Tea

☆ Y.L. Tai, H. Yang, C.L. Wei and X.C. Wan

1. Introduction

Tea is one of the most popular non-alcoholic beverages worldwide because of its bioactive compounds and it is being consumed socially and habitually by people since 3000 B.C. Tea plant, (*Camellia sinensis*) is an evergreen tree or shrub and is native to China (Figures 28.1 and 28.2). Its cultivation later spread to India and Japan, then to Europe and Russia; it finally arrived in the New World in the late 17th century (Sharangi, 2009; Mahmood *et al.*, 2010). The genus *Camellia* has long attracted a great deal of interest due to its great economic values, wide geographic distribution and remarkable species diversity. The main economic value of *C. sinensis* is the production of tea made from the young leaves and it is served as drink for 2/3rd of world population daily (Mondal *et al.*, 2004). Flavonoids (catechins), theanine, caffeine, polysaccharides and pigment among others are the most important metabolites, and these components varied in different development stages, cultivars

Figure 28.1: Tender Shoot of Tea Plant.

Figure 28.2: Six Major Tea Categories in China.

and environments. Most importantly, these compounds have health benefits on human (Hayat *et al.*, 2015; Yen and Chen, 1995). Thus, there is an urgent need for exploring the regulatory mechanism that underlies the production of metabolites and gene expression in tea. Moreover, developing superior tea plant cultivars with high yield, resistance to biotic and abiotic stresses and rich secondary metabolites is another important goal. However, tea plant is a woody and largely self-incompatible species, which has a long juvenile phase and high heterozygosity, which hinder traditional breeding initiatives.

Currently, due to the rapid growth of the next generation sequencing technologies, more research investigating the regulatory mechanism of tea plant metabolic pathways, employing RNA-Seq, are being reported. The main focus of this chapter is to provide a holistic view of tea transcriptomics with a special emphasis on the gene expression (RNA-seq data) and secondary metabolites. In particular, the recent progress in understanding the regulatory mechanism of three major secondary metabolites (catechins, caffeine and theanine) in tea by transcriptomics studies is detailed. The cost-effective and high throughput DNA sequencing technology, such as RNA-seq, is a revolutionary advance in genome-scale sequencing. The comprehensive analysis based on transcriptomics and metabolomics will serve as a crucial information platform to accelerate research of functional genomics and genomics in *C. sinensis*. Recent advances in RNA-seq have provided researchers with a powerful tool for the characterization and quantification of the transcriptome. Transcriptome sequencing using next-generation sequencing technologies is a fast and cost-effective approach to generate genome-scale sequence resources, and thus has increasingly been employed in model organisms and even more in other plants. Despite the significant economic impact of tea and other similar commodities (*i.e.*, coffee, cocoa), a lack of public genomic resources, especially the use of large insert libraries, has resulted in lack of advanced genetic knowledge that could potentially used for modern breeding. In this study, we will discuss the genome size, chloroplast genome from different species of tea plants, and the construction of BAC library which will aid in unravelling the tea plant genome.

2. Characteristics of Genomics of Tea Plant

Tea plant (*C. sinensis*) is characterized with self-incompatibility and polyploidy. Further, research on genomic information and gene expression studies in tea has advanced slowly. During the past 20 years, several genetic maps of tea plant have been generated with various types of molecular markers, *viz.*, randomly amplified polymorphic DNA (RAPD) (Ota and Tanaka, 1999), amplified fragment length polymorphisms (AFLP) (Hackett *et al.*, 2000), inter-simple sequence repeats (ISSR) (Huang *et al.*, 2006) and simple sequence repeats (SSR) (Tan *et al.*, 2013) and single nucleotide polymorphisms (SNPs) (Ma *et al.*, 2015). These genome-wide molecular markers will be a useful tool for genetic diversity and hasten breeding efforts. However, with the advent of whole genome sequencing projects, functional genomics research would gain attention in tea genetics and breeding in the future (Chen *et al.*, 2007). Rapid progress of identification and isolation of important gene(s) from tea plant has been made in the past several years (Ma and Chen, 2007). Previous studies on tea genetics, such as linkage mapping, genetic integrity of somaclonal variants (Hackett *et al.*, 2000; Thomas *et al.*, 2006; Kaundun and Matsumoto, 2003; Huang *et al.*, 2005) and tea plant functional genes and biosynthesis pathways (Park *et al.*, 2004; Chen *et al.*, 2005; Jin *et al.*, 2007) have played an important role in advancing tea genomics. However, progress in tea genomics lags far behind other crop species. Some research efforts have yielded progress in tea genomics, tea chloroplast genome and construction of BAC libraries. Research in this area will provide preliminary knowledge in tea genome, and give a comprehensive analysis of tea genome structure, evolution, and biological characteristic.

2.1. Genome Size

Plant genomes exhibit enormous variation in both their size and structure, which has stimulated speculation regarding the ancestral genome size of these plants and trends in genome evolution (Kellogg and Bennetzen, 2004). The genetic research of *C. sinensis* lags behind other important plants owing to its complex chromosome structures, resulting from numerous hybridization and polyploidization. Very little is known about genome structure and components of tea and other species in genus *Camellia*. Tanaka estimated genome of tea and *Camellia* using flow cytometry (Figure 28.3), and found that the genome sizes of tea and *Camellia* were nearly equal and estimated to 4.0 Gb (Tanaka *et al.*, 2006). In addition, the suitable plant tissues for flow cytometry constitute rapidly dividing cell without any interfering substances. Huang *et al.* (2013) analyzed flowers, leaves and buds by flow cytometry and compared eight species of *Camellia* and found that there was little discrepancy of DNA contents among different tissues. They also detected genome size variation among *Camellia* species from representative sections of the species in the genus *Camellia* and found that the DNA content varied significantly in the genus *Camellia*. Furthermore, the genome size of *C. sinensis var. assamica* was estimated to be 2.94 Gb and the *Camellia sinensis var. sinensis* was 2.84 Gb.

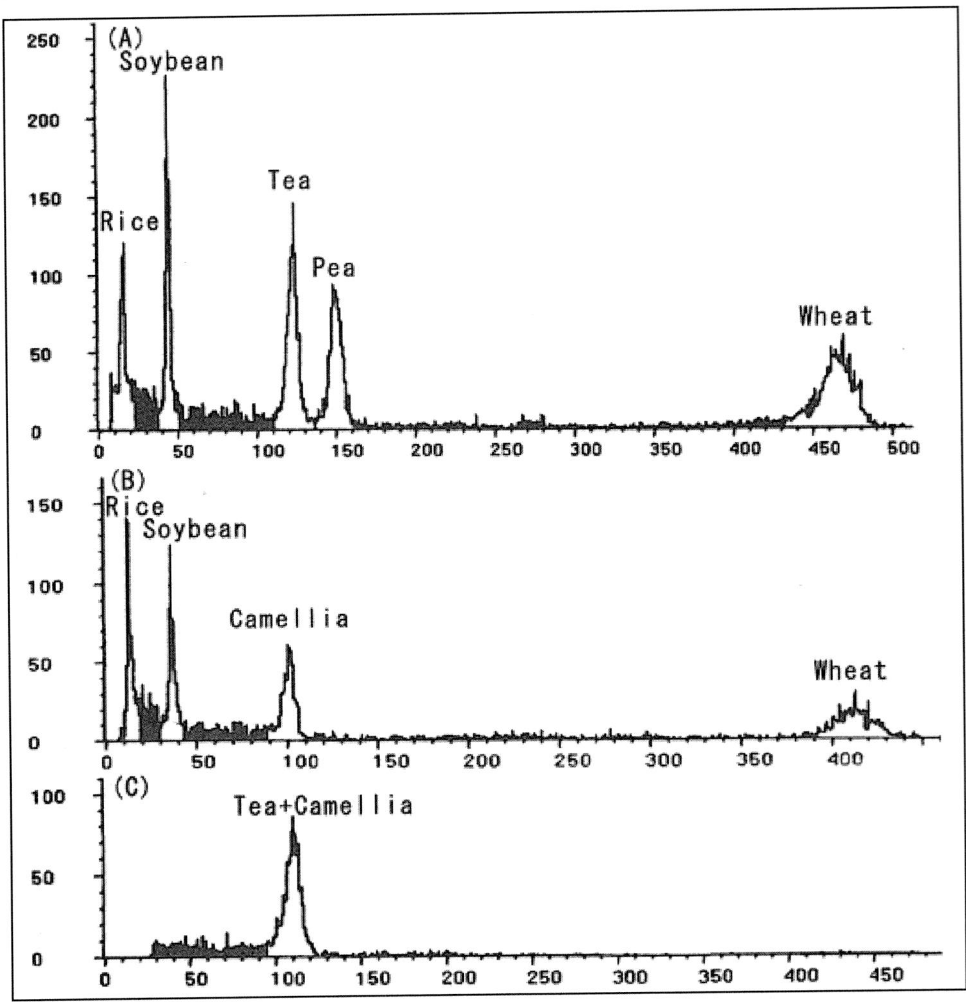

Figure 28.3: Flow Cytometry Diagrams of Tea (Yabukita, (A)), *Camellia* (Nichigetsusei, (B)), Mixture of them (C) and Standards of known Genome Size to Estimate the Genome Size (Tanaka *et al.*, 2006).

2.2. Why Tea Plant has a Large Genome?

An increase in genome size may be caused by the amplification and accumulation of retrotransposons. However, the decrease in genome size may result from a higher overall rate of deletions than insertions, selection against transposable elements, unequal crossing over, and illegitimate recombination (Bennetzen *et al.*, 2005). LTR-retrotransposons are the most significant contributors to large plant genome size, for instance, in maize, LTR-retrotransposons make up over 70 per cent of the nuclear genome (Sanmiguel *et al.*, 1998). Transposable element (TE) comprised the major fraction of repetitive DNA in eukaryotes and appear to be responsible, in large part, for differences in genome sizes among species (Hawkins *et al.*, 2008). The tea genome is found to be highly enriched for LTRs and DNA TEs. Lin *et al.* (2010) evaluated 182 random BAC-end sequences using BLAST searches and queries to the transposable element databases of *Arabidopsis* and rice, and found LTR retrotransposons were the predominant class of repeat elements in *C. sinensis* followed by DNA retroelements.

2.3. Chloroplast Genome of Tea Plant

Chloroplast (cp) genomes in different plants are conservative relatively, but contain enough genetic differences to provide useful phylogenetic information and be useful for researching the taxonomy and phylogenetic relationships. Hence, complete cp genome sequences have been widely used for taxonomic classification, species identification and understanding mechanisms that underlie the evolution of plant species and individuals plants. Fast and convenient techniques (next-generation sequencing) are generating voluminous cp genome sequence data and hence gene-based phylogenetics has been further extended to phylogenomics. Huang *et al.* (2014) sequenced 13 *Camellia* cp genomes using next-generation Illumina genome analyzer platform, and the results showed that, rather than functional constrains, it is the regional constraints that strongly affect sequence evolution of the cp genomes. Despite a high degree of conservation between the *Camellia* cp genomes, sequence variation among species could still be detected, representing a wide phylogenetic diversity in the genus (Table 28.1) (Huang *et al.*, 2014). They also investigated the variation of repeat sequences, SSRs, InDels and substitutions among the five complete *Camellia* cp genomes, representing a wide phylogenetic diversity in the genus *Camellia*. Yang *et al.* (2013a) sequenced seven complete chloroplast genomes from six species representing different subdivisions of the genus *Camellia* using Illumina sequencing technology. They found the length of the *Camellia* cp genome was about 157 kb. The phylogenetic analyses on the complete cp genomes of six *Camellia* species provided enough evidence for unique variations between the different lineages. Xiaoqian *et al.* (2014) carried out sequencing of the cp genome of *Longjing* 43; its cp genome was of the same size as the *Camellia* cp genome (157 kb). There were 15 non-synonymous mutations genes in the coding region and more than 100 polymorphic sites in the non-coding region, which could be the DNA markers for the determination of different *C. sinensis* varieties. These cp genomes may provide useful genetic information of phylogenetics, taxonomy and species identification in the genus *Camellia*.

2.4. BAC Library Construction

BAC libraries are important resources for the construction of genetic and physical maps (Luo and Wing, 2003), gene identification, map-based cloning, comparative genomics analysis and molecular marker development programmes (Shizuya *et al.*, 1992). Construction of a tea genomic BAC library is in progress as part of the tea genome project, and it will provide an important platform for whole-genome sequencing and assembly and annotation during *de novo* sequencing of the tea genome. Lin *et al.* (2010) constructed a BAC library of tea variety *Chin-shin oolong* with an average insert size of 135 kb; it provided 13.54x genome equivalent

Table 28.1: The Sequenced Chloroplast Genome Features of *Camellia* spp. (Huang *et al.*, 2014)

Complete genomes	Matched reads (bp)	Genome size (bp)	Mean coverage	LSC length (bp)	SSC length (bp)	IR length (bp)	GC content (%)
C. grandibracteata	24,127,775	157,127	154	86,657	18,286	26,092	37.29
C. leptophylla	26,635,918	157,102	170	86,648	18,276	26,089	37.30
C. sinensis var. *dehungensis*	24,978,790	157,110	159	86,656	18,276	26,089	37.30
C. sinensis var. *sinensis*	111,673,521	157,117	711	86,663	18,276	26,089	37.29
C. sinensis var. *pubilimba*	7,753,104	157,086	49	86,679	18,267	26,096	37.30
C. petelotii	9,358,318	157,121	60	86,660	18,283	26,089	37.29
C. pubicosta	36,142,305	157,076	230	86,650	18,280	26,073	37.30
C. reticulata	56,357,778	156,971	359	86,606	18,235	26,065	37.30
C. oleifera	8,162,492	157,145	52	86,676	18,291	26,089	37.28
C. sinensis var. *assamica*	2,828,916	157,121	18	86,651	18,286	26,092	37.29
C. taliensis	2,828,754	157,087	18	86,650	18,287	26,075	37.29
Incomplete genomes	**Matched reads (bp)**	**Predicted genome size (bp)**	**Mean coverage**	**Number of gaps**	**Gap length (bp)**		
C. crassicolumna var. *crassicolumna*	6,595,133	157,100	42	180	40,530		
C. fangchengensis	39,446,507	157,364	251	138	30,491		
C. kwangsiensis	8,553,876	156,992	54	207	36,424		
C. ptilophylla	8,426,943	157,057	54	148	22,817		
C. tachangensis	5,504,058	157,009	35	345	57,250		

coverage from a total of 401,280 clones (Figures 28.4 A, B). They analyzed 182 random BAC-end sequences, and found LTR retrotransposons were the most predominant sequence class (86.93 per cent - 87.24 per cent) of repeat elements in *C. sinensis*. They also found 25 simple sequence repeats (SSRs) that could potentially be used as genetic markers. Thus, BAC libraries of *C. sinensis* will help in developing physical maps, and DNA-marker analysis and functional gene cloning, and it would also provide a glimpse into the sequence composition of organization and structure of tea genome.

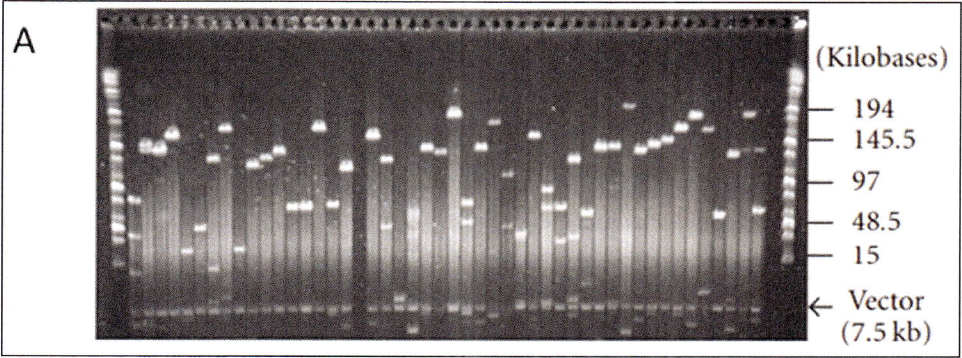

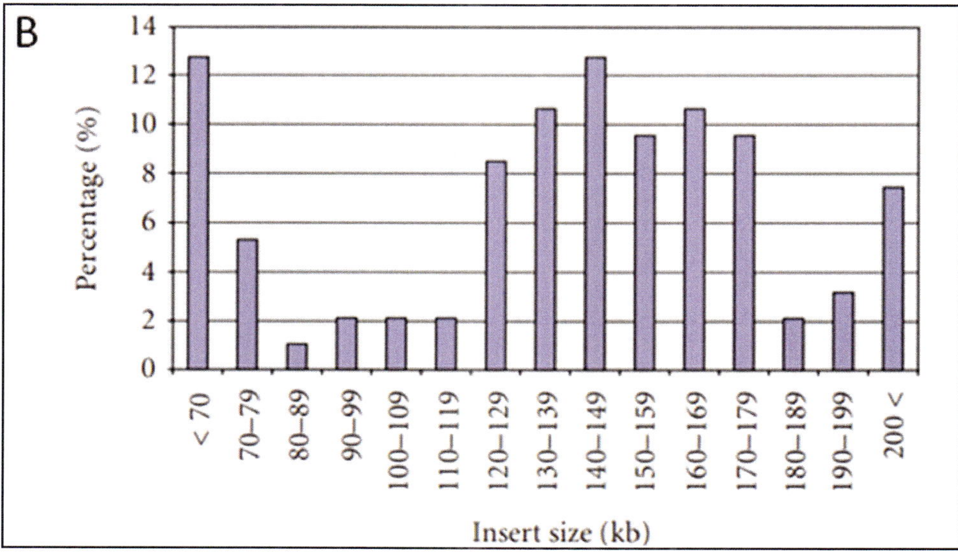

Figure 28.4: A: Insert DNA analysis of random BAC clones from the *Camellia sinensis Hind*III BAC library, CSBCBa, by pulsed-field gel electrophoresis. B: Insert size distribution of *Camellia sinensis* BAC clones from the CSBCBa library (Lin *et al.*, 2010).

3. Transcriptomics of Tea Plant

The transcriptome is the complete set of genes expressed under particular conditions. Most of the sequencing efforts in *C. sinensis* were EST sequencing, with a limited number of tags reported in public databases. Previous EST sequencing-based studies of tea have focused on the analysis of secondary pathway genes. EST sequencing has long been the important technology for transcript(s) discovery (Boguski *et al.,* 1994; Ohlrogge and Benning, 2000). However, EST sequencing has some certain limitations, such as high cost, low throughput, and lack of quantitation of the expressed genes. The high-throughput next-generation DNA sequencing (NGS) technologies have revolutionized transcriptomics by allowing RNA analysis through cDNA sequencing at massive scale (RNA sequencing) (Luciani *et al.,* 2012). It is especially suitable for gene expression profiling in non-model organisms without genomic sequences. The high-throughput of the next generation RNA sequencing (RNA-seq) technologies is a rapid, efficient, and cost-effective technique for genome-wide gene expression analysis and is widely used to define gene structure and expression profiles in more and more organisms. The assembly of *de novo* transcriptomes generated from RNA-seq makes it possible to conduct gene analysis without reference genomes (Yuan *et al.,* 2012; Ge *et al.,* 2011). Studies using this method have already altered our view of the extent and complexity of eukaryotic transcriptomes. RNA-Seq will undoubtedly be valuable for understanding transcriptomic dynamics during development and in the analysis of samples, where it will allow comparison between diseased and normal tissues, as well as the sub-classification of disease states (Wang *et al.,* 2009). In addition, RNA-Seq also can reveal the regulatory mechanism of secondary metabolites in two close species (Yuan *et al.,* 2012). These technologies will continue to help us realize the full potential of genomic information. A paired-end library sequencing strategy is generally applied not only to increase the sequencing depth, but also to improve the efficiency of *de novo* assembly. Efforts have been focused on illuminating the molecular mechanisms of plant growth, development (Ge *et al.,* 2011; Owuor *et al.,* 2008) and secondary metabolite production (Nagar and Sood, 2006) in tea. The secondary metabolites in tea plant include polyphenols (catechins and flavonoids), alkaloids (caffeine, theobromine, theophylline *etc.*), theanine, volatile oils and others. Among them, catechins, caffeine and theanine are the characteristic constituents (Liang *et al.,* 2001) in tea plant, and they are not only the important contributors to the flavour of tea, but also play crucial role in human health. As tea principal flavour substances, catechins which are the main compound of flavonoids usually account for 25 per cent to 30 per cent of the dry weight of fresh tea plant leaves (Singh *et al.,* 1999). In addition, catechins are not only important for tea quality but also related to the growth and metabolism of tea plant, such as antioxidant activity, ultraviolet light protection, and defense against phytopathogens. Caffeine is a purine alkaloid in plants that has been widely used as a stimulant. Tea plant contains 5-6 per cent caffeine that is responsible for the stimulating effect of the beverage (Ramarethinam and Rajalakshmi, 2004). It is accumulated in seeds, buds and young leaves, and serves as an anti-herbivory compound to protect soft tissues from predators (Hewavitharanage *et al.,* 1999). Theanine is a unique non-protein amino acid, and is the predominant amino acid component in tea, accounting for about 50 per cent of the total free amino acids and constitutes between 1 per cent and 2 per cent of the dry weight of tea leaves (Mukai *et al.,* 1992). Theanine acts as a neurotransmitter in the brain and has a relaxation-inducing effect in humans (Kimura *et al.,* 2007;

Kakuda, 2002). It has been shown that theanine not only provides acidic flavor and taste known as "umami" of green tea, but also produces a noticeable relaxation effect in human beings (Yamaguchi and Ninomiya, 2000; Juneja *et al.*, 1999). The genes that are involved in secondary metabolic pathways have been a key focus, and great effort has been made to identify genes related to these pathways in tea plant. To understand the molecular regulatory mechanism of secondary metabolism in tea plant, some researchers have investigated the relationship between gene expression and the secondary metabolites (Nakatsuka *et al.*, 2008; Park *et al.*, 2004; Yang *et al.*, 2012; Rani *et al.*, 2012). However, the lack of genomic information has become an obstacle to explore the molecular regulatory mechanisms underlying secondary metabolite biosynthesis in tea plant. Although unsuitable for the detection of differentially expressed genes with low transcript abundance (Shelton *et al.*, 2002; Davies and Robinson, 2000),transcriptome sequencing represents an efficient approach to obtaining functional genomic information.

3.1. The Variation of Transcriptome at different Developmental Stages

The metabolites, especially the secondary metabolites, are diverse at different developmental stages in tea plant and therefore, it is critical to understand the regulatory mechanism of secondary metabolite biosynthesis at the transcriptional level during development. Yuan *et al.* (2015) used cDNA-amplified fragment length polymorphism (cDNA-AFLP) technique to isolate genes that are differentially expressed during periodic albinism in *AnJiBaiCha*. Differentially expressed genes (DEGs) at three developmental stages (pre-albinistic stage, albinistic stage, re-greening stage) of the tea cultivar *AnJiBaiCha* were identified, which provided insights into the molecular mechanisms of periodic albinism and amino acid accumulation (Figure 28.5).

Tai *et al.* (2015) analyzed the differences in catechins, theanine, and caffeine contents of bud and leaves in tea compared it to oil tea. Further, Illumina RNA-Seq technology was employed to study transcriptomics of buds and second leaves of tea and oil tea (Tai *et al.*, 2015). The chemical composition of metabolites in tea and oil tea were similar but the absolute contents were much lower in oil tea. DEGs were identified and were found to be involved in secondary metabolite pathways, including 594 (14.23 per cent) from TBvsTL (tea buds versus tea leaves), 482 (14.1 per cent) from OTBvsOTL (oil tea buds versus leaves), respectively.

3.2. Transcriptome Analysis of different Organs

Until now, very little was known about the regulatory mechanism of secondary metabolite biosynthesis in different organs/tissues. Shi *et al.* (2011) obtained full-scale transcriptomic information from mixed tissue and leaves (tender shoots, young leaves, mature leaves, stems, young roots, flower buds and immature seeds) by RNA-seq, and characterized the majority of the essential genes in flavonoid, caffeine and theanine biosynthetic pathways. After removal of adaptor sequences, duplication sequences, ambiguous reads and low-quality reads, 30.9 million high-quality clean reads (2.32 Gbp, 89.6 per cent of the raw data) remained. A total of 127,094 unigenes with the N50 length of 506 bp were retained. The results obtained through this study demonstrated that the assembly and mapping of large amount of

Figure 28.5: The three Developmental Stages of *AnJiBaiCha.* A: Pre-albinistic stage leaves; B: Albinistic stage leaves; C: Re-greening stage leaves (Yuan *et al.,* 2015).

short reads in to contigs were successful from the tea plant samples with relatively little redundancy.

In a previous study, four cDNA libraries using different tissues *viz.*, young root, young leaf, subtractive young leaf and drought-stressed root of the *C. sinensis* were generated. Around 1,809, 921, 239 and 419 uniESTs in young root (Shi *et al.*, 2009), tender shoot (Chen *et al.*, 2005), young leaves subtractive cDNA libraries (Park *et al.*, 2004) and in the drought-stressed root SSH (suppression subtractive hybridization) library of *C. sinensis* (Das *et al.*, 2012) were obtained, respectively.

The transcriptome data from full-scale transcriptomic information of tea plant have a high degree of consistency with previous EST data; yet they represent a significant increase in coverage. By comparing transcriptome data with four previously prepared cDNA libraries analyzed by EST sequencing, we found that the number of unigenes from RNA-seq was approximate 20 times more than the existing cDNA libraries. Yet, a small number of genes discovered in the cDNA libraries did not generate BLAST hits in the Illumina transcriptome, which could be resolved by enhancing the accuracy of the assembly, increasing the sequencing depth and perfecting gene annotation strategies.

Genes involved in four primary pathways (glycolysis, citrate cycle, pentose phosphate cycle, and Calvin cycle and photosynthesis pathways) and three secondary metabolic pathways (flavonoid, theanine and caffeine pathway) that were related to tea quality were analyzed using *C. sinensis* unigenes. The majority of the essential genes in these pathways were found, especially the three characteristic metabolic pathways (Figure 28.6). Many of these genes appeared to form multi-gene families that implied the tea genome, like many other higher plants, went through one or more round of genome duplications during evolution.

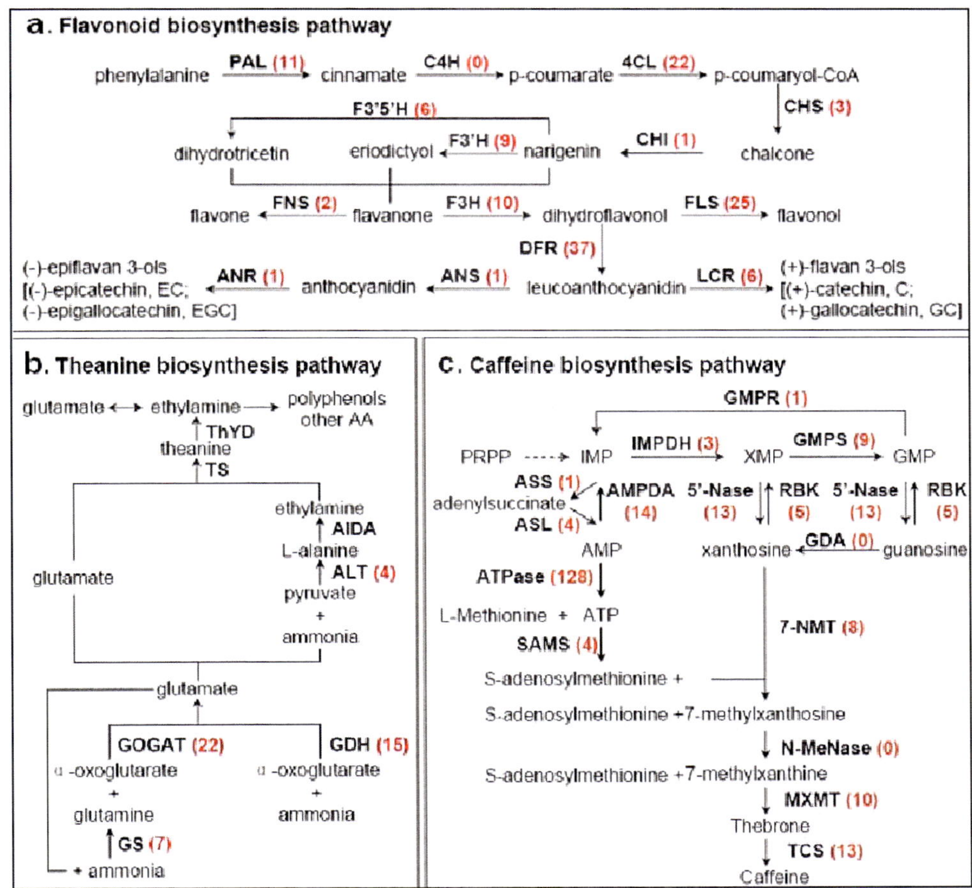

Figure 28.6: C. *sinensis* Unigenes Involved in Three Secondary Metabolic Pathways. (a) *C. sinensis* unigenes involved in the pathway of flavonoid biosynthesis; (b) *C. sinensis* unigenes involved in the pathway of theanine biosynthesis. Putative theanine biosynthesis pathway is based on Sasaoka K; (c) *C. sinensis* unigenes involved in the pathway of caffeine biosynthesis. The red number in the bracket following each gene name indicates the number of corresponding *C. sinensis* unigenes (Shi *et al.*, 2011).

Li *et al.* (2015) analyzed 13 different tissue samples from various organs and developmental stages of [*Camellia sinensis* (L.) O. Kuntze cv. *Longjing 43*], including buds and leaves of different ages, stems, flowers, seeds, and roots. They obtained a total of 43.7 Gb raw data and assembled 347,827 unigenes. In total, 1719 unigenes were identified as being involved in the secondary metabolic pathways and the expression patterns of the genes related to flavonoid, caffeine and theanine pathways were characterized, uncovering regulatory mechanism during plant growth and development. They focused on the biosynthetic pathways of three characteristic metabolites (catechins, caffeine and theanine) and the differential expressions of the related genes in the 13 *C. sinensis* tissues (Figure 28.7). A total of 206 unigenes were annotated and these were found to be associated with the biosynthetic pathways of the three characteristic metabolites based on the KEGG database (Li *et al.*, 2015).

Figure 28.7: Thirteen different Tissues of *C. sinensis* Used in the Study of Li *et al.* (2015). The name of each tissue is shown in yellow.

Transcriptional control is an important mechanism for regulating secondary metabolite production in plant cells. Transcription factors (TFs) are sequence-specific DNA-binding proteins that interact with the promoter regions of target genes and modulate the rate of initiation of mRNA synthesis by RNA polymerase II (Gantet and Memelink, 2002). Some TFs are known to be involved in the regulation of secondary metabolism, such as R2R3-MYB, basic helix-loop-helix (bHLH) proteins, AP2/ERF family proteins, WRKY, NAC, DOF, bZIP, HD-ZIP, and TFIIIA zinc finger TFs (Vom Endt *et al.,* 2002). There were 206 TFs from 33 families, 132 TFs from 30 families and 91 TFs associated with eight were observed to be associated with 36 unigenes involved in flavonoid, caffeine and theanine biosynthesis, respectively. Importantly, many critical biosynthetic genes are associated with a number of TFs from different families, indicating that the transcriptional control of these biosynthetic pathways is complex. The TF regulation network revealed the possible critical links in gene regulation between the flavonoid, caffeine and theanine biosynthesis pathways (Figure 28.8). The TF regulation network revealed the potentially possible critical links in gene regulation among the flavonoid, caffeine, and theanine biosynthesis pathways in *C. sinensis.*

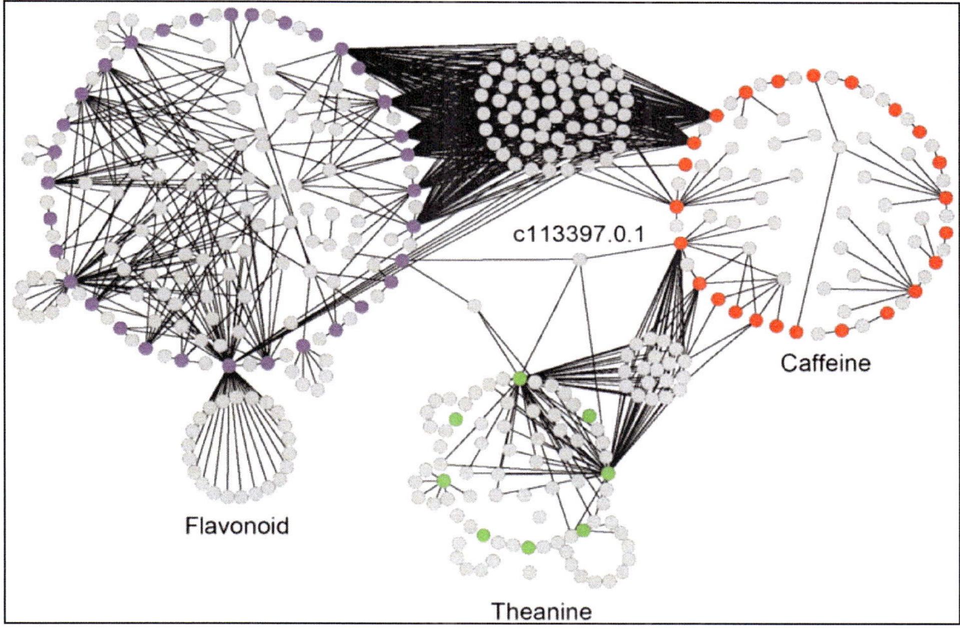

Figure 28.8: Transcription Factor (TF) Regulation Network of the Flavonoid, Caffeine, and Theanine Biosynthesis Pathways.

The purple-, orange-, and green-colored nodes represent unigenes involved in the flavonoid, caffeine, and theanine biosynthesis pathways, respectively. Each gray node represents a TF, and a linked line between a pair of nodes denotes that their expressions are correlated (Li *et al.,* 2015)

3.3. Stress Response Mechanism Based on Transcriptomics

In plants, protective secondary metabolites are usually induced by external stress signals, whereas other secondary metabolites, such as flower pigments, are expressed in a tissue-specific manner. This regulation is coordinated by specific transcription factors (Gantet and Memelink, 2002). Stress resistance mechanisms have very important influence on tea plant, because tea plant suffers from herbivore pressure, its defense responses are highly evolved and tightly regulated involving a series of physical, chemical (such as flavonoid compounds, caffeine and organic acid) and molecular modifications (Kersten *et al.*, 2013). Moreover, it has been reported that the total catechin of tea leaves increased with exposure time to sunlight, suggesting that catechins biosynthesis is also environmentally dependent (Mariya *et al.*, 2003). The role of catechins is generally considered as providing protection of plants from the UV rays damage in sunlight, and catechin production is seriously affected by photosynthesis (Premkumar *et al.*, 2008). Stress resistance also influences the activities of enzymes involved in secondary pathway in tea plant, for example, glutamine synthetase expression and activity was increased by ammonium and nitrate and also by cadmium and salt stress but decreased by copper, aluminum, drought, cold and heat stress (Rana *et al.*, 2008).

3.3.1. Cold Stress

Low temperature is one of the most important environmental factors that temperate plants have to cope with during their life cycles. Low temperature is one of the most critical environmental factors that limits the growth, survival and geographical distribution of tea plant (Wang *et al.*, 2012). Many plants can increase their freezing tolerance after exposure to low temperatures (non-freezing) for some time, a phenomena known as cold acclimation (CA) (Thomashow, 1999). Tea plant has to suffer low temperatures during the wintertime due to the local climate changes. A series of complex regulatory networks were triggered in tea plant during CA. Wang *et al.* (2013) presented a global survey for transcriptome profiles in tea plant during the CA process using RNA-Seq and DGE; genes related to cold signal sensors, stabilization of plasma membranes, osmosensing-responsive, and stress-responsive transcription factors were identified. Furthermore, they illustrated the responses of tea plants to low temperatures during the CA process (Figure 28.9). These results could help to explore the cold-related genes in improving the understanding of low-temperature tolerance ability and plant environment interactions.

Tea plant is an evergreen tree species, and tea leaves do not exhibit the phenomenon of autumnal senescence. Winter dormancy (WD) in tea plant is observed when the day light period becomes shorter than a critical light period of 11 hours 15 minutes and minimum temperature falls below 13°C for at least six weeks (Laycock, 1969). Molecular analyses during WD in tea plant showed down-regulation of genes associated with protein synthesis and cell division leading to diminished growth and developmental activities during winter season (Paul and Kumar, 2011, 2013). Paul *et al.* (2014) dissected the molecular processes operating in the leaves during the period of active growth (PAG) and WD through transcriptome analysis to uncover the mechanism of tea as a non-deciduous species. Several genes associated with molecular functions such as catalytic activity and DNA binding were significantly modulated during WD. Functional and pathway assignments of the differentially expressed unigenes using GO and KEGG classification revealed numerous hormonal, physiological and developmental changes during WD. Whereas genes encoding biotic and abiotic stresses and development related TFs were significantly over-represented during WD (Figure 28.10).

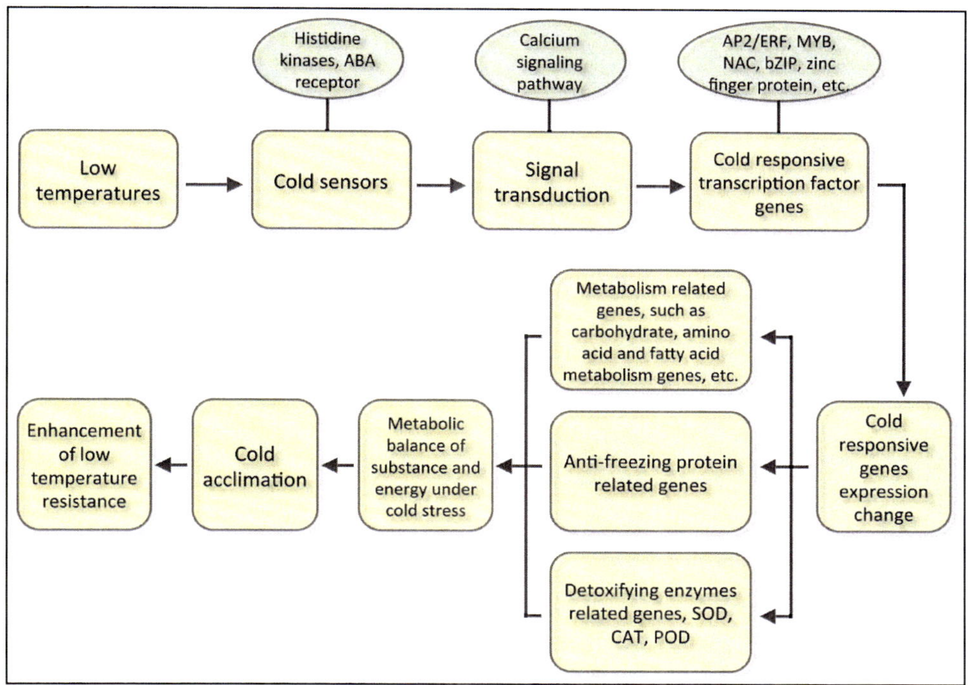

Figure 28.9: A Diagram of Responses of Tea Plants to Low Temperatures during the Cold Acclimatization Process (Wang *et al.*, 2013).

3.3.2. Mechanisms in Response to the *Ectropis oblique*

As an evergreen woody plant, the tea plant is cultivated in typically warm and humid climate areas, thus providing a relatively steady microclimate and food supply for tea geometrid *Ectropis oblique* (Prout) (Chen *et al.*, 2012). Tea plants have evolved a series of mechanisms in response to the two stimuli related to *E. oblique* feeding: mechanical damage and oral secretion stimuli. However, the induced resistance mechanisms are only activated or enhanced when the tea plant is attacked. These mechanisms involve both direct defenses that inhibit the growth or development of herbivorous insects and indirect defenses that induce the release of plant volatiles that attract the parasitoids and predators of the herbivore (Agrawal, 1998). A previous study analyzed the emission of tea plant volatiles induced by insect pests (Cai *et al.*, 2014). After mechanical damage, tea leaves immediately released grass-smelling components. These preferentially expressed genes were mainly involved in plant hormone pathways and herbivore-induced plant volatiles. Jasmonate/ethylene synthesis and signaling transduction appeared to be particularly active in *E. oblique* resistance. Wang *et al.* (2015) performed a comparison of three transcription profiles of CK, mechanically damaged (MD) and geometrid-damaged (GD) tea plant, and identified resistance genes with expression levels related to the FAC inducers of E. *oblique*. A comparison of the results from the analyses of the transcriptomes between MD and GD in tea plant revealed 28,260 unigenes expression changes,

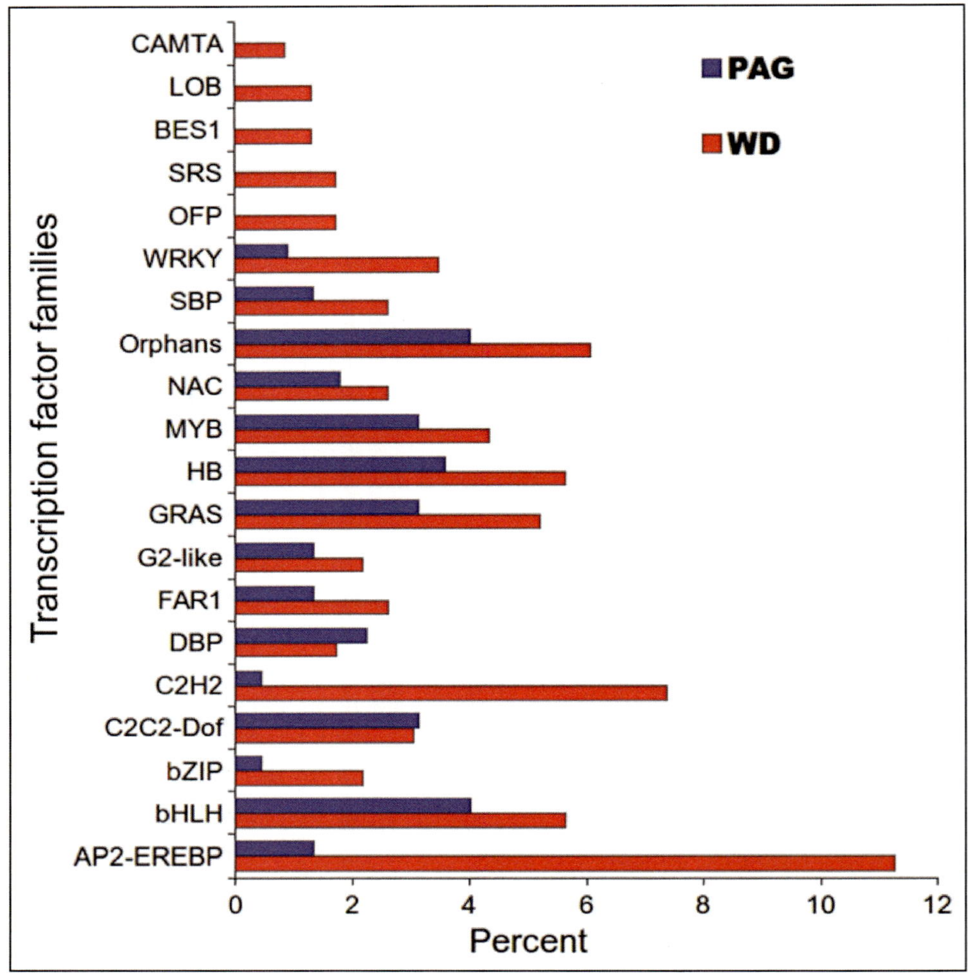

Figure 28.10: Relative Abundance and Distribution of Top 20 Transcription Factor (TF) Families during the Period of Active Growth (PAG) and Winter Dormancy (WD) for Unigenes Exhibiting Significant Differential Expression. "Percent" on X-axis represents percent TF families out of total differentially expressed TF families in the tea transcriptome (Paul *et al.*, 2014).

including 19,950 up regulated and 8310 down regulated unigenes. These unigenes were highly enriched in 15 KEGG pathways. Many genes related to brassinosteroids (BRs) were highly expressed and produced positive resistance effects that may have acted synergistically with jasmonate pathways. Herbivore-induced plant volatiles (HIPVs), particularly terpenoid volatiles and phenylpropanoid, play an important role in tea plant defense responses. Among the transcripts with highly differential expression, an enrichment of transcripts involved in BR biosynthesis was observed. BRs are a class of polyhydroxylated steroids that affect cell elongation and division

(Friedrichsen and Chory, 2001) and regulate many aspects of plant growth and development (Ali *et al.*, 2014). BRs can increase plant resistance to stress, potentially through regulation of the metabolism of ROS and certain secondary metabolites (Ashraf *et al.*, 2010). Wang et al . (2016) identified DGEs involved in plant hormone pathways, JA/ET synthesis, signaling transduction and herbivore-induced plant volatiles after the tea plants were damaged by *E. oblique* feeding; a model of defense mechanism of tea plant induced by *E. oblique* attack was also proposed.

3.3.3. Transcriptional Responses in Methyl Jasmonate Treated Tea Leaves

Tea plant has long been consumed worldwide for its amazing flavor and aroma. The economic importance of tea for its market value and is mostly decided by its taste and aroma. While phenolic compounds are responsible for the color and the taste, volatile compounds are fundamental for tea odor and aroma (Rawat *et al.*, 2007; Schwab *et al.*, 2008; Yang *et al.*, 2013b; Misra *et al.*, 2014). Therefore, it is an important reason that volatile compounds in tea have been investigated since the 1930s. Methyl jasmonate (MeJA), which acts as an effective elicitor, could mostly improve the quality of tea aroma by improving flavor volatiles contents in tea leaves. Up until now, researchers have made progress in determining the main compounds of tea aroma and its formation during tea processing. According to previous studies, the most convenient and efficient methods to identify genes related to secondary metabolic pathways were transcriptome combining metabolic analysis after treatment with stress or exogenous elicitors.

Shi *et al.* (2015) carried out an RNA-Seq analysis of MeJA-elicited transcriptional changes to identify the candidate genes involved in the secondary metabolites pathways in tea leaves, especially the biosynthesis pathway of volatiles (terpenoids and phenylpropanoids pathway). Generally, MeJA could greatly activate secondary metabolism pathways, especially volatiles. The research may show us a full understanding of tea respond to MeJA treatment resulting in the changes of volatile compounds in tea leaves. And these results also represent the massive genetic resource for tea volatile biosynthesis and will provide a new view of the genomic research in the area.

3.4. Comparative Transcriptome Analysis of Genus Camellia

The genus *Camellia* is composed of over 110 taxa (Tien-Lu, 1992), of which *C. sinensis* is often commercially used as a source of the beverage tea. The secondary metabolite compositions vary in different developmental stages, treatments and species of genus *Camellia.* Catechins are special accumulation in tea plant and their contents also have high variability in different tea plant cultivars (Magoma *et al.*, 2000). Comparative transcriptome studies involving different developmental stages, organs, treatments and species of genus *Camellia* were used to reveal the regulatory mechanism, the relationship between gene expression and secondary metabolite biosynthesis pathways. Another widely known member of *Theaceae* is oil tea (*Camellia oleifera* Abel), a tree serving as an important source of edible oil that is grown specifically in China (Lee and Yen, 2006). Although oil tea belongs to the genus *Camellia*, this plant lacks these three characteristic constituents (catechin, caffeine and theanine). Comparative transcriptomic studies have been performed to identify differential gene expression in several organisms (Güimil *et al.*, 2005; Kim *et al.*, 2014; Koenig *et al.*, 2013). By analyzing the transcriptome profiles from four different tea plant cultivars, Wu *et al.* (2014) identified the critical genes that regulate catechins biosynthesis. Tai *et al.* (2015) analyzed of tea and oil tea via RNA-Seq to uncover the genetic components underlying the biosynthesis of characteristic metabolites in tea.

3.4.1. *C. sinensis* and *C. taliensis*

C. taliensis, commonly described as 'wild' tea plant by the local people in its growing areas, is one of the most important wild relatives of the cultivated tea. Yang *et al.* (2008) reported that the compounds of *C. taliensis* rich in tea polyphenols and caffeine were the closest to *C. sinensis* and presented the first transcriptome of *C. taliensis*. *C. sinensis* has high cross-compatibility with most of its allied species in the genus *Camellia*, especially with the *C. taliensis* (Takeda, 1990). The analysis of related pathways identified the majority of candidate genes involved in major secondary metabolic pathways responsible for tea quality in this species. Comparisons with corresponding genes cloned in *C. sinensis* revealed that most of the candidate genes were relatively conserved in *C. taliensis* except FNSII, F4'ST and TCS. Thus, these analysis results may important gene resources that have great potential to enhance genetic improvement of cultivated tea in the future.

3.4.2. *C. oleifera* and *C. sinensis*

Buds and leaves of tea and oil tea were analyzed for differential levels of catechins, theanine and caffeine. These three characteristic compounds in tea are also present in oil tea, but in much lower quantity. Flavonoid, theanine and caffeine metabolism in tea and oil tea may share common pathways, but the expression levels of some of the key genes of these three metabolic pathways might differ between tea and oil tea (Tai *et al.*, 2015) which might lead to differential biosynthesis of catechins, theanine and caffeine (Figure 28.11).

High-performance liquid chromatography (HPLC) analyses were conducted to determine the contents of catechins, theanine and caffeine and related intermediates in buds and five leaves of tea and oil tea. Transcripts of buds and second leaves of tea and oil tea were enumerated by RNA-Seq technology (Figure 28.12). This comparative transcriptomic analysis provides important insights into the molecular mechanisms underlying secondary metabolite biosynthesis in tea, as well as the phytochemical characteristics of its main metabolites. High amino acid sequence identity was found in the homologous genes between tea and oil tea, as 64 per cent of the genes shared over 70 per cent identity.

Moreover, since there were significant differences in the contents of the major compounds from bud and leaves of tea versus oil tea, the comparative transcriptome data were used to search for key genes in these secondary pathways and to uncover molecular mechanisms underlying the biosynthesis of characteristic metabolites in tea. The quality of tea in large part depends on its characteristic secondary metabolites (catechin, theanine and caffeine). Almost all genes in these three secondary metabolic pathways could be detected. Many of these genes appeared to

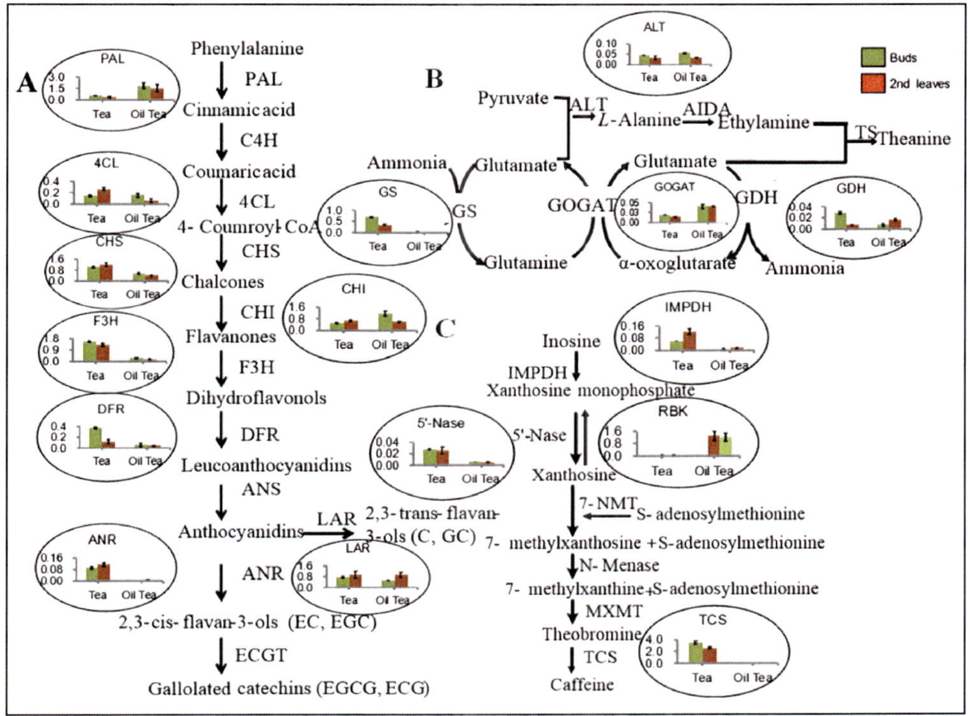

Figure 28.11: Pathways of the Three Main Secondary Metabolites in Tea and Oil Tea. (a) Core reactions of flavonoid biosynthesis involved in the biosynthesis of catechins; (b) Core reactions of the theanine biosynthesis pathway. Compound names are shown below each arrow. Abbreviations beside the arrows indicate the enzymes catalyzing the transfer. The gene expression levels detected by qRT-PCR are shown in the histograms within the corresponding circles. Transcription level is indicated as the mean ($2^{\Delta Ct}$) ± SD. (Tai *et al.*, 2015).

form multigene families which might explain why higher levels of gene expression did not always lead to higher enzyme activity in the present study.

In addition, Xia *et al.* (2014) carried out sequencing and *de novo* assembly of the transcriptome of *C. oleifera* using Roche/454 GS FLX massive parallel pyrosequencing platform, with the main focus on pathways related to lipid metabolism. They have successfully identified transcripts associated with fatty acid metabolism, oil accumulation and breakdown in *C. oleifera*. This research explored the dynamic evolution of orthologous genes between *C. oleifera* and *C. sinensis*, and assessed the natural selection pressure assigning to genes during the *Camellia* evolution.

3.4.3. Different Plant Cultivars

C. sinensis is a perennial cross-pollination plants that has rich genetic diversity of populations (Magoma *et al.*, 2000). The contents and component proportions of catechins in *C. sinensis* are mainly determined by the cultivars of tea and

Figure 28.12: Samples Examined. Buds and five initial leaves were collected from tea and oil tea. Bud, L1, L2, L3, L4 and L5 indicate the bud, first, second, third, fourth and fifth leaf, respectively (Tai *et al.,* 2015).

environmental conditions (Gulati *et al.,* 2009). Research on the regulatory mechanism among different cultivars of tea plant will be a useful and effective tool for improving the tea germplasm. Transcriptomes of four tea cultivars (*Yunnanshilixiang, Chawansanhao, Ruchengmaoyecha* and *Anjibaicha*) from different provinces (Yunnan, Jiangsu, Hunan and Zhejiang province) in China (Figure 28.13) were sequenced using the high-throughput sequencing platform Illumina HiSeq 2000 and were *de novo* assembled (Wu *et al.,* 2014). Analysis of transcriptome profiles and physiological indicators identified putative genes involved in the flavonoid biosynthetic pathway. Results showed that the multi-gene regulation of large-leafed catechins significantly differed compared to other cultivars. The expression levels of genes *ANS, ANR,* and *LAR* may potentially cause differences catechins contents among the cultivars studided. This study provided important insights into the regulatory mechanisms of catechins biosynthesis in tea plant.

4. Conclusions

Tea genomics and transcriptomics studies earlier were mainly concentrated on the development of BACs and EST sequencing so that physical and genetic maps of tea genomes could be constructed. Advent of next generation sequencing technologies has enabled many novel studies related to differential expression of secondary metabolites, such as catechins, that confer characteristic flavor to tea. Despite these recent developments, tea genomics is yet to reap the benefits of genomics and transcriptomics revolution.

Figure 28.13: Four Tea Plant Cultivars Used for Transcriptome Analysis: 'Yunnanshilixiang', 'Chawansanhao', 'Ruchengmaoyecha', and 'Anjibaicha' (Xia *et al.,* 2014).

References

Agrawal, A.A. (1998). Induced responses to herbivory and increased plant performance. *Science*, **279**: 1201-1202.

Ali, S.S., Gunupuru, L.R., Kumar, G.S., Khan, M., Scofield, S., Nicholson, P. and Doohan, F.M. (2014). Plant disease resistance is augmented in uzu barley lines modified in the brassinosteroid receptor BRI1. *BMC Plant Biology*, **14**: 227. doi: 10.1186/s12870-014-0227-1.

Ashraf, M., Akram, N., Arteca, R. and Foolad, M. (2010). The physiological, biochemical and molecular roles of brassinosteroids and salicylic acid in plant processes and salt tolerance. *Critical Reviews in Plant Sciences*, **29**: 162-190.

Bennetzen, J.L., Ma, J. and Devos, K.M. (2005). Mechanisms of recent genome size variation in flowering plants. *Annals of Botany*, **95**: 127-132.

Boguski, M.S., Tolstoshev, C.M. and Bassett Jr, D.E. (1994). Gene discovery in dbEST. *Science*, **265**: 1993-1994.

Cai, X.M., Sun, X.L., Dong, W.X., Wang, G.C. and Chen, Z.M. (2014). Herbivore species, infestation time, and herbivore density affect induced volatiles in tea plants. *Chemoecology*, **24**: 1-14.

Chen, L., Zhao, L.P. and Gao, Q.K. (2005). Generation and analysis of expressed sequence tags from the tender shoots cDNA library of tea plant (*Camellia sinensis*). *Plant Science*, **168**: 359-363.

Chen, L., Zhou, Z.X. and Yang, Y.J. (2007). Genetic improvement and breeding of tea plant (*Camellia sinensis*) in China: From individual selection to hybridization and molecular breeding. *Euphytica*, **154**: 239-248.

Chen, Z.M., Sun, X.L. and Dong, W.X. (2012). Genetics and Chemistry of the Resistance of Tea Plant to Pests. In: *Global Tea Breeding*. Springer, *Berlin Heidelberg*. pp. 343-360.

Das, A., Das, S. and Mondal, T.K. (2012). Identification of differentially expressed gene profiles in young roots of tea [*Camellia sinensis* (L.) O. Kuntze] subjected to drought stress using suppression subtractive hybridization. *Plant Molecular Biology Reporter*, **30**: 1088-1101.

Davies, C. and Robinson, S.P. (2000). Differential screening indicates a dramatic change in mRNA profiles during grape berry ripening: Cloning and characterization of cDNAs encoding putative cell wall and stress response proteins. *Plant Physiology*, **122**: 803-812.

Friedrichsen, D. and Chory, J. (2001). Steroid signaling in plants: from the cell surface to the nucleus. *Bioessays*, **23**: 1028-1036.

Gantet, P. and Memelink, J. (2002). Transcription factors: tools to engineer the production of pharmacologically active plant metabolites. *Trends in pharmacological sciences*, **23**: 563-569.

Ge, G., Xiao, P., Zhang, Y. and Yang, L. (2011). The first insight into the tissue specific taxus transcriptome via Illumina second generation sequencing. *PLoS ONE*, **6(6)**: e21220. doi: 10.1371/*journal.pone*.0021220.

Güimil, S., Chang, H.S., Zhu, T., Sesma, A., Osbourn, A., Roux, C., Ioannidis, V., Oakeley, E.J., Docquier, M. and Descombes, P. (2005). Comparative transcriptomics of rice reveals an ancient pattern of response to microbial colonization. *Proceedings of the National Academy of Sciences*, **102**: 8066-8070.

Gulati, A., Rajkumar, S., Karthigeyan, S., Sud, R., Vijayan, D., Thomas, J., Rajkumar, R., Das, S., Tamuly, P. and Hazarika, M. (2009). Catechin and catechin fractions as biochemical markers to study the diversity of Indian tea (*Camellia sinensis* (L.) O. Kuntze) germplasm. *Chemistry and Biodiversity*, **6**: 1042-1052.

Hackett, C.A., Wachira, F.N., Paul, S., Powell, W. and Waugh, R. (2000). Construction of a genetic linkage map for *Camellia sinensis* (tea). *Heredity*, **85**: 346-355.

Hawkins, J.S., Grover, C.E. and Wendel, J.F. (2008). Repeated big bangs and the expanding universe: Directionality in plant genome size evolution. *Plant Science*, **174**: 557-562.

Hayat, K., Iqbal, H., Malik, U., Bilal, U. and Mushtaq, S. (2015). Tea and its consumption: benefits and risks. *Critical Reviews in Food Science and Nutrition*, **55**: 939-954.

Hewavitharanage, P., Karunaratne, S. and Kumar, N.S. (1999). Effect of caffeine on shot-hole borer beetle (*Xyleborus fornicatus*) of tea (*Camellia sinensis*). *Phytochemistry*, **51**: 35-41.

Huang, F., Liang, Y., Lu, J. and Chen, R. (2006). Genetic mapping of first generation of backcross in tea by RAPD and ISSR markers. *Journal of Tea Science*, **26**: 171-176.

Huang, H., Shi, C., Liu, Y., Mao, S.Y. and Gao, L.Z. (2014). Thirteen *Camellia* chloroplast genome sequences determined by high-throughput sequencing: genome structure and phylogenetic relationships. *BMC Evolutionary Biology*, **14**: 151.

Huang, H., Tong, Y., Zhang, Q.J. and Gao, L.Z. (2013). Genome size variation among and within *Camellia* species by using flow cytometric analysis. *PloS ONE*, **8**: e64981. doi: 10.1371/*journal.pone*.0064981.

Huang, J., Li, J., Huang, Y., Luo, J., Gong, Z. and Liu, Z. (2005). Construction of AFLP molecular markers linkage map in tea plant. *Journal of Tea Science*, **25**: 7-15.

Jin, J., Cui, H., Gong, X., Chen, W. and Xin, Y. (2007). Studies on tea plants (*Camellia sinensis*) germplasms using EST-SSR marker. *Hereditas*, **29**: 103-108.

Juneja, L.R., Chu, D.C., Okubo, T., Nagato, Y. and Yokogoshi, H. (1999). L-theanine—a unique amino acid of green tea and its relaxation effect in humans. *Trends in Food Science and Technology*, **10**: 199-204.

Kakuda, T. (2002). Neuroprotective effects of the green tea components theanine and catechins. *Biological and Pharmaceutical Bulletin*, **25**: 1513-1518.

Kaundun, S.S. and Matsumoto, S. (2003). Identification of processed Japanese green tea based on polymorphisms generated by STS-RFLP analysis. *Journal of Agricultural and Food Chemistry*, **51**: 1765-1770.

Kellogg, E.A. and Bennetzen, J.L. (2004). The evolution of nuclear genome structure in seed plants. *American Journal of Botany*, **91**: 1709-1725.

Kersten, B., Ghirardo, A., Schnitzler, J.P., Kanawati, B., Schmitt-Kopplin, P., Fladung, M. and Schroeder, H. (2013). Integrated transcriptomics and metabolomics decipher differences in the resistance of pedunculate oak to the herbivore *Tortrix viridana* L. *BMC Genomics*, **14**: 737.

Kim, S., Park, M., Yeom, S.I., Kim, Y.M., Lee, J.M., Lee, H.A., Seo, E., Choi, J., Cheong, K. and Kim, K.T. (2014). Genome sequence of the hot pepper provides insights into the evolution of pungency in Capsicum species. *Nature Genetics*, **46**: 270-278.

Kimura, K., Ozeki, M., Juneja, L.R. and Ohira, H. (2007). L-Theanine reduces psychological and physiological stress responses. *Biological Psychology*, **74**: 39-45.

Koenig, D., Jiménez-Gómez, J.M., Kimura, S., Fulop, D., Chitwood, D.H., Headland, L.R., Kumar, R., Covington, M.F., Devisetty, U.K. and Tat, A.V. (2013). Comparative transcriptomics reveals patterns of selection in domesticated and wild tomato. *Proceedings of the National Academy of Sciences*, **110**: E2655-E2662.

Laycock, D. (1969). Latitude, day length and crop distribution. In: *Proceedings of the Tocklai Experiment Station, Jorhat, Assam, India for 1969*. pp. 1-9.

Lee, C. and Yen, G. (2006). Antioxidant activity and bioactive compounds of tea seed (*Camellia oleifera* Abel.) oil. *Journal of Agricultural and Food Chemistry*, **54**: 779-784.

Li, C.F., Zhu, Y., Yu, Y., Zhao, Q.Y., Wang, S.J., Wang, X.C., Yao, M.Z., Luo, D., Li, X. and Chen, L. (2015). Global transcriptome and gene regulation network for secondary metabolite biosynthesis of tea plant (*Camellia sinensis*). *BMC Genomics*, **16**: 560. doi: 10.1186/s12864-015-1773-0.

Liang, Y., Ma, W., Lu, J. and Wu, Y. (2001). Comparison of chemical compositions of Ilex latifolia Thumb and *Camellia sinensis* L. *Food Chemistry*, **75**: 339-343.

Lin, J., Kudrna, D. and Wing, R.A. (2010). Construction, characterization, and preliminary BAC-end sequence analysis of a bacterial artificial chromosome library of the tea plant (*Camellia sinensis*). *BioMed Research International*, doi. org/10.1155/2011/476723

Luciani, F., Bull, R.A. and Lloyd, A.R. (2012). Next generation deep sequencing and vaccine design: today and tomorrow. *Trends in Biotechnology*, **30**: 443-452.

Luo, M. and Wing, R.A. (2003). An improved method for plant BAC library construction. *Plant Functional Genomics*. Springer. pp. 3-19.

Ma, C. and Chen, L. (2007). Research progress on isolation and cloning of functional genes in tea plants. *Frontiers of Agriculture in China*, **1**: 449-455.

Ma, J.Q., Huang, L., Ma, C.L., Jin, J.Q., Li, C.F., Wang, R.K., Zheng, H.K., Yao, M.Z. and Chen, L. (2015). Large-scale SNP discovery and genotyping for constructing a high-density genetic map of tea plant using Specific-Locus Amplified Fragment Sequencing (SLAF-seq). *PloS ONE*, **10**: e0128798. doi: 10.1371/*journal.pone*.0128798.

Magoma, G., Wachira, F., Obanda, M., Imbuga, M. and Agong, S. (2000). The use of catechins as biochemical markers in diversity studies of tea (*Camellia sinensis*). *Genetic Resources and Crop Evolution*, **47**: 107-114.

Mahmood, T., Akhtar, N. and Khan, B.A. (2010). The morphology, characteristics, and medicinal properties of *Camellia sinensis*. *Journal of Medicinal Plants Research*, **4**: 2028-2033.

Mariya, J., Sasikumar, R., Balasubramanian, M., Saravanan, M. and RajKumar, R. (2003). *Influence of light on catechin biosynthesis in tea. Tea*, **24**: 80-86.

Misra, R.C., Maiti, P., Chanotiya, C.S., Shanker, K. and Ghosh, S. (2014). Methyl jasmonate-elicited transcriptional responses and pentacyclic triterpene biosynthesis in sweet basil. *Plant Physiology*, **164**: 1028-1044.

Mondal, T.K., Bhattacharya, A., Laxmikumaran, M. and Ahuja, P.S. (2004). Recent advances of tea (*Camellia sinensis*) biotechnology. *Plant Cell, Tissue and Organ Culture*, **76**: 195-254.

Mukai, T., Horie, H. and Goto, T. (1992). Differences in free amino acids and total nitrogen contents among various prices of green tea. *Tea Research Journal* (Japan), **76**: 45-50.

Nagar, P.K. and Sood, S. (2006). Changes in endogenous auxins during winter dormancy in tea (*Camellia sinensis* L.) O. Kuntze. *Acta Physiologiae Plantarum*, **28**: 165-169.

Nakatsuka, A., Mizuta, D., Kii, Y., Miyajima, I. and Kobayashi, N. (2008). Isolation and expression analysis of flavonoid biosynthesis genes in evergreen azalea. *Scientia Horticulturae*, **118**: 314-320.

Ohlrogge, J. and Benning, C. (2000). Unraveling plant metabolism by EST analysis. *Current Opinion in Plant Biology*, **3**: 224-228.

Ota, S. and Tanaka, J. (1999). RAPD-based linkage mapping using F1 segregating populations derived from crossings between tea cultivar 'Sayamakaori'and strain 'Kana-Ck17'. *Breed Rescue*, **1**: 16.

Owuor, P.O., Obanda, M., Nyirenda, H.E. and Mandala, W.L. (2008). Influence of region of production on clonal black tea chemical characteristics. *Food Chemistry*, **108**: 263-271.

Park, J.S., Kim, J.B., Hahn, B.S., Kim, K.H., Ha, S.H., Kim, J.B. and Kim, Y.H. (2004). EST analysis of genes involved in secondary metabolism in *Camellia sinensis* (tea), using suppression subtractive hybridization. *Plant Science*, **166**: 953-961.

Paul, A. and Kumar, S. (2011). Responses to winter dormancy, temperature, and plant hormones share gene networks. *Functional and Integrative Genomics*, **11**: 659-664.

Paul, A. and Kumar, S. (2013). Dehydrin2 is a stress-inducible, whereas dehydrin1 is constitutively expressed but up-regulated gene under varied cues in tea [*Camellia sinensis* (L.) O. Kuntze]. *Molecular Biology Reports*, **40**: 3859-3863.

Paul, A., Jha, A., Bhardwaj, S., Singh, S., Shankar, R. and Kumar, S. (2014). RNA-seq-mediated transcriptome analysis of actively growing and winter dormant shoots identifies non-deciduous habit of evergreen tree tea during winters. *Scientific Reports*, **4**: 5932. doi:10.1038/srep05932.

Premkumar, R., Ponmurugan, P. and Manian, S. (2008). Growth and photosynthetic and biochemical responses of tea cultivars to blister blight infection. *Photosynthetica*, **46**: 135-138.

Ramarethinam, S. and Rajalakshmi, N. (2004). Caffeine in tea plants [*Camellia sinensis* (L) O. Kuntze]: *In situ* lowering by *Bacillus licheniformis* (Weigmann) Chester. *Indian Journal of Experimental Biology*, **42(6)**: 575-580.

Rana, N., Mohanpuria, P. and Yadav, S. (2008). Expression of tea cytosolic glutamine synthetase is tissue specific and induced by cadmium and salt stress. *Biologia Plantarum*, **52**: 361-364.

Rani, A., Singh, K., Ahuja, P.S. and Kumar, S. (2012). Molecular regulation of catechins biosynthesis in tea [*Camellia sinensis* (L.) O. Kuntze]. *Gene*, **495**: 205-210.

Rawat, R., Gulati, A., Babu, G.K., Acharya, R., Kaul, V.K. and Singh, B. (2007). Characterization of volatile components of Kangra orthodox black tea by gas chromatography-mass spectrometry. *Food Chemistry*, **105**: 229-235.

Sanmiguel, P., Gaut, B., Tikhonov, A., Nakajima, Y. and Bennetzen, J. (1998). The paleontology of intergene retrotransposons of maize. *Nature Genetics*, **20**: 43-45.

Schwab, W., DavidovichRikanati, R. and Lewinsohn, E. (2008). Biosynthesis of plant-derived flavor compounds. *The Plant Journal*, **54**: 712-732.

Sharangi, A. (2009). Medicinal and therapeutic potentialities of tea (*Camellia sinensis* L.): A review. *Food Research International*, **42**: 529-535.

Shelton, D., Leach, D., Baverstock, P. and Henry, R. (2002). Isolation of genes involved in secondary metabolism from *Melaleuca alternifolia* (Cheel) using expressed sequence tags (ESTs). *Plant Science*, **162**: 9-15.

Shi, C.Y., Wan, X.C., Jiang, C.J. and Zhang, Z.Z. (2009). Construction of cDNA library for tender roots of tea plant (*Camellia sinensis* (L.) O. Kuntze) and analysis of its expressed sequence tags [J]. *Journal of Nanjing Agricultural University*, **1**: 025.

Shi, C.Y., Yang, H., Wei, C.L., Yu, O., Zhang, Z.Z., Jiang, C.J., Sun, J., Li, Y.Y., Chen, Q. and Xia, T. (2011). Deep sequencing of the *Camellia sinensis* transcriptome revealed candidate genes for major metabolic pathways of tea-specific compounds. *BMC Genomics*, **12**: 131. doi: 10.1186/1471-2164-12-131

Shi, J., Ma, C., Qi, D., Lv, H., Yang, T., Peng, Q., Chen, Z. and Lin, Z. (2015). Transcriptional responses and flavor volatiles biosynthesis in methyl jasmonate-treated tea leaves. *BMC Plant Biology*, **15**: 233. doi: 10.1186/s12870-015-0609-z.

Shizuya, H., Birren, B., Kim, U.J., Mancino, V., Slepak, T., Tachiiri, Y. and Simon, M. (1992). Cloning and stable maintenance of 300-kilobase-pair fragments of human DNA in *Escherichia coli* using an F-factor-based vector. *Proceedings of the National Academy of Sciences*, **89**: 8794-8797.

Singh, H.P., Ravindranath, S. and Singh, C. (1999). Analysis of tea shoot catechins: spectrophotometric quantitation and selective visualization on two-dimensional paper chromatograms using diazotized sulfanilamide. *Journal of Agricultural and Food Chemistry*, **47**: 1041-1045.

Tai, Y., Wei, C., Yang, H., Zhang, L., Chen, Q., Deng, W., Wei, S., Zhang, J., Fang, C. and Ho, C. (2015). Transcriptomic and phytochemical analysis of the biosynthesis of characteristic constituents in tea (*Camellia sinensis*) compared with oil tea (*Camellia oleifera*). *BMC Plant Biology*, **15**: 190. doi: 10.1186/s12870-015-0574-6.

Takeda, Y. (1990). Cross compatibility of tea (*Camellia sinensis*) and its allied species in the genus *Camellia*. *Japan Agricultural Research Quarterly*, **24(14)**: 111-116.

Tan, L.Q., Wang, L.Y., Wei, K., Zhang, C.C., Wu, L.Y., Qi, G.N., Cheng, H., Zhang, Q., Cui, Q.M. and Liang, J.B. (2013). Floral transcriptome sequencing for SSR marker development and linkage map construction in the tea plant (*Camellia sinensis*). *PLoS ONE*, **8(11)**: e81611. doi: 10.1371/journal.pone.0081611.

Tanaka, J., Taniguchi, F., Hirai, N. and Yamaguchi, S. (2006). Estimation of the genome size of tea (*Camellia sinensis*), camellia (*C. japonica*), and their interspecific hybrids by flow cytometry. *Journal of the Remote Sensing Society of Japan* (Japan). **101**: 1-7.

Thomas, J., Vijayan, D., Joshi, S.D., Lopez, S.J. and Kumar, R.R. (2006). Genetic integrity of somaclonal variants in tea (*Camellia sinensis* (L.) O Kuntze) as revealed by inter simple sequence repeats. *Journal of Biotechnology*, **123**: 149-154.

Thomashow, M.F. (1999). Plant cold acclimation: Freezing tolerance genes and regulatory mechanisms. *Annual Review of Plant Biology*, **50**: 571-599.

Tien-Lu, M. (1992). A revision of *Camellia* sect. thea. *Acta Botanica Yunnanica*, **14**(2): 115-132.

Vom Endt, D., Kijne, J.W. and Memelink, J. (2002). Transcription factors controlling plant secondary metabolism: what regulates the regulators?. *Phytochemistry*, **61**: 107-114.

Wang, D., Li, C.F., Ma, C.L. and Chen, L. (2015). Novel insights into the molecular mechanisms underlying the resistance of *Camellia sinensis* to Ectropis oblique provided by strategic transcriptomic comparisons. *Scientia Horticulturae*, **192**: 429-440.

Wang, X.C., Zhao, Q.Y., Ma, C.L., Zhang, Z.H., Cao, H.L., Kong, Y.M., Yue, C., Hao, X.Y., Chen, L. and Ma, J.Q. (2013). Global transcriptome profiles of *Camellia sinensis* during cold acclimation. *BMC Genomics*, **14**: 415. doi: 10.1186/1471-2164-14-415.

Wang, Y., Jiang, C.J., Li, Y.Y., Wei, C.L. and Deng, W.W. (2012). CsICE1 and CsCBF1: two transcription factors involved in cold responses in *Camellia sinensis*. *Plant Cell Reports*, **31**: 27-34.

Wang, Z., Gerstein, M. and Snyder, M. (2009). RNA-Seq: a revolutionary tool for transcriptomics. *Nature Reviews Genetics*, **10**: 57-63.

Wang, Y.N., Tang, L., Hou, Y., Wang, P., Yang, H. and Wei, C.L. (2016). Differential transcriptome analysis of leaves of tea plant (Camellia sinensis) provides comprehensive insights into the defense responses to Ectropis oblique attack using RNA-Seq. *Functional and Integrative Genomics*, **16(4):** 383-398.

Wu, Z.J., Li, X.H., Liu, Z.W., Xu, Z.S. and Zhuang, J. (2014). *De novo* assembly and transcriptome characterization: Novel insights into catechins biosynthesis in *Camellia sinensis*. *BMC Plant Biology*, **14**: 277.

Xia, E.H., Jiang, J.J., Huang, H., Zhang, L.P., Zhang, H.B. and Gao, L.Z. (2014). Transcriptome analysis of the oil-rich tea plant, *Camellia oleifera*, reveals candidate genes related to lipid metabolism. PLoS ONE, **9(8)**: e104150. doi: 10.1371/journal.pone.0104150.

Xiaoqian, Y., Zhonghui, Z., Quanwu, Z., Yingying, W., Zhangxiang, L., Chuyu, Y., Longjiang, F. and Hairong, X. (2014). Entire chloroplast genome sequence of tea (*Camellia sinensis* cv. Longjing 43): A molecular phylogenetic analysis. *Journal of Zhejiang Universit* (Agriculture and Life Sciences), **40**: 404-412.

Yamaguchi, S. and Ninomiya, K. (2000). Umami and food palatability. *The Journal of Nutrition*, **130**: 921S-926S.

Yang, C.Z, Zhang, Y., Gao, D., Chen, K. and Jiang, H. (2008). The origin of big-leaf tea trees and evaluation of Camellia taliensis germplasm resources. *Tea Science and Technology*, 1-4.

Yang, D., Liu, Y., Sun, M., Zhao, L., Wang, Y., Chen, X., Wei, C., Gao, L. and Xia, T. (2012). Differential gene expression in tea (*Camellia sinensis* L.) calli with different morphologies and catechin contents. *Journal of Plant Physiology*, **169**: 163-175.

Yang, J.B., Yang, S.X., Li, H.T., Yang, J. and Li, D.Z. (2013a) Comparative chloroplast genomes of *Camellia* species. *PLoS ONE*, **8**: e73053. doi: 10.1371/*journal.pone*.0073053.

Yang, Z., Baldermann, S. and Watanabe, N. (2013b). Recent studies of the volatile compounds in tea. *Food Research International*, **53**: 585-599.

Yen, G.C. and Chen, H.Y. (1995). Antioxidant activity of various tea extracts in relation to their antimutagenicity. *Journal of Agricultural and Food Chemistry*, **43**: 27-32.

Yuan, L., Xiong, L.G., Deng, T.T., Wu, Y., Li, J., Liu, S.Q., Huang, J.A. and Liu, Z.H. (2015). Comparative profiling of gene expression in *Camellia sinensis* L. cultivar AnJiBaiCha leaves during periodic albinism. *Gene*, **561**: 23-29.

Yuan, Y., Song, L., Li, M., Liu, G., Chu, Y., Ma, L., Zhou, Y., Wang, X., Gao, W. and Qin, S. (2012). Genetic variation and metabolic pathway intricacy govern the active compound content and quality of the Chinese medicinal plant *Lonicera japonica* thunb. *BMC Genomics*, **13**: 195. doi: 10.1186/1471-2164-13-195.

Zhang, Y.H., Zhang, S.D. and Ling, L.Z. (2015). *De novo* transcriptome analysis to identify flavonoid biosynthesis genes in *Stellera chamaejasme*. *Plant Gene*, 4: 64-68.

Chapter 29

Rubber

☆ *C. Bindu Roy and T. Saha*

1. Introduction

Hevea brasiliensis (Willd. ex Adr. de Juss) Muell. Arg. (rubber tree), belonging to the Euphorbiaceae family, is commercially cultivated for production of natural rubber (NR) (*cis*-1,4-polyisoprene), which is an important strategic raw material for over 40,000 products. It cannot be replaced with synthetic alternatives due to its unique properties, such as resilience, elasticity, impact and abrasion resistance, efficient heat dispersion and flexibility at cold temperature (Cataldo, 2000; Cornish, 2001a). Among over 2,000 plant species producing rubber, *H. brasiliensis* is the only species cultivated commercially for natural rubber. *H. brasiliensis* is a cross-pollinated, diploid (2n = 2x = 36) and perennial plant species with a large genome (~2100 Mb) (Leitch *et al.*, 1998). NR is collected by tapping the bark of rubber tree (Figure 29.1).

India ranks second in productivity and sixth in production of NR in the world. This contributes significantly to the export earnings and agricultural employment. South-east Asia contributes to 92 per cent of NR production followed by Africa (6 per cent) and Latin America (2 per cent). The major rubber producing countries are Thailand, Indonesia, India, Malaysia, China, Vietnam, Côte d'Ivoire, Liberia, Sri Lanka, Brazil, the Philippines, Cameroon, Nigeria, Cambodia, Guatemala, Myanmar, Ghana, DR Congo, Gabon and Papua New Guinea.

Genetic improvement of *Hevea* through breeding is very elaborate and time consuming as in many other perennial species. The major limitations are its narrow genetic base, non-synchronous flowering, low fruit set, long gestation period, heterozygous nature and insufficient availability of land for field experimentation. Therefore, it is essential to identify dependable early selection parameters that can reduce the duration from the beginning of selection of new varieties to the release

Figure 29.1: View of a Healthy Rubber Plantation.

of new recommended clones. A combination of conventional and modern breeding technologies will help in this endeavor.

Transcriptomics is the study of the transcriptome, which is the complete set of transcripts in a cell, and their quantity, for a specific developmental stage or physiological condition. In contrast with the genome, which is characterized by its stability, the transcriptome actively varies depending on many factors, including stage of development and environmental conditions. Because transcriptome includes all mRNA transcripts in the cell, the transcriptome reflects the genes that are being actively expressed at any given time and therefore is also referred to as expression profiling as it examines the expression level of mRNAs in a given cell population. Understanding the transcriptome is essential for interpreting the functional elements of the genome and revealing the molecular constituents of cells and tissues, and also for understanding development and response to biotic and abiotic stresses. The key aims of transcriptomics are: to catalogue all species of transcript, including mRNAs, non-coding RNAs and small RNAs; to determine the transcriptional structure of genes, in terms of their start sites, 5' and 3' ends, splicing patterns and other post-transcriptional modifications; and to quantify the changing expression levels of each transcript during development process and under different conditions.

Various technologies have been developed to deduce and quantify the transcriptome, including hybridization-based (microarray) approach or sequence-based approach. In contrast to microarray methods, sequence-based approaches directly determine the cDNA sequence. Initially, Sanger sequencing of cDNA or EST libraries was used, but this approach is relatively low throughput, expensive and generally not quantitative. Tag-based methods were developed to overcome these limitations, including serial analysis of gene expression (SAGE), cap analysis of gene expression (CAGE) and massively parallel signature sequencing (MPSS). These tag-based sequencing approaches are high throughput and can provide precise, 'digital' gene expression levels. However, they are based on expensive Sanger sequencing technology and a significant portion of the short tags cannot be uniquely mapped to the reference genome. Moreover, only a portion of the transcript is analysed and isoforms are generally indistinguishable from each other. These disadvantages limit the use of traditional sequencing technology in annotating the structure of transcriptomes.

Introduction of Next Generation Sequencing (NGS) technology or high throughput sequencing technologies has opened new doors into the field of DNA sequencing including whole transcriptome sequencing (RNA-seq) for its unprecedented level of sensitivity and high-throughput nature. RNA-seq, combined with appropriate bioinformatic tools, provides a better approach to study gene expression profiling and is also used to construct the complete transcriptome of an organism either by reference-based or *de novo* assembly. Thus, the NGS based RNA-Seq study provides a better approach to gene expression profiling with novel molecular techniques and computational tools allowing us to perform gene expression profiling to reveal transcriptional complexity at a faster pace.

During the past one decade, remarkable progress has been made in studies related to transcriptome sequencing in various crops. RNA-Seq has emerged as a powerful tool for profiling expressed genes in plants and other organisms (Schuster, 2008; Rosenkranz *et al.*, 2008; Ansorge, 2009; Wang *et al.*, 2009; Metzker, 2010; Ozsolak and Milos, 2011; Van Verk *et al.*, 2013) as it has obvious advantages over existing methodologies such as enabling large-scale functional assignment of genes, more thorough qualitative and quantitative analysis of gene expression and more sensitive and accurate profiling of the transcriptome. This review article discusses on various research outcome of studies on transcriptional profiling of different organs and tissues with reference to the rubber trees.

2. Transcriptome Analysis of Latex from Rubber Tree

Latex is the cytoplasmic content of laticifers or latex vessels of *Hevea brasiliensis*. Latex regeneration and duration of latex flow after tapping are the two major factors that determine the rubber yield. Laticifers form a ubiquitous network of tubes in the tree and are the major location of rubber biosynthesis (Gomez and Moir, 1979). The chemical composition of natural rubber is *cis*-polyisoprene, a high-molecular weight polymer formed from sequential condensation of isopentenyl pyrophosphate (IDP) units. Numerous classes of isoprenoids including *Hevea cis*-polyisoprene is produced from the plant isoprenoid biosynthesis pathway *via* IPP as a common intermediate (Kekwick, 1989). The mevalonate (MVA) pathway has been the conventionally studied pathway for isoprenoid biosynthesis since 1950s. In the recent years, the plastidic 1-deoxy-D-xylulose 5-phosphate/2-C-methyl-D-erythritol 4-phosphate (MEP) pathway has been considered a possible alternative route for rubber biosynthesis (Figure 29.2).

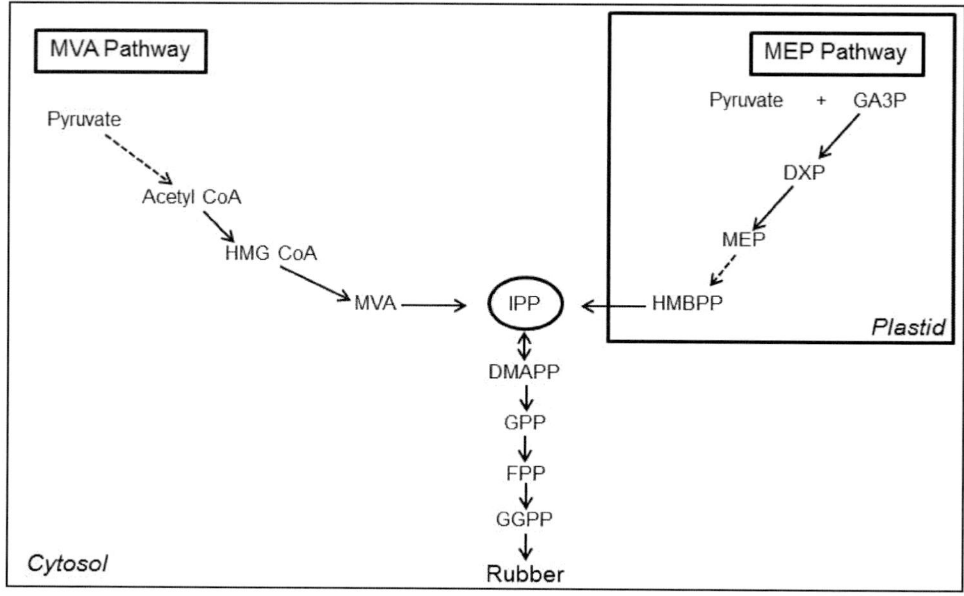

Figure 29.2: Biosynthesis of *Hevea* Rubber.

IPP for the biosynthesis of rubber may be contributed by the MVA and MEP pathways. Broken arrows indicate multiple pathway steps. IPP: isopentenyl pyrophosphate; MVA: Mevalonate; MEP: Methyl erythritol phosphate; HMG CoA: Hydroxymethylglutaryl coenzyme A; DMAPP: Dimethylallyl diphosphate; GPP: Geranyl diphosphate; FPP: Farnesyl diphosphate; GGPP: Geranylgeranyl diphosphate; DXP: Deoxy xylulose phosphate; HMBPP: Hydroxy methyl butenyl pyrophosphate.

In this era of high-throughput technology, expressed sequence tag (EST) sequencing has enabled laticifer gene expression to be surveyed on a large scale. Chow *et al.* (2007) analyzed latex transcriptome based on a collection of 10040 ESTs from the latex of rubber trees, which represents the cytoplasmic content of a single cell type, in order to analyse the latex transcription profile with emphasis on rubber biosynthesis-related genes. A total of 3441 unique transcripts were obtained after quality editing and assembly of EST sequences. Functional classification of unique transcripts according to the Gene Ontology convention showed that 73.8 per cent were related to genes of unknown function. Among highly expressed ESTs, a significant proportion encoded proteins related to rubber biosynthesis and stress or defence responses. Sequences encoding rubber particle membrane proteins (RPMPs) belonging to three protein families accounted for 12 per cent of the ESTs. Characterization of these ESTs revealed nine RPMP variants (7.9–27 kDa) including the 14 kDa REF (rubber elongation factor) and 22 kDa SRPP (small rubber particle protein). The expression of multiple RPMP isoforms in latex was shown using antibodies against REF and SRPP. Both EST and quantitative reverse transcription-PCR (qRT-PCR) analyses demonstrated REF and SRPP to be the most abundant transcripts in latex. Besides rubber biosynthesis, comparative sequence analysis showed that the RPMPs are highly similar to sequences in the plant kingdom having stress-related functions. The EST sequence analysis undertaken by Chow *et al.* (2007), with an aim to gain insights into rubber biosynthesis in *Hevea* reinforces the fact that the entire process of *Hevea* cis-polyisoprene biosynthesis involves the participation of numerous proteins, varying regulatory control of their expression, and interactions between them in the rubber biosynthesis machinery.

In 2015, Chao *et al.* carried out comparative transcriptome analysis of latex from two rubber clones: CATAS8-79 and PR107 with a purpose to globally characterize latex transcriptome. RNA was extracted from CATAS8-79 and PR107 at first tapping and sequenced using Illumina paired-end sequencing technology individually. *De novo* assembly of reads generated 53,571 and 57,806 unigenes in CATAS8-79 and PR107 individually, and finally 51,829 unigenes were integrated. To classify the unigenes, the Blast2GO program was used to get Gene Ontology (GO) annotation based on molecular function, biological process and cellular component. All unigenes were aligned to the Clusters of Orthologous Group (COG) database to predict possible functions and Kyoto Encyclopedia of Genes and Genomes (KEGG) pathway database to perform pathway assignments. Differentially expressed genes between the two DGE libraries (CATAS8-79 versus PR107) were identified. A total of 6726 unigenes with differential expression patterns were detected between CATAS8-79 and PR107. Wei *et al.* (2015), following transcriptome sequencing, reported cysteine and methionine metabolism, energy, oxidative phosphorylation, terpenoid backbone biosynthesis, plant hormone signal transduction and copper and potassium transporters to be significantly enriched during the latex flow.

3. Characterization of Bark Transcriptome of Rubber Tree

In rubber tree, bark is one of important agricultural and biological organs as latex is harvested through tapping of bark tissue. The molecular mechanisms involved in the bark formation and development in rubber tree remain largely unknown, which is at least partially due to lack of bark transcriptomic and genomic information. Li *et al.* (2012) carried out high-throughput transcriptome sequencing of rubber tree bark using Illumina paired-end sequencing technology to generate enormous transcript sequences for the functional characterization and molecular marker development. In total, 22,756 unigenes, with an average length of 485 bp, were obtained with *de novo* assembly. Similarity searches indicated that 16,520 and 12,558 unigenes showed significant similarities to known proteins from NCBI non-redundant and Swiss Prot protein databases, respectively. Among these annotated unigenes, 6,867 and 5,559 unigenes were separately assigned to Gene Ontology (GO) and Clusters of Orthologous Group (COG), suggesting that the assembled unigenes represented a wide diversity of transcripts in the rubber tree genome. Among three GO categories, cell and binding activity were the most abundant classes in cellular component and molecular function, respectively as was reported by Xia *et al.* (2011). Triwitayakorn *et al.* (2011) also indicated that the major category fell into binding activity among molecular function terms. As for biological process, metabolic process was the largest group. Post translational modification, protein turnover, chaperones and transcription were also in the top categories. When 22,756 unigenes were searched against the KEGG database, 12,097 unigenes were assigned to five main categories including 123 KEGG pathways. Among the main KEGG categories, metabolism was the biggest category (9,043; 74.75 per cent), suggesting the active metabolic processes in rubber tree bark. Further, spliceosome, plant pathogen interaction, biosynthesis of plant hormones, biosynthesis of phenylpropanoids and ribosome were other well represented pathways.

Tapping panel dryness (TPD) involves the partial or complete cessation of latex flow thus seriously affecting latex production in rubber tree (Figure 29.3). The annual rubber production loss from TPD accounts for 15-20 per cent with an incidence of 12-50 per cent (Chen *et al.*, 2002) and no effective treatment has been developed for it (Li *et al.*, 2010). Numerous studies have been conducted to define the origin and nature of TPD, but still the molecular nature and mechanism of TPD occurrence remains unknown.

Figure 29.3: A Healthy Rubber Tree with Normal Latex Flow (A) and a Tapping Panel Dryness (TPD) Affected Rubber Tree in which the Tapping Panel has Completely Dried (B).

There have been prolific attempts to understand the nature and molecular mechanisms of TPD. It was initially hypothesized that TPD might be caused by pathogens (Zheng and Chen, 1982; Soyza, 1983), but no further evidence confirmed it (Li, 1982; Wang, 1988; Nandris *et al.*, 1991 a and b). Many researchers proposed that TPD is a physiological anomaly resulting from wounding stress and ethylene over stimulation (Fan and Yang, 1984; Chrestin, 1989; de Fay and Jacob, 1989; Jacob *et al.*, 1994; Faridah *et al.*, 1996). During the development of TPD, many physiological and biochemical changes were observed to occur including decrease of sucrose and dry matter and increase of inorganic phosphorus in the latex (Tupy and Primot, 1976; Pakianathan *et al.*, 1982; Sivakumaran *et al.*, 1984), increase in activities of RNase and proteinase (Tupy, 1969; Fan and Yang, 1995; Zeng, 1997) and decrease in the content of protein, nucleic acid, thiols, ascorbic acid, the levels of variable peroxidase and superoxide dismutase isozyme (SOD) (Xi and Xiao, 1988; Fan and Yang, 1994), and the impaired rubber synthesis (Krishnakumar *et al.*, 2001). Chrestin *et al.* (1984) and Chrestin (1989) put forward that uncompensated oxidative stress within the latex cells might be associated with membrane destabilization that ultimately resulted in bursting of the lutoids and consecutive *in situ* latex coagulation. Several protein markers linked to TPD have also been reported (Darussamin *et al.*, 1995; Dian *et al.*, 1995; Lacrotte *et al.*, 1995; Sookmark *et al.*, 2002), but their functions in TPD onset are unclear.

Biosynthesis of natural rubber is from isopentenyl pyrophosphate (IPP) units, which is formed by the mevalonate (MVA) pathway and the 2C-methyl-d-erythritol-4-phosphate (MEP) pathway (Gronover *et al.*, 2011; Cornish, 2001b). The production of the precursors of rubber biosynthesis through the MVA pathway and the MEP pathway is precisely regulated (Okada, 2011). Genes of these two pathways were cloned and characterized (Sando *et al.*, 2008a; 2008b). HMGR, the key regulatory step for IPP biosynthesis through the MVA pathway, has been extensively studied (Brown *et al.*, 1980; Chang *et al.*, 2013).

Chen *et al.* (2002) cloned and characterized a transcription factor *HbMyb1* with mRNA differential display reverse transcriptase polymerase chain reaction (DDRT-PCR), implying that it may play a role in apoptosis. Venkatachalam *et al.* (2009), using DDRT-PCR, identified translocase of the Outer Mitochondrial Membrane (*HbTOM20*) and suggested that it may be involved in alteration of mitochondrial metabolism. Venkatachalam *et al.* (2010) identified a thymidine kinase gene (*HbTK*) related to TPD by random amplified polymorphic DNA screening. Further, Venkatachalam *et al.* (2007) and Li *et al.* (2010) studied the expression patterns of the TPD responsive genes using suppression subtractive hybridization (SSH) method and found that genes associated with stress/defense response preferentially expressed in the fresh latex samples from rubber tree with the onset of TPD syndrome. Qin *et al.* (2012) identified the genes associated with TPD by oligonucleotide microarrays hybridized with the latex from TPD and healthy rubber trees. Although these approaches are helpful to elucidate the onset of TPD, they still leave many gaps in the knowledge and understanding of the nature and mechanism of TPD.

Mantello *et al.* (2014) performed RNA-Seq of rubber bark on the Illumina GA IIx platform. A total of 50,384 contigs, that were over 400 bp in size, were obtained and subjected to further analyses. A similarity search against the non-redundant protein database returned 32,018 (63 per cent) positive BLASTx hits. The transcriptome analysis was annotated using COG, GO, KEGG and Pfam databases. A search for putative molecular marker identified 17,927 SSRs and 404,114 SNPs. A total of 78 SNPs belonging to the mevalonate (MVA) and 2-C-methyl-D-erythritol 4-phosphate (MEP) pathways, which are involved in rubber biosynthesis, were validated in 36 genotypes of *H. brasiliensis*. They identified that this new dataset represented a powerful information source for rubber tree bark genes and therefore an important tool for the development of microsatellites and SNP markers for use in future genetic analyses such as genetic linkage mapping, quantitative trait loci identification, investigations of linkage disequilibrium and marker-assisted selection.

Liu *et al.* (2015) performed a comparative analysis of the transcriptomes of healthy (H) and TPD-affected (T) rubber trees using Illumina to identify genes and pathways related to TPD. *De novo* assemblies of sequence reads yielded 141,456 and 169,285 contigs, and 96,070 and 112,243 unigenes from H and T libraries, respectively. Among 73597 genes, 22577 genes were identified as differentially expressed genes between H and T library *via* comparative transcript profiling. A majority of genes, involved in natural rubber biosynthesis and jasmonate synthesis with most potential relevance in TPD occurrence, were found to be differentially expressed (Figures 29.4 and 29.5). In TPD-affected trees, the expression of most genes related to the latex biosynthesis and jasmonate synthesis was severely inhibited suggesting as a possible direct cause of TPD. Liu *et al.* (2015) found that the expression of most genes related to latex biosynthesis was severely inhibited and was probably the direct cause of the TPD. It was suspected that latex production dramatically decreased or even completely ceased in TPD affected trees due to inadequate supply or depletion of their immediate substrates (IPPs and FPPs). The expression of the majority of genes involved in JA biosynthesis including 13-lipoxygenase (*LOX*), allene oxide cyclase (*AOC*) and 12-oxophytodienoic acid reductase (*OPR*) were found to be down-regulated. Suppression of *OPR* was the greatest and it suggested that the *OPR* might be a key enzyme for JA biosynthesis pathway. *HMGR1* in the MVA pathway was suppressed. The mevalonate generated from HMG-CoA by HMGR was further

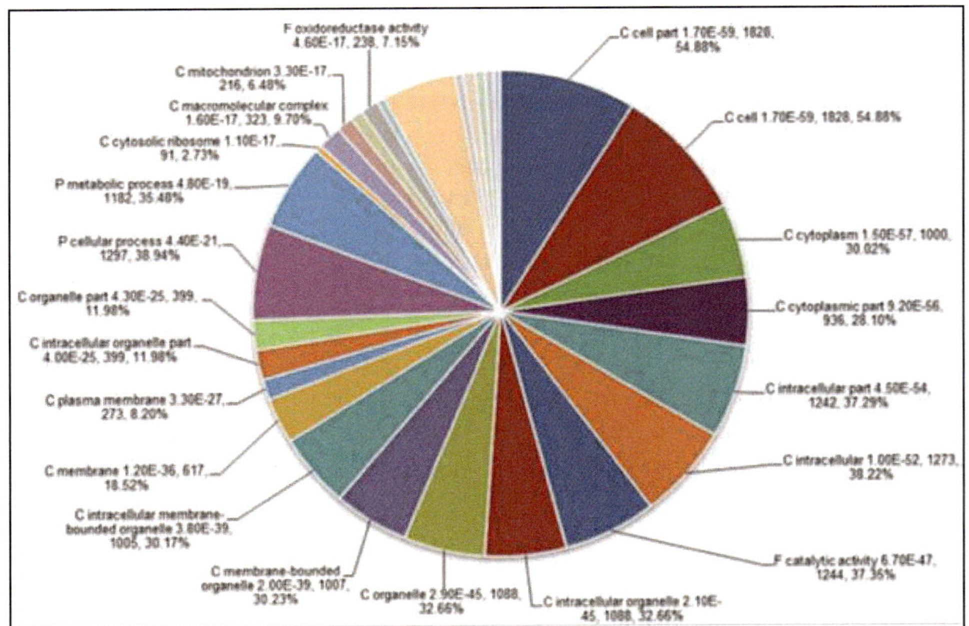

Figure 29.4: Gene Ontology Annotation of Down-Regulated Genes in the TPD-Affected (T) Bark Compared to the Healthy (H) Bark of Rubber Tree. P: The biological process; C: The cellular component; F: The molecular function. The P value, the number of genes and the percentage of genes in the corresponding categories are given after the name of subcategories (Figure adopted from Liu *et al.*, 2015)

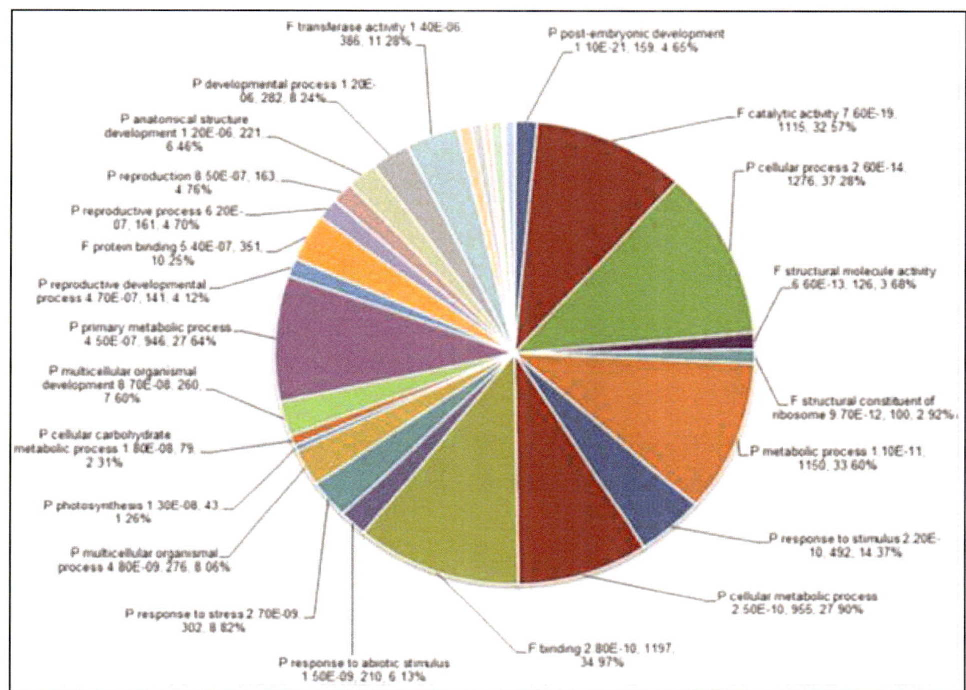

Figure 29.5: Gene Ontology Annotation of Up-regulated Genes in the TPD-affected (T) Bark Compared to the Healthy (H) Bark of Rubber Tree. P: The biological process; C: The cellular component; F: The molecular function. The P value, the number of genes and the percentage of genes in the corresponding categories are given after the name of subcategories (Figure adopted from Liu *et al.*, 2015)

converted to isoprenoid compounds and then to natural rubber. Although there are genes (*HMGR1*, *HMGR2* and *HMGR3*) encoding HMGR in *H. brasiliensis*, but only HMGR1 was likely to be involved in rubber biosynthesis (Suwanmanee *et al.*, 2013). The facts that the expression of rate-limiting *HMGR1* in the MVA pathway was regulated by methyl jasmonate (Burnett *et al.*, 1993; Mehrjerdi *et al.*, 2013; Choi *et al.*, 1994) suggested that the suppression of latex production is possibly achieved by decrease in JA biosynthesis. Moreover, linolenic acid (a precursor of JA biosynthesis) was shown to induce the laticifer differentiation of *H. brasiliensis* (Hao and Wu, 2000; Shi *et al.*, 2012). In some cases, JA induced production of ROS and alterations in mitochondrial dynamics and was found to be involved in local PCD (Zhang and Xing, 2008; Vankova, 2010). Therefore, they presumed that JA may play a central role in latex production as well as in the occurrence and maintenance of TPD.

Li *et al.* (2016) recently compared the transcriptomes of bark between healthy and TPD affected trees to identify TPD-related genes. In total 57,760 assembled genes were obtained and analyzed in detail. In contrast to healthy rubber trees, 5652 and 2485 genes were up- and down-regulated, respectively, in TPD affected trees. Of 115 KEGG pathways associated with the TPD-related genes, metabolic pathway was the most dominant, followed by plant-pathogen interaction, biosynthesis of secondary metabolites, spliceosome, endocytosis, ubiquitin mediated proteolysis, starch and sucrose metabolism, and RNA degradation. These predominant eight KEGG pathways were mainly associated with metabolism, PCD, protein degradation and RNA processing. In addition, the TPD-related genes were significantly enriched in five KEGG pathways including plant-pathogen interaction, ABC transporters, monoterpenoid biosynthesis, base excision repair and caffeine metabolism. Importantly, the TPD-related genes enriched in five KEGG pathways above were also associated with the GO terms such as metabolic process, immune system process, antioxidant activity, response to stimulus, death, *etc.* Studies confirmed that rubber tree TPD is a complex process involving many genes. The observed lower rubber yield from TPD trees might be a result of lower isopentenyl diphosphate (IPP) available for rubber biosynthesis and from downregulation of the genes in post-IPP steps of rubber biosynthesis pathway.

4. Transcriptome Responses of *Hevea brasiliensis* on Interaction with *Microcyclus ulei* Causing South American Leaf Blight (SALB) Disease

SALB, caused by the ascomycete fungus *Microcyclus ulei,* is a major factor limiting cultivation of rubber in South and Central America (Figure 29.6). In the early 20th century, epidemics of SALB led to failure of rubber cultivation in tropical America (Grandin, 2009) and as a result of its potential serious economic consequence, stringent quarantine measures, especially for importation of budwood

Figure 29.6: South American Leaf Blight (SALB) Caused by *Microcyclus ulei* Infected Rubber Plantation (A) and SALB Disease Symptom Observed on Rubber Leaf (B).

material, were adopted to exclude SALB from Asia (FAO, 2007; Lieberei, 2007). However, the continent accounts for only 3 per cent of the global production compared with Southeast Asian countries (IRSG, 2014). Garcia *et al.* (2011) studied the transcriptome of the MDF 180 clone, which is partially resistant to SALB, both in the field and in controlled conditions and compared with the clone PB 314, which is considered highly susceptible to all *M. ulei* isolates. The study was conducted by constructing subtractive libraries (SSH) at three different time points (6-72 hpi, 4-28 dpi and 34-58 dpi) associated with the asexual and sexual development of the fungus after inoculation. Differentially expressed genes varied at different times. The identified ESTs had an average size of 346 bp. The differentially expressed genes involved in the defense response of MDF 180 to *M. ulei* included pathogenesis-related proteins, R genes, proteins involved in the detoxification of reactive oxygen species and phenol metabolism.

Paez *et al.* (2015) carried out studies on a hybrid clone (FX3864) that was resistant to *M. ulei*. This clone FX 3864 was obtained by crossing between clones PB 86 and B 38 (*H. brasiliensis* x *H. brasiliensis*). RNA-Seq technology was used to analyze the differential gene expression of FX 3864 clone at 0 and 48 h post infection with the *M. ulei* isolate GCL012. They identified 86 differentially expressed genes associated with the defense response of FX 3864 to GCL012. Seven putative gene members of the AP2/ERF ethylene (ET)-dependent superfamily were found to be down-regulated. An increase in salicylic acid (SA) was associated with the up-regulation of three genes involved in cell wall synthesis and remodeling, as well as in the down-regulation of the putative gene *CPR5*. They reported that the defense response of FX 3864 against the GCL012 isolate was associated with the antagonistic SA, ET and jasmonic acid (JA) pathways. These responses are characteristic of plant resistance to biotrophic pathogens. The ontological terms found in the differential expression analyses were mainly associated with signal transduction, responses to hormonal stimuli and responses to biotic stress. The results suggest that modulation of plant transcriptional responses was related to the stress caused by the interaction with *M. ulei* within the first 48 hpi. The down-regulation of ET-sensitive genes from the putative superfamily AP2/EFR allowed inferring a decrease in ET and JA within the first 48 hpi of the *H. brasiliensis - M. ulei* interaction. Additionally, the up-regulation of genes associated with cell wall metabolism, such as Hb42704 with homology to a polygalacturonase, and the down-regulation of Hb5539, which is a homologue of CPR-5, suggested an increase in SA within the first 48 hpi of interaction between the resistant clone FX 3864 and the *M. ulei* GCL012 isolate.

5. Genes Conferring Host Tolerance to *Corynespora cassiicola* Causing Corynespora Leaf Disease in Rubber

One of the most economically significant diseases being a major threat to NR production in Asia is Corynespora leaf disease caused by *Corynespora cassiicola*. The pathogen infects young refoliating leaves (Figure 29.7) leading to defoliation thereby extending the immaturity period of rubber trees. At Rubber Research Institute of India, we have been working on deciphering the molecular basis of *Corynespora* disease tolerance to understand genes/factors controlling host tolerance in rubber

Figure 29.7: An Infected Rubber Plantation showing Immature Leaves with Typical Disease Symptoms Caused by *Corynespora cassiicola* on Rubber (A); Leaf Spots Surrounded by Yellow Halo (B); Circular Concentric Spots (C); Infection on Vein Causing Typical Railway Track Symptom (D).

trees (Roy *et al.*, 2008; 2009; 2012; 2013; Saha *et al.*, 2010). Plant resistance to diseases is associated with a number of defense responses, activated by the host after contact with the pathogens. Identification of host genes involved in defense response, is important to understand plant resistance mechanisms against phytopathogens.

In an effort to identify the disease responsive genes in response to *C. cassiicola* infection and clone genes of interest *via* candidate gene approach, we adopted differential display reverse transcriptase polymerase chain reaction (DD-RT-PCR) technique (Roy *et al.*, 2008; 2009; 2012; 2013). For this, the pathogen was isolated and conidial suspension was sprayed on two healthy rubber clones: RRII 105 and GT 1, a *Corynespora* susceptible and tolerant clone respectively. Leaf samples were collected after 6, 12 and 24 hours following challenge inoculation for RNA isolation. Differential display of cDNA was performed using RNAimage Kit (GenHunter Corporation). Changes in mRNA transcripts between challenged and control samples were recorded for each of 24 primer combinations. The DD-RT-PCR technique resulted in equal detection of most cDNA bands in RNA samples from both control and challenged plants (Figure 29.8). Comparison of band intensity on the autoradiogram revealed that some of the bands were of lesser intensity in challenged samples as compared to their levels in the control samples. This indicated that pathogen infection suppressed mRNA synthesis quantitatively reducing overall protein synthesis. Similarly, infection also activated several new transcripts, as revealed by the production of new cDNA bands in the infected samples with light to medium intensity, which were absent in the uninoculated samples. Some bands, however, were unique to either challenged or control samples. Based on the comparative intensity, the cDNA bands expressed following pathogen infection were grouped into three classes: induced (up-regulated), suppressed (down-regulated), and newly expressed (activated following pathogen inoculation). Since our interest was only to identify the genes over-expressed/induced during the symptom development due to pathogen infection, we selected 170 distinctly up-regulated bands only from the challenged sample, cloned and sequenced. Out

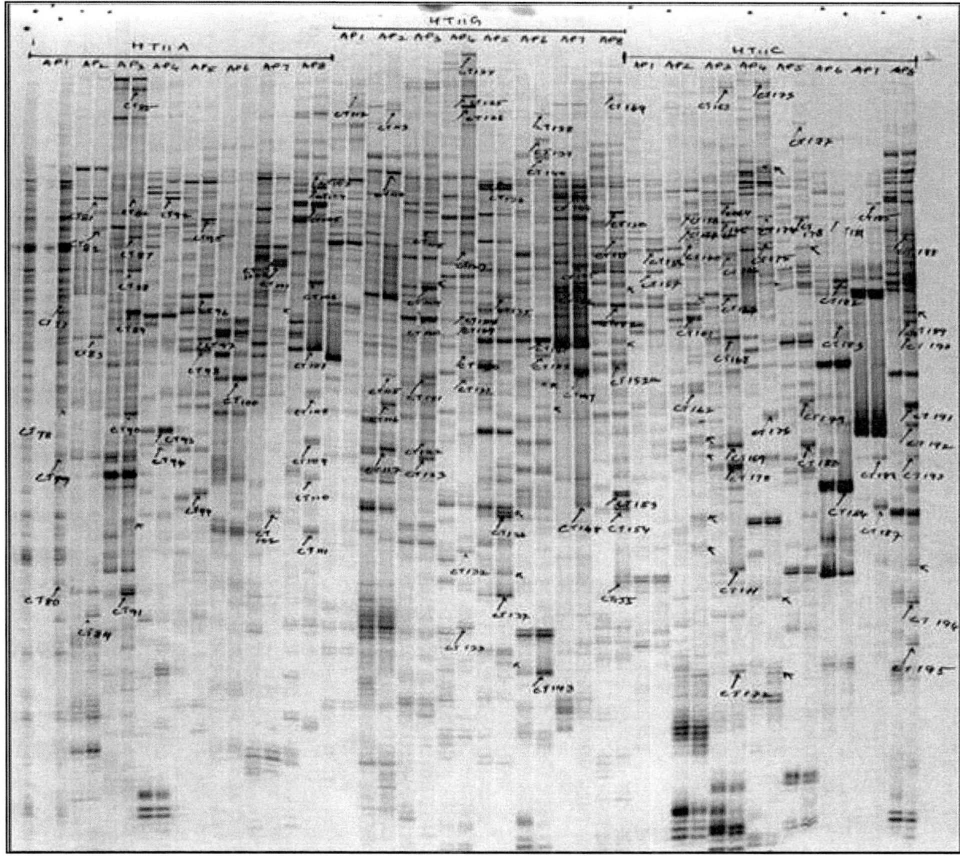

Figure 29.8: Representative Electrophorogram Showing Differentially Expressed Genes in Response to *Corynespora cassiicola* Infection in Two Clones of Rubber: RRII 105 (Highly susceptible) and GT 1 (Moderately resistant).

of those, only sixteen clones showed homology with existing GenBank sequences. Among them two of the clones, DDCT7 and DDCT12, were most significant in terms of disease resistance/tolerance, which showed homology with GRAS transcription factor (E value: 1^{e-15}) and anthocyanidin 3-O-glucosyl transferase (E-value: 4^{e-48}) respectively. The importance of these two genes in plant disease resistance has been well established in other plant species. Besides these two genes, regulation of cellulose synthase catalytic subunit, NADH dehydrogenase and ABL Interactor like protein 2 were also identified during *Corynespora* infection. All the cDNA clones were subjected to 'Reverse Northern' analysis followed by Northern analysis to detect transcript accumulation in rubber leaf tissue during *Corynespora* infection. A few cDNA clones were found to be over-expressed compared to control within 24 h of infection in tolerant clone GT1. Among the over-expressed clones, some clones showed almost similar response (up-regulation) in susceptible clone RRII 105 while a few others showed opposite trend in expression level between susceptible and tolerant genotypes. These clones down were regulated in RRII 105 and showed considerable levels of up regulation in GT1 at 24 h of infection. Some of the over-expressed clones at 24 h of infection in GT1 showed reduced levels of expression at 48 h. Although DD-RT-PCR is a laborious and time consuming process, it helped to identify a few genes involved in Corynespora disease resistance.

6. RNA-Seq for Identification of Genes/Transcripts in Response to *Corynespora cassiicola* Infection in Rubber

In continuation with the DD-RT-PCR technique adopted to discover genes involved in host tolerance during Corynespora leaf disease development, we undertook a Next Generation Sequencing (NGS) based approach to sequence the transcriptome of rubber from a susceptible (RRII 105) and resistant clone (GT 1) in both healthy and pathogen challenged condition. RNA isolated from the control plants of RRII 105 and GT 1 were designated as C1 and C2 respectively. Equal quantity of RNA from the three treatments (6, 12 and 24 h following challenge inoculation with *C. cassiicola*) for each clone were pooled together to ensure that all the genes expressed at different time points in response to pathogen infection would be represented in a single transcriptome. Thus, the treated pooled RNA samples were named as T1 and T2 for RRII 105 and GT 1 respectively. Accordingly, four samples were processed for library preparation according to the Illumina TruSeq RNA library protocol and sequencing was performed on Illumina HiSeq 2000.

The number of transcripts generated was 89017 for C1, 83395 for C2, 122507 for T1 and 126702 for T2 with a mean transcript length of 1373 bases (Figure 29.9). Analysis of gene expression through transcriptome profiling of RRII 105 and GT 1 in response to *Corynespora* infection revealed that they responded differently upon infection. A comparison of gene expression between C1 and T1 identified genes specifically triggered only in susceptible clone (RRII 105) in response to pathogen infection. Similarly, expression study between C2 and T2 revealed the genes conferring tolerance in resistant clone (GT 1). Genes showing differential regulation between the susceptible and resistant clone upon challenge inoculation was obtained by comparing T1 and T2, which was a reflection of the clonal response to *C. cassiicola*.

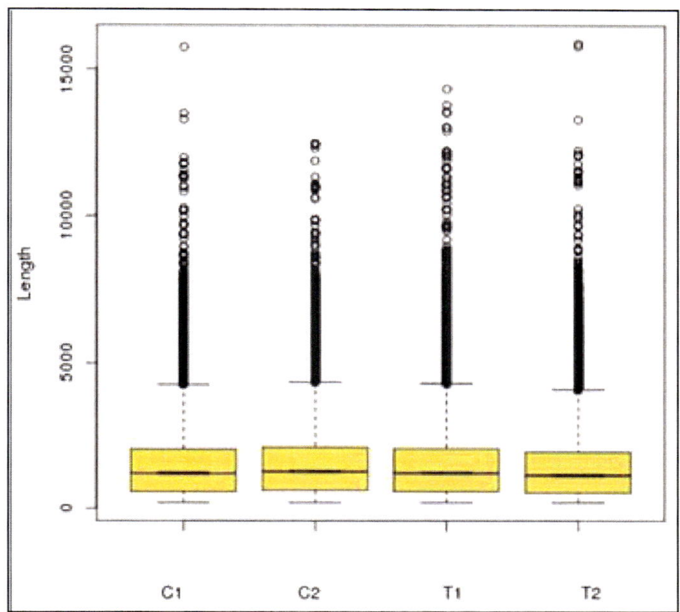

Figure 29.9: Box Plot Showing Length Distribution of Assembled Transcripts.

The box and whiskers show that the transcript lengths are uniformly distributed in control (C1, C2) and treated (T1, T2) assemblies. Data points in C1 and T2 indicate that the maximum transcript size is over 15 Kb.

Sharing of differentially regulated transcripts between the control and treated samples (C1C2, T1T2, C1T1 and C2T2) for both susceptible and resistant clones are depicted in Venn diagrams (Figure 29.10). It was observed that in the resistant clone GT1 there was up regulation of 15809 and down regulation of 15187 transcripts in

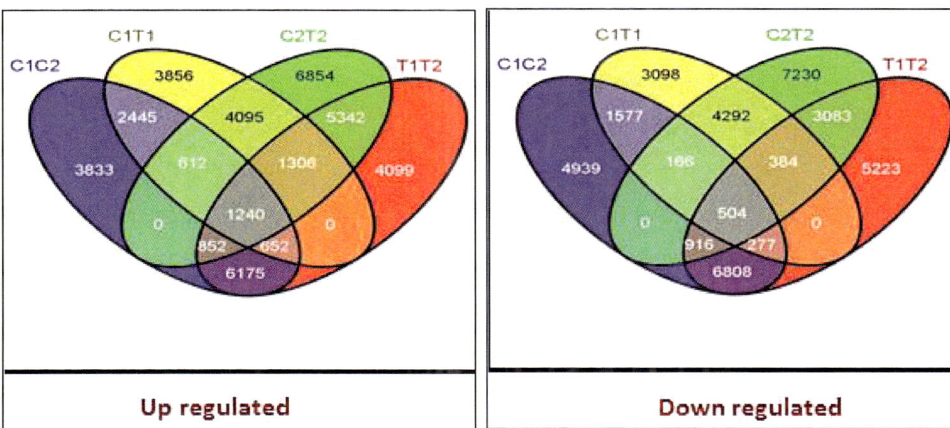

Figure 29.10: Sharing of differentially Expressed Transcripts in RRII 105 (Susceptible) and GT 1 (Tolerant) Clones in Control (C1 and C2) and following Challenge Inoculation with *C. cassiicola* (T1 and T2).

the control (unchallenged) condition. However, in the challenged condition, there were 19666 and 17195 transcripts up regulated and down regulated respectively in the resistant clone GT1. The transcripts with least expression values could be novel genes of interest with very low copy numbers.

The annotated genes were functionally classified according to the gene ontology. In control samples (C1 vs. C2), it was observed that in C2 a significant number of transcripts were grouped under defense response followed by proteolysis, signal transduction and carbohydrate metabolic process, which were annotated under biological process. Most represented transcripts involved in ATP and ADP binding activities were assigned under molecular functions. Similarly, significant number of transcripts related to membrane proteins was grouped under cellular components in C2. Differential gene expression observed between resistant and susceptible clone in the absence of pathogen indicated that resistance mechanism had already

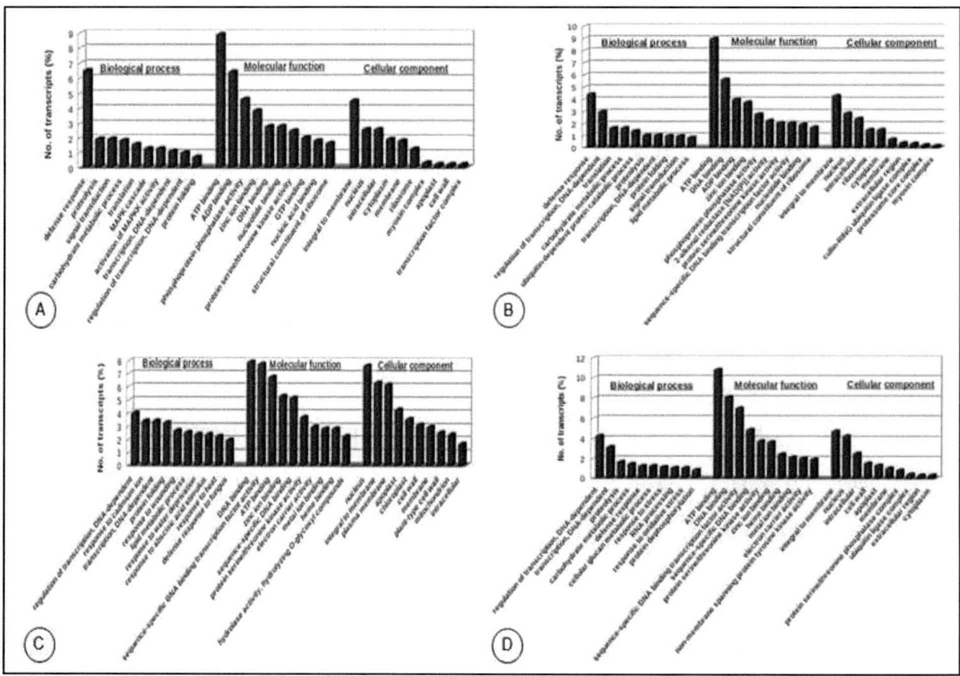

Figure 29.11: Comparison of the Functionally Classified (Biological process, molecular function and cellular component) Transcripts/Genes Belonging to RRII 105 (Susceptible) and GT 1 (Tolerant) Clones in Control (C1 and C2) and following Challenge Inoculation with *C. cassiicola* (T1 and T2).

Bar diagram demonstrate comparisons between the two clones in control condition (C1 Vs. C2) (A); in pathogen challenged condition (T1 Vs. T2) (B); control Vs. challenged samples from RRII 105 (C1 Vs. T1) (C) and control Vs. challenged samples from GT 1 (C2 Vs. T2) (D). Defence response genes are over expressed in GT 1 in both control (C2) and treated (T2) conditions compared to RRII 105 (T1 and T2).

been built-up in the resistant clone GT 1 as a priming mechanism. Therefore, GT1 appears inherently tolerant to *C. cassiicola* compared to RRII 105 (Figure 29.11).

Upon pathogen infection in GT1 (T2) the genes involved in defense response and transcription regulation were most represented under biological process. Similarly ATP binding and DNA binding activities were represented the most under molecular functions. Nucleotide-binding site (NBS), Leucine-rich repeat (LRR) and a putative amino-terminal signaling domain is an integral part of plant disease resistance (R) proteins, commonly known as NBS-LRR (Belkhadir *et al.*, 2004). In GT1, 19 unique NBS-LRR transcripts were found to over express following challenge inoculation, which could be grouped under TIR-NBS-LRR and CC-NBS-LRR subfamilies of disease resistance genes prevailing in plant species having active role in pathogen detection. Conformational changes in the amino-terminal and LRR domains of plant NBS-LRR proteins due to interaction with either a modified host protein or a pathogen protein, might promote the exchange of ADP for ATP by the NBS domain, which could activate downstream signaling, leading to pathogen resistance (DeYoung and Innes, 2006). This could be a reason that large number of ATP binding proteins was up regulated in resistant clone GT1 upon infection (T2). Further, transcripts encoding transcription factors were identified through GO annotations falling under the same category. Most of the transcripts encoding membrane protein were categorized under cellular components. Cullin ring ubiquitin ligase complex was uniquely observed in resistant clone GT1 upon infection. This complex has been reported as key mediators of post-translational protein regulation (Petroski and Deshaies, 2005) involved in disease resistance caused by microorganisms (Gray, 2002). A comparison of the differential gene expression data between C1 and T1 revealed upregulation of DNA binding and transcription factors (TF) (WRKY, NAC, GATA and other putative TF proteins), zinc finger protein and putative uncharacterized proteins. It is presumed that these TFs being regulatory proteins, triggered during disease development in T1, play a crucial role in resistance mechanism in the susceptible clone RRII 105. It has been reported that adaptability of plants to various stressful environmental conditions is achieved by reprogramming their transcriptome in a dynamic and temporal manner, through enforcement of a network of various transcription factors (Pandey and Somssich, 2009).

A comparison of the gene expression pattern in the tolerant clone GT1 revealed many of the major transcription factor genes (WRKY, NAC, GATA) were triggered and up regulated also in resistant clone GT1 during disease development, which could be associated with enhanced immunity against the pathogen. Comparison of expression pattern between T1 and T2 revealed the highly enriched gene ontology terms which commonly fell into (i) defense response, (ii) response to stress and (iii) response to stimulus. Besides these, other GO categories containing cellular catabolic process, RNA biosynthetic process, carbohydrate metabolism *etc.* were enriched only in T2. Disease resistant proteins and leucine rich repeat containing protein are generally classified under defense response in plants. Significantly enriched gene ontology terms specifically in T2 were grouped into the major categories (i) heterocycle catabolic process (ii) carbohydrate metabolic process and (iii) RNA biosynthetic process.

Some of the differentially induced genes in T2 encoded pathogenesis-related (PR) proteins including chitinases and glucanases that are capable of degrading cell wall components of microbial pathogens. The differential expression of this enzyme is highly enhanced by fungi, bacteria, viruses and other biotic and abiotic stresses (Punja and Zhang 1993). The plant susceptibility towards pathogen is highly reduced by chitinase in combination with several other pathogen resistance proteins (Punja and Zhang 1993), especially when the pathogen has chitinaceous cell wall, as in the case of *C. cassiicola.*

The study of disease transcriptome of rubber in response to *C. cassiicola* infection identified molecular signatures providing vital clues related to disease resistance in rubber. Transcriptome sequencing in rubber using NGS platform was performed by several researchers for identification of tissue-specific genes involved in various metabolic pathways, development of SSR and SNP markers for linkage map construction *etc.* However, very few reports are available on transcriptome sequencing of rubber in response to both biotic and abiotic stresses.

Traditional rubber tree improvement is a lengthy and difficult process. The availability of the transcriptome will hopefully enable the development of a large database of molecular markers that can be used to identify genes involved in disease tolerance thereby providing scope for improvement of the rubber tree with disease resistance. This study is a significant step towards development of genomic resources for rubber and will accelerate functional genomic studies thereby facilitating marker-assisted selection breeding in rubber tree.

7. Transcriptome Sequencing for Development of Microsatellite Markers and Construction of Genetic Linkage Map in Rubber

Molecular markers have been used in several studies to assess genetic diversity among rubber tree genotypes (Roy *et al.,* 2004; Saha *et al.,* 2005). Microsatellites or simple sequence repeats (SSRs), combine codominance, high polymorphism, good genome coverage and Mendelian inheritance. They have also been capable of discriminating closely related individuals. Successful use and application of SSRs in rubber tree breeding (Feng *et al.,* 2009; Le Guen *et al.,* 2009, 2011; Gouvêa *et al.,* 2010) have also been reported. The presence of SSRs in the transcripts of genes indicates that they might have a role in gene expression or function; however, it remains to be seen whether any unusual phenotypic variation might be associated with the length of SSRs in coding regions (Varshney *et al.,* 2005). This fact, together with the increasing availability of sequences in databases, mainly expressed sequence tags (ESTs), has enabled the development of functional microsatellite markers or EST-SSRs and the use of these markers has been reported in several species, including *Hevea* sp. (Feng *et al.,* 2009). The fact that these markers are derived from genes and that, consequently, their sequences are more conserved, make it extremely robust to identify polymorphisms in closely related species (Varshney *et al.,* 2005). Functional EST-SSR microsatellites are efficient in evaluating the genetic diversity among clones, are used to translate the genetic differences among cultivars and to fingerprint closely related materials. Perseguini *et al.* (2012) studied the genetic diversity of cultivated accessions and wild species of rubber tree using EST-SSR markers with an objective to evaluate the efficiency of EST-SSR markers in the assessment of genetic diversity of rubber tree genotypes and to verify the transferability of these markers for wild species of *Hevea*.

Triwitayakorn *et al.* (2011) performed transcriptome sequencing from the vegetative shoot apical tissue, which is a highly dynamic structure, to discover genes, expand the EST database and develop EST-SSR markers that can be used for assessing genetic diversity, constructing linkage maps and identifying traits of commercial interest. The transcriptome sequencing yielded 2311497 reads. Clustering and assembly of the reads produced a total of 113313 unique sequences, comprising 28387 isotigs and 84926 singletons. A total of 17819 EST-SSRs were identified from the data set. To demonstrate the use of these EST resources for marker development, primers were designed for 430 of the EST-SSRs. Three hundred and twenty-three primer pairs were amplifiable in *H. brasiliensis* clones. Polymorphic information content values of selected 47 SSRs among 20 *H. brasiliensis* clones ranged from 0.13 to 0.71, with an average of 0.51. A dendrogram of genetic similarities between the 20 *H. brasiliensis* clones using these 47 EST-SSRs suggested two distinct groups that correlated well with clone pedigree. These novel EST-SSRs, together with the published SSRs, were used for the construction of an integrated parental linkage map of *H. brasiliensis* based on 81 lines of an F1 mapping population. The map consisted of 97 loci, consisting of 37 novel EST-SSRs and 60 published SSRs, distributed on 23 linkage groups and covered 842.9 cM with a mean interval of 11.9 cM and ~"4 loci per linkage group. Although the numbers of linkage groups exceeded the haploid number (18) of *H. brasiliensis*, several common markers between homologous linkage groups with the previous map indicated that the map was appropriate for further study in marker-assisted selection.

8. Conclusion

Transcriptome sequencing in rubber was initiated only during the end of the last decade resulting in generation of enormous data for both tissue specific and stress related (both biotic and abiotic) genes/transcripts. These genomic resources need to be utilized effectively for developing clones suitable for cultivation in non-traditional (cold/drought-prone) areas in India. Crop improvement in traditional rubber growing regions requires greater attention towards increasing latex production, which is a very complex trait to understand as there is involvement of several genes both from bark as well as latex. At Rubber Research Institute of India, we are making an effort to identify key genes involved in latex biosynthesis and also the genes influencing latex yield through regulation of gene expression. This is being done by comparing the transcriptome of different high yielding as well as low yielding clones and also wild *Hevea* accessions. We also been working on deciphering the molecular basis of disease tolerance using RNA-Seq technique to understand genes/factors controlling host tolerance in rubber trees. Enormous genomic data has been generated through transcriptome sequencing in rubber clones in response to biotic and abiotic stresses. The biological information generated has to be extracted properly from this data set. There are a large number of genes annotated where the gene designations are known but even more number of transcripts/ genes for which the gene designations are unannotated, which also needs to be studied in detail. The transcriptome data from all these studies on rubber provide a significant resource for discovery of genes related to various biological processes, thereby helping in developing rubber clones with greater yielding potential as well as possessing desirable secondary traits.

References

Ansorge, W.J. (2009). Next-generation DNA sequencing techniques. *Nature Biotechnology*, **25(4)**: 195-203.

Brown, M.S. and Goldstein, J.L. (1980). Multivalent feedback regulation of HMG CoA reductase, a control mechanism coordinating isoprenoid synthesis and cell growth. *Journal of Lipid Research*, **21**: 505-517.

Burnett, R.J., Maldonado-Mendoza, *I.E.*, McKnight, T.D. and Nessler, C.L. (1993). Expression of a 3-hydroxy-3-methylglutaryl coenzyme a reductase gene from *Camptotheca acuminate*: Is differentially regulated by wounding and methyl Jasmonate. *Plant Physiology*, **103**: 41-8.

Belkhadir, Y., Subramaniam, R. and Dangl, J.L. (2004). Plant disease resistance protein signaling: NBS-LRR proteins and their partners. *Current Opinion in Plant Biology*, **7**(4): 391-399.

Cataldo, F. (2000). Guayule rubber: A new possible world scenario for the production of natural rubber. *Progress in Rubber Plastics and Recycling Technology*, **16**: 31-59.

Chang, W.C., Song, H., Liu, H.W. and Liu, PH. (2013). Current development in isoprenoid precursor biosynthesis and regulation. *Current Opinion in Chemical Biology*, **17(4):** 571-579.

Chao, J., Chen, Y., Wu, S. and Tian, W.M. (2015). Comparative transcriptome analysis of latex from rubber tree clone CATAS8-79 and PR107. *Genomics Data*, **5**: 120-121.

Chen, S.C., Peng, S.Q., Huang, G.X., Wu, K., Fu, X. and Chen, Z. (2002). Association of decreased expression of a Myb transcription factor with the TPD (tapping panel dryness) syndrome in *Hevea brasiliensis*. *Plant Molecular Biology*, **51**: 51-58.

Choi, D., Bostock, R.M., Avdiushko, S. and Hildebrand, D.F. (1994). Lipid-derived signals that discriminate wound- and pathogen-responsive isoprenoid pathways in plants: methyl jasmonate and the fungal elicitor arachidonic acid induce different 3-hydroxy-3-methylglutaryl-coenzyme A reductase genes and antimicrobial isoprenoids in *Solanum tuberosum* L. *Proceedings of the National Academy of Sciences*, USA, **91(6)**: 2329-2333.

Chow, K.S., Wan, K.L., Mohd. Isa, N.M., Bahari, A., Tan, S.H., Harikrishna, K. and Yeang, H.Y. (2007). Insights into rubber biosynthesis from transcriptome analysis of *Hevea brasiliensis* latex. *Journal of Experimental Botany*, **58(10)**: 2429-2440.

Chrestin, H., Bangratz, J., d'Auzac, J. and Jacob, J.L. (1984). Role of the lutoidic tonoplast in the senescence and degeneration of the laticifers of *Hevea brasiliensis*. *Zeitschrift für Pflanzenphysiologie* **114**: 261.

Chrestin, H. (1989). Biochemical aspects of bark dryness induced by overstimulation of rubber trees with ethrel. In: D'Auzac, J., Jacob, J.L., Chrestin, H. (Eds.), *Physiology of Rubber Tree Latex*. CRC Press, Boca Raton, Florida, USA. pp. 431-442.

Cornish, K. (2001a). Biochemistry of natural rubber, a vital raw material, emphasizing biosynthetic rate, molecular weight and compartmentalization, in evolutionarily divergent plant species. *Nature Products Reports*, **18**: 182-189.

Cornish, K. (2001b). Similarities and differences in rubber biochemistry among plant species. *Phytochemistry*, **57**: 1123-1134.

Darussamin, A., Suharyanto, S. and Chaidamsari, T. (1995). Change in the chemical composition and electrophoretic profile of latex and bark protein related to tapping panel dryness incidence in *Hevea brasiliensis*. *Menara Perkebunan*, **63**: 52-59.

de Fay, E. and Jacob, J.L. (1989). Symptomatological, histological and cytological aspects. In: D'Auzac J., Jacob J.L. and Chrestin H. (Eds.), *Physiology of Rubber Tree Latex*. CRC Press, *Boca Raton*, Florida, USA. pp. 407-428.

DeYoung, B.J. and Innes, R.W. (2006). Plant NBS-LRR proteins in pathogen sensing and host defense. *Nature Immunology*, **7**: 1243-1249.

Dian, K., Sangare, A. and Diopoh, J.K. (1995). Evidence for specific variations of protein pattern during tapping panel dryness condition development in *Hevea brasiliensis*. *Plant Science* **105**: 207-216.

Fan, S.W. and Yang, S.Q. (1984). Cause of disease and hypothesis on tapping panel dryness of *Hevea brasiliensis*. *Chinese Journal of Tropical Crops Research*, **18**: 43-48.

Fan, S.W. and Yang, S.Q. (1994).) Cause of disease and hypothesis on tapping panel dryness of *Hevea brasiliensis*. *Chinese Journal of Tropical Crops Research*, **18**: 43-48.

Fan, S.W. and Yang, S.Q. (1995). Tapping panel dryness induced by excessive tapping is a local senescence phenomenon. *Chinese Journal of Tropical Crops Research* **19**: 15-22.

FAO. (2007). *Pest risk analysis for South American leaf blight (SALB) of rubber (Hevea)*. http://www.fao.org/docrep/010/ai003e/AI003E25.htm. Accessed 13/11/15.

Faridah, Y., Siti Arija, M.A. and Ghandimathi, H. (1996). Changes in some physiological latex parameters in relation to over exploitation and onset of induced tapping panel dryness. *Journal of Natural Rubber Research*, **10**: 182-186.

Feng, S.P., Li, W.G., Huang, H.S., Wang, Y. and Wu, Y.T. (2009). Development, characterization and cross-species/genera transferability of EST-SSR markers for rubber tree (*Hevea brasiliensis*). *Molecular Breeding*, **23**: 85-97.

Garcia, D., Carels, N., Koop, D.M., de Sousa, L.A., Junior, A. and Renaud, P.V. (2011). EST profiling of resistant and susceptible *Hevea* infected by *Microcyclus ulei*. *Physiological and Molecular Plant Pathology*, **76(2)**: 126–36.

Gomez, J.B. and Moir, G.F.J. (1978).*The ultracytology of latex vessels in Hevea brasiliensis*. Monograph No. 4. *Malaysian Rubber Research and Development Board*, Kuala Lumpur.

Gouvêa, L.R.L., Rubiano, L.B., Chioratto, A.F., Zucchi, M.I. and Gonçalves, P.S. (2010). Genetic divergence of rubber tree estimated by multivariate techniques and microsatellite markers. *Genetics and Molecular Biology*, **33**(2): 308-318.

Grandin, G. (2009). Fordlandia: The Rise and Fall of Henry Ford's Forgotten Jungle City. *Metropolitan Books*, New York.

Gray, W.M. (2002). Plant Defence: A New Weapon in the arsenal. *Current Biology*, **12(10)**: 352-354.

Gronover, C.S., Wahler, D. and Prüfer, D. (2011). Natural rubber biosynthesis and physicochemical studies on plant derived latex. In: Elnashar, M. (Ed.), *Biotechnology of Biopolymers*. *Intech Open Access*, Croatia. pp. 75-88.

Hao, B.Z. and Wu, J.L. (2000). Laticifer differentiation in *Hevea brasiliensis*: Induction by exogenous jasmonic acid and linolenic acid. *Annals of Botany*, **85(1)**: 37-43.

IRSG. (2014). *International Rubber Statistical Group*. Rubber Stastistical Bulletin. http://www.rubberstudygroup.com/publications.aspx. Accessed 5 July 2014: 68.

Jacob, J.L, Prevot, J.C. and Laccrotte, R. (1994). Tapping panel dryness in *Hevea brasiliensis*. *Plantations Recherche Développement*, **2**: 15-21.

Kekwick, R.G.O. (1989). The formation of isoprenoids in *Hevea* latex. In: d'Auzac, J., Jacob, J.L. and Chrestin, L. (Eds.), *Physiology of Rubber Tree Latex*. CRC Press, Boca Raton, Florida. pp. 145-164.

Krishnakumar, R., Cornish, K. and Jacob, J. (2001). Rubber biosynthesis in tapping panel dryness affected *Hevea* trees. *Journal of Natural Rubber Research*, **4**: 131-139.

Lacrotte, R., Gidrol, X., Vichitcholchai, N., Pujade-Renaud, V., Narangajavana, J. and Chrestin, H. (1995). Protein markers of tapping panel dryness. *Plantations Recherche Développement*, **2**: 40-45.

Le Guen, V., Doare, F., Weber, C. and Seguin, M. (2009). Genetic structure of Amazonian populations of *Hevea brasiliensis* is shaped by hydrographical network and isolation by distance. *Tree Genetics and Genomes*, **5**: 673-683.

Le Guen, V., Gay, C., Xiong, T.C., Souza, L.M., Rodier-Goud, M. and Seguin, M. (2011). Development and characterization of 296 new polymorphic microsatellite markers for rubber tree (*Hevea brasiliensis*). *Plant Breeding*, **130**: 294-296.

Leitch, A.R., Lim, K.Y., Leitch, I.J., O'Neill, M., Chye, M.L. and Low, F.C. (1998). Molecular cytogenetic studies in rubber, *Hevea brasiliensis* Muell. Arg. (Euphorbiaceae). *Genome*, **41**: 464-467.

Li, D.J., Deng, Z., Chen, C.L., Xia, Z.H., Wu, M., He, P. and Chen, S.C. (2010). Identification and characterization of genes associated with tapping panel dryness from *Hevea brasiliensis* latex using suppression subtractive hybridization. *BMC Plant Biology*, **10**: 140. doi: 10.1186/1471-2229-10-140.

Li, D. J., Deng, Z., Qin, B. and Meng, Z. H. (2012). *De novo* assembly and characterization of bark transcriptome using Illumina sequencing and development of EST-SSR markers in rubber tree (*Hevea brasiliensis* Muell. Arg.). *BMC Genomics*, **13**: 192. doi: 10.1186/1471-2164-13-192.

Li, D., Wang, X., Deng, Z., Liu, H., Yang, H. and He, G. (2016). Transcriptome analyses reveal molecular mechanism underlying tapping panel dryness of rubber tree (*Hevea brasiliensis*). *Scientific Reports*, **6:** 23540. doi: 10.1038/srep23540.

Li, Z.Y. (1982). The relationship of brown blast of rubber trees and over exploitation. *Chinese Journal of Tropical Agriculture*, **5**: 12-6.

Lieberei, R. (2007). South American leaf blight of the rubber tree (*Hevea* spp.): New steps in plant domestication using physiological features and molecular markers. *Annals of Botany*, **100**: 1-18.

Liu, J.P., Xia, Z.Q., Tian, X.Y. and Li, Y.J. (2015). Transcriptome sequencing and analysis of rubber tree (*Hevea brasiliensis* Muell.) to discover putative genes associated with tapping panel dryness (TPD). *BMC Genomics*, **16**: 398-410.

Mantello, C.C., Cardoso-Silva, C.B., da Silva, C.C., Moura de Souza, L., Scaloppi Junior, E.J., Gonalves, P.S., Vicentini, R. and de Souza, A.P. (2014). *De novo* assembly and transcriptome analysis of the rubber tree (*Hevea brasiliensis*) and SNP markers development for rubber biosynthesis pathways. *PLoS ONE*, **9(7)**: e102665. doi:10.1371/*journal.pone*.0102665.

Mehrjerdi, M.Z., Bihamta, M.R., Omidi, M., Naghavi, M.R., Soltanloo, H. and Ranjbar, M. (2013). Effects of exogenous methyl jasmonate and 2-isopentenyladenine on artemisinin production and gene expression in Artemisia annua. *Turkish Journal of Botany*, **37**: 499–505.

Metzker, M.L. (2010). Sequencing technologies-the next generation. *Nature Review Genetics*, **11**: 31–46.

Nandris, D., Chrestin, H., Noirot, M., Nicole, M., Thouvenel, J.C. and Geiger, J.P. (1991a) The phloem necrosis of the trunk of rubber tree in Ivory Coast. (1) Symptomatology and biochemical characteristics. *European Journal of Forest Pathology*, **21**: 325-39.

Nandris, D., Thouvenel, J.C., Nicole, M., Chrestin, H., Rio, B. and Noirot, M. (1991b) The phloem necrosis of the trunk of the rubber tree in Ivory Coast. (2) Etiology of the disease. *European Journal of Forest Pathology*, **21**: 340-353.

Okada, K. (2011). The biosynthesis of isoprenoids and the mechanisms regulating it in plants. *Bioscience Biotechnology and Biochemistry*, **75(7)**: 1219-1225.

Ozsolak, F. and Milos, P.M. (2011). RNA sequencing: Advances, challenges and opportunities. *Nature Review Genetics*, **12**: 87-98.

Páez, H.U.A., Romero, G.I.A., Restrepo, S.R., Gutiérrez, A.F.A. and Castano, M.D. (2015). Assembly and analysis of differential transcriptome responses of *Hevea brasiliensis* on interaction with *Microcyclus ulei*. PLoS ONE, **10(8)**: e0134837. doi:10.1371/*journal. pone*.0134837

Pakianathan, S.W., Samsidar, H., Sivakumaran, S. and Gomez, J.B. (1982). Physiological and anatomical investigation on long term ethephon-stimulated trees. *Journal of Rubber Research Institute of Malaysia*, **30**: 63-79.

Pandey, S.P. and Somssich, I.E. (2009). The Role of WRKY transcription factors in plant immunity. *Plant Physiology* **150(4)**: 1648-1655.

Perseguini, J., Romao, L., Brinez, B., Scaloppi, E.J.R., Gonçalves, P. and Benchimol, L.L. (2012). Genetic diversity of cultivated accessions and wild species of rubber tree using EST-SSR markers. *Pesquisa Agropecuária Brasileira*, **47(8)**: 1087-1094.

Petroski, M.D. and Deshaies, R.J. (2005). Function and regulation of cullin–ring ubiquitin ligases. *Nature Reviews Molecular Cell Biology*, **6**: 9-20.

Punja, Z.K. and Zhang, Y.Y. (1993). Plant chitinases and their roles in resistance to fungal diseases. *Journal of Nematology*, **25(4)**: 526-540.

Qin, B., Liu, X.H., Deng, Z. and Li, D.J. (2012). Identification of genes associated with tapping panel dryness in *Hevea brasiliensis* using oligonucleotide microarrays. *Chinese Journal of Tropical Crops*, **33(2)**: 296-301.

Rosenkranz, R., Borodina, T., Lehrach, H. and Himmelbauer, H. (2008). Characterizing the mouse ES cell transcriptome with Illumina sequencing. *Genomics*, **92**: 187-94.

Roy, C.B., Nazeer, M.A. and Saha, T. (2004). Identification of simple sequence repeats in rubber (*Hevea brasiliensis*). *Current Science*, **87**: 807-811.

Roy, C.B., Jacob, C.K. and Saha, T. (2008). Differential gene expression during disease development in rubber (*Hevea brasiliensis*) by the fungal pathogen *Corynespora cassiicola*. *Proceedings of the International Rubber Research and Development Board (IRRDB) Natural Rubber Conference*, Selangor, Malaysia.

Roy, C.B., Jacob, C.K. and Saha, T. (2009). Towards identification of genes involved in host tolerance to Corynespora leaf disease in rubber (*Hevea brasiliensis*). *Proceedings of 5th International Conference on Plant Pathology in the Globalized Era. Indian Agricultural Research Institute*, New Delhi.

Roy, C.B. and Saha, T. (2012). Generation of transcriptome resources in rubber (*Hevea brasiliensis*) in response to *Corynespora cassiicola* causing Corynespora leaf disease for gene discovery and marker identification using NGS platform. In: *International Rubber Conference IRC 2012. Tiruvananthapuram*, Kerala. India.

Roy, C.B. and Saha, T. (2013). Gene discovery using next generation sequencing (NGS) technology in response to *Corynespora cassiicola* causing Corynespora leaf disease in rubber (*Hevea brasiliensis*). In: First International and Third National Conference on Biotechnology, *Bioinformatics and Bioengineering*, Tirupati, India.

Saha, T., Roy, C.B. and Nazeer, M.A. (2005). Microsatellite variability and its use in the characterization of cultivated clones of *Hevea brasiliensis*. *Plant Breeding*, **124**: 86-92.

Saha, T., Roy, C.B. and Ravindran, M. (2010). Characterization of a family of disease resistant gene analogues (RGAs) in rubber (*Hevea brasiliensis*) and their relationship with functional RGAs in response to *Corynespora* infection. In: National Symposium on Molecular approaches for management of fungal diseases of crop plants. *Indian Institute of Horticultural Research*, Bangalore, idnia.

Sando, T., Takaoka, C., Mukai, Y., Yamashita, A. and Hattori, M. (2008a). Cloning and characterization of mevalonate pathway genes in natural rubber producing plant *Hevea brasiliensis*. *Bioscience Biotechnology and Biochemistry*, **72(8)**: 2049-2060.

Sando, T., Takeno, S., Watanabe, N. and Okumoto, H. (2008b). Cloning and characterization of the 2-C-methyl-D-erythritol 4-phosphate (MEP) pathway genes of a natural-rubber producing plant *Hevea brasiliensis*. *Bioscience Biotechnology and Biochemistry* **72(11)**: 2903-17.

Schuster, S.C. (2008). Next-generation sequencing transforms today's biology. *Nature Methods*, **5(1)**: 16–8.

Shi, M.J. and Tian, W.M. (2012). Effect on the induction of the secondary laticifer differentiation by the transportation of exogenous JA in *Hevea brasiliensis*. *Chinese Journal of Tropical Crops*, **33(9)**: 1647-1653.

Sivakumaran, S., Pakianathan, S.W. and Abraham, P.D. (1984). Continuous yield stimulation. Plausible cause of yield decline. *Journal of Rubber Research Institute of Malaysia*, **32**: 119-143.

Sookmark, U., Pujade-Renaud, V., Chrestin, H., Lacotem, R., Naiyanetr, C. and Seguin, M. (2002). Characterization of polypeptides accumulated in the latex cytosol of rubber trees affected by the tapping panel dryness syndrome. *Plant Cell Physiology*, **43**: 1323-1333.

Soyza, A.G. (1983). The investigation of the occurring rule and distributing pattern of brown blast disease of rubber tree in Sri Lanka. *Journal of Rubber Research Institute of Sri Lanka*, **61**: 1-6.

Suwanmanee, P., Sirinupong, N. and Suvachittanont, W. (2013). Regulation of 3-hydroxy-3-methylglutaryl-CoA synthase and 3-hydroxy-3-methylglutaryl-CoA reductase and rubber biosynthesis of *Hevea brasiliensis* (B.H.K.) Mull. Arg. In: Bach, T.J. and Rohmer, M. (Eds.), *Isoprenoid Synthesis in Plants and Microorganisms: New Concepts and Experimental Approaches*. Springer-Verlag, Berlin/Heidelberg/New York. pp. 315–328.

Tupy, J. (1969). Nucleic acid in latex production of rubber in *Hevea brasiliensis*. *Journal of Rubber Research Institute of Malaysia*, **21**: 468-476.

Tupy, J. and Primot, L. (1976). Control of carbohydrate metabolism by ethylene in latex vessels in *Hevea brasiliensis* in relation to rubber production. *Biologia Plantarum*, **18**: 373-384.

Triwitayakorn, K., Chatkulkawin, P., Kanjanawattanawong, S., Sraphet, S., Yoocha, T., Sangsrakru, D., Chanprasert, J., Ngamphiw, C., Jomchai, N., Therawattanasuk, K. and Tangphatsornruang, S. (2011). Transcriptome sequencing of *Hevea brasiliensis* for development of microsatellite markers and construction of a genetic linkage map. *DNA Research*, **18**: 471-482.

Vankova, R. (2010). Plant hormone functions in abiotic and biotic stress responses. In: Pessarakli, M. (Ed.), *Handbook of Plant and Crop Stress*, 3rd Edition. CRC Press, Boca Raton, Florida, USA. pp. 191-211.

Van Verk, M.C., Hickman, R., Pieterse, C.M.J. and Van Wees, S.C.M. (2013). RNA-Seq: Revelation of the messengers. *Trends in Plant Science*, **18**: 175-179.

Varshney, R.K., Graner, A. and Sorrells, M.E. (2005). Genic microsatellite markers in plants: features and applications. *Trends Biotechnology*, **23**: 48-55.

Venkatachalam, P., Thulaseedharan, A. and Raghothama, K. (2007). Identification of expression profiles of tapping panel dryness (TPD) associated genes from the latex of rubber tree (*Hevea brasiliensis* Muell. Arg.). *Planta*, **226(2)**: 499-515.

Venkatachalam, P., Thulaseedharan, A. and Raghothama, K. (2009). Molecular identification and characterization of a gene associated with the onset of tapping panel dryness (TPD) syndrome in rubber tree (*Hevea brasiliensis* Muell.) by mRNA differential display. *Molecular Biotechnology*, **41(1)**: 42–52.

Venkatachalam, P., Geetha, N., Priya, P. and Thulaseedharan, A. (2010). Identification of a differentially expressed thymidine kinase gene related to tapping panel dryness syndrome in the rubber tree (*Hevea brasiliensis* Muell.Arg.) by random amplified polymorphic DNA screening. *International Journal of Plant Biology*, **1(1)**: e7. doi: 10.4081/pb.2010.e7.

Wang, C.Z. (1988). The report about bark of TPD *Hevea brasiliensis* inoculating. Chinese Journal of Tropical Agriculture, **11**: 25-30.

Wang, C.Z., Gerstein, M. and Snyder, M. (2009). RNA-Seq: a revolutionary tool for transcriptomics. *Nature Review Genetics*, **10**: 57-63.

Wei, F., Luo, S., Zheng, Q., Qiu, J., Yang, W., Wu, M. and Xiao, X. (2015). Transcriptome sequencing and comparative analysis reveal long-term flowing mechanisms in *Hevea brasiliensis* latex. *Gene*, **556**: 153-162.

Xia, Z.H., Xu, H.M., Zhai, J.L., Li, D.J., He, C.Z. and Huang, X. (2011). RNA-Seq analysis and *de novo* transcriptome assembly of *Hevea brasiliensis*. *Plant Molecular Biology*, **77**: 299-308.

Xi, W.X. and Xiao, X.Z. (1988). Study on peroxidase isozyme and superoxide dismutase isozyme of TPD *Hevea* trees. *Chinese Journal of Tropical Crops*, **9**: 31-36.

Zeng, R.Z. (1997). The relation between contents of nucleic acid and tapping panel dryness in latex from *Hevea brasiliensis*. *Chinese Journal of Tropical Crops Research*, **18**: 10-15.

Zheng, G.B. and Chen, M.R. (1982). Study of the cause for brown blast disease. *Chinese Journal of Tropical Crops*, **3**: 57-61.

Zhang, L. and Xing, D. (2008). Methyl jasmonate induces production of reactive oxygen species and alterations in mitochondrial dynamics that precede photosynthetic dysfunction and subsequent cell death. *Plant Cell Physiology*, **49**: 1092-1110.

Chapter 30

Spices

☆ *D. Prasath*

1. Introduction

Advances in DNA sequencing technology during the last decade have dramatically impacted genome sequencing and transcriptome analysis. Techniques such as microarrays and serial analysis of gene expression (SAGE) have facilitated transcriptome analysis at large scale from numerous plants. However, those techniques could be used only for model plants with known genome sequences. The expressed sequence tag (EST) sequencing has been successfully used to analyze the transcriptome in non-model plants. However, deep EST sequencing using capillary sequencing, which requires cDNA cloning and individual DNA preparations for each clone, is time consuming and very expensive.

Breeding for durable crop resistance is a difficult task due to both the high degree of pathogenic variability. The related/wild species is an invaluable source for breeding durable resistant cultivars. Utilization of the wild species depends largely upon the compatibility with the cultivated species and the molecular mechanism of resistance against particular pathogen. Transcriptome analysis using Next Generation Sequencing (NGS) is one of the most popular tools to unravel host-pathogen interaction mechanism, and it has been applied recently to several non-model species that lack genomic sequence information.

2. Transcriptome Sequencing/RNA-seq

The transcriptome encompasses the set of transcripts from a cell or a population of cells, which include protein-coding mRNAs and non-coding small RNAs (*e.g.* ribosomal, tRNA, miRNA). RNA-seq, also called whole-transcriptome shotgun sequencing, refers to the use of high-throughput sequencing technologies for characterizing the RNA content and composition of a given sample. Traditionally, transcriptome profiling, or transcriptomics, has focused on quantifying gene expression. With the advent of ultra high-throughput sequencing (UHTS) technologies, it is now possible to obtain highly resolved structural information of RNA populations on a high-throughput platform. This includes mapping transcript initiation and termination sites, splice junctions and post-transcriptional modifications. Such information will lead to a better understanding of the functional elements within the genome and the discovery of novel developmental or environmental regulatory networks. Whole-genome or whole transcriptome analyses have become a realistic option for genetic non-model organisms, even for individual laboratories, and will soon be standard practice in molecular studies.

2.1. RNA-seq: Applications

- ✰ Transcriptome sequence constitutes a meaningful resource to develop a large number of popular molecular markers such as single-nucleotide polymorphisms and microsatellites. In situations where full sequencing cannot be afforded, but the application requires the use of many markers (*e.g.* genome scans), the transcriptome provides a useful functionally relevant subset of the genome.
- ✰ The great advantage of RNA-seq data is that it allows users to investigate differences in gene expression patterns between populations.
- ✰ It is a more comprehensive and efficient way to measure transcriptome composition, obtain RNA expression patterns, and discover new genes. In addition, this approach is very sensitive, and thus allows the detection of low-abundance transcripts

2.2. Overview of RNA-seq Experimental Procedures

For a typical RNA-seq experiment, mRNA is isolated and reverse-transcribed (RT) into cDNA libraries with homogeneous lengths and sequenced using NGS platforms. The most commonly used sequencing platforms are the pyrosequencing-based 454 system by Roche, the sequencing-by-synthesis-based HiSeq/MiSeq machines from Illumina and the sequencing by-ligation SOLiD system (Figure 30.1).

2.3. RNA Sequencing Analysis

Although the volume of data from RNA-seq experiments is often burdensome, it can provide enormous insights. RNA-seq analysis tools generally fall into three categories:

(i) Those for read alignment;

(ii) Those for transcript assembly or genome annotation; and

(iii) Those for transcript and gene quantification.

2.4. Differential Gene and Transcript Expression Analysis

In the field of molecular biology, gene expression profiling is the measurement of the activity of thousands of genes at once, to create a global picture of cellular function. The volume and complexity of data from RNA-seq experiments necessitate

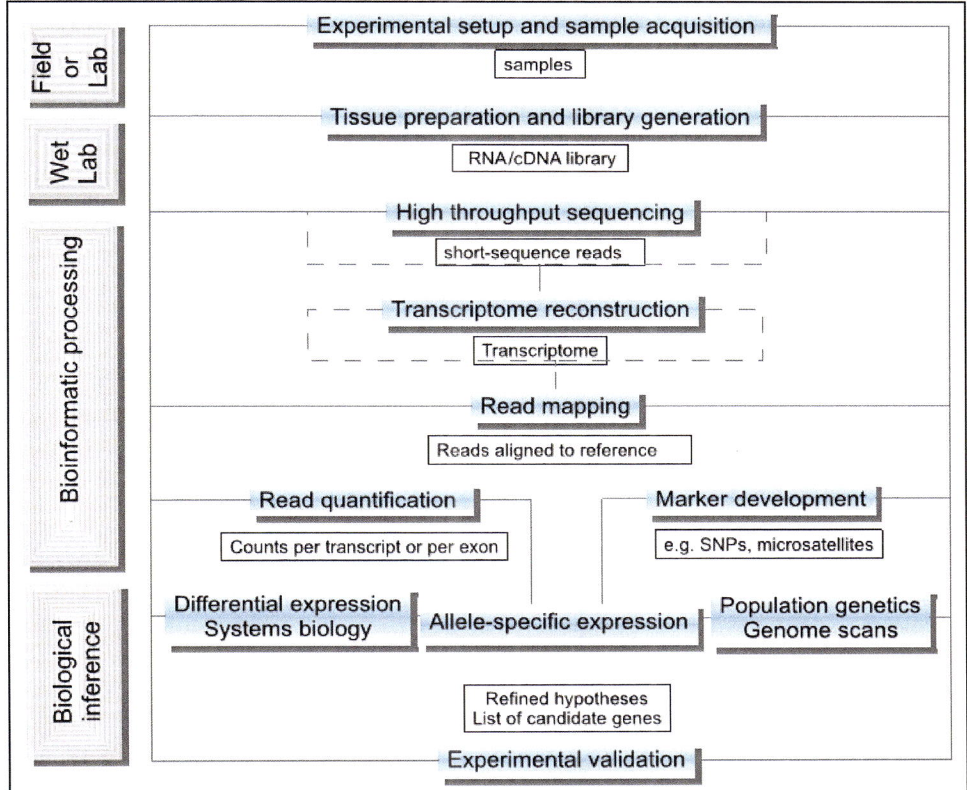

Figure 30.1: Flow Chart of RNA-seq Experiment (Wolf, 2013).

scalable, fast and mathematically principled analysis software. TopHat and Cufflinks are free, open-source software tools for gene discovery and comprehensive expression analysis of high-throughput mRNA sequencing (RNA-seq) data. Together, they allow biologists to identify new genes and new splice variants of known ones, as well as compare gene and transcript expression under two or more conditions. TopHat and Cufflinks performs such analyses, also covers several accessory tools and utilities that aid in managing data, including CummeRbund, a tool for visualizing RNA-seq analysis results.

2.5. Marker Discovery

Molecular genetic markers have found wide application in plants. Advances in DNA sequencing provide tools for efficient large-scale discovery of markers for use in plants. Discovery options include transcriptome sequencing, gene-enriched genome sequencing and whole genome sequencing. Transcriptome sequencing allows genome-wide analysis of large, complex plant genomes and the potential to identify biologically significant microsatellite markers Single Nucleotide Polymorphism markers (SNPs).

2.6. Genotyping-by-Sequencing and Restriction Site Associated DNA Sequencing

The advent of next-generation DNA sequencing (NGS) technologies has led to the development of rapid genome-wide Single Nucleotide Polymorphism (SNP) detection applications in various plant species. Recent improvements in high throughput sequencing, combined with an overall decrease in costs per gigabase of sequence, is allowing NGS to be applied to not only the evaluation of small subsets of parental inbred lines, but also the mapping and characterization of traits of interest in much larger populations. Such an approach, where sequences are used simultaneously to detect and score SNPs, therefore bypassing the entire marker assay development stage, is known as genotyping-by-sequencing (GBS) (Deschamps *et al.* (2012). The GBS approach also uses data directly from the populations being genotyped, thus removing ascertainment bias towards a particular population. Genetic maps generated using GBS-based sequencing information then can be used subsequently for identifying loci of interest from different sets of individuals, including segregating populations or mutant pools. Construction of a low-density GBS linkage map using the reduced representation sequence-based marker discovery technique known as restriction site associated DNA sequencing (RAD) has been reported in barley (Chutimanitsakun *et al.*, 2011).

3. Transcriptome Studies in Spices using NGS

NGS is being used routinely for transcript profiling from non-model spice crops. The primary goal of transcriptome analysis is to learn about how changes in transcript abundance control growth and development and plant-pathogen interactions of an organism and its response to the environment. The current state of NGS based transcriptome studies in spices are summarized below.

3.1. Chillies (*Capsicum annuum* L.)

RNA-Seq was used to obtain transcriptomes of whole Serrano-type chili pepper fruits collected at 10, 20, 40 and 60 days after anthesis (DAA). More profound changes in the chili fruit transcriptome were observed in the intervals between 10 to 20 and 40 to 60 DAA. The last interval, between 40 to 60 DAA, included 49 per cent of all significant changes detected, and was characterized predominantly by a global decrease in gene expression. This period signals the end of maturation and the beginning of senescence of chili pepper fruit. The transcriptome at 60 DAA was the most specialized and least diverse of the four states sampled (Martínez-López *et al.*, 2014).

The use of cytoplasmic male sterility (CMS) in F_1 hybrid seed production of chili pepper is increasingly becoming popular. However, the molecular mechanisms of cytoplasmic male sterility and fertility restoration remain poorly understood due to limited transcriptomic and genomic data. The difference between a CMS line 121A and its near-isogenic restorer line 121C in transcriptome level, aiming to find out critical genes and pathways associated with the male sterility was reported by Liu *et al.* (2013) using next generation sequencing technology (NGS). Many of the differentially expressed unigenes represent a set of potential candidate genes associated with the formation or abortion of pollen. Global transcriptome analysis revealed common regulons for biotic/abiotic stresses, with some of these regulons encoding signaling components in both stresses (Lee and Choi, 2013).

Studies on transcriptome analysis of symptomatic and recovered leaves of geminivirus-infected pepper revealed a total of 309 differentially expressed genes between healthy (mock) and symptomatic or recovered tissues. Computational prediction of differential expression was validated using quantitative reverse-transcription PCR confirming the robustness of the bioinformatic methods. Within the set of differentially expressed genes associated with the recovery process were genes involved in defense responses including pathogenesis-related proteins, reactive oxygen species, systemic acquired resistance, jasmonic acid biosynthesis, and ethylene signaling. Upon comparison, no major differences were found among the differentially expressed genes in symptomatic and recovered tissues. On the other hand, a set of genes with novel roles in defense responses was identified including genes involved in histone modification. This latter result suggested that post-transcriptional and transcriptional gene silencing may be one of the major mechanisms involved in the recovery process. Genes orthologous to the *C. annuum* proteins involved in the pepper-PepGMV recovery response were identified in both *Solanum lycopersicum* and *Solanum tuberosum* suggesting conservation of components of the viral recovery response in the Solanaceae (Góngora-Castillo *et al.*, 2012).

High-throughput transcriptome profiling of two pepper cultivars *viz.*, Mandarin and Blackcluster, carried out using 454 GS-FLX pyrosequencing, resulted in the identification of a total of 9701 and 12,741 potential SNPs which eventually resulted in 1025 and 1059 genotype specific SNPs respectively, after examining SNP frequency distribution for each mapped unigenes. These markers for pepper will be highly valuable for marker-assisted breeding and other genetic studies (Ahn *et al.*, 2014).

Pepper microarray-based transcriptome revealed that above ground *Bemisia tabaci* infestation may lead to the accumulation of nutrient molecules (*i.e.*, minerals, phosphate, and peptides) and plant hormone in roots from above ground plant parts and/or from the soil environment by strongly inducing the over-expression of transporters, leading to an increase in root biomass (Park and Ryu, 2014).

3.2. Black Pepper (*Piper nigrum* L.)

De novo sequencing using Illumina HiSeq™ 2000 to generate leaf transcriptome of black pepper was carried out by Joy *et al.* (2013). The catalogue of transcripts identified, together with the demonstration of reliable existence of SSRs in the miRNA precursors, permits future opportunities for understanding the genetic mechanism of black pepper and likely functions of 'tandem repeats' in miRNAs (Figure 30.2). Further, Asha *et al.* (2016) reported microRNA-mediated gene regulation in black pepper through analysis of high-throughput small RNA deep sequencing data, in combination with transcriptome sequences.

Hu *et al.* (2015) studied transcriptome of black pepper berries using Illumina RNA-Seq technology. The study highlighted the potential of RNA-seq for functional genomics researches on different species for which genomic sequence data are not available. The root transcriptome of black pepper was sequenced by the NGS

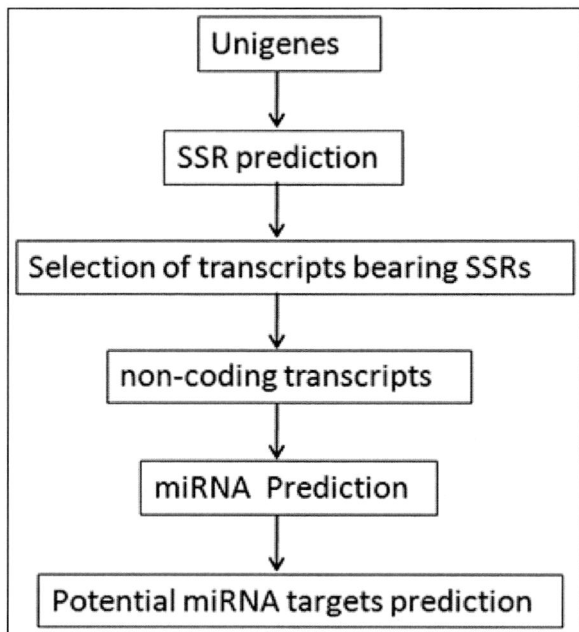

Figure 30.2: Flow Diagram of Whole Transcriptome Analysis and Prediction of Potential miRNA in Black Pepper (Joy *et al.*, 2013).

SOLiD platform and assembled using the multiple-k method (Gordo *et al.*, 2012). The 4472 predicted proteins showed about 52 per cent homology with the *Arabidopsis* proteome. A total of 615 proteins were identified from two root proteomes, which seem to define the plant's root pattern

Foot rot, caused by *Phytophthora capsici*, is the most devastating disease of black pepper (*Piper nigrum* L.). In black pepper-*P. capsici* disease system, the sources for high level of resistance are scarce but, it is found in one of the distant relative of black pepper, *Piper colubrinum*. *De novo* sequencing and transcriptome characterization of the leaves of *Piper nigrum* and *Piper colubrinum* challenge inoculated with *Phytophthora capsici* were undertaken and detailed analysis of the Illumina (Genome AnalyzerIIx) derived sequences led to the identification of large number genes, especially those associated with resistance to the biotic stress factors. *Piper colubrinum* and *Piper nigrum* transcripts showed maximum hit with *Vitis vinifera* (wine grape) sequences, followed by *Populus trichocarpa* (poplar) sequences indicating closer relationship of magnolids (clade to which *Piper* belong) with eudicots. The genes identified also include those involved in pathogen recognition and signalling, transcription factors besides NBS-LRR type resistance genes. The transcriptome data analysis also revealed identification of many defense related genes differentially expressed in these two different species of *Piper* (Johnson, 2012).

The molecular mechanisms underlying foot rot susceptibility were studied by comparing transcriptome analysis between resistant (*Piper flaviflorum*) and susceptible (*Piper nigrum* cv. Reyin-1) black pepper species. A comparison of susceptible and resistant black pepper lines by RNA-Seq, followed by qRT-PCR of phenylpropanoid genes led to the suggestion that disease resistance is accompanied by elevated transcript levels of some phenylpropanoid genes in the resistant *Piper flaviflorum*, compared to the susceptible *P. nigrum* (Hao *et al.*, 2016).

3.3. Ginger (*Zingiber officinale* Rosc.)

Ginger is affected by many diseases of economic importance and among them, bacterial wilt (*Ralstonia solanacearum*) is one of the most important production constraints in tropical, sub tropical and warm temperature regions of the world. In spite of extensive search, no resistance source could be identified in ginger. However, mango ginger (*Curcuma amada*), which is resistant to *R. solanacearum*, is a potential donor if the exact mechanism of resistance can be deciphered. Genomic tools are now being developed to accelerate the identification of resistance genes and the development of bacterial wilt resistant ginger. The study by Prasath *et al.* (2014). describes the global analysis of ginger and mango ginger-*Ralstonia solanacearum* challenge inoculated regimes, which would serve as a blueprint of gene expression profile. The transcriptomes were sequenced from rhizome tissue samples after challenge inoculation with *R. solanacearum* using Illumina platform and several candidate genes were identified that may underline the difference in resistance to *R. solanacearum* between ginger and mango ginger.

3.4. Ginger lily (*Hedychium coronarium*)

The molecular mechanism underlying floral scent production in *Hedychium coronarium* was studied by Yue *et al.* (2015) using Illumina platform. Thirty five and 33 candidate genes were reported to be possibly involved in the biosynthesis of floral volatile terpenes and benzenoids, respectively. Among them, flower-specific HcDXS2A, HcGPPS, HcTPSs, HcCNL and HcBCMT1 might play critical roles in regulating the formation of floral fragrance through DGE profiling coupled with floral volatile profiling analyses.

3.5. Turmeric (*Curcuma longa* L.)

The rhizome transcriptome of three varieties of *turmeric* using Illumina reversible dye terminator sequencing followed by *de novo* transcriptome assembly was compared by Annadurai *et al.* (2013). The presence of transcripts related to biosynthetic pathways of several anti-cancer compounds like taxol, curcumin, and vinblastine in addition to anti-malarial compounds like artemisinin and acridone alkaloids, emphasizes turmeric's importance as a storehouse of highly potent phytochemicals. Later, Sheeja *et al.* (2015) used Illumina sequencing platform and generated a substantial amount of expressed sequence tag (EST) dataset from two species *viz., C. longa* and its wild relative *Curcuma aromatica* Salisb. contrasting in curcumin content. The candidate genes for enzymes involved in curcuminoid biosynthesis were identified from both the species.

3.6. Seed Spices

To investigate essential oil metabolism, the transcriptome of coriander mericarps, at three developmental stages (early, mid, late) was sequenced via

Illumina technology and a transcript library was produced (Galata *et al.*, 2014) and CscTRPS and CsLINS which account for the majority of essential oil constituents in coriander mericarps, were identified. *De novo* transcriptome analysis along with the diosgenin pathway in fenugreek was studied using SOLiD 4 Genome Analyzer and the genes responsible for diosgenin biosynthesis have been identified (Vaidya *et al.*, 2013).

4. Conclusion

Despite these challenges, ultra high-throughput sequencing based transcriptomics approaches promise 'never-before' opportunities to explore plant transcriptomes. As improvements to the sequencing chemistry, sequencing hardware and software and statistical methods of analysis continue to progress, the expectations for transcriptomics studies will continue to increase. In summary, NGS based transcriptome approaches have clearly demonstrated their advantages over previously developed methods and are becoming the new standard for transcriptomics studies.

References

Ahn, Y.K., Tripathi, S., Kim, J.H., Cho, Y.I., Lee, H.E., Kim, D.S., Woo, J.G. and Cho, M.C. (2014). Transcriptome analysis of *Capsicum annuum* varieties Mandarin and Black cluster: assembly, annotation and molecular marker discovery. *Gene*, **533**: 494–499.

Annadurai, R.S., Neethiraj, R., Jayakumar, V., Damodaran, A.C., Rao, S.N., Katta, M.A., Gopinathan, S., Sarma, S.P., Senthilkumar, V., Niranjan, V., Gopinath, A. and Mugasimangalam, R.C. (2013). *De Novo* transcriptome assembly (NGS) of *Curcuma longa* L. rhizome reveals novel transcripts related to anticancer and antimalarial terpenoids. *PLoS ONE*, **8**(2): e56217. doi:10.1371/*journal. pone*.0056217

Asha, S., Sweda, S. and Soniya, E.V. (2016). Unravelling the complexity of microRNA-mediated gene regulation in black pepper (*Piper nigrum* L.) using high-throughput small RNA profiling. *Plant Cell Reports*, **35**(1): 53-63.

Chutimanitsakun, Y., Nipper, R.W., Cuesta-Marcos, A., Cistue, L., Corey, A., Filichkina, T., Johnson, E.A. and Hayes, P.M. (2011). Construction and application for QTL analysis of a restriction site associated DNA (RAD) linkage map in barley. *BMC Genomics*, **12**: 1-13.

Deschamps, S., Llaca, V. and Gregory, D.M. (2012). Genotyping-by-Sequencing in Plants. *Biology (Basel)*, **1(3)**: 460-483.

Galata, M., Sarker, L.S. and Mahmoud, S.S. (2014). Transcriptome profiling, and cloning and characterization of the main monoterpene synthases of *Coriandrum sativum* L. *Phytochemistry*, **102**: 64–73.

Góngora-Castillo, E., Ibarra-Laclette, E., Trejo-Saavedra, D.L. and Rivera-Bustamante, R.F. (2012). Transcriptome analysis of symptomatic and recovered leaves of geminivirus-infected pepper (*Capsicum annuum*). *Virology Journal*, **9**: 295. doi: 10.1186/1743-422X-9-295.

Gordo, S.M., Pinheiro, D.G., Moreira, E.C., Rodrigues, S.M., Poltronieri, M.C. and Lemos, O.F. (2012). High-throughput sequencing of black pepper root transcriptome. *BMC Plant Biology*, **12**: 168. doi: 10.1186/1471-2229-12-168.

Hao, C., Xia, Z., Fan, R., Tan, L., Hu, L., Wu, B. and Wu, H. (2016). *De novo* transcriptome sequencing of black pepper (*Piper nigrum* L.) and an analysis of genes involved in phenylpropanoid metabolism in response to *Phytophthora capsici*. *BMC Genomics*, **17**: 822. doi: 10.1186/s12864-016-3155-7.

Hu, L., Hao, C., Fan, R., Wu, B. and Tan, L. (2015). *De novo* assembly and characterization of fruit transcriptome in black pepper (*Piper nigrum*). *PLoS ONE*, **10**(8): e0136028. doi: 10.1371/*journal.pone*.0136028.

Johnson, K.G. (2012). Transcriptomics approaches for gene discovery in Plants - a case study in *Piper*. International Conference on Agricultural and Horticultural Sciences, September 14-15, 2012 *Hyderabad International Convention Centre*, Hyderabad, India.

Joy, N., Asha, S., Mallika, V. and Soniya, E.V. (2013). *De novo* Transcriptome sequencing reveals a considerable bias in the incidence of simple sequence repeats towards the downstream of 'pre-miRNAs' of black pepper. *PLoS ONE*, **8**(3): e56694. doi:10.1371/*journal.pone*.0056694.

Liu, C., Ma, N., Wang, P.Y., Fu, N. and Shen, H.L. (2013). Transcriptome sequencing and de novo analysis of a cytoplasmic male sterile line and its near-isogenic restorer line in chili pepper (*Capsicum annuum* L.). *PLoS ONE*, **8**(6): e65209. doi:10.1371/*journal.pone*.0065209.

Lee, S. and Choi, D. (2013). Comparative transcriptome analysis of pepper (*Capsicum annuum*) revealed common regulons in multiple stress conditions and hormone treatments. *Plant Cell Reports*, **32(9)**: 1351–1359.

Martínez-López, L.A., Ochoa-Alejo, N. and Martínez, O. (2014). Dynamics of the chili pepper transcriptome during fruit development. *BMC Genomics*, **15**: 143. doi: 10.1186/1471-2164-15-143.

Park, Y.S. and Ryu, C.M. (2014). Understanding cross-communication between aboveground and belowground tissues via transcriptome analysis of a sucking insect whitefly-infested pepper plants. *Biochemical and Biophysical Research Communications*, **443**: 272–277.

Prasath, D., Karthika, R., Habeeba, N.T., Suraby, E.J., Rosana, O.B., Shaji, A., Eapen, S.J., Deshpande, U. and Anandaraj, M. (2014). Comparison of the transcriptomes of ginger (*Zingiber officinale* Rosc.) and mango ginger (*Curcuma amada* Roxb.) in response to the bacterial wilt infection. PLoS One, **9**(6): e99731. doi:10.1371/*journal.pone*.0099731.

Sheeja, T.E., Deepa, K., Santhi, R. and Sasikumar, B. (2015). Comparative transcriptome analysis of two species of *Curcuma* contrasting in a high-value compound curcumin: Insights into genetic basis and regulation of biosynthesis. *Plant Molecular Biology Reporter*, **33(6):** 1825-1836.

Yue, Y., Yu, R. and Fan, Y. (2015). Transcriptome profiling provides new insights into the formation of floral scent in *Hedychium coronarium*. *BMC Genomics*, **16**: 470. doi: 10.1186/s12864-015-1653-7.

Vaidya, K., Arpita, G., Vinay, K., Spandan, C., Navin, S., Kalpesh, K., Tanushree, T. and Surendra, K.C. (2013). *De novo* transcriptome sequencing in *Trigonella foenum-graecum* L. to identify genes involved in the biosynthesis of diosgenin. *Plant Genome*, **6**: 1–11.

Wolf, B. and Jochen, W. (2013). Principles of transcriptome analysis and gene expression quantification: An RNA-seq tutorial. *Molecular Ecology Resources*, **13**(4): 559-572.

Chapter 31

Coconut

☆ *M. Neema, M.K. Rajesh, S.V. Ramesh and P. Chowdappa*

1. Introduction

Coconut is one of the versatile and indispensable food items for millions of inhabitants in South, South-East Asia, and Pacific islands. Coconut (*Cocos nucifera*) is a monospecific member of the monocotyledonous family Arecaceae. Several hundred varieties of the coconut palms are grown all over the tropics, possessing high morphological, physiological and biochemical variations. The variability extents from tall to dwarf, large to small sized nuts, susceptibility and resistance to drought and salinity, high to low oil and fiber content *etc.* The uses of different parts of coconut tree (both edible and non-edible) from the leaves down to roots for many ages on tropical countries have given it the name 'Tree of Life'. The antimicrobial property of coconut husk has also been proved (Jose *et al.*, 2014). Coconut water is rich in potassium and is recommended as a health drink (Manna *et al.*, 2014). Despite the importance of coconut for humans and tropical ecosystems, the potential of the crop has not been utilized to the full extent. A number of barriers like non-availability of quality planting material and true to type seed nuts, susceptibility to biotic and abiotic stresses and fluctuating prices have considerably affected the productivity of coconut palms (CPCRI, 2015). Apart from conventional breeding techniques which are constrained due to the long life cycle of the palm, non-conventional techniques of tissue culture, molecular markers and 'omics' approaches are gaining importance. The genome sequence information of coconut is relatively scarce compared to other economically important palms like oil palm and date palm, where whole genome sequence has been made available in the public domain (Al-Dous *et al.*, 2011; Al-Mssallem *et al.*, 2013; Singh *et al.*, 2013; Dussert *et al.*, 2013).

Genome sequencing provides perspectives to tackle genetic problems, discovering regulatory signals, determining coding regions and the novel genes involved in growth and functioning of the system. Non-model plants like coconut, where the species possesses characters such as long life cycle, difficulty to grow in the laboratory or poor fecundity, have been schemed out of sequencing projects earlier due to high running cost of Sanger sequencing; consequently, the information about their genomics and key biological processes are inadequate. However, the advent of fast and cost effective next generation sequencing (NGS) platforms in the recent past has enabled the unearthing of certain characteristic gene structures unique to these species (Unamba *et al.*, 2015). The chloroplast and mitochondrial genomes of coconut have been sequenced (Huang *et al.*, 2013; Aljohi *et al.*, 2016), but the whole genome sequence is not available till date in the public domain. Transcriptome analysis, using RNA-Seq and assembly, could be utilized to gain global overview of the transcriptome of a plant, thus providing a wider outlook into the genetic mechanisms of growth, varietal characterization, and resistance to biotic and abiotic stresses. It has also aided in gaining insights about mechanisms underlying processes of secondary metabolism as well as facilitated development of genomic resources for evolutionary analysis and marker assisted breeding even without prior availability of genomic sequence information (Unamba *et al.*, 2015). This chapter discusses the recent research advances in genomics and transcriptomics of coconut.

2. Coconut Genome

2.1. Genome Size

The C-value (haploid genome size) of a species corresponds to the DNA amount in its unreplicated haploid or gametic nucleus (pollen or sperm), regardless of its ploidy level (Greilhuber *et al.*, 2005) and it is measured in picograms (pg) or base pairs (bp). Determination of the genome sizes and ploidy levels of cultivated coconuts are essential prerequisite for the sequencing of the coconut genome as it would provide the precise information regarding optimal depth of reads required for the accurate assembly and coconut genome annotation (Gunn *et al.*, 2015). The genome size of a species is associated with growth, development and adaptation to environmental changes. Genome size variation among related palm species is of evolutionary significance because changes characterize clads and may be associated with polyploidy, transposon amplification, deletion or rearrangements. Gunn *et al.* (2015) have conducted analysis, through flow cytometry, to estimate the genome sizes of 23 coconut cultivars which included dwarf and tall cultivars and hybrids. They found that the coconut genome is large, averaging 5.757 Gbp. The value was larger than the report of Zonneveld *et al.* (2005), but was consistent with data published by Sandoval *et al.* (2003) but differs from the results of Röser *et al.* (1997). Their results also suggest that intraspecific variation was associated with domestication and the variation among tall coconut was significantly greater in comparison to dwarfs.

2.2. Chloroplast Genome

Chloroplast is the photosynthetic organelle that has a genome of its own. Huang *et al.* (2013) reported the chloroplast genome sequence of a dwarf coconut plant. Gene content and organization, inverted repeat fluctuations, repeated sequence structure, and occurrence of RNA editing in coconut chloroplast genome were described in detail. Analysis of the data obtained from high-throughput sequencing demonstrated that the chloroplast genome of coconut is a typical quadripartite molecule within which a pair of inverted repeats is separated by a large single copy region and a small single copy region. The chloroplastic genome was 154,731 bp in length and was predicted to encode 130 genes and four pseudogenes. Although the chloroplast genome of coconut was the smallest found so far among palms, it shared the same overall organization, gene content and repeat structures that have been observed with chloroplast DNA sequenced from other palm species (Huang *et al.*, 2013). Some of the unique features found in the coconut chloroplast genome included pseudogenization of *rps19*-like gene and an unusually high number of RNA editing sites.

2.3. Mitochondrial Genome

Mitochondria are semi-autonomous bodies that contribute to energy production, metabolism and cell homeostasis. The mitochondrial genome (mt genome) codes for important polypeptides that build up complexes for the oxidative phosphorylation chain, together with nuclear-encoded subunits. In most plant species, the mt DNA gene sequences evolve very slowly and point mutations are rare because plant mitochondria contain an active DNA recombination system that allows copy correction of mutations (Gualberto *et al.*, 2014) The plant mitochondrial DNA (mtDNA) is large and variable in size (200 to 2,500 kb), contains many introns and repeated elements (typically 90 per cent of the total sequence), and experiences frequent gene gain/loss/transfer/duplication, and genome rearrangements (Galtier, 2011).

The coconut mitochondrial genome of an Oman Local Tall cultivar was sequenced by Aljohi *et al.* (2016). This was the second whole genome mitochondria assembly from Arecaceae family after that of date palm (Fang *et al.*, 2012). The coconut mitochondrial genome is around 679 kb in length with GC content of 45.5 per cent. Within the genome, 72 proteins, nine pseudogenes, 23 tRNAs, and three ribosomal RNAs were encoded. The chloroplast (cp) derived regions accounted for 5.07 per cent of the total assembly length in contrast to the date palm mt genome, where the 93.5 per cent of the genome sequence is cp derived (Fang *et al.*, 2012). In coconut, the cp-derived regions included 13 proteins, two pseudogenes, and 11 tRNAs. The mt genome of coconut has a relatively large fraction of repeat content (17.26 per cent), including both forward (tandem) and inverted (palindromic) repeats whereas in date palm there are 0.33 per cent tandem and 2.3 per cent long repeats. In coconut, sequence variation analysis shows that the transition transversion ratio of 0.3 in the mt genome was much lower (2-2.1) when compared to that of the nuclear genome (Aljohi *et al.*, 2016).

2.4. Whole Genome Sequencing of Coconut

Alsaihati *et al.* (2014) had reported a draft genome of coconut by generating seven libraries for pair-end and mate-pair genome sequencing using HiSeq. The estimated size of coconut genome was around 2.6 Gb, with repetitive sequences between 50 to 70 per cent based on draft genome data analysis. Since the genome assembly using ALLPATHS-LG and SOAPdenovo2 was unsatisfactory (Table 31.1), a custom pipeline was developed using SOAP assembly and gap closing modules, SSPACE scaffolder and intermediate correction scripts to produce quality genome draft with coverage of 94.5 per cent.

Table 31.1: Comparison of Coconut Draft Genome using different Assembly Pipelines (Alsaihati *et al.*, 2014)

Draft	*AllPathsLG*	*SoapDenovo2*	*Custom Pipeline*
Scaffolds	80,145	212,688	88,400
Total size (est. coverage)	0.97Gb (37 per cent)	2.6Gb (~100 per cent)	2.47Gb (94.5 per cent)
Longest scaffold	771Kb	891Kb	1246Kb
N50	48Kb	66.6Kb	114Kb
N90	8Kb	4Kb	14.8Kb
Total gap (per cent of draft size)	255Mb (26 per cent)	749Mb (29 per cent)	251Mb (10 per cent)

It is imperative to obtain whole genome sequence of coconut which is essential for precise understanding of various cellular processes, signaling pathways, defense mechanisms and host pathogen interactions of the palm.

3. Coconut Transcriptome

The transcriptome represents the whole complement of RNA transcripts in cells or tissues and reflects the expressed genes at various life stages, tissue types, physiological states, and environmental conditions. In crops where whole genome sequence is not available, transcriptome analysis using RNA-Seq technology enables the precise measurement of level of transcripts. Knowledge of the transcripts would be advantageous in determining the various genetic factors involved in various morphological and physiological processes in the plant. Next-generation transcriptome sequencing (RNA-seq) has been validated as an efficient methodology for production of genomic sequences by assembling short- read sequences (Xia *et al.*, 2011) with higher sensitivity and broader dynamic range. Studies based on large scale transcriptome profiling have given a fundamental insights into the aspects of co-expressing genes and their roles in metabolic pathways (Wickramasuriya and Dunwell, 2015), candidate genes associated with tolerance/resistance to different stresses as well as genes associated with several agronomic traits. Until 2012, the coconut transcriptome data made available publicly were very scarce; only 774 sequences were available in National Centre for Biotechnology Information database (Fan *et al.*, 2013).

3.1. Fatty Acid Biosynthesis and Metabolism

The first genome-wide study of coconut using transcriptome analysis was carried out by Fan *et al.* (2013). Total RNA was isolated from spear leaves, young leaves and fruit flesh of Hainan Tall cultivar. Illumina RNA-Seq technology was used to generate 54,931,406 short reads containing a total of 4,943,826,540 nucleotide bases. *De novo* assembly using the short read assembly programme 'Trinity' was applied to combine these short reads to 57,304 unigenes, with an average length of 752 bp. Unigene sequences were annotated against NCBI Nr database with E-value, 0.00001. They were also aligned by BLASTX to protein databases such as Swiss-Prot, KEGG and COG, in order to retrieve proteins with the highest sequence similarity. ESTScan, Blast2GO program, WEGO software and Blast All software's were used for annotations. Ten putative non-coding RNA genes were identified when the coconut transcriptome was compared to whole genome of date palm. A total of 347 unigenes involved in the biosynthesis and metabolism of fatty acid was also identified. These unigenes could be assigned to five steps of the fatty acid biosynthesis pathway, providing a means of elucidating the molecular mechanisms for fatty acid biosynthesis in coconut palm. Of these, 20 unigenes were predicted to be related to fatty acyl-ACP thioesterase, which is a crucial enzyme for terminating the elongation of carbon chains and therefore regulating the length of fatty acids. Therefore, the study suggested that the expression of fatty acyl-ACP thioesterase might be correlated with the observed accumulation of medium chain fatty acids (*i.e.* lauric acid) in coconut.

3.2. Embryogenesis

Somatic embryogenesis (SE), where a single or a group of somatic cells differentiate to form embryonic cells under suitable *in vitro* conditions (Zimmerman, 1993), is an ideal system to explore gene expression patterns associated with initial stages of embryo development. The formation of embryos from somatic cells closely resembles the developmental pathway of zygotic embryos (ZEs) and hence, the molecular information generated for the SE pathway could be used to explain the dynamic molecular interactions that take place during early embryogenesis (Zimmerman, 1993). Rajesh *et al.* (2015a) carried out *de novo* assembly and characterization of global transcriptome of coconut embryogenic calli using Illumina paired end sequencing. Here transcriptome analysis of coconut embryogenic calli derived from plumular explants of West Coast tall cultivar was undertaken on Illumina Hi Seq 2000 platform. The reads obtained (50,839,994 paired end reads; 7.73 Gb) were trimmed using a Perl script (46.2 million reads; 5.42GB), assembled using SOAPdenovo assembler and the trimmed reads were aligned using Bowtie programme. The assembled reads were subjected to annotation, classification and ontology analysis using BLASTx, BLAST2GO and KEGG programmes. Genes known to be involved in SE, namely protein kinases like receptor-like kinases [somatic embryogenesis receptor kinase (*SERK*) and CLAVATA 1 (*CLV1*)], mitogen-activated protein kinase (*MAPK*), transcription factors [WUSCHEL (*WUS*), APETALA2/ Ethylene-responsive factor (*AP2/ERF*), PICKLE (*PKL*), AINTEGUMENTA (*ANT*) and *WRKY*], extracellular proteins [arabinogalactan protein (*AGP*), Germin-like protein (*GLP*), embryogenic cell protein (*ECP*), and late embryogenesis-abundant protein (*LEA*)] and glutathione S-transferase (*GST*) were identified. Gene ontology (GO) annotations identified 8300 transcripts associated with biological processes majority having transcription and regulatory function, 13193 transcripts with molecular functions involved in ATP, zinc ion and metal ion binding processes and 6076 transcripts with cellular components; the highly expressed one were components integral to membrane followed by nucleus. Gene validation and expression analysis was carried out by real time quantitative PCR (RT-qPCR). For these, three stages of developmental pathway of somatic embryogenesis, namely initial calli, embryogenic calli and somatic embryos, were selected. The results indicated that *CLV* was upregulated in the initial stage of callogenesis. Transcripts of *GLP*, *GST*, *PKL*, *WUS* and *WRKY* were expressed more in somatic embryo stage. The expression of *SERK*, *MAPK*, *AP2*, *SAUR*, *ECP*, *AGP*, *LEA* and *ANT* were higher in embryogenic callus stage compared to initial culture and somatic embryo stages. This study was an effort to aid in the development of an efficient *in vitro* production protocol for coconut which is otherwise recalcitrant to *in vitro* culture.

Bandupriya *et al.* (2015) carried out a detailed analysis of ESTs by examining the transcriptome data of the different embryo tissue types together with one somatic tissue in order to identify important embryo-specific genes as well as other functional genes in different biochemical pathways in coconut. Four cDNA libraries were generated from immature embryo, mature embryo, microspore derived embryo and mature leaves and cDNA was sequenced by the Roche-454 GS-FLX system.

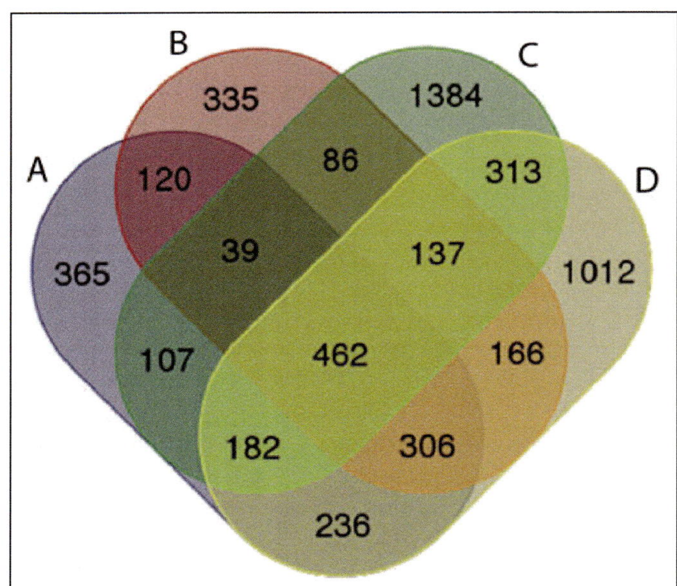

Figure 31.1: Venn Chart Showing Unique and Shared Unigenes found in Four Coconut Transcriptome Libraries.

(A) Immature embryos at the age of nine months after pollination, (B) Mature embryos at the age of 12 months after pollination, (C) Microspore-derived embryos and (D) Developing leaf.

The unigene sequences were annotated in BLAST; 462 unigenes were shared by all four libraries and the numbers of tissue specific unigenes were 365 in immature embryos at the age of nine months after pollination, 335 in mature embryos at the age of 12 months after pollination, 1012 in microspore-derived embryos and 1384 in developing leaf (Figure 31.1).

Comparative expression analysis demonstrated that the embryo tissue libraries shared certain degree of similarities while they were different from leaf abundant ESTs. The result emphasizes on the effectiveness of 454 sequencing approach for the identification of transcripts for a particular organ of coconut. In this study, a list of putative transcripts such as chitinase, beta-1,3-glucanase, ATP synthase CF0 subunit, thaumatin-like protein and metallothionein-like protein, which may be involved in various biological processes during embryogenesis were identified, which can further be utilized for future research.

3.3. Host-pathogen Interactions

The root (wilt) disease is a serious debilitating disease causing severe yield loss in coconut palms in Southern India (Figure 31.2). RNA-seq was utilized to characterize global transcriptome responsive to root (wilt) disease in coconut (Rajesh *et al.*, 2013). To explore the molecular mechanisms involved in compatible and incompatible interactions, transcriptome profiling of diseased and healthy Chowghat Green Dwarf (CGD) palms was conducted. The flow diagram of different steps followed is provided in Figure 31.3. RNA-Seq analysis generated more than 102 million 86 bp paired-end reads, which were assembled into 2,54,302 contigs, with a mean size of 488 bp. Based on sequence similarity searches, 1,66,429 (~ 65 per

Figure 31.2: Chowghat Green Dwarf Palm Affected by Root (Wilt) Disease.

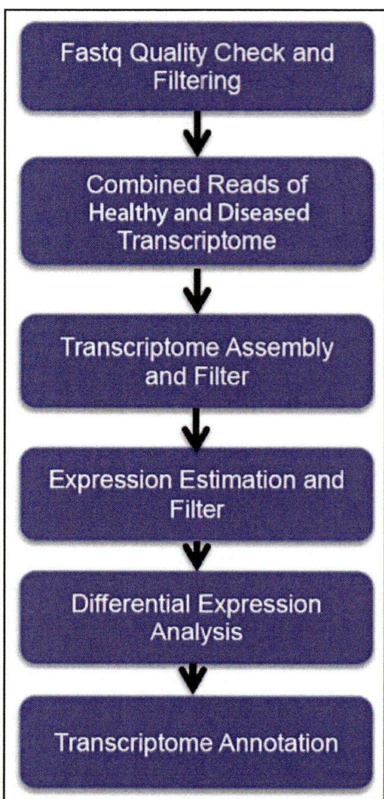

Figure 31.3: Flow Diagram of Whole Transcriptome Analysis from Healthy and Root (Wilt) Diseased Leaf Samples of Coconut (Rajesh *et al.*, 2013).

cent) of assembled transcripts had at least one significant hit (E^{-20}) in NCBI database. Differential gene expression analysis was performed statistically using DESeq program. Overall, 3,033 transcripts differentially expressed transcripts (p-value < 0.01 and at least 100 reads in one of the sample) could be detected between healthy and root (wilt) diseased samples. Of the total, 1266 transcripts were up-regulated in healthy samples. Many of these differentially expressed transcripts were primarily involved in defense responses, signalling pathways, cellular transport and other metabolic processes.

The gene ontology (GO) terms for transcripts were extracted wherever possible. The top 25 GO terms identified in molecular function, biological process and cellular component category are provided in Figures 31.4–31.6.

Transcriptome analysis to identify complex host pathogen interactions, with respect to coconut yellow decline disease, was carried out by Nejat *et al.* (2015). Whole transcriptome profiles of naturally infected leaves of Malayan Red Dwarf in response to yellow decline phytoplasma, and healthy leaves were generated through RNA-Seq technique. Illumina sequencing yielded a total of 72,019,264 and 70,935,896 reads from healthy and infected leaves transcriptome respectively which

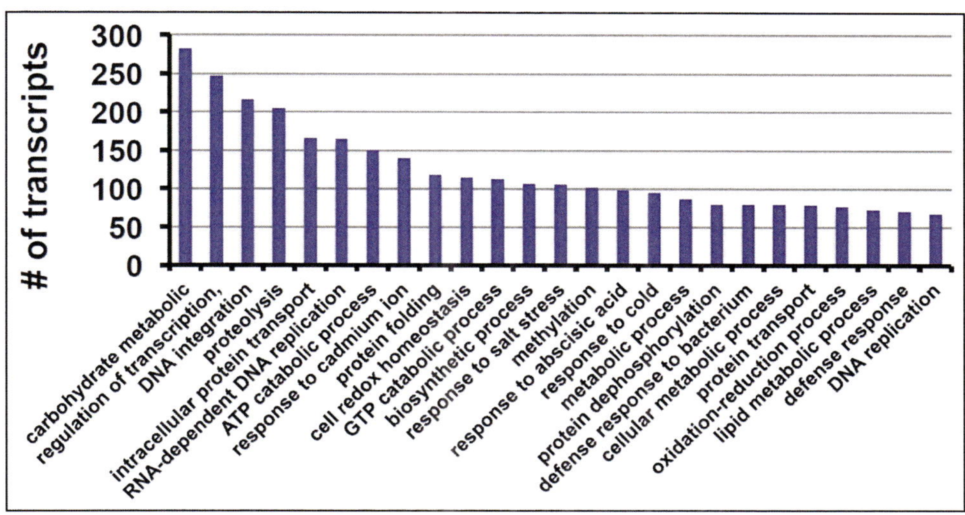

Figure 31.4: Top 25 Terms in Biological Process Category from GO Annotation.

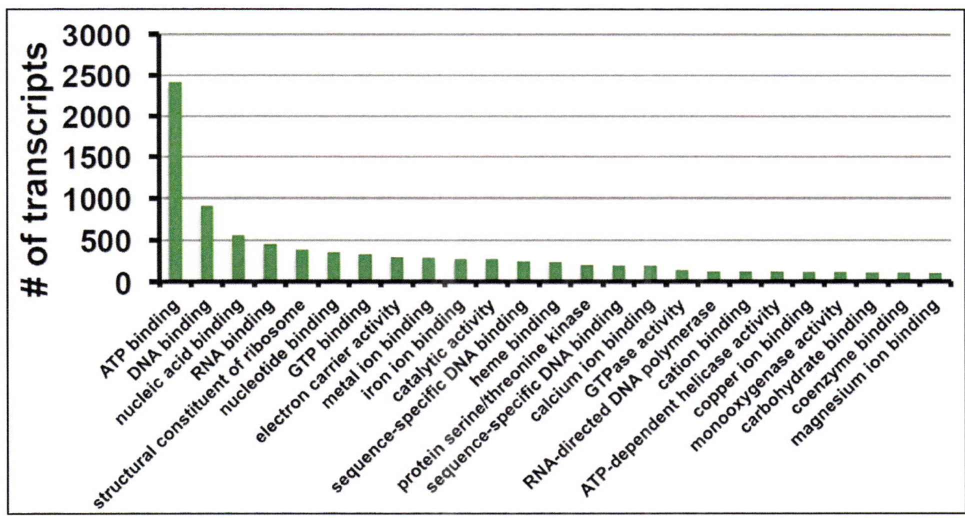

Figure 31.5: Top 25 Terms in Molecular Function Category from GO Annotation.

were assembled into 108,994 contigs for healthy and 148,264 contigs for diseased coconut palm. An analysis to determine the differentially expressed genes in the healthy as well as infected revealed that 18,013 transcripts were upregulated and 21,860 transcripts were down-regulated in infected leaves compared to healthy ones. This transcriptome analysis showed that many of the biological and cellular processes were reprogrammed in coconut due to phytoplasma infection. A number of genes associated with the production of defense related proteins, reactive oxygen species, ABC transport protein family, hydrolase and kinases, flavanol synthesis, auxin induced protein, no apical meristem (NAM) gene family, and ethylene were

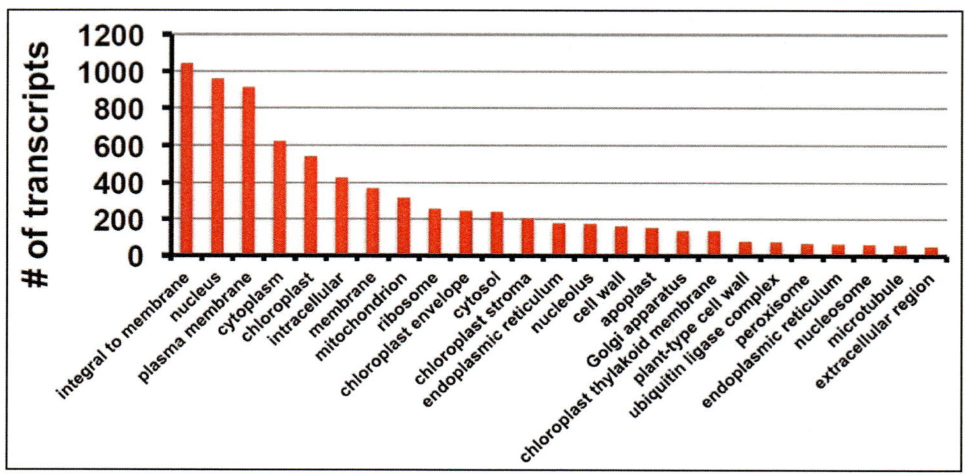

Figure 31.6: Top 25 Terms in Cellular Component Category from GO Annotation.

upregulated as a result of phytoplasmal infection. A total of 35 *WRKY* genes were induced, whereas 21 were suppressed in infected coconut leaves. Transcripts of cytokinin dehydrogenase were down regulated. Up-regulation of gibberellin-2-oxidase was presumed to down-regulate levels of gibberellins in infected coconut palms and result in stunting, inflorescence necrosis and premature nut fall, which are the characteristics of phytoplasma infection. Wide range of genes associated with photosynthesis was down regulated, which is part of the defense mechanism in response to biotic stress. It was also suggested that phytoplasma might alter the expression of genes involved in carbohydrate metabolism in infected palms to import sugars, which are their main source of energy, through ABC transporter system, which might explain the up-regulation of ABC protein family transcripts in phytoplasma infection.

Rajesh *et al.* (2015b) carried out comprehensive bioinformatics analysis of transcriptome data of leaf samples of coconut root (wilt) disease-resistant cultivar Chowghat Green Dwarf, generated through RNA-Seq, and identified 243 resistance gene analog (RGA) sequences, comprising six classes of RGAs. Domain and conserved motif predictions of clusters were performed to analyze the architectural diversity. Phylogenetic analysis of deduced amino acid sequences revealed that coconut NBS-LRR type RGAs were classified into distinct groups based on the presence of TIR or CC motifs in the N-terminal regions.

3.4. RNA Directed DNA Methylation

Huang *et al.* (2014) carried out transcriptome analysis to classify gene expression in seed tissues and leaves of dwarf coconut palm with a focus on the identification of factors involved in RNA-directed DNA methylation. In the study, RNA-Seq data were collected from maturing gelatinous endosperm, mature embryo and young leaf of fragrant dwarf green coconut. Of the three tissues, the leaf transcriptome had the maximum total number of sequencing reads of 121,151,552 with total unigenes of 33,446 and embryo transcriptome had total sequencing reads 81,128,552 with total unigenes of 86254. Total transcripts in endosperm were 229,866. The study was focused on factors involved in RNA directed DNA methylation (RdDM) and suggests that the relative abundance of most of these factors is highest in endosperm tissue compared with leaves and embryos. The research results suggest that small RNA-mediated silencing pathways were active in coconut seeds, particularly maturing endosperm.

3.5. Aroma

The liquid endosperm of Aromatic Green Dwarf coconut of Thailand has a peculiar 'pandan-like' aroma due to the presence of 2-acetyl-1-pyrroline (2AP) (Saensuk *et al.*, 2016), which is also present in rice accessions (Grimm *et al.*, 2011; Arikit *et al.*, 2011). *De novo* assembly of transcriptome from the aromatic coconut endosperm was undertaken by Saensuk *et al.* (2016) to identify the gene(s) responsible for 2AP biosynthesis. The gene, *CnAMADH*2, which is an ortholog of rice aromatic gene, was present in all Aromatic Green Dwarf accessions. But compared to other aromatic plants, in the case of coconut, a G-to-C substitution found in exon 14 and was associated exclusively with 2AP content. This base substitution resulted in an amino acid change of alanine to proline at position 442, which might have resulted in unstabilized dimer conformation that could lower AMADH enzyme activity. The authors confirmed the substitution by PCR based on this sequence variation.

4. Conclusion

The precise understanding of key processes and pathways are important for the genetic manipulations of the plant, thereby making them more productive and tolerant to biotic and abiotic stresses. The whole genome sequence of coconut is not yet published except for a draft genome. Due to the reduced size and complexity of the transcriptome relative to the genome, transcriptome sequencing provides a rapid, inexpensive approach to access gene sequences, gene expression abundances, and gene expression patterns in any species, including those that lack a reference genome sequence. To date, successful applications of RNA sequencing in conjunction with *de novo* transcriptome assembly has enabled identification of new genes in an array of biochemical pathways in plants (Gongora-Castillo *et al.*, 2013). Transcriptome analysis till date in coconut has provided insight into the genetic mechanism and factors influencing host pathogen interaction, somatic embryogenesis, RNA directed DNA methylation and other novel genes. These studies have increased transcriptomics resources for coconut and provide a foundation for further functional and molecular studies that will improve coconut through molecular breeding and genetic engineering technologies. Still there is a need for improving the sequencing platforms to obtain long and high quality output reads and better algorithms and bioinformatic softwares to handle this palm which possess a large genome. Finally, unraveling of genome sequence of coconut and large scale transcriptome studies would provide molecular insights and help provide break-through in successful coconut regeneration and genetic transformation that are essential for accelerating crop improvement programmes.

References

Al-Dous, E.K., George, B., Al-Mahmoud, M.E., Al-Jaber, M.Y., Wang, H., Salameh, Y.M., Al-Azwani, E.K., Chaluvadi, S., Pontaroli, A.C., DeBarry, J. and Arondel, V. (2011). *De novo* genome sequencing and comparative genomics of date palm (*Phoenix dactylifera* L.). *Nature Biotechnology*, **29:** 521-527.

Aljohi, H. A., Liu, W., Lin, Q., Zhao, Y., Zeng, J., Alamer, A., Alanazi, I.O., Alawad, A.O., Al-Sadi, A.M., Hu, S. and Yu, J. (2016). Complete sequence and analysis of coconut palm (*Cocos nucifera*) mitochondrial genome. *PLoS ONE*, **11:** 0163990. doi: 10.1371/*journal.pone*.0163990.

Al-Mssallem, I.S., Hu, S., Zhang, X., Lin, Q., Liu, W., Tan, J., Yu, X., Liu, J., Pan, L., Zhang, T. and Yin, Y. (2013). Genome sequence of the date palm (*Phoenix dactylifera*). *Nature Communications*, **6:** 4.

Alsaihati, B., W. Liu, Q. Lin, and Al-Mssallem, I. S. (2014). Coconut genome *de novo* sequencing. Genome sequencing and ESTs session, January 13, 2014. *Plant and Animal Genome XXII (International Plant and Animal Genome Conference)*, San Diego, California, USA. Poster PO46.

Arikit, S., Yoshihashi, T., Wanchana, S., Uyen, T. T., Huong, N.T., Wongpornchai, S., and Vanavichit, A. (2011). Deficiency in the amino aldehyde dehydrogenase encoded by *GmAMADH2*, the homologue of rice *Os2AP*, enhances 2acetyl1pyrroline biosynthesis in soybeans (*Glycine max* L.). *Plant Biotechnology* Journal, **9**: 75-87.

Bandupriya, H.D. and Dunwell, J.M. (2015). Transcriptome analysis for discovering candidate genes involve in embryogenesis in coconut (*Cocos nucifera* L.) through 454 pyrosequencing. *Journal of the National Science Foundation of Sri Lanka*, **43: 319-336.**

CPCRI (2015). Vision 2050. *Indian Council of Agricultural Research*, New Delhi. 71p.

Dussert, S., Guerin, C., Andersson, M., Joët, T., Tranbarger, T.J., Pizot, M., Sarah, G., Omore, A., Durand-Gasselin, T. and Morcillo, F. (2013). Comparative transcriptome analysis of three oil palm fruit and seed tissues that differ in oil content and fatty acid composition. *Plant Physiology*, **162:** 1337-1358.

Fan, H., Xiao, Y., Yang, Y., Xia, W., Mason, A.S., Xia, Z., Qiao, F., Zhao, S. and Tang, H. (2013). RNA-Seq analysis of *Cocos nucifera*: Transcriptome sequencing and *de novo* assembly for subsequent functional genomics approaches. *PLoS ONE*, **8:** 59997. doi: 10.1371/*journal.pone*.0059997.

Fang, Y., Wu, H., Zhang, T., Yang, M., Yin, Y., Pan, L., Yu, X., Zhang, X., Hu, S., Al-Mssallem, I.S. and Yu, J. (2012). A complete sequence and transcriptomic analyses of date palm (*Phoenix dactylifera* L.) mitochondrial genome. *PloS one*, **7:** 37164.

Galtier, N. (2011). The intriguing evolutionary dynamics of plant mitochondrial DNA. *BMC Biology*, **9:** 1. doi: 10.1186/1741-7007-9-61.

Gongora-Castillo, E. and Buell, C.R. (2013). Bioinformatics challenges in *de novo* transcriptome assembly using short read sequences in the absence of a reference genome sequence. *Natural Product Reports*, **30:** 490-500.

Greilhuber, J., Dolezel, J., Lysak, M.A. and Bennett, M.D. (2005). The origin, evolution and proposed stabilization of the terms 'genome size' and 'C-value' to describe nuclear DNA contents. *Annals of Botany*, **95:** 255-260.

Grimm, C.C., Champagne, E.T., Lloyd, S.W., Easson, M., Condon, B. and McClung, A. (2011). Analysis of 2-acetyl-1-pyrroline in rice by HSSE/GC/MS. *Cereal Chemistry*, **88**: 271-277.

Gualberto, J.M., Mileshina, D., Wallet, C., Niazi, A.K., Weber-Lotfi, F. and Dietrich, A. (2014). *The plant mitochondrial genome: Dynamics and maintenance. Biochimie*, **100**: 107-120.

Gunn, B.F., Baudouin, L., Beulé, T., Ilbert, P., Duperray, C., Crisp, M., Issali, A., Konan, J.L. and Rival, A. (2015). Ploidy and domestication are associated with genome size variation in Palms. *American Journal of Botany*, **102:** 1625-1633.

Huang, Y.Y., Lee, C.P., Fu, J.L., Chang, B.C.H., Matzke, A.J. and Matzke, M. (2014). *De novo* transcriptome sequence assembly from coconut leaves and seeds with a focus on factors involved in RNA-directed DNA methylation. G3: *Genes Genomes Genetics*, **4:** 2147-2157.

Huang, Y.Y., Matzke, A.J. and Matzke, M. (2013). Complete sequence and comparative analysis of the chloroplast genome of coconut palm (*Cocos nucifera*). *PLoS ONE*, **8**: 74736. doi: 10.1371/*journal.pone*.0074736.

Jose, M., Cyriac, M.B., Pai, V., Varghese, I. and Shantaram, M. (2014). Anti-microbial properties of *Cocos nucifera* (coconut) husk: An extrapolation to oral health. *Journal of Natural Science Biology and Medicine*, **5:** 359-364.

Manna, K., Khan, A., Das, D.K., Kesh, S.B., Das, U., Ghosh, S., Dey, R.S., Saha, K.D., Chakraborty, A., Chattopadhyay, S. and Dey, S. (2014). Protective effect of coconut water concentrate and its active component shikimic acid against hydroperoxide mediated oxidative stress through suppression of NF-êB and activation of Nrf2 pathway. *Journal of Ethnopharmacology*, **155:** 132-146.

Nejat, N., Cahill, D.M., Vadamalai, G., Ziemann, M., Rookes, J. and Naderali, N. (2015). Transcriptomics-based analysis using RNA-Seq of the coconut (*Cocos nucifera*) leaf in response to yellow decline phytoplasma infection. *Molecular Genetics and Genomics*, **290:** 1899-1910.

Rajesh, M. K., Rachana, K. E., Babu, M., Thomas, R. J. and Anitha Karun (2013). Characterization of the global transcriptome responsive to root (wilt) disease in coconut using RNA-seq. In: *National Symposium on 'Pathogenomics for Diagnosis and Management of Plant Diseases. CTCRI, Thiruvananthapuram*, India, 24-25 October, 2013.

Rajesh, M.K., Fayas, T.P., Naganeeswaran, S., Rachana, K.E., Bhavyashree, U., Sajini, K.K. and Karun, A. (2015a). *De novo* assembly and characterization of global transcriptome of coconut palm (*Cocos nucifera* L.) embryogenic calli using Illumina paired-end sequencing. *Protoplasma*, **253:** 913-928.

Rajesh, M.K., Rachana, K.E., Naganeeswaran, S., Shafeeq, R., Thomas, R. J., Shareefa, M., Merin, B. and Anitha Karun (2015b). Identification of expressed resistance gene analog sequences in coconut leaf transcriptome and their evolutionary analysis. *Turkish Journal of Agriculture and Forestry*, **39**: 489-502.

Röser, M., Johnson, M.A.T. and Hanson, L. (1997). Nuclear DNA amounts in palms (Arecaceae). *Acta Botanica,* **110:** 79-89.

Saensuk, C., Wanchana, S., Choowongkomon, K., Wongpornchai, S., Kraithong, T., Imsabai, W., Chaichoompu, E., Ruanjaichon, V., Toojinda, T., Vanavichit, A. and Arikit, S. (2016). De novo transcriptome assembly and identification of the gene conferring a "pandan-like" aroma in coconut (*Cocos nucifera* L.). *Plant Science,* **252:** 324-334.

Sandoval, A., Hocher, V. and Verdeil, J.L. (2003). Flow cytometric analysis of the cell cycle in different coconut palm (*Cocos nucifera* L.) tissues cultured *in vitro*. *Plant Cell Reports,* **22:** 25-31.

Singh, R., Ong-Abdullah, M., Low, E.T.L., Manaf, M.A.A., Rosli, R., Nookiah, R., Ooi, L.C.L., Ooi, S.E., Chan, K.L., Halim, M.A. and Azizi, N. (2013). Oil palm genome sequence reveals divergence of interfertile species in Old and New worlds. *Nature,* **500:** 335-339.

Unamba, C.I.N., Nag, A. and Sharma, R.K. (2015). Next generation sequencing technologies: The doorway to the unexplored genomics of non-model plants. *Frontiers in Plant Science,* **6:** 1074. doi: 10.3389/fpls.2015.01074.

Wickramasuriya, A.M. and Dunwell, J.M. (2015). Global scale transcriptome analysis of *Arabidopsis* embryogenesis *in vitro*. *BMC Genomics,* **16:** 301.

Xia, Z., Xu, H., Zhai, J., Li, D., Luo, H., He, C. and Huang, X. (2011). RNA-Seq analysis and *de novo* transcriptome assembly of *Hevea brasiliensis*. *Plant Molecular Biology,* **77:** 299-308.

Zimmerman, J.L. (1993). Somatic embryogenesis: A model for early development in higher plants. *The Plant Cell,* **5:** 1411.

Zonneveld, B.J.M., Leitch, I.J. and Bennett, M.D. (2005). First nuclear DNA amounts in more than 300 angiosperms. *Annals of Botany,* **96:** 229-244.

Chapter 32

Date Palm

☆ *Arvind K. Yadav, Rhitu Rai, Prasanta K. Dash and S.V. Ramesh*

1. Introduction

Date palm (*Phoenix dactylifera* L.) is a multipurpose tree that provides fibre, carbohydrates, minerals and vitamins besides being source of an array of medicinal compounds. The tree is a long-living monocotyledonous plant belonging to the family of Arecaceae. Cultivated for their edible sweet fruit. Because of their long cultivation history their place of origin is unknown but it is speculated to have originated from lands around Iraq (Morton, 1987). However, the plant is widely cultivated and is naturalized in many tropical and subtropical regions worldwide. The date palm trees typically reach about 75-80 feet (23-25 meter) height and the leaves are approximately 20 feet (6 meter) long. The full span of the crown of plants ranges from 6–10 meter (Robinson *et al.*, 2012). It is one of the oldest cultivated plants in the world and is the most important subsistence crop in Northern Africa and the Middle East countries. Nearly 2000 cultivars of date palm are known in the world. Among them, only genotypes some have been evaluated for their performance and fruit quality (Parvin *et al.*, 2015). Because of its high nutritional value and long-life the date palm has been mentioned as the 'tree of life' (Augstburger *et al.*, 2002). Date pulp hold easily digestible sugars (70%), mostly glucose, sucrose and fructose; dietary fibres and less proteins and fats. It also comprises vitamins like riboflavin, biotin, thiamine, ascorbic and folic acid that are essential for the body. The pulp of the fruit is rich in calcium, iron, copper, cobalt, magnesium, fluorine, manganese, phosphorus, potassium, copper, sodium, boron, sulfur, zinc and selenium (Al-Farsi and Lee, 2008). Dates are rich in nutrients and are a good source of rapid energy due to high carbohydrate content (70–80%). Biochemically, date fruits contain fat (0.20–0.50%), protein (2.30– 5.60%), dietary fibre (6.40–11.50%), minerals (0.10– 916 mg/100 g dry weight) and vitamins with very little or no starch (Al-Shahib and Marshal, 2003). The fruit is also a good source of important phytochemicals, including carotenoids, phenolics and flavonoids. Date fruits not only provide antioxidant, anti-mutagenic and immune-modulatory benefits to health but also have diverse medicinal values, including anti-hyperlipidemic, anticancer, gastroprotective, hepatoprotective and nephroprotective properties (Tang *et al.*, 2013).

2. Genetic Map of Date Palm

Date palm is a diploid crop with estimated size of 685Mb. The exact number of date palm chromosomes has been uncertain with some publications reporting 14, 16, or 18 pairs with most evidence pointing to 18 pairs (Al-Salih and Al-Rawi, 1987). Despite its importance from antiquity, few genetic resources are available for improving the productivity and development of the dioecious date palm. Analysis of the genetic maps combined with genomic data will assist in the improvements of this commercially important palm. In order to overcome the challenges Mathew *et al.* (2014), presented the first genetic map of date palm and identified the putative date palm sex chromosome based on a modified gentoyping-by-sequencing (GBS) approach. The total genetic map length was 1293 cM. Assuming a genome size of approximately 670 Mb, genetic to physical distance conversion in date palm is approximately 1.93 cM/Mb of sequence or 520 kb/cM. Inter-marker distances were evenly distributed though some biases were observed. The largest gap between any two markers was 9.7 cM on linkage group (LG) 6. Both LG5 and LG10 had an excess of gaps greater than three cM with a total of nine and seven respectively. Distribution of marker types among the linkage groups showed pronounced marker type bias in LG5, 6, 9 and 10. Analysis of the date palm sex-determination region suggests it is telomeric on linkage group 12 and recombination is not suppressed in the full chromosome.

3. Date Palm Genome

A major initiative in the genomics of date palm was published by Zhang *et al.*, 2012, who generated large scale gene sequences from 14 cDNA libraries representing various tissues derived from the cultivar Khalas. Large scale annotation of gene sequences based on plant database and functional domain annotation based on Pfam database were performed (Zhang *et al.*, 2012). Interestingly, comparative analysis revealed that 70.6 %, 69.4 %, 68.4 % and 69.3 % of date palm gene models were shown to be shared with rice, sorghum (monocots), Arabidopsis and grapevine (dicots) respectively.

The genome sequence of date palm has been made available by sequencing an elite cultivar Khalas (Al-Mssallem *et al.*, 2013). The size of genome is 605.4 Mb and it covers >90% of the genome (~671 Mb) and >96% of its genes (~41,660 genes). The sequencing effort yielded 41,660 gene models (42,957 isoforms) in 10,363 scaffolds (472,329,057 bp in length; 84.6% of the total length). Proteome comparison of date palm to Arabidopsis, Rice, Sorghum and Grape revealed that 8,093 gene families are shared among all five plant genomes and 1,127 gene families are unique to date palm. These unique gene families are mostly related to DNA/RNA metabolic processes and ion binding. In date palm the abundant biologically-defined repeats that accounted for 21.99% of the genome, of which 14.03% and 4.17% are Ty1/Copia and Ty3/Gypsy, respectively. The genome harbours 38.41% repetitive sequences and was observed that Ty1/Copia elements have a much higher copy number than Ty3/Gypsy elements. The most abundant Ty1/Copia elements in *P. dactylifera* show the highest homology to the rice retrotransposon element 1 in the conserved region of the reverse transcriptase genes.

Genome-wide duplication (GWD) provides essential genetic material for the creation of novel functions for adapting new environment and tolerating biotic and abiotic stresses (Freeling, 2009). Using 4,215 paralogous gene pairs in 411 collinear regions of the date palm genome assembly, the distribution of Ks or 4DTv was assessed and it showed two distinct peaks: Ks ~0.314 (4DTv~0.107) and Ks ~0.833 (4DTv~0.332). The first peak corresponds to a GWD event shared among all angiosperms and the second peak derives from either a single more ancient GWD or massive consecutive segmental duplications when the slow substitution rate of the palm family. Significant macro-synteny between date palm and other monocotyledons were also found but macro-synteny is non-existent between date palm and any dicotyledons. The biggest scaffold, pdS00001 (~4.5 Mb in length), appears highly conserved and is part of the 'concentric circles' of monocotyledons.

Whole genome re-sequencing of 62 cultivars of data palm has paved way for generation of a comprehensive registry of approximately seven million single nucleotide polymorphisms (SNPs) that has a greater potential for date palm crop improvement programmes. Population and genetic diversity analysis of date palm cultivars using those SNPs again confirmed a genetic dichotomy between North Africa and the Middle East/South Asian date palms. Furthermore, selection pressure analysis among the population of date palms based on SNP genotyping revealed that geographic adaptation of these cultivars was correlated to preferential selection in few genomic regions. Thus, the analysis is a comprehensive genomics resource for this important crop that captures diversity of date palm post domestication (Hazzouri *et al.*, 2015).

Genotyping-by-sequencing (GBS) is a potential strategy to rapidly describe genetic composition of plant cultivars. A collection of 13,000-65,000 SNPs were utilized in genotyping 70 date palm cultivars using leaf and fruit samples to gain knowledge on the origin of date palm cultivation. SNP based genotyping also provided further credence to the earlier findings that North African and Arabian Gulf regions are the two main earliest centres of date palm domestication. Furthermore, the study found genomic regions of date palm that display high degree of geographic segregation and higher proportion of allele fixation in sex chromosomes (Mathew *et al.*, 2014).

4. Chloroplast Genome

A complete sequence of the date palm chloroplast (cp) genome based on pyrosequencing was reported by Yang *et al.*, (2010). The date palm chloroplast genome is 158,462 bp in length and has a typical circular double-stranded DNA molecule. It shares a common quadripartite structure with the vast majority of other angiosperms: a pair of inverted repeats (IRs, 27,276 bp) separated by the large single-copy (LSC, 86,198 bp) and small single-copy (SSC, 17,712 bp) regions. It encodes 131 predicted functional genes; 112 are unique and 19 are duplicated in the IR regions. Among the 112 unique genes, 79 protein-coding, 29 transfer RNA and 4 ribosomal RNA genes were identified. 50.93%, 1.79%, and 5.71% of the genome sequence encode proteins, tRNAs and rRNAs, respectively, whereas the remaining 41.57% are non coding and filled with introns, intergenic spacers, or pseudogenes. The date palm chloroplast genome has 18 intron-containing genes among the 112 unique genes. Almost all are single-intron except two genes, *ycf*3 and *clp*P, whose exons are separated by two introns. A total of 22,950 codons represent the coding capacity of all protein-coding genes of date palm chloroplast genome. Among these codons, 2001 (8.72%) encode for isoleucine and 271 (1.18%) for cysteine, which were the most and the least amino acids, respectively. Similar to other chloroplast genomes (Raubeson *et al.*, 2007), the date palm cp genome is also AT-rich (62.77%), and the values vary slightly among defined sequences of non-coding, protein-coding, tRNA and rRNA, where their A+T contents are 66.60%, 61.03%, 57.94% and 52.19%, respectively. 78 SNPs were identified as major intra-varietal polymorphisms within the population of a specific chloroplast genome, most of which were located in genes with vital functions. Based on RNA-sequencing data, 18 polycistronic transcription units and three highly expression-biased genes—*atpF*, *trnA-UGC* and *rrn23* were also found (Yang *et al.*, 2010) in date palm.

5. Mitochondrial Genome

Based on next-generation sequencing strategy,data from pyrosequencing and ligation-based sequencing, date palm mitochondrial genome (cultivar Khalas, Al-Hasa Oasis, Saudi Arabia) was deciphered (Fang *et al.*, 2012). Analysis of the mitochondrial genome sequence and transcriptomic data are of importance in revealing mechanisms underlying mitochondrial genome evolution and the unique evolutionary status of date palm among angiosperms. The mitochondrial genome chromosome is made of 715,001 bp circular molecule with an average GC content of 45.1%. Date palm mitochondrial genome represents the fourth largest mitochondrial genome sequenced after cucumber with 1,555,935 bp, melon with 982,833 bp and grape with 773,279 bp. The complete sequence of the date palm mitochondrial genome for analysis is available at GenBank (accession number JN375330). Its protein coding sequence is composed of only 6.5% of the mitochondrial genome (46,770 bp) and this gene content is similar to other angiosperm genomes. The mitochondrial genome contains at least 38 protein-coding genes, 30 tRNAs, three ribosomal RNAs and five complete ORFs. Most of these genes encode proteins of the electron transport chain. The majority (93.5%) of the genome sequence is comprised of chloroplast-derived (10.3% with respect to the whole genome length) sequences and are non-coding which harbours 0.33% tandem and 2.3% long repeats. This is the second highest proportion (10.3%) of chloroplast-derived sequences among the sequenced mitochondrial genomes to date, of which several intact genes, such as petA, petG, petL, psaJ, psbT, rpl20, rpl33 and rps8 are identified (Fang *et al.*, 2012).

Chloroplast and mitochondrial genomes are known to share sequences due to frequent gene transfer events (Stern and Lonsdale, 1982). Frequent DNA transfer from chloroplast DNA to mitochondrial DNA occurs as far back as the common ancestor of the extant gymnosperms and angiosperms, about 300 MYA (million-years-ago) (Wang *et al.* 2007). Fang *et al.* (2012) reported that mitochondrial genome of date palm contains more than 100 fragments of chloroplast origin (over 80% identity) ranges from 50 to 6,521 bp in length. The total fraction of chloroplast DNA sequences of date palm present in its genome is 73,691 bp, corresponding to 10.3% of the whole mitochondrial genome, and 46.5% of date palm chloroplast genome. These finding suggest that chloroplast DNA sequence insertion is an important mechanism for plant mitochondrial genome size expansion and sequence diversity.

6. Transcriptomics

Transcriptome refers to the complete set of all RNA (coding and non-coding), that are transcribed from the genome. Transcriptomics studies using next generation sequencing technologies has widely been adapted to unravel gene expression mechanism in many crops. In date palm transcriptional changes during development of fruit has been studied. Using RNA-Seq data 4,134 differentially expressed genes (DEGs) were identified in date palm whose expressions significantly vary among seven fruit developmental stages. When DEGs were clustered into different groups (up-regulated, down-regulated and not-regulated) it was found that different enrichments of DEGs such as gluconeogenesis, cellular carbohydrate metabolism and small molecule biosynthesis were up-regulated and biological regulation, transcription and regulation of RNA metabolic process were found in the down-regulated group. This large-scale genomic data is a basis for further genomic studies not only on date palm but also in other *Arecaceae* plants (Xin *et al.*, 2015).

Among the non-coding RNAs, microRNAs (miRNAs) are class of small RNAs that play major role in gene regulation thereby decide the growth and development of an organism. Expression profiling of date palm identified 276 novel fruit-development associated (FDA) miRNAs and their targets. Expectedly, most of the miRNA target genes were encoding transcripts involved in starch/ sucrose metabolism (Xin *et al.*, 2015). Deep sequencing of leaves and roots of date palm treated with NaCl stress identified majority of conserved miRNA homologs (153) and novel miRNAs (180). Among the salt stress responsive miRNAs it was demonstrated that 54 and 25 miRNAs derived from leaves and roots respectively were upregulated. The target transcripts for miRNAs were identified to be potassium channel AKT2-like proteins, vacuolar protein sorting-associated protein, calcium-dependent and mitogen-activated proteins (Yaish and Kumar, 2015).

7. Proteomics

Date palm (cultivar 'Barhi') proteome was studied to understand the fruit development biology at molecular level and provide the first detailed comparative proteome of the date fruit (Marondedze *et al.*, 2014). Altogether 189 significantly altered proteins were observed of which 171 were positively identified by using LC MS/MS. Of the 193 identified unique proteins that were significantly differentially expressed 82 proteins and 96 proteins were up or down regulated at different times. Further 36 proteins showed differential accumulation throughout development. It was noted that 29 proteins were specific to date fruit and belong to ten functional categories and these proteins have not been identified in any other fleshy fruits (Bevan *et al.*, 1998). 64 differentially expressed date proteins showed contrasting expression patterns in other fruits. Additionally, five proteins, proline iminopeptidase, aspartyl tRNA synthetase, GDP dissociation inhibitor, disproportionating enzyme and ornithine carbamoyl transferase have been detected in other fleshy fruits, but not identified in this fruit. The identified proteins were classified into 14 functional categories as, 'disease and defense' (16.5%), 'metabolism' (15.4%), 'unclassified' (15.1%), 'protein destination and storage' (10.7%), 'energy' (9.9%), 'cellular structure' (7.0%), 'secondary metabolism' (5.5%), 'signal transduction' (5.1%), 'protein synthesis' (5.1%), 'unclear classification' (2.6%), 'transporters' (2.6%), 'transcription' (2.2%), 'cell growth/division' (1.8%) and 'intracellular traffic' (0.4%). In response to abiotic stress, some proteins have dual roles. In this category, nine protein were identified as date hypanthium-specific, 29 proteins showed contrasting accumulation patterns and seven had similar patterns. Two date-specific proteins involved in protein synthesis, group antigen polymerase (Gag-pol) polyprotein (spot 249) and aspartyl tRNA synthetase were upregulated during ripening. Further, five stress-responsive proteins classified in the category 'protein destination and storage' showed differential accumulation during development. Three proteins were detected as upregulated and two other proteins decreased at all stages. Three heat shock proteins (HSP82) were identified as down regulated at NTR and RIPE, and these were date response specific proteins. Other stress-related proteins were detected as differentially expressed and included universal-stress protein (USP) and E3 ubiquitin ligase. USPs are involved in ethylene-mediated stress adaptation (Sauter *et al.*, 2002) and Ubiquitination plays a crucial role in abiotic and biotic stress responses (Mazzucotelli *et al.*, 2006). A total of 20 proteins involved in energy generation were identified. This proteomics analysis provides insights into physiological processes during date fruit development and ripening, and offers a reference proteome for the study of regulatory mechanisms that can help to improvements of horticultural traits including fruit quality and yield.

In addition, comparative proteome analysis of date palm subjected to salt stress, drought (PEG induced and non-irrigated) stress using protein 2D gel electrophoresis technique showed differential protein expression pattern. Under all the three conditions of stress, ATP synthase CF1 alpha chains were found to be upregulated. Abundance in Rubisco activase and oxygen-evolving enhancer protein 2 was observed in salt and drought stress induced under no-irrigation conditions whereas expression levels of transketolase was significantly altered in PEG-induced drought stress (El-Rabey *et al.*, 2016). Thus proteome analysis has provided a comprehensive insight into drought and salt induced stress in date palm. Most importantly, protein biomarker linked to gender of date palm received a fillip with a proteome study by Dakhlaoui-Dkhil *et al.* (2013). Comparative proteome maps of male and female date palms (cultivar Deglet Nour) identified ABC superfamily ATP binding cassette transporter as male specific protein. The relevance of this protein spot was further confirmed by analysing proteomes of other cultivars such as Aligue, Khouet Aligue, Kentichi and Kenta (Dakhlaoui-Dkhil *et al.*, 2013).

8. Conclusion

Date palm, due to its invaluable nutritious fruit, has invited attention of the biotechnologists worldwide. Developments in the field of sequencing technologies, generation of SNP based markers, has played a greater role in refining genetic and linkage map of date palm. The advent of next generation sequencing (NGS) technologies enabled deciphering transcriptional changes during fruit development and helped identifying gene regulatory networks under the control of small non-coding RNAs such as miRNAs. Thus, it is certain that developments in the field of date palm genomics and transcriptomics would lead to deeper understanding of molecular phenomenon that underlie various important aspects of date palm cultivation, and aid in crop improvement programmes.

References

Al Farsi, M.A. and Lee, C.Y. (2008). Nutritional and functional properties of dates: a review. *Critical Reviews in Food Science and Nutrition*, **48**: 877–887.

Al-Mssallem, I.S., Hu, S., Zhang, X., Lin, Q., Liu, W., Tan, J., Yu, X., Liu, J., Pan, L., Zhang, T. and Yin, Y. (2013). Genome sequence of the date palm *Phoenix dactylifera* L. *Nature Communications*, **4**: 2274. doi:10.1038/ncomms3274.

Al-Salih, A.A. and Al-Rawi, A.M.A. (1987). A study of the cytology of two female cultivars of date palm. *Date Palm Journal*, **5(2)**: 123–142.

Al-Shahib, W. and Marshal, R.J. (2003). The fruit of the date palm: it's possible use as the best food for the future. *Food Science and Nutrition*, **54**: 247–259.

Augstburger, F., Berger, J., Censkowsky, U., Heid, P., Milz, J. and Streit, C. (2002). Date Palm. Naturland, Germany.

Bevan, M., Bancroft, I., Bent, E., Love, K., Goodman, H., Dean, C., Bergkamp, R., Dirske, W., Van Staveren, M. and Stiekema, W. (1998). Analysis of 1.9 Mb of contiguous sequence from chromosome 4 of *Arabidopsis thaliana*. *Nature*, **391**: 485– 488.

Dakhlaoui-Dkhil, S.S., Coquet, L.L., Cosette, P., Elkahoui, S., Song, P.C.T., Vaudry, D., Hannachi, A.S., Trifi, N., Limam, F. and Jouenne, T. (2013). The date palm (*Phoenix dactylifera* L.) leaf proteome: identification of a gender biomarker to screen male parents. *Plant Omics*, **6(1)**: 18-23.

El-Rabey, H.A., Al-Malki, A.L. and Abulnaja, K.O. (2016). Proteome analysis of date palm (*Phoenix dactylifera* L.) under severe drought and salt stress. *International Journal of Genomics*, **2016**: 7840759. doi:10.1155/2016/7840759.

Fang, Y., Wu, H., Zhang, T., Yang, M., Yin, Y., Pan, L., Yu, X., Zhang, X., Hu, S., Al-Mssallem, I.S. and Yu, J. (2012). A Complete sequence and transcriptomic analyses of date palm (*Phoenix dactylifera* L.) mitochondrial genome. *PLoS ONE*, **7(5)**: e37164. doi:10.1371/journal.pone.0037164.

Freeling, M. (2009). Bias in plant gene content following different sorts of duplication: tandem, whole-genome, segmental, or by transposition. *Annual Review of Plant Biology*, **60**: 433–453.

Hazzouri, K.M., Flowers, J.M., Visser, H.J., Khierallah, H.S., Rosas, U., Pham, G.M. and Haider, N. (2015). Whole genome re-sequencing of date palms yields insights into diversification of a fruit tree crop. *Nature Communications*, 6. doi:10.1038/ncomms9824.

Marondedze, C., Gehring, C. and Thomas, L. (2014). Dynamic changes in the date palm fruit proteome during development and ripening. *Horticulture Research*, **1**: 1–15.

Mathew, L.S., Spannagl, M., Al-Malki, A., George, B., Torres, M.F., Al-Dous, E.K., Al-Azwani, E.K., Hussein, E., Mathew, S., Mayer, K.F., Mohamoud, Y.A., Suhre, K. and Malek, J.A. (2014). A first genetic map of date palm (*Phoenix dactylifera*) reveals long-range genome structure conservation in the palms. *BMC Genomics*, **15**: 285. doi:10.1186/1471-2164-15-285.

Mazzucotelli, E., Belloni, S., Marone, D., De Leonardis, A.M., Guerra, D., Di Fonzo, N., Cattivelli, L. and Mastrangelo, A.M. (2006). The E3 ubiquitin ligase gene family in plants: regulation by degradation. *Current Genomics*, **7**: 509–522.

Morton, J.F. (1987). Date. *Fruits of warm climates*. JF Morton. pp. 5-11.

Parvin, S., Easmin, D., Sheikh, A., Biswas, M., Sharma, S.C.D., Jahan, M.G.S., Islam, M.A., Roy, N. and Shovon, M.S. (2015). Nutritional analysis of date fruits (*Phoenix dactylifera* L.) in perspective of Bangladesh. *American Journal of Life Sciences*, **3(4)**: 274-278.

Raubeson, L.A., Peery, R., Chumley, T.W., Dziubek, C., Fourcade, H.M., Boore, J.L. and Jansen, R.K. (2007). Comparative chloroplast genomics: analyses including new sequences from the angiosperms *Nuphar advena* and *Ranunculus macranthus*. *BMC Genomics*, **8**: 174. doi:10.1186/1471-2164-8-174.

Robinson, M.L., Brown, B. and Williams, C. (2012). The date palm in Southern Nevada. *Nevada:* The University of Nevada, 23p.

Sauter, M., Rzewuski, G., Marwedel, T. and Lorbiecke, R. (2002). The novel ethylene-regulated gene OsUsp1 from rice encodes a member of a plant protein family related to prokaryotic universal stress proteins. *Journal of Experimental Botany*, **53**: 2325-2331.

Stern, D.B. and Lonsdale, D.M. (1982). Mitochondrial and chloroplast genomes of maize have a 12-kilobase DNA sequence in common. *Nature*, **299**: 698–702.

Tang, Z.X., Shi, L.E. and Aleid, S.M. (2013). Date fruit: chemical composition, nutritional and medicinal values, products. *Journal of the Science of Food and Agriculture*, **93**: 2351–2361.

Wang, D., Wu, Y.W., Shih, A.C., Wu, C.S., Wang, Y.N. and Chaw, S.M. (2007). Transfer of chloroplast genomic DNA to mitochondrial genome occurred at least 300 MYA. *Molecular Biology and Evolution*, **24**: 2040–2048.

Xin, C., Liu, W., Lin, Q., Zhang, X., Cui, P., Li, F., Zhang, G., Pan, L., Al-Amer, A., Mei, H. and Al-Mssallem, I.S. (2015). Profiling microRNA expression during multi-staged date palm (*Phoenix dactylifera* L.) fruit development. *Genomics*, **105(4)**: 242-251.

Yaish, M.W. and Kumar, P.P. (2015). Salt tolerance research in date palm tree (*Phoenix dactylifera* L.), past, present, and future perspectives. *Frontiers in Plant Science*, **6**: 348. doi:10.3389/fpls.2015.00348.

Yang, M., Zhang, X., Liu, G., Yin, Y., Chen, K., Yun, Q., Zhao, D., Al-Mssallem, I.S. and Yu, J. (2010). The complete chloroplast genome sequence of date palm (*Phoenix dactylifera* L.). *PLoS ONE*, **5(9)**: e12762. doi:10.1371/journal.pone.0012762.

Zhang, G., Pan, L., Yin, Y., Liu, W., Huang, D., Zhang, T., Wang, L., Xin, C., Lin, Q., Sun, G. and Abdullah, M.M. (2012). Large-scale collection and annotation of gene models for date palm (*Phoenix dactylifera* L.). *Plant Molecular Biology*, **79(6)**: 521-536.

Chapter 33

Cocoa

☆ *V. Arunachalam*

1. Introduction

Cocoa (*Theobroma cacao* L.) is an important perennial tree of the tropics and forms a major ingredient of chocolates and confectionery recipes. It is a shade loving tree commonly grown as inter/mixed crop in orchards and plantation crops. Cocoa is a diploid plant (2n = 2x = 20) and belongs to family *Malvaceae* with an estimated genome size of 380 Mbp (Figueira *et al.*, 1992). Genetic improvement of cocoa is aimed at improving traits such as yield, bean quality and resistance to diseases. Three groups of cacao cultivars are known traditionally worldwide: Criollo, Forastero and the hybrid between the two, Trinitario. Criollo is a high quality variety, but with poor yield and susceptible to fungal diseases. Currently, microsatellite markers have greatly aided in grouping the cacao cultivars into ten major groups (Motamayor *et al.*, 2008). Cocoa is vegetatively propagated by grafting method- hence desirable single plants can be multiplied clonally. Clonal selection among progenies helps in the further improvement in the traits of interest. Recent developments in genomics and transcriptomics research have paved the way for understanding of genes, enzymes, and pathways of important traits of cocoa crop. Genomic science has provided novel molecular markers hence quick strategies are available to cocoa breeders. Readers are advised to refer review articles on cocoa agronomy and general aspects (Badrie *et al.*, 2015); genomics (Bennett, 2003; Guiltinan *et al.*, 2008), linkage analysis and QTL mapping (Lanuad *et al.*, 2009), Witches' broom disease interactions (Teixeira *et al.*, 2015) for further details. The objective of the current chapter is to highlight the major developments in the areas of cocoa genomics and transcriptomics research.

2. Cocoa Genomics

The cacao plant is characterised by small chromosomes, single secondary constriction, and lack of C-banding due to small size (2.01×10^8) of its genome (Couch *et al.*, 1993). Besides these physical attributes of chromosomes, extent of methylation in the cacao genome has also been studied in detail. A massive transcriptome resource was generated by sequencing 56 cDNA libraries representing various tissues, genotypes and environments, which yielded 149,650 sequences linked to genes governing important traits in cocoa (Argout *et al.*, 2008). Bacterial Artificial Chromosome (BAC) resources were developed for the first time by Clement *et al.* (2004) to aid physical mapping of resistance genes. Generated from the genotype 'SCA6', the BAC resource contains approximately 11 genome equivalents, with an average insert size of 120 kb. Soon, a second BAC library was created; the library obtained from 'LCT-EEN37' genotype, collected in Ecuador, represents approximately 11 genome equivalents with an average insert size of 120 kb. BAC libraries were also generated from Criollo genotype 'B97-61', collected from Belize and a Forastero genotype, 'Matina 1-6', collected from lower Amazon. The main aim of construction of these BAC resources was to support genome sequencing projects. BAC-based approach was also followed to sequence and assemble a QTL-rich region, of ~3 Mbp cacao genome which revealed several important genes governing the traits such as black pod disease resistance, bean shape index, and pod weight (Feltus *et al.*, 2011).

The International Cacao Genome Sequencing (ICGS) consortium was formally organized in 2006 at the International Cocoa Research Conference held in Costa Rica. Two years later, in 2008, the Cocoa Genome Consortium (CGS)—an industry-funded partnership—was formed. The objectives of these two consortiums were to undertake sequencing of the genomes of two quite distinct genetic groups of cacao: Criollo and Amelonado, respectively. The rationale of sequencing two different genetic types of cocoa was that the comparison of these two quite distinct types would provide a far deeper understanding of the structure and function of the cocoa genome compared to a single genome (Guiltinan and Maximova, 2015). Both these types were chosen because of their highly homozygous genomes which would immensely facilitate final genome assemblies.

The ICGS sequenced the genome of a Criollo type (B97-61/B2) collected from Maya mountains of Belize (Argout *et al.*, 2011), whereas the CGS sequenced Matina 1-6, a Costa Rican variety from the Matina river valley (Motamayor *et al.*, 2013). The sequencing data revealed a difference of about 3.4% (430 Mbp and 445 Mbp for B97-61/B2 and Matina 1-6, respectively) between the genomes of these two sequenced types of cacao; most of the difference was accounted for by the increased amount of repetitive DNA and transposons found in the Matina 1-6 genome (Motamayor *et al.*, 2013). Approximately 29, 000 genes have been predicted in cacao genome, similar to number of genes found in the model plant *Arabidopsis*. In addition, over 700 novel genes were found; these genes have been suggested to be involved in specializations within the cacao lineage (family Malvaceae).The global organization of the genomes of the two cocoa types was found to be quite similar; however, 12 relatively small regions were found to be located on different chromosomes in the two types, which could have possibly have arisen as a result of transposon activity (Motamayor *et al.*, 2013). The release of these two genomes is a major milestone in the genomics of cocoa as it offered interesting conclusions and useful information for cocoa breeders. Argout *et al.* (2011) have also listed the genes responsible for fungal and oomycetes resistance and flavour quality of cocoa.

3. Omics of Fungal Disease Resistance

3.1. Witches' Broom Disease

Witches' broom disease (WBD), caused by the pathogenic hemibiotrophic fungus *Moniliophthora perniciosa*, is a major disease of *Theobroma cacao* and can cause up to 90% yield loss. The disease has severely affected the cocoa industry in Brazil and is a major factor of yield loss in cacao (Teixeira *et al.*, 2014). Gesteira *et al.* (2007) generated cDNA libraries from meristem of resistant cacao genotype, TSH1188, and the susceptible genotype, Catongo, after inoculation with *M. perniciosa*. From these two libraries, a total of 6884 ESTs could be obtained, which corresponded to 2926 non-redundant sequences (2585 singletons plus 341 contigs). Putative functional categories could be assigned to 54% of these sequences. Even though the overall distribution of sequences in functional categories between the two libraries was quite similar, differences could be observed with respect to genes encoding PR proteins in TSH188, the resistant genotype, and genes involved in programmed cell death (PCD) in Catongo, the susceptible genotype.

Two SSH (subtractive suppressive hybridization) libraries were constructed from meristems, collected from the resistant genotype CAB 214 and the susceptible genotype ICS39, after inoculation with *M. perniciosa* by Leal *et al.* (2007), subtracting common transcripts in both directions. A total of 104 and 187 unique sequences were obtained respectively, from each of these two libraries. Out of the 23 genes evaluated by RT-qPCR, only 16 were induced in the susceptible genotype, while 21 were induced in the resistant genotype.

Gene expression analyses in the disease resistant variety 'TSH1188' and the susceptible variety 'Catongo', revealed the production of ROS (reactive oxygen species) and elicitor molecules during infection followed by detoxification of ROS in the resistant variety (da Hora Junior *et al.*, 2012). The report also identified 154 and 227 genes from TSH118 and Catongo, respectively, which were differentially expressed during fungal infection. One hundred and fifty-three genes, potentially related to plant-pathogen interaction, were identified from infected plant tissue libraries along with 71 putative SNPs (Lima *et al.*, 2009).

Lopes *et al.* (2010) focused on cacao transcription factors (TFs) by developing a macroarray with 88 TF cDNA from interaction libraries (Gesteira *et al.*, 2007). Seventy-two TFs were found differentially expressed between the susceptible (Catongo) and resistant (TSH1188) genotypes and/or during the disease time course—from 24 to 30 days after infection. Most of the TFs differentially expressed belonged to bZIP, MYB and WRKY families, and presented opposite expression patterns in susceptible and resistant cacao–*M. perniciosa* interactions. The results of the macroarray were confirmed by RT-qPCR for bZIP and WRKY TFs (Lopes *et al.*,

2010). On the other hand, SVP (short vegetative phase), which shared similarity with a *Populus tomentosa* MAD-Box transcription factor, was up-regulated in resistant CAB plants (Leal *et al.*, 2007). Dual RNA-seq analysis was performed to characterize transcriptional changes of both cacao and *M. perniciosa* during infection of pathogen in biotrophic stage. The normal transcriptional machinery of cacao was found to be heavily disturbed during the infection of *M. perniciosa* leading to hormonal imbalances in host (Teixera *et al.*, 2014). In addition, infection due to *M.perniciosa* causes carbon deprivation status in the cacao plants leading to premature senescence. Thus the phenotypic symptoms associated with fungal infection and transition from biotropic to necrotic stage has been correlated with transcriptional changes of cacao plants. Transcriptomics studies of fungus in necrotrophic phase revealed upregulation of the secreted proteins that are actively involved in pathogenesis by degradation of plant cell wall (Meinhardt *et al.*, 2014).

Analysis of segregating mapping population of cocoa (derived from a cross between the resistant 'TSH 1188' and the tolerant 'CCN 51') for Witches' broom disease (WBD) identified seven QTLs spread over five chromosomes that confer resistant to WBD. The study not only identified potential candidate disease resistance genes in the QTL regions, but a few SNP-based molecular markers were also proposed to aid breeding for resistance against WBD (Royaert *et al.*, 2016). Ultimately, greater understanding of molecular mechanism underlying WBD has led to development of WBD Transcriptome Atlas initiative (http://www.lge.ibi.unicamp.br/wbdatlas), which is a repository of all sequence libraries generated from cacao-*M. perniciosa* interactions.

3.2. Black Pod /Pod Rot Disease

Black pod, caused by *Phytophthora megakarya,* is a major disease of cocoa in West African and Asian countries. The genetic resistance of cacao to three pathogens of pod rot (*Phytophthora palmivora, P. megakarya* and *P. capsici*) were studied by linkage analysis. A total of 13 QTLs, spread over six chromosomes, were identified. Among these, a major one was qPsp-5 governing resistance to five races and three species of pathogens (Risterucci *et al.*, 2003). Three QTLs for black pod resistance were found on LG 4, 8, and 10, with the most favourable alleles coming from the cacao genotype Pound 7 (Brown *et al.*, 2007). A meta-QTL study on cocoa combining QTL and linkage mapping experiments revealed only 13 major and consensus QTLs (Clement *et al.*, 2003; Brown *et al.*, 2007) from the linkage studies (Review: Lanuad *et al.*, 2009). The study lists unique and common QTLs for pod rot and disease resistance. *In silico* analysis of transcriptomics data from cocoa infected by black pod disease, found 272 enzymes corresponding to 114 metabolic pathways. The annotated enzymes from the study were involved in amino acid biosynthesis and phenylpropanoid biosynthesis. The study has implications in understanding the biotic stress response pathway in cacao (Naganeeswaran *et al.*, 2012). A major QTL region spanning 3 Mb size on LG5 covering the resistance to black pod disease was sequenced from Matina 1–6 cacao clone to identify SNP markers on COS (conserved orthologous sequence) genes (Kuhn *et al.*, 2012). Unravelling molecular mechanism underlying resistance to pod rot pathogen (*Phytophthora tropicalis*) in two cacao genotypes with contrasting disease resistance traits Scania6 (Sca6-resistant) and Imperial College Selection 1

(ICS1-susceptible), revealed that salicylic acid treatment enhanced production of reactive oxygen species (ROS) through upregulation of organeller genes involved in ROS production in Sca6 whereas ICS1 produced pathogenesis related proteins (Fister et al., 2015). Furthermore, it was also deduced that transient overexpression of *TcNPR*1- a transcriptional regulator involved in salicylic acid dependent immune system also- enhanced resistance to pod rot pathogen (Fister *et al.*, 2015).

3.3. Ceratocystis Wilt Resistance

Ceratocystis wilt is a lethal wilt disease of cacao present in Caribbean and Central and South America. A mapping population involving Scavina 6 (Sca 6; resistant genotype) × Imperial College Selections 1 (ICS 1; susceptible genotype) and a set of EST-SSR markers was employed by Santos *et al.* (2013) to identify additional SSR markers (CEPEC13, CEPEC 14, CEPEC 28 and CEPEC 17) and to tag resistance genes of the disease.

4. Genomics of Flooding Stress Tolerance

Studies conducted regarding flooding stress tolerance (40-day flooding) on 35 elite cacao genotypes, also identified polymorphism for 248 alleles of 18 microsatellite loci (Bertolde *et al.*, 2010). Gene expression pattern in cocoa plants suffering from soil anoxia caused by flooding found three major genes (alcohol / lactate hydrogenases and pyruvate carboxylase in leaves and roots. Activity of these enzymes also differed during flooding in TSA-792 and TSH-774 the stress tolerant and susceptible genotypes respectively (Bertolde *et al.*, 2014).

5. Omics of Cacao *in vitro* Culture

Deciphering molecular mechanisms underlying cocoa somatic embryogenesis (SE) would help devise better tools for regeneration. A gene governing somatic embryogenesis in cocoa called as leafy cotyledon1 (*TcLEC1*) was characterised by Alemanno *et al.* (2008). Later, a candidate gene *TcBBM*, known as baby boom transcription factor orthologous to that of *Arabidopsis*, was identified as a biomarker for embryogenesis in cacao tissue (Florez *et al.*, 2015) and was confirmed by over-expression in *Arabidopsis* and cocoa transgenic systems. Transient expression studies of *TcLEC1* and *TcBBM* and enhanced somatic embryogenesis observed in cacao underlined the importance of transcriptional factors (TFs) induced embryogenesis and identification of functional biomarkers associated with SE in cacao (Zhang *et al.*, 2014; Florez *et al.*, 2015). Most importantly, studies on cacao TFs have opened new avenues for development of efficient regeneration system thereby making cocoa transgenics a possibility.

6. Genomics of Pod Colour

The genome sequence of Matina 1-6 was analysed *via* haplotype, association mapping and gene expression studies to identify candidate gene(s) that governs cacao pod colour (Motomayor *et al.*, 2013). The R2R3 MYB class transcription factor *TcMYB113* has been found to be involved in red colour pigmentation of cocoa pods. Furthermore, SNP identified in the *TcMYB113* has been shown to affect the activity of *trans*-acting siRNAs (siRNAs) targeting *TcMYB113* and hence cause pod colour variation (Motomayor et al., 2013).

7. Omics of Cacao Flavour

Flavour of cacao is an important criterion for chocolate industry and it has also been demonstrated that flavour is a genetically controlled character besides flavour components depend on post-harvest processing conditions (Clapperton *et al.*, 1994). Relative expression levels of *TcLIS* (cacao linalool synthase) from cotyledons during fermentation were found to increase in 'ICS1' and 'Nacional' seeds (Sabau *et al.*, 2006). Similarly, to decipher other quality parameters, cDNA macroarray based expression analysis was carried out to delineate genes involved in terpenes and polyphenol biosynthesis pathways (Sabau *et al.*, 2012).

8. Bioinformatics and Databases

Storage and retrieval of the genomics data on user-friendly online databases and web servers is important so as to use the information for crop breeding. Table 33.1 lists GenBank accessions that house major genomic and transcriptomic resources of cacao. Cacao genome database (http:// www.cacaogenomedb.org) provides several services of visualization and browsing of the DNA sequence, genes on all the 10 chromosomes of cocoa plant (Zheng, 2012). TropGeneDB (http://tropgenedb.cirad.fr) holds updated information regarding cocoa linkage maps, QTLs, molecular markers and genes (Hamelin *et al.*, 2013).

Table 33.1: Major Genomic and Transcriptomic Resources of Cacao

GenBank Id(s)	*Type*	*Materials*	*Reference*
CU469588 to CU633156	56 EST libraries	Genotypes differing self compatibility, disease resistance, flavour / quality of bean and corresponding tissues	Argout *et al.* (2008)
CACC01000001–CACC01025912	Whole genome sequence of 10 chromosomes	Criollo cultivar	Argout *et al.* (2011)
ALXC01000000	Genome assembly of whole genome	Matina 1-6 clone	Motamayor *et al.* (2013)

With the advent of cost and time effective sequencing technologies, use of computational approaches becomes essential for analysis of voluminous data from genomics and transcriptomics projects (review by Arunachalam, 2014). Gene index provides curated sequence information and details of the known genes in an organism. TIGR gene index was developed for cocoa plant (Quackenbush *et al.*, 2001). Few attempts were made using bioinformatics tools to mine the SSR (Riju *et al.*, 2009) and SNP (Lima *et al.*, 2009; Riju and Arunachalam, 2010) markers. The SNPs mined from cocoa ESTs are available in the online database (http://www.riju.byethost31.com/cocoa/ccsnp.html). Of the 6578 EST sequences from seven tissues/libraries, a density of one SNP/166 bp and one Indel/360 bp were found in cocoa. Similarly, Cacao EST sequences are mined for simple sequence repeats by performing computational analysis . Identified SSRs and primers designed were made available as online database as http://riju.byethost31.com/cocoa/. Putative function of the simple sequence repeat containing sequences were analysed *in silico* and found to be PGK (phosphoglycerate kinase) and gibberellin 20-oxidase 1, among others (Riju *et al.*, 2009).

A bioinformatics tool has been developed to locate RAPD/ISSR primers and design *in silico* SCAR primers (PremKrishnan and Arunachalam, 2012). The software was used to mine the complete genome, EST and core nucleotide sequences of cacao for RAPD/ISSR priming sites and iSCAR (*in silico* Sequence Characterised Amplified Region) Markers (http://www.bioinfoindia.org/fv-iscardb/) have been designed. The database current holds 9123 predicted SCAR markers for use in cacao breeding programs (PremKrishnan and Arunachalam, in press).

Conclusion

The availability of the cacao genome sequence will accelerate the discovery of candidate genes underlying important QTLs identified in genetic studies and in relation to functional genomics. The genome sequencing of both contrasting Criollo and Forastero genotypes, originated from distinct genetic groups, will allow to produce a wide SNP resource useful for all genetic and genomics studies. Developments in the field of cacao 'omics' especially in transcriptomics have unraveled role of molecular components that play significant role in somatic embryogenesis, conferring resistance to diseases such as WBD, pod rot, identification of biomarkers associated with flooding tolerance etc would aid in effective designing future crop improvement programmes.

References

Alemanno, L., Devic, M., Niemenak, N., Sanier, C., Guilleminot, J., Rio, M. and Montoro, P. (2008). Characterization of leafy cotyledon1-like during embryogenesis in *Theobroma cacao* L. *Planta*, **227(4)**: 853-866.

Argout, X., Fouet, O., Wincker, P., Gramacho, K., Legavre, T., Sabau, X., Risterucci, A.M, Da Silva, C., Cascardo, J., Allegre, M. and Kuhn, D. (2008). Towards the understanding of the cocoa transcriptome: Production and analysis of an exhaustive dataset of ESTs of *Theobroma cacao* L. generated from various tissues and under various conditions. *BMC Genomics*, **9(1)**: 512.

Argout, X., Salse, J., Aury, J. M., Guiltinan, M. J., Droc, G., Gouzy, J. Allegre, M., Chaparro, C., Legavre, T., Maximova, S.N. and Abrouk, M. (2011). The genome of *Theobroma cacao*. *Nature Genetics*, **43**: 101-108.

Arunachalam, V. (2014). Computational genomics of plants. CAB Reviews. Perspectives in Agriculture, Veterinary Science, Nutrition and Natural Resources.*CAB Reviews*: (CABI Press): **9(036)**: 1-9.

Badrie, N., Bekele, F., Sikora, E. and Sikora, M. (2015). Cocoa agronomy, quality, nutritional, and health aspects. *Critical Reviews in Food science and Nutrition*, **55**(5): 620-659.

Bennett, A. B. (2003). Out of the Amazon: *Theobroma cacao* enters the genomic era. *Trends in Plant Sciences*, **8(12)**: 561-563.

Bertolde, F.Z., Almeida, A.A.F. and Pirovani, C.P. (2014). Analysis of gene expression and proteomic profiles of clonal genotypes from *Theobroma cacao* subjected to soil flooding. *PLoS ONE*, **9**(10): e108705. doi: 10.1371/journal.pone.0108705.

Bertolde, F.Z., Almeida, A-AF, Correa, R.X., Gomes, F.P., Gaiotto, F.A., Baligar, V.C. and Loguercio, L. L. (2010). Molecular, physiological and morphological analysis of water logging tolerance in clonal genotypes of *Theobroma cacao* L. *Tree Physiology*, **30**: 56–67.

Brown, J. S., Phillips-Mora, W., Power, E. J., Krol, C., Cervantes-Martinez, C., Motamayor, J. C. and Schnell, R. J. (2007). Mapping QTLs for resistance to frosty pod and black pod diseases and horticultural traits in *Theobroma cacao* L. *Crop Science*, **47**(5): 1851-1858.

Clapperton, J. F., Yow, S. T. K., Chan, J. and Lim, D. H. K. (1994). Effects of planting materials on flavour. *Cocoa Growers' Bulletin*, 48: 47–59.

Clement, D., Lanaud, C., Sabau, X., Fouet, O., Le Cunff, L., Ruiz, E., Risterucci, A.M., Glaszmann, J.C. and Piffanelli, P. (2004). Creation of BAC genomic resources for cocoa (*Theobroma cacao* L.) for physical mapping of RGA containing BAC clones. *Theoretical and Applied Genetics*, **108**(8): 1627-1634.

Clement, D., Risterucci, A. M., Motamayor, J. C., N'goran, J. and Lanaud, C. (2003). Mapping QTL for yield components, vigour and resistance to *Phytophthora palmivora* in *Theobroma cacao* L. *Genome*, **46(2)**: 204-212.

Couch, J. A., Zintel, H. A. and Fritz, P. J. (1993). The genome of the tropical tree *Theobroma cacao* L. *Molecular and General Genetics*, **237(1-2)**: 123-128.

Feltus, F.A., Saski, C.A., Mockaitis, K., Haiminen, N., Parida, L., Smith, Z., Ford, J.,Staton, M.E., Ficklin, S.P., Blackmon, B.P., Cheng, C.H. (2011). Sequencing of a QTL-rich region of the Theobroma cacao genome using pooled BACs and the identification of trait specific candidate genes. *BMC Genomics*, **27**(12): 379.

Figueira, A., Janik, J. and Goldsbrough, P. (1992). Genome size and DNA polymorphism in *Theobroma cacao*. *Journal of the American Society for Horticultural Science*, **117:** 673-677.

Fister, A.S., O'Neil, S.T., Shi, Z., Zhang, Y., Tyler, B.M., Guiltinan, M.J. and Maximova, S.N. (2015). Two *Theobroma cacao* genotypes with contrasting pathogen tolerance show aberrant transcriptional and ROS responses after salicylic acid treatment. *Journal of Experimental Botany*, doi:10.1093/jxb/erv334.

Florez, S. L., Erwin, R. L., Maximova, S. N., Guiltinan, M. J. and Curtis, W. R. (2015). Enhanced somatic embryogenesis in *Theobroma cacao* using the homologous BABYBOOM transcription factor. *BMC Plant biology*, **15**(1): 121.

Gesteira, A.S., Micheli, F., Carels, N., da Silva, A.C., Gramacho, K.P., Schuster, I., Macedo, J., Pereira, G.A.G. and Cascardo, J.C.M., 2007. Comparative analysis of expressed genes from cacao meristems infected by *Moniliophthora perniciosa*. *Annals of Botany*, **100**: 129–140

Guiltinan, M. J., Verica, J., Zhang, D. and Figueira, A. (2008). Genomics of Theobroma cacao, "The Food of the Gods". In: *Genomics of Tropical Crop Plants*. Springer, New York. pp. 145-170.

Hamelin, C., Sempere, G., Jouffe, V. and Ruiz, M. (2013), TropGeneDB, the multi-tropical crop information system updated andextended. *Nucleic Acids Research*, **41**: D1172–1175.

Kuhn, D. N., Livingstone III, D., Main, D., Zheng, P., Saski, C., Feltus, F. A., Mockaitis, K., Farmer, A.D., May, G.D., Schnell, R.J. and Motamayor, J.C. (2012). Identification and mapping of conserved ortholog set (COS) II sequences of cacao and their conversion to SNP markers for marker-assisted selection in Theobroma cacao and comparative genomics studies. Tree Genetics & Genomes, **8(1):** 97-111.

Lanaud, C., Fouet, O., Clément, D., Boccara, M., Risterucci, A. M., Surujdeo-Maharaj, S. and Argout, X. (2009). A meta–QTL analysis of disease resistance traits of *Theobroma cacao* L. *Molecular Breeding*, **24**(4): 361-374.

Leal, Jr, G.A., Albuquerque, P.S.B., Figueira A. (2007). Genes differentially expressed in Theobroma cacao associated with resistance to witches' broom disease caused by *Crinipellis perniciosa*. *Molecular Plant Pathology*, **8**: 279-292.

Lima, L.S., Gramacho, K.P, Carels, N., Novais, R., Gaiotto, F.A., Lopes, U.V., Gesteira, A.S., Zaidan, H.A., Cascardo, J.C.M., Pires, J.L. and Micheli, F. (2009). Single nucleotide polymorphisms from *Theobroma cacao* expressed sequence tags associated with witches' broom disease in cacao. *Genetics and MolecularResearch*, **8(3)**: 799-808.

López, C.M.R., Wetten, A.C. and Wilkinson, M.J. (2010). Progressive erosion of genetic and epigenetic variation in callus-derived cocoa (*Theobroma cacao*) plants. *New Phytologist*, **186**: 856-868.

Meinhardt, L.W., Costa, G.G., Thomazella, D.P., Teixeira, P.J., Carazzolle, M.F., Schuster, S.C., Carlson, J.E., Guiltinan, M.J., Mieczkowski, P., Farmer, A. and Ramaraj, T. (2014). Genome and secretome analysis of the hemibiotrophic fungal pathogen,Moniliophthoraroreri, which causes frosty pod rot disease of cacao: mechanisms of the biotrophic and necrotrophic phases. *BMC Genomics*, **27(15)**:164.

Motamayor, J. C., Lachenaud, P., Mota, J. W. D. S., Loor, R., Kuhn, D. N., Brown, J. S. and Schnell, R. J. (2008), Geographic and genetic population differentiation of the Amazonian chocolate tree (*Theobroma cacao* L). *PLoS ONE*, **3**(10): e3311.

Motamayor, J.C., Mockaitis, K., Schmutz, J., Haiminen, N., Livingstone, D. III, Cornejo, O., Findley, S.D., Zheng, P., Utro, F., Royaert, S., Saski, C., Jenkins, J., Podicheti, R., Zhao, M., Scheffler, B.E., Stack, J.C., Feltus, F.A., Mustiga, G.M., Amores, F., Phillips, W., Marelli, J.P., May, G.D., Shapiro, H., Ma, J., Bustamante, C.D., Schnell, R.J., Main, D., Gilbert, D., Parida, L. and Kuhn, D.N. (2013). The genome sequence of the most widely cultivated cacao type and its use to identify candidate genes regulating pod colour. *Genome Biology*, **14**: 1–24.

Naganeeswaran, S.A., Subbian, E.A. and Ramaswamy, M. (2012). Analysis of expressed sequence tags (ESTs) from cocoa (*Theobroma cacao* L.) upon infection with *Phytophthora megacarya*. **8**: 65-69.

Premkrishnan, B.V. and Arunachalam, V. (2012). *In Silico* RAPD priming sites in expressed sequences and iSCAR markers for oil palm. *Comparative and Functional Genomics*, doi: 10.1155/2012/913709.

Premkrishnan, B.V. and Arunachalam, V. (In Press). Database of predicted SCAR markers in five fruit and three vegetable crops. *J Genet*. Unedited article online http://www.ias.ac.in/public/Resources/General/jgen/jgen-14-703-ue.pdf.

Quackenbush J., Cho D., Lee F. L., Hott I., Karamychera S. And Parizi B. (2001). The *TIGR* gene indices: Analysis of gene transient sequences in highly sample eukaryotic species. *Nucleic Acids Research*, **29**: 159–164.

Riju, A. and Arunachalam, V. (2010). Electronic sorting of SNP/Indel Sites in expressed sequence tag libraries of cocoa (*Theobroma cacao* L.). *Genes, Genomes and Genomics*, **4**: 37-40.

Riju, A., Rajesh, M.K.,Sherin, P.T.P.F., Chandrasekar, A., Apshara, S.E., and Arunachalam, V. (2009). Mining of expressed sequence tag libraries of cacao for microsatellite markers using five computational tools. *Journal of Genetics*, **88**: 217–225.

Risterucci, A. M., Paulin, D., Ducamp, M., N'goran, J. A. K. and Lanaud, C. (2003). Identification of QTLs related to cocoa resistance to three species of *Phytophthora*. *Theoretical and Applied Genetics*, **108(1)**: 168-174.

Royaert, S., Jansen, J., da Silva, D.V., de Jesus Branco, S.M., Livingstone, D.S., Mustiga, G., Marelli, J.P., Araújo, I.S., Corrêa, R.X. and Motamayor, J.C. (2016) .Identification of candidate genes involved in Witches' broom disease resistance in a segregating mapping population of *Theobroma cacao* L. in Brazil. *BMC Genomics*, **17(1)**: 1.

Sabau, X., LoorSolorzano, R. G., Boccara, M., Fouet, O., Jeanneau, M., Argout, X., Legavre, T., Risterucci, A. M., Vincker, P., Da Silva, C., Jimenez, J. C. and Cros, E. (2006). Preliminary results on linalool synthase expression during seed development and fermentation of nacional and trinitario clones. *15th International Cacao Research Conference*, San Jose, Costa Rica.

Santos L.L.L., Legravre, T., Lanaud, C. Schnell, R.J., Pires, J.L., Lopes, U.V., Micheli. F. and Gramacho, K.P. (2013). Identification, characterization and mapping of EST-derived SSRs from the cacao–*Ceratocystis cacaofunesta* interaction *Tree Genetics & Genomes*, **9**: 117–127.

Teixeira, P.J., Thomazella, D.P., Reis, O., do Prado, P.F., do Rio, M.C., Fiorin, G.L., José , J., Costa, G.G., Negri, V.A., Mondego, J.M. and Mieczkowski P. (2014). High-resolution transcript profiling of the atypical biotrophic interaction between *Theobroma cacao* and the fungal pathogen *Moniliophthora perniciosa*. *Plant Cell*, **26(11)**: 4245-4269.

Teixeira, P.J.P.L., Thomazella, D.P.dT. and Pereira, G.A.G. (2015). Time for Chocolate: Current understanding and new perspectives on cacao witches' broom disease research. *PLoS Pathology*, **11**(10): e1005130. doi:10.1371/journal.ppat.1005130.

Zhang, Y., Clemens, A., Maximova, S.N. and Guiltinan, M.J. (2014). The *Theobroma cacao* B3 domain transcription factor *TcLEC2* plays a duel role in control of embryo development and maturation. *BMC Plant Biology*, **14**: 106. doi: 10.1186/1471-2229-14-106

Zheng, P. (2012). Cacao Genome Database. In: *Plant and Animal Genome XX Conference*, January 14-18, 2012.

Index

C

S

T

U

V

W

Z

Figures 1.1a-f: Plantlet Regeneration from Immature Anthers and (g) RAPD Analysis with Primer OPB20. (p. 9)

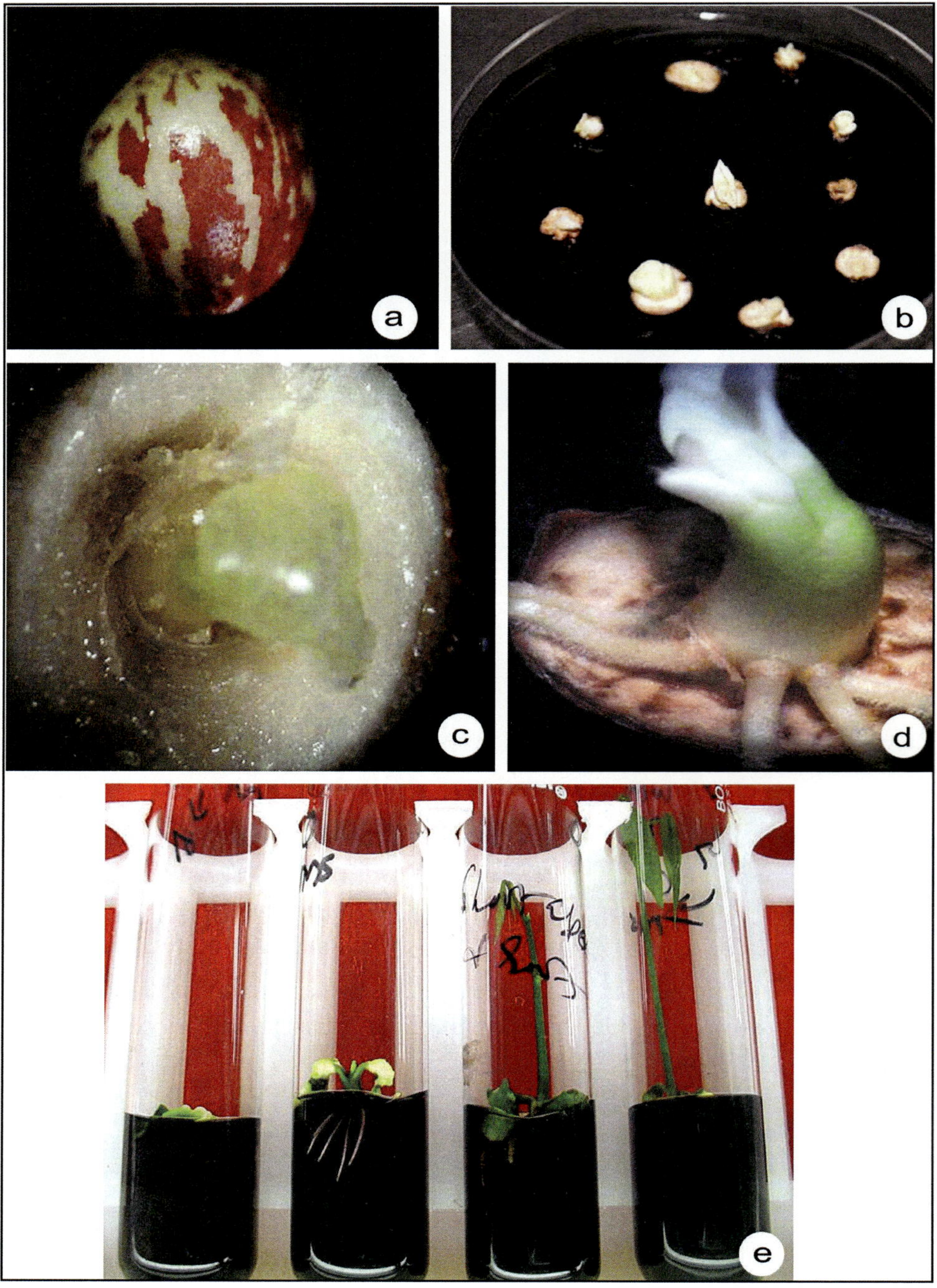

Figure 1.2: Different Stages of Plant Regeneration via Embryo Rescue: (a) Seed development, (b) Ovule culture, (c) Developing embryo, (d) Root differentiation and (e) Plantlet regeneration. (p. 16)

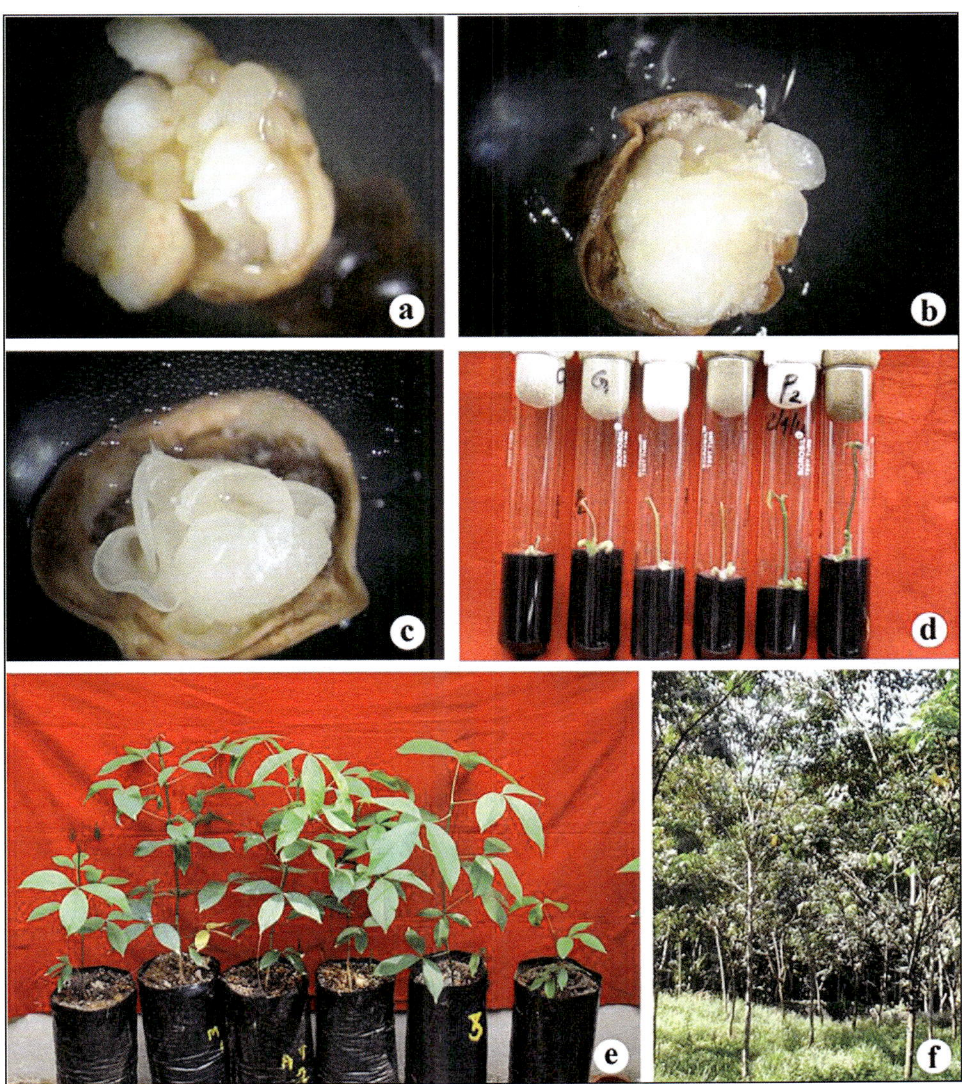

Figure 1.3: Induction of Polyembrony and Plantlet Regeneration. (p. 17)

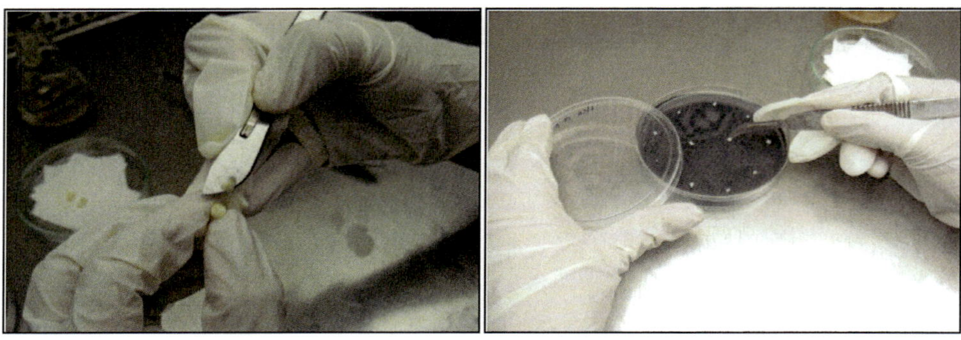

Figure 2.1: Excision of Plumules from Mature Zygotic Embryos of Coconut and Inoculation on to Culture Medium. (p. 35)

Figure 2.3: Regeneration of Plantlet from Coconut Plumule. (p. 35)

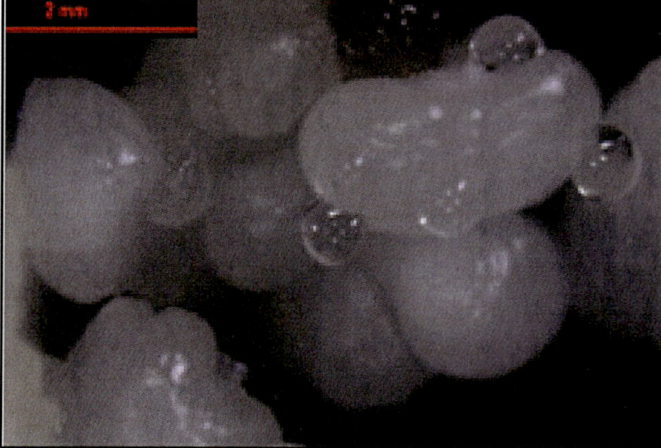

Figure 2.2: Development of Embryogenic Calli from Plumular Explants of Coconut.(p. 35)

Figure 2.4: Different Growth Stages in Plumule Culture. (p. 35)

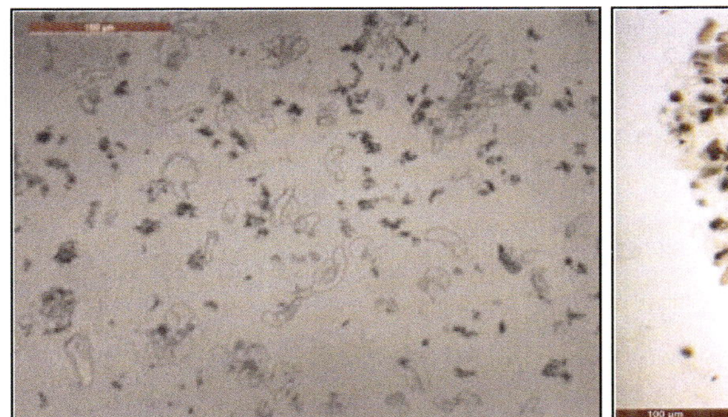

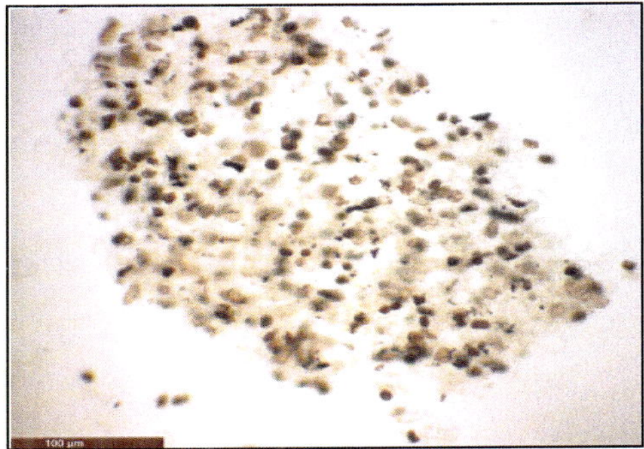

Figure 2.5: Initiation of Cell Suspension from Coconut Embryogenic Calli Derived from Plumular Explants. (p. 39)

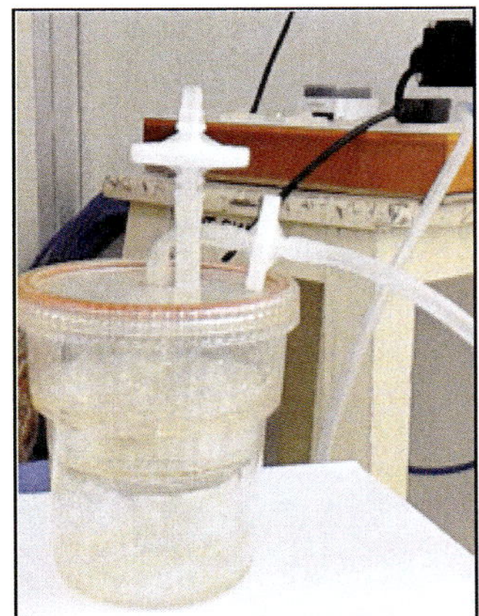

Figure 2.6: Use of Temporary Immersion Systems in Coconut Plumule Culture Experiments. (p. 41)

Figure 4.1: Various Stages of Somatic Embryogenesis Based Technology Developed at AAU, Anand. (p. 82)

Figure 5.1: Arecanut Clonal Propagation through Somatic Embryogenesis: (A) Inflorescence explants, (B) Callusing and formation of somatic embryos, (C) Germination of somatic embryos, (D) and (E) Plantlet development. (p. 96)

Figure 7.1: RITA Vessels Based Temporary Immersion Bioreactor System Developed at CIRAD, France and Adopted for Commercial Scale Micropropagation of F_1 Hybrids of Arabica. Top: Overview of the system; Bottom: Close up view of a few RITA vessels (Etienne *et al.,* 2012). (p. 132)

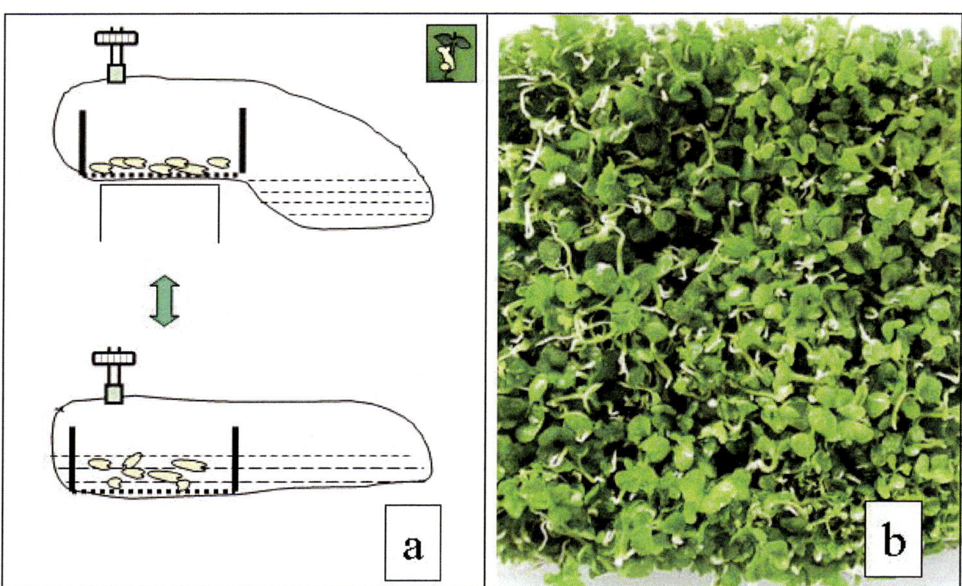

Figure 7.2: Flexible Plastic Based Disposable Temporary Immersion Bioreactors Developed by Nestle R and D Centre for Micropropagation of Robusta Coffee Clones. Top: Overview of a culture room with disposable bioreactors; Bottom left (a): Diagram of a disposable bioreactor; right (b): Pregerminated somatic embryos at the end of the phase (Ducos *et al.,* 2011). (p. 133)

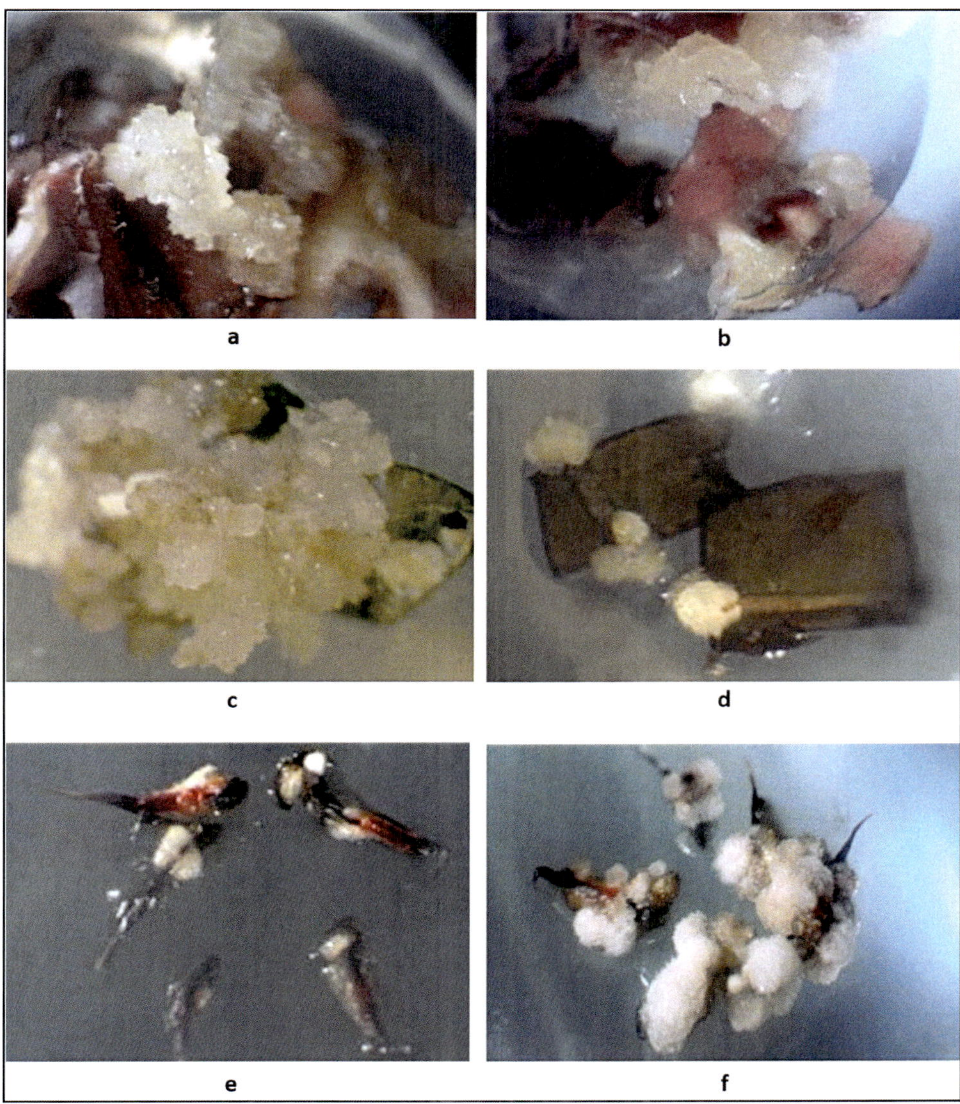

Figure 9.1: Callus Initiation from Different Explants of Cocoa. Callogenesis in cotyledonary explants (a, b), in tender leaves (c, d) and staminodal (e, f) explants. (p. 176)

Figure 16.1: Cacao Germplasm, in the International Cocoa Genebank Trinidad, Maintained as Living Trees. (p. 316)

Figure 17.3: Flowering (Left) and Bearing Pattern (Right) in Wild Species ***Psilanthus bengalensis*** Endemic to India. The fruits are less in number at each node, mostly single seeded, and beans are known to contain low caffeine or no caffeine. (p. 361)

Figure 17.4: Sln.13 (Chandragiri), an Improved Variety of Arabica Coffee. (p. 364)

Figure 17.5: Sln.3R (CxR), an Improved Variety of Robusta Coffee. (p. 367)

Figure 17.8: A New Breeding Line of Arabica (S.4817) with Integration of SH3 Gene by MAS. (p. 385)

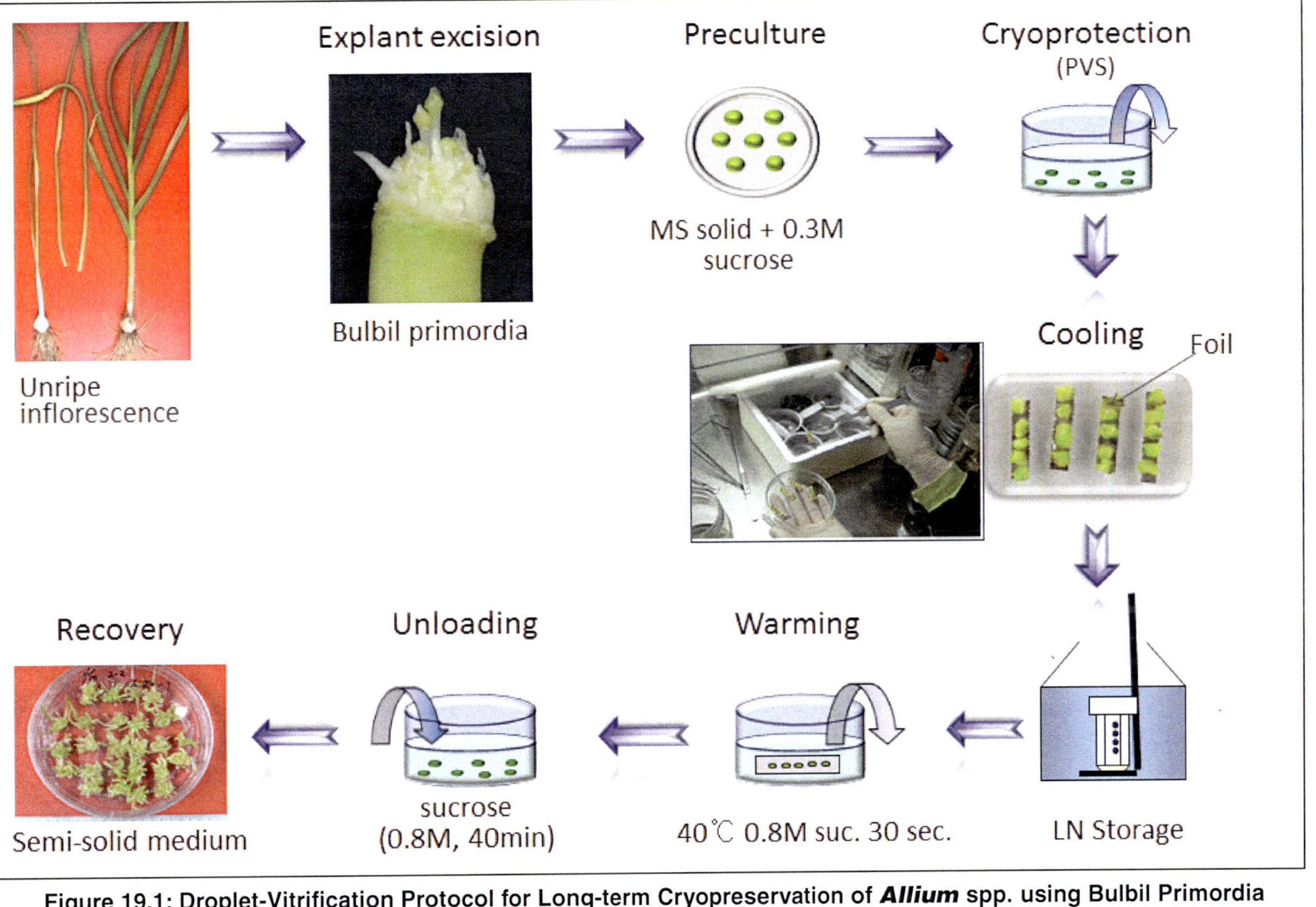

Figure 19.1: Droplet-Vitrification Protocol for Long-term Cryopreservation of ***Allium*** spp. using Bulbil Primordia from Immature Inflorescences (Modified from Kim ***et al.***, 2012a). (p. 442)

Figure 20.1: Plantlet Formation from Coconut Zygotic Embryo following Cryopreservation in PVS3. (p. 465)

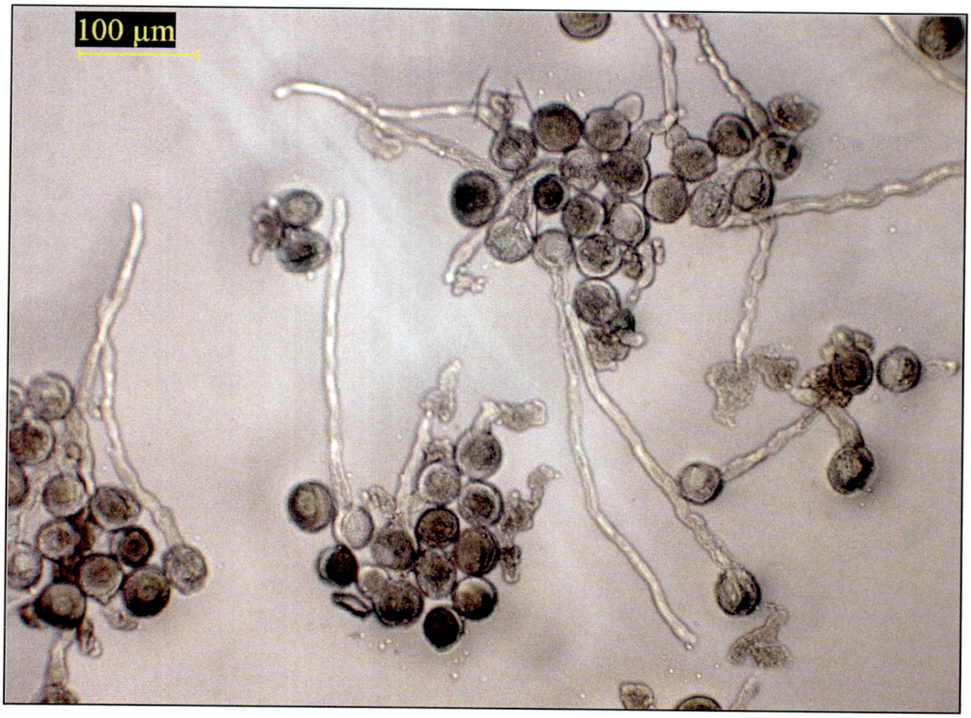

Figure 20.2: Germination in WCT Pollen Cryostored for Four Years. (p. 469)

Figure 20.3: Nut Set in COD Palms Hybridized with WCT Pollen Cryostored for Six Years. (p. 470)

Figure 22.2: Shoot Buds of Ginger Encapsulated in Na-alginate (*Source*: Nirmal Babu *et al.*, 2012). (p. 493)

Figure 22.4. Shoot Tips of ***Piper barberi*** Encapsulated in Na-alginate, arrow Indicates Shoot Tip Used as Explants (Source: Nirmal Babu ***et al.,*** 2012). (p. 496)

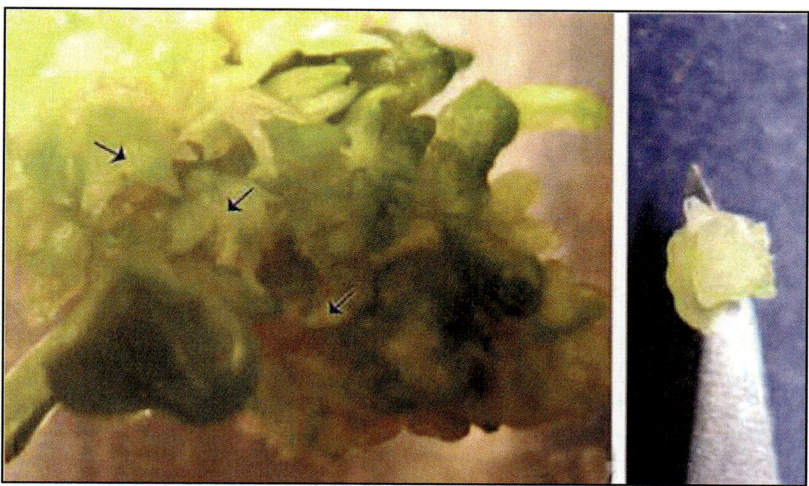

Figure 22.5: Plant Regeneration from Cryopreserved Miniature Shoots of Cardamom by Vitrification. Left: cardamom culture with miniature shoots; Right: excised meristematic clumps used for cryopreservation (Source: Nirmal Babu ***et al.,*** 2012). (p. 497)

Figure 22.6: Plantlets Regenerating from Cryopreserved Shoot Bud of Ginger (Source: Nirmal Babu ***et al.,*** 2012). (p. 498)

Figure 22.7: Germination of Cryopreserved Vanilla Pollen (Source: Nirmal Babu ***et al.,*** 2012). (p. 500)

Figure 24.1: A Mature Rubber Plantation Showing an Enlarged View of the Tapping Area at the Inset. (p. 527)

Figure 24.3: Development of *Hevea* Transgenic Plants Integrated with *Hb. mn SOD* Gene using Immature Anther-Derived Callus as initial Explants.

(A) emergence of transgenic callus lines on selection medium containing kanamycin (300 mg/l) and cefotaxime (500 mg/l) after *Agrobacterium* infection; (B) globular embryos from transgenic callus; (C) transformed embryos showing histochemical GUS expression; (D) regenerated transgenic plantlets; (E) PCR confirmation of transgenic plants using SOD gene specific primers: the upper bands indicates amplification of native SOD and lower bands indicates transgene (cDNA); (F) Southern blot analysis with ***nptII*** specific probe and (G) hardened transgenic plants growing in containment facility. (p. 532)

Figure 25.1: Different Stages of Transgenesis in Black Pepper.

(a) Co-cultivated embryogenic mass, (b) Growing points under kanamycin selection, (c) Embryo clusters formed under kanamycin selection, (d) Embryogenic mass in liquid SH medium, (e) Fully developed plantlets in liquid SH medium, (f) Hardened plants maintained in green house (Source: Jiby and Bhat, 2011). (p. 558)

Figure 25.2: *Agrobacterium* Mediated Genetic Transformation in Vanilla.

(a) Protocorm like bodies (PLBs) in selection medium after 30 days of co-cultivation (b, c, d, e and f) after 40, 50, 60, 90 and 120 days of culture in regeneration medium (g) hardened transgenic plants in green house. (p. 573)

C. *eugenioides* 0,67 Gb ♀

C. *canephora* 0,71 Gb ♂

C. *arabica* 1,3 Gb

Figure 27.1: Origin of *Coffea arabica*. (p. 592)

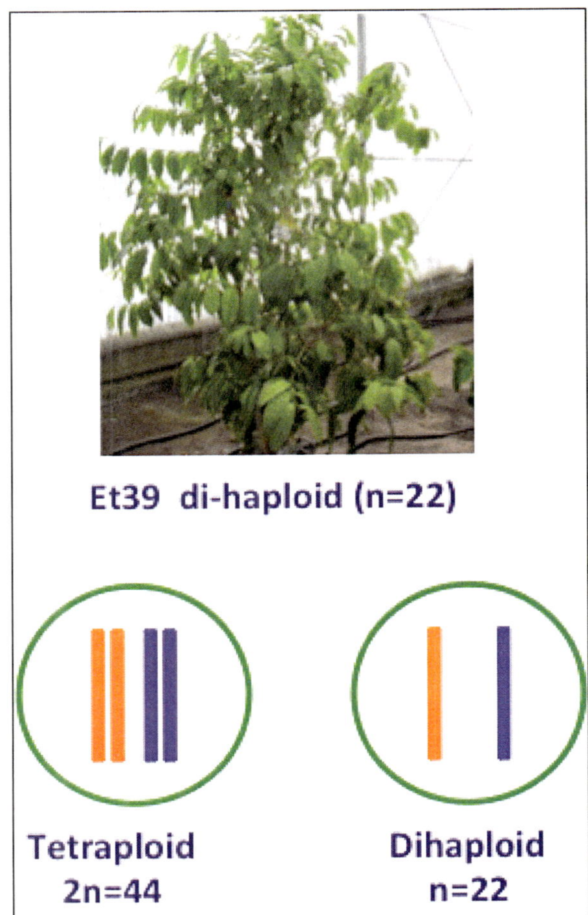

Figure 27.2: Et39 the *C. arabica* Genotype Sequenced. A dihaploid has only one set of each parental chromosomes, the C. arabica dihaploid has 22 chromosomes instead of 44, 11 chromosomes from each parental species. (p. 599)

Figure 28.1: Tender Shoot of Tea Plant. (p. 609)

Figure 28.2: Six Major Tea Categories in China. (p. 610)

Figure 29.1: View of a Healthy Rubber Plantation. (p. 638)

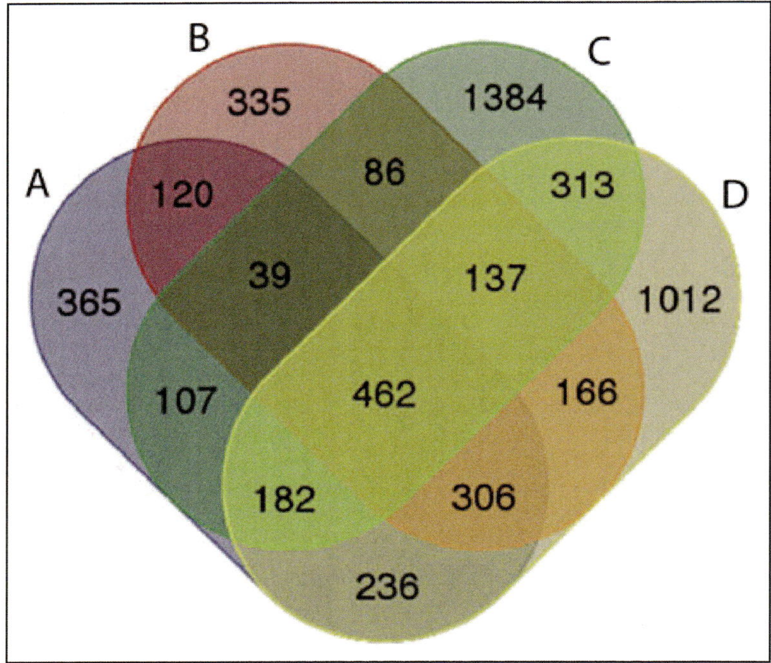

Figure 31.1: Venn Chart Showing Unique and Shared Unigenes found in Four Coconut Transcriptome Libraries. (p. 680)

Figure 31.2: Chowghat Green Dwarf Palm Affected by Root (Wilt) Disease. (p. 681)

Figure 28.2: Six Major Tea Categories in China. (p. 610)

Figure 29.1: View of a Healthy Rubber Plantation. (p. 638)

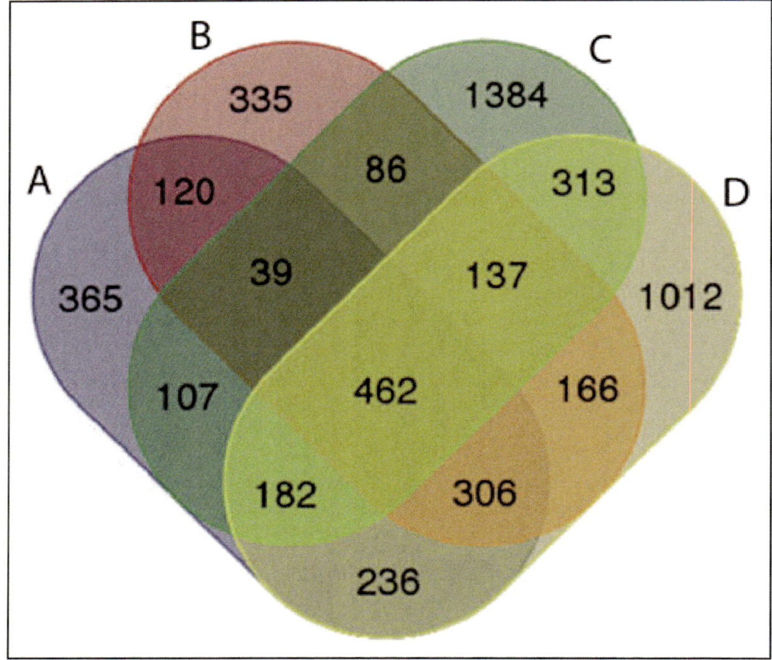

Figure 31.1: Venn Chart Showing Unique and Shared Unigenes found in Four Coconut Transcriptome Libraries. (p. 680)

Figure 31.2: Chowghat Green Dwarf Palm Affected by Root (Wilt) Disease. (p. 681)